D1640893

Public Transport Optimization

Konstantinos Gkiotsalitis

Public Transport Optimization

 Springer

Konstantinos Gkiotsalitis ⓘ
Department of Civil Engineering
University of Twente
Enschede, The Netherlands

ISBN 978-3-031-12443-3   ISBN 978-3-031-12444-0 (eBook)
https://doi.org/10.1007/978-3-031-12444-0

This Springer imprint is published by the registered company Springer Nature Switzerland AG
The registered company address is: Gewerbestrasse 11, 6330 Cham, Switzerland

To my colleagues at the University of Twente and the National Technical University of Athens for their support, my wife, and my son.

Preface

Climate change, pandemics, service disruptions, shared mobility services, and mobility-as-a-service schemes change rapidly the landscape in public transport planning. In this ever-changing environment, professional engineers and students of engineering schools should attain the required mathematical modeling skillset to meet the public transport planning needs of today and tomorrow.

This book serves as a guide for an aspiring public transport modeler. It systematically builds the skillset required by a public transport modeler covering topics ranging from the mathematical modeling of public transport systems to developing optimization solution methods, evaluating the complexity of problems, programming mathematical models, creating decision support tools, and evaluating the system's performance in the face of uncertainty. It offers a detailed description of exact optimization approaches, metaheuristics, and multi-objective optimization covering the vast majority of real-world problems in the public transport optimization field.

After studying this book, the engineering student, the professional engineer, or the public transport practitioner will build the necessary skillset to create new public transport models that meet the future challenges of the public transport sector. More specifically, the reader will be able to

- Formulate public transport problems into mathematical programs.
- Analyze the computational complexity of mathematical programs.
- Select appropriate solution methods.
- Implement one or more solution methods and return a solution to the problem.
- Develop decision support tools for public transport planners and service providers.
- Evaluate the performance of a decision support tool in a public transport system.

In terms of content, this book covers the topic of public transport optimization at the strategic, tactical, and operational stages. It also covers the topic of planning the services of shared mobility services, on-demand services, and feeder modes.

This book grew out of a set of course notes used for two postgraduate courses at the University of Twente, the Netherlands. The book is designed as a text for the following classroom environments:

- A primary text for an intensive one-semester course on Public Transport Optimization.
- A primary text for an intensive one-semester course on Mathematical Modeling of Urban Transport Systems.
- A primary text for two trimester courses on Mathematical Optimization in Transport and Public Transport Planning.
- A supporting text in a course on operations research and mathematical programming techniques, demonstrating the applications of the methodology.
- A supporting text on Rail Transport and Mobility-as-a-Service modeling courses.

Enschede, The Netherlands   Dr. Konstantinos Gkiotsalitis
June 2022

Acknowledgements

I would like to acknowledge the help of fellow professors, colleagues, supporting staff, and MSc. students at the University of Twente for providing me with the time and feedback that was required to complete the textbook and strengthen its learning objectives by testing it in real classroom settings.

I would also like to acknowledge the support of my family and my wife, Alexandra Saran, throughout this journey.

I am also indebted to Prof. Emeritus Antony Stathopoulos from the National Technical University of Athens and Prof. John Polak from Imperial College London for their early guidance in mathematical programming, and my former colleagues at NEC Laboratories Europe.

Contents

Acronyms

AI	Artificial Intelligence
BFS	Basic Feasible Solution
CG	Conjugate Gradient
CNF	Conjuctive Normal Form
CPP	Chinese Postman Problem
CSP	Crew Scheduling Problem
CVRP	Capacitated Vehicle Routing Problem
DE	Differential Evolution
DNF	Disjunctive Normal Form
FONC	First-order Necessary Conditions
FSP	Frequency Settings Problem
GA	Genetic Algorithm
GMI	Gomory Mixed Integer cuts
GSEC	Generalized Sub-tour Elimination Constraints
IP	Integer Program
IQP	Inequality-constrained Quadratic Program
KKT	Karush-Kuhn-Tucker conditions
LICQ	Linear Independence Constraint Qualification
LP	Linear Program
LPP	Line Planning Problem
LSCP	Location Set Covering Problem
MCLP	Maximal Covering Location Problem
mCP	m-Center Problem
MDP	Model Predictive Control
MDVSP	Multi-depot Vehicle Scheduling Problem
MDVSPTW	Multi-depot Vehicle Scheduling Problem with Time Windows
MILP	Mixed-integer Linear Program
MINLP	Mixed-integer Nonlinear Program
MIP	Mixed-integer Program
MOEA	Multi-objective Evolutionary Algorithms
MOOP	Multi-objective Optimization Problem

NSGA	Non-dominated Sorting Genetic Algorithm
PDP	Pickup and Delivery Problem
PESP	Periodic Event Scheduling Problem
PSO	Particle Swarm Optimization
QP	Quadratic Program
SAT	Boolean Satisfiability Problem
SD-VSP	Single-depot Vehicle Scheduling Problem
SLSQP	Sequential Least Squares Quadratic Programming
SONC	Second-order Necessary Conditions
SOSC	Second-order Sufficient Conditions
SPEA	Strength-Pareto EA
SPP	Shortest Path Problem
SPSPP	Single-Pair Shortest Path Problem
SQP	Sequential Quadratic Programming
TNDFSP	Transit Network Design and Frequency Setting Problem
TNDP	Transit Network Design Problem
TNT	Transit Network Timetabling
TRNDP	Transit Route Network Design Problem
TSP	Traveling Salesman Problem
VC	Vertex Cover
VRP	Vehicle Routing Problem
VSP	Vehicle Scheduling Problem

Part I
Mathematical Programming of Public Transport Problems

Problems in public transportation, as problems in many other sectors, change rapidly from year to year. The introduction of in-vehicle sensors and smart payment systems, the introduction of electric and autonomous vehicles, the use of demand-responsive services, and the integration of shared modes in public transport services have rapidly changed the sector over the past years. Undoubtedly, many revolutionary changes lie ahead in the years to come. This requires from a public transport modeler to be equipped with knowledge on how to translate new problem requirements into comprehensive mathematical programs that can be analyzed and solved to provide decision support to the relevant stakeholders. Problems ranging from optimally scheduling the trips of a fleet of electric vehicles to selecting the stations of a public transit network with autonomous vehicles are examples of future-looking problems that require knowledge in mathematical modeling to be able to formulate and solve them.

The first part of this handbook focuses on equipping the reader with the required knowledge in mathematical programming, linear algebra, numerical optimization methods, and complexity theory. After completing the first part of the handbook, the reader is expected to be able to do the following:

1. Translate problem descriptions of potential future problems into mathematical formulations that can be analyzed and solved to provide decision support to the relevant stakeholders.
2. Analyze the complexity of the developed mathematical formulations, propose efficient reformulations, and develop or select the appropriate solution methods to solve these formulations in a computer machine.
3. Understand the space and time complexities of solution methods used to solve mathematical programs.

Importantly, there is a detailed description of open-source optimization tools that can be used to solve mathematically formulated problems with a computer machine. After finishing this part of the handbook, the reader will be aware of which tools are suitable depending on the formulation of the problem and will be able to apply them to practical problems. Problems covered are continuous and discrete optimization problems of convex and non-convex origin.

Chapter 1
Introduction to Mathematical Programming

Abstract This chapter presents general information on mathematical programming and introduces the basics of linear algebra and propositional logic. Furthermore, it introduces mathematical formulations for common transport problems and discusses possible linearizations that can result in deriving globally optimal solutions. It also offers the basis for common terminologies used in mathematical programming, such as convexity, linearity, feasibility, global and local optimality.

1.1 Mathematical Modeling

1.1.1 Introductory Definitions

Mathematical modeling is the process of translating specific problems from an application area into mathematical formulations. Mathematical models are used in engineering, economics and in many other industries. They can take different forms, including statistical models, logical models and many more. This book will specifically focus on mathematical programming models that can be used for the representation and optimization of a public transport problem.

> Mathematical programming translates a real-world problem concerning the allocation of limited resources among competitive activities under a set of constraints into mathematical formulations that capture the problem's objectives and constraints.

Mathematical programming is used at several different levels in public transport planning. It can be applied at the strategic level to determine the stations and the routes of a public transport network, at the tactical planning level to determine the frequencies and timetables of the service routes, or at the operational level to make real-time control decisions [1].

Mathematical programming generally refers to the optimization of a decision problem. The word 'programming' appeared in the 1940s because decision problems were programmed in computer machines. Mathematical programming has an abstract meaning and oftentimes might not be associated with the solution methods deployed to solve an optimization problem. Naud et al. [2] state that mathematical programming contrasts with computer programming which solves decision problems by implementing algorithms that may be designed specifically for a given problem. The main difference is that mathematical programming considers declarative approaches which separate the representation of the problem through a *mathematical model* and its *solving*, where the *solving* may be done through various solution methods.

Before proceeding with the general representation of a mathematical program, Tables 1.1 and 1.2 present a list with common mathematical notations that will be used in the remainder of the book. In addition to that, if not mentioned otherwise, this book will use the following conventions:

- Small letters to represent *scalars*. For instance, $x = 5$.
- Small letters in bold to represent *vectors*. For instance, $\mathbf{x} = \begin{bmatrix} x_1 \\ x_2 \\ \dots \\ x_n \end{bmatrix}$.
- Capital letters to represent *sets*. For instance, S can represent a set with elements $S = \{1, 2, ..., |S|\}$.
- Capital letters in bold to represent *matrices*. For instance, a matrix $\mathbf{S} = \{s_{ij}\}$ where each element s_{ij} of the matrix is a scalar.
- $\mathbb{R}, \mathbb{Q}, \mathbb{Z}, \mathbb{N}$ for the sets of real numbers, rational numbers, integer numbers, and natural numbers, respectively. Note that $\mathbb{N} \subset \mathbb{Z} \subset \mathbb{Q} \subset \mathbb{R}$.

1.1.2 Basics of Linear Algebra

In mathematical modeling we make use of vectors, scalars, and matrices. Herein is provided a series of brief definitions concerning the basics of linear algebra.

Definition 1.1 A *scalar* is a quantity that is completely described by its magnitude. In the real coordinate space, a scalar can be seen as a real number $x \in \mathbb{R}$.

Definition 1.2 A *vector* is a list of at least two scalars. A vector can be interpreted as a point in space. In n dimensions, a point is described by column vector $\mathbf{x} = \begin{bmatrix} x_1 \\ x_2 \\ \dots \\ x_n \end{bmatrix}$ where $\mathbf{x} \in \mathbb{R}^n$ meaning that $x_1, ..., x_n$ are scalars in $\mathbb{R}$. By convention, when we refer to a vector we mean a *column* vector which has *row* vector $\mathbf{x}^{\mathsf{T}} = [x_1, x_2, ..., x_n]$ as its transpose.

Table 1.1 Common mathematical symbols (1/2)

Symbol	Name	Meaning
$\pm$	plus-minus	4 ± 2 means both $4 + 2$ and $4 - 2$
$\mp$	minus-plus	$6 + (3 \mp 5)$ means both $6 + (3 - 5)$ and $6 + (3 + 5)$
$\cdot$	dot	$\mathbf{u} \cdot \mathbf{v}$ is the dot product of vectors $\mathbf{u}$ and $\mathbf{v}$
$\therefore$	therefore	$x > y$ and $y > z \therefore x > z$
$\because$	because	if $x > y$, then $x > z \because y > z$
$\neg$	logical negation	$x \neq y \Leftrightarrow \neg(x = y)$
$\propto$	proportional to	if $y = 2x$, then $y \propto x$
$\approx$	approximately equal	$\pi \approx 3.14$
$\sim$	similar to	$x \sim D$ can mean that x is similar to D, or that the random variable x follows the probability distribution D
$:=$ or $\triangleq$ or $\doteq$	is defined as	$x \triangleq 52 - y$
$\equiv$	equivalent	$x \neq y \equiv \neg(x = y)$
$\ll$	much less	$5 \ll 100$
$\leqq$	less than or equal	$\mathbf{x} \leqq \mathbf{y}$: each element x_i of vector $\mathbf{x}$ is less than or equal to each corresponding element y_i of vector $\mathbf{y}$
$\subseteq$	subset	$A \subseteq B$ means that each element of A is also an element of B
$\mapsto$	maps to	$f : x \mapsto y$ means that the function f maps x to y.
$\models$	entails	$A \models B$ means that each sentence in A entails the sentence in B
$\vdash$	infers	$A \to B \vdash \neg B \to \neg A$
$\lfloor ... \rfloor$	floor (lowest integer)	$\lfloor 2.9 \rfloor = 2$
$\lceil ... \rceil$	ceil	$\lceil -2.6 \rceil = -2$, $\lceil 3.1 \rceil = 4$
$\lfloor ... \rceil$	nearest integer	$\lfloor 2.6 \rceil = 3$, $\lfloor 4.49 \rceil = 4$
$:$	such that	$x \in \mathbb{R} : a < x < b$

Definition 1.3 A *matrix* $\mathbf{C}$ is an array that has a fixed number of rows and columns. Each element of the matrix is a scalar,

$$\mathbf{C} = \begin{bmatrix} c_{11} & c_{12} & \cdots & c_{1n} \\ c_{21} & c_{22} & \cdots & c_{2n} \\ \vdots & \vdots & \ddots & \vdots \\ c_{m1} & c_{m2} & \cdots & c_{mn} \end{bmatrix}$$

where each column of the matrix can be seen as a column vector and each row as a row vector.

Table 1.2 Common mathematical symbols (2/2)

Symbol	Name	Meaning
$\setminus$	minus set	$A \setminus B$ means that we consider all elements of set A except the ones belonging to set B
$\oplus$	exclusive or	$(x > y) \oplus (z > y)$ is true when $x > y$ or $z > y$, but not both
$\forall$	for all	$n^2 \geq n, \ \forall n \in \mathbb{R}$
∂	partial derivative	$\partial f(x, y)/\partial x$
$\mathbb{E}$	Expected value (average)	$\mathbb{E}[X] = 52$
$\exists$	there exists	$\exists \, x \in \mathbb{R} : 5x > 17$
$\exists!$	there exists exactly one	$\exists! \, n \in \mathbb{N} : n + 5 = 10$
$\nexists$	there does not exist	$\nexists \, n \in \mathbb{N} : n + 5 = -10$
$\in$	belongs to	$x \in \mathbb{R}$
$\circ$	composition of functions	$(f \circ g)(x) = f(g(x))$
$\perp$	$x \perp y$ means that x is independent of y	If $x \perp y$, then $P(x\|y) = P(x)$
$\vee$	logical or	$n > 4 \vee n < 2$ means that n should be either greater than 4 of less than 2
$\wedge$	logical and	$n > 4 \wedge n < 12$ means that n should be greater than 4 and less than 12
∇	gradient of	$\nabla_{\mathbf{x}} f(x_1, x_2, ..., x_n)$
$\prod$	product	$\prod_{j=0}^{2} (j + 1) = 1 \cdot (1 + 1) \cdot (2 + 1)$
$\cup$	union of sets	$\bigcup_i A_i$
$\cap$	intersection of sets	$A \bigcap B$

Definition 1.4 The *diagonal* matrix is a matrix in which the entries outside the main diagonal are all zero. For instance,

$$\mathrm{diag}(\mathbf{C}) = \begin{bmatrix} c_{11} & 0 & \dots & 0 \\ 0 & c_{22} & \dots & 0 \\ \vdots & \vdots & \ddots & \vdots \\ 0 & 0 & \dots & c_{nn} \end{bmatrix}$$

for an $n \times n$ matrix.

Definition 1.5 The *identity* matrix $\mathbf{I}$ of size n is the $n \times n$ matrix with ones in the main diagonal and zeros outside of it.

Definition 1.6 The *transpose* $\mathbf{C}^\mathsf{T}$ of a matrix $\mathbf{C}$ is a matrix for which the i-th row, j-th column element of $\mathbf{C}^\mathsf{T}$ is the j-th column, i-th row element of $\mathbf{C}$. It follows that if $\mathbf{C}$ is an $n \times m$ matrix, its transpose $\mathbf{C}^\mathsf{T}$ is an $m \times n$ matrix. In addition, the transpose of a vector $\mathbf{x} = \begin{bmatrix} x_1 \\ x_2 \\ ... \\ x_n \end{bmatrix}$ is the row vector $\mathbf{x}^\mathsf{T} = [x_1, x_2, ...x_n]$.

Definition 1.7 The *inverse* $\mathbf{C}^{-1}$ of matrix $\mathbf{C}$ is a matrix for which $\mathbf{C}^{-1}\mathbf{C} = \mathbf{I}$. If matrix $\mathbf{C}$ does not have an inverse, it is called *singular* matrix. If it is invertible, it is called *nonsingular*. It follows that only square matrices $\mathbf{C} \in \mathbb{R}^{n \times n}$ can be nonsingular.

Definition 1.8 A square matrix $\mathbf{C} \in \mathbb{R}^{n \times n}$ is called *symmetric* if for every i, j we have $c_{ij} = c_{ji}$. It follows that if $\mathbf{C}$ is symmetric, then $\mathbf{C} = \mathbf{C}^\mathsf{T}$.

Based on the above definitions, we can present the mathematical operations among scalars, vectors and matrices. Let us focus first on vectors.

Vector operations

- Scalar addition, subtraction, multiplication or division of scalar b and vector $\mathbf{x}$ results in $b + \mathbf{x} = [b + x_1, b + x_2, ..., b + x_n]^\mathsf{T}$ where the plus symbol can be replaced by the symbol for subtraction, multiplication or division.
- Vector addition or subtraction of same size vectors $\mathbf{x}$ and $\mathbf{y}$:

$$\mathbf{x} + \mathbf{y} = [x_1 + y_1, x_2 + y_2, ..., x_n + y_n]^\mathsf{T}$$

Note that if the added/subtracted vectors are not of the same size, the vector addition/subtraction is *undefined*.
- Inner product (also known as dot product in the Euclidean space, which is the real n-space $\mathbb{R}^n$) of same size vectors $\mathbf{x}$ and $\mathbf{y}$:

$$\langle \mathbf{x}, \mathbf{y} \rangle = \mathbf{x} \cdot \mathbf{y} = \mathbf{x}^\mathsf{T}\mathbf{y} = [x_1, x_2, ..., x_n] \cdot \begin{bmatrix} y_1 \\ y_2 \\ ... \\ y_n \end{bmatrix} = x_1 y_1 + x_2 y_2 + \cdots + x_n y_n = \sum_{i=1}^{n} x_i y_i$$

The outcome of the inner product of two vectors is always a scalar. If the two vectors are not of the same size, the dot product is *undefined*.
- Outer product of two vectors $\mathbf{x}$ and $\mathbf{y}$:

$$\mathbf{x} \otimes \mathbf{y} = \begin{bmatrix} x_1 \\ x_2 \\ ... \\ x_n \end{bmatrix} \otimes \begin{bmatrix} y_1 \\ y_2 \\ ... \\ y_m \end{bmatrix} = \begin{bmatrix} x_1 y_1 & x_1 y_2 & ... & x_1 y_m \\ x_2 y_1 & x_2 y_2 & ... & x_2 y_m \\ \vdots & \vdots & \ddots & \vdots \\ x_n y_1 & x_n y_2 & ... & x_n y_m \end{bmatrix}$$

- l_2-norm, known also as Euclidean norm or 2-norm in the Euclidean space $\mathbb{R}^n$:

$$\|\mathbf{x}\|_2 = \sqrt{x_1^2 + x_2^2 + \cdots + x_n^2}$$

  gives the ordinary distance from the origin to point $\mathbf{x}$
- l_p-norm for $p \geq 1$, known also as p-norm, is the norm:

$$\|\mathbf{x}\|_p = \left(\sum_{i=1}^{n} |x_i|^p \right)^{1/p}$$

  note that for $p = 2$ the p-norm is the Euclidean norm. In addition, the absolute-value l_1 norm $\|x_i\|$ of a scalar is the absolute value $|x_i|$ and the absolute-value l_1 norm of a vector $\mathbf{x}$ is:

$$\|\mathbf{x}\| = |x_1| + |x_2| + \cdots + |x_n|$$

- l_∞-norm, known as l-infinity norm, is the largest magnitude among all elements of a vector:

$$\|\mathbf{x}\|_{+\infty} = \max\{|x_1|, |x_2|, ..., |x_n|\}$$

We can now proceed with the operations of matrices.

Matrix operations

- Scalar addition, subtraction, multiplication or division of scalar b and matrix $\mathbf{C}$ results in

$$\mathbf{C} + b = b + \mathbf{C} = \begin{bmatrix} b + c_{11} & b + c_{12} & \dots & b + c_{1n} \\ b + c_{21} & b + c_{22} & \dots & b + c_{2n} \\ \vdots & \vdots & \ddots & \vdots \\ b + c_{m1} & b + c_{m2} & \dots & b + c_{mn} \end{bmatrix}$$

  where the plus symbol can be replaced by the symbol for subtraction, multiplication or division, respectively.
- Matrix addition or subtraction of same size matrices $\mathbf{C}$ and $\mathbf{D}$ is similar to vector addition/subtraction, i.e., element-wise addition/subtraction of corresponding elements in two matrices with the same dimensions. Note that if the added/subtracted matrices are not of the same size, the matrix addition/subtraction is *undefined*.
- Matrix multiplication of two matrices is defined if their inner dimensions agree, meaning that the number of columns in the first matrix is equal to the number of rows in the second matrix. An $m \times n$ matrix $\mathbf{C}$ multiplied by an $n \times p$ matrix $\mathbf{D}$ results in an $m \times p$ matrix:

$$
\underbrace{\mathbf{C}}_{m\times n}\ \underbrace{\mathbf{D}}_{n\times p} = \underbrace{\begin{bmatrix} \sum\limits_{i=1}^{n} c_{1i}d_{i1} & \sum\limits_{i=1}^{n} c_{1i}d_{i2} & \dots & \sum\limits_{i=1}^{n} c_{1i}d_{ip} \\ \sum\limits_{i=1}^{n} c_{2i}d_{i1} & \sum\limits_{i=1}^{n} c_{2i}d_{i2} & \dots & \sum\limits_{i=1}^{n} c_{2i}d_{ip} \\ \vdots & \vdots & \ddots & \vdots \\ \sum\limits_{i=1}^{n} c_{mi}d_{i1} & \sum\limits_{i=1}^{n} c_{mi}d_{i2} & \dots & \sum\limits_{i=1}^{n} c_{mi}d_{ip} \end{bmatrix}}_{m\times p}
$$

Note that if an $m \times n$ matrix $\mathbf{C}$ is multiplied by an n-valued vector $\mathbf{x}$ this will result in an m-valued vector:

$$
\underbrace{\mathbf{C}}_{m\times n}\ \underbrace{\mathbf{x}}_{n\times 1} = \underbrace{\mathbf{b}}_{m\times 1}
$$

In addition, an n-valued transpose vector $\mathbf{x}^\mathsf{T}$ multiplied by an $n \times m$ matrix $\mathbf{C}$ results in the $1 \times m$-valued matrix $\mathbf{b} = \mathbf{x}^\mathsf{T}\mathbf{C}$, which is a row vector with m values.

Matrix multiplication has the following properties:

1. matrix multiplication is associative: $(\mathbf{AB})\mathbf{C} = \mathbf{A}(\mathbf{BC})$
2. matrix multiplication is distributive: $\mathbf{A}(\mathbf{B} + \mathbf{C}) = \mathbf{AB} + \mathbf{AC}$
3. matrix multiplication obeys the transpose property: $(\mathbf{AB})^\mathsf{T} = \mathbf{B}^\mathsf{T}\mathbf{A}^\mathsf{T}$
4. matrix multiplication is not commutative: $\mathbf{AB} \neq \mathbf{BA}$ for any $\mathbf{A}$ and $\mathbf{B}$

The correspondence between matrix notation and calculus is provided in Table 1.3.

In addition to the basics of linear algebra, in the next section we present the basics of differentiation.

Table 1.3 Correspondence between matrix notation and calculus

Matrix notation	Calculus	Outcome
$c = \underset{1\times n\ n\times 1}{\mathbf{x}^\mathsf{T}\ \mathbf{y}}$	$c = \sum_{i=1}^{n} x_i y_i$	scalar
$c = \underset{1\times n\ n\times 1}{\mathbf{y}^\mathsf{T}\ \mathbf{x}}$	$c = \sum_{i=1}^{n} y_i x_i$	scalar
$\mathbf{c} = \underset{m\times n\ n\times 1}{\mathbf{A}\ \mathbf{x}}$	$c_i = \sum_{j=1}^{n} a_{ij}x_j$	m-valued vector
$\mathbf{C} = \underset{m\times n\ n\times p}{\mathbf{A}\ \mathbf{B}}$	$c_{ij} = \sum_{k=1}^{n} a_{ik}b_{kj}$	$m \times p$-valued matrix
$\mathbf{c}^\mathsf{T} = \underset{1\times m\ m\times n}{\mathbf{y}^\mathsf{T}\ \mathbf{A}}$	$c_i = \sum_{k=1}^{m} y_k a_{ki}$	n-valued row vector $\mathbf{c}^\mathsf{T}$
$\mathbf{c}^\mathsf{T} = \underset{1\times m\ m\times n\ n\times 1}{\mathbf{y}^\mathsf{T}\ \mathbf{A}\ \mathbf{x}}$	$c = \sum_{j=1}^{m}\sum_{k=1}^{n} y_j a_{jk}x_k$	scalar

1.1.3   Basics of Differentiation

1.1.3.1   Scalar-by-Scalar

Scalar-by-scalar refers to the case when we seek to find the derivative of a scalar function f by a scalar x. The derivative of a single-variable, scalar-valued function $f : \mathbb{R} \to \mathbb{R}$ is the slope of the tangent line of the function at a particular point. Using Leibniz's notation, the derivative of $f(x)$ is represented by $\frac{df(x)}{dx}$. Using Euler's notation, it is represented by $D_x f(x)$. The derivative of $f(x)$ is expressed as:

$$\frac{df(x)}{dx} = \lim_{h \to 0} \frac{f(x+h) - f(x)}{h}$$

Differentiating combined functions is based on the following basic rules.

Rules of differentiation for single-variable, scalar-valued functions

- sum rule: $\frac{d(af(x)+\beta g(x))}{dx} = a\frac{df(x)}{dx} + \beta\frac{dg(x)}{dx}$
- product rule: $\frac{d(f(x)g(x))}{dx} = g(x)\frac{df(x)}{dx} + f(x)\frac{dg(x)}{dx}$
- chain rule: for $(f \circ g)(x) = f(g(x))$ we have $\frac{d(f(g(x)))}{dx} = \frac{d(f(g(x)))}{dg(x)}\frac{dg(x)}{dx}$

Reckon some basic derivatives:

$$\frac{dx^a}{dx} = ax^{a-1}$$

$$\frac{de^x}{dx} = e^x$$

$$\frac{da^x}{dx} = a^x \ln a \text{ for } a > 0$$

$$\frac{d\ln x}{dx} = \frac{1}{x} \text{ for } x > 0$$

Let us consider an example to demonstrate these rules. The derivative of $\ln(x)e^x$ with respect to x can be considered as the derivative of $f(x)g(x)$ where $f(x) = \ln(x)$ and $g(x) = e^x$. Using the product rule:

$$\frac{d(f(x)g(x))}{dx} = g(x)\frac{df(x)}{dx} + f(x)\frac{dg(x)}{dx} = e^x\frac{1}{x} + \ln(x)e^x$$

Consider now the derivative of $\sin(x^2)$ with respect to x. We can write this as $(f \circ g)(x) = f(g(x))$ where $g(x) = x^2$. Using the chain rule we have:

$$\frac{\mathrm{d}f(g(x))}{\mathrm{d}x} = \frac{\mathrm{d}f(g(x))}{\mathrm{d}g(x)} \frac{\mathrm{d}g(x)}{\mathrm{d}x} = \cos(g(x))2x = 2x \cos x^2$$

1.1.3.2 Directional and Partial Derivatives

Definition 1.9 Directional Derivative: consider a scalar-valued function $f : \mathbb{R}^n \to \mathbb{R}$ with multiple variables, expressed in a column vector $\mathbf{x} = [x_1, x_2, ..., x_n]^\mathsf{T}$. If we are given a direction unit vector $\mathbf{u} = [u_1, u_2, ..., u_n]^\mathsf{T}$ with $u_1^2 + u_2^2 + \cdots + u_n^2 = 1$, we can define the directional derivative of $f(\mathbf{x})$ in the direction of $\mathbf{u}$ by:

$$\mathrm{D}_\mathbf{u} f(\mathbf{x}) = \lim_{h \to 0} \frac{f(\mathbf{x} + h\mathbf{u}) - f(\mathbf{x})}{h}$$

Definition 1.10 Partial Derivative: consider direction unit vector $\mathbf{u}$. If $\mathbf{u} = [1, 0, ..., 0]^\mathsf{T}$, then the direction vector is parallel to the x_1 axis and $\mathrm{D}_\mathbf{u} f(\mathbf{x})$ is the partial derivative with respect to x_1. That is,

$$\frac{\partial f(\mathbf{x})}{\partial x_1} = \mathrm{D}_\mathbf{u} f(\mathbf{x}) \ \text{ for } \mathbf{u} = [1, 0, ..., 0]^\mathsf{T}$$

Similarly, the partial derivative with respect to x_i is:

$$\frac{\partial f(\mathbf{x})}{\partial x_i} = \mathrm{D}_\mathbf{u} f(\mathbf{x}) \ \text{ for } u_i = 1 \text{ and } u_j = 0 \ \forall u_j \in \mathbf{u} \setminus \{u_i\}$$

1.1.3.3 Scalar-by-Vector (Gradient Vector)

Consider a scalar-valued function f with multiple variables, expressed in a column vector $\mathbf{x} \in \mathbb{R}^n$. Function $f : \mathbb{R}^n \to \mathbb{R}$. In this case, the derivative of a scalar function f by a vector $\mathbf{x} = [x_1, x_2, ..., x_n]^\mathsf{T}$ is:

$$\frac{\partial f(\mathbf{x})}{\partial \mathbf{x}}$$

There are two main notations for laying out systems of partial derivatives:

- the *numerator* layout notation
- the *denominator* layout notation

In the *numerator* layout notation:

$$\frac{\partial f(\mathbf{x})}{\partial \mathbf{x}} = \left[\frac{\partial f(\mathbf{x})}{\partial x_1}, \frac{\partial f(\mathbf{x})}{\partial x_2}, ..., \frac{\partial f(\mathbf{x})}{\partial x_n} \right]$$

In the *denominator* layout notation:

$$\frac{\partial f(\mathbf{x})}{\partial \mathbf{x}} = \begin{bmatrix} \frac{\partial f(\mathbf{x})}{\partial x_1} \\ \frac{\partial f(\mathbf{x})}{\partial x_2} \\ \vdots \\ \frac{\partial f(\mathbf{x})}{\partial x_n} \end{bmatrix}$$

It is common to call *gradient* vector the derivative of a scalar by a vector when using the denominator layout notation:

$$\nabla_{\mathbf{x}} f(\mathbf{x}) = \begin{bmatrix} \frac{\partial f(\mathbf{x})}{\partial x_1} \\ \frac{\partial f(\mathbf{x})}{\partial x_2} \\ \vdots \\ \frac{\partial f(\mathbf{x})}{\partial x_n} \end{bmatrix}$$

Adopting the denominator layout notation, the gradient is an n-valued column vector. From the definitions of the directional derivative and the gradient, it follows that the directional derivative is a scalar:

$$D_{\mathbf{u}} f(\mathbf{x}) = \nabla_{\mathbf{x}} f(\mathbf{x}) \cdot \mathbf{u} = \mathbf{u}^{\mathsf{T}} \nabla_{\mathbf{x}} f(\mathbf{x}) = [u_1, u_2, ..., u_n] \begin{bmatrix} \frac{\partial f(\mathbf{x})}{\partial x_1} \\ \frac{\partial f(\mathbf{x})}{\partial x_2} \\ \vdots \\ \frac{\partial f(\mathbf{x})}{\partial x_n} \end{bmatrix} = \sum_{i=1}^{n} u_i \frac{\partial f(\mathbf{x})}{\partial x_i}$$

Consider, for instance, function $f(\mathbf{x}) = 3x_1 + 5x_2^2 + e^{x_3}$. Note that $f : \mathbb{R}^3 \rightarrow \mathbb{R}$ and its gradient is:

$$\nabla_{\mathbf{x}} f(\mathbf{x}) = \begin{bmatrix} \frac{\partial(3x_1 + 5x_2^2 + e^{x_3})}{\partial x_1} \\ \frac{\partial(3x_1 + 5x_2^2 + e^{x_3})}{\partial x_2} \\ \frac{\partial(3x_1 + 5x_2^2 + e^{x_3})}{\partial x_n} \end{bmatrix} = \begin{bmatrix} 3 \\ 10x_2 \\ e^{x_3} \end{bmatrix}$$

Notice that the same rules apply for scalar-by-vector derivatives of combined functions.

Rules of differentiation for multi-variable, scalar-valued functions

Consider $g : \mathbb{R}^n \rightarrow \mathbb{R}$ and $\mathbf{x} \in \mathbb{R}^n$.

- sum rule: $\frac{\partial(af(\mathbf{x}) + \beta g(\mathbf{x}))}{\partial \mathbf{x}} = a\frac{\partial f(\mathbf{x})}{\partial \mathbf{x}} + \beta\frac{\partial g(\mathbf{x})}{\partial \mathbf{x}}$
- product rule: $\frac{\partial f(\mathbf{x})g(\mathbf{x})}{\partial \mathbf{x}} = g(\mathbf{x})\frac{\partial f(\mathbf{x})}{\partial \mathbf{x}} + f(\mathbf{x})\frac{\partial g(\mathbf{x})}{\partial \mathbf{x}}$
- chain rule: $\frac{\partial f(g(\mathbf{x}))}{\partial \mathbf{x}} = \frac{\partial f(g(\mathbf{x}))}{\partial g(\mathbf{x})}\frac{\partial g(\mathbf{x})}{\partial \mathbf{x}}$

Table 1.4 Examples of scalar-by-vector derivatives

Scalar-by-vector	Outcome with numerator layout	Outcome with denominator layout
$\frac{\partial a}{\partial \mathbf{x}}$	$\mathbf{0}^{\mathsf{T}}$	$\mathbf{0}$
$\frac{\partial a f(\mathbf{x})}{\partial \mathbf{x}}$	$a \frac{\partial f(\mathbf{x})}{\partial \mathbf{x}}$	$a \frac{\partial f(\mathbf{x})}{\partial \mathbf{x}}$
$\frac{\partial \mathbf{a}^{\mathsf{T}} \mathbf{x}}{\partial \mathbf{x}}$	$\mathbf{a}^{\mathsf{T}}$	$\mathbf{a}$

Using the numerator layout notation and the denominator layout notation has an impact on the outcome of the scalar-by-vector derivative (see Table 1.4). For this reason, the use of numerator layout notation or denominator layout notation should be consistent when solving a specific problem (i.e., use consistently only one of the two).

1.1.3.4 Vector-by-Scalar

In this case we have a vector-valued function $\mathbf{f}(x) = [f_1(x), f_1(x), ..., f_m(x)]^{\mathsf{T}}$ and a scalar variable x. That is, $\mathbf{f} : \mathbb{R} \to \mathbb{R}^m$. Using the numerator layout notation, we have:

$$\frac{\partial \mathbf{f}(x)}{\partial x} = \begin{bmatrix} \frac{\partial f_1(x)}{\partial x} \\ \frac{\partial f_2(x)}{\partial x} \\ \vdots \\ \frac{\partial f_m(x)}{\partial x} \end{bmatrix}$$

with chain rule:

$$\frac{\partial \mathbf{g}(\mathbf{f}(x))}{\partial x} = \frac{\partial \mathbf{g}(\mathbf{f}(x))}{\partial \mathbf{f}(x)} \frac{\partial \mathbf{f}(x)}{\partial x}$$

Using the denominator layout notation, we have:

$$\frac{\partial \mathbf{f}(x)}{\partial x} = \left[\frac{\partial f_1(x)}{\partial x}, \frac{\partial f_2(x)}{\partial x}, ..., \frac{\partial f_m(x)}{\partial x} \right]$$

with chain rule:

$$\frac{\partial \mathbf{g}(\mathbf{f}(x))}{\partial x} = \frac{\partial \mathbf{f}(x)}{\partial x} \frac{\partial \mathbf{g}(\mathbf{f}(x))}{\partial \mathbf{f}(x)}$$

1.1.3.5   Vector-by-Vector (Jacobian Matrix)

Let us now consider a multivariate vector-valued function $\mathbf{f}(\mathbf{x})$, where $\mathbf{x} = [x_1, ..., x_n]^\mathsf{T}$ and $\mathbf{f} : \mathbb{R}^n \to \mathbb{R}^m$. Using the numerator layout notation, the vector-by-vector derivative, known as the Jacobian matrix, is:

$$\frac{\partial \mathbf{f}(\mathbf{x})}{\partial \mathbf{x}} = \begin{bmatrix} \frac{\partial f_1(\mathbf{x})}{\partial x_1} & \frac{\partial f_1(\mathbf{x})}{\partial x_2} & \cdots & \frac{\partial f_1(\mathbf{x})}{\partial x_n} \\ \frac{\partial f_2(\mathbf{x})}{\partial x_1} & \frac{\partial f_2(\mathbf{x})}{\partial x_2} & \cdots & \frac{\partial f_2(\mathbf{x})}{\partial x_n} \\ \vdots & \vdots & \ddots & \vdots \\ \frac{\partial f_m(\mathbf{x})}{\partial x_1} & \frac{\partial f_m(\mathbf{x})}{\partial x_2} & \cdots & \frac{\partial f_m(\mathbf{x})}{\partial x_n} \end{bmatrix}$$

which is an $m \times n$ matrix with chain rule:

$$\frac{\partial \mathbf{g}(\mathbf{f}(\mathbf{x}))}{\partial \mathbf{x}} = \frac{\partial \mathbf{g}(\mathbf{f}(\mathbf{x}))}{\partial \mathbf{f}(\mathbf{x})} \frac{\partial \mathbf{f}(\mathbf{x})}{\partial \mathbf{x}}$$

and the product rule of the dot product of two vector-valued functions becomes:

$$\frac{\partial (\mathbf{f}(\mathbf{x}) \cdot \mathbf{g}(\mathbf{x}))}{\partial \mathbf{x}} = (\mathbf{f}(\mathbf{x}))^\mathsf{T} \frac{\partial \mathbf{g}(\mathbf{x})}{\partial \mathbf{x}} + (\mathbf{g}(\mathbf{x}))^\mathsf{T} \frac{\partial \mathbf{f}(\mathbf{x})}{\partial \mathbf{x}}$$

Using the denominator layout notation:

$$\frac{\partial \mathbf{f}(\mathbf{x})}{\partial \mathbf{x}} = \begin{bmatrix} \frac{\partial f_1(\mathbf{x})}{\partial x_1} & \frac{\partial f_2(\mathbf{x})}{\partial x_1} & \cdots & \frac{\partial f_m(\mathbf{x})}{\partial x_1} \\ \frac{\partial f_1(\mathbf{x})}{\partial x_2} & \frac{\partial f_2(\mathbf{x})}{\partial x_2} & \cdots & \frac{\partial f_m(\mathbf{x})}{\partial x_2} \\ \vdots & \vdots & \ddots & \vdots \\ \frac{\partial f_1(\mathbf{x})}{\partial x_n} & \frac{\partial f_2(\mathbf{x})}{\partial x_n} & \cdots & \frac{\partial f_m(\mathbf{x})}{\partial x_n} \end{bmatrix}$$

which is an $n \times m$ matrix with chain rule:

$$\frac{\partial \mathbf{g}(\mathbf{f}(\mathbf{x}))}{\partial \mathbf{x}} = \frac{\partial \mathbf{f}(\mathbf{x})}{\partial \mathbf{x}} \frac{\partial \mathbf{g}(\mathbf{f}(\mathbf{x}))}{\partial \mathbf{f}(\mathbf{x})}$$

and the product rule of the dot product of two vector-valued functions becomes:

$$\frac{\partial (\mathbf{f}(\mathbf{x}) \cdot \mathbf{g}(\mathbf{x}))}{\partial \mathbf{x}} = \frac{\partial \mathbf{f}(\mathbf{x})}{\partial \mathbf{x}} \mathbf{g}(\mathbf{x}) + \frac{\partial \mathbf{g}(\mathbf{x})}{\partial \mathbf{x}} \mathbf{f}(\mathbf{x})$$

In addition to the basics of differentiation, in the next section we present the basics of propositional logic that are used to translate logical statements into mathematical models.

1.1.4 Basics of Propositional Logic

Many mathematical symbols are used to express fundamental problems in propositional logic. For instance, we typically use boolean connectives $\vee, \wedge, \vdash, \oplus, \neg$ in proposition logic. Propositional variables can evaluate to either *True* or *False*. For instance, if propositional variable P is true, the logical negation $\neg P$ is false. That is, $P \vdash \neg(\neg P)$. If, for example, P refers to the movement of a train and it has the value *true*, this entails ($\vdash$) that the 'no movement' $\neg P$ cannot be true $\neg(\neg P)$.

In propositional logic, formulas are built up from propositional variables through linear systems [3]. Consider, for example, the satisfiability of the logic formula:

$$x \vee \neg y$$

Considering that x and y can take the values *True* or *False*, the logical formula is satisfied if x is true or $\neg y$ is true (that is, y is false). Based on that, we have the four possible assignments of truth values presented in Table 1.5.

Table 1.5 shows that the disjunction $x \vee \neg y$ is not true when x and y are false and true, respectively (in that case, neither x nor $\neg y$ are true). If we now associate the number 0 to the symbolic value False and the number 1 to the symbolic value True, then the logical formula is satisfied if, and only if,

$$x + (1 - y) \geq 1 \quad \text{where } x \in \{0, 1\}, \ y \in \{0, 1\}$$

Indeed, if x is false and y is true, then $x + (1 - y) = 0 + (1 - 1) = 0 < 1$. In any other case, $x + (1 - y) \geq 1$. That is, $x \vee \neg y \Leftrightarrow x + (1 - y) \geq 1$. With this we have translated a logical formula into a linear inequality constraint.

Consider now the logical formula $x \wedge y$. This is a conjunction which is satisfied when both x and y are true. This can be translated into the linear inequality constraint:

$$xy \geq 1 \quad \text{where } x \in \{0, 1\}, \ y \in \{0, 1\}$$

which is satisfied only if $x = 1$ and $y = 1$ (that is, they are both true).

Table 1.5 Possible assignment of values to the logic formula $x \vee \neg y$

x	y	$x \vee \neg y$
True	True	True
True	False	True
False	True	False
False	False	True

To generalize, let $+x$ abbreviate the atom x and $-x$ abbreviate the negation $\neg x$. Let also $X := \{\pm x_j, \ \forall j \in \{1, 2, ..., n\}\}$ be a finite set of n logical variables. Each logical variable (literal) can either be an atom x_j or a negation $\neg x_j$ depending on the $\pm$ sign of x_j. Then, we can proceed with the following definitions:

Definition 1.11 *Empty clause* is an empty set of literals (typically symbolized as $\square$).

Definition 1.12 *Disjunctive clause* is a proposition of the form $\bigvee_{j=1}^{n} \pm x_j$. For instance, $x_1 \vee x_2 \vee \neg x_3$.

Definition 1.13 *Conjunctive clause* is a proposition of the form $\bigwedge_{j=1}^{n} \pm x_j$. For instance, $x_1 \wedge x_2 \wedge \neg x_3$.

Definition 1.14 *Disjunctive Normal Form (DNF)* is a disjunction of conjunctive clauses (e.g., $(x_1 \wedge x_2) \vee (x_3 \wedge \neg x_4)$).

Definition 1.15 *Conjunctive Normal Form (CNF)* is a conjunction of disjunctive clauses (e.g., $(x_1 \vee x_2 \vee \neg x_3) \wedge (x_4 \vee \neg x_5)$).

These definitions can be used to generalize the translation of logical expressions to linear inequality constraints. For example, consider a $0-1$ variable y_j which takes the 0 value when atom x_j is false and the 1 value when atom x_j is true. Then, the disjunctive clause $\bigvee_{j=1}^{n} \pm x_j$ can be written as [4]:

$$\sum_{j=1}^{n} \pm y_j \geq 1 \quad \text{where } y_j \in \{0, 1\} \ \forall j \in \{1, 2, ..., n\}$$

It follows that if $\pm x_j$ is an atom, we can replace $\pm y_j$ by y_j. If, however, $\pm x_j$ is a negation, $\neg x_j$, then we should replace $\pm y_j$ by $1 - y_j$. For instance, the disjunctive clause $x_1 \vee x_2 \vee \neg x_3$ is expressed as:

$$y_1 + y_2 + (1 - y_3) \geq 1 \quad y_1, y_2, y_3 \in \{0, 1\}$$

Let us now consider the conjunctive clause $\bigwedge_{j=1}^{n} \pm x_j$. This logical formula is satisfied if, and only if,

$$\prod_{j=1}^{n} \pm y_j \geq 1 \quad \text{where } y_j \in \{0, 1\} \ \forall j \in \{1, 2, ..., n\}$$

Notice that the main difference here is that the left-hand side (lhs) is a product. This is requested because all logical variables $\pm x_{ij}$ should be true (even if one of them is false, the product will be equal to zero and the constraint will not be satisfied). For example,

$$x_1 \wedge \neg x_2 \wedge x_3 \Leftrightarrow y_1(1 - y_2)y_3 \geq 1 \quad \text{where } y_1, y_2, y_3 \in \{0, 1\}$$

Consider now the CNF which is a conjunction of disjunctive clauses. Let, for instance, $(x_1 \vee x_2 \vee \neg x_3) \wedge (x_4 \vee \neg x_5) \wedge (x_5 \vee \neg x_6)$. This can be expressed as:

$$(x_1 \vee x_2 \vee \neg x_3) \wedge (x_4 \vee \neg x_5) \wedge (x_5 \vee \neg x_6) \quad \Leftrightarrow \quad \begin{array}{l} y_1 + y_2 + (1 - y_3) \geq 1 \\ y_4 + (1 - y_5) \geq 1 \\ y_5 + (1 - y_6) \geq 1 \\ y_1, y_2, ..., y_6 \in \{0, 1\} \end{array}$$

where every conjunction of disjunctive clauses results in an additional inequality constraint.

Consider also the DNF which is a disjunction of conjunctive clauses. The DNF can be described as a sum of products. Let, for instance, $(x_1 \wedge x_2 \wedge \neg x_3) \vee (x_4 \wedge \neg x_5) \vee (x_5 \wedge \neg x_6)$. This is satisfied if, and only if, the following inequality constraint is satisfied:

$$y_1 y_2(1 - y_3) + y_4(1 - y_5) + y_5(1 - y_6) \geq 1 \quad \text{where } y_1, ..., y_6 \in \{0, 1\}$$

To summarize, Table 1.6 presents possible assignments of values of logic formulas depending on the values of the literals. Note that $x_1 \wedge x_2$ is equivalent to $y_1 y_2 \geq 1$, $x_1 \vee x_2$ is equivalent to $y_1 + y_2 \geq 1$, $x_1 \oplus x_2$ is equivalent to $y_1 + y_2 = 1$, $x_1 \rightarrow x_2$ is equivalent to $y_1 \leq y_2$, and $x_1 \leftrightarrow x_2$ is equivalent to $y_1 = y_2$, where $y_1, y_2 \in \{0, 1\}$.

The basic laws that govern the most common equivalences of logical statements are presented in Table 1.7.

Finally, an important topic on logic and mathematical programming are the necessary and sufficient conditions. These are explained below.

Definition 1.16 *Necessary Condition:* is a condition (or set of conditions) that must be present for another condition to occur. The assertion that A is necessary for B is equivalent to B cannot be true unless A is true. This indicates that if A is false, then B is also false. Note though that B might be false even if A is true.

Table 1.6 Possible assignment of values for compound propositions

x_1	x_2	$x_1 \wedge x_2$	$x_1 \vee x_2$	$x_1 \oplus x_2$	$x_1 \rightarrow x_2$	$x_1 \leftrightarrow x_2$
True	True	True	True	False	True	True
True	False	False	True	True	False	False
False	True	False	True	True	True	False
False	False	False	False	False	True	True

Table 1.7 Basic laws of propositional logic

Law	Logical statement equivalency
Symmetry	$A \wedge B = B \wedge A$
Symmetry	$A \vee B = B \vee A$
Associativity	$A \wedge (B \wedge C) = (A \wedge B) \wedge C$
Associativity	$A \vee (B \vee C) = (A \vee B) \vee C$
De Morgan's Law	$\neg(A \wedge B) = \neg A \vee \neg B$
De Morgan's Law	$\neg(A \vee B) = \neg A \wedge \neg B$
Distributive Law	$A \wedge (B \vee C) = (A \wedge B) \vee (A \wedge C)$
Distributive Law	$A \vee (B \wedge C) = (A \vee B) \wedge (A \vee C)$

Definition 1.17 *Sufficient Condition:* is a condition (or set of conditions) that produces another condition. That is, a condition A is sufficient for a condition B, if (and only if) the truth of A guarantees the truth of B. However, knowing A is false does not meet a minimal need to conclude that B is also false.

The key difference between the two is that a necessary condition must be there for an event to occur, but it alone does not provide sufficient cause for the occurrence of the event.

1.2 General Representation of an Optimization Problem

An optimization problem consists of a well-defined objective (e.g., goal, aim) expressed in the form of an objective function, parameters and sets which represent the input of the optimization problem, and variables that are the unknowns of the optimization problem. Except for the objective function, equality and inequality equations (also known as constraints) might restrict the values of different problem variables.

An optimization problem seeks to find the optimal values of the problem variables such that all constraints are satisfied and the objective function is minimized (or maximized) depending on whether the optimization problem is a minimization or a maximization one. These optimal values of the variables are the *optimal solution* (sometimes simply called *solution*) of the optimization problem. The definitions of the *feasible solution*, the *feasible region*, the *globally optimal* solution and the *locally optimal* solution are provided below.

Important Definitions:

Feasible solution: A solution is feasible if the values of the variables of this solution satisfy all problem constraints.

Feasible region: The feasible region is the set of all possible sets of values of the variables that satisfy all problem constraints. The feasible region is the set of all feasible solutions. It is also known as *feasible set* [5].

Domain of a function: The set of inputs accepted by a function. For instance, the domain of $f : \mathbb{N} \to \mathbb{R}$ is the set of natural numbers, denoted as $\text{dom}(f) = \mathbb{N}$.

Globally optimal solution: When solving a minimization problem, a solution is globally optimal if its objective function value is smaller than or equal to the objective function values of all other feasible solutions. For a maximization problem, the objective function value of the globally optimal solution is larger than or equal to all other feasible solutions.

Locally optimal solution: A locally optimal solution is a solution that performs better than or equal to all other feasible solutions in a specific region (neighborhood) of the feasible region.

Any mathematical optimization problem can be written in its *standard form*. A single-objective optimization problem can be a maximization or a minimization problem, where a maximization problem with objective function f can be transformed into a minimization problem by minimizing objective function $-f$. The standard form of a continuous minimization problem is provided below.

! Standard form of a minimization problem with scalar-valued objective function [6]

$$\min\ f(\mathbf{x}) \tag{1.1}$$
$$\text{subject to:}\ h_i(\mathbf{x}) = 0, \qquad i = 1, ..., l \tag{1.2}$$
$$g_j(\mathbf{x}) \leq 0, \qquad j = 1, ..., m \tag{1.3}$$
$$\mathbf{x} \in \mathbb{R}^n \tag{1.4}$$

---

where $f : \mathbb{R}^n \to \mathbb{R}$ is the objective function, $h_i(\mathbf{x})$ are *finitely* many equality constraint functions and $g_j(\mathbf{x})$ are *finitely* many inequality constraint functions of the minimization problem. The objective function $f(\mathbf{x})$ and the constraint functions $h_i(\mathbf{x}), g_j(\mathbf{x})$ for $i = 1, ..., l$ and $j = 1, ..., m$ are scalar, real-valued functions of an n-dimensional column vector $\mathbf{x} = [x_1, ..., x_n]^\mathsf{T}$. The feasible region of the optimization problem can be expressed as:

$$\mathcal{F} := \{\mathbf{x} \in \mathbb{R}^n\ :\ \mathbf{h}(\mathbf{x}) = \mathbf{0} \wedge \mathbf{g}(\mathbf{x}) \leq \mathbf{0}\}$$

where $\mathbf{h}(\mathbf{x}) = [h_1(\mathbf{x}), ..., h_l(\mathbf{x})]^\mathsf{T}$ and $\mathbf{g}(\mathbf{x}) = [g_1(\mathbf{x}), ..., g_m(\mathbf{x})]^\mathsf{T}$.

Fig. 1.1 Feasible region $\mathcal{F}$
of the minimization problem
in Eqs. (1.9)–(1.13)

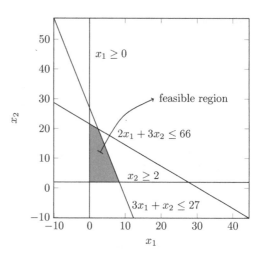

This minimization problem can be written more succinctly if we replace each equality constraint $h_i(x)$, $i = 1, ..., l$ by two inequality constraints:

$$h_i(\mathbf{x}) = 0 \quad \Leftrightarrow \quad \begin{array}{l} h_i(\mathbf{x}) \leq 0 \\ h_i(\mathbf{x}) \geq 0 \end{array} \quad \forall i = 1, ..., l \qquad (1.5)$$

In addition, any $h_i(\mathbf{x}) \geq 0$ can be written as $-h_i(\mathbf{x}) \leq 0$ and, by transforming all equality constraints $h_i(\mathbf{x}) = 0$ to the form $g_j(\mathbf{x}) \leq 0$, we have the more succinct, yet equivalent, inequality-constrained mathematical program:

$$\min \ f(\mathbf{x}) \qquad (1.6)$$
$$\text{subject to: } g_j(\mathbf{x}) \leq 0, \qquad\qquad j = 1, ..., m + 2l \qquad (1.7)$$
$$\mathbf{x} \in \mathbb{R}^n \qquad (1.8)$$

If the objective function, f, and the constraint functions are linear or affine functions, and $\mathbf{x}$ are continuous variables, then the problem is a linear programming problem. Consider, for example, the minimization problem:

$$\min \ 5x_1 + 3x_2 \qquad (1.9)$$
$$\text{subject to: } 3x_1 + x_2 \leq 27 \qquad (1.10)$$
$$2x_1 + 3x_2 \leq 66 \qquad (1.11)$$
$$x_2 \geq 2 \qquad (1.12)$$
$$x_1, x_2 \in \mathbb{R}_{\geq 0} \qquad (1.13)$$

Given these constraint functions, the feasible region $\mathcal{F}$ of this minimization problem is provided in Fig. 1.1.

The minimization problem in Eqs. (1.9)–(1.13) is a linear program because its objective function is linear, its variables x_1, x_2 are continuous, and all inequality constraints are affine functions. Continuous minimization problems with linear/affine equality and inequality constraints can be written as:

! Minimization problem with linear constraints

$$\min_{\mathbf{x} \in \mathbb{R}^n} f(\mathbf{x}) \tag{1.14}$$

$$\text{subject to: } \mathbf{C}\,\mathbf{x} = \mathbf{d} \tag{1.15}$$

$$\mathbf{A}\,\mathbf{x} \leq \mathbf{b} \tag{1.16}$$

where $\mathbf{x} \in \mathbb{R}^n$ is a column vector $\mathbf{x} = [x_1, ..., x_n]^\mathsf{T}$. In addition, $\mathbf{C}$ is a $l \times n$ matrix with elements:

$$\mathbf{C} = \begin{bmatrix} c_{11} & c_{12} & \cdots & c_{1n} \\ c_{21} & c_{22} & \cdots & c_{2n} \\ \vdots & \vdots & \ddots & \vdots \\ c_{l1} & c_{l2} & \cdots & c_{ln} \end{bmatrix} \tag{1.17}$$

where $1, 2, ..., l$ are the indexes of the equality constraints. Similarly, $\mathbf{A}$ is an $m \times n$ matrix, where $1, 2, ..., m$ are the indexes of the inequality constraints, $\mathbf{d}$ a column vector $l \times 1$, and $\mathbf{b}$ a column vector $m \times 1$.

Consider, for example, the minimization problem with affine constraints and a nonlinear objective function:

$$\min_{\mathbf{x} \in \mathbb{R}^2} 5x_1 + 3x_2 + 12x_1 x_2 \tag{1.18}$$

$$\text{subject to: } 3x_1 = 5x_2 \tag{1.19}$$

$$2x_1 \leq 7 \tag{1.20}$$

$$4x_2 \leq 29 \tag{1.21}$$

$$x_1 \geq 1 \tag{1.22}$$

The objective function is $f(\mathbf{x}) = 5x_1 + 3x_2 + 12x_1 x_2$. In addition, $d = 0$, $\mathbf{b} = [7, 29, -1]^\mathsf{T}$, $\mathbf{C} = [3, -5]$ and

$$\mathbf{A} = \begin{bmatrix} 2 & 0 \\ 0 & 4 \\ -1 & 0 \end{bmatrix}$$

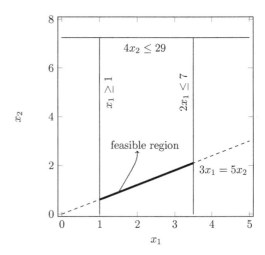

Fig. 1.2 Feasible region $\mathcal{F}$ of the minimization problem in Eqs. (1.18)–(1.22). Because of the equality constraint $3x_1 = 5x_2$ the feasible region lies in the line $3x_1 - 5x_2 = 0$

resulting in:

$$\min_{\mathbf{x} \in \mathbb{R}^2} 5x_1 + 3x_2 + 12x_1x_2$$

$$\text{subject to: } [3 \quad -5] \begin{bmatrix} x_1 \\ x_2 \end{bmatrix} = 0$$

$$\begin{bmatrix} 2 & 0 \\ 0 & 4 \\ -1 & 0 \end{bmatrix} \begin{bmatrix} x_1 \\ x_2 \end{bmatrix} \leq \begin{bmatrix} 7 \\ 29 \\ -1 \end{bmatrix}$$

The feasible region of this optimization problem is provided in Fig. 1.2 and it lies in the line $3x_1 - 5x_2 = 0$ for $x_1 \in [1, 3.5]$.

1.2.1 Sets, Parameters and Variables

To explain the representation of an optimization problem, consider the transportation problem of Hitchcock [7] that has the following problem description.

Hitchcock's Transportation Problem
Suppose there are several factories that supply a product to a number of cities. Due to freight rates, the cost of supplying a unit of the product to a particular city varies according to which factory supplies it, and it also varies from city to city. These costs are known in advance. In addition, the total product shipped from any factory and the total product that must be supplied to every city are

known. The supply should also match the demand, therefore the total shipped product from all factories is equal to the total supplied product to the cities. All products of a factory should be shipped. Minimize the total shipping costs.

When translating a problem description into a mathematical program, one needs to determine the sets, parameters and variables of the problem before formulating the objective function(s) and the constraints. It is typical to determine the *sets* of the problem first. In the transportation problem of Hitchcock, the sets of the problem are:

- the factories
- the cities

When solving a specific instance of the aforementioned transportation problem the exact number of factories and the number of cities will be provided. In an abstract formulation though, one can assume that the number of factories is m and the number of cities is n, where both m and n are positive integer numbers. The sets of the factories and the cities can then be defined as $M = \{1, ..., m\}$ and $N = \{1, ..., n\}$, respectively.

After the sets, one can define the problem's *parameters*. Parameters refer to the fixed values (provided inputs) that do not change their values during the optimization process. In the transportation problem of Hitchcock, the parameters are:

- the cost of transporting a unit of the product from a factory $i \in M$ to a city $j \in N$
- the amount of products produced at factory $i \in M$
- the amount of products that need to be supplied to a city $j \in N$

The first set of parameters can be denoted by an $m \times n$ matrix, $\mathbf{C}$. Each element c_{ij} of matrix $\mathbf{C}$ indicates the cost of transporting a unit of the product from a factory $i \in M$ to a city $j \in N$ and it is a non-negative number. The second set of parameters can be denoted by an m-dimensional vector, $\mathbf{a}$. Each element a_i of the vector indicates the amount of products produced at factory $i \in M$ and it is a non-negative number. Finally, the third set of parameters can be denoted by an n-dimensional vector, $\mathbf{b}$. Each element b_j of vector $\mathbf{b}$ indicates the amount of products that need to be supplied to a city $j \in N$ and it is a non-negative number.

In addition to the above, one should declare the *variables* of the optimization problem. In this problem, the variables refer to the amount of product that will be actually transported from any factory $i \in M$ to any city $j \in Z$. The variables can be denoted by an $m \times n$ matrix, $\mathbf{X} = \{x_{ij}\}$, where x_{ij} indicates the amount of product transported from factory $i \in M$ to city $j \in N$ and it is a non-negative number.

1.2.2 Objectives

The objective function of a single-objective problem is a scalar function that changes its value based on the values of the problem's variables. In the transportation problem of Hitchcock, the objective is to minimize the total shipping costs. That is,

$$\min \sum_{i \in M} \sum_{j \in N} c_{ij} x_{ij} \tag{1.23}$$

More generally, if $\mathbf{x}$ is the set of variables of a single-objective optimization problem, then $\mathbf{x} \mapsto f(\mathbf{x})$ where $f(\mathbf{x})$ is the objective function. In the case of the examined problem, $\mathbf{X} \mapsto \sum_{i \in M} \sum_{j \in N} c_{ij} x_{ij}$. Notice that if we have 3 products transported from factory $i \in M$ to city $j \in N$ and 2 products transported from factory $i' \in M$ to city $j' \in N$, then the objective function value is equal to $3c_{ij} + 2c_{i'j'}$.

1.2.3 Constraints

The constraints of an optimization problem are equality and inequality constraints that restrict the values of the problem variables. In some problem instances, there might not be any possible variable values that can satisfy all problem constraints resulting in an infeasible optimization problem.

In the discussed transportation problem, the amount of products shipped from a factory to a city should be a non-negative number. This results in a first set of inequality constraints:

$$x_{ij} \geq 0 \quad \forall i \in M, j \in N \tag{1.24}$$

In addition, each factory i produces exactly a_i product units, resulting in the equality constraints:

$$\sum_{j \in N} x_{ij} = a_i \quad \forall i \in M \tag{1.25}$$

Constraints (1.25) state that all transported product units from any factory $i \in M$, $\sum_{j \in N} x_{ij}$, should be equal to the amount of product units produced by that factory, a_i.

Finally, the last constraints refer to the number of product units supplied to each city $j \in N$, which should be equal to b_j:

$$\sum_{i \in M} x_{ij} = b_j \quad \forall j \in N \tag{1.26}$$

After defining the sets, the parameters, the variables, the objective and the constraints of Hitchcock's transportation problem, we have a complete mathematical program. The mathematical program is the problem formulation of the examined optimization problem that can be cast as follows.

Hitchcock's Transportation Problem

$$\min_{\mathbf{X} \in \mathbb{R}^{m \times n}} \sum_{i \in M} \sum_{j \in N} c_{ij} x_{ij} \tag{1.27}$$

$$\text{subject to: } \sum_{j \in N} x_{ij} = a_i \qquad \forall i \in M \tag{1.28}$$

$$\sum_{i \in M} x_{ij} = b_j \qquad \forall j \in N \tag{1.29}$$

$$x_{ij} \geq 0 \qquad \forall i \in M, j \in N \tag{1.30}$$

This mathematical program is a linear program because its objective function is a linear function and its constraints consist of affine equalities and inequalities. If we assume that the variables x_{ij} take natural values because we are allowed to ship items in whole number of batches, then the problem is an integer linear program because the variables cannot take continuous values in $\mathbb{R}$.

1.2.4 Modeling Example of a Public Transport Problem

Many public transport problems can be formulated as mathematical programs. An example of such problems is the periodic event scheduling problem (PESP) introduced by Serafini [8] in 1989. In the problem definition of [9], a periodic timetable of a specific time period $[0, t)$ should assign points of time to any relevant event, where an event is either the arrival or departure of a directed public transport line at a station. In this problem, any two events i, j should have a pre-defined minimum and maximum time difference (a lower and upper bound). These time differences might bound, for instance, the allowed elapsed time from event i to event j. Events are connected to each other by introducing a set of arcs that represents all admissible pairs of events (i, j). For instance, an arc might represent the travel of a trip from the departure from the stop related to event i to the arrival to the stop related to event j. In its classic formulation, PESP does not have a specific objective function and the only requirement is to find the event-related departure and/or arrival times that meet the constraints.

Starting with the sets of the problem, we have a set $E = \{1, ..., |E|\}$ that represents all planned events within the time period $[0, t)$. We also have a set of arcs A that represents the transitions among events, i.e., $(i, j) \in A$ if there is a connection between them. The time $t \geq 0$ is a parameter of the problem and it is provided as input. In addition, for any two events $(i, j) \in A$, we have a pre-defined minimum allowed time difference, $l_{ij} \geq 0$, and a pre-defined maximum allowed time difference, $l_{ij} \leq u_{ij} < l_{ij} + t$. Both l_{ij} and u_{ij} are problem parameters because their values are known. Finally, we introduce vector $\boldsymbol{\pi} = [\pi_1, ..., \pi_{|E|}]^\mathsf{T}$ to represent the variables of the problem, where π_i refers to the time point of event $i \in E$. Note that $|E|$ is the cardinality of set E and indicates the number of elements in set E. We can state that vector $\boldsymbol{\pi} \in [0, t)^{|E|}$. This results in the following mathematical program.

PESP formulation

Find a vector $\boldsymbol{\pi}$ that fulfills:

$$l_{ij} \leq \pi_j - \pi_i \leq u_{ij} \qquad \forall(i, j) \in A \qquad (1.31)$$

$$0 \leq \pi_i < t \qquad \forall i \in E \qquad (1.32)$$

By solving the aforementioned problem, one seeks to find one of the potential feasible solutions. As long as $\pi_i < t - l_{ij}$ it is possible to find a $\pi_j \in [0, t)$ such that $\pi_j - \pi_i > l_{ij}$. If, however, $\pi_i \geq t - l_{ij}$ there is no $\pi_j \in [0, t)$ satisfying constraint (1.31) because this constraint would require that $\pi_j \geq t$. To rectify this, one can introduce integer variables $z_{ij} \in \mathbb{Z}_{\geq 0}$, known in the literature as *periodic offsets* [10], that transform constraints (1.31) into:

$$l_{ij} \leq \pi_j - \pi_i + z_{ij}t \leq u_{ij} \qquad \forall(i, j) \in A$$

A final note is that this classic formulation assumes that all feasible timetable solutions are equally good. One can easily expand this formulation by adding an objective function that strives to minimize the slack time between two events [11]. If the selected time points of events i and j are such that $\pi_j - \pi_i + z_{ij}t = l_{ij}$ we have no slack because the time difference between the two events is the minimum possible. Seeking such time points, yields the extended PESP formulation presented below.

PESP formulation with minimized slack times

$$\min \sum_{(i,j)\in A} w_{ij}(\pi_j - \pi_i + z_{ij}t - l_{ij}) \qquad (1.33)$$

$$\text{subject to: } l_{ij} \leq \pi_j - \pi_i + z_{ij}t \leq u_{ij} \qquad \forall(i, j) \in A \qquad (1.34)$$

$$0 \leq \pi_i < t \qquad \forall i \in E \qquad (1.35)$$

$$z_{ij} \in \mathbb{Z}_{\geq 0} \qquad \forall(i, j) \in A \qquad (1.36)$$

where each w_{ij} is a parameter representing the penalty that is to be applied to any time unit of slack between events i and j.

1.3 Continuous Optimization

1.3.1 Introduction

Optimization problems can be characterized based on the nature of their variables. As opposed to discrete optimization, continuous optimization problems refer to problems where the problem variables can be chosen from a set of real values between which there are no gaps.

Continuous optimization problems can be either *unconstrained* or *constrained*. In the former case, there are no restrictions on the variables. In the latter case, there are explicit constraints on the variables. It is important to note that unconstrained optimization problems are less common in public transport problems because they typically require planning specific services under resource constraints. Notwithstanding this, unconstrained optimization problems are very useful because they arise in the reformulation of constrained optimization problems where the problem constraints can be replaced by penalty terms that are added to the objective function of the problem.

To provide a tangible example, consider the uni-dimensional optimization problem of the form:

$$\min 5x^2 \tag{1.37}$$

$$\text{subject to: } x \in \mathbb{R} \tag{1.38}$$

This is a continuous optimization problem since the problem's variable (x) can be chosen from a set of infinitely many real values (see Fig. 1.3).

Fig. 1.3 Example of the objective function values of the continuous optimization problem of the form $\min 5x^2$ s.t. $x \in \mathbb{R}$

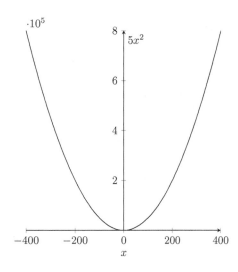

Fig. 1.4 Example of the continuous constrained optimization problem $\min 5x^2$ subject to $-200 \leq x \leq 200$

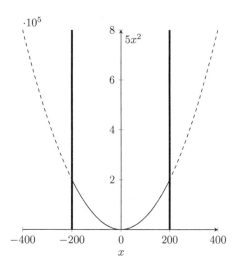

Adding problem constraints, such as $-200 \leq x \leq 200$, will result in a continuous optimization problem with feasible region $\mathcal{F} := \{x \in \mathbb{R} : -200 \leq x \leq 200\}$, as presented in Fig. 1.4.

In discrete optimization, some or all variables are drawn from sets with finitely many elements. If, for instance, x is only allowed to take values from the set $x = \{-400, -350, ..., +350, +400\}$ this will result in the discrete optimization problem:

$$\min \ 5x^2 \tag{1.39}$$

$$\text{subject to: } x \in \{-400, -350, ..., +350, +400\} \tag{1.40}$$

which is presented in Fig. 1.5.

The discrete optimization problem in Fig. 1.5 is an *integer optimization* problem (IP) because $\{-400, -350, ..., +350, +400\}$ contains only integer values. *Integer* programming and *binary* programming are subcategories of discrete optimization problems where, in the former case, variables can take values from an integer set and, in the latter case, variables can take values from a binary set, meaning that $x \in \{0, 1\}$. Since binary programming can be perceived as a subcategory of integer programming, it holds that *discrete* programming $\supseteq$ *integer* programming $\supseteq$ *binary* programming. Integer programming is a subcategory of discrete programming since a discrete program can allow its variables to take values from finite sets that might include non-integers (i.e., real or rational numbers).

Fig. 1.5 Example of the discrete optimization problem min $5x^2$ subject to $x \in \{-400, -350, ..., +350, +400\}$

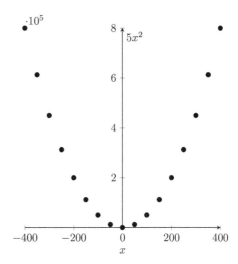

1.3.2 Example of a Continuous Public Transport Optimization Problem

Continuous optimization is common in public transport problems. One of the common continuous optimization problems is the vehicle holding problem. Vehicle holding is a real-time control strategy that is used to reduce the difference between the *scheduled* headway of two successive public transport vehicles and the *actual* one. In public transport, the scheduled time headway of a service line indicates the time difference between two successive trip arrivals at the same station. Because of mixed traffic conditions and passenger demand variations, the actual headway of two vehicles $i, i+1$ of a service line might differ from the scheduled headway when departing from a specific station.

Let d_{i+1} be the departure time of trip $i+1$ that arrives to the station before trip i. Let also h be the scheduled time headway between trips $i+1$ and i. If trip i is ready to depart from the station at time t, then there is an opportunity to hold that trip at the station and let it depart at time $d_i \geq t$ in order to try to meet the scheduled headway. This is known as the one-headway-based control strategy [12]. In practice, the variable of the problem is the departure time of trip i, d_i, from the station since h, t and d_{i+1} are already known (parameters). Finding the optimal value of d_i can be a result of solving the following continuous constrained optimization model.

Fig. 1.6 Optimal solution $d_i = 4$ min of the vehicle holding problem for $t = 3$ min, $d_{i+1} = 0$ min, and $h = 4$ min

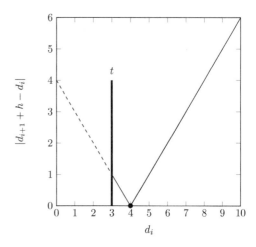

One-headway-based vehicle holding

$$\min |d_{i+1} + h - d_i| \tag{1.41}$$
$$\text{subject to: } d_i \geq t \tag{1.42}$$
$$d_i \in \mathbb{R}_{\geq 0} \tag{1.43}$$

The objective function $|d_{i+1} + h - d_i|$ strives to minimize the difference between the scheduled time headway h and the actual time headway $d_{i+1} - d_i$ upon the departure of trips $i, i + 1$ from the same station. There are also the additional constraints that d_i should be non-negative and d_i should be greater than or equal to t since t is the earliest possible departure time of trip i if we do not apply vehicle holding.

Let, for instance, d_{i+1} be equal to 0 min, t to 3 min, and h to 4 min. Then, the optimal value of d_i can be obtained according to Fig. 1.6.

In the aforementioned example, trip i will depart from the station at time $d_i = 4$ min and it will maintain the scheduled headway $h = 4$ min. Note, however, that if $t > d_{i+1} + h$ the scheduled time headway cannot be met (Fig. 1.7).

1.4 Discrete Optimization

1.4.1 Introduction

In discrete optimization, some or all the variables of a problem should be drawn from sets with finitely many elements. This is achieved by adding the so-called *integrality*

Fig. 1.7 Optimal solution
$d_i = 5$ min of the vehicle
holding problem for $t = 5$
min, $d_{i+1} = 0$ min, and
$h = 4$ min

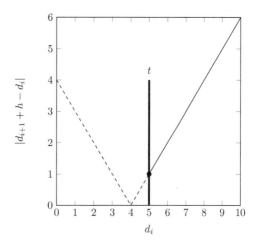

constraints to the optimization problem. Integrality constraints have the form $x_i \in \mathcal{P}$,
where x_i is a variable and $\mathcal{P}$ a set with finitely many elements. If $\mathcal{P}$ is a set of integers,
$\mathcal{P} \subseteq \mathbb{Z}$, the optimization problem is referred to as an integer program.

Common categorizations of discrete optimization problems include:

- (Purely) Integer optimization problems where the variables can take values from
  integer or binary sets (the latter are also called binary optimization problems).
- Mixed-integer optimization problems where we have a nonempty subset of integer
  variables and a subset of real-valued (continuous) variables [13].

Another important category is the combinatorial optimization problems. Combinatorial optimization problems are discrete optimization problems with a combinatorial origin, meaning that the problem is one of graph theory, or arranging objects in
a particular way. Note that a combinatorial optimization problem can be formulated
as an integer optimization or a mixed-integer optimization problem, depending on
its nature.

Two important terms in discrete optimization problems are the *compactness* and
tightness. These are defined below.

Definition 1.18 *Tightness* of a discrete optimization problem refers to the distance
between the discrete solution and the associated solution of its continuous relaxation
that does not consider any integrality constraints.

Definition 1.19 *Compactness* of an optimization problem refers to its size (i.e.,
number of constraints and variables) and impacts the searching speed of the solution
method.

Although tightness and compactness are preferable properties, on many occasions
a tighter formulation results in a less compact one, and vice versa. For instance,
well-known public transportation problems, such as the vehicle routing problem,

can become more tight by adding many constraints that will reduce considerably the search space resulting in a lower distance between the discrete solution and the solution of its continuous relaxation. This addition of constraints, however, will increase the size of the problem making it less compact and thus increasing the difficulty of solving its continuous relaxation. In reverse, more compact formulations result in weak lower bounds and it might take way more time to find a discrete solution which is equivalent to the solution of its continuous relaxation. To summarize, creating computationally efficient models is not a trivial task since very weak (not tight) formulations will need more time to close the distance between the solutions of the discrete and the continuously relaxed problems, whereas tight formulations are typically less compact resulting in increased computational times when solving each continuous relaxation [14].

1.4.2 Combinatorial Optimization

Combinatorial optimization problems are very common in public transport. Combinatorial optimization problems appear in public transport scheduling, routing, network design, service pattern design and many more. In practice, the classic topics in combinatorial optimization include shortest path problems, spanning trees, matchings and matroids. A combinatorial optimization problem can be typically represented using graph theory.

Classic combinatorial optimization problems include the shortest path problem (SPP) [15], the Chinese postman problem (CPP) [16], the traveling salesman problem (TSP) and the vehicle routing problem (VRP). To provide an example, let us focus on the TSP which has been used to solve the school bus routing problem [17]. The objective of the TSP is to find the shortest tour of a salesman who visits different cities. The traveling salesman wishes to visit the cities and return to the starting point after visiting each city exactly once.

Considering the mathematical modeling part, we have a set of cities $N = \{1, 2, ..., n\}$ and the cost of traveling between two cities $i \in N$ and $j \in N$ is known, i.e., c_{ij}, where c_{ij} is non-negative. After defining set N and parameters $\mathbf{C} = \{c_{ij}\}$, we can represent the problem using a graph (Fig. 1.8 shows the case with three cities that need to be visited exactly once). The TSP in Fig. 1.8 is asymmetric because we allow $c_{ij} \neq c_{ji}$.

Concerning the variables of this combinatorial optimization problem, every link between two cities i, j can be either traversed or not. For this, binary variables $\mathbf{X} = \{x_{ij}\}$, where $x_{ij} \in \{0, 1\}$ can be used to indicate whether the link i, j is traversed by the traveling salesman ($x_{ij} = 1$) or not ($x_{ij} = 0$). The objective function of the TSP can be formulated as:

$$\min \sum_{i \in N} \sum_{j \in N \setminus \{i\}} c_{ij} x_{ij} \tag{1.44}$$

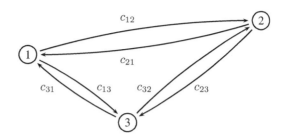

Fig. 1.8 Asymmetric TSP with three cities

and it seeks to find the links i, j that will be part of the traveling salesman's tour such that the total traveling cost is minimized.

Concerning the constraints, each city should have exactly one successor in the shortest tour. That is,

$$\sum_{j \in N \setminus \{i\}} x_{ij} = 1 \quad \forall i \in N \tag{1.45}$$

This means that from a city i there should only be a city $j \neq i$ which is the successor of i in the tour. Similarly, any city j should have only one predecessor in the tour. That is,

$$\sum_{i \in N \setminus \{j\}} x_{ij} = 1 \quad \forall j \in N \tag{1.46}$$

Note though that the constraints of having exactly one successor and exactly one predecessor for every city do not ensure that we can build a complete tour that ends at the starting location of the traveling salesman. For instance, consider the example of the symmetric TSP with four cities presented in Fig. 1.9. This TSP is symmetric because the cost of traveling from i to j is equivalent to the cost of traveling from j to i ($c_{ij} = c_{ji} \ \forall i \in N, j \in N$) resulting in a symmetric $N \times N$ matrix $\mathbf{C} = \{c_{ij}\}$. A possible solution that meets the aforementioned constraints is $x_{12} = x_{21} = x_{34} = x_{43} = 1$. However, this solution will not create a tour that starts from city 1 and returns back to city 1. Instead, it forms two sub-tours as presented in Fig. 1.9.

In Fig. 1.9 cities 1, 2, 3 and 4 have exactly one predecessor and successor city, but we have two sub-tours (1-2-1) and (3-4-3) that do not result in a tour which serves all cities and returns back to the 1st one. For this reason, additional constraints are needed to eliminate sub-tours. These constraints are called *sub-tour elimination constraints* and they grow with the number of cities in the network resulting in a very complex problem that has a *non-compact* formulation according to the following definition.

Definition 1.20 A *compact* formulation is a mathematical programming formulation featuring a polynomial increase in the number of constraints when the problem's size increases, whereas a *non-compact* formulation is a formulation featuring an exponential increase in the number of constraints when the problem's size increases.

Fig. 1.9 Solution of the TSP that satisfies constraints (1.45)–(1.46) but does not result in a complete tour because there is no tour that returns back to city 1

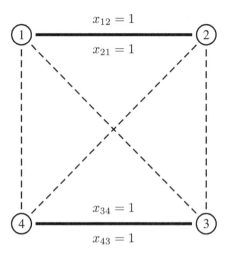

There are different approaches to model the sub-tour elimination constraints in a way that allows us to attain a compact formulation. A classic one is the Miller-Tucker-Zemlin (MTZ) formulation introduced in 1960 [18].

Following the MTZ formulation of the sub-tour elimination constraints, let u_i be an integer variable indicating the position of city i in the tour, where $i \in \{2, 3, ..., n\}$. Note that city 1 is the starting/ending point in the tour and i does not start from 1. To avoid forming a sub-tour, we need to impose that if $x_{ij} = 1$ for $i \neq j \neq 1$, then the position of city j is greater than the position of i, $u_j > u_i$. This will not allow visiting i after j if $x_{ij} = 1$. That is, we seek to force the logical expression $x_{ij} = 1 \Rightarrow u_j > u_i$ for $i \neq j \neq 1$. This logical expression is translated into the following inequality constraints:

$$(n - 1)x_{ij} + u_i - u_j \leq n - 2 \quad \forall i, j \in \{2, 3, ..., n\} \mid i \neq j \qquad (1.47)$$

Indeed if $x_{ij} = 1$, then $n - 1 + u_i - u_j \leq n - 2 \Rightarrow u_i + 1 \leq u_j$ forcing u_j to take a value greater than u_i since j is the next city of i in the tour. That is, this inequality constraint is equivalent to the logical expression $x_{ij} = 1 \Rightarrow u_j > u_i$ for $i \neq j \neq 1$. Note that if $x_{ij} = 0$ then it is not necessary that $u_j < u_i$ because the inequality constraint will result in $u_i - u_j \leq n - 2$ which is always true if we bound the integer values of any u_i to take values within the range $[1, n - 1]$:

$$1 \leq u_i \leq n - 1 \quad \forall i \in \{2, 3, ..., n\} \qquad (1.48)$$

Even in the extreme case where $u_1 = n - 1$ and $u_j = 1$ we still have that $u_i - u_j \leq n - 1$. To summarize, the TSP with the MTZ sub-tour elimination constraints is formulated as follows.

Traveling Salesman Problem with MTZ sub-tour elimination constraints

$$\min \sum_{i \in N} \sum_{j \in N \setminus \{i\}} c_{ij} x_{ij} \tag{1.49}$$

$$\text{subject to: } \sum_{j \in N \setminus \{i\}} x_{ij} = 1 \qquad \forall i \in N \tag{1.50}$$

$$\sum_{i \in N \setminus \{j\}} x_{ij} = 1 \qquad \forall j \in N \tag{1.51}$$

$$(n-1)x_{ij} + u_i - u_j \leq n - 2 \quad \forall i, j \in \{2, 3, ..., n\} \mid i \neq j \tag{1.52}$$

$$1 \leq u_i \leq n - 1 \qquad \forall i \in \{2, 3, ..., n\} \tag{1.53}$$

$$u_i \in \mathbb{Z} \qquad \forall i \in \{2, 3, ..., n\} \tag{1.54}$$

$$x_{ij} \in \{0, 1\} \qquad \forall i \in N, j \in N \tag{1.55}$$

The combinatorial optimization problem formulated in Eqs. (1.49)–(1.55) can be characterized as an integer programming problem because the variables can receive values from integer sets.

1.4.3 Mixed-Integer Problems

As already discussed, mixed-integer optimization problems (MIPs) are problems where we have a nonempty subset of integer variables and a subset of real-valued (continuous) variables. Unlike pure integer optimization problems (IPs), MIPs do not require all variables to be integer.

In some cases, MIPs can emerge from continuous problems with optional inequality constraints or disjunctive constraints. These standard cases are described below.

1.4.3.1 Optional Inequality Constraints in MIPs

Let us consider the case where one of the constraints of the problem is optional. Let, for example, a constraint $c(\mathbf{x})$ that needs to be greater than or equal to $b \in \mathbb{R}$ in some cases (when, for example, $z = 1$) and it does not need to be satisfied in some other cases (when $z = 0$). This can be expressed as:

$$\begin{cases} c(\mathbf{x}) \geq b & \text{for } z = 1 \\ c(\mathbf{x}) \geq b \vee c(\mathbf{x}) \leq b & \text{for } z = 0 \end{cases} \tag{1.56}$$

Note that when $z = 0$ the constraint is inactive because $c(\mathbf{x}) \geq b \vee c(\mathbf{x}) \leq b$ can be satisfied for any value of $\mathbf{x}$ since we request $c(\mathbf{x}) \geq b$ or $c(\mathbf{x}) \leq b$. Now, if the constraint function $c(\mathbf{x})$ has a lower and an upper bound, l_c and u_c, respectively, we can replace the optional constraint with the following inequality constraint that uses the binary variable z:

$$c(\mathbf{x}) \geq l_c + (b - l_c)z \tag{1.57}$$

Note that if $z = 0$ the constraint should hold for any $\mathbf{x}$ because it should be inactive. Indeed, for $z_c = 0$ we have $c(\mathbf{x}) \geq l_c$ which is always true since l_c is the lower bound of function $c(\mathbf{x})$. If, however, $z = 1$ then $c(\mathbf{x}) \geq b$. This is indeed the case because the right-hand-side (rhs) of Eq. (1.57) becomes $l_c + (b - l_c) = b$.

The optional constraint could have also been of the form:

$$\begin{cases} c(\mathbf{x}) \leq b & \text{for } z = 1 \\ c(\mathbf{x}) \leq b \vee c(\mathbf{x}) \leq b & \text{for } z = 0 \end{cases} \tag{1.58}$$

In that case, we can replace this constraint by introducing the binary variable z as part of the following linear inequality constraint:

$$c(\mathbf{x}) \leq u_c + (b - u_c)z \tag{1.59}$$

Indeed, when $z = 0$ we have $c(\mathbf{x}) \leq u_c$ which is always true because u_c is the upper bound of $c(\mathbf{x})$ and when $z = 1$ we have $c(\mathbf{x}) \leq u_c + b - u_c$ which is equivalent to $c(\mathbf{x}) \leq u_c + (b - u_c)z$.

1.4.3.2 Disjunctive Constraints in MIPs

Another example is the case of disjunctive constraints where one of them must be verified, but not necessarily both. This can be expressed as:

$$c_1(\mathbf{x}) \geq b_1 \quad \vee \quad c_2(\mathbf{x}) \geq b_2 \tag{1.60}$$

where c_1, c_2 are constraint functions and $b_1, b_2 \in \mathbb{R}$ fixed-value scalars. Note that if one of the disjunctive constraints is satisfied, Eq. (1.60) is satisfied as well. Let again l_{c1}, l_{c2} and u_{c1}, u_{c2} be the lower and upper bounds of functions c_1 and c_2, respectively. Then, we can introduce binary variable $z \in \{0, 1\}$ and replace the disjunctive constraints by:

$$c_1(\mathbf{x}) \geq l_{c1} + (b_1 - l_{c1})z \tag{1.61}$$

$$c_2(\mathbf{x}) \geq l_{c2} + (b_2 - l_{c2})(1 - z) \tag{1.62}$$

$$z \in \{0, 1\} \tag{1.63}$$

From the above system of equations, if $z = 0$ we only request that $c_2(\mathbf{x}) \geq b_2$ since $c_1(\mathbf{x})$ is always greater than or equal to l_{c1}. For the same reason, when $z = 1$ we only request that $c_1(\mathbf{x}) \geq b$. That is to say, $z \in \{0, 1\}$ will require the satisfaction of at least one of the two constraints.

The same idea can be applied when $c_1(\mathbf{x}) \leq b_1 \vee c_2(\mathbf{x}) \leq b_2$ resulting in:

$$c_1(\mathbf{x}) \leq u_{c1} + (b_1 - u_{c1})z \tag{1.64}$$

$$c_2(\mathbf{x}) \leq u_{c2} + (b_2 - u_{c2})(1 - z) \tag{1.65}$$

$$z \in \{0, 1\} \tag{1.66}$$

1.4.3.3   MIP Example: The Joint Stop-Skipping and Vehicle Holding Problem

A MIP example is the joint problem of stop-skipping and vehicle holding. Consider that in the one-headway-based vehicle holding problem described in Eq. (1.41)–(1.43) we are also allowed to let trip i skip the station if the vehicle arrives there with a delay. Trip i was ready to depart at time t after completing its passenger boardings and alightings at the station. Trip i will actually depart at time $d_i \geq t$ after deciding about the potential holding time at the station. Let now $x_i \in \mathbb{R}_{\geq 0}$ be a continuous variable that indicates the holding time decision. That is,

$$d_i = t + x_i \tag{1.67}$$

If we consider that trip i arrives at the station at time $t_0 < t$, it can depart immediately without making a stop there if we decide to skip the station. Deciding to skip the station can be formulated with the use of a binary variable z_i which can take the value of 1 if the station is served and the value of 0 if it is skipped. Clearly, if $z_i = 0$, then $d_i = t_0$ because the trip will not stop at the station. If, however, $z_i = 1$, then $d_i = t + x_i$. These logical conditions can be expressed as:

$$d_i = z_i(t + x_i) + (1 - z_i)t_0 \tag{1.68}$$

Note that if $z_i = 1$, then $d_i = t + x_i$. In addition, if $z_i = 0$, then $d_i = t_0$. That is, Eq. (1.68) incorporates the logical conditions. If the objective of the joint stop-skipping and vehicle holding problem is still to reduce the difference between two successive trip arrivals and the scheduled time headway, h, the joint decision about holding trip i at the station and skipping that station can be expressed in the following mixed-integer linear program (MILP).

Joint stop-skipping and one-headway-based vehicle holding problem

$$\min \ |d_{i+1} + h - d_i| \qquad\qquad (1.69)$$

$$\text{subject to: } d_i = z_i(t + x_i) + (1 - z_i)t_0 \qquad\qquad (1.70)$$

$$x_i \in \mathbb{R}_{\geq 0} \qquad\qquad (1.71)$$

$$z_i \in \{0, 1\} \qquad\qquad (1.72)$$

The aforementioned problem is a MIP because the x_i variable can take real values in the set $[0, +\infty)$ that has infinitely many elements, whilst the z_i variable can take values from the discrete set $\{0, 1\}$ that has a finite number of elements (0 and 1). To provide an example, if trip i is behind schedule (e.g., $t_0 = 3$ min, $t = 4$ min, $d_{i+1} = 0$ min, and $h = 2$ min), then the solution of the optimization problem is $(x_i, z_i) = (0, 0)$ resulting in $d_i = t_0 = 2$ min.

Closing, it is worth noting that the aforementioned optimization problem would have been a binary programming problem if vehicle holding was not considered. In that case, $d_i = z_i t + (1 - z_i)t_0$ since we exclude the holding variable x_i from the mathematical program.

1.5   Global and Local Optimum

1.5.1   Local Optimum

A locally optimal solution to an optimization problem is a solution that performs better than or equal to all other feasible solutions in a specific region (*neighborhood*) of the feasible region. If this solution performs better than all other feasible solutions in the specific region (neighborhood), then it is called a *strict* local optimum.

Formally,

> **Local Optimum**

Considering a minimization problem, a solution $\mathbf{x}^* = [x_1^*, x_2^*, ...]^\mathsf{T}$ is a local optimum in the neighborhood N of $\mathbf{x}^*$ if $f(\mathbf{x}^*) \leq f(\mathbf{x}) \ \ \forall \mathbf{x} \in N$, where set N contains all feasible solutions in the neighborhood. In a maximization problem, $\mathbf{x}^*$ is a local optimum if $f(\mathbf{x}^*) \geq f(\mathbf{x}) \ \ \forall \mathbf{x} \in N$.

In the aforementioned definition f is the objective function of the optimization problem. This local optimum is also called *weak* local optimum to distinguish it from the *strict* local optimum that is described below.

Fig. 1.10 Example of strict local minima in neighborhoods [0,3] and [3,7]

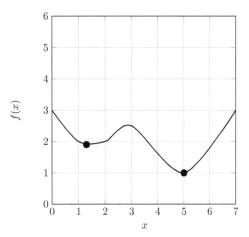

Considering a minimization problem, a solution $\mathbf{x}^*$ is a strict local optimum in the neighborhood N of $\mathbf{x}^*$ if $f(\mathbf{x}^*) < f(\mathbf{x})$ $\forall \mathbf{x} \in N$ with $\mathbf{x} \neq \mathbf{x}^*$, where set N contains all feasible solutions in the neighborhood. In a maximization problem, $\mathbf{x}^*$ is a strict local optimum if $f(\mathbf{x}^*) > f(\mathbf{x})$ $\forall \mathbf{x} \in N$ with $\mathbf{x} \neq \mathbf{x}^*$.

Consider, for example, the function presented in Fig. 1.10 that needs to be minimized. In the closed interval [0,3], the strict local optimum is $x^* = 1.3$ with $f(x^*) = 1.9 < f(x)$ $\forall x \in [0, 3]$ for $x^* \neq x$. Similarly, in the neighborhood [3,7] the strict local minimum is $x^* = 5$ with $f(x^*) = 1 < f(x)$ $\forall x \in [3, 7]$ for $x^* \neq x$.

In contrast, the objective function presented in Fig. 1.11 has no strict local optima in the neighborhood [0,7]. If we seek to minimize this function, any point $x^* \in [2, 3]$ is a weak local optimum since $f(x^*) \leq f(x)$ $\forall x \in [0, 7]$.

1.5.2 Global Optimum

A solution is globally optimal if the resulting objective function value is smaller than or equal to the objective function values of all other feasible solutions in case of solving a minimization problem (for a maximization problem, it should be greater than or equal to all other feasible solutions). Formally,

Fig. 1.11 Example of weak
local optima $x^* \in [2, 3]$

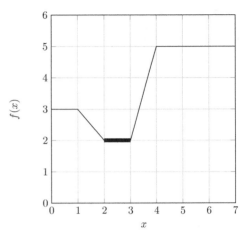

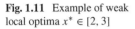

> **Global Optimum**

Considering a minimization problem, a solution $\mathbf{x}^*$ is a global optimum if $f(\mathbf{x}^*) \leq f(\mathbf{x}) \; \forall \mathbf{x} \in \mathcal{F}$, where $\mathcal{F}$ is the feasible region of the problem. In a maximization problem, $\mathbf{x}^*$ is a global optimum if $f(\mathbf{x}^*) \geq f(\mathbf{x}) \; \forall \mathbf{x} \in \mathcal{F}$.

A globally optimal solution might not be unique. For instance, there might be several globally optimal solutions with similar performance in the feasible region $\mathcal{F}$ of the optimization problem. A globally optimal solution is a *unique* global optimum if it fulfills the following condition.

> **Unique Global Optimum**

Considering a minimization problem, a solution $\mathbf{x}^*$ is a *unique* global optimum if $f(\mathbf{x}^*) < f(\mathbf{x}) \; \forall \mathbf{x} \in \mathcal{F}$ with $\mathbf{x} \neq \mathbf{x}^*$, where $\mathcal{F}$ is the problem's feasible region.

Consider, for example, a minimization problem with the objective function in Fig. 1.12 and constraints $x \geq 0$, $x \leq 10$, resulting in the feasible region:

$$\mathcal{F} := \{x \in \mathbb{R} : 0 \leq x \leq 10\}$$

In this example, $x^* = 5$ is a *unique* globally optimal solution in $\mathcal{F}$, whereas $x_1 = 1.3$ and $x_2 = 9$ are strict locally optimal solutions in their respective neighborhoods.

An optimization problem can also have multiple globally optimal solutions within its feasible region. Consider, for instance, a minimization problem with the objective function in Fig. 1.13. In this problem, we have two globally optimal solutions, $x_1 = 5$ and $x_2 = 9$. Naturally, a decision maker would be equally satisfied with any of the

Fig. 1.12 Example of *unique* globally optimal solution $x^* = 5$ in $[0,10]$

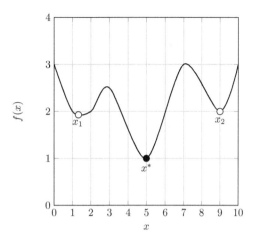

Fig. 1.13 Example of multiple globally optimal solutions: $x_1 = 5$ and $x_2 = 9$ in $\mathcal{F} := \{x \in \mathbb{R} : 0 \le x \le 10\}$

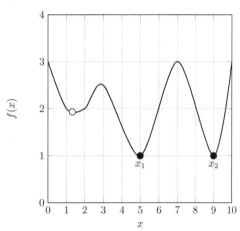

two solutions since $f(x_1) = f(x_2) \le f(x) \; \forall x \in \mathcal{F}$, where $\mathcal{F}$ is the feasible region $[0, 10]$.

1.5.3 Convexity

When solving an optimization problem we might have numerous locally optimal solutions. The objective is to find one of the globally optimal solutions from the set of locally optimal ones. If the optimization problem has several locally optimal solutions, finding all of them and selecting a globally optimal solution becomes a very complicated task.

Convexity plays an important role in finding a globally optimal solution when solving a minimization problem because if the mathematical program is convex, then any local minimizer will be a global minimizer. This is a key property of a mathematical optimization problem because it makes it easy to solve. It is well-argued that the distinction between convex and non-convex mathematical programs is even more important than the distinction between linear and non-linear programs because convexity affects significantly the ability to solve an optimization problem to global optimality [19].

! Convex Minimization Problem

For a minimization problem to be convex, its *feasible* region $\mathcal{F}$ should be a convex set and its *objective* function $f : \mathcal{F} \to \mathbb{R}$ should be a convex function.

If the above conditions hold, any local minimizer $\mathbf{x}^*$ is also a global minimizer in the feasible region $\mathcal{F}$. We note that if our optimization problem is a maximization one, any local maximizer $\mathbf{x}^*$ is also a global maximizer in the feasible region $\mathcal{F}$ if the objective function is concave and the feasible region is convex.

In addition, we call a locally optimal solution x^* a *unique* global minimizer if the following conditions are fulfilled.

! Unique Global Minimizer of a Minimization Problem

If the *feasible* region $\mathcal{F}$ of a minimization problem is a convex set and its *objective* function $f : \mathcal{F} \to \mathbb{R}$ is a *strictly* convex function, a local minimizer $\mathbf{x}^*$ is the unique global minimizer of this minimization problem.

Examples of a convex and a non-convex mathematical program are provided in Fig. 1.14. The convex mathematical program has a convex feasible region and a strictly convex objective function resulting in a local minimizer which is also a global minimizer. In contrast, the non-convex mathematical program has a convex feasible region and a non-convex objective function resulting in multiple local minimizers. Finding a global minimizer in the latter case is not a trivial task for mathematical optimization solvers.

To determine the convexity of a mathematical program, we need to establish the conditions under which a set (feasible region) and a function (objective function) are convex. Before doing that, it is important to stress again that these conditions hold for minimization problems. If we seek to solve a maximization problem, then a local maximizer is a global maximizer if the feasible region is a convex set and the objective function is a *concave* function. Similarly, a local minimizer is the *unique* global maximizer of a maximization problem if the feasible region is a convex set and the objective function is a *strictly concave* function, where function f is concave if function $-f$ is convex.

Convex Mathematical Program:

$$\min x^2$$
$$\text{s.t. } -5 \le x \le 5$$

Non-convex Mathematical Program:

$$\min 0.01x + \sin x$$
$$\text{s.t. } -5 \le x \le 38$$

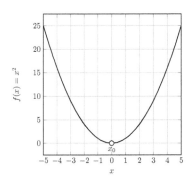
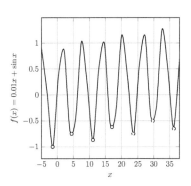

Fig. 1.14 Examples of convex and non-convex mathematical programs

1.5.3.1 Convex Sets

Before defining the convex set, we define first the *line* and the *affine* sets.

Definition 1.21 *Line* in the Euclidean space $\mathbb{R}^n$ is a set of points $\mathcal{L} \subseteq \mathbb{R}^n$ that connect two points $\mathbf{x}_1 \ne \mathbf{x}_2 \in \mathbb{R}^n$:

$$\mathcal{L} = \{\mathbf{y} \in \mathbb{R}^n \mid \mathbf{y} = \beta\mathbf{x}_1 + (1 - \beta)\mathbf{x}_2\}$$

where β is a scalar in $\mathbb{R}$.

Based on the above definition, the line segment between points $\mathbf{x}_1$ and $\mathbf{x}_2$ is:

$$\mathcal{L} = \{\mathbf{y} \in \mathbb{R}^n \mid \mathbf{y} = \beta\mathbf{x}_1 + (1 - \beta)\mathbf{x}_2\} \quad 0 \le \beta \le 1$$

Intuitively, for parameter value $\beta = 0$ we have the point of the line $\mathbf{y} = \mathbf{x}_2$ and for $\beta = 1$ we have the point of the line $\mathbf{y} = \mathbf{x}_1$. All other line points for $0 < \beta < 1$ will lie in the line segment between $\mathbf{x}_1$ and $\mathbf{x}_2$. Consider, for example, two points in the 3-dimensional space: $\mathbf{x}_1 = [3, 5, 7]^\mathsf{T}$ and $\mathbf{x}_2 = [7, 4, 1]^\mathsf{T}$. The line segment between these two points is a set of points $\mathbf{y} \in \mathbb{R}^3$ such that $\mathbf{y} = \beta\mathbf{x}_1 + (1 - \beta)\mathbf{x}_2$ where $0 \le \beta \le 1$ (see Fig. 1.15).

If we have a number of points $\mathbf{x}_1, \mathbf{x}_2, ..., \mathbf{x}_k \in \mathbb{R}^n$, a *linear combination* of these points is:

$$\beta_1\mathbf{x}_1 + \beta_2\mathbf{x}_2 + \cdots + \beta_k\mathbf{x}_k, \text{ where } \beta_1, ..., \beta_k \in \mathbb{R}$$

Let us now proceed with the definition of an affine set.

Fig. 1.15 Line segment
$\mathcal{L} = \{\mathbf{y} \in \mathbb{R}^3 \mid \mathbf{y} = \beta\mathbf{x}_1 + (1 - \beta)\mathbf{x}_2\}$ with
$0 \leq \beta \leq 1$ between points
$\mathbf{x}_1 = [3, 5, 7]^\mathsf{T}$ and
$\mathbf{x}_2 = [7, 4, 1]^\mathsf{T}$

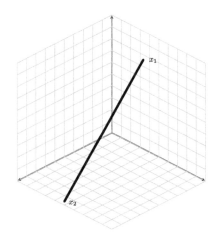

Definition 1.22 *Affine set* is a set $\mathcal{A} \subseteq \mathbb{R}^n$ if for any two distinct points $\mathbf{x}_1, \mathbf{x}_2$ in $\mathcal{A}$, the line passing through these points lies in the set $\mathcal{A}$.

From the definition of affine sets follows the theorem:

Theorem 1.1 *The solution set of linear equations, $\mathcal{A} = \{\mathbf{x} \in \mathbb{R}^n \mid A\mathbf{x} = \mathbf{b}\}$, where $A \in \mathbb{R}^{m \times n}$ and $\mathbf{b}$ is a column vector $\mathbf{b} \in \mathbb{R}^m$, is an affine set.*

Proof If set $\mathcal{A}$ is affine, then any two points $\mathbf{x}_1, \mathbf{x}_2 \in \mathcal{A}$ should be connected by a line that lies in the set $\mathcal{A}$. Let two points $\mathbf{x}_1, \mathbf{x}_2 \in \mathcal{A}$. Then, $A\mathbf{x}_1 = \mathbf{b}$ and $A\mathbf{x}_2 = \mathbf{b}$. For any $\beta \in \mathbb{R}$ we have:

$$\beta A\mathbf{x}_1 + (1 - \beta)A\mathbf{x}_2 = \beta\mathbf{b} + (1 - \beta)\mathbf{b} = \mathbf{b}$$

Thus, any two points of set $\mathcal{A}$ are connected by a line that lies in the set $\mathcal{A}$ proving that $\mathcal{A}$ is affine. □

If we have a number of points $\mathbf{x}_1, \mathbf{x}_2, ..., \mathbf{x}_k \in \mathbb{R}^n$, the *affine combination* of these points is:

$$\beta_1\mathbf{x}_1 + \beta_2\mathbf{x}_2 + \cdots + \beta_k\mathbf{x}_k, \text{ where } \beta_1, ..., \beta_k \in \mathbb{R}, \sum_{j=1}^{k} \beta_j = 1$$

The difference between the affine and the linear combination is that in the former one we require that $\sum_{j=1}^{k} \beta_j = 1$. Consider, for instance, the case where $k = 2$ and we have points $\mathbf{x}_1, \mathbf{x}_2 \in \mathbb{R}^3$ in the 3-dimensional space. Then, the linear combination is $\beta_1\mathbf{x}_1 + \beta_2\mathbf{x}_2$ and the affine combination is $\beta_1\mathbf{x}_1 + (1 - \beta_1)\mathbf{x}_2$. It becomes evident that the linear combination of points $\mathbf{x}_1$ and $\mathbf{x}_2$ is a plane and the affine combination is a line that passes through $\mathbf{x}_1$ and $\mathbf{x}_2$. The set of all affine combinations of points

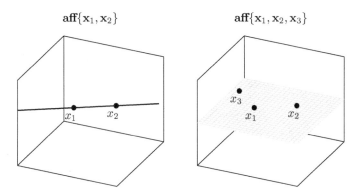

Fig. 1.16 Affine hulls for sets $\{\mathbf{x}_1, \mathbf{x}_2\}$ and $\{\mathbf{x}_1, \mathbf{x}_2, \mathbf{x}_3\}$ containing two and three points, respectively. The affine hull of a set with two points is a line passing through them and for a set with three noncolinear points a plane that contains them

$\mathbf{x}_1, ..., \mathbf{x}_k$ is called the *affine hull* or the *affine span* of these points. The affine hull is denoted as:

$$\mathbf{aff}\{\mathbf{x}_1, ..., \mathbf{x}_k\} := \left\{ \sum_{j=1}^{k} \beta_j \mathbf{x}_j \; : \; \begin{array}{c} \beta_1, ..., \beta_k \in \mathbb{R} \\ \sum_{j=1}^{k} \beta_j = 1 \end{array} \right\}$$

The affine hull is the smallest affine set containing points $\mathbf{x}_1, ..., \mathbf{x}_k$. The affine hull of two different points is the line passing through them. The affine hull of three different points that are not colinear is a plane that includes these three points. Figure 1.16 presents the affine hull of a set of two points $\mathbf{x}_1 = [3, 5, 0]^\mathsf{T}$, $\mathbf{x}_2 = [4, 7, 0]^\mathsf{T}$ and the affine hull of a set of three points $\mathbf{x}_1 = [3, 5, 0]^\mathsf{T}$, $\mathbf{x}_2 = [4, 7, 0]^\mathsf{T}$, $\mathbf{x}_3 = [2.2, 7.5, 0]^\mathsf{T}$.

We now define the convex set:

> **Convex Set**

A convex set is a set $\mathcal{F} \subseteq \mathbb{R}^n$ if for any two points $\mathbf{x}, \mathbf{y} \in \mathcal{F}$ and any $0 \leq \beta \leq 1$ it holds:

$$\beta \mathbf{x} + (1 - \beta)\mathbf{y} \in \mathcal{F}$$

The expression $\beta \mathbf{x} + (1 - \beta)\mathbf{y} \in \mathcal{F}$ implies that the line segment connecting any two points $\mathbf{x}, \mathbf{y} \in \mathcal{F}$ should lie entirely inside $\mathcal{F}$ in order for set $\mathcal{F}$ to be convex. For instance, Fig. 1.17 presents examples of convex and non-convex sets. In the latter, there is at least a pair of points $\mathbf{x}, \mathbf{y}$ for which the line segment that connects them does not lie entirely within the set $\mathcal{F}$.

If we have a number of points $\mathbf{x}_1, \mathbf{x}_2, ..., \mathbf{x}_k \in \mathbb{R}^n$, the *convex combination* of these points is:

Fig. 1.17 Example of convex and non-convex sets

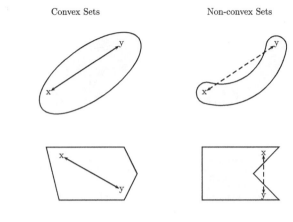

$$\beta_1 x_1 + \beta_2 x_2 + \cdots + \beta_k x_k, \text{ where } \quad \begin{array}{l} \beta_1, ..., \beta_k \in \mathbb{R} \\ \sum_{j=1}^{k} \beta_j = 1 \\ \beta_1, ..., \beta_k \geq 0 \end{array}$$

Note that the convex combination differs from the affine combination because the former also requests that $\beta_j \geq 0 \; \forall j \in \{1, ..., k\}$. It becomes clear that an affine set is a convex set, whereas the opposite is not necessarily true. The set of all convex combinations of points $x_1, ..., x_k$ is called the *convex hull* or the *convex span* of these points. The convex hull is denoted as:

$$\mathbf{conv}\{x_1, ..., x_k\} := \left\{ \sum_{j=1}^{k} \beta_j x_j \; : \; \begin{array}{l} \beta_1, ..., \beta_k \in \mathbb{R} \\ \sum_{j=1}^{k} \beta_j = 1 \\ \beta_j \geq 0 \; \forall j \in \{1, ..., k\} \end{array} \right\}$$

To examine the difference between an affine and a convex hull, let us consider the convex hulls of two points $x_1 = [3, 5, 0]^\mathsf{T}$, $x_2 = [4, 7, 0]^\mathsf{T}$ and four points $x_1 = [3, 5, 0]^\mathsf{T}$, $x_2 = [4, 9, 0]^\mathsf{T}$, $x_3 = [2.1, 10, 0]^\mathsf{T}$, $x_4 = [2.7, 8, 0]^\mathsf{T}$ in the 3-dimensional space. Figure 1.18 presents these convex hulls.

Practitioner's Corner
Convex Hull in Python 3

To determine the convex hull of multiple points in the Euclidean plane, one can use the Scipy library in Python 3. For instance, an example of determining the convex hull of six points in the 2-dimensional space can be found below (Fig. 1.19).

```
import numpy as np
from scipy.spatial import ConvexHull
import matplotlib.pyplot as plt
points = np.array([[1, 4],[2, 4],[3, 1],[2, 2],[2.5,
    3],[4, 5]])
hull = ConvexHull(points)
```

```
hull_points = hull.simplices
plt.scatter(points[:,0], points[:,1])
for simplex in hull_points:
  plt.plot(points[simplex,0], points[simplex,1], 'k-')
plt.show()
```

This results in the convex hull plot presented in Fig. 1.19.

It is important to note that the intersection of k convex sets $\bigcap_{i=1}^{k} \mathcal{F}_i$ is always convex. This property can be used to examine whether the feasible region of a mathematical program that results from the intersection of many convex sets is convex or not. Common convex sets are the *halfspaces* and *hyperplanes*.

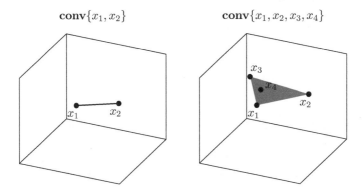

Fig. 1.18 Convex hulls for sets $\{x_1, x_2\}$ and $\{x_1, x_2, x_3, x_4\}$ containing two and four points in the 3-dimensional space, respectively. The convex hull of a set with two points is a line segment between the two points

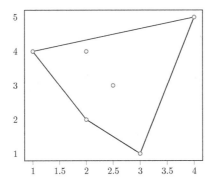

Fig. 1.19 Convex hull plot of points $[1, 4]^T$, $[2, 4]^T$, $[3, 1]^T$, $[2, 2]^T$, $[2.5, 3]^T$, $[4, 5]^T$

Definition 1.23 *Hyperplane* is a set of the form:

$$\{\mathbf{x} \in \mathbb{R}^n : \mathbf{c}^\mathsf{T}\mathbf{x} = b\}$$

where $\mathbf{c}$ and $\mathbf{x}$ are column vectors of n dimensions, $\mathbf{c}$ is additionally a nonzero vector, and b a scalar, $b \in \mathbb{R}$.

A hyperplane in the 2-dimensional space is a line. Consider, for example, the hyperplane:

$$\{\mathbf{x} \in \mathbb{R}^2 : [3, 5]\begin{bmatrix} x_1 \\ x_2 \end{bmatrix} = 7\}$$

in the 2-dimensional space. This hyperplane is the line in Fig. 1.20 that splits the 2-dimensional space into two halfspaces.

In a 3-dimensional space, a hyperplane is a plane that splits the 3-dimensional place into two halfspaces. Consider, for example, the hyperplane:

$$\{\mathbf{x} \in \mathbb{R}^3 : [3, 5, 7]\begin{bmatrix} x_1 \\ x_2 \\ x_3 \end{bmatrix} = 7\}$$

in the 3-dimensional space. An example of a plane in the 3-dimensional space is presented in Fig. 1.21.

The hyperplane in the n-dimensional space, where $n > 3$, is the space $\mathbf{x} \in \mathbb{R}^n : \mathbf{c}^\mathsf{T}\mathbf{x} = b$ which divides the n-dimensional space into two open halfspaces:

$$c_1 x_1 + c_2 x_2 + \cdots + c_n x_n < b$$

and

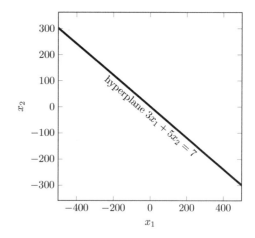

Fig. 1.20 Hyperplane $3x_1 + 5x_2 = 7$ of the 2-dimensional space that divides it into two halfspaces. Note that a hyperplane in the 2-dimensional space is a line

Fig. 1.21 Example of a plane in the 3-dimensional space. This hyperplane divides the 3-dimensional space into two halfspaces

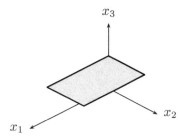

$$c_1 x_1 + c_2 x_2 + \cdots + c_n x_n > b$$

Each instance of a hyperplane is an $n-1$ dimensional affine sub-space of the n-dimensional space and a hyperplane can be seen as the boundary of two halfspaces [20]. As affine sets, hyperplanes are also convex sets. It is important to note here that a *closed* halfspace is defined as:

$$\{\mathbf{x} \in \mathbb{R}^n \mid \mathbf{c}^\mathsf{T}\mathbf{x} \leq b\}$$

and an *open* halfspace as:

$$\{\mathbf{x} \in \mathbb{R}^n \mid \mathbf{c}^\mathsf{T}\mathbf{x} < b\}$$

From the above definition of halfspaces, it becomes clear that a closed halfspace is a convex set since any two points $\mathbf{x}, \mathbf{y}$ in the halfspace can be connected with a line which lies entirely inside the halfspace.

Let us now define the polyhedron:

> **Polyhedron**

A polyhedron is the intersection of a finite number of (closed) halfspaces. That is, a polyhedron is a subset $\mathcal{P} \subseteq \mathbb{R}^n$ of the form $\mathcal{P} := \{\mathbf{x} \in \mathbb{R}^n \mid \mathbf{A}\mathbf{x} \leq \mathbf{b}\}$ where $\mathbf{A}$ is an $m \times n$-dimensional matrix and $\mathbf{b}$ a column vector $\mathbf{b} \in \mathbb{R}^m$.

Because the intersection of convex sets is convex, the intersection of halfspaces is a convex set. That is, a *polyhedron* is a convex set. This is also formally presented in the following proof.

Theorem 1.2 *Any polyhedron is a convex set.*

Proof Formally, a polyhedron is the set $\mathcal{P} := \{\mathbf{x} \in \mathbb{R}^n \mid \mathbf{A}\mathbf{x} \leq \mathbf{b}\}$. If $\mathbf{x}, \mathbf{y} \in \mathcal{P}$, then $\mathbf{A}\mathbf{x} \leq \mathbf{b}$ and $\mathbf{A}\mathbf{y} \leq \mathbf{b}$. $\mathcal{P}$ is not a convex set if there exist two points $\mathbf{x}, \mathbf{y} \in \mathcal{P}$ for some $0 \leq \beta \leq 1$ such that:

$$\beta\mathbf{x} + (1-\beta)\mathbf{y} \notin \mathcal{P}$$

Fig. 1.22 Example of polyhedron in the 2-dimensional space. Note that this polyhedron is bounded and thus it is also a polytope. If we remove constraint $x_2 \leq 20$ the polyhedron becomes unbounded and it is not a polytope

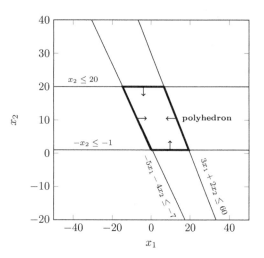

That is, we need to prove that $\mathbf{A}(\beta \mathbf{x} + (1 - \beta)\mathbf{y}) > \mathbf{b}$, and thus $\beta \mathbf{x} + (1 - \beta)\mathbf{y} \notin \mathcal{P}$, for some $\mathbf{x}, \mathbf{y}$. However, $\mathbf{A}(\beta \mathbf{x} + (1 - \beta)\mathbf{y}) = \beta \mathbf{A}\mathbf{x} + (1 - \beta)\mathbf{A}\mathbf{y} \leq \beta \mathbf{b} + (1 - \beta)\mathbf{b}$. Consequently, $\mathbf{A}(\beta \mathbf{x} + (1 - \beta)\mathbf{y}) \leq \mathbf{b}$ for any points $\mathbf{x}, \mathbf{y}$ and we thus reached a contradiction.   □

Another important definition is the *polytope*.

Definition 1.24 A *polytope* is a *bounded* polyhedron $\mathcal{P} := \{\mathbf{x} \in \mathbb{R}^n \mid \mathbf{A}\mathbf{x} \leq \mathbf{b}\}$ that does not contain a line or a halfline. Polyhedron $\mathcal{P}$ is bounded if $\exists M > 0$ such that $\|\mathbf{x}\| \leq M \ \forall \mathbf{x} \in \mathcal{P}$.

Figure 1.22 presents an example of a polyhedron in the 2-dimensional space that is the intersection of closed halfspaces $3x_1 + 2x_2 \leq 60, -5x_1 - 4x_2 \leq -7, x_2 \leq 20$, and $-x_2 \leq -1$. This polyhedron could have been expressed as:

$$\begin{bmatrix} 3 & 2 \\ -5 & -4 \\ 0 & 1 \\ 0 & -1 \end{bmatrix} \begin{bmatrix} x_1 \\ x_2 \end{bmatrix} \leq \begin{bmatrix} 60 \\ -7 \\ 20 \\ -1 \end{bmatrix}$$

To satisfy all inequality constraints, x_2 cannot receive values outside of the range $[0, 20]$ and x_1 cannot receive values outside of the range $[-73/5, 58/3]$. Thus, there exists a finite number $M > 0$ such that $|x_1| + |x_2| \leq M$ for any values of x_1 and x_2 that satisfy the inequality constraints.

After establishing the convexity of polyhedra, it becomes evident that the feasible region of continuous optimization problems with linear and/or affine equality and inequality constraints is a convex set. This is a valuable observation because many continuous optimization problems have only linear and/or affine constraint functions.

Following the standard form notation of a continuous minimization problem with linear constraints from Eqs. (1.14)–(1.16), the feasible region of such problems is:

$$\mathcal{F} = \{\mathbf{x} \in \mathbb{R}^n \mid \mathbf{Cx} = \mathbf{d} \wedge \mathbf{Ax} \leq \mathbf{b}\}$$

Note that the equality constraints $\mathbf{Cx} = \mathbf{d}$ are hyperplanes that can be expressed as closed halfspaces by replacing them with

$$\mathbf{Cx} \leq \mathbf{d} \text{ and } -\mathbf{Cx} \leq -\mathbf{d}$$

Thus, the feasible region of problems with linear equality and inequality constraints becomes:

$$\mathcal{F} = \{\mathbf{x} \in \mathbb{R}^n \mid \mathbf{Cx} \leq \mathbf{d} \wedge -\mathbf{Cx} \leq -\mathbf{d} \wedge \mathbf{Ax} \leq \mathbf{b}\}$$

which is a polyhedron, and thus a convex set.

Another important definition is the polyhedral cone:

> **Polyhedral Cone**

A cone is polyhedral if it is the intersection of a finite number of half-spaces that contain the origin [21]. That is, a polyhedral cone is a subset $C \subseteq \mathbb{R}^n$ of the form $C := \{\mathbf{x} \in \mathbb{R}^n \mid \mathbf{Ax} \leq 0\}$ where $\mathbf{A}$ is an $m \times n$-dimensional matrix.

Other well-known sets that are convex are provided below.

Well-known convex sets

- Euclidean balls $B(\mathbf{x}_0, \epsilon) := \{\mathbf{x} \in \mathbb{R}^n : \|\mathbf{x} - \mathbf{x}_0\|_2 \leq \epsilon\}$ where $\epsilon \in \mathbb{R}_{\geq 0}$ is a scalar denoting the radius of the ball and $\mathbf{x}_0 \in \mathbb{R}^n$ the center of the ball.
- The empty set $\emptyset$ and single points.
- The ellipsoid $\mathcal{E} := \{\mathbf{x} \in \mathbb{R}^n : (\mathbf{x} - \mathbf{x}_0)^\mathsf{T} \mathbf{P}^{-1} (\mathbf{x} - \mathbf{x}_0) \leq 1\}$ where $\mathbf{x}_0 \in \mathbb{R}^n$ is the center of the ellipsoid and $\mathbf{P} \in \mathbb{R}^{n \times n}$ is a symmetric and positive definite matrix.

There are four main ways to determine whether a set (e.g., the feasible region of an optimization problem) $\mathcal{F}$ is a convex set:

1. Applying the definition of the convex set $\mathcal{F}$.
2. Writing $\mathcal{F}$ as the convex hull of a set of points $\mathbf{x}_1, \mathbf{x}_2, ..., \mathbf{x}_k$.
3. Proving that set $\mathcal{F}$ is one of the well-known convex sets.
4. Proving that $\mathcal{F}$ can be built from convex sets after performing convexity preserving operations.

The main convexity preserving operations are presented below.

Convexity preserving operations of sets

- **Intersection** – a set $\mathcal{F}$ is convex if it consists of the intersection of convex sets $\mathcal{F}_1, \mathcal{F}_2, ..., \mathcal{F}_n$. That is, $\mathcal{F} = \bigcap_{i=1}^{n} \mathcal{F}_i$ is convex if $\mathcal{F}_1, \mathcal{F}_2, ..., \mathcal{F}_n$ are convex sets. For instance, a polyhedron is the intersection of halfspaces (which are convex), and therefore is convex.
- **Affine Transform** – if set $\mathcal{F} \subseteq \mathbb{R}^n$ is convex, then for matrix $\mathbf{A} \in \mathbb{R}^{m \times n}$ and vector $\mathbf{b} \in \mathbb{R}^m$ the set $\mathcal{S} \subseteq \mathbb{R}^m$ such that:

$$\mathcal{S} := \mathbf{A}\mathcal{F} + \mathbf{b} = \{\mathbf{A}\mathbf{x} + \mathbf{b} \mid \mathbf{x} \in \mathcal{F}\}$$

  is also convex. Examples of affine transformations of convex sets include:

  - Scaling: if $\mathcal{F}$ is a convex set, $a\mathcal{F}$ is also convex
  - Translation: if $\mathcal{F}$ is a convex set, $\mathcal{F} + b$ is also convex
  - The projection of a convex set onto some of its coordinates
  - Set sum: $\mathcal{F} = \mathcal{F}_1 + \mathcal{F}_2$ is convex if $\mathcal{F}_1$ and $\mathcal{F}_2$ are convex sets

- **Perspective transform** – if $\mathcal{F}$ is a convex set, then the perspective transform $P(\mathbf{x}) = \frac{\mathbf{x}}{t} = [x_1/t, x_2/t, ..., x_n/t]^\mathsf{T}$ for $\mathbf{x} \in \mathcal{F}$ and $t \in \mathbb{R} \mid t > 0$ is also a convex set.

A prominent example of an operation that does not preserve convexity is the set union. In particular, if $\mathcal{F}_1, \mathcal{F}_2, ..., \mathcal{F}_n$ are convex sets, $\mathcal{F} = \bigcup_{i=1}^{n} \mathcal{F}_i$ is not necessarily a convex set.

1.5.3.2 Convex Functions

The definition of a convex function is provided below.

> **Convex Function**

Formally, a function $f : \mathcal{F} \to \mathbb{R}$ is a convex function if $\mathcal{F} \subseteq \mathbb{R}^n$ is a convex set and for any two points $\mathbf{x} \neq \mathbf{y}$ in $\mathcal{F}$, the following property is satisfied:

$$f(\theta\mathbf{x} + (1 - \theta)\mathbf{y}) \leq \theta f(\mathbf{x}) + (1 - \theta) f(\mathbf{y}) \quad \text{for any } 0 < \theta < 1$$

If the objective function and the feasible region in a minimization problem are both convex, then any local solution of the problem is also a globally optimal solution. A strictly convex function is also defined as follows.

> **Strictly Convex Function**

Formally, a function $f : \mathcal{F} \to \mathbb{R}$ is a strictly convex function if $\mathcal{F}$ is a convex set and for any two points $\mathbf{x} \neq \mathbf{y}$ in $\mathcal{F}$, the following property is satisfied:

$$f(\theta\mathbf{x} + (1-\theta)\mathbf{y}) < \theta f(\mathbf{x}) + (1-\theta)f(\mathbf{y}) \quad \text{for any } 0 < \theta < 1$$

We can also easily determine that a function f is concave, since f is concave if $-f$ is convex. This can be used to determine whether a maximization problem has a concave objective function.

Following the definition of a convex function, let us consider a single-dimensional function f. Geometrically, f is convex function if a line segment drawn from any point $(x, f(x))$ to another point $(y, f(y))$, where $x, y \in \mathcal{F} \subseteq \mathbb{R}$, lies on or above the graph of f (Fig. 1.23). The set of all points lying on or upon the graph of f is called *epigraph*, and an alternative way to prove that function f is convex is to show that its epigraph is a convex set.

A typical example of a convex function is an affine function of the form $f(\mathbf{x}) = \mathbf{c}^\mathsf{T}\mathbf{x} + \beta$ in a convex feasible region $\mathcal{F}$. In fact, this function is both convex and concave, as shown below.

Theorem 1.3 *An affine function $f(x) = \mathbf{c}^\mathsf{T}x + \beta$ is both convex and concave.*

Proof For any $\mathbf{x}, \mathbf{y} \in \mathcal{F}$ where $\mathbf{x} \neq \mathbf{y}$, we have $\mathbf{c}^\mathsf{T}(\theta\mathbf{x} + (1-\theta)\mathbf{y}) + \beta \leq \theta (\mathbf{c}^\mathsf{T}\mathbf{x} + \beta) + (1-\theta)(\mathbf{c}^\mathsf{T}\mathbf{y} + \beta)$ for any $\theta \in (0, 1)$. The left hand side is exactly equal to the right hand side of the inequality expression. This indicates that a segment drawn from any point to any other point of an affine function will lie on the graph of f, and thus f is convex. In addition, an affine function is also concave because $-f$ is still convex. □

As linear functions are a sub-category of affine functions, linear functions are also both convex and concave. This property of affine functions is very useful when

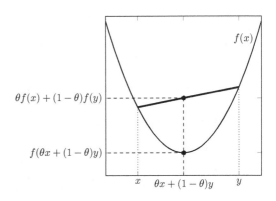

Fig. 1.23 The line segment between any two points of a convex function f does not lie below its graph

solving minmax problems since we can compute a globally optimal solution when solving both the min and max problems if the objective function is affine.

Considering multi-dimensional functions, $f(\mathbf{x})$, one can prove that f is convex if its domain $\mathcal{F}$ is convex and it is possible to prove one of the following:

1. **Line segment:** for any two points $\mathbf{x} \neq \mathbf{y}$ in $\mathcal{F}$ and any $0 < \theta < 1$ we have $f(\theta\mathbf{x} + (1 - \theta)\mathbf{y}) \leq \theta f(\mathbf{x}) + (1 - \theta)f(\mathbf{y})$.
2. **First order condition:** if $f(\mathbf{x})$ is differentiable, then

$$f(\mathbf{y}) \geq f(\mathbf{x}) + \nabla f(\mathbf{x})^\mathsf{T}(\mathbf{y} - \mathbf{x}) \ \ \forall(\mathbf{x}, \mathbf{y}) \in \mathcal{F}$$

   using the denominator layout notation [22].
3. **Hessian matrix test:** if its Hessian matrix $\mathbf{H}(f)$ exists everywhere on $\mathcal{F}$ and it is positive semi-definite.

Proving one of the three aforementioned conditions suffices to prove that a function is convex. We note here that a function f is differentiable if the gradient

$$\nabla f(\mathbf{x}) = \left[\frac{\partial f(\mathbf{x})}{\partial x_1}, \frac{\partial f(\mathbf{x})}{\partial x_2}, ..., \frac{\partial f(\mathbf{x})}{\partial x_n}\right]^\mathsf{T}$$

exists for any $\mathbf{x} = [x_1, x_2, ..., x_n]^\mathsf{T} \in \mathcal{F}$.

Regarding the Hessian matrix test, if the Hessian matrix (or, simply, the Hessian) of a function is positive semi-definite, then the function is convex. If the Hessian is positive definite, then the function is *strictly* convex. The Hessian of a function f with n variables $\mathbf{x} = [x_1, x_2, ..., x_n]^\mathsf{T}$ is an $n \times n$ matrix that is defined as follows.

Definition 1.25 Hessian matrix:

$$\mathbf{H}(f) = \begin{bmatrix} \frac{\partial^2 f(\mathbf{x})}{\partial x_1^2} & \frac{\partial^2 f(\mathbf{x})}{\partial x_1 x_2} & \cdots & \frac{\partial^2 f(\mathbf{x})}{\partial x_1 x_n} \\ \frac{\partial^2 f(\mathbf{x})}{\partial x_2 x_1} & \frac{\partial^2 f(\mathbf{x})}{\partial x_2^2} & \cdots & \frac{\partial^2 f(\mathbf{x})}{\partial x_2 x_n} \\ \vdots & \vdots & \ddots & \vdots \\ \frac{\partial^2 f(\mathbf{x})}{\partial x_n x_1} & \frac{\partial^2 f(\mathbf{x})}{\partial x_n x_2} & \cdots & \frac{\partial^2 f(\mathbf{x})}{\partial x_n^2} \end{bmatrix}$$

If a Hessian $n \times n$ matrix $\mathbf{H}$ is symmetric (which is always true when the second partial derivatives are continuous according to Schwarz's theorem [23]), we can prove its definiteness by performing one of the following three tests (note that implementing one of the three tests suffices to prove the definiteness).

Matrix Definiteness tests in the Euclidean space

1. **Test of vector multiplication** – a symmetric matrix $n \times n$-dimensional $\mathbf{H}$ is:

   - positive semi-definite if $\mathbf{c}^\mathsf{T}\mathbf{H}\mathbf{c} \geq 0$ for any $\mathbf{c} \in \mathbb{R}^n$;
   - positive definite if $\mathbf{c}^\mathsf{T}\mathbf{H}\mathbf{c} > 0$ for any $\mathbf{c} \in \mathbb{R}^n \mid \mathbf{c} \neq \mathbf{0}$;

- negative semi-definite if $\mathbf{c}^T \mathbf{H} \mathbf{c} \leq 0$ for any $\mathbf{c} \in \mathbb{R}^n$;
- negative definite if $\mathbf{c}^T \mathbf{H} \mathbf{c} < 0$ for any $\mathbf{c} \in \mathbb{R}^n \mid \mathbf{c} \neq \mathbf{0}$;
- indefinite if $\mathbf{c}^T \mathbf{H} \mathbf{c} > 0$ for some $\mathbf{c} \in \mathbb{R}^n$ and $\mathbf{c}^T \mathbf{H} \mathbf{c} < 0$ for some other $\mathbf{c} \in \mathbb{R}^n$.

2. **Test of principal minors** – a symmetric matrix $n \times n$-dimensional $\mathbf{H}$ is:

- positive semi-definite if all of its n principal minors are greater than or equal to 0;
- positive definite if all of its *leading* principal minors are greater than 0;
- negative semi-definite if all principal minors alternate in sign;
- negative definite if all *leading* principal minors alternate in sign;
- indefinite if they do not fit to any of the previous.

3. **Test of eigenvalues** – a symmetric matrix $n \times n$-dimensional $\mathbf{H}$ is:

- positive semi-definite if all its eigenvalues are non-negative;
- positive definite if all its eigenvalues are positive;
- negative semi-definite if all its eigenvalues are non-positive;
- negative definite if all its eigenvalues are negative;
- indefinite if it has at least one positive eigenvalue and at least one negative eigenvalue.

To perform the aforementioned tests we need to consider the following definitions from linear algebra.

Definition 1.26 *Eigenvalue.* The eigenvalue λ of an $n \times n$ matrix $\mathbf{H}$ is a value such that for an n-dimensional eigenvector $\mathbf{u}$ we have $\mathbf{H} \mathbf{u} = \lambda \mathbf{u}$.

For eigenvector $\mathbf{u}$,

$$\mathbf{H} \mathbf{u} = \lambda \mathbf{u} \Rightarrow (\mathbf{H} - \lambda \mathbf{I}) \mathbf{u} = \mathbf{0}$$

which has a nonzero solution if, and only if, the determinant of matrix $(\mathbf{H} - \lambda \mathbf{I})$ is zero:

$$\det(\lambda \mathbf{I} - \mathbf{H}) = 0$$

This is known as the *characteristic polynomial* of $\mathbf{H}$ and it is used to calculate the values of the eigenvalues.

Definition 1.27 *Minor.* Let $\mathbf{H}$ be an $n \times n$ matrix. Let $\mathbf{H}_{i,j}$ be a submatrix obtained by deleting row i and column j from $\mathbf{H}$. Then, the scalar $M_{ij} = \det(\mathbf{H}_{ij})$ is the (i, j) *minor* of $\mathbf{H}$, $C_{ij} = (-1)^{i+j} M_{ij}$ the (i, j)th *cofactor* of $\mathbf{H}$ and $\det(\mathbf{H}) = \sum_{j=1}^{n} h_{1j} C_{1j}$ the *determinant* of $\mathbf{H}$.

Definition 1.28 *Principal Minor.* Let $\mathbf{H}$ be an $n \times n$ matrix. The $k \times k$ submatrix of $\mathbf{H}$ obtained by deleting $n - k$ rows and the same $n - k$ columns of $\mathbf{H}$ is called principal submatrix of $\mathbf{H}$. The determinant of a principal submatrix is called a *principal minor*.

Definition 1.29 *Leading Principal Minor.* Let $\mathbf{H}$ be an $n \times n$ matrix. The k-th order principal submatrix of $\mathbf{H}$ obtained by deleting the last $n - k$ rows and columns of $\mathbf{H}$ is called the k-th order leading principal submatrix of $\mathbf{H}$, and its determinant is called the k-th order *leading principal minor*.

As an example, we investigate the definiteness of the following Hessian matrix.

$$\mathbf{H} = \begin{bmatrix} 2 & -1 & -1 \\ -1 & 2 & -1 \\ -1 & -1 & 2 \end{bmatrix}$$

Based on the first method that uses vector multiplication, for $\mathbf{c} \in \mathbb{R}^3$ we have

$$\mathbf{c}^\mathsf{T}\mathbf{H}\mathbf{c} = [c_1, c_2, c_3] \begin{bmatrix} 2 & -1 & -1 \\ -1 & 2 & -1 \\ -1 & -1 & 2 \end{bmatrix} \begin{bmatrix} c_1 \\ c_2 \\ c_3 \end{bmatrix} = [2c_1 - c_2 - c_3, \ -c_1 + 2c_2 - c_3, \ -c_1 - c_2 + 2c_3] \begin{bmatrix} c_1 \\ c_2 \\ c_3 \end{bmatrix} =$$

$$c_1(2c_1 - c_2 - c_3) + c_2(-c_1 + 2c_2 - c_3) + c_3(-c_1 - c_2 + 2c_3) =$$

$$2c_1^2 + 2c_2^2 + 2c_3^2 - 2c_1c_2 - 2c_1c_3 - c_2c_3 =$$

$$(c_1 - c_2)^2 + (c_1 - c_3)^2 + (c_2 - c_3)^2$$

which is greater than or equal to 0 for any $c_1, c_2, c_3 \in \mathbb{R}$. Thus, the matrix is positive semi-definite based on the test of vector multiplication.

Consider now the second test that uses principal minors. The leading principal minors of $\mathbf{H}$ are:

- first order leading principal minor: $D_1 = 2 > 0$
- second order leading principal minor: $D_2 = \begin{vmatrix} 2 & -1 \\ -1 & 2 \end{vmatrix} = 2 \cdot 2 - (-1)(-1) = 3 > 0$

- third order leading principal minor: $D_3 = \sum_{j=1}^3 h_{1j}C_{1j} = 2(-1)^2 \begin{vmatrix} 2 & -1 \\ -1 & 2 \end{vmatrix} +$

$$(-1)(-1)^3 \begin{vmatrix} -1 & -1 \\ -1 & 2 \end{vmatrix} + (-1)(-1)^4 \begin{vmatrix} -1 & 2 \\ -1 & -1 \end{vmatrix} = 6 - 3 - 3 = 0.$$

The first and second order leading principal minors are greater than 0. However, the third order one is equal to 0. Therefore, $\mathbf{H}$ is not positive definite. To investigate if it is positive semi-definite, we need to calculate the values of all principal minors. $\mathbf{H}$ has:

- one third order principal minor which is equal to $D_3 = 0$.
- three second order principal minors, which are all of the form $\begin{vmatrix} 2 & -1 \\ -1 & 2 \end{vmatrix} = 2 \cdot 2 - (-1)(-1) = 3$.
- three first order principal minors: $h_{11} = 2$, $h_{22} = 2$, $h_{33} = 2$.

Thus, all principal minors are greater than or equal to 0 and the Hessian matrix is positive semi-definite.

Consider now the third method that conducts the eigenvalues' test. The eigenvalues λ of matrix $\mathbf{H}$ can be found by solving $\det(\lambda \mathbf{I} - \mathbf{H}) = 0$. We have:

$$\lambda \mathbf{I} - \mathbf{H} = \lambda \begin{bmatrix} 1 & 0 & 0 \\ 0 & 1 & 0 \\ 0 & 0 & 1 \end{bmatrix} - \begin{bmatrix} 2 & -1 & -1 \\ -1 & 2 & -1 \\ -1 & -1 & 2 \end{bmatrix} = \begin{bmatrix} \lambda - 2 & 1 & 1 \\ 1 & \lambda - 2 & 1 \\ 1 & 1 & \lambda - 2 \end{bmatrix}$$

Therefore,

$$\det(\lambda \mathbf{I} - \mathbf{H}) = (\lambda - 2)(-1)^2 \begin{vmatrix} \lambda - 2 & 1 \\ 1 & \lambda - 2 \end{vmatrix} + 1(-1)^3 \begin{vmatrix} 1 & 1 \\ 1 & \lambda - 2 \end{vmatrix} + 1(-1)^4 \begin{vmatrix} 1 & \lambda - 2 \\ 1 & 1 \end{vmatrix} =$$

$$(\lambda - 2)[(\lambda - 2)^2 - 1] - (\lambda - 2 - 1) + [1 - (\lambda - 2)] = (\lambda - 2)^3 - 3\lambda + 8$$

Setting now

$$\det(\lambda \mathbf{I} - \mathbf{H}) = 0$$

we get

$$(\lambda - 2)^3 - 3\lambda + 8 = 0$$

that results in eigenvalues $\lambda = 0$ and $\lambda = 3$ when solving the equation. Note that both eigenvalues are non-negative, thus the Hessian matrix is positive semi-definite. At this point, it is important to state again that one does not have to perform all three tests to determine the definiteness of a matrix—one of them suffices.

Practitioner's Corner
Eigenvalues in Python 3

To determine the definiteness of a large matrix, we can compute its eigenvalues using Python 3. For instance, the eigenvalues of our example matrix are computed as follows:

```
import numpy as np
from numpy import linalg as LA
H=np.array([[2,-1,-1],[-1,2,-1],[-1,-1,2]])
eigenvalues,eigenvectors=LA.eig(H)
print(eigenvalues)
```

To simplify the process of investigating the definiteness of a matrix, one can use operations that preserve convexity and use matrices of lower order for the definiteness tests. The main operations of convex functions that preserve convexity are presented below.

Common operations that preserve convexity of functions

- **Non-negative weighted sums** – if $f_1, ..., f_m : \mathbb{R}^n \to \mathbb{R}$ are convex functions and $\alpha_1, ..., \alpha_m$ are non-negative scalars, then their nonlinear combination:

$$f(\mathbf{x}) = \alpha_1 f_1(\mathbf{x}) + ... + \alpha_m f_m(\mathbf{x})$$

  is also convex. This property extends to infinite sums and integrals.
- **Composition with an affine mapping** – if $f(\mathbf{x}) : \mathbb{R}^n \to \mathbb{R}$ is convex, then for matrix $\mathbf{A} \in \mathbb{R}^{m \times n}$ and vector $\mathbf{b} \in \mathbb{R}^m$ function $g : \mathbb{R}^m \to \mathbb{R}$ such that $g(\mathbf{y}) = f(\mathbf{Ax} + \mathbf{b})$ is also convex.
- **Pointwise maximum** – if $f_1, f_2, ..., f_m$ are convex functions, then their pointwise maximum $f(\mathbf{x}) = \max\{f_1(\mathbf{x}), f_2(\mathbf{x}), ..., f_m(\mathbf{x})\}$ is also convex.
- **General Composition** – if $f : \mathbb{R}^n \to \mathbb{R}$ and $g : \mathbb{R} \to \mathbb{R}$ are convex functions and g is non-decreasing, then $h(\mathbf{x}) = g(f(\mathbf{x}))$ is convex. For instance, $e^{f(\mathbf{x})}$ is convex if f is convex, because $g(\mathbf{y}) = e^{\mathbf{y}}$ is convex and monotonically increasing.
- **Perspective** – if f is convex in $\mathrm{dom}(f) \subseteq \mathbb{R}^n$, then its perspective

$$g(\mathbf{x}, t) = t f\left(\frac{\mathbf{x}}{t}\right) \qquad \mathbf{x} \in \mathrm{dom}(f), \ t \in \mathbb{R}_{>0}$$

is also convex with $\mathrm{dom}(g) = \{(\mathbf{x}, t) \mid \frac{x}{t} \in \mathrm{dom}(f), \ t > 0\}$. For example, $f(x) = -\log x$ is convex on the domain of positive real numbers; hence, $g(x, t) = -t \log(x/t) = -t \log x + t \log t$ is also convex for $t > 0$.

Using the definitions and the operations that preserve convexity, simple examples of convex functions include:

- affine functions $f(\mathbf{x}) = \mathbf{c}^\mathsf{T}\mathbf{x} + \beta$ and linear functions $f(\mathbf{x}) = \mathbf{c}^\mathsf{T}\mathbf{x}$ which are simultaneously convex and concave.
- quadratic functions $f(\mathbf{x}) = \mathbf{x}^\mathsf{T}\mathbf{H}\mathbf{x}$, if $\mathbf{H}$ is a symmetric positive semi-definite matrix.
- norms $||\mathbf{x}||_p = (\sum_{i=1}^{n} |x_i|^p)^{1/p}$ for $p \geq 1$
- the LogSumExp (LSE) function, also known as multivariable softplus, $f(x_1, x_2, ..., x_m) = \log(\exp(x_1) + \cdots + \exp(x_m))$

1.5.4 Example of a Convex Transport Problem

Let us now investigate the convexity of a transport problem. Consider again Hitchcock's Transportation Problem by replacing all equality constraints by inequality ones:

$$\min_{\mathbf{X}\in\mathbb{R}^{m\times n}} \quad f(\mathbf{X}) := \sum_{i\in M}\sum_{j\in N} c_{ij}x_{ij} \tag{1.73}$$

$$\text{subject to: } \sum_{j\in N} x_{ij} \leq a_i \qquad\qquad \forall i \in M \tag{1.74}$$

$$\sum_{j\in N} x_{ij} \geq a_i \qquad\qquad \forall i \in M \tag{1.75}$$

$$\sum_{i\in M} x_{ij} \leq b_j \qquad\qquad \forall j \in N \tag{1.76}$$

$$\sum_{i\in M} x_{ij} \geq b_j \qquad\qquad \forall j \in N \tag{1.77}$$

$$x_{ij} \geq 0 \qquad\qquad \forall i \in M, j \in N \tag{1.78}$$

All constraints are closed halfspaces forming a polyhedron $\mathcal{P}$. Thus, the feasible region of the problem is a convex set. The objective function is a linear function, which is both convex and concave (note that its Hessian is a zero-valued matrix). Thus, Hitchcock's Transportation Problem is convex and any locally optimal solution is also a globally optimal one.

1.6 Linear Programming Reformulations

A mathematical program is linear if it has a linear or affine objective function and linear and/or affine constraints. Nonlinear mathematical programs are programs that have a nonlinear objective function or at least one nonlinear constraint.

Linear programs have favorable properties, one of which is convexity. In more detail, their feasible region is a set defined as the intersection of finitely many halfspaces because of the sole presence of affine (or linear) constraints. Thus, the feasible region is convex. In addition, the objective function of linear programs is both convex and concave allowing to solve both a maximization or a minimization version of a linear program to global optimality. Because of these properties, a crucial task of a modeler is to transform nonlinear programs into linear programs provided that this is possible given the characteristics of the problem. A linear program is said to be in standard form if:

- it is a maximization program;
- all constraints are described as equality constraints;
- all variables are restricted to be non-negative.

Standard form of Linear Program

$$\max \ \mathbf{c}^\mathsf{T}\mathbf{x} \tag{1.79}$$

$$\text{s.t. } \mathbf{Ax} = \mathbf{b} \tag{1.80}$$

$$\mathbf{x} \in \mathbb{R}^n_{\geq 0} \tag{1.81}$$

where $\mathbf{A}$ is an $m \times n$ matrix, $\mathbf{c}$ a vector with n values and $\mathbf{b}$ a vector with m values. Note that a minimization problem can be transformed to the standard form by multiplying the objective function by -1. In addition, inequality constraints can be transformed into equality constraints by adding slack variables. If we transform all equality constraints into inequality ones, then the linear program is cast in its canonical form:

Canonical form of Linear Program

$$\max \ \mathbf{c}^\mathsf{T}\mathbf{x} \tag{1.82}$$

$$\text{s.t. } \mathbf{Ax} \leq \mathbf{b} \tag{1.83}$$

$$\mathbf{x} \in \mathbb{R}^n_{\geq 0} \tag{1.84}$$

Linear programs and their solution methods will be examined in more detail in the next chapters. Because linear programs have favorable properties and a translation of a nonlinear program to a linear one can typically provide significant benefits in terms of attaining a globally optimal solution, we present a number of common transformations of nonlinear programs to linear ones. These transformations are commonly known as linearizations.

1.6.1 Linearizations of Max and Absolute Terms

Mathematical programs can include max terms in their objective function or their constraints. A function $\max\{x, y\}$ can be linearized by introducing continuous variable $r \in \mathbb{R}$, a very large positive number M (big-M), and binary variables $\delta \in \{0, 1\}$ that replace $\max\{x, y\}$ with r which is restricted by the following inequality constraints:

$$r \geq x \tag{1.85}$$

$$r \geq y \tag{1.86}$$

$$r \leq x + \delta M \tag{1.87}$$

$$r \leq y + (1 - \delta)M \tag{1.88}$$

Note that r can now replace $\max\{x, y\}$ because r will always be equal to $\max\{x, y\}$ when forced to satisfy the inequality constraints (1.85)–(1.88). In more detail, constraints (1.85) and (1.86) ensure that r cannot receive a value lower than x or y. That is, $r \geq \max\{x, y\}$. To proceed, there are three cases: $x > y$ or $y > x$ or $y = x$.

- If $x > y$, then from constraints (1.85) and (1.86) we have $r \geq x$. Because of that, constraint (1.88) can only be satisfied if $\delta = 0$ since r cannot be less than or equal to y. If $\delta = 0$, then constraint (1.87) imposes that $r \leq x$. That is, $x \leq r \leq r$ and r receives the value of $\max\{x, y\} = x$.
- If $y > x$, then from constraints (1.85) and (1.86) we have $r \geq y$. Because of this, constraint (1.87) is satisfied only if $\delta = 1$ since r cannot be less than or equal to x. For $\delta = 1$, constraint (1.88) results in $r \leq y$. Thus, $y \leq r \leq y$ ensuring again that $r = \max\{x, y\} = y$.
- Finally, if $x = y$ then r should be forced to be equal to either x or y since there is no difference between them. In this case δ can be either 0 or 1. If $\delta = 0$, then $r \leq x$ and $r \leq y + M$. In addition, $r \geq x$ from (1.85) and thus $r = x$. If $\delta = 1$, then $r \leq x + M$ and $r \leq y$. In addition, $r \geq y$ from (1.86) and thus $r = y$. Summarizing, r is also equal to $\max\{x, y\}$ when $x = y$

This linearization is useful in many applications. One of them is if we need to linearize the absolute term $|x|$. In this case, $|x|$ can be written as $\max\{-x, x\}$ which is linearized by the following system of inequality constraints:

$$r \geq x \tag{1.89}$$

$$r \geq -x \tag{1.90}$$

$$r \leq x + \delta M \tag{1.91}$$

$$r \leq -x + (1 - \delta)M \tag{1.92}$$

In a similar way, the term $\min\{x, y\}$ can be linearized by linearizing the $-\max\{-x, -y\}$.

A particular easy case occurs when the max term is in the objective function of a minimization problem (minimax). In this case,

$$\min\{\max\{c_1 x_1, ..., c_i x_i, ..., c_{|I|} x_{|I|}\}\}$$

can be linearized by introducing continuous variable z such that:

$$\min z$$
$$\text{s.t. } z \geq c_i x_i \ \ \forall i \in I$$

The same can be applied for a maximin problem, namely $\max\{\min\{c_1 x_1, \ldots, c_i x_i, \ldots c_{|I|} x_{|I|}\}\}$, where c_i are parameter values.

1.6.2 Linearization of a Fractional Program

We consider the case of a fractional function $\frac{1}{f(\mathbf{x})}$ where $f(\mathbf{x}) : \mathbb{R}^n \to \mathbb{R}$ is linear (or affine). In practice, there are many linear-fractional programs of the following form in public transport applications.

Linear-fractional program

$$\max_{\mathbf{x}} \; \frac{\mathbf{c}^\mathsf{T}\mathbf{x} + a}{\mathbf{d}^\mathsf{T}\mathbf{x} + \beta} \qquad (1.93)$$

$$\text{s.t. } \mathbf{Ax} \leq \mathbf{b} \qquad (1.94)$$

where the denominator is an affine function and the feasible region is a polyhedron. Note that $\mathbf{c}, \mathbf{d}, \mathbf{x} \in \mathbb{R}^n$, $a, \beta \in \mathbb{R}$, $\mathbf{A} \in \mathbb{R}^{m \times n}$ and $\mathbf{b} \in \mathbb{R}^m$.

A well-known transformation for this problem is the Charnes-Cooper transformation [24]. First, the constraints have to ensure that the denominator is always positive. That is $\{\mathbf{x} \mid \mathbf{d}^\mathsf{T}\mathbf{x} + \beta > 0\}$. Then, we introduce variables $\mathbf{y} \in \mathbb{R}^n$ and $t \in \mathbb{R}$ such that:

$$\mathbf{y} = \frac{1}{\mathbf{d}^\mathsf{T}\mathbf{x} + \beta}\mathbf{x}$$

and

$$t = \frac{1}{\mathbf{d}^\mathsf{T}\mathbf{x} + \beta}$$

Then, the linear-fractional program is translated to the equivalent linear program:

$$\max_{\mathbf{y},t} \; \mathbf{c}^\mathsf{T}\mathbf{y} + at \qquad (1.95)$$

$$\text{s.t. } \mathbf{Ay} \leq \mathbf{b}t \qquad (1.96)$$

$$\mathbf{d}^\mathsf{T}\mathbf{y} + \beta t = 1 \qquad (1.97)$$

$$t \geq 0 \qquad (1.98)$$

and the solution of the original problem is computed as $\mathbf{x} = \frac{1}{t}\mathbf{y}$.

1.6.3 Product Function to a Separable Function

Let us consider the nonlinear function $f(\mathbf{x}) = x_1 x_2$ where x_1 and x_2 are variables. To make this function separable, we introduce variables y_1, y_2 and r such that:

$$r = y_1^2 - y_2^2 \tag{1.99}$$

$$y_1 = \frac{1}{2}(x_1 + x_2) \tag{1.100}$$

$$y_2 = \frac{1}{2}(x_1 - x_2) \tag{1.101}$$

Then, $x_1 x_2$ can be replaced by r because r will always take the value of $x_1 x_2$. In more detail,

$$r = y_1^2 - y_2^2 = 1/4(x_1^2 + 2x_1 x_2 + x_2^2) - 1/4(x_1^2 - 2x_1 x_2 + x_2^2) = x_1 x_2$$

Note that r is still nonlinear, but it is a separable function of y_1, y_2 compared to f.

1.6.4 Linearization of Conditional Expressions

In many cases we have a binary variable y that takes values based on the value of the inequality constraint of another continuous variable x. For instance, we have to use $y = 1$ public transport vehicle if the passenger demand is more that a threshold value $x > 0$. This can be written as:

$$y = \begin{cases} 1 & \text{if } x > 0 \\ 0 & \text{if } x < 0 \end{cases} \tag{1.102}$$

Introducing a large positive number M this conditional can be linearized as:

$$x \leq My \tag{1.103}$$
$$x \geq (y - 1)M \tag{1.104}$$
$$y \in \{0, 1\} \tag{1.105}$$

Note that if $x > 0$ then $y = 1$ to satisfy both constraints. If $x < 0$ then y is forced to be equal to 0 to satisfy both constraints.

In reverse, if we require that:

$$y = \begin{cases} 1 & \text{if } x < 0 \\ 0 & \text{if } x > 0 \end{cases} \tag{1.106}$$

Table 1.8 Linearization of conditional expressions for linear functions f, g

Conditional expression	Proposed linearization with $M \ggg 0$
$f(\mathbf{x}) \leq 0$ or $g(\mathbf{x}) \leq 0$	$f(\mathbf{x}) \leq My$ $g(\mathbf{x}) \leq M(1 - y)$ $y \in \{0, 1\}$
if $f(\mathbf{x}) > 0$ then $g(\mathbf{x}) \geq 0$	$-g(\mathbf{x}) \leq My$ $f(\mathbf{x}) \leq M(1 - y)$ $y \in \{0, 1\}$
$y = \begin{cases} 1 & \text{if } f(\mathbf{x}) > 0 \\ 0 & \text{if } f(\mathbf{x}) < 0 \end{cases}$	$f(\mathbf{x}) \leq My$ $f(\mathbf{x}) \geq M(y - 1)$ $y \in \{0, 1\}$
$y = \begin{cases} 1 & \text{if } f(\mathbf{x}) < 0 \\ 0 & \text{if } f(\mathbf{x}) > 0 \end{cases}$	$f(\mathbf{x}) \leq M(1 - y)$ $f(\mathbf{x}) \geq -My$ $y \in \{0, 1\}$
$y = \begin{cases} 1 & \text{then } g(\mathbf{x}) = f(\mathbf{x}) \\ 0 & \text{then } g(\mathbf{x}) \text{ might not meet the constraint} \end{cases}$	$g(\mathbf{x}) + z = f(\mathbf{x})$ $z \leq M(1 - y)$ $z \geq -M(1 - y)$ $z \in \mathbb{R}$
$xy \leq 0$ where $x, y \in \mathbb{R}$	$x + z \leq 0$ $z \leq M(1 - y)$ $z \geq -M(1 - y)$ $z \in \mathbb{R}$
$z = xy$ where $x \in \{0, 1\}, y \in \mathbb{R}$	$z \leq Mx$ $z \geq -Mx$ $z \leq y + M(1 - x)$ $z \geq y - M(1 - x)$ $z \in \mathbb{R}$

then after introducing a large positive number M this conditional can be linearized as:

$$x \leq M(1 - y) \tag{1.107}$$

$$x \geq -My \tag{1.108}$$

$$y \in \{0, 1\} \tag{1.109}$$

where y is forced to be equal to 0 for $x > 0$ and $y = 1$ for $x < 0$. Note if $x = 0$, then y can be either 0 or 1.

Table 1.8 offers a useful summary of typical linearizations with the use of binary variables.

1.6.5 Linearization of Constraints on a Collection of Variables

1.6.5.1 Special Ordered Sets of Type 1 (SOS1)

A set of variables is a Special Ordered Set of type 1 (SOS1) if at most one of them can take a non-zero value and all others must be equal to 0. Typically, they are applied in problems where one has to choose at most one option from a set of options (they are referred to as multiple choice). This might arise, for instance, when we need to decide which station to close at a particular public transport service line.

Commercial optimization solvers have routines to account for SOS1 sets. For instance, consider continuous variables $(y_1, y_2, ..., y_n)$, where $y_i \in [0, \beta_i]$. That is, each variable y_i is non-negative and has a fixed upper bound β_i. If these variables are SOS1, i.e., $SOS1(y_1, y_2, ..., y_n)$, then the usual approach to formulate them is to introduce a binary variable x_i for each continuous variable y_i such that:

$$y_i \leq x_i \beta_i \qquad\qquad \forall i \in \{1, 2, ..., n\} \qquad (1.110)$$

$$\sum_{i=1}^{n} x_i \leq 1 \qquad\qquad\qquad\qquad\qquad\qquad (1.111)$$

$$x_i \in \{0, 1\}, y_i \in \mathbb{R}_{\geq 0} \qquad \forall i \in \{1, 2, ..., n\} \qquad (1.112)$$

Constraints (1.111)–(1.112) force at most one x_i to be equal to 1. Because of this, and given constraints (1.110), only its corresponding variable y_i can be greater than 0 and take any value in the range $[0, \beta_i]$. This will force the decision maker to select at most one option with cost y_i. Constraints (1.110)–(1.112) can be simply written as $SOS1(y_1, y_2, ..., y_n)$. Note that if the upper bound β_i for each variable i is $+\infty$, we can use a very large positive value for β_i to be able to solve this problem with an optimization solver.

1.6.5.2 Special Ordered Sets of Type 2 (SOS2)

A set of variables is Special Ordered Set of type 2 (SOS2) if [25]:

- the variables are non-negative,
- at most two variables can be non-zero,
- if two variables are non-zero, they must be adjacent to each other based on their pre-defined ordering.

A typical application occurs when a non-linear function is approximated by a piecewise linear function. They are typically used in separable functions, defined below.

Definition 1.30 A *separable* function $f(\mathbf{x})$ is a function that can be written as the sum of functions of scalar variables, $\sum_{i \in I} f_i(x_i)$.

For instance, $f(\mathbf{x}) = x_1^2 + 5x_2 + 1/x_3$ can be seen as a separable function $f_1(x_1) + f_2(x_2) + f_3(x_3)$, whereas the function $x_1x_2^2 + 5/(x_3 + x_4)$ is not separable.

Following the definition of SOS2, let us consider ordered continuous variables $y_1, y_2, ..., y_n$, where $y_i \in [0, \beta_i]$. Then, SOS2$(y_1, y_2, ..., y_n)$ can be formulated by introducing binary variables $x_1, x_2, ..., x_n$ such that [26]:

$$y_i \leq x_i \beta_i \qquad\qquad \forall i \in \{1, 2, ..., n\} \qquad\qquad (1.113)$$

$$\sum_{i=1}^{n} x_i \leq 2 \qquad\qquad\qquad\qquad\qquad\qquad (1.114)$$

$$x_i + x_j \leq 1 \qquad\qquad \forall i \in \{1, 2, ..., n-2\}, j \in \{i+2, ..., n\} \qquad (1.115)$$

$$x_i \in \{0, 1\}, y \in \mathbb{R}_{\geq 0} \qquad \forall i \in \{1, 2, ..., n\} \qquad\qquad\qquad (1.116)$$

1.6.6 Linearizations of Nonlinear Functions with Piecewise Linear Functions

In several cases, piecewise linear functions are used to linearize nonlinear functions that appear in the objective function or the constraints of an optimization problem. These linearizations are subject to approximation errors, as can be seen in Fig. 1.24.

If we want to approximate a nonlinear function with a piecewise linear function, we need first to introduce breakpoints that lie in the nonlinear function and draw lines that connect these breakpoints. If the breakpoints along the x-axis are $x_1, x_2, ..., x_n$, then we can derive the values of the nonlinear function at the locations of these breakpoints $f(x_1), f(x_2), ..., f(x_n)$. The piecewise linear function at a set of breakpoints $x_1, ..., x_n$ has the general form:

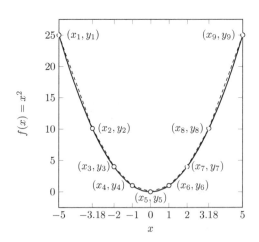

Fig. 1.24 Approximating the quadratic function $f(x) = x^2$ (solid) in the region $[-5, +5]$ with a piecewise linear function (trimmed) with breaking points $(x_1, y_1), ..., (x_9, y_9)$

$$\tilde{f}(x) = \begin{cases} f(x_1) + (x - x_1)\frac{f(x_2)-f(x_1)}{x_2-x_1} & \text{for } x_1 \leq x \leq x_2 \\ f(x_2) + (x - x_2)\frac{f(x_3)-f(x_2)}{x_3-x_2} & \text{for } x_2 \leq x \leq x_3 \\ \cdots \\ f(x_{n-1}) + (x - x_{n-1})\frac{f(x_n)-f(x_{n-1})}{x_n-x_{n-1}} & \text{for } x_{n-1} \leq x \leq x_n \end{cases} \quad (1.117)$$

If we introduce nonnegative weights $\lambda_1, \lambda_2, ..., \lambda_n$ we can rewrite the conditional $\tilde{f}(x)$ as a linearized function:

$$\tilde{f}(x) = \lambda_1 f(x_1) + \lambda_2 f(x_2) + \cdots + \lambda_n f(x_n) \quad (1.118)$$

$$x = \lambda_1 x_1 + \lambda_2 x_2 + \cdots + \lambda_n x_n \quad (1.119)$$

$$\sum_{i=1}^{n} \lambda_i = 1 \quad (1.120)$$

$$SOS2(\lambda_1, \lambda_2, ..., \lambda_n) \quad (1.121)$$

Note that for a value x the SOS2 constraints will force the program to evaluate $\tilde{f}(x)$ within the two breakpoints that x belongs to. Consider, for instance, the piecewise linear function of Fig. 1.24. Equations (1.118)–(1.121) for $\lambda_1, ..., \lambda_9 \in \mathbb{R}_{\geq 0}$ and $x = 3.14$ will give us $\tilde{f}(3.14) = 9.905$, which is slightly higher that the actual value of the nonlinear function x^2 at that point, $f(3.18) = 9.860$. This difference is the approximation error between $\tilde{f}$ and f which is equal to zero at the locations of the breakpoints. An important observation is that if we introduce more breakpoints, the approximation error between $\tilde{f}$ and the nonlinear function f is reducing. Notwithstanding this, if we have too many breakpoints it will take more computation time to compute the value of $\tilde{f}$ because the number of variables and constraints increases in Eqs. (1.118)–(1.121).

Exercises

1.1 Derivatives
Compute the derivative of $\sin(a^x)$ with respect to x.

1.2 Gradient Vector
Compute the gradient of $f(\mathbf{x}) = 5x_1^2 + 3x_2^3 + 6^{x_3} + x_1 \sin x_4$.

1.3 Propositional Logic
Write the disjunctive clause $x_1 \vee \neg x_2 \vee x_3 \vee \neg x_4$ as an inequality constraint. Do the same for the conjunctive clause $x_1 \wedge \neg x_2 \wedge x_3 \wedge \neg x_4$.

1.4 Linearity
Is the following program a linear programming problem? Justify your answer.

$$\min_{\mathbf{x}\in\mathbb{R}^2} 5x_1x_2 + 3x_2 + 12x_1x_2$$

$$\text{subject to: } 3x_1 = 5x_2$$
$$2x_1 \leq 7$$
$$4x_2 \leq 29$$
$$x_1 \geq 1$$

1.5 Periodic Event Scheduling

Consider the PESP formulation with minimized slack times in (1.33)–(1.36). Is this formulation a continuous, integer or mixed-integer program?

1.6 Affine Hull

What is the shape of the affine hull of:

- two different points
- three different points that are not colinear

1.7 Global Optimality

Has the minimization problem:

$$\min_{\mathbf{x}\in\mathbb{R}^2} x_1^2 + 6x_2^2 + 5$$

a unique globally optimal solution? Justify your answer.

1.8 Polyhedron

Prove that the following inequality constraints form a polyhedron. If yes, is this polyhedron a polytope?

$$3x_1 + 2x_2 \leq 60$$
$$x_2 \leq 40$$
$$x_2 \geq 1$$
$$-5x_1 - 4x_2 \leq -7$$

1.9 Convex/Concave

Prove that the objective function $\mathbf{c}^\mathsf{T}\mathbf{x}$ where $\mathbf{c}$ is a parameter vector is both convex and concave.

1.10 Convexity

Prove that the following minimization problem is convex.

$$\min_{x\in\mathbb{R}} x^2 + 5x$$

$$\text{s.t. } x \leq 2$$
$$x \geq -5$$

1.11 Convexity preservation
Prove that the following functions are convex by using operations that preserve convexity:

(a) $f(\mathbf{x}) = (15x_1 - 12x_2 + 7)^4 + e^{x_2}$
(b) $f(x) = e^{5x^2}$

1.12 Linearizations
Reformulate the vehicle holding problem presented in Eqs. (1.41)–(1.43) as a linear programming problem.

1.13 SOS1 constraints
Solve the mathematical program:

$$\max \sum_{i \in I} y_i$$
$$\text{s.t. } \text{SOS1}(y_1, y_2, ..., y_n)$$

for $I = \{1, 2, 3, 4, 5\}$ and $y_i \in [0, \beta_i]$ for all $i \in I$, where $\beta_1 = 5, \beta_2 = 7$, $\beta_3 = 8, \beta_4 = 6$ and $\beta_5 = 12$.

1.14 Piecewise linear function
Consider the nonlinear function $f(x) = \frac{1}{x^2} + 0.05x$. Approximate the nonlinear function in the region [1,15] with a piecewise linear function $\tilde{f}(x)$ with breaking points at the x-axis $(x_1, x_2, ..., x_{15}) = (1, 2, ..., 15)$.

1.15 Eigenvalues
Compute the eigenvalues of the following matrix. Is this matrix positive definite?

$$\mathbf{H} = \begin{bmatrix} 10 & 2 & 0 \\ 2 & 0 & 0 \\ 0 & 0 & -6 \end{bmatrix}$$

References

1. O. Ibarra-Rojas, F. Delgado, R. Giesen, J. Muñoz, Transp. Res. Part B: Methodol. **77**, 38 (2015)
2. O. Naud, J. Taylor, L. Colizzi, R. Giroudeau, S. Guillaume, E. Bourreau, T. Crestey, B. Tisseyre, in *Agricultural Internet of Things and Decision Support for Precision Smart Farming*, ed. by A. Castrignan, G. Buttafuoco, R. Khosla, A.M. Mouazen, D. Moshou, O. Naud (Academic Press, 2020), pp. 183–224. https://doi.org/10.1016/B978-0-12-818373-1.00004-4. https://www.sciencedirect.com/science/article/pii/B9780128183731000044
3. S.K. Mitter, V. Borkar, V. Chandru, D. Micciancio, in *Control Using Logic-Based Switching* (Springer, 1997), pp. 79–91
4. in *Logic-Based Decision Support, Annals of Discrete Mathematics*, vol. 40, ed. by R.G. Jeroslow (Elsevier, 1989), pp. 79–102

5. B. Beavis, I. Dobbs, *Optimisation and Stability Theory for Economic Analysis* (Cambridge University Press, 1990)
6. A. Ben-Tal, A. Nemirovski, *Lectures on Modern Convex Optimization: Analysis, Algorithms, and Engineering Applications* (SIAM, 2001)
7. F.L. Hitchcock, J. Math. Phys. **20**(1–4), 224 (1941)
8. P. Serafini, W. Ukovich, SIAM J. Discrete Math. **2**(4), 550 (1989)
9. C. Liebchen, R.H. Möhring, Electron. Notes Theor. Comput. Sci. **66**(6), 18 (2002)
10. R. Borndörfer, N. Lindner, S. Roth, J. Rail Transp. Plan. Manag. **15**, 100175 (2020)
11. M. Goerigk, C. Liebchen, in *17th Workshop on Algorithmic Approaches for Transportation Modelling, Optimization, and Systems (ATMOS 2017)* (Schloss Dagstuhl-Leibniz-Zentrum fuer Informatik, 2017)
12. L. Fu, X. Yang, Transp. Res. Record: J. Transp. Res. Board **1791**(1), 6 (2002)
13. L.A. Wolsey, *Wiley Encyclopedia of Computer Science and Engineering* (2007), pp. 1–10
14. L.A. Wolsey, *Integer Programming* (Wiley, 2020)
15. E.W. Dijkstra et al., Numerische mathematik **1**(1), 269 (1959)
16. M. Guan, Chin. Math. **1**, 237 (1962)
17. J. Park, B.I. Kim, Eur. J. Oper. Res. **202**(2), 311 (2010)
18. C.E. Miller, A.W. Tucker, R.A. Zemlin, J. ACM (JACM) **7**(4), 326 (1960)
19. R.T. Rockafellar, *Convex Analysis*, vol. 36 (Princeton University Press, 1997)
20. U. Bondhugula, A. Hartono, J. Ramanujam, P. Sadayappan, in *Proceedings of the ACM SIG-PLAN 2008 Conference on Programming Language Design and Implementation (PLDI 08), Tucson, AZ (June 2008)* (Citeseer, 2008)
21. H. Weyl, Commentarii Mathematici Helvetici **7**(1), 290 (1935)
22. S. Boyd, S.P. Boyd, L. Vandenberghe, *Convex Optimization* (Cambridge University Press, 2004)
23. H.A. Schwarz, *Gesammelte mathematische abhandlungen*, vol. 260 (American Mathematical Society, 1972)
24. A. Charnes, W.W. Cooper, Naval Res. Logist. Q. **9**(3–4), 181 (1962)
25. E.M.L. Beale, J.A. Tomlin, OR **69**(447–454), 99 (1970)
26. G.B. Dantzig, Econometrica, J. Econom. Soc. 30–44 (1960)

Chapter 2
Introduction to Computational Complexity

Abstract This chapter introduces the Big O, Big Omega, and Big Theta notations which characterize the scalability of algorithms in terms of space requirements and computational times. It also introduces the Polynomial and Non-deterministic Polynomial problem classes which are used to categorize decision and optimization problems. Furthermore, it presents decision problems that are NP-complete and optimization problems that are NP-Hard. Finally, it introduces reduction strategies which can be used to prove that a decision problem is NP-complete.

2.1 Big O Notation

With complexity theory, we investigate how algorithms scale with an increase in the input size of a decision problem. The main complexity types are:

- **Best-case complexity:** the complexity of solving the problem for the best possible input with respect to complexity.
- **Average-case complexity:** the complexity of solving the problem averaged over all possible inputs.
- **Worst-case complexity:** the complexity of solving the problem for the worst possible input with respect to complexity.

Typically, when examining the complexity of an algorithm we are interested in its *worst-case* complexity because it guarantees that the algorithm will finish within the indicated period of time. To examine the complexity of an algorithm, one should devise rules that describe how the algorithm scales when its input size increases. For this purpose, one can use asymptotic notation to describe the bounds on asymptotic growth rates of input sizes. The most common of these notations are O (known as Big O or Big Oh), Ω (known as Big Omega), and Θ (known as Big Theta).

Big O notation is a mathematical notation that describes the asymptotic upper bound of a function when the argument tends towards a particular value or infinity. In computational efficiency, Big O is used to classify algorithms according to how their run time or space requirements grow as the input size grows (see Definition 2.1). It is important to stretch that it can be applied to describe both the run time and the space requirements of an algorithm.

© The Author(s), under exclusive license to Springer Nature Switzerland AG 2022

K. Gkiotsalitis, *Public Transport Optimization*,

https://doi.org/10.1007/978-3-031-12444-0_2

Fig. 2.1 $O(n^2)$ is the Big O
of $T(n) = 6n^2 + 200n$
because for some c, i.e.,
$c = 24$, there exists n_0
such that $0 \le T(n) \le cn^2$
$\forall n \ge n_0$

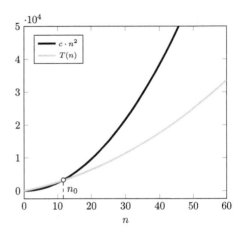

Definition 2.1 If $T(n)$ is the number of required steps for solving a problem or the problem's space requirements, then $O(g(n))$ is the Big O of $T(n)$ if there exists positive constant c for which we can find constant n_0 such that $0 \le T(n) \le cg(n)$ $\forall n \ge n_0$.

The Big O notation is asymptotic and it refers to the upper bound of $T(n)$ for very large problem sizes. Let, for instance, the number of steps needed to solve a problem of size n be $T(n) = 6n^2 + 200n$. A first observation is that $O(n^3)$ is an upper bound of $T(n)$ for large values of n. Indeed $6n^2 + 200n \le n^3$ because $\exists\, n_0 : 6n^2 + 200n \le n^3$ $\forall n \ge n_0$. Based on the definition of Big O, there is also a tighter upper bound of $T(n)$, namely, $O(n^2)$. In more detail, there exists c such that $0 \le 6n^2 + 200n \le cn^2$ for large values of n (see Fig. 2.1). That is, there exists $c > 0$ and n_0 such that $g(n) = n^2$ can keep function $T(n)$ *below it* for all $n \ge n_0$.

In general, as $n \to +\infty$, we seek $g(n)$ such that $O(g(n))$ is close to $T(n)$ in order to achieve a tight upper bound. In more detail, for $n \to +\infty$, the term $6n^2$ dominates $200n$ and, as n^2 grows larger, the effect of 6 is minimal. Hence, if a problem requires $T(n) = 6n^2 + 200n$ steps to be solved, then $g(n)$ can be equal to n^2 and its complexity can be characterized as $O(n^2)$ using the Big O notation. As demonstrated in this example, the following rules are applied when using Big O.

Big O rules for tight upper bounds:
1. If $T(n)$ is a sum of several terms, we can keep the term with the largest growth rate and omit all others.
2. If $T(n)$ is a product of several factors, any constants that do not depend on n can be omitted.

Let us consider an example to clarify the meaning of $T(n)$ and $O(n)$. Consider that we have a set $S = \{s_1, s_2, \ldots, s_n\}$ of n real numbers and we want to find:

- If the number of elements in the set is lower than a number k
- Which is the element with the smallest value.

  To do this, one can use the following algorithm.

Algorithm 1 Find the smallest number in a set and check if set size $< k$

1: given $S = \{s_1, s_2, \ldots, s_n\}$ and some integer $k > 0$
2: **if** $n \geq k$ **then**
3:    **return:** size n is not smaller than k
4: **end if**
5: initialize smallest_number=$+\infty$
6: **for** $i \in S$ **do**
7:    **if** $i \leq$ smallest_number **then**
8:       smallest_number=i
9:    **end if**
10: **end for**
11: **return:** smallest_number

Lines 2 and 3 of the algorithm require 2 steps at most (one to check if $n \geq k$ and at most one to declare that size n is not smaller than k). Line 5 requires one assignment step to initialize the value of smallest_number. In lines 6–10, each time we go around the for loop we count 1 step for selecting another i (line 6), 1 step to perform the check $i \leq$ smallest_number, and (at most) 1 step to perform the assignment statement smallest_number=i. In total, we need to execute a maximum of 3 steps each time we go around the for loop, and, because we go around it n times, we need $3n$ steps. Finally, we need 1 step in line 11 to return the value of the smallest_number. In total, we need $T(n) = 4 + 3n$ steps. Using the Big O notation, the upper bound of $T(n)$ for very large n is $O(n)$ indicating that the algorithm is of linear complexity.

With the Big O notation, algorithms can be characterized based on their computational complexity and/or their space requirements at worst case scenarios. Examples of the main Big O complexities are presented in Table 2.1 in ascending order.

The increase in the number of operations based on the input size n of a problem is presented in Fig. 2.2 considering a list of the most common complexities.

To summarize, Big O offers an asymptotic upper bound to a function. However, this bound can be loose (for instance, $O(n^5)$ is a valid upper bound for $T(n) = n^2$). We will later see that we can achieve tighter bounds by using the Big Theta notation.

Table 2.1 Big O complexities in ascending order

Big O	Description	Example
$O(1)$	Constant—not related to the problem's size	$T(72)$
$O(\log n)$	Logarithmic	$T(250 + \log 5n)$
$O(n^c)$ with $0 < c < 1$	Fractional power	$T(7n^{0.8})$
$O(n)$	Linear	$T(32n + 17n + 3n)$
$O(n \log n)$	Loglinear	$T(52n \log n)$
$O(n^c)$ with $c > 1$	Polynomial	$T(57n^4 + 82n^3 + 71n)$
$O(c^n)$ with $c > 1$	Exponential	$T(2^n + 6n^7)$
$O(n!)$	Factorial	$T(n! + 5n^2 + 6)$

Fig. 2.2 Computational complexity showing the number of operations based on the input size n

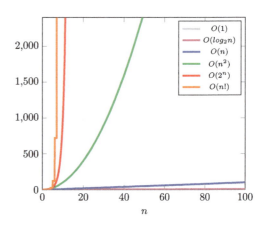

2.2   Big Omega (Ω) and Big Theta (Θ) Notations

To understand the Big Theta notation, we will briefly introduce the less common Big Omega (Ω) notation which is used to identify the best case complexity. Ω is an asymptotic lower bound to function $T(n)$, where $T(n)$ represents the actual computation time or storage space requirements of an algorithm. It is used if we want to say that an algorithm will take at least a certain amount of time or space.

Definition 2.2 *Big Omega* (Ω). If $T(n)$ is the number of required steps for solving a problem or the problem's space requirements, $\Omega(g(n))$ is the Big Omega of $T(n)$ if there exists positive constant c for which we can find constant n_0 such that $0 \leq cg(n) \leq T(n) \ \forall n \geq n_0$.

We can take again the example of $T(n) = 6n^2 + 200n$. We note that $g(n) = n$ is Big Omega of $T(n)$ because there exists $c > 0$ such that $0 \leq cn \leq 6n^2 + 200n \ \forall n \geq n_0$ for some n_0. This asymptotic lower bound is valid, but it can be tightened. For instance, n^2 is also Big Omega of $T(n)$. Indeed, as shown in Fig. 2.3, there exists a small enough $c > 0$ such that $0 \leq cn^2 \leq 6n^2 + 200n \ \forall n \geq n_0$ for some n_0.

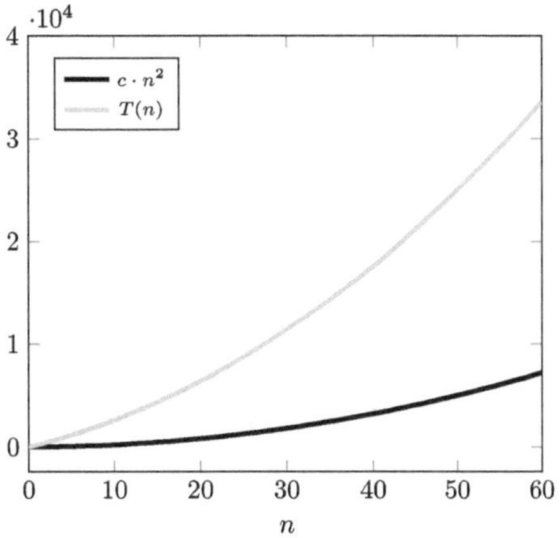

Fig. 2.3 $\Omega(n^2)$ is the Big Omega of $T(n) = 6n^2 + 200n$ because for some c, i.e., $c = 2$, there exists n_0 such that $0 \leq cn^2 \leq T(n) \ \forall n \geq n_0$

Fig. 2.4 $\Theta(n^2)$ is the Big Theta of $T(n) = 6n^2 + 200n$ because for some c_1, i.e., $c_1 = 2$, and c_2, i.e., $c_2 = 24$, there exists n_0 such that $0 \leq c_1 n^2 \leq T(n) \leq c_2 n^2 \ \forall n \geq n_0$

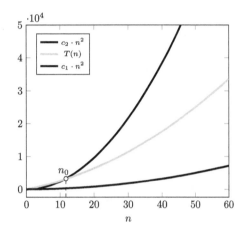

With the definition of Ω, we can now proceed to the definition of Big Theta. Big Theta is the exact (tight) asymptotic bound of a function $T(n)$. In more detail, for a function $T(n)$ we seek to find a function $g(n)$ which after multiplied with the right constants bounds it from below and from above for large values of n. Formally, Big Theta is defined as follows.

Definition 2.3 *Big Theta* (Θ). If $T(n)$ is the number of required steps for solving a problem or the problem's space requirements, $\Theta(g(n))$ is the Big Theta of $T(n)$ if there exist positive constants c_1 and c_2 for which we can find constant n_0 such that $0 \leq c_1 g(n) \leq T(n) \leq c_2 g(n) \ \forall n \geq n_0$.

For our example function $T(n) = 6n^2 + 200n$ it follows that its Big Theta is $\Theta(n^2)$ because there exist positive c_1 and c_2 such that $0 \leq c_1 g(n) \leq T(n) \leq c_2 g(n) \ \forall n \geq n_0$, as presented in Fig. 2.4.

Summarizing the asymptotic notations, Big O provides an upper asymptotic bound and ensures that, at the worst case, the running time or the memory space required from an algorithm will be below that upper bound. Big Omega provides a lower asymptotic bound indicating that the running time or the memory space required from an algorithm will be at least the one suggested from the Big Omega notation. Finally, Big Theta is both an upper and lower asymptotic bound, thus providing an exact bound.

2.3 P vs NP

After the introduction to algorithmic complexities, we divert our attention to polynomial (P) vs Non-deterministic polynomial (NP) problem classes. The P versus NP problem is one of the most important unsolved problems in computer science. It asks if P=NP, or, equivalently, if a decision problem whose solution can be quickly *checked* in polynomial time can be also *solved* in polynomial time [1]. To understand more about P vs NP and the importance of this problem, we will define the P and NP problem classes. Before doing so, we introduce TIME and NTIME.

Definition 2.4 TIME($f(n)$) is the class of all decision problems (or languages) that can be decided by a *deterministic* Turing machine whose maximum running time is $O(f(n))$, where n is the size of the problem that needs to be decided.

Definition 2.5 NTIME($f(n)$) is the class of all decision problems (or languages) that can be decided by a *nondeterministic* Turing machine whose maximum running time is $O(f(n))$, where n is the size of the problem that needs to be decided.

One can imagine a Deterministic Turing Machine (DTM) as a simple computer with a deterministic set of rules that prescribe at most one action to be performed for any given situation. In contrast, a Nondeterministic Turing Machine (NTM) is a theoretical model of computation where its set of rules may prescribe more than one action to be performed for any given situation [2].

A decision problem is a problem that has a yes or no (or, equivalently, True or False) answer. We thus say that a decision problem can be decided in TIME($f(n)$) if a DTM can decide whether an instance of this problem is True or False in time $O(f(n))$. To provide an example, a decision problem can be *"is t a prime number?"* Given the value of t and a set of prescribed rules (algorithm), a DTM can decide whether this statement is true or false.

At the definitions of TIME and NTIME we used the terms decision problems (or languages). We briefly note that a formal language is a set of strings whose letters are taken from an alphabet and are well-formed according to a specific set of rules [3]. The connection between a *decision problem* and a *formal language* is that a decision problem can be encoded as a set X of strings for which the answer to the decision problem is "yes". That is, X includes all strings that satisfy the decision problem. For example, for the decision problem *"is t a prime number?"* we can define a formal language:

$X = \{$list of strings for which the decision problem *"is t a prime number?"* has a "yes" answer.$\}$

An instance of the decision problem x can be also encoded as a string. Using the string encoding, x is a "yes" answer of the decision problem if, and only if, $x \in X$. Then, asking an algorithm[1] to decide whether an instance x of a decision problem is true or false is equivalent to checking whether this instance is a member of X, that is $x \in X$.

We now proceed to the definition of P.

P (Polynomial)

P (Polynomial) is the class (set) of all decision problems that can be decided by a deterministic Turing machine in an amount of time that is polynomial in the size of the input, or can be upper bounded by a polynomial. Formally,

$$P = \bigcup_k \text{TIME}(n^k)$$

where k can be any positive constant and $\text{TIME}(n^k)$ is the complexity class of problems that can be decided by a DTM in time $O(n^k)$.

The definition implies that a decision problem is in P if there exists an algorithm (called also decision procedure) that runs in (at most) polynomial time for deciding this problem. The required steps, S, of this algorithm when solving the decision problem should be $S = f(n)$, where n is the size of the decision problem and f a polynomial function, i.e., $f(n) = n^k$ for $k \geq 0$. Using the Big O notation, these problems can be solved in $O(n^k)$, where k is a constant.

Decision problems that can be decided by a polynomial-time algorithm are called computationally *tractable* problems. These problems can be solved quickly since the running times of polynomial-time algorithms increase polynomially with the size of the decision problem (as opposed to, for example, exponentially—see Fig. 2.2). Typical examples of problems that are in P are the sorting problem [4] or Dijkstra's algorithm for the shortest path decision problem [5].

Let us consider again the P vs NP problem which asks if P=NP. This problem is very important because if P≠NP, then NP decision problems are intractable [6]. The debate of P vs NP stems from the fact that for many problems that were previously thought to be NP, there have been introduced efficient polynomial-time algorithms. An example is linear programming that was thought to be an NP problem before the introduction of the Ellipsoid method, which was the first worst-case polynomial-time algorithm [7].

To underline the difference between P and NP, we hereby define the NP problem class. The NP class has two equivalent definitions. We start from the definition that is closer to the definition of P.

---

[1] Algorithm A of a decision problem is called the algorithm for which $A(x) = yes \Leftrightarrow x \in X$, where x refers to any instance of the decision problem and X to the set of strings that have a 'yes' answer to the decision problem.

Fig. 2.5 Relation between P
and NP problems assuming
that P≠NP, and thus
NP-P≠ ∅. Note that over
time the size of set P
increases because there are
new research discoveries of
polynomial-time algorithms
for previously thought NP
problems

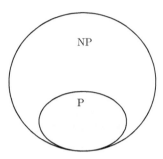

NP (Non-deterministic Polynomial)—NTM definition

NP (Non-deterministic Polynomial) is the class (set) of all decision problems that
can be decided by a *nondeterministic* Turing machine in an amount of time that
is polynomial in the size of the input, or can be upper bounded by a polynomial.
Formally,

$$NP = \bigcup_k NTIME(n^k)$$

where k can be any positive constant.

This definition implies directly that P is a subset of NP if P≠NP because a non-
deterministic Turing machine can decide all problems in P in time $O(n^k)$, where k
is any positive constant (see Fig. 2.5).

Let us now proceed to the second, equivalent, definition of the NP class that uses
a deterministic Turing machine instead of a nondeterministic one.

NP (Non-deterministic Polynomial)—DTM definition

NP (Non-deterministic Polynomial) is the class (set) of all decision problems for
which for every problem instance which is decided as True, if an oracle machine
provides a *certificate*,[2] the certificate for this instance can be verified by a *certifier*[3]
in polynomial time using a deterministic Turing machine.

It is important to note that the second definition does not require to *decide* the
instance of the decision problem. Instead, we are provided already with a *certificate*,
which is a solution of this instance, and we are required to prove that there exists an
algorithm (called certifier) that can verify whether this certificate is True or False in
polynomial time. Finding a certificate is a much more complicated task, but here it
is assumed that this certificate is provided by an oracle machine and we just have to

---

[2] Also known as *witness* or *proof*.

[3] Also known as *verifier*.

verify it. The task of our certifier C is to prove in polynomial time that certificate c is a "yes" or "no" solution for the instance x of the decision problem. If the answer is yes, then the certificate c is a proof that instance x is True. Note that here we use the terms "yes" and "true" interchangeably.

Using the concept of languages, NP can be seen as the class of problems that are *efficiently checkable*. Efficiently checkable means that for any string x in the set X of all strings that satisfy the decision problem:

- There is a certificate c
- The length of the certificate when it is encoded as a string is $|c| \leq \text{poly}(|x|)$, where poly stands for polynomial
- Given c, x we can efficiently check with a polynomial time certifier C whether $C(c, x) = \text{yes}$, meaning that x is a yes instance of the decision problem ($x \in X$)

To understand this better, consider the *subset sum* decision problem that asks whether given a set S of integers $\{s_1, s_2, \ldots, s_n\}$, there is a subset $T \subseteq S$ such that the integers in that subset have a sum equal to zero. To answer this question for a specific set S, we need to create all possible subsets of that set and sum their integer elements to check if there is a subset with integers that sum to zero. As the size n of set S increases, the number of possible subsets increases exponentially and there is no polynomial-time algorithm to perform this task. Thus, the subset sum problem is not in P. It is important here to make the following categorization to describe the decision problem of our example in a more structured way:

- **Decision Problem:** the *subset sum* is our decision problem that has a "yes" or "no" answer for a specific problem instance. We can consider set X that contains all strings that satisfy this decision problem.
- **Instance x:** the specific set S based on which we should return a "yes" or "no" answer to our decision problem. I.e., does set S contain a subset such that its integers have a sum equal to zero? This is equivalent to ask whether $x \in X$, where x is the encoded string of instance $x = \{\text{set } S\}$.
- **Certificate c:** a certificate (or witness or proof) is a solution of this problem instance provided by an oracle. In particular, a subset T with integers that have a sum equal to zero.

Note that this categorization holds if instance x has a "yes" answer (equivalently, string $x \in X$) because, if $x \notin X$, a certificate does not exist and it cannot be provided by an oracle. This can be generalized as $\exists$ certificate c if, and only if, $x \in X$. This is why in the definition of NP with DTM we ask that the considered problem instances have a "yes" answer (decided as being True). The process is illustrated in Fig. 2.6.

For a problem instance $x = \{\text{set } S\}$ and a certificate $c = \{\text{subset } T\}$ provided by an oracle, the certifier $C(x, c)$ is an algorithm which verifies if the certificate is indeed a "yes" answer for instance x. This certifier is provided in the following algorithm.

Fig. 2.6 If for any instance
x with 'yes' answer to
decision problem X there is
a polynomial time certifier
$C(x, c)$ that can verify
whether the certificate c is a
yes or no answer to X for
instance x, then X is in NP

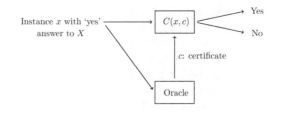

Algorithm 2 Certifier of Subset Sum solution

1: given certificate/witness T
2: initialize $sum = 0$
3: **for** $i \in T$ **do**
4:     $sum = sum + i$
5: **end for**
6: **if** $sum \neq 0$ **then**
7:     **return:** FALSE
8: **else**
9:     **return:** TRUE
10: **end if**

The certifier in Algorithm 2 requires at most $4 + 2|T|$ steps and it is of linear complexity $O(|T|)$, where $|T|$ is the cardinality of subset T indicating the number of integers in the subset. Therefore, we have a polynomial time certifier that can verify our certificate, and thus the subset sum decision problem is in NP.

To summarize, NP problems can be seen as decision problems for which if an oracle machine provides a solution (certificate) for an instance that has a yes solution, we can verify it (i.e., check if it satisfies our decision problem) in at most polynomial time. That is, there exists a certifier that can determine in polynomial time $O(n^k)$ whether the solution from the oracle satisfies (i.e., returns a yes answer) to our decision problem. This is the critical difference between P and NP since:

- P class decision problems can be *decided* in polynomial time.
- NP class decision problems can *verify whether an externally provided certificate is a yes answer to an instance of our decision problem that has a yes answer* in polynomial time.

Essentially, if P = NP that would mean that being able to quickly verify whether a solution is correct is equivalent to being able to find a quick solution. Although it is argued that there are no polynomial-time algorithms for problems in the set NP-P, this has not been proven yet. In fact, there have been several discoveries of polynomial time algorithms for previously thought NP problems. As previously discussed, linear programming was considered to be NP before the introduction of the Ellipsoid algorithm in 1979 [7]. In 2004, the Agrawal-Kayal-Saxena (AKS) polynomial-time algorithm solved the decision problem of determining whether an integer n is prime moving it from NP-P to P [8].

Closing this section, we provide some decision problem examples with their *instances* and their *certificates*. For brevity, we define these problems in undirected graphs although some of them can also be defined in directed ones. One can use these certificates to prove that these problems are in NP.

1. *Hamiltonian Cycle*: given an undirected graph, does it have a cycle that visits each vertex exactly once?

   - **Instance** x: An undirected graph $G = \{V, E\}$ with V vertices and E edges.
   - **Certificate** c: An ordered set $P = \{v_1, v_2, \ldots, v_n\}$ of vertices that contains each vertex $v \in V$ exactly once and for any two adjacent vertices v_i, v_{i+1} there exists an edge, $(v_i, v_{i+1}) \in E$.

2. *3SAT*: given a CNF where each clause is a disjunctive clause with exactly 3 literals, is there a satisfying assignment?

   - **Instance** x: A CNF $(l_1 \vee l_2 \vee l_3) \wedge \ldots \wedge (l_{n-2} \vee l_{n-1} \vee l_n)$.
   - **Certificate** c: A CNF with n literals that has at least one true literal in each disjunctive clause.

3. *CLIQUE*: given an undirected graph, does it have a subgraph with k adjacent vertices?

   - **Instance** x: A graph $G = \{V, E\}$ of vertices and edges and positive integer k.
   - **Certificate** c: An ordered set $P = \{v_1, v_2, \ldots, v_k\}$ that contains k vertices where each vertex $v \in V$ is included at most once. For any two vertices $v, u \in P$ there exists an edge, $(v, u) \in E$.

4. *Traveling Salesman Problem (TSP)*: given an undirected graph and distances for the edges of the graph, is there a possible tour with cost less than or equal to a prescribed cost?

   - **Instance** x: A graph $G = \{V, E\}$ of vertices and edges, a set W of the distances of the edges, and a prescribed cost k.
   - **Certificate** c: An ordered set of vertices $P = \{v_1, v_2, \ldots, v_n\}$ that contains each vertex $v \in V$ exactly once. For any two adjacent vertices v_i, v_{i+1} there exists an edge, $(v_i, v_{i+1}) \in E$, and the sum of distances of all these edges is less than or equal to k.

Notice that the decision problem of TSP is the Hamiltonian Cycle decision problem with the extra consideration of edge costs.

2.4   NP-complete, NP-Hard and Other Classes

We already discussed the P vs NP problem. Another interesting class is the NP-complete class which is a subclass of the NP set and it contains the hardest decision problems in NP. Formally [9]:

NP-complete

NP-complete is the class (set) of all decision problems $X \in$ NP-complete, where X:

1. is in NP;
2. and every problem k in NP is reducible to X, ($k \leq_p X$), in polynomial time.

In the definition of NP-complete decision problems, we do not require from an NP-complete decision problem X to be just in NP, but we also require that every problem k in NP is reducible to X in polynomial time. We will later see that the term *reduction* is used to describe the conversion of one decision problem into a second decision problem in order to measure their relative computational difficulty. It is said that decision problem k is polynomial time reducible to X ($k \leq_p X$) if X is at least as hard to solve as k. That is, a decision problem X is NP-complete if it is NP and any problem k in NP is reduced to X in polynomial time, meaning that X is at least as hard as any other NP problem k.

It is important to note that there are no discoveries of algorithms that can decide NP-complete decision problems in polynomial time. The same is not true for problems in NP that are not NP-complete, i.e., we already showed that there have been discoveries of polynomial-time algorithms for previously thought NP problems. The first decision problem that was proven to be NP-complete based on the definition of NP-complete problems was the SAT (Boolean satisfiability problem) in the early 1970s by Cook [10] and Levin [11] (known as the Cook-Levin Theorem). A SAT problem asks whether the variables of a given Boolean formula can be replaced by the values TRUE or FALSE in such a way that the formula evaluates to TRUE. SAT is clearly in NP because any assignment of Boolean values to Boolean variables that is claimed to satisfy the given expression can be verified in polynomial time by a deterministic Turing machine.

In 1978, Schaefer [12] identified five classes of Boolean relations for which SAT is in P (known as Schaefer's dichotomy theorem):

- Setting all variables true or all variables false satisfies all clauses;
- All relations are equivalent to a conjunction of binary clauses (2-CNF);
- All relations are equivalent to a conjunction of Horn clauses, where a Horn clause is a disjunction of literals with at most one unnegated literal;
- All relations are equivalent to a conjunction of dual-Horn clauses, where a dual-Horn clause is a disjunction of literals with at most one negated literal;
- All relations are equivalent to a conjunction of affine formulas.

All other SAT decision problems are NP-complete. We can now use the definition of NP-complete problems to present also the NP-Hard class of problems.

NP-Hard

NP-Hard is the class (set) of all decision problems $X \in$ NP-Hard for which every problem k in NP is reducible to X (that is, $k \leq_p X$) in polynomial time.

It follows that a decision problem is NP-Hard if it satisfies the second requirement of the NP-complete definition. That is, every NP-complete problem is NP-Hard, but the reverse is not always true since an NP-Hard problem might not be in NP. That is, an NP-Hard problem should be as hard to solve as any other problem in NP but it is not required to be in NP itself. In contrast, an NP-complete problem must be as hard as any other problem in NP, and, at the same time, it should be in NP. Formally, NP-complete=NP-Hard $\cap$ NP (see Fig. 2.7). Considering the implications of the P vs NP problem, if it is proven that P=NP, then NP-complete=P=NP.

Aside from P and NP, there are also other important classes of decision problems, such as NEXP (known also as NEXPTIME) for which verifying the certificate with a deterministic Turing machine takes exponential time (or, equivalently, a non-deterministic Turing machine can solve them in exponential time). Formally,

$$\text{NEXP} = \bigcup_k \text{NTIME}(2^{n^k})$$

Fig. 2.7 Relation between P, NP, NP-complete and NP-Hard if P$\neq$NP

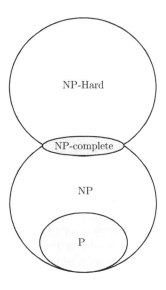

where n is the size of the decision problem and k any possible positive constant. Following the definitions of NP-complete and NP-Hard, there are similar definitions for NEXP-complete and NEXP-Hard.

2.5 Decision vs Optimization Problems

It is worth noting that the above complexity classes were defined for decision problems. The analysis was primarily focused on decision problems because it is easier to determine their difficulty compared to optimization problems. In reality, however, being able to solve a decision problem in a specific time class will provide an indication of the class of its corresponding optimization problem (if such a problem exists). Because of this, determining the difficulty of a decision problem can be used to determine the difficulty of its corresponding optimization problem. To understand the difference between a decision and an optimization problem, the definition of a decision problem is provided below.

Decision Problem

A decision problem is a problem that has only two possible outputs (yes or no) on any input.

An example of a decision problem is deciding whether a given number is prime. The answer to that decision problem can be a yes or a no for a given input. An algorithm A that solves a decision problem is called a *decision procedure*.

Many optimization problems can be translated to decision problems to determine their complexity. Unlike decision problems, for which there is only one correct answer for each input, optimization problems seek to find the best answer to a particular input. For example, the Traveling Salesman Problem, which seeks to find the shortest tour of a salesman who visits different cities and it was expressed in Eqs. (1.49–1.55), can be posed as a decision problem as follows:

"for a given cost k, decide whether the graph has any tour with cost less than or equal to k".

By repeatedly answering this yes or no decision problem, it is possible to find the shortest tour. Generally, in an optimization problem we ask to find the best possible solution, whereas in a decision problem we provide input k and we ask if there is a satisfiable solution for k (*yes* or *no* answer). That is, an optimization problem can be turned into a decision problem by introducing a threshold value k and asking whether there exists a feasible solution (yer or no question) for value k. Then, by repeatedly answering the yes or no decision problem with different input values k, we can find a solution for its corresponding optimization problem.

In another example, consider the shortest path optimization problem from an origin A to a destination B. This can be posed as a decision problem as follows: *"for a given cost k, decide whether there is a path from A to B with cost less than or equal to k"*. Again, by repeatedly answering this problem for different values of k we can find the shortest (minimum cost) path from A to B.

It becomes clear from the definition of decision problems that an optimization problem is at least as hard as its corresponding decision problem because it requires to perform also a search. Therefore, we can say that:

- If an optimization problem is solved in *polynomial time*, then its corresponding decision problem (if it has one) is in P since it is not harder to solve than the optimization problem. Consider, for example, the shortest path optimization problem. A polynomial time algorithm (Dijkstra) is known for that problem. Thus, its corresponding decision problem is also in P.
- If a decision problem is NP-complete, then its corresponding optimization problem (if it has one) is NP-Hard because it is at least as hard as the decision problem (but it can be harder). Consider, for example, the traveling salesman optimization problem. This is an NP-Hard problem since its corresponding decision problem is NP-complete.

The latter observation is very important because it allows us to prove that an optimization problem is NP-Hard by knowing that its corresponding decision problem is NP-complete. As we will later see, there is a list of known NP-complete problems that can be used to categorize our decision problem of interest with the use of poly-time reductions. This classification of the difficulty of the optimization problem is crucial because if a problem is characterized as NP-Hard it effectively means that the size of problem which can be solved optimally using mathematical programming or combinatorial optimization is limited (*intractable* problem).

2.6 Reductions

We discussed the main classes of decision problems. It is often difficult, however, to categorize a problem into one of these classes. To this end, it is very useful to use reductions that can reduce a problem with known complexity to the examined problem. By doing so, we can prove that the examined decision problem belongs to a particular class.

Definition 2.6 A *poly-time many-one reduction* of problem A to problem B ($A \leq_P B$) is a polynomial time algorithm R that converts any instance of one decision problem A into an instance of a decision problem B such that the instance of problem A is true if, and only if, the produced instance of problem B is also true.

The subscript p in the reduction $\leq_p$ indicates that the translation algorithm R is of polynomial complexity. From the definition of poly-time reduction (where poly-time stands for polynomial time), it follows that if we know the class of B and A is reduced to B, then if B is decidable, A is also decidable. In addition, if A is undecidable, then B is also undecidable. We will later show that:

- If B is in P and A$\leq_p$B, then A is also in P.
- If A is NP-complete and A$\leq_p$B and B is in NP, then B is also NP-complete.
- If A is NP-complete or NP-Hard and A$\leq_p$B, then B is NP-Hard.

Let us discuss first the poly-time many-one reduction. Consider an instance x of a decision problem A. This instance is evaluated by an algorithm a and it returns a decision (True or False). For example, for the TSP decision problem an instance x can be "for a given cost k, does the graph $G = \{V, E\}$ have any tour that costs less than k?". This can be either True or False and it is checked by an algorithm a. Now, consider a decision problem B. An instance y of a decision problem B can be decided by an algorithm β. For illustration purposes, this is presented in Fig. 2.8. We have a poly-time many-one reduction $A \leq_p B$ if there exists an algorithm R that:

1. Takes instance x of A and translates it to instance y of B, namely, $R(x) = y$, so that:
$$a(x) = \text{TRUE if, and only if, } \beta(R(x)) = \text{TRUE}$$

   That is, an instance x of problem A is true if, and only if, an instance $R(x)$ of problem B is true (and thus x of A is false if, and only if, $R(x)$ of B is false).
2. The algorithm R transforms x to $R(x)$ in polynomial time $O(n^c)$, where n is the input size of x and c a positive constant.

Let us now show that if B is in P and A$\leq_p$B, then A is also in P. If B is in P, there exists algorithm β that determines whether an instance y of B is True or False in polynomial time $O(n^c)$, where n is the input size of y. Any input x of A can be translated to input $y = R(x)$ of B in polynomial time with algorithm R. That is, we would require some polynomial time $O(n^c)$ to make the translation from x to $R(x)$, where n is the size of instance x, and an additional polynomial time to solve instance $y = R(x)$ of B with algorithm β. Note that algorithm β is of polynomial time and can be called a polynomial number of times, thus the computational time of deciding any instance x of A when using algorithms R and β is bounded from above by a

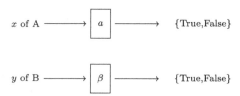

Fig. 2.8 Evaluating instances x and y of decision problems A and B, respectively, with algorithms a and β. A$\leq_p$B if there exists a polynomial time algorithm R that translates x to $y = R(x)$ such that A(x) is true if, and only if, B($R(x)$) is true

polynomial. Hence, A is also in P. From this, it follows that if $A \leq_p B$ and A is not in P, then B is also not in P because if B was in P then A would have also been in P.

Now we will show how one can use poly-time reductions to prove that a decision problem is NP-complete. Reckon that a problem B is NP-complete if it belongs to NP and all other problems A in NP can be reduced to B in polynomial time. This means that all A problems are at most as complex as B, or, equivalently, B is at least as hard as any other problem in NP. Using reductions, we can use decision problems for which we know their class to determine the class of other problems. An important class categorization for a problem B is whether it is NP-complete or NP-Hard because this will mean that it is computationally intractable provided that $P \neq NP$. Using reductions, we can use the following proof strategies.

Prove that a problem B is NP-complete

To prove that a problem B is NP-complete it suffices to prove that:

1. B is in NP
2. There is a known NP-complete problem A that can be reduced to B in polynomial time ($A \leq_p B$).

If one can prove both conditions, then B is NP-complete because it is at least as hard as the NP-complete problem A and it also belongs to NP. Note that since B is at least as hard as the NP-complete A, it follows that any other problem Y in NP is poly-time reducible to B. Formally, any problem Y in NP is $Y \leq_p A$ because A is NP-complete. Because the poly-time reductions are *transitive*, $A \leq_p B$ means that $Y \leq_p B$ for any problem Y in NP.

To prove that a problem B is NP-Hard:

We should prove that there is a known NP-complete or NP-Hard problem A that can be reduced to B.

Indeed if there is a known NP-complete or NP-Hard problem A that can be reduced to B, then B is at least as hard as the hardest problems in NP.

In order to prove the class of a problem with reductions, we should use the classes of known NP-complete problems. As of now, there are many problems for which we know their classes and they can be used for reduction purposes. For instance, in 1972 Karp provided a list of 21 NP-complete problems [13] and in 1979 Garey and Johnson presented more than 300 NP-complete problems [14]. Since then, many more problems have been characterized as NP-complete or NP-Hard. In Fig. 2.9 we present Karp's 21 NP-complete problems with the reduction hierarchy that was used in Karp's proofs (i.e., the satisfiability problem was reduced to clique, clique was reduced to vertex cover, etc.). All 21 decision problems were proven to be NP-complete with poly-time reductions starting from the boolean satisfiability (SAT)

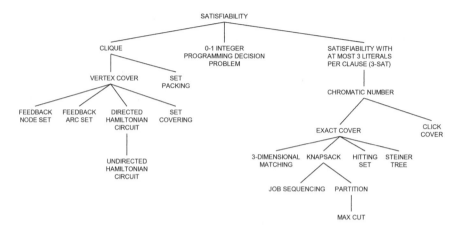

Fig. 2.9 Karp's 21 NP-complete problems

problem that was proven to be NP-complete in Cook's (also known as Cook-Levin) theorem. Note that many public transport problems, such as PESP, are proven to be NP-complete using a poly-time reduction and the fact that SAT is NP-complete [15].

The description of each one of the 21 NP-complete problems according to Karp is provided in Table 2.2.

2.6.1 Reduction Example I: SAT $\leq_p$ 3SAT

Let us discuss one of Karp's proofs of NP-completeness. 3SAT is NP-complete because the NP-complete boolean satisfiability problem (SAT) reduces to 3SAT and 3SAT is in NP. SAT can have a number of clauses $\sigma_1 \vee \sigma_2 \vee \ldots \vee \sigma_m$. 3SAT is a CNF SAT where each clause is a disjunctive clause of exactly three literals. That is, 3SAT = $\{\phi \mid \phi$ is a satisfiable 3CNF-formula$\}$. An example of 3SAT with two clauses is $(\sigma_1 \vee \sigma_2 \vee \sigma_3) \wedge (\sigma_4, \neg\sigma_5, \sigma_6)$. To prove that unrestricted SAT with at least 3 literals $\leq_p$ 3SAT, we use algorithm R that takes a SAT instance x and for each clause $\sigma_1 \vee \ldots \vee \sigma_m$:

- If the clause has one literal ($m = 1$), then introduces two new variables u_1 and u_2 and replaces σ by the equisatisfiable $(\sigma \vee u_1 \vee u_2) \wedge (\sigma \vee \neg u_1 \vee u_2) \wedge (\sigma \vee u_1 \vee \neg u_2) \wedge (\sigma \vee \neg u_1 \vee \neg u_2)$
- If the clause has two literals ($m = 2$), then introduces new variable u and replaces $\sigma_1 \vee \sigma_2$ by the equisatisfiable $(\sigma_1 \vee \sigma_2 \vee u) \wedge (\sigma_1 \vee \neg\sigma_2 \vee \neg u)$
- If the clause has three literals ($m = 3$), then this clause remains unchanged
- If the clause has $m > 3$ literals, it transforms it into clauses $(\sigma_1 \vee \sigma_2 \vee u_1) \wedge (\sigma_3 \vee \ldots \vee \sigma_m \vee \neg\sigma_1) \wedge (\neg\sigma_3 \vee u_1) \wedge \ldots \wedge (\neg\sigma_m \vee u_1)$ where u_1 is a new variable and repeats this transformation until each of the above clauses has no more than three literals

Table 2.2 Karp's 21 NP-complete decision problems. Karp's notation in [13] is mostly preserved

Decision problem	Description
0–1 Integer Programming decision problem	There exists a 0–1 vector $\mathbf{x}$ such that $\mathbf{Cx} = \mathbf{d}$ for an integer matrix $\mathbf{C}$ and vector $\mathbf{d}$
Clique	Graph $G = \{V, E\}$ has k mutually adjacent nodes
Set Packing	Family of sets $\{S_j\}$ contains l mutually disjoint sets
Vertex Cover (also known as Node Cover)	There is a set $C \subseteq V$ in graph $G = \{V, E\}$, such that $|C| \leq k$ for a $k > 0$ and every edge $(i, j) \in E$ has at least one endpoint at some vertex in set C
Set Covering	For a finite family of sets $\{S_j\}$ and $k > 0$, there is a sub-family $\{T_h\} \subseteq \{S_j\}$ containing $\leq k$ sets such that $\cup T_h = \cup S_j$
Feedback Node Set	There is a set $R \subseteq V$ such that every directed cycle of digraph H contains a node in R
Feedback Arc Set	There is a set $S \subseteq E$ such that every cycle of digraph H contains an arc in S
Directed Hamilton Circuit	A digraph H has a directed cycle which includes each node exactly once.
Undirected Hamilton Circuit	A graph G has a cycle which includes each node exactly once.
3SAT	Conjunction of clauses $C_1, \ldots, C_3$ each containing at most 3 literals per clause is satisfiable
Chromatic Number	For graph G and given positive integer k there is a function $\phi : N \mapsto Z_k$ such that, if u and v are adjacent, then $\phi(u) \neq \phi(v)$
Clique Cover	For graph G' and given positive integer l, N' is the union on l or fewer cliques.
Exact Cover	For family $\{S_j\}$ of subsets of a set $\{u_i, \ i = 1, 2, \ldots, t\}$, there is a subfamily $\{T_h\} \subseteq \{S_j\}$ such that the sets T_h are disjoint and $\cup T_h = \sum S_j = \{u_i, \ i = 1, 2, \ldots, r\}$.
Hitting Set	For a family $\{U_i\}$ of subsets of $s_j, \ j = \{1, 2, \ldots, r\}$ there is a set W such that, for each i, $|W \cap U_i| = 1$.
Steiner Tree	For graph G, $R \subseteq N$, weighting functions $w : A \mapsto Z$, and positive integer k, G has a subtree of weight $\leq k$ containing the set of nodes in R.
3-Dimensional Matching	For set $U \subseteq T \times T \times T$, where T a finite set, there is a set $W \subseteq U$ such that $|W| = |T|$ and no two elements of W agree in any coordinate.
Knapsack	For input $(a_1, \ldots, a_r, b) \in \mathbb{Z}^{r+1}$, $\sum_{j=1}^{r} a_j x_r = b$ has a 0–1 solution.
Job Sequencing	For execution vector $T = (T_1, \ldots, T_p) \in \mathbb{Z}^p$, deadline vector $D = (D_1, \ldots, D_p) \in \mathbb{Z}^p$, penalty vector $P = (P_1, \ldots, P_p) \in \mathbb{Z}^p$ and positive integer k there is a permutation π of $\{1, 2, \ldots, p\}$ such that $\sum_{j=1}^{p} [\text{if } T_{\pi(1)} + \cdots + T_{\pi(j)} > D_{\pi(j)} \text{ then } P_{\pi(j)} \text{ else } 0] \leq k$.
Partition	For $(c_1, \ldots, c_S) \in \mathbb{Z}^s$ there is a set $I \subseteq \{1, 2, \ldots, s\}$ such that $\sum_{h \in I} c_h = \sum_{i \notin I} c_h$.
Max Cut	For graph G, weighting function $w : A \mapsto Z$, and positive integer W, there is a set $S \subseteq N$ such that $\sum_{\{u,v\} \in A : u \in S, v \neq S} w(u, v) \geq W$.

Algorithm R took the instance x of SAT and transformed it to instance $R(x)$ of 3SAT. Note that if we use algorithm β of 3SAT to decide if $R(x)$ is True or False, then $R(x)$ will be True if, and only if, x is True for the original SAT problem. This is because the two formulas are equisatisfiable (the second formula is satisfiable if, and only if, the first formula is satisfiable—and vice versa). In addition, the transformation of input x to $R(x)$ is done by polynomial-time algorithm R. Algorithm R has polynomial complexity because when replacing a clause with a shorter clause the total number of iterations (steps) is at most equal to the total number of literals of the initial formula.

Summarizing, we found poly-time algorithm R such that SAT(x) is translated to 3SAT$(R(x))$ and instance x of SAT is true if, and only if, instance $R(x)$ of 3SAT is true. Thus, unrestricted SAT$\leq_p$3SAT. In addition, if an oracle provides a certificate c for a 3SAT instance x that has a 'yes' answer, a certifier $C(x, c)$ can determine whether it is true or not in polynomial time by checking if there is a variable with True value in every clause (this takes linear time $O(|x|)$, where $|x|$ is the number of variables in all clauses). Thus, 3SAT is NP-complete given that 3SAT is in NP and unrestricted SAT$\leq_p$3SAT.

2.6.2   Reduction Example II: Vertex Cover $\leq_p$ Independent Set

Let us prove that the Independent Set decision problem is NP-complete with the use of poly-time reduction. For this, we will reduce the NP-complete Vertex Cover (VC) decision problem (also known as Node Cover) to the Independent Set (IS) decision problem in polynomial time. Let us first define the two decision problems.

Definition 2.7 *Vertex Cover (VC)* decision problem: Given an undirected graph $G = \{V, E\}$ and a number $k > 0$, does G contain a vertex cover C of size at most k ($|C| \leq k$)? Note that a vertex cover C is a subset of V ($C \subseteq V$) such that for every edge $(i, j) \in E$, we have $i \in C$ or $j \in C$ (or both). That is, a vertex cover C is a set of vertices such that every edge of graph G has at least one endpoint in it.

From the definition of the vertex cover, it follows that all vertices of a graph form a vertex cover and there might be multiple vertex covers for a specific graph. This is why the corresponding optimization problem strives to find the *minimum* vertex cover of a graph. In the example of Fig. 2.10 the set of vertices $C = \{1, 3, 4, 6\}$ form a vertex cover. There are more vertex covers than $\{1, 3, 4, 6\}$. For instance, the set $\{2, 3, 4\}$ is also a vertex cover (in fact, it is the minimum vertex cover for this graph because it requires only 3 vertices). Let us now define the independent set decision problem.

Definition 2.8 *Independent Set (IS) decision problem*: Given an undirected graph $G = \{V, E\}$ and a number k, does G contain a set $S \subseteq V$ of at least k independent vertices ($|S| \geq k$)? Note that this means that for every two vertices $i, j \in S$ there is no edge connecting them: $(i, j) \neq E$.

Fig. 2.10 Example graph $G = \{V, E\}$

Clearly, a set with just one vertex is an independent set and we might have many independent sets for a particular graph G. For instance, in the graph of Fig. 2.10 set $S = \{2, 5\}$ is an independent set, but also $\{1, 5, 6\}$ is an independent set. Therefore, the corresponding optimization problem is to find the maximum independent set of a graph.

Our objective is to prove that the IS decision problem is NP-complete. To do so, we need first to show that IS is in NP. Consider an instance x of the IS decision problem with a graph $G = \{V, E\}$ and a required number of at least k independent vertices that has a 'yes' answer. For problem instance x, an oracle provides a certificate c which is a set $S = \{s_1, s_2, \ldots, s_n\}$ with n vertices. Given instance x and certificate c we construct the certifier $C(x, c)$ presented in Algorithm 3.

Algorithm 3 Poly-time certifier of Independent Set solution

1: given certificate $S = \{s_1, s_2, \ldots, s_n\}$ of instance $\{G = \{V, E\}, k\}$
2: **if** $|S| < k$ **then**
3:     **return:** FALSE
4:     **break** and **terminate**
5: **end if**
6: **for** $i \in S$ **do**
7:     **for** $j \in S - \{i\}$ **do**
8:       **if** $i = j$ **then**
9:         **return:** FALSE
10:         **break** and **terminate**
11:       **end if**
12:       **for** $(l, m) \in E$ **do**
13:         **if** $i = l$ and $j = m$ **then**
14:           **return:** FALSE
15:           **break** and **terminate**
16:         **end if**
17:       **end for**
18:     **end for**
19: **end for**
20: **return:** TRUE

In line 2 of the certifier we ask if the size of the certificate $|S|$ is smaller than the minimum allowed number of independent vertices in the set, k. If this is the case, the certifier terminates with a False decision. This verification requires at most 3 computational steps. In lines 8-11 the certifier checks if any two vertices i and $j \neq i$ in the S are the same vertex in graph G. If yes, the certifier terminates and returns false because set S cannot have the same vertex twice. Lines 8-11 require at most 3 steps and because we have to go around two for loops with sizes $|S|$ and $|S| - 1$ when executing them, the required steps are at most $3|S|(|S| - 1)$. If one wants to be precise, the "return false" and "terminate" can occur only once, thus the required steps of lines 8-11 can be considered to be at most $|S|(|S| - 1) + 2$. From line 12 until 17 the certifier checks if there exists an edge in the graph whose starting vertex and ending vertex belong to set S. If this is the case, the certifier returns false and terminates. This requires at most $5|E||S|(|S| - 1)$ steps. If the certifier does not return false in the above checks, it decides that the certificate is a yes answer to this instance of the decision problem and terminates by performing 1 more step. The maximum possible number of required steps is $5|E||S|(|S| - 1) + 3|S|(|S| - 1) + 4$ which results in a polynomial time complexity, $O(|S|^2|E|)$. That is, the Independent Set problem is in NP.

In addition, we need to prove that the NP-complete Vertex Cover is poly-time reducible to the Independent Set: $VC \leq_p IS$. To do so, we need to find a polynomial time algorithm R that converts instances of decision problem VC to instances of decision problem IS such that the VC instance x is TRUE if, and only if, the IS instance $y = R(x)$ is true. To do the above, we use the following property.

Lemma 2.1 *If S is an independent set, then its complement set $V - S$ is a vertex cover C and vice versa, e.g., if $V - S$ is a vertex cover, its complement set S is an independent set.*

Proof Let S be an independent set. Then, every edge $(i, j) \in E$ either goes from a vertex in S to a vertex in $V - S$ or from a vertex in $V - S$ to another vertex in $V - S$. Either way, every edge $(i, j) \in E$ has at least one vertex i or j that is not in S and all these vertices that are not in S form a vertex cover $C = V - S$ since they offer at least one endpoint to every edge.

Let us now prove the reverse. Suppose $V - S$ is a vertex cover and $(i, j) \in S$. Note that $(i, j) \neq E$ because if $(i, j) \in E$ either i or j (or both) should have been in the vertex cover $V - S$. Thus, all $(i, j) \in S$ are independent vertices and S is an independent set.   □

Based on the aforementioned property, there is an algorithm R that:

- Takes an instance $x = \{G = \{V, E\}, k\}$ of the vertex cover decision problem that asks if there exists a vertex cover C such that $|C| \leq k$ in a graph G, where $|C|$ is the size of C.

- Creates an instance $y = R(x) = \{G = \{V, E\}, |V| - k\}$ of the IS decision problem that asks if there exists an independent set S in G such that the size of $|S| \geq |V| - k$.

The reduction algorithm R is polynomial since its only operation was to replace k with $|V| - k$. In addition, instance $y = \{G = \{V, E\}, |V| - k\}$ has a yes answer if, and only if, instance $x = \{G = \{V, E\}, k\}$ has a yes answer. Consequently, there exists independent set S in G such that $|S| \geq |V| - k$ if, and only if, there exists vertex cover C in G such that $|C| \leq k$. Thus, the Independent Set decision problem is NP-complete.

2.6.3 Reduction Example III: Vertex Cover $\leq_p$ Set Cover

We first define the Set Cover decision problem.

Definition 2.9 *Set Cover (SC)* decision problem: if there is a set U of elements $\{u_1, u_2, \ldots\}$ and $S_1, \ldots, S_i, \ldots, S_m$ a collection of subsets $S_i \subseteq U$, is there a collection of at most k subsets whose union is equal to U?

If an instance of the set cover decision problem has a yes answer and an oracle provides a solution (certificate c) with subsets $S = S_1, \ldots, S_m$, it can be easily verified in polynomial time whether this solution has at most k subsets and its union is U. In more detail, to check whether $m \leq k$ we need one computational step. To check if $S_1 \cup \cdots \cup S_m = U$ requires at most $O(r|U|)$ time, where r is the number of elements in all subsets $S_1, \ldots, S_m$ and $|U|$ the number of elements in U. Thus the SC decision problem is in NP.

Now, we prove that SC is NP-complete because there is a poly-time algorithm R that reduces the Vertex Cover decision problem to Set Cover. In more detail, algorithm R:

- Takes an instance $x = \{G = \{V, E\}, k\}$ of the vertex cover decision problem that asks if there exists a vertex cover C such that $|C| < k$ in the graph $G = \{V, E\}$.
- Creates an instance $y = R(x) = \{U, (S_1, \ldots, S_m), k\}$ of the set cover decision problem where U contains all vertices in V and there is a subset S_u for every vertex $u \in V$, where S_u contains the edges adjacent to u.

Note that with this translation U can be covered by less than or equal to k subsets if, and only if, G has a vertex cover C of size less than or equal to k. The reason is that if at most k sets $S_{u_1}, \ldots, S_{u_k}$ cover U, then every edge is adjacent to at least one of the vertices $u_1, \ldots, u_k$ yielding a vertex cover of size k. The opposite is also true, namely, if $u_1, \ldots, u_k$ is a vertex cover, then subsets $S_{u1}, \ldots, S_{u_k}$ cover U. Thus, instance $x = \{G = \{V, E\}, k\}$ of the vertex cover decision problem has a yes answer if, and only if, instance $y = R(x)$ of the set cover has a yes answer. Note that algorithm R has a polynomial time complexity $O(|V||E|)$ as can be seen below.

Algorithm 4 Poly-time algorithm R reducing Vertex Cover to Set Cover

---
1: given instance $x = \{G = \{V, E\}, k\}$ of the vertex cover
2: initialize $U = \emptyset$
3: **for** $u \in V$ **do**
4:    $U \leftarrow U + \{u\}$
5: **end for**
6: **for** $u \in V$ **do**
7:    initialize $S_u = \emptyset$
8:    **for** $(i, j) \in E$ **do**
9:      **if** $u = i$ **then**
10:        $S_u = S_u \cup \{j\}$
11:     **end if**
12:    **end for**
13: **end for**
14: **return:** instance $y = \{U, (S_1, \ldots, S_{|V|}), k\}$ of the set cover

---

Thus, Set Cover is NP-complete.

2.6.4 2SAT and Problems in P

Now, let us consider the case of 2SAT which is solved in polynomial time. 2SAT is a CNF SAT where each clause is a disjunctive clause of two literals, i.e., $(x_1 \vee x_2) \wedge (x_3 \vee x_4) \wedge \ldots \wedge (x_{n-1} \vee x_n)$. 2SAT is in P because each clause in a 2-CNF formula is logically equivalent to an implication from one variable or negated variable to the other. That is, $(x_1 \vee x_2) \equiv \neg x_1 \implies x_2$. This results in a simple implication (if we ever set x_1 to be false, x_2 must be true for our expression to be true). Every clause can be transformed into such an implication and we can solve 2SAT by selecting one variable x_i in each clause, setting it to false, and following all implications to see whether we get a contradiction (i.e., there is some x_{i+1} which is true). If there is not, our expression is satisfiable. Other prominent decision problems in P are presented in Table 2.3.

Table 2.3 Examples of decision problems in P presented in [13]

Decision problem	Description
2SAT	Satisfiability of conjunction of clauses containing at most 2 literals
Minimum Spanning Tree	There is a spanning tree of weight $\leq$ a given threshold k
Shortest Path	There is a path between an origin/destination pair of weight $\leq$ a given threshold k
Minimum Cut	There is an s, t cut of weight $\leq$ a given threshold k
Arc Cover	There is a set $Y \subseteq A$ such that $|Y| \leq k$ and every node is incident with an arc in Y
Arc Deletion	There is a set of k arcs whose deletion breaks all cycles
Bipartite Matching	There are p elements of $S \subseteq Z_p \times Z_p$, no two of which are equal in either component
Sequencing with Deadlines	Starting at time 0, one can execute jobs $1, 2, ..., n$ with execution times T_i and deadlines D_i in some order such that k jobs miss their deadlines
Solvability of Linear Equations	There exists a vector y_i such that, for each i, $\sum_j c_{ij} y_j = a_i$

Exercises

2.1 Asymptotic Notation

An algorithm requires $T(n) = 3^n + 125n^5 + 4n$ steps. Characterize its complexity using the Big Theta notation.

2.2 Asymptotic Notation

Write an algorithm that finds two numbers from a set of integers from 1 to n, where n is a finite number. These two numbers should have a product of 15 and a sum of 8. Define the computational complexity of the algorithm using the Big O notation.

2.3 Asymptotic Notation

Determine the computational complexity of the Subset Sum problem by using the Big O notation.

2.4 NP

Prove that the Traveling Salesman decision problem is in NP.

2.5 Hamiltonian Cycle—NP-complete

Prove that the Hamiltonian Cycle decision problem is NP-complete using the Vertex Cover decision problem.

2.6 Clique—NP-complete

Prove that the Clique decision problem is NP-complete using the 3SAT decision problem.

2.7 Periodic Event Scheduling Problem—NP-complete

Prove that the PESP decision problem is NP-complete.

2.8 Complexity

Are all NP-complete and NP-Hard problems equally hard?

References

1. M.A. Harrison, *Proceedings of the Third Annual ACM Symposium on Theory of Computing* (ACM, 1971)
2. J.R. Sampson, in *Adaptive Information Processing* (Springer, 1976), pp. 40–55
3. D.E. Knuth, Inf. Control **8**(6), 607 (1965)
4. R.C. Bose, R.J. Nelson, J. ACM (JACM) **9**(2), 282 (1962)
5. E.W. Dijkstra et al., Numer. Math. **1**(1), 269 (1959)
6. L. Fortnow, Commun. ACM **52**(9), 78 (2009)
7. L.G. Khachiyan, in *Doklady Akademii Nauk*, vol. 244 (Russian Academy of Sciences, 1979), pp. 1093–1096
8. M. Agrawal, N. Kayal, N. Saxena, *Annals of Mathematics* (2004), pp. 781–793
9. J. Van Leeuwen, *Handbook of Theoretical Computer Science (vol. A) Algorithms and Complexity* (Mit Press, 1991)
10. S.A. Cook, in *Proceedings of the Third Annual ACM Symposium on Theory of Computing* (1971), pp. 151–158
11. L.A. Levin, Probl. Peredachi Inf. **9**(3), 115 (1973)
12. T.J. Schaefer, in *Proceedings of the Tenth Annual ACM Symposium on Theory of Computing* (1978), pp. 216–226
13. R.M. Karp, in *Complexity of Computer Computations* (Springer, 1972), pp. 85–103
14. M.R. Garey, D.S. Johnson, *Computers and Intractability: A Guide to the Theory of NP-Completeness* (W. H. Freeman, 1979)
15. P. Großmann, S. Hölldobler, N. Manthey, K. Nachtigall, J. Opitz, P. Steinke, in *International Conference on Industrial, Engineering and other Applications of Applied Intelligent Systems* (Springer, 2012), pp. 166–175

Chapter 3
Continuous Unconstrained Optimization

Abstract This chapter presents the basics of continuous unconstrained optimization for single-dimensional and multi-dimensional problems with a scalar-valued objective function. It also introduces the two most important strategies for solving unconstrained nonlinear optimization problems (*line search* and *trust region*) and presents in detail the main line search-based solution algorithms. The order and range of convergence of solution algorithms are also presented.

3.1 Single-Dimensional Problems

A single-dimensional optimization problem can be written in the form:

$$\min\ f(x) \tag{3.1}$$

$$\text{subject to: } g_j(x) \leq 0, \qquad j = 1, \ldots, m \tag{3.2}$$

$$x \in \mathbb{R} \tag{3.3}$$

where the feasible region is the set of all possible values of variable x that satisfy the m constraints of the problem. Notice that the problem is defined in a single dimension, i.e., $x \in \mathbb{R}$, resulting in a single-dimensional constrained optimization problem. If we ignore all constraints, i.e., x is a continuous variable that belongs to $\mathbb{R}$, then we solve a continuous, single-dimensional, unconstrained optimization problem:

$$\min\ f(x) \tag{3.4}$$

$$\text{subject to: } x \in \mathbb{R} \tag{3.5}$$

To solve single-dimensional unconstrained optimization problems, one should examine the properties of the objective function. First, it is desirable that the objective function is well-defined:

Definition 3.1 A function $f : \mathbb{R}^n \to \mathbb{R}$ with domain $\text{dom}(f)$ is *well-defined* or *unambiguous* if for every point $\mathbf{x} \in \text{dom}(f)$ there is a *unique* element $y \in \mathbb{R}$

satisfying $f(\mathbf{x}) = y$ and there is an image for every point in $\mathrm{dom}(f)$. A function which is not well-defined is called ill-defined or ambiguous.

Another significant property is the differentiability class of the objective function. The definition of differentiability classes is provided below.

Definition 3.2 Differentiability classes: a function f is said to be of differentiability class C^k if the derivatives f', f'', $\ldots$, $f^{(k)}$ exist and are continuous.

We note that a function is continuous if an infinitely small increment a of the independent variable x always produces an infinitely small change $f(x + a) - f(x)$. In case a function has abrupt changes in value, these changes are known as discontinuities. To be more precise, we present below the definition of continuity using limits.

Continuity test based on limits

Let $f : \mathbb{R}^n \to \mathbb{R}$ and $\mathbf{x}_0 \in \mathbb{R}^n$. f is continuous at $\mathbf{x}_0$ if:

- There exist $f(\mathbf{x}_0)$ and $\lim_{\mathbf{x} \to \mathbf{x}_0} f(\mathbf{x})$
- $\lim_{\mathbf{x} \to \mathbf{x}_0} f(\mathbf{x}) = f(\mathbf{x}_0)$

Reckon that the limit L of a function f at a point $\mathbf{x}_0 \in \mathbb{R}^n$ is defined as the real value L for which for every $\epsilon > 0$ there exists some $\delta > 0$ such that:

$$0 < \|\mathbf{x} - \mathbf{x}_0\|_2 < \delta \implies |f(\mathbf{x}) - L| < \epsilon$$

Another, equivalent, continuity test is the *epsilon delta* continuity test [1] provided below.

Epsilon delta continuity test

Let $f : \mathbb{R}^n \to \mathbb{R}$ and $\mathbf{x}_0 \in \mathbb{R}^n$. f is continuous at $\mathbf{x}_0$ if for every $\epsilon > 0$ there exists $\delta > 0$ such that: $\|\mathbf{x} - \mathbf{x}_0\|_2 < \delta$ implies that $|f(\mathbf{x}) - f(\mathbf{x}_0)| < \epsilon$.

For instance, function $f(\mathbf{x}) = x_1 x_2$ is continuous at $\mathbf{x}_0 = [0, 0]^\mathsf{T}$ because $f(\mathbf{x}_0) = 0$ and $\lim_{\mathbf{x} \to \mathbf{x}_0} f(\mathbf{x}) = 0$.

Let us now consider function:

$$f(x) = \begin{cases} x & \text{if } x \geq 0 \\ 0 & \text{if } x < 0 \end{cases}$$

This function is also continuous at $x_0 = 0$ (see Fig. 3.1) because

Fig. 3.1 Example of graph
$f(x) = x$ if $x \geq 0$ and
$f(x) = 0$ if $x < 0$

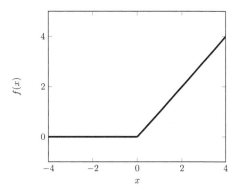

$$\lim_{x \to x_0^+} f(x) = \lim_{x \to x_0^-} f(x) = f(x_0) = 0.$$

To provide a counterexample, let us consider the function:

$$f(x) = \begin{cases} x + 5 & \text{if } x \geq 0 \\ 0 & \text{if } x < 0 \end{cases}$$

Function f is *not* continuous at $x_0 = 0$ because, even though there exists $f(x_0) = 5$, we have:

$$\lim_{x \to x_0^+} f(x) \neq \lim_{x \to x_0^-} f(x)$$

Another important definition of continuity is the Lipschitz continuous function.

Lipschitz continuous

A function $f : \mathbb{R}^n \to \mathbb{R}^m$ is called Lipschitz continuous if there exists Lipschitz constant $L > 0$ such that $\| f(\mathbf{x}) - f(\mathbf{y}) \| \leq L \| \mathbf{x} - \mathbf{y} \|$ for all $\mathbf{x}, \mathbf{y}$.

Because $\| f(\mathbf{x}) - f(\mathbf{y}) \| \leq L \| \mathbf{x} - \mathbf{y} \|$ for all $\mathbf{x}, \mathbf{y}$, a Lipschitz continuous function is limited in how fast it can change by L. It follows that every function that has bounded first derivatives is Lipschitz continuous. This does not mean, however, that any Lipschitz continuous function should be differentiable (i.e., $f(x) = |x|$ is Lipschitz continuous because $||x| - |y|| \leq |x - y|$ for all x, y, but not differentiable).

Another important property of a function is its smoothness, described below.

Definition 3.3 A function f is said to be *smooth* over some domain if it has continuous derivatives up to some desired order.

The *smoothness* of a function is a property measured by the number of continuous derivatives it has over some domain. To explore the smoothness of a function, it is useful to use the following definition of the differentiability class.

Definition 3.4 A function f is of *Differentiability Class* C^n over some domain, if, and only if, its derivatives f', f'', ..., $f^{(n)}$ exist and are continuous functions.

From the definition of the differentiability classes, it follows that the requirement for a function to be considered smooth over some domain is to be C^n, where $n \in \mathbb{N} \setminus \{0\}$ is the desired order of smoothness. At the very least, a function f should be C^1 to be considered smooth. Functions that are C^1 are also known as *continuously differentiable* functions. In general, though, one might require from a function f to be of a higher differentiability class to characterize it as smooth. It is not uncommon to reserve the characterization *smooth function* to describe functions in class $C^{+\infty}$ that have continuous derivatives of all orders $n \in \mathbb{N}$. Such functions (i.e., polynomials) are referred to as C-infinity functions [2]. In general, we have the following hierarchy:

Continuous function $\subseteq$ Continuously Differentiable function $\subseteq$ Smooth function

For instance, function:

$$f(x) = \begin{cases} x & \text{if } x \geq 0 \\ 0 & \text{if } x < 0 \end{cases}$$

is C^0 because it is continuous at any point in $x \in \mathbb{R}$ (even at $x_0 = 0$ since $\lim_{x \to x_0^+} f(x) = \lim_{x \to x_0^-} f(x) = f(x_0) = 0$). However, it is not C^1 because it is not differentiable at $x_0 = 0$ since:

$$\lim_{h \to 0^-} \frac{f(0+h) - f(0)}{h} \neq \lim_{h \to 0^+} \frac{f(0+h) - f(0)}{h}$$

Thus, f is not smooth because it does not even meet the minimum requirement of being C^1. Generally, a function f needs to be continuous in order to be smooth over some domain (that is, continuity is a *necessary* condition for smoothness). Continuity is not a sufficient condition though. On the other side, if function f is smooth over some domain, then it is also continuous (smoothness is a *sufficient* condition for continuity). In Fig. 3.2 we provide a function which is not continuous (and thus not smooth) at $x_0 = -3$ and continuous, yet not smooth at $x_1 = 2$. Geometrically, the graph of f has a break at $x_0 = 3$ indicating non-continuity and a sharp turn at $x_1 = 2$ indicating non-smoothness.

Following the aforementioned definitions, we will now discuss the optimality conditions of single-dimensional unconstrained optimization problems. In single-dimensional optimization problems, there are necessary and sufficient conditions for a solution to be a locally optimal one. This is discussed in the next sections.

Fig. 3.2 Function f is not continuous (and thus not smooth) at $x_0 = -3$ and not smooth (yet continuous) at $x_1 = 2$

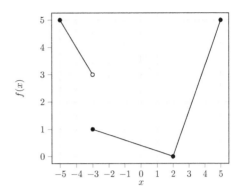

3.1.1 Necessary Conditions for Local Optimality

Before presenting the necessary conditions of local optimality for single-variable optimization problems, we provide the following key definitions of stationary, critical, and inflection points.

Definition 3.5 *Stationary point:* consider a continuous function $f(x)$, differentiable at x^*. Point x^* is a stationary point of f if $\frac{df(x^*)}{dx} = 0$.

Definition 3.6 *Critical point:* consider a function f, well-defined at x^*. Point x^* is called critical point if the derivative of f does not exist at that point or it is equal to zero (stationary point). That is, $f'(x^*) = 0$ or $\nexists\, f'(x^*)$.

Definition 3.7 *Inflection point:* is a point on a smooth plane curve at which the curvature changes sign. At the inflection point the function changes from being concave to convex, or vice versa. That is, x^* is an inflection point of an at least C^2 function f if $f''(x^*) = 0$ and $f''(x^* + h)$ has opposite sign to $f''(x^* - h)$ where h is a very small positive number in the neighborhood of x^*. Formally, a point x^* is an inflection point of a C^k function f if k is odd and $k \geq 3$ such that $f^{(k)}(x^*) \neq 0$ and $f^{(n)}(x^*) = 0$ for $n = 2, \ldots, k - 1$.

From the above definitions, it follows that for any continuous, single-variable function f:

$$\text{Inflection points} \subseteq \text{Stationary points} \subseteq \text{Critical points}$$

The *necessary* condition for a differentiable function in one variable, $f(x)$, to have a local minimum at x^* is that the derivative at x^* is equal to zero:

$$\frac{df(x^*)}{dx} = 0 \tag{3.6}$$

Fig. 3.3 Inflection point
$x^* = 0$ of function
$f(x) = x^3$. This point is not
a locally optimal solution
despite being a stationary
point with $f'(x^*) = 0$

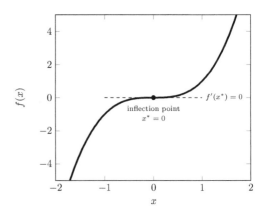

We can formalize this as follows:

First-order Necessary condition of single-variable optimization

If x^* is a locally optimal solution of single-variable function $f(x)$ and f is C^1 in the
vicinity of x^*, then $\frac{df(x^*)}{dx} = 0$.

Note that the necessary condition is also called first-order condition because it
uses the first-order derivative of f. The necessary condition of local optimality is not
a sufficient one. It only indicates that a point x^* cannot be a locally optimal solution
of f if $\frac{df(x^*)}{dx} \neq 0$ and f is C^1 in the neighborhood of x^*. That is, satisfying this
condition does not necessarily mean that x^* is a local optimum of f. This is why the
condition is called *necessary* and not *sufficient*.

To understand better the necessary condition, consider the function $f(x) = x^3$ in
Fig. 3.3. This function has $f'(x) = 3x^2$ and $f'(x^*) = 0$ for $x^* = 0$. Although f is
C^1 in the vicinity of x^*, x^* is not a locally optimal solution because it is an inflection
point. Indeed, $f''(0) = 0$ and $f'''(0) = 6$ meaning that it is an inflection point. This
is also evident in Fig. 3.3 where the curvature of f changes from concave to convex
at $x^* = 0$.

If we know that a point is a local optimum, we can use the first derivative rule to
determine whether a local optimum is a local minimum or local maximum:

- x^* is a local minimum if the derivative goes from negative to positive at x^*.
- x^* is a local maximum if the derivative goes from positive to negative at x^*.

If a function f is C^2 at the vicinity of x^*, then we can apply the second-order
necessary condition of local optimality. Note that this condition is still necessary
(that is, it does not ensure the x^* is a local optimum).

Second-order Necessary condition of single-variable minimization

If x^* is a local minimizer of $f(x)$ and f is C^2 at the vicinity of x^*, then $\frac{\mathrm{d}f(x^*)}{\mathrm{d}x} = 0$ and $\frac{\mathrm{d}^2 f(x^*)}{\mathrm{d}x^2} \geq 0$.

The above condition refers to x^* being a local minimizer. The same condition applies for x^* being a local maximizer with the only difference that $\frac{\mathrm{d}^2 f(x^*)}{\mathrm{d}x^2} \geq 0$ is replaced by $\frac{\mathrm{d}^2 f(x^*)}{\mathrm{d}x^2} \leq 0$. Note that $\frac{\mathrm{d}^2 f(x^*)}{\mathrm{d}x^2} \geq 0$ indicates that f is locally convex at x^*, and $\frac{\mathrm{d}^2 f(x^*)}{\mathrm{d}x^2} \geq 0$ indicates that f is locally concave at x^*.

3.1.2 Sufficient Conditions for Local Optimality

Previously, we discussed that a locally optimal solution x^* is a stationary point with $f'(x^*) = 0$ if f is C^1 in the vicinity of x^*. A *sufficient condition* for a stationary point x^* to be a local minimum, is for the function to be strictly convex in the vicinity of the stationary point. If a function f is C^2 in the vicinity of x^*, meaning that f is twice differentiable, the strict convexity condition in the vicinity of x^* is equivalent to:

$$\frac{\mathrm{d}^2 f(x^*)}{\mathrm{d}x^2} > 0$$

Summarizing,

Second-order Sufficient conditions of single-variable optimization

Let continuous function f which is C^2 in the vicinity of x^*. Then, x^* is a local minimum (maximum) of f if:

- $\frac{\mathrm{d}f(x^*)}{\mathrm{d}x} = 0$
- $\frac{\mathrm{d}^2 f(x^*)}{\mathrm{d}x^2} > 0$ $\left(\frac{\mathrm{d}^2 f(x^*)}{\mathrm{d}x^2} < 0 \text{ for maximum} \right)$.

Consider the example of $f(x) = x^3 - 3x$ restricted in the interval $[-3, 3]$, which is part of its domain $\mathrm{dom}(f)$. The function has first-order derivative $f'(x) = 3x^2 - 3$ and stationary points $x_a = -1$ and $x_b = 1$. Its second-order derivative is $f''(x) = 6x$.

At the vicinity of x_a function f is C^2 and $f'(x_a) = -6 < 0$. Thus, x_a satisfies the second-order sufficient conditions for being a local maximizer of f. In addition, at the vicinity of x_b function f is C^2 and $f'(x_b) = 6 > 0$. Thus, x_b satisfies the second-order sufficient conditions for being a local minimizer of f. The local minimizer and the local maximizer of the function are presented in Fig. 3.4.

Fig. 3.4 $f(x) = x^3 - 3x$
with local minimizer $x_b = 1$
and local maximizer
$x_a = -1$

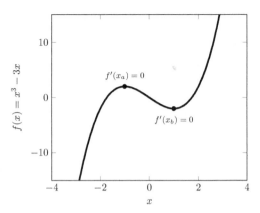

Reckon that function f is allowed to take values in the closed set $[-3, 3]$. From Fig. 3.4 it becomes apparent that even if x_a and x_b are indeed a local maximizer and a local minimizer of f, they are not its global maximizer/minimizer. In particular, its global maximizer is the endpoint $x = 3$ and its global minimizer is the endpoint $x = -3$. This highlights the fact that if a point satisfies the second-order sufficient conditions for local optimality, we only know that it is a local—and not a global—optimum.

3.1.3 Global Optimality

We have already provided the sufficient conditions for a solution x^* to be a locally optimal one. We showed, however, that a locally optimal solution is not necessarily a globally optimal solution. Finding a globally optimal solution becomes an easier task if the minimization (maximization) problem is convex (concave). Consider, for instance, the previous example in Fig. 3.4. If function f was restricted to the interval $[0, 3]$ instead of $[-3, 3]$, then we would have had a convex function and x_b would have been a global minimizer (see Fig. 3.5).

Because f is single-dimensional and twice differentiable, we can formally prove the convexity of f in $[0,3]$ by showing that:

$$f''(x) \geq 0 \quad \forall x \in [0, 3]$$

Indeed, $f''(x) = 6x \geq 0$ for any $x \in [0, 3]$.

Similarly, if f was allowed to take values in the interval $[-3, 0]$ it would have been a concave function and $x_a = -1$ would have been a global maximizer. As already discussed, convexity (concavity) is a very important property in mathematical optimization because it ensures that a locally optimal solution is also a globally optimal one. This is formally presented below.

Fig. 3.5 $f(x) = x^3 - 3x$ with local minimizer $x_b = 1$, which is also a global minimizer in the interval $x \in [0, 3]$

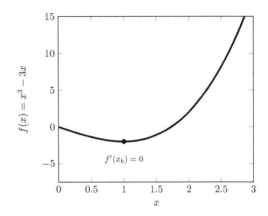

Fig. 3.6 Relationship between critical points, end points and local extrema

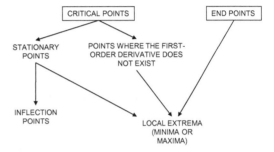

Sufficient Condition for Global Optimality

Let f be C^1 and convex (concave) in some interval of $\mathrm{dom}(f)$. If $\frac{\mathrm{d}f(x^*)}{\mathrm{d}x} = 0$ then x^* is a global minimizer (maximizer) of f in this interval. In addition, if f is strictly convex (concave) then x^* is a unique global minimizer (maximizer) of f.

Interestingly, the *first-order necessary* condition for x^* to be a local minimizer is also a *sufficient* condition for x^* to be a global minimizer if f is *convex*. That is, any stationary point becomes a global minimizer.

If the objective function is not convex in the interval of interest, then finding a globally optimal solution becomes a complicated task. In theory, one should find all local extrema and select the one with the best objective function value as the globally optimal solution. The local extrema of a function f are illustrated in Fig. 3.6.

In essence, to determine the global minimizer of a single-variable minimization problem with a non-convex objective function over some interval $\mathcal{F}$ one should:

- Compute all stationary points x^* for which $f'(x^*) = 0$ and exclude the ones that are inflection points;
- Find local extrema at other critical points where the derivative of f does not exist;
- Find all end points;

- Compare the objective function values of all the above points and select the one with the lowest value as the global minimizer.

In many cases, performing the aforementioned steps to find all local extrema is not easy and finding a globally optimal solution for a non-convex problem can be a computationally intractable task.

3.2 Multivariate Problems

A multi-dimensional unconstrained minimization problem can be written in the form:

$$\min f(\mathbf{x}) \tag{3.7}$$

$$\mathbf{x} \in \text{dom}(f) \subseteq \mathbb{R}^n \tag{3.8}$$

where $\text{dom}(f)$ is the set of all possible inputs of f and n the number of independent variables (dimensions). In multivariate optimization problems we use the following definitions.

Definition 3.8 *Stationary point:* consider a continuous function $f(\mathbf{x})$, differentiable at an open neighborhood of $\mathbf{x}^*$. The vector $\mathbf{x}^* = [x_1, \ldots, x_n]^\mathsf{T}$ is a stationary point of f if $\|\nabla f(\mathbf{x}^*)\|_2 = 0$, where $\nabla f(\mathbf{x}^*) = \left[\frac{\partial f(\mathbf{x}^*)}{\partial x_1}, \frac{\partial f(\mathbf{x}^*)}{\partial x_2}, \ldots, \frac{\partial f(\mathbf{x}^*)}{\partial x_n} \right]^\mathsf{T}$.

Definition 3.9 *Critical point:* a critical point is a value in the domain of function f where the gradient is undefined or is equal to zero (stationary point).

Definition 3.10 *Saddle point:* consider a continuous function $f(\mathbf{x})$, differentiable at an open neighborhood of $\mathbf{x}^*$. The vector $\mathbf{x}^* = [x_1, \ldots, x_n]^\mathsf{T}$ is a saddle point of f if $\|\nabla f(\mathbf{x}^*)\|_2 = 0$ and the Hessian matrix of f at $\mathbf{x}^*$ is indefinite (has both positive and negative eigenvalues). In the special case where there are only two independent variables $\mathbf{x} = [x_1, x_2]^\mathsf{T}$, then $\mathbf{x}^*$ is a saddle point if $\frac{\partial f(\mathbf{x}^*)}{\partial x_1} = \frac{\partial f(\mathbf{x}^*)}{\partial x_2} = 0$ and $\frac{\partial^2 f(\mathbf{x}^*)}{\partial x_1^2} \frac{\partial^2 f(\mathbf{x}^*)}{\partial x_2^2} - \left[\frac{\partial^2 f(\mathbf{x}^*)}{\partial x_1 \partial x_2} \right]^2 < 0$.

The saddle point is neither a local maximum nor a local minimum, as was the case with the inflection point in single-variable optimization problems.

3.2.1 Necessary Conditions for Local Optimality

There are two forms of necessary conditions that apply to a locally optimal solution $\mathbf{x}^*$ in the case of multivariate (also called multidimensional or multivariable) optimization. Both of them are presented below.

First-order Necessary condition of multi-variable optimization

If $\mathbf{x}^*$ is a local minimizer (maximizer) of f and $f(\mathbf{x})$ is C^1 in an open neighborhood of $\mathbf{x}^*$, then $\|\nabla f(\mathbf{x}^*)\|_2 = 0$.

Second-order Necessary condition of multi-variable optimization

If $\mathbf{x}^*$ is a local minimizer (maximizer) of f and $f(\mathbf{x})$ is C^2 in an open neighborhood of $\mathbf{x}^*$, then:

- $\|\nabla f(\mathbf{x}^*)\|_2 = 0$
- The Hessian of f at $\mathbf{x}^*$ is positive (negative) semi-definite.

Note that both necessary conditions refer to a locally optimal solution and they are not sufficient.

3.2.2 Sufficient Conditions for Local Optimality

We now proceed to the sufficient condition for a solution $\mathbf{x}^*$ to be a strict local minimizer of a multivariable unconstrained minimization problem. To show this, it is sufficient to demonstrate that f is strictly convex in the vicinity of $\mathbf{x}^*$, as was the case with the single-dimensional problems. Function f can be shown to be strictly convex in the vicinity of $\mathbf{x}^*$ if the Hessian matrix of f at $\mathbf{x}^*$ is positive definite, as described below.

Second-order Sufficient condition of multi-variable optimization

If f is C^2 in an open neighborhood of $\mathbf{x}^*$ and:

- $\|\nabla f(\mathbf{x}^*)\|_2 = 0$
- The Hessian of f at $\mathbf{x}^*$ is positive (negative) definite.

then $\mathbf{x}^*$ is a strict local minimizer (maximizer) of f.

The sufficient condition guarantees that $\mathbf{x}^*$ is a strict local minimizer and requests that the Hessian of f at $\mathbf{x}^*$ is positive definite, instead of positive semi-definite. It is important to note that the sufficient condition refers to points $\mathbf{x}^*$ for which f is C^2 in their vicinity. There can be other strict local minimizers at points where f is not C^2.

3.2.3 Global Optimality

As in single-variable optimization, a special case arises when the objective function is convex in its domain. In that case, it is easy to characterize a globally optimal solution as it only needs to satisfy the first-order necessary conditions for local optimality.

Sufficient Condition for Global Optimality

Let f be C^1 and convex (concave) in $\text{dom}(f) \subseteq \mathbb{R}^n$. If $\mathbf{x}^*$ is a stationary point of f, then it is a global minimizer (maximizer) of f.

Let us consider the example $\min f(\mathbf{x})$ with $f(\mathbf{x}) := x_1^2 + 2x_2^2$ and $\text{dom}(f) = \mathbb{R}^2$. Function f is $C^{+\infty}$ and we have stationary point $\mathbf{x}^* = [0, 0]^\mathsf{T}$ for which $\|\nabla f(\mathbf{x}^*)\|_2 = 0$. In addition, the Hessian of f in $\text{dom}(f)$ is:

$$\mathbf{H} = \begin{bmatrix} 2 & 0 \\ 0 & 4 \end{bmatrix}$$

Because the Hessian of f in $\text{dom}(f)$ is positive definite, f is strictly convex and stationary point $\mathbf{x}^* = [0, 0]^\mathsf{T}$ is a strict global minimizer of f.

3.3 Optimization Algorithms

We discussed the sufficient conditions to derive locally optimal solutions for single-variable and multi-variable unconstrained minimization problems. It is important to note that these conditions require the computation of gradients and Hessian matrices. In this section we will present optimization algorithms that can be implemented in computing machines to compute locally optimal solutions.

The two most important strategies for solving unconstrained nonlinear optimization problems are the *line search* and the *trust-region*. Both line search and trust-region strategies start from an initial solution guess $\mathbf{x}_0$ and they perform iterations until converging to a (hopefully) locally optimal solution. Line search methods pick first an improving search direction. Then, they pick the step size (length) based on which they update the incumbent solution. In contrast, Trust-region methods choose the search direction and the step size simultaneously when updating the incumbent solution.

3.3.1 Order of Convergence

The methods used for solving unconstrained nonlinear optimization problems are iterative methods that start from an initial solution guess $\mathbf{x}_{k=0} \in \mathbb{R}^n$ and intend to converge to a stationary point $\mathbf{x}^*$ which is (hopefully) a local optimum. Among others, these methods are analyzed based on their order of convergence and convergence rate which provide an indication of their computation times. Suppose that we have a sequence of points in $\mathbb{R}^n$ that converge to some point $\mathbf{x}^*$:

$$\mathbf{x}_{k=0}, \mathbf{x}_{k=1}, \cdots \rightarrow \mathbf{x}^*$$

The error between each iterate $\mathbf{x}_k$ and $\mathbf{x}^*$ is $\|\boldsymbol{\epsilon}_k\|_2 = \|\mathbf{x}_k - \mathbf{x}^*\|_2$, where $\|\boldsymbol{\epsilon}_k\|_2$ is the Euclidean norm,

$$\|\boldsymbol{\epsilon}_k\|_2 := \sqrt{\epsilon_{k1}^2 + \epsilon_{k2}^2 \cdots + \epsilon_{kn}^2}$$

and n are the dimensions of vector $\boldsymbol{\epsilon}_k = [\epsilon_{k1}, \epsilon_{k2}, \ldots, \epsilon_{kn}]^\mathsf{T}$.

We would like to know how quickly this error converges to zero. If there exist $q \geq 1$ and $\mu > 0$ for which

$$\lim_{k \to +\infty} \frac{\|\boldsymbol{\epsilon}_{k+1}\|_2}{\|\boldsymbol{\epsilon}_k\|_2^q} = \lim_{k \to +\infty} \frac{\|\mathbf{x}_{k+1} - \mathbf{x}^*\|_2}{\|\mathbf{x}_k - \mathbf{x}^*\|_2^q} = \mu$$

we say that the order of converge of the sequence to $\mathbf{x}^*$ is q and the convergence rate (known also as asymptotic error) is μ. The higher the q value, the faster the convergence. For the same q, the lower the μ value, the faster the convergence. This expression can be understood more easily if it is stated as:

$$\|\boldsymbol{\epsilon}_{k+1}\|_2 = \mu \|\boldsymbol{\epsilon}_k\|_2^q \quad \text{for } k \to +\infty$$

which is equivalent to:

$$\|\mathbf{x}_{k+1} - \mathbf{x}^*\|_2 = \mu \|\mathbf{x}_k - \mathbf{x}^*\|_2^q \quad \text{for } k \to +\infty$$

As an example, consider the sequence $1, 1/2, 1/4, 1/8, 1/16, \cdots \rightarrow 0$ that can be written as $x_k = 1/2^k$. This sequence converges to $x^* = 0$ and has:

$$\lim_{k \to +\infty} \frac{\epsilon_{k+1}}{\epsilon_k^q} = \lim_{k \to +\infty} \frac{1/2^{k+1}}{1/2^{kq}} = \lim_{k \to +\infty} \frac{2^{kq}}{2^{k+1}}$$

which is equal to $+\infty$ for $q = 2$ (does not converge) and $1/2$ for $q = 1$. Thus, $x_k = 1/2^k$ has order of convergence $q = 1$ and convergence rate $\mu = 1/2$ (converges linearly with asymptotic error $\mu = 1/2$).

In another example, consider the bisection method [3] applied to find the root $x^* \mid f(x^*) = 0$ of single-variable problems with a continuous objective function f

defined in interval $[a, b]$ for which $f(a)f(b) < 0$. The bisection method iterates until finding a root $f(x^*) = 0$ that lies in the interval (a, b) since there is at least one root based on Bolzano's theorem [4]. Bolzano's theorem is a corollary of the intermediate value theorem stating that a continuous function f where $\text{dom}(f)$ contains the interval $[a, b]$ will take any given value between $f(a)$ and $f(b)$ at some point within the interval. The process of the bisection algorithm is described in the following algorithm. Notice that the algorithm terminates when finding a point c for which $|f(c)| < \tau$, where τ is a very small convergence tolerance indicating that $f(c) \approx 0$, and thus c is the root.

Algorithm 5 Bisection algorithm for finding the root of single-variable function f

---

1: given interval $[a, b]$ for which $f(a)f(b) < 0$ and convergence tolerance τ
2: for midpoint $c = \frac{a+b}{2}$ compute $f(c)$
3: **repeat**
4: **if** $f(a)$ and $f(c)$ have the same sign **then**
5:    set $a = c$ and $f(a) = f(c)$
6: **else**
7:    set $b = c$ and $f(b) = f(c)$
8: **end if**
9: compute $c = \frac{a+b}{2}$ and $f(c)$
10: **until** $|f(c)| < \tau$
11: **return:** c

It is clear that at each iteration of the bisection method $|\epsilon_{x_{k+1}}| = \frac{1}{2}|\epsilon_{x_k}|$. That is, the order of convergence is $q = 1$ (linear convergence) and the asymptotic error $\mu = \frac{1}{2}$. Consider, for instance, the case where $f(x) = x^2 - 1$ and $[a, b] = [0, 3]$. This will result in root $c \approx 1$ after using the bisection method with convergence tolerance $\tau = 10^{-12}$.

We provide below the basic definitions for estimating the convergence rates of iterative algorithms [5].

Convergence types (from slower to faster)

1. **Sublinear Convergence** (which is slower than linear) is the convergence to a target point $\mathbf{x}^*$ if
$$\lim_{k \to +\infty} \frac{\|\mathbf{x}_{k+1} - \mathbf{x}^*\|_2}{\|\mathbf{x}_k - \mathbf{x}^*\|_2} = 1$$

2. **Linear Convergence** is called the convergence to a target point $\mathbf{x}^*$ if there exists $\mu \in (0, 1)$ such that
$$\lim_{k \to +\infty} \frac{\|\mathbf{x}_{k+1} - \mathbf{x}^*\|_2}{\|\mathbf{x}_k - \mathbf{x}^*\|_2} = \mu$$

3. **Superlinear Convergence** (which is faster than linear) is the convergence to a target point $\mathbf{x}^*$ if

$$\lim_{k \to +\infty} \frac{\|\mathbf{x}_{k+1} - \mathbf{x}^*\|_2}{\|\mathbf{x}_k - \mathbf{x}^*\|_2} = 0$$

4. **Quadratic Convergence** is called the convergence to a target point $\mathbf{x}^*$ if there exists $\mu \in (0, \infty)$ such that

$$\lim_{k \to +\infty} \frac{\|\mathbf{x}_{k+1} - \mathbf{x}^*\|_2}{\|\mathbf{x}_k - \mathbf{x}^*\|_2^2} = \mu$$

5. **Cubic Convergence** is called the convergence to a target point $\mathbf{x}^*$ if there exists $\mu \in (0, +\infty)$ such that

$$\lim_{k \to +\infty} \frac{\|\mathbf{x}_{k+1} - \mathbf{x}^*\|_2}{\|\mathbf{x}_k - \mathbf{x}^*\|_2^3} = \mu$$

We also present a practical approach to calculate the order of convergence. Reckon that $\|\epsilon_{k+1}\|_2 = \mu \|\epsilon_k\|_2^q$. We also have that $\|\epsilon_k\|_2 = \mu \|\epsilon_{k-1}\|_2^q$. By dividing them, we get:

$$\frac{\|\epsilon_{k+1}\|_2}{\|\epsilon_k\|_2} = \frac{\mu \|\epsilon_k\|_2^q}{\mu \|\epsilon_{k-1}\|_2^q} = \frac{\|\epsilon_k\|_2^q}{\|\epsilon_{k-1}\|_2^q}$$

which, with the use of $\log_{10}$, can be expressed as:

$$\log \left(\frac{\|\epsilon_{k+1}\|_2}{\|\epsilon_k\|_2} \right) = \log \left(\frac{\|\epsilon_k\|_2^q}{\|\epsilon_{k-1}\|_2^q} \right)$$

and, by using the logarithmic properties,

$$q = \frac{\log \left(\frac{\|\epsilon_{k+1}\|_2}{\|\epsilon_k\|_2} \right)}{\log \left(\frac{\|\epsilon_k\|_2}{\|\epsilon_{k-1}\|_2} \right)} = \frac{\log \left(\frac{\|\mathbf{x}_{k+1} - \mathbf{x}^*\|_2}{\|\mathbf{x}_k - \mathbf{x}^*\|_2} \right)}{\log \left(\frac{\|\mathbf{x}_k - \mathbf{x}^*\|_2}{\|\mathbf{x}_{k-1} - \mathbf{x}^*\|_2} \right)}$$

Let us consider an example of applying this method for calculating the order of convergence. We seek to find the root $x^* \mid f(x^*) = 0$ of the single-variable function $x^2 - x - 1$ with the iterative Newton-Raphson method that updates $x_k = x_{k-1} - \frac{f(x_{k-1})}{f'(x_{k-1})}$ starting from an initial solution guess $x_{k=0} = 10$. This root is $x^* = \frac{1+\sqrt{5}}{2}$. If we terminate the algorithm when $|f(x_k)| < \tau$, where $\tau = 1.00E - 6$ is a small precision error, then we get the results of Table 3.1.

From the last column of Table 3.1 we see that the order of convergence $q \approx 2$ for large values of k (in particular, for $k > 4$). That is, we have quadratic convergence close to the root. We should note though that the algorithm started rather slowly with

Table 3.1 Iterations of Newton-Raphson method to find the root x^* of $f(x) = x^2 - x - 1$

k	x_k	$f(x_k)$	$f'(x_k)$	$\begin{aligned}&\vert\epsilon_k\vert =\\&\vert x_k - x^*\vert\end{aligned}$	$\dfrac{\log\left(\frac{\vert\epsilon_{k+1}\vert}{\vert\epsilon_k\vert}\right)}{\log\left(\frac{\vert\epsilon_k\vert}{\vert\epsilon_{k-1}\vert}\right)}$
0	10	89	19	8.381966	–
1	5.315789	21.94183	9.631579	3.697755	1.169809
2	3.037676	5.1898	5.075352	1.419642	1.330788
3	2.015126	1.045608	3.030253	0.397092	1.59517
4	1.67007	0.119064	2.34014	0.052036	1.872832
5	1.619191	2.59E-03	2.238382	1.16E-03	1.988319
6	1.618035	1.34E-06	2.236069	5.98E-07	1.999967
7	1.618034	3.57E-13	2.236068	1.60E-13	–

Fig. 3.7 Log-log plot of the Newton-Raphson method when finding the root of $x^2 - x - 1$ starting from $x_{k=0} = 10$

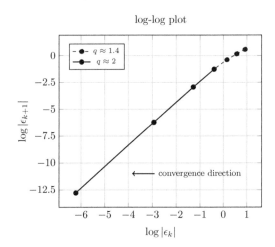

a relatively linear order of convergence which was close to 1. Note that with this convergence test we can conclude that the Newton-Raphson converges faster (in fewer iterations) than the Bisection method which was converging linearly. This convergence test will be used later on when examining more complex iterative algorithms that solve unconstrained minimization problems in multiple dimensions.

The order of convergence q and the asymptotic error μ can be also obtained graphically using 2-dimensional log-log plots. Notice that $\Vert\epsilon_{k+1}\Vert_2 = \mu\Vert\epsilon_k\Vert_2^q$. If we take the logarithms with base 10 we have:

$$\log\Vert\epsilon_{k+1}\Vert_2 = \log\mu + q\log\Vert\epsilon_k\Vert_2$$

That is, we can plot $\log\Vert\epsilon_k\Vert_2$ in the x-axis and $\log\Vert\epsilon_{k+1}\Vert_2$ in the y-axis to derive $\log\mu$ and the slope q (see Fig. 3.7 for the case presented in Table 3.1).

Figure 3.7 shows a function which appears to be piecewise linear with one part from iteration $k = 0$ to $k = 3$ and another part from $k = 4$ to $k = 6$. At $k = 0$ we have $\log |\epsilon_k| \approx 0.923$ and $\log |\epsilon_{k+1}| \approx 0.568$. This gives us an order of convergence $q \approx 1.4$ (note that in Table 3.1 this order varied from 1.169809 to 1.59517). After iteration $k = 3$ this order of convergence changes to $q \approx 2$ and we converge in iteration $k = 6$ with $\log |\epsilon_k| = -6.22320069$ and $\log |\epsilon_{k+1}| = -12.79622728$. Using $\log |\epsilon_{k+1}| = \log \mu + q \log |\epsilon_k|$ at the last iteration we can compute the value of the asymptotic error $\mu = 10^{\log |\epsilon_{k+1}| - q \log |\epsilon_k|} \simeq 0.44686$.

A final note is that the order of convergence can be linked to the Big O notation. Consider, for instance, the linearly converging case $\|\epsilon_{k+1}\|_2 = \mu \|\epsilon_k\|_2$. Let $\|\epsilon_0\|_2$ be the initial error for $k = 0$ and $\tau > 0$ a very small precision error which is close to 0. Then,

$$\|\epsilon_0\|_2 \rightarrow \|\epsilon_1\|_2 \rightarrow \|\epsilon_2\|_2 \rightarrow \cdots \rightarrow \tau$$

can be written as:

$$\|\epsilon_0\|_2 \rightarrow \mu \|\epsilon_0\|_2 \rightarrow \mu^2 \|\epsilon_0\|_2 \cdots \rightarrow \mu^k \|\epsilon_0\|_2$$

where $\mu^k \|\epsilon_0\|_2 \leq \tau$ and triggers the termination of the algorithm. Note that in this case k is the total number of required iterations until convergence and we can compute it based on the error of our initial solution guess, $\|\epsilon_0\|_2$, and the selected precision error τ. Because the algorithm will terminate when $\mu^k \|\epsilon_0\|_2 \leq \tau$, we seek to find the k-th iteration for which $\mu^k = \tau/\|\epsilon_0\|_2$. Taking the logarithms with base 10 we have:

$$k \log \mu = \log \tau - \log \|\epsilon_0\|_2 \Leftrightarrow k = \frac{\log \tau - \log \|\epsilon_0\|_2}{\log \mu}$$

That is, the total number of required iterations k is a function $T(\mu, \tau, \|\epsilon_0\|_2) = \frac{\log \tau - \log \|\epsilon_0\|_2}{\log \mu}$ which can be asymptotically bounded resulting in $O\left(\frac{\log \tau - \log \|\epsilon_0\|_2}{\log \mu}\right)$ (logarithmic complexity). This is very useful information because we can estimate the required iterations for solving a problem. For instance, if we solve the problem presented in Table 3.1 with the Bisection method that converges linearly with $\mu = 1/2$, we would need $k \approx \frac{\log (1.00E-6) - \log 8.381966}{\log 0.5} = 23$ iterations. Indeed, for $\tau = 1.00E - 6$ and $[a, b] = [0, 10]$ we have the 23 iterations presented in Table 3.2.

3.3.2 Line Search Strategy

The line search strategy is one of two basic iterative approaches to find a local minimum $\mathbf{x}^*$ of an objective function $f : \mathbb{R}^n \rightarrow \mathbb{R}$ (the other is the trust-region strategy).

The line search approach first finds a direction along which the objective function f will be reduced and then computes a step size that determines how far the incumbent

Table 3.2 Iterations of Bisection method to find the root x^* of $f(x) = x^2 - x - 1$

k	c	$f(c)$
1	5	19
2	2.5	2.75
3	1.25	−0.6875
4	1.875	0.640625
5	1.5625	−0.12109
6	1.71875	0.235352
7	1.640625	0.051025
8	1.601563	−0.03656
9	1.621094	0.006851
10	1.611328	−0.01495
11	1.616211	−0.00407
12	1.618652	0.001383
13	1.617432	−0.00135
14	1.618042	1.79E-05
15	1.617737	−0.00066
16	1.617889	−0.00032
17	1.617966	−0.00015
18	1.618004	−6.7E-05
19	1.618023	−2.5E-05
20	1.618032	−3.4E-06
21	1.618037	7.23E-06
22	1.618035	1.9E-06
23	1.618034	−7.6E-07

solution $\mathbf{x}_k$ should move along that direction. Each iteration of a line search method computes a search direction, $\mathbf{p}_k$ (known as *step direction*), and then decides how far to move along that direction, a_k (known as *step size* or *step length*). The iteration from point $\mathbf{x}_k = [x_{k1}, x_{k2}, \ldots, x_{kn}]^\mathsf{T}$ to $\mathbf{x}_{k+1}$ is given by

$$\mathbf{x}_{k+1} = \mathbf{x}_k + a_k \mathbf{p}_k \qquad (3.9)$$

where the step length a_k is a positive scalar and $\mathbf{p}_k$ a column vector that has the same number of elements as vector $\mathbf{x}_k$. Line search is a strategy that strives to update $\mathbf{x}_k$ by updating the values of a_k and $\mathbf{p}_k$ in each iteration with the objective to converge at some point, i.e., when $\|\nabla f(\mathbf{x}_{k+1})\|_2 \leq \tau$ where $\tau \geq 0$ is a small convergence tolerance (known also as precision error). Note that because $\|\nabla f(\mathbf{x}_{k+1})\|_2 \approx 0$, point $\mathbf{x}_{k+1}$ is a stationary point provided that f is differentiable in an open neighborhood of $\mathbf{x}_{k+1}$.

The line search strategy can be written in an algorithmic form as follows.

Line search strategy in an algorithmic form

1: set $k = 0$ and make initial solution guess $\mathbf{x}_k$
2: **while** $\|\nabla f(\mathbf{x}_k)\|_2 > \tau$ **do**
3:    compute a *step direction* $\mathbf{p}_k$
4:    compute a *step size* a_k to minimize $z(a) = f(\mathbf{x}_k + a\mathbf{p}_k)$ over $a \in \mathbb{R}_{\geq 0}$
5:    set $\mathbf{x}_{k+1} = \mathbf{x}_k + a_k\mathbf{p}_k$
6:    set $k = k + 1$
7: **end while**

It is important to note that we always need to start with an initial solution guess $\mathbf{x}_{k=0}$. As we will discuss later, the selection of the initial solution guess $\mathbf{x}_{k=0}$ can affect the computed $\mathbf{x}_k$ for which $\|\nabla f(\mathbf{x}_k)\|_2 \approx 0$.

In every iteration k one needs to update the step direction $\mathbf{p}_k$ and the step size a_k. The step direction $\mathbf{p}_k$ can be updated by various methods that have specific advantages and disadvantages. Some of the most common ones that will be presented in the following sections are:

- Steepest/Gradient descent
- Newton method
- Quasi-Newton methods
- Conjugate gradient

Note that the aforementioned step direction methods can be applied to trust-region methods as well, except for the conjugate gradient method. In addition, Newton's method requires the computation of second-order derivatives whereas all other methods require the computation of first-order ones.

3.3.3 Exact and Inexact Methods for Determining the Step Length

We first focus on exact and inexact methods for updating the step size (length) a_k. Line search methods derive their name from the fact that they determine the step size by performing a line search. Exact methods update the step size a_k at iteration k by solving the following single-variable minimization problem:

$$a_k := \underset{a \in \mathbb{R}_{\geq 0}}{\operatorname{argmin}} \phi(a)$$

where $\phi(a) = f(\mathbf{x}_k + a\mathbf{p}_k)$.

Fig. 3.8 2-dimensional contour plot of $f(\mathbf{x}) = (x_1 - 1)^2 + (2x_2 - 1)^2$. The values of f for specific combinations of x_1, x_2 appear in the contours. With step length $a_k = 0.275$ and direction $\mathbf{p}_k = [4, 6]^\mathsf{T}$ we move from $\mathbf{x}_k$ with $f(\mathbf{x}_k) = 13$ to $\mathbf{x}_{k+1}$ with $f(\mathbf{x}_{k+1}) = 0.9$

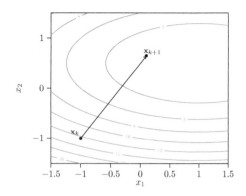

Inexact methods seek to find an approximate solution a_k that meets specific conditions (i.e., Wolfe conditions, Goldstein conditions, Armijo condition) without explicitly solving a minimization problem.

The most accurate approach to compute the step length a_k is to solve the minimization problem $\min_{a \geq 0} \phi(a)$ to global optimality. Consider, for example, the two-dimensional function $f(\mathbf{x}) = (x_1 - 1)^2 + (2x_2 - 1)^2$ and assume that we are at point $\mathbf{x}_k = [-1, -1]^\mathsf{T}$ with step direction $\mathbf{p}_k = [4, 6]^\mathsf{T}$. To find step size a_k would require to solve the single-variable minimization problem:

$$\min_{a \geq 0} \phi(a) =$$

$$\min_{a \geq 0} (x_1 + ap_1 - 1)^2 + (2x_2 + 2ap_2 - 1)^2 =$$

$$\min_{a \geq 0} (-2 + 4a)^2 + (-3 + 12a)^2$$

for which $\phi'(a) = 320a - 88$. Setting $\phi'(a) = 0$, we get $a = 0.275$. Thus, $a_k = 0.275$ is the selected step length when using exact line search. This will move us from point $\mathbf{x}_k = [-1, -1]^\mathsf{T}$ to point $\mathbf{x}_{k+1} = [-1, -1]^\mathsf{T} + [0.275 \cdot 4, 0.275 \cdot 6]^\mathsf{T} = [0.1, 0.65]^\mathsf{T}$ (see the contour plot in Fig. 3.8, where a contour plot is a graphical technique to plot a 3-dimensional surface on a 2-dimensional format by plotting slices of $f(\mathbf{x})$, called contours, for connecting the (x_1, x_2) coordinates where the value of $f(x_1, x_2)$ occurs).

Generally, solving minimization problems in each iteration of k can be very costly. For this reason, many approaches use an inexact line search to identify a step length a_k that achieves adequate reductions in f at a reasonable computational cost. Below we discuss these inexact methods in more detail.

The first inexact line search method for determining the step size a_k is known as the Wolfe conditions, which were published in 1969 and 1971 by Philip Wolfe [6, 7]. Wolfe conditions are a set of inequalities for performing inexact line search. Their objective is to find a_k that approximates the solution of $\min_{a \geq 0} f(\mathbf{x}_k + a\mathbf{p}_k)$. The conditions are presented below.

Fig. 3.9 To satisfy Armijo's rule, we can select $a_k \in [0, \bar{a}_k]$ because $f(\mathbf{x}_k + a\mathbf{p}_k)$ remains below or intersects $f(\mathbf{x}_k) + c_1 a \mathbf{p}_k^\mathsf{T} \nabla f(\mathbf{x}_k)$ at this interval. Note that the selected step length when using exact line search is $a_k = 0.275$

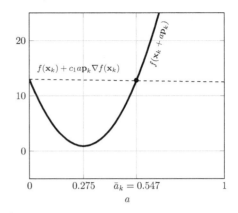

Wolfe conditions for inexact selection of step length a_k

A step length $a_k \geq 0$ is said to satisfy the Wolfe conditions, restricted to the search direction $\mathbf{p}_k$, if the following inequalities hold for some $0 < c_1 < c_2 < 1$:

1. Armijo rule: $f(\mathbf{x}_k + a_k \mathbf{p}_k) \leq f(\mathbf{x}_k) + c_1 a_k \mathbf{p}_k^\mathsf{T} \nabla f(\mathbf{x}_k)$
2. Curvature condition: $\mathbf{p}_k^\mathsf{T} \nabla f(\mathbf{x}_k + a_k \mathbf{p}_k) \geq c_2 \mathbf{p}_k^\mathsf{T} \nabla f(\mathbf{x}_k)$

In particular, c_1 is proposed to take very small values, i.e., $c_1 = 10^{-4}$, and $c_2 = 0.9$ when using Newton or quadi-Newton methods and $c_2 = 0.1$ when using the conjugate gradient method for determining the step direction $\mathbf{p}_k$ (see [8]). The Armijo rule inequality aims to give a sufficient decrease in the objective function f which is proportional to the step length a_k and the directional derivative. The curvature condition inequality aims to ensure that the slope is reduced sufficiently. The selected step length a_k can be any value that satisfies both inequality constraints. That is, the inequality constraints can be perceived as a lower and upper bound that form an admissible set of step length values.

To understand more the Wolfe conditions and the relation between a step length that satisfies the Wolfe conditions and a step length computed by exact line search, let us consider again the example of the two-dimensional function $f(\mathbf{x}) = (x_1 - 1)^2 + (2x_2 - 1)^2$ and assume that we are at point $\mathbf{x}_k = [-1, -1]^\mathsf{T}$ with step direction $\mathbf{p}_k = [4, 6]^\mathsf{T}$. We previously showed that the selected step length when applying exact line search is $a_k = 0.275$. This step length is the global minimizer of function $\phi(a) := f(\mathbf{x}_k + a\mathbf{p}_k)$, as presented in Fig. 3.9.

Let us consider the Wolfe conditions with $c_1 = 10^{-2}, c_2 = 0.1$. The Armijo condition requires that the selected step length a_k satisfies the inequality constraint $f(\mathbf{x}_k + a_k \mathbf{p}_k) \leq f(\mathbf{x}_k) + c_1 a_k \mathbf{p}_k^\mathsf{T} \nabla f(\mathbf{x}_k)$. Because $f(\mathbf{x}_k + a\mathbf{p}_k) = f(\mathbf{x}_k) + c_1 a \mathbf{p}_k^\mathsf{T} \nabla f(\mathbf{x}_k)$ for $\bar{a}_k = 0.547$, any value of a_k in $[0, \bar{a}_k]$ satisfies the Armijo condition (see Fig. 3.9).

Fig. 3.10 To satisfy Wolfe conditions, we should select a step length a_k within the interval $[\hat{a}_k, \bar{a}_k]$, where $\hat{a}_k$ is obtained by the curvature condition and $\bar{a}_k$ by the Armijo condition

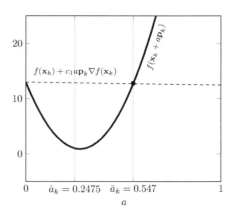

Satisfying Armijo's rule does not necesarily mean that we make reasonable progress. For instance, some values $a_k \in [0, \bar{a}_k]$, i.e., values close to 0.275, result in a far greater reduction of $f(\mathbf{x}_k + a_k\mathbf{p}_k)$ compared to others, i.e., values close to 0. To rule out unacceptably short steps, Wolfe imposed also the curvature condition. There, we require from the step length to satisfy the inequality constraint $\mathbf{p}_k^\mathsf{T}\nabla f(\mathbf{x}_k + a_k\mathbf{p}_k) \geq c_2\mathbf{p}_k^\mathsf{T}\nabla f(\mathbf{x}_k)$. We have $c_2\mathbf{p}_k^\mathsf{T}\nabla f(\mathbf{x}_k) = 0.1[4, 6]^\mathsf{T}[-4, -12] = -8.8$. Because $\mathbf{p}_k^\mathsf{T}\nabla f(\mathbf{x}_k + a\mathbf{p}_k) = c_2\mathbf{p}_k^\mathsf{T}\nabla f(\mathbf{x}_k)$ for $\hat{a}_k = 0.2475$, the allowed step length according to Wolfe's conditions is any value in the range $[\hat{a}_k, \bar{a}_k]$. This is shown graphically in Fig. 3.10 where $[\hat{a}_k, \bar{a}_k]$ is the interval of a possibly selected step length when applying the Wolfe conditions.

It is evident from Fig. 3.10 that the curvature condition rules out unacceptably short steps that lie in the interval $[0, 0.2475]$. Even when satisfying the curvature condition though, a step length might not be particularly close to a minimizer of $\phi(a) = f(\mathbf{x}_k + a\mathbf{p}_k)$, i.e., it can be close to $\bar{a}_k = 0.547$. For this reason, the curvature condition can be modified to force a_k to lie in the neighborhood of a stationary point of $\phi(a)$. This modification results in the strong Wolfe conditions that are presented below and exclude step lengths that are far from the stationary points of ϕ.

Strong Wolfe conditions for inexact selection of step length a_k

A step length a_k is said to satisfy the strong Wolfe conditions, restricted to the search direction $\mathbf{p}_k$, if the following inequalities hold:

1. Armijo rule: $f(\mathbf{x}_k + a_k\mathbf{p}_k) \leq f(\mathbf{x}_k) + c_1 a_k \mathbf{p}_k^\mathsf{T}\nabla f(\mathbf{x}_k)$
2. Curvature condition: $|\mathbf{p}_k^\mathsf{T}\nabla f(\mathbf{x}_k + a_k\mathbf{p}_k)| \leq c_2|\mathbf{p}_k^\mathsf{T}\nabla f(\mathbf{x}_k)|$

Implementing the updated curvature condition imposed by the strong Wolfe conditions, we request that:

$$-c_2|\mathbf{p}_k^\mathsf{T}\nabla f(\mathbf{x}_k)| \leq \mathbf{p}_k^\mathsf{T}\nabla f(\mathbf{x}_k + a_k\mathbf{p}_k) \leq c_2|\mathbf{p}_k^\mathsf{T}\nabla f(\mathbf{x}_k)|$$

Fig. 3.11 To satisfy the strong Wolfe conditions, we should select a step length a_k within the interval $[0.2475, 0.3025]$, obtained by the curvature condition. Notice that a selection of step length a_k in this interval meets also the Armijo condition which requires that $a_k \leq 0.547$

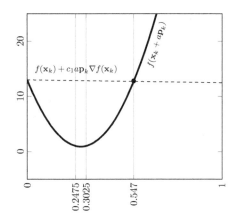

In our example, we have $c_2 |\mathbf{p}_k^{\mathsf{T}} \nabla f(\mathbf{x}_k)| = 8.8$ and the aforementioned curvature condition is satisfied for any $a_k \in [0.2475, 0.3025]$. This is advantageous because the selected step length is forced to be closer to the optimal step length 0.275 that was computed by the exact line search (see Fig. 3.11).

Another inexact line search method for the selection of the step length a_k is based on the Goldstein conditions. Like the Wolfe conditions, the Goldstein conditions ensure that the step length a_k achieves sufficient decrease but is not too short.

Goldstein conditions for inexact selection of step length a_k

1. $f(\mathbf{x}_k) + (1 - c)a_k \mathbf{p}_k^{\mathsf{T}} \nabla f(\mathbf{x}_k) \leq f(\mathbf{x}_k + a_k \mathbf{p}_k)$
2. $f(\mathbf{x}_k + a_k \mathbf{p}_k) \leq f(\mathbf{x}_k) + c a_k \mathbf{p}_k^{\mathsf{T}} \nabla f(\mathbf{x}_k)$

The Goldstein conditions are often used in Newton-type methods but are not well-suited for quasi-Newton methods.

Finally, one can only use the *Armijo condition* to perform an inexact line search for the selection of the step length [9]. For completeness, the Armijo condition is re-stated below. Reckon, however, that meeting the Armijo condition does not ensure the selection of a step length which is close to the optimal step length computed via exact line search.

Armijo condition for inexact selection of step length a_k

$$f(\mathbf{x}_k + a_k \mathbf{p}_k) \leq f(\mathbf{x}_k) + c a_k \mathbf{p}_k^{\mathsf{T}} \nabla f(\mathbf{x}_k)$$

To satisfy the Armijo, Goldstein, or (strong) Wolfe conditions we should find a step length that satisfies the respective inequality constraints. A typical approach to do so is to start from an initial solution guess of the step length a and reduce its value iteratively (a process called *backtracking*) until we find a value a_k that meets the conditions' requirements.

Let us consider the implementation of backtracking to find a step length a_k that meets the Armijo rule. This process is called Armijo backtracking line search and starts with a large estimate of the step size, a, and, iteratively, reduces the step size (backtracking) until a decrease of the objective function adequately corresponds to the expected decrease based on the local gradient of the objective function. Armijo backtracking line search performs well for determining a_k when combined with the gradient descent method or Newton's method with positive definite Hessian for the search direction $\mathbf{p}_k$. In more detail, given search direction $\mathbf{p}_k$, an initially large value of the step length estimate $a > 0$ and search control parameters $\rho \in (0, 1)$ and $c \in (0, 1)$, we start from a and we repeatedly update $a = \rho a$ until we satisfy the Armijo condition $f(\mathbf{x}_k + a\mathbf{p}_k) \leq f(\mathbf{x}_k) + ca\mathbf{p}_k^\mathsf{T} \nabla f(\mathbf{x}_k)$. This is formalized in the following algorithm.

Algorithm 6 Armijo Backtracking Line search

1: select $a > 0$, $\rho \in (0, 1)$, and $c \in (0, 1)$
2: **repeat**
3:     set $a = \rho a$
4: **until** $f(\mathbf{x}_k + a\mathbf{p}_k) \leq f(\mathbf{x}_k) + ca\mathbf{p}_k^\mathsf{T} \nabla f(\mathbf{x}_k)$
5: Return step length $a_k = a$

The backtracking algorithm can be used to find a_k values that meet the Wolfe, strong Wolfe, or Goldstein conditions. For example, let us consider that we seek to find a_k that satisfies the strong Wolfe conditions:

1. Armijo rule: $f(\mathbf{x}_k + a_k\mathbf{p}_k) \leq f(\mathbf{x}_k) + c_1 a_k \mathbf{p}_k^\mathsf{T} \nabla f(\mathbf{x}_k)$
2. Curvature condition: $|\mathbf{p}_k^\mathsf{T} \nabla f(\mathbf{x}_k + a_k\mathbf{p}_k)| \leq c_2 |\mathbf{p}_k^\mathsf{T} \nabla f(\mathbf{x}_k)|$

We can find such a_k by starting from a large value of a and performing the steps of the following algorithm.

Algorithm 7 Strong Wolfe conditions with backtracking

1: select $a > 0$ and $\rho \in (0, 1)$
2: **repeat**
3:     set $a = \rho a$
4: **until** $f(\mathbf{x}_k + a\mathbf{p}_k) \leq f(\mathbf{x}_k) + c_1 a\mathbf{p}_k^\mathsf{T} \nabla f(\mathbf{x}_k)$ and $|\mathbf{p}_k^\mathsf{T} \nabla f(\mathbf{x}_k + a\mathbf{p}_k)| \leq c_2 |\mathbf{p}_k^\mathsf{T} \nabla f(\mathbf{x}_k)|$
5: Return step length $a_k = a$

3.3.4 Steepest/Gradient Descent Line Search Method

We already discussed methods for deciding the step length a_k of a line search strategy (exact line search, strong Wolfe conditions, Goldstein conditions, Armijo rule). As already discussed, there are several different methods for determining the step direction $\mathbf{p}_k$, including steepest/gradient descent, Newton methods, quasi-Newton methods, and conjugate gradient. We will now discuss the gradient descent method for determining the search direction $\mathbf{p}_k$ at each iteration $\mathbf{x}_{k+1} = \mathbf{x}_k + a_k \mathbf{p}_k$. Reckon that the objective of a line search strategy is to iteratively move from an initial solution guess $\mathbf{x}_0$ to a stationary point $\mathbf{x}_k$ for which $\|\nabla f(\mathbf{x}_k)\|_2 \approx 0$. We initially define the *descent* direction, which is the search direction that moves us closer to a stationary point.

Definition 3.11 A search direction $\mathbf{p}_k$ at step k is a *descent* direction if $\langle \mathbf{p}_k, \nabla f(\mathbf{x}_k) \rangle < 0$, where $\langle \mathbf{p}_k, \nabla f(\mathbf{x}_k) \rangle$ is the inner product of $\mathbf{p}_k$ and $\nabla f(\mathbf{x}_k)$.

Reckon that the inner product $\langle \mathbf{p}_k, \nabla f(\mathbf{x}_k) \rangle$ of two vectors in the $\mathbb{R}^n$ space is the dot product $\mathbf{p}_k^\top \nabla f(\mathbf{x}_k)$. From this definition, the negative of a non-zero gradient is always a descent direction. That is, if $\mathbf{p}_k = -\nabla f(\mathbf{x}_k)$ we guarantee a descent direction because $\mathbf{p}_k^\top \nabla f(\mathbf{x}_k) = -(\nabla f(\mathbf{x}_k))^\top \nabla f(\mathbf{x}_k) < 0$. Consider, for instance,

$$\nabla f(\mathbf{x}_k) = \begin{bmatrix} a \\ b \\ \dots \\ n \end{bmatrix}$$

Then, $-(\nabla f(\mathbf{x}_k))^\top \nabla f(\mathbf{x}_k) = -(a^2 + b^2 - \dots + n^2) < 0$ for any non-zero $\nabla f(\mathbf{x}_k)$. Gradient descent uses this natural choice for the search direction, $\mathbf{p}_k$:

$$\mathbf{p}_k := -\nabla f(\mathbf{x}_k) \quad \text{(step direction in gradient descent)}$$

which generates iterates:

$$f(\mathbf{x}_{k+1}) < f(\mathbf{x}_k) \quad \forall k \in \mathbb{N}$$

Gradient descent is based on the observation that if the multi-variable function f is defined and differentiable in a neighborhood of $\mathbf{x}_k$, then $f(\mathbf{x}_k)$ decreases the fastest if one goes from $\mathbf{x}_k$ in the direction of the negative gradient of f in $\mathbf{x}_k$, $\nabla f(\mathbf{x}_k)$. The gradient descent line search method is provided below. We note here that the difference between *gradient* and *steepest* descent is that the latter descents in the direction of the largest directional derivative. The step direction in steepest descent is [10]:

$$\mathbf{p}_k := -\frac{\nabla f(\mathbf{x}_k)}{\|\nabla f(\mathbf{x}_k)\|_2} \quad \text{(step direction in steepest descent)}$$

The gradient descent line search algorithm is presented in Algorithm 8.

Algorithm 8 Gradient Descent Line Search Algorithm

1: set convergence tolerance τ, iteration $k = 0$, and make initial solution guess $\mathbf{x}_k$
2: **repeat**
3:     compute *step direction* $\mathbf{p}_k = -\nabla f(\mathbf{x}_k)$
4:     compute *step size* a_k to minimize $f(\mathbf{x}_k + a\mathbf{p}_k)$ with an exact or inexact method
5:     set $\mathbf{x}_{k+1} = \mathbf{x}_k + a_k\mathbf{p}_k$
6:     set $k = k + 1$
7: **until** $\|\nabla f(\mathbf{x}_k)\|_2 \leq \tau$

The gradient descent line search algorithm follows the steps of the line search strategy with the change that the search direction in line 3 of the algorithm is set to $\mathbf{p}_k = -\nabla f(\mathbf{x}_k)$. The step length in line 4 of the algorithm can be determined exactly by solving the minimization problem to global optimality or inexactly (i.e., strong Wolfe conditions, Goldstein conditions or Armijo backtracking line search). The algorithm stops when $\|\nabla f(\mathbf{x}_k)\|_2 \leq \tau$, where:

$$\|\nabla f(\mathbf{x}_k)\|_2 := \sqrt{\left(\frac{\partial f(\mathbf{x}_k)}{\partial x_1}\right)^2 + \left(\frac{\partial f(\mathbf{x}_k)}{\partial x_2}\right)^2 + \cdots + \left(\frac{\partial f(\mathbf{x}_k)}{\partial x_n}\right)^2}$$

for an n-dimensional vector $\mathbf{x}_k$. For a small convergence tolerance τ, this will ensure that $\|\nabla f(\mathbf{x}_k)\|_2 \approx 0$.

For demonstration purposes, let us consider the following single-variable unconstrained minimization example:

$$\min_{x \in \mathbb{R}} f(x) := x^2 - 5x$$

We apply the gradient descent for determining the step direction and Armijo backtracking line search for determining the step length. We start with initial solution guess $x_{k=0} = 5$ and we set $a = 1, c = 0.1, \rho = 0.7$ for the backtracking line search. In addition, we set precision error $\tau = 0.0001$ for the termination of the algorithm. In the first iteration, the computed step direction is $p_k = -f'(x_k) = -5$. Based on that, the resulting step size a_k after implementing the Armijo backtracking line search is the first one that satisfies the Armijo rule:

$$f(\mathbf{x}_k + a\mathbf{p}_k) \leq f(\mathbf{x}_k) + ca\mathbf{p}_k^\mathsf{T}\nabla f(\mathbf{x}_k)$$

This rule is satisfied for any $a \in [0, 0.9]$ and, because we use $\rho = 0.7$ for backtracking, the selected a_k is equal to 0.7. Therefore, $x_{k+1} = x_k + a_k p_k = 5 - 0.7 \cdot 5 = 1.5$. We update $k = k + 1$ and we proceed to the next iteration until the algorithm converges after 12 iterations with $x_k = 2.500042$ and $f(x_k) \simeq -6.25$ for which $|f'(x_k)| = 0.000084 \leq \tau$. The iterations are presented in Fig. 3.12.

Fig. 3.12 Convergence of gradient descent line search with Armijo backtracking in twelve iterations to a stationary point $x^* \simeq 2.5$ for which $f(x^*) \simeq -6.25$ and $f'(x^*) \simeq 0$

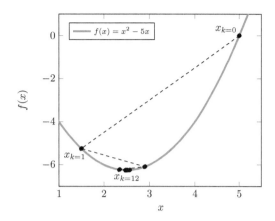

Table 3.3 Gradient descent iterations with Armijo backtracking line search ($c = 0.1$) when minimizing the unconstrained function $f(x) = x^2 - 5x$

k	x_k	$f(x_k)$	$f'(x_k)$	p_k	a_k	x_{k+1}	$\frac{|x_{k+1}-x^*|}{|x_k-x^*|}$
0	5	0	5	−5	0.7	1.5	0.4
1	1.5	−5.25	−2	2	0.7	2.9	0.4
2	2.9	−6.09	0.8	−0.8	0.7	2.34	0.4
3	2.34	−6.2244	−0.32	0.32	0.7	2.564	0.4
4	2.564	−6.2459	0.128	−0.128	0.7	2.4744	0.4
5	2.4744	−6.24934	−0.0512	0.0512	0.7	2.51024	0.4
6	2.51024	−6.2499	0.02048	−0.02048	0.7	2.495904	0.4
7	2.495904	−6.24998	−0.00819	0.008192	0.7	2.501638	0.399902
8	2.501638	−6.25	0.003277	−0.00328	0.7	2.499345	0.399878
9	2.499345	−6.25	−0.00131	0.001311	0.7	2.500262	0.4
10	2.500262	−6.25	0.000524	−0.00052	0.7	2.499895	0.400763
11	2.499895	−6.25	−0.00021	2.10E-04	0.7	2.500042	0.4
12	2.500042	−6.25	8.39E-05	–	–	–	–

The detailed results in each iteration of the gradient descent with Armijo backtracking from Fig. 3.12 are presented in Table 3.3. Note that all directions are descent directions because $p_k^\mathsf{T} \nabla f(x_k) < 0$ for all k. In addition, the computed stationary point $x^* \simeq 2.5$ is a local minimum because $f''(x^*) = 2 < 0$. This local minimum is also a global minimum because f is convex.

From the last column of Table 3.3 one can note that the gradient descent method converges linearly. In particular, there is a $0 < \mu < 1$ which is equal to, approximately, 0.4 for which $|x_{k+1} - x^*| \leq \mu |x_k - x^*|$. This is not a fast convergence and might result in computational issues when solving more complicated problems.

Table 3.4 Gradient descent iterations with strong Wolfe conditions ($c_1 = 0.1, c_2 = 0.4$) when minimizing the unconstrained function $f(x) = x^2 - 5x$

k	x_k	$f(x_k)$	$f'(x_k)$	p_k	a_k	x_{k+1}	$\frac{|x_{k+1}-x^*|}{|x_k-x^*|}$
0	5	0	5	−5	0.7	1.5	0.4
1	1.5	−5.25	−2	2	0.7	2.9	0.4
2	2.9	−6.09	0.8	−0.8	0.49	2.508	0.02
3	2.508	−6.249936	0.016	−0.016	0.49	2.50016	0.02
4	2.50016	−6.25	0.00032	−0.00032	0.7	2.499936	0.4
5	2.499936	−6.25	−0.00013	0.000128	0.7	2.500026	0.4
6	2.500026	−6.25	5.12E-05	0.000128	0.7	–	–

It is also important to note that the step length selection method plays an important role. When using the Armijo rule, we required 12 iterations. We note though that if we use exact line search the algorithm terminates in only 1 iteration. In particular, for $k = 0$ we seek $a \geq 0$ to minimize $f(x_k + ap_k)$. For $f(x_k + ap_k) = (x_k + ap_k)^2 - 5(x_k + ap_k)$ we have $f'(x_k + ap_k) = 2p_k(x_k + p_k a) - 5p_k$. Setting $f'(x_k + ap_k) = 0$ we get $a = \frac{5}{2p_k} - \frac{x_k}{p_k}$ which results in $a_k = 0.5$ for $p_k = -5$ and $x_k = 5$. Thus, the updated solution $x_{k+1} = x_k + a_k p_k = 5 + 0.5(-5) = 2.5$ and the algorithm terminates in 1 iteration. Similarly, if we use the strong Wolfe conditions that impose the curvature condition in addition to the Armijo rule the algorithm terminates in 6 iterations instead of 12 (see Table 3.4).

Consider now an example of a multivariable minimization problem. We seek to minimize the nonlinear Rosenbrock's function in two dimensions:

$$\min f(\mathbf{x}) = 100(x_2 - x_1^2)^2 + (1 - x_1)^2$$

that has gradient:

$$\nabla f(\mathbf{x}) = [-400x_1(-x_1^2 + x_2) + 2x_1 - 2, -200x_1^2 + 200x_2]^\mathsf{T}$$

The 2-dimensional Rosenbrock's function is presented in Fig. 3.13. At the bottom layer of this figure is also projected its 2D contour plot that provides a view from the top. As it can be seen graphically, it has a strict local minimum of $\mathbf{x}^* = [x_1, x_2]^\mathsf{T} = [1, 1]^\mathsf{T}$ with $f(\mathbf{x}^*) = 0$.

We minimize this nonlinear function with gradient descent line search. In particular, if we use Armijo backtracking for the step length a_k with $a = 1$, $c = 0.1$ and $\rho = 0.7$, initial solution guess $\mathbf{x}_k = [2, 2]^\mathsf{T}$ for $k = 0$ and termination threshold $\tau = 0.0001$, we will receive solution $\mathbf{x}^k = [1.00007198, 1.00014408]^\mathsf{T}$ after $k = 12806$ iterations with $f(\mathbf{x}_k) = 5.18E - 09 \simeq 0$ and $\|\nabla f(\mathbf{x}_k)\|_2 \simeq \tau = 0.0001$. It is important to stretch that the solution $\mathbf{x}^k$ after $k = 12806$ iterations is very close

Fig. 3.13 Two-dimensional Rosenbrock function. The marker (black dot) in the 3D figure and the projected 2D contour plot presents the locally optimal solution $\mathbf{x}^* = [1, 1]^\mathsf{T}$ with $f(\mathbf{x}^*) = 0$

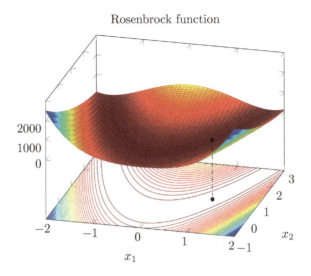

Rosenbrock function

Fig. 3.14 Log-log plot of the gradient descent method when finding stationary point $\mathbf{x}^* = [1, 1]^\mathsf{T}$ of Rosenbrock's function starting from $\mathbf{x}_{k=0} = [2, 2]^\mathsf{T}$

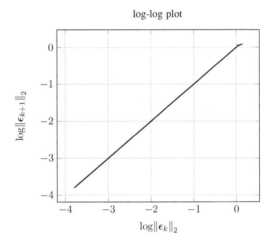

log-log plot

to $\mathbf{x}^* = [1, 1]^\mathsf{T}$ but not exactly equal to it because we allow for a precision error τ in the termination criterion of the algorithm. It is also important to note that the convergence of the algorithm is slow, as presented in Fig. 3.14. Note that this does not change significantly if we use strong Wolfe conditions instead of Armijo back-tracking line search. With strong Wolfe conditions the algorithm still required around 11420 iterations (a slight reduction from the 12806 iterations required when selecting the step length based on the Armijo rule).

The slope of Fig. 3.14 is almost equal to 1, and thus the order of convergence is $q = 1$ (linear convergence). At the last iteration we also have $\mu = \frac{\|\mathbf{x}_{k+1}-\mathbf{x}^*\|_2}{\|\mathbf{x}_k-\mathbf{x}^*\|_2} = \frac{0.000152818}{0.000153027} \approx 0.998634$. This is also true for all iterations except the very first ones.

The provided solution $\mathbf{x}_k = [1.00007198, 1.00014408]^\mathsf{T}$ is a stationary point with $\nabla f(\mathbf{x}_k) \simeq [0,0]^\mathsf{T}$. The Hessian matrix at this point is

$$\mathbf{H}(f(\mathbf{x}_k)) \simeq \begin{bmatrix} 802.115 & -400.029 \\ -400.029 & 200 \end{bmatrix}$$

which is positive definite because its eigenvalues 0.399 and 1001.715 are both positive. Thus, this stationary point is a *local minimum* and not a saddle point based on the second-order sufficient conditions for local optimality.

To summarize, the gradient descent line search method has the following main advantages and disadvantages:

- It is easy to implement and its iterations are fast because we just need to compute a gradient to determine the step direction.
- It converges slowly when problems are not strongly convex because moving in the direction of the negative of the gradient can lead to a lot of *zigzagging*.
- It cannot handle non-differentiable functions.

Closing, it is important to note that numerical optimization methods compute only a stationary point (they terminate when $\|f(\mathbf{x}_k)\|_2 \approx 0$). It is up to us to further characterize this stationary point (i.e., determine whether it is a local or global minimum/maximum). In addition, the selection of the initial solution guess $\mathbf{x}_0$ can result in the computation of different stationary points. Let us consider, for instance, the non-convex function $f(x) := \sin x + \cos 2x + \sin 3x$ in the interval $x \in [-2, 2]$. The graph of f is provided in Fig. 3.15. Applying gradient descent with initial solution guess $x_0 = -1.5$ will return stationary point $x^* \simeq -0.8751$ with $f(x^*) = -1.4397$ and $f'(x^*) \simeq 0$. If we use, however, another initial solution guess, i.e., $x_0 = 1.3$, gradient descent will converge to stationary point $\hat{x}^* \simeq 1.5708$ with $f(\hat{x}^*) = -1$ and $f'(\hat{x}^*) \simeq 0$.

Fig. 3.15 Plot $f(x) :=$ $\sin x + \cos 2x + \sin 3x$ for $x \in [-2, 2]$. For initial solution guess $x_0 = -1.5$ gradient descent converges to stationary point $x^* \simeq -0.8751$, which is a global minimum in $[-2, -2]$. For initial solution guess $x_0 = 1.3$ gradient descent converges to stationary point $\hat{x}^* \simeq 1.5708$, which is a local minimum

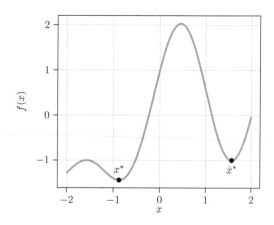

3.3.5 Newton Line Search Method

Newton's method (called also Newton-Raphson method) for computing the step direction $\mathbf{p}_k$ can be applied if the multi-variable function f is defined and is C^2 (twice differentiable) in a neighborhood of $\mathbf{x}_k$. The requirement that f is C^2 might limit its application to specific problems, but when this requirement is met the use of Newton's method is typically much faster than gradient descent. In Newton's method the objective function $f : \mathbb{R}^n \rightarrow \mathbb{R}$ around the point $\mathbf{x}_k$ is approximated by a second-order Taylor series:

$$f(\mathbf{x}_k + \Delta\mathbf{x}) \approx f(\mathbf{x}_k) + \Delta\mathbf{x}^{\mathsf{T}}\nabla f(\mathbf{x}_k) + \frac{1}{2}\Delta\mathbf{x}^{\mathsf{T}}\nabla^2 f(\mathbf{x}_k)\Delta\mathbf{x} \qquad (3.10)$$

where $\Delta\mathbf{x} = \mathbf{x} - \mathbf{x}_k$ for a $\mathbf{x}$ in the neighborhood of $\mathbf{x}_k$. Newton's method minimizes the second-order Taylor series approximation of f at $\mathbf{x}_k$, which is a quadratic function, to compute the step direction and find the next iterate $\mathbf{x}_{k+1}$. The gradient of quadratic function (3.10) considering our column vector variable $\Delta\mathbf{x}$ is:

$$\nabla f(\mathbf{x}_k + \Delta\mathbf{x}) = \nabla f(\mathbf{x}_k) + \nabla^2 f(\mathbf{x}_k)\Delta\mathbf{x}$$

To minimize $f(\mathbf{x}_k + \Delta\mathbf{x})$ near $\mathbf{x}_k$, we set $\nabla f(\mathbf{x}_k + \Delta\mathbf{x}) = [0, 0, \ldots, 0]^{\mathsf{T}}$ to find a stationary point. This gives us:

$$\nabla f(\mathbf{x}_k) + \nabla^2 f(\mathbf{x}_k)\Delta\mathbf{x} = \mathbf{0} \implies \Delta\mathbf{x} = -(\nabla^2 f(\mathbf{x}_k))^{-1}\nabla f(\mathbf{x}_k)$$

Thus, we select the search direction $\mathbf{p}_k = -(\nabla^2 f(\mathbf{x}_k))^{-1}\nabla f(\mathbf{x}_k)$, known as the *Newton direction*.

Algorithm 9 Newton Line Search Algorithm

1: set convergence tolerance τ, iteration $k = 0$, and make initial solution guess $\mathbf{x}_k$
2: **repeat**
3:     compute a *Newton direction* $\mathbf{p}_k = -(\nabla^2 f(\mathbf{x}_k))^{-1}\nabla f(\mathbf{x}_k)$
4:     compute *step size* $a_k > 0$ to minimize $f(\mathbf{x}_k + a\mathbf{p}_k)$ with an exact or inexact method
5:     set $\mathbf{x}_{k+1} = \mathbf{x}_k + a_k\mathbf{p}_k$
6:     set $k = k + 1$
7: **until** $\|\nabla f(\mathbf{x}_k)\|_2 \leq \tau$

A key distinction between Newton methods is the Pure vs Damped Newton, provided below. Note that the *Pure* Newton method often does not converge if the initial point is not close enough to a local minimum.

Fig. 3.16 Plot of $f(x)$ and
its second-order Taylor
series approximation $q(x)$ at
$x_k = -1$. The optimal value
of q results in Newton step
$p_k = 1$ and next iterate
$x_{k+1} = x_k + p_k = 0$ which
is closer to the local
minimizer $x^* \simeq 1.26$ of f
than x_k

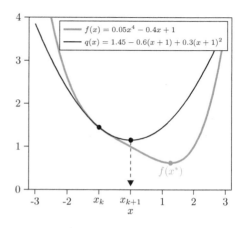

Pure vs Damped Newton Method

- If the step size is not determined from line search but it is set equal to $a_k = 1$, then
  the method is called *Pure* Newton method.
- If the step length a_k is not set equal to 1, but it is determined by other methods,
  i.e., (in)exact line search, then it is called *Damped* Newton method.

The main idea behind Newton's method is that at each iteration k we minimize
the second-order Taylor series approximation of f at $\mathbf{x}_k$. This approximates f by
an easier-to-solve quadratic polynomial q which is minimized to compute the step
direction of iteration k. Consider, for instance, the implementation of Pure Newton
in function $f(x) := 0.05x^4 - 0.4x + 1$ at iteration k, where $x_k = -1$. At iteration
k, function f is approximated by its second-order Taylor series approximation q:

$$q(x) = f(x_k) + (x - x_k)f'(x_k) + \frac{1}{2}(x - x_k)^2 f''(x_k)$$
$$= 1.45 - 0.6(x + 1) + 0.3(x + 1)^2$$

Function f and its aforementioned Taylor series approximation q at $x_k = -1$ are
presented in Fig. 3.16. By minimizing the Taylor-series approximation q instead of f,
we compute the Newton direction $p_k = -f'(x_k)/f''(x_k) = 1$ which will lead us to
the next iterate $x_{k+1} = x_k + p_k = -1 + 1 = 0$ (note that the step length a_k is omitted
because $a_k = 1$ in Pure Newton).

Newton's line search algorithm converges very fast. In practice, Newton's method
is typically slower at the beginning (linear convergence) and it turns to quadratic

Fig. 3.17 Log-log error plot
of Newton's method with
Armijo backtracking line
search when finding
stationary point
$\mathbf{x}^* = [2, -1]^\mathsf{T}$ of $f(\mathbf{x}) =$
$100(x_2 - x_1^2)^2 + (1 - x_1)^2$
starting from $\mathbf{x}_{k=0} = [1, 1]^\mathsf{T}$

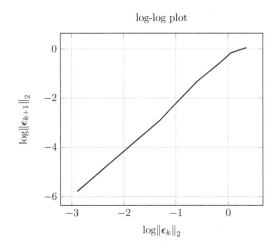

convergence as we approach a stationary point $\mathbf{x}^*$. In fact, if f is C^3 and the initial
solution guess $\mathbf{x}_k$ is sufficiently close to $\mathbf{x}^*$, where $\nabla f(\mathbf{x}^*) = \mathbf{0}$ and $\nabla^2 f(\mathbf{x}^*)$ is
nonsingular, then the sequence of iterates $\mathbf{x}_k, \mathbf{x}_{k+1}, \mathbf{x}_{k+2}, \ldots$ converges *quadratically*
to $\mathbf{x}^*$. Consider, for instance, the minimization of Rosenbrock's function $f(\mathbf{x}) =$
$100(x_2 - x_1^2)^2 + (1 - x_1)^2$. Using Newton's line search with Armijo backtracking
and initial solution guess $\mathbf{x} = [1, 1]^\mathsf{T}$ the convergence with convergence tolerance
$\tau = 0.0001$ at stationary point $\mathbf{x}^* = [2, -1]^\mathsf{T}$ is presented at the log-log plot of
Fig. 3.17. Note that this convergence starts slow and becomes rapidly quadratic with
$q = 2$.

Despite its fast convergence when starting close to a stationary point, Newton's
method is more complex compared to gradient descent because it requires the computation and the storage of the inverse Hessian matrix $(\nabla^2 f(\mathbf{x}_k))^{-1}$ at each iteration.
In addition, unlike gradient/steepest descent, it does not guarantee that $\mathbf{p}_k$ is always a
descent direction. That is, its steps may not even be descent directions if the algorithm
starts far away from a stationary point. This is described in the following lemma.

Lemma 3.1 *Newton direction $\boldsymbol{p}_k$ is a descent direction if, and only if, Hessian*
$\nabla^2 f(\boldsymbol{x}_k)$ *is positive definite.*

Proof A descent direction requires that the inner product $\langle \mathbf{p}_k, \nabla f(\mathbf{x}_k) \rangle$, which is
equal to $\mathbf{p}_k^\mathsf{T} \nabla f(\mathbf{x}_k)$ in the Euclidean space, is negative. Since $\mathbf{p}_k = -(\nabla^2 f(\mathbf{x}_k))^{-1} \nabla$
$f(\mathbf{x}_k)$, this is possible if, and only if, the inverse Hessian $(\nabla^2 f(\mathbf{x}_k))^{-1}$ is positive
definite. Because the inverse of a positive definite matrix is also a positive definite
matrix, Hessian $\nabla^2 f(\mathbf{x}_k)$ should also be positive definite. $\square$

To ensure positive definiteness, one can modify the Hessian $\nabla^2 f(\mathbf{x}_k)$ by replacing it with positive definite matrix $\mathbf{B}_k$, where:

$$
\mathbf{B}_k = \begin{cases} \nabla^2 f(\mathbf{x}_k) & \text{if } \nabla^2 f(\mathbf{x}_k) \text{ is sufficiently positive definite} \\ \nabla^2 f(\mathbf{x}_k) + \mathbf{E}_k & \text{if } \nabla^2 f(\mathbf{x}_k) \text{ is not sufficiently positive definite} \end{cases} \quad (3.11)
$$

where $\mathbf{E}_k$ is a matrix such that $\nabla^2 f(\mathbf{x}_k) + \mathbf{E}_k$ is sufficiently positive definite. A general description of Newton's line search with Hessian modification is provided below.

Algorithm 10 Newton Line Search Algorithm with Hessian Modification

---
1: set convergence tolerance τ, iteration $k = 0$, and make initial solution guess $\mathbf{x}_k$
2: **repeat**
3:   **if** $\nabla^2 f(\mathbf{x}_k)$ is positive definite **then**
4:     set $\mathbf{B}_k = \nabla^2 f(\mathbf{x}_k)$
5:   **else**
6:     set $\mathbf{B}_k = \nabla^2 f(\mathbf{x}_k) + \mathbf{E}_k$ such that $\mathbf{B}_k$ is positive definite
7:   **end if**
8:   compute *step direction* $\mathbf{p}_k = -\mathbf{B}_k^{-1} \nabla f(\mathbf{x}_k)$
9:   compute *step size* a_k with an exact or inexact method
10:    set $\mathbf{x}_{k+1} = \mathbf{x}_k + a_k \mathbf{p}_k$
11:    set $k = k + 1$
12: **until** $\|\nabla f(\mathbf{x}_k)\|_2 \leq \tau$

In the above algorithm, $\mathbf{E}_k$ can be set equal to $\theta\mathbf{I}$, where $\mathbf{I}$ is the identity matrix and $\theta > 0$ a scalar multiplier that is calculated such that $\nabla^2 f(\mathbf{x}_k) + \theta\mathbf{I}$ is sufficiently positive definite. The question is how to compute θ such that $\nabla^2 f(\mathbf{x}_k) + \theta\mathbf{I}$ is sufficiently positive definite. This can be performed in an iterative manner by starting with $\theta = 0$ and increasing its value until $\nabla^2 f(\mathbf{x}_k) + \theta\mathbf{I}$ becomes sufficiently positive definite. To do this, we would need a test for evaluating whether $\nabla^2 f(\mathbf{x}_k) + \theta\mathbf{I}$ is positive definite. For this, we can use Cholesky's decomposition (also known as Cholesky factorization [11]) which, in the space of real numbers, converts a symmetric positive definite matrix $\nabla^2 f(\mathbf{x}_k) + \theta\mathbf{I}$ to the product of a lower triangular matrix $\mathbf{L}$ and its transpose $\mathbf{L}^\mathsf{T}$, i.e., $\nabla^2 f(\mathbf{x}_k) + \theta\mathbf{I} = \mathbf{L}\mathbf{L}^\mathsf{T}$. The idea behind this is that if $\nabla^2 f(\mathbf{x}_k) + \tau\mathbf{I}$ is not positive definite, Cholesky's decomposition is not successful and we should increase the value of θ to $\theta + \beta$ before trying again (a value of $\beta = 0.001$ was proposed by [8]). This practical algorithm can be expressed as:

Algorithm 11 Newton Line Search Algorithm with Hessian Modification via Cholesky Decomposition

---

1: set convergence tolerance τ, iteration $k = 0$, and make initial solution guess $\mathbf{x}_k$
2: **repeat**
3:    **if** $\nabla^2 f(\mathbf{x}_k)$ is positive definite **then**
4:      set $\mathbf{B}_k = \nabla^2 f(\mathbf{x}_k)$
5:    **else**
6:      initialize θ, i.e., set $\theta = 0$
7:      **repeat**
8:        set $\theta = \theta + \beta$, where $\beta > 0$
9:      **until** Cholesky's Decomposition $\nabla^2 f(\mathbf{x}_k) + \theta \mathbf{I} = \mathbf{LL}^{\mathsf{T}}$ is successful
10:     set $\mathbf{B}_k = \nabla^2 f(\mathbf{x}_k) + \theta \mathbf{I}$
11:    **end if**
12:    compute *step direction* $\mathbf{p}_k = -\mathbf{B}_k^{-1} \nabla f(\mathbf{x}_k)$
13:    compute *step size* a_k with an exact or inexact method
14:    set $\mathbf{x}_{k+1} = \mathbf{x}_k + a_k \mathbf{p}_k$
15:    set $k = k + 1$
16: **until** $\|\nabla f(\mathbf{x}_k)\|_2 \leq \tau$

Practitioner's Corner
Cholesky's decomposition in Python 3

Cholesky's decomposition is a standard algorithm of linear algebra in many programming languages. Let, for example, try to find positive definite $\mathbf{B}_k$ of $f(\mathbf{x}) := 0.5x_1^2 + x_1 \cos x_2$ at $\mathbf{x}_k = [1, 1]^{\mathsf{T}}$. In Python 3 this is achieved as follows.

```python
import numpy as np
import numdifftools as nd
import math
x=[1,1];theta=0;beta=0.001
def f(x):
 return(0.5*x[0]**2+x[0]*math.cos(x[1]))
while True:
 try:
  Bk = np.array(nd.Hessian(f)(x) + theta * np.identity
    (len(x)))
  L = np.linalg.cholesky(Bk)
  break
 except:
  theta = theta + beta
print(Bk,theta)
```

resulting in $\mathbf{B}_k = \begin{bmatrix} 1.911 & -0.84147098 \\ -0.84147098 & 0.37069769 \end{bmatrix}$ for $\theta = 0.911$.

Let us apply Newton's line search algorithm with Armijo backtracking line search and $a = 1$, $c = 10^{-4}$ and $\rho = 0.99$ to minimize Rosenbrock's function in two dimensions, $f(\mathbf{x}) = 100(x_2 - x_1^2)^2 + (1 - x_1)^2$ (see Fig. 3.13).

We start with an initial solution guess $\mathbf{x}_k = [2, 2]^\mathsf{T}$ for $k = 0$. We first compute the gradient $\nabla f(\mathbf{x}) = [-400x_1(x_2 - x_1^2) - 2(1 - x_1), 200(x_2 - x_1^2)]^\mathsf{T}$ and then the Hessian:

$$\nabla^2 f(\mathbf{x}) = \begin{bmatrix} 1200x_1^2 - 400x_2 + 2 & -400x_1 \\ -400x_1 & 200 \end{bmatrix}$$

For the initial solution guess $\mathbf{x}_k = [2, 2]^\mathsf{T}$ we have $\nabla f(\mathbf{x}_k) = [1602, -400]^\mathsf{T}$, and

$$\nabla^2 f(\mathbf{x}_k) = \begin{bmatrix} 4002 & -800 \\ -800 & 200 \end{bmatrix}$$

The inverse Hessian $(\nabla^2 f(\mathbf{x}))^{-1}$ is such that $\nabla^2 f(\mathbf{x})(\nabla^2 f(\mathbf{x}))^{-1} = \mathbf{I}$. Using Gaussian elimination, we compute its values:

$$(\nabla^2 f(\mathbf{x}_k))^{-1} = \begin{bmatrix} 0.00124688 & 0.00498753 \\ 0.00498753 & 0.02495012 \end{bmatrix}$$

We can now compute the Newton direction:

$$\mathbf{p}_k = -(\nabla^2 f(\mathbf{x}_k))^{-1}\nabla f(\mathbf{x}_k) = [-0.00249377, 1.99002494]^\mathsf{T}$$

For this step direction, the Armijo backtracking line search results in a step length of $a_k \simeq 0.9950187$ and $\mathbf{x}_{k+1} = \mathbf{x}_k + a_k\mathbf{p}_k \simeq [1.997506, 3.990025]^\mathsf{T}$. Updating $k = k + 1$ we proceed to the next iterations.

As shown in Table 3.5, Newton's line search terminates after $k = 13$ iterations. It convergences to $\mathbf{x}_k = [1, 1]^\mathsf{T}$ with $f(\mathbf{x}_k) = 1.6E - 21 \simeq 0$ and $\|\nabla f(\mathbf{x}_k)\|_2 = 1.19E - 10 < \tau$, where the convergence tolerance is $\tau = 0.0001$. It is remarkable that Newton's line search required only 13 iterations compared to more than 11 thousand iterations required by the gradient descent for solving the same unconstrained minimization problem. The iteration steps of Newton's method are presented in Fig. 3.18.

Note that in Table 3.5 the inner product $\langle \mathbf{p}_k, \nabla f(\mathbf{x}_k) \rangle = \mathbf{p}_k^\mathsf{T}\nabla f(\mathbf{x}_k)$ is negative for any iteration (column 4). That is, all search directions were descent directions. We have already discussed, however, that if the Hessian $\nabla^2 f(\mathbf{x}_k)$ is not positive definite for some iteration k, then the step direction $\mathbf{p}_k$ in this iteration is not a descent direction because $\mathbf{p}_k^\mathsf{T}\nabla f(\mathbf{x}_k) \geq 0$. Consider, for example, the unconstrained minimization of the following function:

$$f(\mathbf{x}) = \frac{1}{2}x_1^2 + x_1 \cos x_2$$

Applying Newton's line search with Armijo backtracking for $a = 5$ and a starting point $\mathbf{x}_{k=0} = [1, 1]^\mathsf{T}$ will result in the convergence to stationary point

Fig. 3.18 Newton's line search steps until convergence (black dots), starting from $\mathbf{x}_0 = [2, 2]^\mathsf{T}$ and terminating at $\mathbf{x}_k = [1, 1]^\mathsf{T}$, for $k = 13$

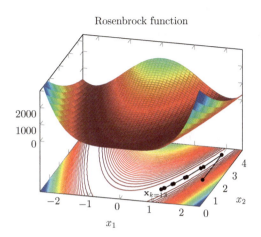

Table 3.5 Newton's iterations with Armijo backtracking line search when minimizing Rosenbrock's function

k	$\mathbf{x}_k^\mathsf{T}$	$f(\mathbf{x}_k)$	$\mathbf{p}_k^\mathsf{T} \nabla f(\mathbf{x}_k)$	a_k	$\|\nabla f(\mathbf{x}_k)\|_2$
0	[2,2]	401	-800.005	1	1651.183
1	[1.9975,3.9900]	0.995018692	-1.987565	0.257485	1.999982
2	[1.7409,2.9652]	0.982136558	-0.943703	1	49.1075
3	[1.6886,2.8488]	0.475002176	-0.614413	0.494839	3.272644
4	[1.4684,2.1064]	0.468186126	-0.537465	1	31.83473
5	[1.4257,2.0309]	0.181611606	-0.266392	0.592966	1.925439
6	[1.2407,1.5044]	0.18031798	-0.259228	1	19.16487
7	[1.2106,1.4647]	0.044445058	-0.075275	0.785678	0.879044
8	[1.0705,1.1262]	0.04425574	-0.080566	1	9.494956
9	[1.0563,1.1156]	0.003176671	-0.006107	1	0.202046
10	[1.0021,1.0014]	0.000863902	-0.001724	1	1.317025
11	[1.0008,1.0016]	6.53E-07	-1.31E-06	1	0.002407
12	[1.0000,0.9999]	4.27E-11	-8.53E-11	1	0.000292
13	[1.0000,1.0000]	1.60E-21	$-$	$-$	1.19E-10

$\mathbf{x}^* \approx [0, 1.5708]^\mathsf{T}$ with $\nabla f(\mathbf{x}^*) \simeq [0, 0]^\mathsf{T}$ and $f(\mathbf{x}^*) = 4.97E - 09 \simeq 0$. However, this stationary point is not a local minimum (it is just a saddle point) because the Hessian $\nabla^2 f(\mathbf{x}^*) \simeq \begin{bmatrix} 1 & -1 \\ -1 & 0 \end{bmatrix}$ is indefinite (its eigenvalues $- 0.61803399$ and 1.61803399 are both positive and negative).

Let us solve this problem with Hessian modification using Cholesky's decomposition (Algorithm 11). For this, we set $\beta = 0.075$ when updating θ in order to maintain always a descent direction. At the first iteration of the algorithm, for $\mathbf{x}_{k=0} = [1, 1]^\mathsf{T}$, we have Hessian matrix

Fig. 3.19 Line search for $0.5x_1^2 + x_1 \cos x_2$ with Newton's line search (gray) that leads to saddle point $[0, 1.5708]^\mathsf{T}$ and Newton's line search with Hessian modification (black) that leads to local minimum $[1, 3.1416]^\mathsf{T}$. Both searches start from initial solution guess $[1, 1]^\mathsf{T}$

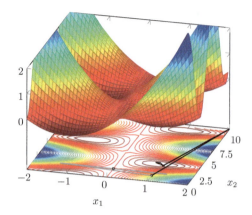

$$\nabla^2 f(\mathbf{x}_k) = \begin{bmatrix} 1 & -0.84147098 \\ -0.84147098 & -0.54030231 \end{bmatrix}$$

which is indefinite with eigenvalues -0.91085 and 1.37055. Starting from $\theta = 0$, Cholesky's decomposition is not successful until $\theta = 0.975$ which results in a modified Hessian $\mathbf{B}_k = \nabla^2 f(\mathbf{x}_k) + \theta \mathbf{I} = \begin{bmatrix} 1.975 & -0.84147098 \\ -0.84147098 & 0.43469769 \end{bmatrix}$ that is positive definite guaranteeing a descent direction $\mathbf{p}_k = [0.2559415, 2.43120299]^\mathsf{T}$. In the next iterations, θ is always equal to zero and we do not need to modify the Hessian. The modified Newton's line search converges to stationary point $\mathbf{x}^* = [1, 3.1416]^\mathsf{T}$ with $f(\mathbf{x}^*) \simeq -0.5$ and $\|\nabla^2 f(\mathbf{x}^*)\|_2 \simeq 0$ which is positive definite, meaning that the stationary point is a local minimum. For illustration purposes, the saddle point produced by Newton's method and the local minimum produced by Newton's method with Hessian modification are presented in Fig. 3.19.

Despite the rapid (quadratic) convergence when close to a stationary point, Newton's line search has three significant disadvantages:

- Its step direction might not always be a descent direction;
- It requires that f is C^2 in the neighborhood of all $\mathbf{x}_k$'s to derive the Hessian;
- It requires to compute and store the inverse Hessian matrix in each iteration, a task that is computationally intensive even for medium-size problems, and totally impractical for large ones.

3.3.6 Quasi-Newton Line Search Methods (DFP, BFGS, SR1, Broyden)

Newton's method assumes that the objective function can be locally approximated by a quadratic function, and uses the gradient and Hessian to find a stationary point. Quasi-Newton methods try to mitigate the two main disadvantages of Newton's method by not requiring that f is C^2 and not requiring to compute and store the inverse Hessian matrix at each iteration. Instead, they approximate the Hessian and inverse Hessian matrices by analyzing successive gradient vectors $\nabla f(\mathbf{x}_{k+1})$ and $\nabla f(\mathbf{x}_k)$. Because they approximate the inverse Hessian instead of computing it, these methods are known as *quasi-Newton*.

Their main difference is that the step size $\mathbf{p}_k$ is updated as follows:

$$\mathbf{p}_k = -\mathbf{B}_k^{-1} \nabla f(\mathbf{x}_k)$$

or, equivalently,

$$\mathbf{p}_k = -\mathbf{H}_k \nabla f(\mathbf{x}_k)$$

where $\mathbf{B}_k$ is a matrix that approximates the Hessian and $\mathbf{H}_k = \mathbf{B}_k^{-1}$ approximates the inverse Hessian $(\nabla^2 f(\mathbf{x}_k))^{-1}$. The overall goal is to benefit from the rapid convergence of Newton's method without the need of inverting the Hessian, which is typically implemented by solving a system of linear equations and is quite costly even for medium-scale problems.

The main idea behind several quasi-Newton algorithms is that the gradients of successive iterates $\nabla f(\mathbf{x}_k)$ and $\nabla f(\mathbf{x}_{k+1})$ yield curvature information. That is,

$$\nabla f(\mathbf{x}_{k+1}) - \nabla f(\mathbf{x}_k) \approx \nabla^2 f(\mathbf{x}_{k+1})$$

Hence, $\mathbf{y}_k = \nabla f(\mathbf{x}_{k+1}) - \nabla f(\mathbf{x}_k)$ can be used to update the approximate Hessian $\mathbf{B}_{k+1}$ and its inverse $\mathbf{H}_{k+1} = \mathbf{B}_{k+1}^{-1}$. One of the first known quasi-Newton method was the method of Davidon-Fletcher-Powell (DFP) introduced by Davidon in 1959 [12] and Fletcher and Powell [13] in 1963. In the DFP method, successive inverse Hessian approximations, $\mathbf{H}_{k+1} = \mathbf{B}_{k+1}^{-1}$, are constructed by the formula:

$$\mathbf{H}_{k+1} = \mathbf{B}_{k+1}^{-1} = \mathbf{H}_k + \frac{\mathbf{d}_k \mathbf{d}_k^{\mathsf{T}}}{\mathbf{y}_k^{\mathsf{T}} \mathbf{d}_k} - \frac{\mathbf{H}_k \mathbf{y}_k \mathbf{y}_k^{\mathsf{T}} \mathbf{H}_k}{\mathbf{y}_k^{\mathsf{T}} \mathbf{H}_k \mathbf{y}_k} \quad \text{(DFP)}$$

where $\mathbf{d}_k = \mathbf{x}_{k+1} - \mathbf{x}_k = a_k \mathbf{p}_k$ and $\mathbf{H}_k$ is initialized as any positive definite symmetric matrix, i.e., the identity matrix.

Another quasi-Newton method, and one of the most common ones, is the Broyden, Fletcher, Goldfarb and Shanno (BFGS) introduced in 1970 [14–17]. In BFGS, the successive Hessian approximations are constructed by the formula:

$$\mathbf{B}_{k+1} = \mathbf{B}_k + \frac{\mathbf{y}_k \mathbf{y}_k^{\mathsf{T}}}{\mathbf{y}_k^{\mathsf{T}} \mathbf{d}_k} - \frac{\mathbf{B}_k \mathbf{d}_k \mathbf{d}_k^{\mathsf{T}} \mathbf{B}_k}{\mathbf{d}_k^{\mathsf{T}} \mathbf{B}_k \mathbf{d}_k} \quad (3.12)$$

and the successive inverse Hessian approximations are constructed by the formula:

$$\mathbf{H}_{k+1} = \left(\mathbf{I} - \frac{\mathbf{d}_k \mathbf{y}_k^{\mathsf{T}}}{\mathbf{y}_k^{\mathsf{T}} \mathbf{d}_k} \right) \mathbf{H}_k \left(\mathbf{I} - \frac{\mathbf{y}_k \mathbf{d}_k^{\mathsf{T}}}{\mathbf{y}_k^{\mathsf{T}} \mathbf{d}_k} \right) + \frac{\mathbf{d}_k \mathbf{d}_k^{\mathsf{T}}}{\mathbf{y}_k^{\mathsf{T}} \mathbf{d}_k} \quad \text{(BFGS)}$$

The BFGS method is summarized in the following algorithm. This algorithm can be used for other quasi-Newton methods by modifying the computation of $\mathbf{H}_{k+1}$ in line 8. Note that the curvature condition

$$\mathbf{d}_k^\mathsf{T} \mathbf{y}_k > 0 \quad \text{(curvature condition)}$$

must be satisfied. This is ensured if we apply Wolfe or *strong* Wolfe conditions in the line search.

Algorithm 12 BFGS Line Search Algorithm

1: set convergence tolerance τ, iteration $k = 0$, $\mathbf{H}_k = \mathbf{I}$, and make initial solution guess $\mathbf{x}_k$
2: **repeat**
3:     compute *step direction* $\mathbf{p}_k = -\mathbf{H}_k \nabla f(\mathbf{x}_k)$
4:     compute *step size* a_k with an exact method or (strong) Wolfe conditions
5:     set $\mathbf{x}_{k+1} = \mathbf{x}_k + a_k \mathbf{p}_k$
6:     set $\mathbf{d}_k = a_k \mathbf{p}_k$ and $\mathbf{y}_k = \nabla f(\mathbf{x}_{k+1}) - \nabla f(\mathbf{x}_k)$
7:     set $\mathbf{H}_{k+1} = \left(\mathbf{I} - \frac{\mathbf{d}_k \mathbf{y}_k^\mathsf{T}}{\mathbf{y}_k^\mathsf{T} \mathbf{d}_k} \right) \mathbf{H}_k \left(\mathbf{I} - \frac{\mathbf{y}_k \mathbf{d}_k^\mathsf{T}}{\mathbf{y}_k^\mathsf{T} \mathbf{d}_k} \right) + \frac{\mathbf{d}_k \mathbf{d}_k^\mathsf{T}}{\mathbf{y}_k^\mathsf{T} \mathbf{d}_k}$
8:     set $k = k + 1$
9: **until** $\|\nabla f(\mathbf{x}_k)\|_2 \leq \tau$

Consider the minimization of $f(\mathbf{x}) = x_1^4 - 2x_2 x_1^2 + x_1^2 + x_2^2 - 2x_1 + 5$. Applying the BFGS quasi-Newton method to find the minimum of $f(x) = x_1^4 - 2x_2 x_1^2 + x_1^2 + x_2^2 - 2x_1 + 5$ with starting point $\mathbf{x} = [1, 2]^\mathsf{T}$, convergence tolerance $\tau = 0.0001$, initial $\mathbf{H}_k = \mathbf{I}$ and strong Wolfe conditions with $c_1 = 10^{-3}$, $c_2 = 0.1$, $a = 1000$, $\rho = 0.99$ results in solution $\mathbf{x}^* \simeq [1, 1]^\mathsf{T}$ with $f(\mathbf{x}^*) \simeq 4$ after 7 iterations (Table 3.6).

The rate of convergence of the BFGS method is usually superlinear. This is evident in Table 3.6 where $\mu = \frac{\|\mathbf{x}_{k+1} - \mathbf{x}^*\|_2}{\|\mathbf{x}_k - \mathbf{x}^*\|_2} \simeq 0$ for large values of k. This is also made explicit in the log-log error plot of Fig. 3.20. Note that the slope of Fig. 3.20 is greater than 1 but less than 2, indicating a superlinear convergence.

Practitioner's Corner
BFGS in Python 3

BFGS can be implemented in Python using the `scipy.optimize` library. Let, for example, try to solve the example problem in Table 3.6 in Python 3 using the default parameters of the BFGS algorithm in scipy. An implementation example follows.

```
import numpy as np
from scipy.optimize import minimize
import math
t=0.0001
x0=np.array([1,2])
def f(x):
```

Fig. 3.20 Log-log error plot of the BFGS method with strong Wolfe conditions when finding stationary point $\mathbf{x}^* = [1, 1]^\mathsf{T}$ of $f(\mathbf{x}) = x_1^4 - 2x_2x_1^2 + x_1^2 + x_2^2 - 2x_1 + 5$ starting from $\mathbf{x}_{k=0} = [1, 2]^\mathsf{T}$

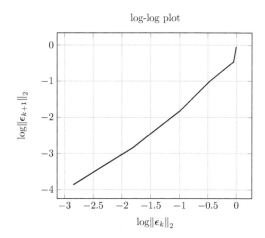

```
return(x[0]**4-2*x[0]**2*x[1]+x[0]**2+x[1]**2-2*x
    [0]+5)
print(minimize(f, x0, method='BFGS', tol=t))
```

resulting in solution $\mathbf{x}^* \simeq [1, 1]^\mathsf{T}$ with $f(\mathbf{x}^*) \simeq 4$ after 7 iterations.

For completeness, we also report the inverse Hessian approximations when using the Broyden and SR1 quasi-Newton methods.

Table 3.6 BFGS iterations with strong Wolfe conditions when minimizing $f(\mathbf{x}) = x_1^4 - 2x_2x_1^2 + x_1^2 + x_2^2 - 2x_1 + 5$

k	$f(\mathbf{x})$	$\mathbf{p}_k^\mathsf{T} \nabla f(\mathbf{x}_k)$	a_k	$\|\nabla f(\mathbf{x}_k)\|_2$	$\|\mathbf{x}_k - \mathbf{x}^*\|_2$	$\frac{\|\mathbf{x}_{k+1}-\mathbf{x}^*\|_2}{\|\mathbf{x}_k-\mathbf{x}^*\|_2}$
0	5	−20	0.084652	4.472136	1	0.897057
1	4.116162506	−0.077706	2.197017	0.475703	0.897057	0.384385
2	4.05381159	−0.065313	1.658131	0.952069	0.344815	0.294295
3	4.002120453	−0.004252	1.076297	0.090353	0.101477	0.149842
4	4.000071611	−0.000131	1.190092	0.028293	0.015206	0.093686
5	4.000000597	−1.2E−06	1.087169	0.002489	0.001425	0.097087
6	4.000000005	−1.01E−08	1.098151	0.000212	0.000138	–
7	4	–	–	1.36E-05	–	–

$$\mathbf{H}_{k+1} = \mathbf{H}_k + \frac{(\mathbf{d}_k - \mathbf{H}_k \mathbf{y}_k)\mathbf{d}_k^\mathsf{T}\mathbf{H}_k}{\mathbf{d}_k^\mathsf{T}\mathbf{H}_k \mathbf{y}_k} \qquad \text{(Broyden)}$$

$$\mathbf{H}_{k+1} = \mathbf{H}_k + \frac{(\mathbf{d}_k - \mathbf{H}_k \mathbf{y}_k)(\mathbf{d}_k - \mathbf{H}_k \mathbf{y}_k)^\mathsf{T}}{(\mathbf{d}_k - \mathbf{H}_k \mathbf{y}_k)^\mathsf{T}\mathbf{y}_k} \qquad \text{(SR1)}$$

3.3.7 Conjugate Gradient Line Search Method

The earliest conjugate gradient (CG) method was devised by Fletcher and Reeves in 1964 [18]. Conjugate gradient line search improves significantly the gradient descent line search without the need to compute second-order derivatives that are required in Newton's method.

CG combines the current gradient vector with the gradient vector from the previous iteration to obtain the new search direction $\mathbf{p}_k$ via a linear combination of the current and the previous search direction. It updates initial solution guess $\mathbf{x}_{k=0}$ by

$$\mathbf{x}_{k+1} = \mathbf{x}_k + a_k \mathbf{p}_k$$

where the step direction $\mathbf{p}_k$ is calculated by

$$\mathbf{p}_k = -\nabla f(\mathbf{x}_k) + \mathbf{p}_{k-1}\frac{(\nabla f(\mathbf{x}_k))^\mathsf{T}\nabla f(\mathbf{x}_k)}{(\nabla f(\mathbf{x}_{k-1}))^\mathsf{T}\nabla f(\mathbf{x}_{k-1})}$$

and the step length a_k is determined with line search. Initially, for $k = 0$ the step direction can be set to $\mathbf{p}_{k=0} = -\nabla f(\mathbf{x}_k)$. The algorithm is presented in Algorithm 13.

Algorithm 13 Conjugate Gradient Line Search Algorithm of Fletcher-Reeves

---

1: set convergence tolerance τ, initialize iteration $k = 0$ and make initial solution guess $\mathbf{x}_k$
2: initialize search direction $\mathbf{p}_k = -\nabla f(\mathbf{x}_k)$
3: **repeat**
4:    compute a *step size* a_k to minimize $f(x_k + a\mathbf{p}_k)$ with an exact line search or strong Wolfe conditions
5:    set $\mathbf{x}_{k+1} = \mathbf{x}_k + a_k \mathbf{p}_k$
6:    update the *step direction* $\mathbf{p}_{k+1} = -\nabla f(\mathbf{x}_{k+1}) + \mathbf{p}_k \frac{(\nabla f(\mathbf{x}_{k+1}))^\mathsf{T}\nabla f(\mathbf{x}_{k+1})}{(\nabla f(\mathbf{x}_k))^\mathsf{T}\nabla f(\mathbf{x}_k)}$
7:    set $k = k + 1$
8: **until** $\|\nabla f(\mathbf{x}_k)\|_2 \le \tau$

To ensure descent step directions for which $\mathbf{p}_k^\mathsf{T}\nabla f(\mathbf{x}_k) < 0$ at each iteration k, one should perform exact line search or apply the *strong* Wolfe conditions. Applying the CG line search of Fletcher-Reeves to $f(\mathbf{x}) = x_1^4 - 2x_2 x_1^2 + x_1^2 + x_2^2 - 2x_1 + 5$ with starting point $\mathbf{x} = [1, 2]^\mathsf{T}$, convergence tolerance $\tau = 0.0001$, and strong

Table 3.7 CG Fletcher-Reeves iterations with strong Wolfe conditions when minimizing $f(\mathbf{x}) = x_1^4 - 2x_2x_1^2 + x_1^2 + x_2^2 - 2x_1 + 5$

k	$f(\mathbf{x})$	$\mathbf{p}_k^\mathsf{T} \nabla f(\mathbf{x}_k)$	$\|\nabla f(\mathbf{x}_k)\|_2$	a_k	$\|\mathbf{x}_k - \mathbf{x}^*\|_2$	$\frac{\|\mathbf{x}_{k+1}-\mathbf{x}^*\|_2}{\|\mathbf{x}_k-\mathbf{x}^*\|_2}$
0	5	−20	4.472136	0.085354	1	0.896822
1	4.117458579	−0.24926	0.525834	0.091575	0.896822	0.974436
2	4.107362051	−0.07593	0.288666	1.0115	0.873896	0.667452
3	4.072336025	−0.68928	0.86889	0.235542	0.583284	0.139451
4	4.001670515	−0.01163	0.112966	0.090659	0.08134	1.013936
5	4.001194660	−0.0008	0.029647	2.329404	0.082473	0.227603
6	4.000354071	−0.00636	0.083408	0.116556	0.018771	0.488267
7	4.000014533	−1.24E-05	0.003685	0.260446	0.009165	0.924311
8	4.000013066	−2.49E-05	0.005217	1.129741	0.008472	0.106975
9	4.000000242	−2.32E-06	0.001588	0.087087	0.000906	1.033298
10	4.000000151	−9.62E-08	0.000324	2.791331	0.000936	0.122959
11	4.000000029	−5.82E-07	0.000797	0.10228	0.000115	0.844311
12	4.000000002	−1.19E-09	3.60E-05	–	9.72E-05	–

Wolfe conditions with $c_1 = 10^{-4}$, $c_2 = 0.1$, $a = 5$, $\rho = 0.99$ results in solution $\mathbf{x}^* \simeq [1, 1]^\mathsf{T}$ with $f(\mathbf{x}^*) \simeq 4$ after 12 iterations (Table 3.7). Besides the Fletcher-Reeves CG, there are more variants of the Conjugate Gradient line search method, such as the Polak and Ribière CG line search that changes slightly the formula of updating the step direction.

Practitioner's Corner

Conjugate Gradient in Python 3

The Polak and Ribière CG [19], which is a variant of the Fletcher-Reeves CG for which $\mathbf{p}_{k+1} = -\nabla f(\mathbf{x}_{k+1}) + \mathbf{p}_k \frac{(\nabla f(\mathbf{x}_{k+1}))^\mathsf{T}(\nabla f(\mathbf{x}_{k+1})-\nabla f(\mathbf{x}_k))}{\|\nabla f(\mathbf{x}_k)\|_2^2}$, can be implemented in Python using the `scipy.optimize` library. Let, for example, try to solve the example problem in Table 3.7 in Python 3 using the default parameters of the CG algorithm in scipy. An implementation example follows.

```python
import numpy as np
from scipy.optimize import minimize
t=0.0001
x0=np.array([1,2])
def f(x):
    return(x[0]**4-2*x[0]**2*x[1]+x[0]**2+x[1]**2-2*x
    [0]+5)
print(minimize(f, x0, method='CG', tol=t))
```

resulting in solution $\mathbf{x}^* \simeq [1, 1]^\mathsf{T}$ with $f(\mathbf{x}^*) \simeq 4$.

3.3.8   Trust-Region Strategy

We previously discussed the main line search algorithms for unconstrained optimization. We explored:

- Algorithms that require the computation of gradients (gradient/steepest descent, Fletcher-Reeves conjugate gradient).
- Algorithms that require the computation of Hessian matrices (Newton's method).
- Algorithms that require the computation of gradients and approximated values of Hessian matrices (Quasi-Newton methods including DFP, BFGS, Broyden, SR1).

The aforementioned methods employed the line search strategy to define a step length by solving a 1-dimensional problem and proceed to the next iterate. Unlike line search methods, trust-region methods define a region around the current iterate $\mathbf{x}_k$ within which they trust a quadratic model to be an adequate representation of the objective function. Then, they choose the step length to be the approximate minimizer of the model in this region. This means that they choose the step length and the step direction simultaneously. If a step length is not acceptable, the length is reduced and the step direction is also updated. In brief, both line search and trust-region are iterative optimization strategies that solve a much easier version of the original optimization problem in each iteration:

- Line search methods compute a step direction based on gradients/Hessians and then solve (in)exactly a simple 1-dimensional problem at each iteration to determine the step length.
- Trust-region methods solve an n-dimensional problem at each iteration, where n is the number of dimensions of the original problem, but this problem is based on a simpler objective function m_k (linear or quadratic) which is trusted to be a good approximation of the original objective function in a simple region (a ball of specified radius).

Consider, for example, minimization problem

$$\min_{\mathbf{x} \in \mathbb{R}^n} \ f(\mathbf{x})$$

where $f : \mathbb{R}^n \to \mathbb{R}$. Starting from an initial solution guess $\mathbf{x}_0$, at each iteration k the trust-region strategy solves the following constrained optimization sub-problem:

$$\min_{\mathbf{p}} \ m_k(\mathbf{p})$$
$$\text{s.t. } \|\mathbf{p}\|_2 \leq \Delta_k \ \text{(trust region)} \tag{3.13}$$

where $\mathbf{p}$ is the step that will create the next iterate $\mathbf{x}_{k+1} = \mathbf{x}_k + \mathbf{p}$ and $\Delta_k > 0$ a parameter called the *trust-region radius* that prohibits making a step greater than Δ_k. Although we use the l_2 norm in the constraints, other norms can also be applied. The two main questions when applying the trust-region strategy are the selection of Δ_k and the selection of a simple (i.e., linear or quadratic) function $m_k(\mathbf{p})$ that is a good approximation of f. To highlight the differences between the line search and the trust-region strategies, the general steps of both iterative strategies at each iteration k are presented below.

Line search: pick step direction $\mathbf{p}_k$ and solve the single-dimensional problem $\min_{a \geq 0} f(\mathbf{x}_k + a\mathbf{p}_k)$ to pick step size a_k. Then, set $\mathbf{x}_{k+1} = \mathbf{x}_k + a_k\mathbf{p}_k$.

Trust-region: create model $m_k(\mathbf{p})$ and pick $\mathbf{p}$ as the optimal solution of the constrained optimization problem (3.13). Then, set $\mathbf{x}_{k+1} = \mathbf{x}_k + \mathbf{p}$.

Typically, function m_k at each iteration k is either modeled as a linear function:

$$m_k(\mathbf{p}) := f(\mathbf{x}_k) + \mathbf{p}^\mathsf{T}\nabla f(\mathbf{x}_k) \tag{3.14}$$

or as a quadratic function:

$$m_k(\mathbf{p}) := f(\mathbf{x}_k) + \mathbf{p}^\mathsf{T}\nabla f(\mathbf{x}_k) + \frac{1}{2}\mathbf{p}^\mathsf{T}\mathbf{B}_k\mathbf{p} \tag{3.15}$$

where $\mathbf{B}_k$ can be the Hessian of f at $\mathbf{x}_k$ or a modified positive definite matrix. After selecting a model function, we need to minimize it at each step subject to $\|\mathbf{p}\|_2 \leq \Delta_k$. If we choose a linear function m_k, the optimal value of $\mathbf{p}$ is:

$$\mathbf{p} = -\Delta_k \frac{\nabla f(\mathbf{x}_k)}{\|\nabla f(\mathbf{x}_k)\|_2}$$

leading to

$$\mathbf{x}_{k+1} = \mathbf{x}_k - \Delta_k \frac{\nabla f(\mathbf{x}_k)}{\|\nabla f(\mathbf{x}_k)\|_2}$$

The question is whether this step is acceptable or we need to modify the value of the trust-region radius Δ_k to limit ourselves to a trusted region where $m_k(\mathbf{p})$ offers a good approximation of $f(\mathbf{x}_k + \mathbf{p})$. To check that, we compute the actual and predicted reduction of f when making step $\mathbf{p}$:

$$\rho_k = \frac{f(\mathbf{x}_k) - f(\mathbf{x}_k + \mathbf{p})}{m_k(\mathbf{0}) - m_k(\mathbf{p})} \tag{3.16}$$

where if ρ_k is close to 1 there is a good approximation of f in this trust region and we can even expand the trust region radius, whereas if $\rho \leq 0$ or it is very close to 0 we need to shrink Δ_k and solve the problem again because step $\mathbf{p}$ is not accepted. This iterative update of the trust-region radius Δ_k and the fact that when using a linear m_k the step is always $-\Delta_k \frac{\nabla f(\mathbf{x}_k)}{\|\nabla f(\mathbf{x}_k)\|_2}$ highlights the analogy between the *line search steepest descent algorithm* for which $\mathbf{x}_{k+1} = \mathbf{x}_k - a_k\frac{\nabla f(\mathbf{x}_k)}{\|\nabla f(\mathbf{x}_k)\|_2}$ and the *trust-region strategy* with a *linear model function* m_k for which $\mathbf{x}_{k+1} = \mathbf{x}_k - \Delta_k \frac{\nabla f(\mathbf{x}_k)}{\|\nabla f(\mathbf{x}_k)\|_2}$.

Similarly, there is an analogy between the line search Newton method and the trust-region strategy when using a quadratic function m_k with $\mathbf{B}_k$ equal to the Hessian

$\nabla^2 f(\mathbf{x}_k)$. For this reason, the trust-region strategy with:

$$m_k(\mathbf{p}) := f(\mathbf{x}_k) + \mathbf{p}^\mathsf{T} \nabla f(\mathbf{x}_k) + \frac{1}{2} \mathbf{p}^\mathsf{T} \nabla^2 f(\mathbf{x}_k) \mathbf{p}$$

is known as the *trust-region Newton method*.

To summarize, the main steps of the trust-region strategy are presented below.

Trust-region strategy in an algorithmic form

1: set convergence tolerance τ, iteration $k = 0$ and make initial solution guess $\mathbf{x}_k$
2: choose model function m_k (linear or quadratic)
3: choose $\eta_1 > 0$ close to 1 indicating a good model fit in the trust region
4: choose $\eta_2 > 0$ close to 0 indicating a bad model fit in the trust region
5: choose $\Delta_k > 0$ as trust-region radius
6: **while** $\|\nabla f(\mathbf{x}_k)\|_2 > \tau$ **do**
7:     find $\mathbf{p} := \mathrm{argmin}_{\mathbf{p}}\, m_k(\mathbf{p})$ s.t. $\|\mathbf{p}\|_2 \leq \Delta_k$
8:     set $\rho_k = \frac{f(\mathbf{x}_k) - f(\mathbf{x}_k + \mathbf{p})}{m_k(\mathbf{0}) - m_k(\mathbf{p})}$
9:     **if** $\rho_k < \eta_2$ **then**
10:        set $\mathbf{x}_{k+1} = \mathbf{x}_k$ (unsuccessful step) and shrink Δ_k
11:     **else if** $\eta_2 \leq \rho_k < \eta_1$ **then**
12:        set $\mathbf{x}_{k+1} = \mathbf{x}_k + \mathbf{p}$ (successful step)
13:     **else if** $\rho_k \geq \eta_1$ **then**
14:        set $\mathbf{x}_{k+1} = \mathbf{x}_k + \mathbf{p}$ (very successful step) and expand Δ_k
15:     **end if**
16:     set $k = k + 1$
17: **end while**

The above algorithm indicates that there are three available options in each iteration:

- Our model m_k is not a good fit of f within the trust-region radius Δ_k and we need to shrink it (possibly by setting $\Delta_k = \beta \Delta_k$ where $\beta \in (0, 1)$, i.e., $\beta = 0.5$).
- Our model m_k is a good fit of f within the trust-region radius Δ_k and we accept the step $\mathbf{p}$ resulting in $\mathbf{x}_{k+1} = \mathbf{x}_k + \mathbf{p}$.
- Our model m_k is a very good fit of f within the trust-region radius Δ_k and we can accept step $\mathbf{p}$ while also expanding $\Delta_k = \beta \Delta_k$ where $\beta > 1$ to allow taking a larger step in the next iteration and increase the speed of convergence.

One final note is that at each iteration we need to solve the constrained minimization problem:

$$\mathbf{p} := \mathrm{argmin}_{\mathbf{p}}\, m_k(\mathbf{p}) \quad \text{s.t.} \quad \|\mathbf{p}\|_2 \leq \Delta_k \tag{3.17}$$

As we already discussed, if m_k is defined as the linear function (3.14), then its global minimizer is the closed-form expression:

$$\mathbf{p} = -\Delta_k \frac{\nabla f(\mathbf{x}_k)}{\|\nabla f(\mathbf{x}_k)\|_2}$$

When m_k is a quadratic function, we have two options. The first option is to solve (3.17) exactly, a procedure that might be computationally intensive. The second option is to use an approximate solution of (3.17) instead of solving it. In the latter case, we need to ensure that the selected approximate solution will guarantee the convergence of the overall method. This is ensured if the selected approximate solution achieves as much reduction in m_k as would a steepest descent step constrained by the trust-region. Let $\mathbf{p}^s = -\Delta_k \frac{\nabla f(\mathbf{x}_k)}{\|\nabla f(\mathbf{x}_k)\|_2}$ denote the steepest descent step in the trusted region. Then, we seek to find $\mathbf{p}^c = t_k \mathbf{p}^s$, known as the *Cauchy* point, where:

$$t_k := \operatorname{argmin}_{t \geq 0} m_k(t\mathbf{p}^s) \text{ s.t. } \|t\mathbf{p}^s\| \leq \Delta_k$$

that has solution:

$$t_k = \begin{cases} 1 & \text{if } \nabla f(\mathbf{x}_k)^\mathsf{T} \mathbf{B}_k \nabla f(\mathbf{x}_k) \leq 0 \\ \min\left(\frac{\|\nabla f(\mathbf{x}_k)\|_2^3}{\Delta_k \nabla f(\mathbf{x}_k)^\mathsf{T} \mathbf{B}_k \nabla f(\mathbf{x}_k)}, 1 \right) & \text{otherwise.} \end{cases}$$

Any approximate solution $\hat{\mathbf{p}}$ of (3.17) that results in a reduction of m_k not less than the Cauchy point, that is $m_k(\hat{\mathbf{p}}) \leq m_k(\mathbf{p}^c)$, is an acceptable approximate solution that guarantees convergence of the overall iterative process. In practice, we request from the approximate solution $\hat{\mathbf{p}}$ of (3.17) to satisfy two inequality constraints:

$$m_k(\hat{\mathbf{p}}) \leq m_k(\mathbf{p}^c) \text{ and } \|\hat{\mathbf{p}}\| \leq \Delta_k$$

Closing this chapter, it is important to note that the proposed solution methods refer to continuous unconstrained optimization problems. In the next chapter, we will focus on optimization methods for problems with constraints.

Exercises

3.1 Continuity
Prove that
$$f(x) = \begin{cases} x^2 - 6 & \text{if } x < 1 \\ x - 1 & \text{if } x \geq 1 \end{cases}$$

is discontinuous at $x_0 = 1$.

3.2 Differentiability class
Provide the differentability class of $f(x) = |x|$. Is this function continuous?

3.3 Lipschitz continuous
Prove that $f(\mathbf{x}) = |x_1| + 5|x_2|$ is Lipschitz continuous.

3.4 Inflection point
Prove that $x^* = 0$ is an inflection point of $f(x) = 12x^3 + 6$.

3.5 Local extrema
Find the local extrema of $f(x) = x^3 - 3x$ in the closed set $[-20, +20]$. Which is the global minimum of f in this interval?

3.6 Convexity
Consider the function $g(x_1, x_2) = 2x_1^2 + x_2^2 + 2x_2 + 3$.

(a) Is g convex on $\mathbb{R}^2$? Provide calculations and explanations for your answer.
(b) Does g have local minimizers? If so, compute them and argue if they are also global minimizers.

3.7 Newton vs Bisection
Is Newton's method preferable compared to the Bisection method in terms of computational complexity? Justify your answer.

3.8 Gradient Descent line search
Solve

$$\min_{\mathbf{x} \in \mathbb{R}} x^2 - 12x$$

with gradient descent using the *Armijo* and *strong Wolfe* conditions considering initial solution guess $x_{k=0} = 15$. Consider also $c = c_1 = 0.1$, $c_2 = 0.4$, precision error $\tau = 0.0001$, and backtracking with $a = 1$, $\rho = 0.7$.

3.9 Damped Newton line search
Considering initial solution guess $\mathbf{x}_{k=0} = [0.5, 0]^{\mathsf{T}}$, solve:

$$\min_{\mathbf{x} \in \mathbb{R}^2} f(\mathbf{x}) = x_1^3 + 2x_1 x_2 + 3x_2^2$$

with Newton's line search with Hessian modification. Use precision error $\tau = 0.0001$ and $\beta = 0.075$ in Cholesky's decomposition. In addition, use inexact line search (Armijo's condition) with $a = 1$, $\rho = 0.99$, and $c = 10^{-4}$.

3.10 Conjugate Gradient line search
Considering initial solution guess $\mathbf{x}_{k=0} = [1, 2]^{\mathsf{T}}$, solve the minimization problem:

$$\min_{\mathbf{x} \in \mathbb{R}^2} f(\mathbf{x}) = x_1^4 - 2x_2 x_1^2 + x_1^2 + x_2^2 - 2x_1 + 5$$

with the Conjugate Gradient using inexact line search with strong Wolfe conditions for which $a = 5$, $\rho = 0.99$, $c_1 = 10^{-4}$, $c_2 = 0.1$. Use precision error $\tau = 0.0001$.

References

1. M.Ó. Searcóid, *Metric Spaces* (2007). pp. 125–146
2. F.W. Warner, *Foundations of Differentiable Manifolds and Lie Groups*, vol. 94. (Springer Science & Business Media, 1983)
3. G.R. Wood, Math. Program. **55**(1), 319 (1992)
4. S.B. Russ, Hist. Math. **7**(2), 156 (1980)
5. R.L. Burden, J.D. Faires, Cole Thomson Learn. Inc. **14**, 190 (1997)
6. P. Wolfe, SIAM Rev. **11**(2), 226 (1969)
7. P. Wolfe, SIAM Rev. **13**(2), 185 (1971)
8. J. Nocedal, S. Wright, *Numerical Optimization* (Springer Science & Business Media, 2006)
9. L. Armijo, Pac. J. Math. **16**(1), 1 (1966)
10. W.L. Winston, J.B. Goldberg, *Operations Research: Applications and Algorithms*, vol. 3. (Thomson Brooks/Cole Belmont, 2004)
11. G.H. Golub, C.F. Van Loan, *Matrix Computations*, vol. 3. (JHU Press, 2013)
12. W.C. Davidon, in *Technical Report, Argonne National Laboratories*, vol. Ill (1959)
13. R. Fletcher, M.J. Powell, Comput. J. **6**(2), 163 (1963)
14. C.G. Broyden, IMA J. Appl. Math. **6**(1), 76 (1970)
15. R. Fletcher, Comput. J. **13**(3), 317 (1970)
16. D. Goldfarb, Math. Comput. **24**(109), 23 (1970)
17. D.F. Shanno, Math. Comput. **24**(111), 647 (1970)
18. R. Fletcher, C.M. Reeves, Comput. J. **7**(2), 149 (1964)
19. E. Polak, G. Ribiere, ESAIM: Math. Model. Numer. Anal. **3**(R1), 35 (1969)

Chapter 4
Continuous Constrained Optimization

Abstract This chapter presents the necessary and sufficient conditions for the local and global optimality of continuous, constrained optimization problems with scalar-valued objective functions. It also presents analytic and iterative solution methods for solving linear and nonlinear optimization problems that are subject to equality and inequality constraints. Solution methods include simplex, interior point, active set, sequential quadratic programming, penalty and augmented Lagrangian methods, among others.

4.1 Necessary Conditions for Local Optimality

A general formulation of a constrained minimization problem is:

$$\min_{\mathbf{x} \in \mathbb{R}^n} f(\mathbf{x}) \tag{4.1}$$

$$\text{subject to: } h_i(\mathbf{x}) = 0, \qquad\qquad i = 1, \ldots, l \tag{4.2}$$

$$g_i(\mathbf{x}) \leq 0, \qquad\qquad i = 1, \ldots, m \tag{4.3}$$

where $L = \{1, \ldots, l\}$ are finitely many equality constraints and $M = \{1, \ldots, m\}$ are finitely many inequality constraints. One can define the feasible region (set) $\mathcal{F}$ as the set of points that satisfy all constraints:

$$\mathcal{F} = \{\mathbf{x} \in \mathbb{R}^n \mid \mathbf{h}(\mathbf{x}) = \mathbf{0} \wedge \mathbf{g}(\mathbf{x}) \leq \mathbf{0}\}$$

where $\mathbf{h}(\mathbf{x}) = [h_1(\mathbf{x}), \ldots, h_l(\mathbf{x})]^\mathsf{T}$ and $\mathbf{g}(\mathbf{x}) = [g_1(\mathbf{x}), \ldots, g_m(\mathbf{x})]^\mathsf{T}$ are vector-valued functions. Reckon that the gradient of a scalar-valued function $f : \mathbb{R}^n \to \mathbb{R}$ when using the *denominator* layout notation is an n-valued vector:

$$\nabla f(\mathbf{x}) = \frac{\partial f(\mathbf{x})}{\partial \mathbf{x}} = \left[\frac{\partial f(\mathbf{x})}{\partial x_1}, \dots, \frac{\partial f(\mathbf{x})}{\partial x_n} \right]^\mathsf{T},$$

whereas for a vector-valued function $\mathbf{h} : \mathbb{R}^n \to \mathbb{R}^l$, we have to compute its *Jacobian matrix*, which is a vector-by-vector derivation.

Definition 4.1 The Jacobian matrix of a vector-valued function $\mathbf{h} : \mathbb{R}^n \to \mathbb{R}^l$ when using the *numerator* layout notation is the $l \times n$ matrix:

$$\nabla \mathbf{h}(\mathbf{x}) = \frac{\partial \mathbf{h}(\mathbf{x})}{\partial \mathbf{x}} = \begin{bmatrix} \frac{\partial h_1(\mathbf{x})}{\partial x_1} & \frac{\partial h_1(\mathbf{x})}{\partial x_2} & \cdots & \frac{\partial h_1(\mathbf{x})}{\partial x_n} \\ \frac{\partial h_2(\mathbf{x})}{\partial x_1} & \frac{\partial h_2(\mathbf{x})}{\partial x_2} & \cdots & \frac{\partial h_2(\mathbf{x})}{\partial x_n} \\ \vdots & \vdots & \ddots & \vdots \\ \frac{\partial h_l(\mathbf{x})}{\partial x_1} & \frac{\partial h_l(\mathbf{x})}{\partial x_2} & \cdots & \frac{\partial h_l(\mathbf{x})}{\partial x_n} \end{bmatrix}$$

If we now use the definition of the feasible set $\mathcal{F}$, we can restate the minimization problem in (4.1)–(4.3) as:

$$\min_{\mathbf{x} \in \mathcal{F}} f(\mathbf{x}) \tag{4.4}$$

Reckon that a locally optimal solution is defined as:

> **Local Optimum**

Considering a minimization problem, a solution $\mathbf{x}^* = [x_1^*, x_2^*, \dots, x_n^*]^\mathsf{T}$ is a local optimum in the neighborhood N of $\mathbf{x}^*$ if $f(\mathbf{x}^*) \leq f(\mathbf{x}) \ \forall \mathbf{x} \in \mathcal{F} \cap N$, where set $\mathcal{F}$ contains all feasible solutions.

> **Strict Local Optimum**

Considering a minimization problem, a solution $\mathbf{x}^*$ is a strict local optimum in the neighborhood N of $\mathbf{x}^*$ if $f(\mathbf{x}^*) < f(\mathbf{x}) \ \forall \mathbf{x} \in \mathcal{F} \cap N$ with $\mathbf{x} \neq \mathbf{x}^*$, where set $\mathcal{F}$ contains all feasible solutions.

As in the case of unconstrained optimization, constrained optimization problems have *necessary* and *sufficient* conditions. Reckon that the necessary and sufficient conditions for a local minimizer/maximizer of an unconstrained optimization problem with objective function f were:

- *First-order Necessary* conditions (FONC): If $\mathbf{x}^*$ is a local minimizer (maximizer) of f and $f(\mathbf{x})$ is C^1 in an open neighborhood of $\mathbf{x}^*$, then $\|\nabla f(\mathbf{x}^*)\|_2 = 0$.

- *Second-order Necessary* conditions (SONC): If $\mathbf{x}^*$ is a local minimizer (maximizer) of f and $f(\mathbf{x})$ is C^2 in an open neighborhood of $\mathbf{x}^*$, then $\|\nabla f(\mathbf{x}^*)\|_2 = 0$ and the Hessian of f at $\mathbf{x}^*$ is positive (negative) semi-definite.
- *Second-order Sufficient* conditions (SOSC): If f is C^2 in an open neighborhood of $\mathbf{x}^*$, $\|\nabla f(\mathbf{x}^*)\|_2 = 0$ and the Hessian of f at $\mathbf{x}^*$ is positive (negative) definite, then $\mathbf{x}^*$ is a strict local minimizer (maximizer) of f.

To expand these conditions to the case of constrained optimization problems, we first present the concept of *duality*.

4.1.1 Duality

Consider the general constrained minimization problem of Eqs. (4.1)–(4.3). An intuitive approach to minimize this problem is to introduce function:

$$J(\mathbf{x}) = \begin{cases} f(\mathbf{x}) & \text{if } \mathbf{x} \in \mathcal{F} := \{\mathbf{x} \in \mathbb{R}^n \mid \mathbf{h}(\mathbf{x}) = \mathbf{0} \wedge \mathbf{g}(\mathbf{x}) \leq \mathbf{0}\} \\ +\infty & \text{otherwise.} \end{cases}$$

and turn the constrained minimization problem into the following unconstrained minimization problem:

$$\min_{\mathbf{x} \in \mathbb{R}^n} J(\mathbf{x})$$

Notice that function $J(\mathbf{x})$ is equal to $f(\mathbf{x})$ if $\mathbf{x}$ is feasible, that is $\mathbf{x} \in \mathcal{F}$, and $+\infty$ if it is not. This will force a solution method to disregard all infeasible points given their $+\infty$ costs in $J(\mathbf{x})$. We can achieve the same effect by defining $J(\mathbf{x})$ as $J(\mathbf{x}) = f(\mathbf{x}) + I(\mathbf{x})$, where:

$$I(\mathbf{x}) = \begin{cases} 0 & \text{if } \mathbf{h}(\mathbf{x}) = \mathbf{0} \text{ and } \mathbf{g}(\mathbf{x}) \leq \mathbf{0} \\ +\infty & \text{otherwise.} \end{cases}$$

Since the violation of only one constraint is enough to turn $\mathbf{x}$ into an infeasible point and $J(\mathbf{x})$ to $+\infty$, we can write $J(\mathbf{x})$ as:

$$J(\mathbf{x}) = f(\mathbf{x}) + \sum_{i=1}^{l} I_1(h_i(\mathbf{x})) + \sum_{i=1}^{m} I_2(g_i(\mathbf{x}))$$

where I_1 and I_2 are functions such that:

$$I_1(h_i(\mathbf{x})) = \begin{cases} 0 & \text{if } h_i(\mathbf{x}) = 0 \\ +\infty & \text{otherwise.} \end{cases}$$

$$I_2(g_i(\mathbf{x})) = \begin{cases} 0 & \text{if } g_i(\mathbf{x}) \leq 0 \\ +\infty & \text{otherwise.} \end{cases}$$

The problem with this structure is that functions I_1 and I_2 are not differentiable and not continuous since their graphs break by moving from $f(\mathbf{x})$ to $+\infty$ when moving from a feasible to an infeasible point that are next to each other. Notice that if we use $\lambda_i h_i(\mathbf{x})$ instead of $I_1(h_i(\mathbf{x}))$ for any $\lambda_i \in \mathbb{R}$ the following inequality holds:

$$\lambda_i h_i(\mathbf{x}) \leq I_1(h_i(\mathbf{x}))$$

because:

- When $h_i(\mathbf{x}) = 0$ we have that $\lambda_i h_i(\mathbf{x}) = I_1(h_i(\mathbf{x})) = 0$.
- When $h_i(\mathbf{x}) \neq 0$ we have that $\lambda_i h_i(\mathbf{x}) < I_1(h_i(\mathbf{x}))$ since $I_1(h_i(\mathbf{x})) = +\infty$.

Thus, $\lambda_i h_i(\mathbf{x})$ is a lower bound of $I_1(h_i(\mathbf{x}))$ for any equality constraint $i \in L$. Notice also that if we use $\mu_i g(\mathbf{x})$ instead of $I_2(h_i(\mathbf{x}))$ for any $\mu_i \in \mathbb{R}_{\geq 0}$ we have:

$$\mu_i g_i(\mathbf{x}) \leq I_2(g_i(\mathbf{x}))$$

because:

- When $g_i(\mathbf{x}) = 0$ we have that $\mu_i g_i(\mathbf{x}) = I_2(g_i(\mathbf{x})) = 0$,
- When $g_i(\mathbf{x}) > 0$ we have that $\mu_i g_i(\mathbf{x}) < I_2(g_i(\mathbf{x}))$ since $I_2(g_i(\mathbf{x})) = +\infty$,
- When $g_i(\mathbf{x}) < 0$ we have that $\mu_i g_i(\mathbf{x}) \leq 0$ since $\mu_i \geq 0$ and $I_2(g_i(\mathbf{x})) = 0$; thus, $\mu_i g_i(\mathbf{x}) \leq I_2(g_i(\mathbf{x}))$.

Therefore, $\mu_i g_i(\mathbf{x})$ is a lower bound of $I_2(g_i(\mathbf{x}))$ for any inequality constraint $i \in M$. The resulting objective function $\mathcal{L} : \mathbb{R}^n \times \mathbb{R}^l \times \mathbb{R}^m \to \mathbb{R}$:

$$\mathcal{L}(\mathbf{x}, \boldsymbol{\lambda}, \boldsymbol{\mu}) := f(\mathbf{x}) + \sum_{i=1}^{l} \lambda_i h_i(\mathbf{x}) + \sum_{i=1}^{m} \mu_i g_i(\mathbf{x})$$

$$= f(\mathbf{x}) + \boldsymbol{\lambda}^{\mathsf{T}}\mathbf{h}(\mathbf{x}) + \boldsymbol{\mu}^{\mathsf{T}}\mathbf{g}(\mathbf{x})$$

where $\boldsymbol{\lambda} \in \mathbb{R}^l$ and $\boldsymbol{\mu} \in \mathbb{R}^m_+$ is called the *Lagrangian function*. The Lagrangian function is a lower bound of $J(\mathbf{x})$ at any point $\mathbf{x}$, i.e., $\mathcal{L}(\mathbf{x}, \boldsymbol{\lambda}, \boldsymbol{\mu}) \leq J(\mathbf{x})$. Although $\mathcal{L}(\mathbf{x}, \boldsymbol{\lambda}, \boldsymbol{\mu})$ has favorable properties compared to the nondifferentiable and noncontinuous $J(\mathbf{x})$, it cannot directly replace $J(\mathbf{x})$ in the minimization problem because their values are not always equal. To find a function that is always equal to $J(\mathbf{x})$, we present first the definitions of infima and suprema.

Definition 4.2 *Infinum* of a set of numbers A is the greatest lower bound of that set, inf A. *Supremum* of a set of numbers A is the smallest upper bound of that set, sup A.

For instance, $\inf_{x}\{x \in \mathbb{R} \mid 0 < x < 10\} = 0$ and $\sup_{x}\{x \in \mathbb{R} \mid 0 < x < 10\} = 10$. It is clear that if the supremum and infinum of a set A exist, then $\inf A \leq x \leq \sup A$ for every $x \in A$. Notice that in the example of set $A = \{x \in \mathbb{R} \mid 0 < x < 10\}$ the infinum and supremum do not belong to set A. If, however, $A = \{x \in \mathbb{R} \mid 0 \leq x \leq 10\}$ then the supremum and infinum belong to that set and $\inf A = \min A$, $\sup A = \max A$. It follows that if a function f is bounded from above for $x \in A$, then $\sup_{x \in A} f(x)$ is finite (not $+\infty$), and if f is bounded from below $\inf_{x \in A} f(x)$ is also finite (not $-\infty$). Other common properties are:

- If $A \subseteq B$ then $\inf A \geq \inf B$ and $\sup A \leq \sup B$
- $\inf(A + B) = \inf A + \inf B$ and $\sup(A + B) = \sup A + \sup B$
- If $r > 0$ then $\inf rA = r \inf A$ and $\sup rA = r \sup A$
- If $r \leq 0$ then $\inf rA = r \sup A$ and $\sup rA = r \inf A$

Using the definition of suprema, we now have:

$$\sup_{\mu \geq 0, \lambda} \mathcal{L}(\mathbf{x}, \lambda, \mu) = J(\mathbf{x})$$

where $\lambda \in \mathbb{R}^{l}$ and $\mu \in \mathbb{R}^{m}_{+}$. Let us see why this is the case. In general, for a given $\mathbf{x}$ we have two possible cases: $\mathbf{x}$ satisfies all constraints resulting in $J(\mathbf{x}) = f(\mathbf{x})$, $\mathbf{x}$ violates at least one (and possibly more) constraints resulting in $J(\mathbf{x}) = +\infty$.

In the first case, $\mathbf{x}$ is a feasible solution, thus $\mathbf{h}(\mathbf{x}) = \mathbf{0}$ and $\mathbf{g}(\mathbf{x}) \leq \mathbf{0}$. This means that:

$$\sup_{\mu \geq 0, \lambda} \mathcal{L}(\mathbf{x}, \lambda, \mu) = \sup_{\mu \geq 0, \lambda} f(\mathbf{x}) + \sup_{\mu \geq 0, \lambda} \lambda^{\mathsf{T}} \mathbf{h}(\mathbf{x}) + \sup_{\mu \geq 0, \lambda} \mu^{\mathsf{T}} \mathbf{g}(\mathbf{x}) = f(\mathbf{x})$$

because $\sup_{\mu \geq 0, \lambda} f(\mathbf{x}) = f(\mathbf{x})$, $\sup_{\mu \geq 0, \lambda} \lambda^{\mathsf{T}} \mathbf{h}(\mathbf{x}) = 0$ since $\mathbf{h}(\mathbf{x}) = \mathbf{0}$, and $\sup_{\mu \geq 0, \lambda} \mu^{\mathsf{T}} \mathbf{g}(\mathbf{x}) = 0$ because $\mu^{\mathsf{T}} \mathbf{g}(\mathbf{x}) \in (-\infty, 0]$ for $\mathbf{g}(\mathbf{x}) \leq \mathbf{0}$ and $\mu \geq \mathbf{0}$.

Let us now consider the second case where $\mathbf{x}$ violates at least one (and possibly more) constraints. In that case,

$$\sup_{\mu \geq 0, \lambda} \mathcal{L}(\mathbf{x}, \lambda, \mu) = \sup_{\mu \geq 0, \lambda} f(\mathbf{x}) + \sup_{\mu \geq 0, \lambda} \lambda^{\mathsf{T}} \mathbf{h}(\mathbf{x}) + \sup_{\mu \geq 0, \lambda} \mu^{\mathsf{T}} \mathbf{g}(\mathbf{x}) = +\infty$$

because:

- If an equality constraint $i \in \{1, \ldots, l\}$ is violated, then $h_i(\mathbf{x}) \neq 0$ and $\lambda_i h_i(\mathbf{x}) \in (-\infty, +\infty)$. Thus, $\sup_{\mu \geq 0, \lambda} \lambda_i h_i(\mathbf{x}) = +\infty$ resulting in $\sup_{\mu \geq 0, \lambda} \mathcal{L}(\mathbf{x}, \lambda, \mu) = +\infty$

- If an inequality constraint $i \in \{1, \ldots, m\}$ is violated, then $g_i(\mathbf{x}) > 0$ and $\mu_i g_i(\mathbf{x}) \in [0, +\infty)$ given that $\mu_i \geq 0$. Thus, $\sup_{\mu \geq 0, \lambda} \mu_i g_i(\mathbf{x}) = +\infty$ resulting in

$$\sup_{\mu \geq 0, \lambda} \mathcal{L}(\mathbf{x}, \lambda, \mu) = +\infty$$

Obviously, if several equality and/or inequality constraints are violated $\sup_{\mu \geq 0, \lambda} \mathcal{L}(\mathbf{x}, \lambda, \mu)$ will continue to be equal to $+\infty$, and thus we proved that:

$$\sup_{\mu \geq 0, \lambda} \mathcal{L}(\mathbf{x}, \lambda, \mu) = J(\mathbf{x})$$

In fact, we could write that:

$$\sup_{\mu \geq 0, \lambda} \mathcal{L}(\mathbf{x}, \lambda, \mu) = \begin{cases} f(\mathbf{x}) & \text{if } \mathbf{h}(\mathbf{x}) = \mathbf{0} \text{ and } \mathbf{g}(\mathbf{x}) \leq \mathbf{0} \\ +\infty & \text{otherwise.} \end{cases}$$

Then, solving the constrained minimization problem of Eqs. (4.1)–(4.3) is equivalent to solving:

$$\min_{\mathbf{x}} \sup_{\mu \geq 0, \lambda} \mathcal{L}(\mathbf{x}, \lambda, \mu) \qquad (4.5)$$

This problem cannot be easily solved. Instead, we can solve the so-called *Lagrangian Dual Problem* presented below.

Lagrangian Dual Problem

Consider the constrained minimization problem $\min_{\mathbf{x} \in \mathbb{R}^n} f(\mathbf{x})$ s.t. $\mathbf{h}(\mathbf{x}) = \mathbf{0} \wedge \mathbf{g}(\mathbf{x}) \leq \mathbf{0}$ with Lagrangian function $\mathcal{L}(\mathbf{x}, \lambda, \mu) := f(\mathbf{x}) + \lambda^{\mathsf{T}} \mathbf{h}(\mathbf{x}) + \mu^{\mathsf{T}} \mathbf{g}(\mathbf{x})$. Its Lagrangian Dual Problem is:

$$\max_{\lambda, \mu} \inf_{\mathbf{x}} \mathcal{L}(\mathbf{x}, \lambda, \mu) \qquad (4.6)$$

$$\text{subject to: } \mu \geq \mathbf{0} \qquad (4.7)$$

The Lagrangian Dual Problem can be written as:

$$\max_{\lambda, \mu} q(\lambda, \mu) \qquad (4.8)$$

$$\text{subject to: } \mu \geq \mathbf{0} \qquad (4.9)$$

where function $q(\lambda, \mu) = \inf_{\mathbf{x}} \mathcal{L}(\mathbf{x}, \lambda, \mu)$ is known as the *Lagrangian dual function* and it does not depend on $\mathbf{x}$. This is a very important property because the resulting $\max_{\lambda, \mu} q(\lambda, \mu)$ s.t. $\mu \geq \mathbf{0}$ is an easy problem that can be solved to global optimality since $q(\lambda, \mu)$ is pointwise affine, and thus concave for a given x, and the feasible region $\mu \geq 0$ is convex. The optimal value of this dual problem is:

$$q(\lambda^*, \mu^*) = \sup_{\mu \geq 0, \lambda} \inf_x \mathcal{L}(x, \lambda, \mu)$$

The fact that the dual has a concave objective function and a convex feasible region is very important because even if our original problem in Eqs. (4.1)–(4.3) does not have a convex objective function and/or feasible region, we will still be able to solve problem $\max_{\lambda, \mu} q(\lambda, \mu)$ s.t. $\mu \geq 0$ to global optimality. A solution λ^*, μ^* is dual feasible if $\mu^* \geq 0$ and $q(\lambda^*, \mu^*) > -\infty$. Below we summarize what we have presented so far to make the relation between the primal and the dual problem more explicit.

Primal and Dual problems for the case of minimization

Consider original constrained minimization problem, called the *Primal Problem*:

PRIMAL PROBLEM:
$$\begin{aligned} &\min_{x \in \mathbb{R}^n} f(x) \\ &\text{s.t.: } h_i(x) = 0, \quad i = 1, \dots, l \\ &\phantom{\text{s.t.: }} g_i(x) \leq 0, \quad i = 1, \dots, m \end{aligned} \tag{4.10}$$

This problem has Lagrangian function:

$$\mathcal{L}(x, \lambda, \mu) := f(x) + \sum_{i=1}^{l} \lambda_i h_i(x) + \sum_{i=1}^{m} \mu_i g_i(x)$$

where $\lambda \in \mathbb{R}^l$ and $\mu \in \mathbb{R}^m_+$. The primal problem in Eqs. (4.10) is equivalent to:

$$\min_x \sup_{\mu \geq 0, \lambda} \mathcal{L}(x, \lambda, \mu) \tag{4.11}$$

which has the *Dual Problem*:

DUAL PROBLEM:
$$\begin{aligned} &\max_{\lambda, \mu} \inf_x \mathcal{L}(x, \lambda, \mu) \\ &\text{s.t.: } \mu \geq 0 \end{aligned} \tag{4.12}$$

with solution:

$$q(\lambda^*, \mu^*) = \sup_{\mu \geq 0, \lambda} \inf_x \mathcal{L}(x, \lambda, \mu)$$

where $\lambda \in \mathbb{R}^l$ and $\mu \in \mathbb{R}^m_+$ are known as the *dual variables* or, as we will later see, *Lagrange/KKT multipliers*, and $q(\lambda, \mu) = \inf_x \mathcal{L}(x, \lambda, \mu)$ is known as the *Lagrangian dual function*.

The main issue here is that the Primal and its Dual are not equivalent since the optimal objective function value of program (4.11) is not necessarily equal to the optimal objective function value of program (4.12). There is, however, a relationship between the two that can help us to use the easy-to-solve dual for solving the primal. This is provided below in the *weak duality* theorem.

Theorem 4.1 *Weak Duality: when solving a constrained minimization problem,*

$$\inf_{x} \sup_{\substack{\lambda \in \mathbb{R}^l \\ \mu \in \mathbb{R}^m_+}} \mathcal{L}(x, \lambda, \mu) \geq \sup_{\substack{\lambda \in \mathbb{R}^l \\ \mu \in \mathbb{R}^m_+}} \inf_{x} \mathcal{L}(x, \lambda, \mu)$$

meaning that the optimal objective function value of the dual is a lower bound to the optimal objective function value of the primal.

Proof Let us denote the optimal objective function value of the primal as $J(\mathbf{x}^*) = \min_{\mathbf{x} \in \mathbb{R}^n} J(\mathbf{x})$. Because $J(\mathbf{x}) = \sup_{\substack{\lambda \in \mathbb{R}^l \\ \mu \in \mathbb{R}^m_+}} \mathcal{L}(\mathbf{x}, \lambda, \mu)$ we have that the optimal solution $\mathbf{x}^*$ must satisfy:

$$J(\mathbf{x}^*) = \inf_{\mathbf{x} \in \mathbb{R}^n} \sup_{\substack{\lambda \in \mathbb{R}^l \\ \mu \in \mathbb{R}^m_+}} \mathcal{L}(\mathbf{x}, \lambda, \mu)$$

We have already shown that for any $\lambda \in \mathbb{R}^l$ and $\mu \in \mathbb{R}^m_+$:

$$\mathcal{L}(\mathbf{x}, \lambda, \mu) \leq J(\mathbf{x}) \Rightarrow \mathcal{L}(\mathbf{x}, \lambda, \mu) \leq \sup_{\substack{\lambda \in \mathbb{R}^l \\ \mu \in \mathbb{R}^m_+}} \mathcal{L}(\mathbf{x}, \lambda, \mu)$$

Thus,

$$\inf_{\mathbf{x}} \mathcal{L}(\mathbf{x}, \lambda, \mu) \leq \inf_{\mathbf{x}} \sup_{\substack{\lambda \in \mathbb{R}^l \\ \mu \in \mathbb{R}^m_+}} \mathcal{L}(\mathbf{x}, \lambda, \mu)$$

for any $\lambda \in \mathbb{R}^l$ and $\mu \in \mathbb{R}^m_+$. This means that $\inf_{\mathbf{x}} \mathcal{L}(\mathbf{x}, \lambda, \mu)$ is a lower bound of the primal's optimal objective function value $J(\mathbf{x}^*)$. Notice that the above inequality holds for any $\lambda \in \mathbb{R}^l$ and $\mu \in \mathbb{R}^m_+$, even for $\sup_{\substack{\lambda \in \mathbb{R}^l \\ \mu \in \mathbb{R}^m_+}} \inf_{\mathbf{x}} \mathcal{L}(\mathbf{x}, \lambda, \mu)$ which corresponds to the tightest lower bound of $J(\mathbf{x}^*)$. Thus,

$$\sup_{\substack{\lambda \in \mathbb{R}^l \\ \mu \in \mathbb{R}^m_+}} \inf_{\mathbf{x}} \mathcal{L}(\mathbf{x}, \lambda, \mu) \leq \inf_{\mathbf{x}} \sup_{\substack{\lambda \in \mathbb{R}^l \\ \mu \in \mathbb{R}^m_+}} \mathcal{L}(\mathbf{x}, \lambda, \mu)$$

and this completes the weak duality proof.   □

Notice from the weak duality proof that the solution λ^*, μ^* of the dual which satisfies:

$$q(\boldsymbol{\lambda}^*, \boldsymbol{\mu}^*) = \sup_{\substack{\boldsymbol{\lambda} \in \mathbb{R}^l \\ \boldsymbol{\mu} \in \mathbb{R}^m_+}} \inf_{\mathbf{x}} \mathcal{L}(\mathbf{x}, \boldsymbol{\lambda}, \boldsymbol{\mu})$$

is not just a lower bound to the primal, it is the *tightest possible* lower bound. The difference between the optimal objective function of the primal and the optimal objective function of the dual, denoted as $J(\mathbf{x}^*) - q(\boldsymbol{\lambda}^*, \boldsymbol{\mu}^*)$, is called *Duality Gap*. It follows that:

- If the primal problem is *unbounded from below*, then $J(\mathbf{x}^*) = -\infty$ and thus $q(\boldsymbol{\lambda}^*, \boldsymbol{\mu}^*) = -\infty$ meaning that the dual problem is *infeasible*.
- If the dual problem is *unbounded from above*, then $q(\boldsymbol{\lambda}^*, \boldsymbol{\mu}^*) = +\infty$ and thus $J(\mathbf{x}^*) = +\infty$ meaning that the primal problem is *infeasible*.

The weak duality theorem is very important because solving the dual provides a lower bound to the primal minimization problem, even if the primal problem is not convex. Under certain conditions, known as *Strong Duality*, the duality gap between the primal and the dual is 0 meaning that solving the dual returns also an optimal solution for the primal. Typical conditions that are used to examine whether a primal problem satisfies the Strong Duality condition are the *refined Slater's constraint qualification* conditions provided below.

Theorem 4.2 *Refined Slater's constraint qualification conditions: if a minimization problem is convex and there exists a feasible point* $\mathbf{x}$ *which is strictly feasible for all inequality constraints that are not affine, then strong duality holds.*

In more detail, the refined Slater's constraint qualification conditions require that the problem is convex and $\exists\, \mathbf{x} \in \mathbb{R}^n$ such that $h_i(\mathbf{x}) = 0 \;\forall i \in L$, $g_i(\mathbf{x}) \leq 0$ for any $i \in M$ for which g_i is an affine function, and $g_i(\mathbf{x}) < 0$ for any $i \in M$ for which g_i is not affine.

Duality gives us an option of trying to solve our original constrained optimization problem in another way. If strong duality holds, we have potentially found an easier solution approach to our original problem. If not, then we still have a lower bound which may be of use. Duality also let us formulate optimality conditions for constrained optimization problems, as we will discuss in the following sections. Closing, we provide the primal-dual relationship in case the primal is a maximization problem.

Primal and Dual problems for the case of maximization

Consider original constrained maximization problem:

PRIMAL PROBLEM:

$$\max_{\mathbf{x} \in \mathbb{R}^n} f(\mathbf{x})$$
$$\text{s.t.:}\; h_i(\mathbf{x}) = 0, \quad i = 1, \ldots, l \tag{4.13}$$
$$g_i(\mathbf{x}) \leq 0, \quad i = 1, \ldots, m$$

This problem has Lagrangian function:

$$\mathcal{L}(\mathbf{x}, \boldsymbol{\lambda}, \boldsymbol{\mu}) := f(\mathbf{x}) - \sum_{i=1}^{l} \lambda_i h_i(\mathbf{x}) - \sum_{i=1}^{m} \mu_i g_i(\mathbf{x})$$

where $\boldsymbol{\lambda} \in \mathbb{R}^l$ and $\boldsymbol{\mu} \in \mathbb{R}_+^m$. The primal problem is equivalent to:

$$\max_{\mathbf{x}} \inf_{\boldsymbol{\mu} \geq 0, \boldsymbol{\lambda}} \mathcal{L}(\mathbf{x}, \boldsymbol{\lambda}, \boldsymbol{\mu}) \tag{4.14}$$

which has the *Dual Problem*:

DUAL PROBLEM: $\quad \begin{aligned} &\min_{\boldsymbol{\lambda}, \boldsymbol{\mu}} \sup_{\mathbf{x}} \mathcal{L}(\mathbf{x}, \boldsymbol{\lambda}, \boldsymbol{\mu}) \\ &\text{s.t.: } \boldsymbol{\mu} \geq \mathbf{0} \end{aligned}$ $\qquad$ (4.15)

with solution:

$$q(\boldsymbol{\lambda}^*, \boldsymbol{\mu}^*) = \inf_{\boldsymbol{\mu} \geq 0, \boldsymbol{\lambda}} \sup_{\mathbf{x}} \mathcal{L}(\mathbf{x}, \boldsymbol{\lambda}, \boldsymbol{\mu})$$

4.1.2 Necessary Conditions for Opimization Problems with Equality Constraints (Lagrange Multipliers)

We first start with the simpler case of a minimization problem that has only equality constraints:

$$\min_{\mathbf{x} \in \mathbb{R}^n} f(\mathbf{x}) \tag{4.16}$$

$$\text{subject to: } \mathbf{c}(\mathbf{x}) = \mathbf{0} \tag{4.17}$$

where $f(\mathbf{x})$ is a scalar-valued function $\mathbb{R}^n \to \mathbb{R}$ and $\mathbf{c}(\mathbf{x})$ is a vector-valued function $\mathbb{R}^n \to \mathbb{R}^l$, where l is the number of equality constraints. Given an equality-constrained minimization problem, one can introduce the *Lagrangian* function $\mathcal{L}$:

$$\mathcal{L}(\mathbf{x}, \boldsymbol{\lambda}) := f(\mathbf{x}) + \boldsymbol{\lambda}^\mathsf{T} \mathbf{c}(\mathbf{x}) \tag{4.18}$$

where $\boldsymbol{\lambda} = [\lambda_1, \ldots, \lambda_l]^\mathsf{T}$ are the dual variables, known as *Lagrange multipliers*. This function can be also written as:

$$\mathcal{L}(\mathbf{x}, \lambda_1, \ldots, \lambda_l) := f(\mathbf{x}) + \sum_{i=1}^{l} \lambda_i c_i(\mathbf{x}) \tag{4.19}$$

It is important to note that if we were solving a maximization instead of a minimization problem, then $\mathcal{L}(\mathbf{x}, \boldsymbol{\lambda}) := f(\mathbf{x}) - \boldsymbol{\lambda}^\mathsf{T} \mathbf{c}(\mathbf{x})$. Now, reckon that the first-order necessary conditions for local optimality in unconstrained optimization state that if $\mathbf{x}^*$ is a local optimum of f and f is C^1 in an open neighborhood of $\mathbf{x}^*$, then $\nabla f(\mathbf{x}^*) = \mathbf{0}$, or, equivalently, $\|\nabla f(\mathbf{x}^*)\|_2 = 0$. In the case of equality-constrained optimization, we have to use the Lagrangian function instead:

$$\nabla \mathcal{L}(\mathbf{x}^*, \boldsymbol{\lambda}^*) = \mathbf{0} \tag{4.20}$$

where, when using the *numerator* layout notation,

$$\nabla \mathcal{L}(\mathbf{x}, \boldsymbol{\lambda}) = \mathbf{0}^\mathsf{T} \Leftrightarrow \begin{cases} \nabla_\mathbf{x} \mathcal{L}(\mathbf{x}, \boldsymbol{\lambda}) = \mathbf{0}^\mathsf{T} \Rightarrow \underbrace{\nabla_\mathbf{x} f(\mathbf{x})}_{1 \times n} + \underbrace{\boldsymbol{\lambda}^\mathsf{T}}_{1 \times l} \underbrace{\nabla_\mathbf{x} \mathbf{c}(\mathbf{x})}_{l \times n} = \mathbf{0}^\mathsf{T} \Leftrightarrow \underbrace{\nabla_\mathbf{x} f(\mathbf{x})}_{1 \times n} + \underbrace{\sum_{i=1}^{l} \lambda_i \nabla_\mathbf{x} c_i(\mathbf{x})}_{1 \times n} = \mathbf{0}^\mathsf{T} \\ \nabla_\lambda \mathcal{L}(\mathbf{x}, \boldsymbol{\lambda}) = \mathbf{0}^\mathsf{T} \Rightarrow \underbrace{\mathbf{c}^\mathsf{T}(\mathbf{x})}_{1 \times l} = \underbrace{\mathbf{0}^\mathsf{T}}_{1 \times l} \end{cases}$$

$$(4.21)$$

and when using the *denominator* layout notation,

$$\nabla \mathcal{L}(\mathbf{x}, \boldsymbol{\lambda}) = \mathbf{0} \Leftrightarrow \begin{cases} \nabla_\mathbf{x} \mathcal{L}(\mathbf{x}, \boldsymbol{\lambda}) = \mathbf{0} \Rightarrow \underbrace{\nabla_\mathbf{x} f(\mathbf{x})}_{n \times 1} + \underbrace{\nabla_\mathbf{x} \mathbf{c}(\mathbf{x})}_{n \times l} \underbrace{\boldsymbol{\lambda}}_{l \times 1} = \mathbf{0} \Leftrightarrow \underbrace{\nabla_\mathbf{x} f(\mathbf{x})}_{n \times 1} + \underbrace{\sum_{i=1}^{l} \lambda_i \nabla_\mathbf{x} c_i(\mathbf{x})}_{n \times 1} = \mathbf{0} \\ \nabla_\lambda \mathcal{L}(\mathbf{x}, \boldsymbol{\lambda}) = \mathbf{0} \Rightarrow \underbrace{\mathbf{c}(\mathbf{x})}_{l \times 1} = \underbrace{\mathbf{0}}_{l \times 1} \Leftrightarrow c_i(\mathbf{x}) = 0 \; \forall i \in 1, \dots, l \end{cases}$$

$$(4.22)$$

Notice that the outcome when using the numerator and the denominator layout notation is similar (both sides in the equations are just transposed). The important part is to be consistent with the use of a layout notation when analyzing a problem. The solution of this system with $n + l$ equations will give us the values of $\mathbf{x}^*$ and $\boldsymbol{\lambda}^*$.

To demonstrate this, consider the single-variable minimization problem $\min f(x) := x^2$ subject to $c(x) = 0$, where $c(x) := 8x - 6$. The unconstrained version of this problem has solution $x' = 0$ with $f(x') = 0$. In Fig. 4.1 we present the solution of the unconstrained problem, x'. We also present constraint $c(x)$. Any solution of the constrained optimization problem should lie in the graph of c and the graph of the objective function f. Both graphs intersect at two points, and the local minimizer $x^* = 0.75$ is the one where $f(x^*)$ has a lower score. Notice that f and c are C^1 (in fact, they are $C^{+\infty}$) and for point x^* there exists some λ^* such that $\nabla \mathcal{L}(x^*, \lambda^*) = \mathbf{0}$, where $\mathcal{L}(x, \lambda) = f(x) + \lambda c(x)$. Indeed,

$$\nabla_x \mathcal{L}(x, \lambda) = 0 \Rightarrow 2x + 8\lambda = 0 \Rightarrow \lambda = -0.1875$$
$$\nabla_\lambda \mathcal{L}(x, \lambda) = 0 \Rightarrow 8x = 6 \Rightarrow x = 0.75$$

Fig. 4.1 Plot of $f(x) := x^2$
and $c(x) := 8x - 6$ in
interval $[-6, 10]$. Point
$x' = 0$ is the minimizer of
the unconstrained problem
min $f(x)$. Point x^* is the
minimizer of the constrained
problem min $f(x)$ s.t.
$c(x) = 0$

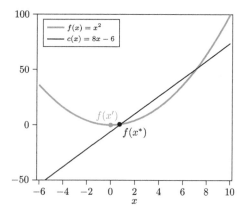

We now state the necessary optimality condition of a constrained optimization problem with equality constraints.

First-order Necessary conditions of multi-variable optimization problems with equality constraints

Assume that functions $f(\mathbf{x})$ and $c_i(\mathbf{x})$ where $i = 1, \dots, l$ are C^1 in an open neighborhood of $\mathbf{x}^*$. If $\mathbf{x}^*$ is a local optimum and a *regular point*, then $\nabla\mathcal{L}(\mathbf{x}^*, \boldsymbol{\lambda}^*) = \mathbf{0}$ for some Lagrange multiplier $\boldsymbol{\lambda}^* \in \mathbb{R}^l$.

In the first-order necessary conditions we use the term *regular point* for $\mathbf{x}^*$. In particular, $\mathbf{x}^*$ is a regular point if it satisfies any constraint qualification condition. We have already discussed about the refined *Slater's Constraint Qualification* condition, but we can also use more generally applicable ones that do not require convexity, such as the *Linear Independence Constraint Qualification (LICQ)*. LICQ requires that $\nabla c_i(\mathbf{x}^*)$ are linearly independent vectors for $i = 1, \dots, l$ for $\mathbf{x}^*$ to be a regular point. We can generally investigate if the regularity conditions are satisfied by checking if LICQ is satisfied.

Definition 4.3 *Linear Independence Constraint Qualification (LICQ)* requests that the gradients of the active inequality constraints and the gradients of the equality constraints are linearly independent at $\mathbf{x}^*$.

Because we consider only equality constraints, LICQ requests that all $\nabla c_1(\mathbf{x}^*), \dots, \nabla c_l(\mathbf{x}^*)$ are linearly independent at $\mathbf{x}^*$. These gradients are linearly independent if the following is true:

$$a_1 \nabla c_1(\mathbf{x}^*) + \cdots + a_l \nabla c_l(\mathbf{x}^*) = \mathbf{0} \text{ if, and only if, } a_i = 0 \; \forall i \in 1, \dots, l$$

where $a_1, \dots, a_l$ are scalars. To summarize, if f and $c_1, \dots, c_l$ are C^1 in an open neighborhood of $\mathbf{x}^*$ and $\mathbf{x}^*$ is a local optimum, there are two cases:

- Vectors $\nabla c_1(\mathbf{x}^*), \ldots, \nabla c_l(\mathbf{x}^*)$ are linearly dependent.
- There exists Lagrange multipler $\boldsymbol{\lambda}^* \in \mathbb{R}^l$ such that $\nabla \mathcal{L}(\mathbf{x}^*, \boldsymbol{\lambda}^*) = \mathbf{0}$.

Let us consider the example of the minimization problem:

$$\min_{\mathbf{x} \in \mathbb{R}^n} 3x_1^2 + 6x_2 + x_3$$

$$\text{subject to: } x_1 = 5x_2 \tag{4.23}$$

$$x_1 = 7x_3$$

Its Lagrangian function is:

$$\mathcal{L}(x_1, x_2, x_3, \lambda_1, \lambda_2) := 3x_1^2 + 6x_2 + x_3 + \lambda_1(x_1 - 5x_2) + \lambda_2(x_1 - 7x_3)$$

According to the first-order necessary conditions, if $\mathbf{x}^*$ is a local minimizer and a regular point, then there exist $\lambda_1 \in \mathbb{R}$ and $\lambda_2 \in \mathbb{R}$ such that $\nabla \mathcal{L}(x_1, x_2, x_3, \lambda_1, \lambda_2) = \mathbf{0}$.

This gives us the system of equations:

$$\nabla_{\mathbf{x}} \mathcal{L}(\mathbf{x}, \boldsymbol{\lambda}) = \mathbf{0} \Rightarrow \begin{bmatrix} \lambda_1 + \lambda_2 + 6x_1 \\ 6 - 5\lambda_1 \\ 1 - 7\lambda_2 \end{bmatrix} = \begin{bmatrix} 0 \\ 0 \\ 0 \end{bmatrix} \tag{4.24}$$

$$\nabla_{\boldsymbol{\lambda}} \mathcal{L}(\mathbf{x}, \boldsymbol{\lambda}) = \mathbf{0} \Rightarrow \begin{bmatrix} x_1 - 5x_2 \\ x_1 - 7x_3 \end{bmatrix} = \begin{bmatrix} 0 \\ 0 \end{bmatrix} \tag{4.25}$$

That has solution $\mathbf{x}^* = [-47/210, -47/1050, -47/1470]^{\mathsf{T}}$ and $\boldsymbol{\lambda}^* = [6/5, 1/7]^{\mathsf{T}}$. Thus, there exists $\boldsymbol{\lambda}^*$ such that $\nabla \mathcal{L}(\mathbf{x}^*, \boldsymbol{\lambda}^*) = \mathbf{0}$. Note that $\mathbf{x}^*$ is also a regular point because:

$$a_1 \nabla c_1(\mathbf{x}^*) + a_2 \nabla c_2(\mathbf{x}^*) = a_1 \begin{bmatrix} 1 \\ -5 \\ 0 \end{bmatrix} + a_2 \begin{bmatrix} 1 \\ 0 \\ -7 \end{bmatrix} = \begin{bmatrix} a_1 + a_2 \\ -5a_1 \\ -7a_2 \end{bmatrix}$$

is equal to $[0, 0, 0]^{\mathsf{T}}$ if, and only if, $a_1 = a_2 = 0$.

Tips

Note that $a_1 \nabla c_1(\mathbf{x}^*) + a_2 \nabla c_2(\mathbf{x}^*)$ can be written as $\nabla \mathbf{c}(\mathbf{x}^*)^{\mathsf{T}} \mathbf{a}$, where

$$\nabla \mathbf{c}(\mathbf{x}) = \frac{\partial \mathbf{c}(\mathbf{x})}{\partial \mathbf{x}} = \begin{bmatrix} \frac{\partial c_1(\mathbf{x})}{\partial x_1} & \frac{\partial c_1(\mathbf{x})}{\partial x_2} & \cdots & \frac{\partial c_1(\mathbf{x})}{\partial x_n} \\ \frac{\partial c_2(\mathbf{x})}{\partial x_1} & \frac{\partial c_2(\mathbf{x})}{\partial x_2} & \cdots & \frac{\partial c_2(\mathbf{x})}{\partial x_n} \\ \vdots & \vdots & \ddots & \vdots \\ \frac{\partial c_l(\mathbf{x})}{\partial x_1} & \frac{\partial c_l(\mathbf{x})}{\partial x_2} & \cdots & \frac{\partial c_l(\mathbf{x})}{\partial x_n} \end{bmatrix}$$

is the $l \times n$ Jacobian matrix according to the numerator layout notation and $\mathbf{a}$ an $l \times 1$ column vector.

Let us now consider a counterexample where $\mathbf{x}^*$ is a local optimum and f, $c_1, \ldots, c_l$ are C^1 in an open neighborhood of $\mathbf{x}^*$ but $\nabla \mathcal{L}(\mathbf{x}^*, \boldsymbol{\lambda}^*) \neq \mathbf{0}$ because the gradients of $c_1, \ldots, c_l$ are not linearly independent. Suppose that $f(\mathbf{x}) := x_1 + x_2 + x_3^2$ and $c_1(\mathbf{x}) := x_1 - 1, c_2(\mathbf{x}) = x_1^2 + x_2^2 - 1$. Minimizing f subject to constraints c_1, c_2 has solution $\mathbf{x}^* = [1, 0, 0]^\mathsf{T}$. For this $\mathbf{x}^*$:

$$
a_1 \nabla c_1(\mathbf{x}^*) + a_2 \nabla c_2(\mathbf{x}^*) = a_1 \begin{bmatrix} 1 \\ 0 \\ 0 \end{bmatrix} + a_2 \begin{bmatrix} 2 \\ 0 \\ 0 \end{bmatrix} = \begin{bmatrix} a_1 + 2a_2 \\ 0 \\ 0 \end{bmatrix}
$$

which is equal to $[0, 0, 0]^\mathsf{T}$ even if $a_1, a_2 \neq 0$, i.e., $a_1 = -2a_2$. Therefore, $\mathbf{x}^*$ is not a regular point and, despite being a solution to the constrained problem with f, c_1, c_2 being C^1 in its open neighborhood, $\nabla \mathcal{L}(\mathbf{x}^*, \boldsymbol{\lambda}^*) \neq \mathbf{0}$. In more detail,

$$
\nabla \mathcal{L}(\mathbf{x}, \boldsymbol{\lambda}) = \begin{bmatrix} 1 + \lambda_1 + 2\lambda_2 x_1 \\ 1 + 2\lambda_2 x_2 \\ 2x_3 \\ x_1 - 1 = 0 \\ x_1^2 + x_2^2 - 1 = 0 \end{bmatrix}
$$

which cannot be equal to $[0, 0, 0, 0, 0]^\mathsf{T}$ for $\mathbf{x}^* = [1, 0, 0]^\mathsf{T}$. For instance, in the second row we have $1 + 2\lambda x_2^* = 1 + 0 \neq 0$.

Finally, the second-order necessary conditions of multi-variable minimization problems with equality constraints are presented below. In the second-order necessary conditions, the Lagrangian function must be *locally* convex in the constrained area of interest. This is achieved by requesting that the Hessian of the Lagrangian function, $\nabla_{xx}^2 \mathcal{L}(\mathbf{x}^*, \boldsymbol{\lambda}^*)$, is positive semi-definite. That is, $\mathbf{d}^\mathsf{T} \nabla_{xx}^2 \mathcal{L}(\mathbf{x}^*, \boldsymbol{\lambda}^*)\mathbf{d} \geq 0$ where $\mathbf{d}$ is an n-valued vector. There is one additional step though. Unlike unconstrained optimization, we are not interested in proving that this relationship holds for any $\mathbf{d} \in \mathbb{R}^n$ since our problem is constrained and we cannot make a step in any direction in the vicinity of $\mathbf{x}^*$. Instead, we are only interested in proving that $\mathbf{d}^\mathsf{T} \nabla_{xx}^2 \mathcal{L}(\mathbf{x}^*, \boldsymbol{\lambda}^*)\mathbf{d} \geq 0$ holds for all vectors $\mathbf{d}$ that are feasible directions, and thus lie in the tangent space meaning that the gradients of the constraints are orthogonal to $\mathbf{d}$. That is, $\mathbf{d} \in \mathbb{R}^n$ such that $(\nabla c_i(\mathbf{x}^*))^\mathsf{T}\mathbf{d} = 0$ for all $i \in 1, \ldots, l$.

Second-order Necessary conditions of multi-variable minimization problem with equality constraints

Assume that functions $f(\mathbf{x})$ and $c_i(\mathbf{x})$ where $i = 1, \ldots, l$ are C^2 in an open neighborhood of $\mathbf{x}^*$. If $\mathbf{x}^*$ is a local optimum and a *regular point*, then $\mathcal{L}(\mathbf{x}^*, \boldsymbol{\lambda}^*) = \mathbf{0}$ for some Lagrange multiplier $\boldsymbol{\lambda}^* \in \mathbb{R}^l$ and $\mathbf{d}^\mathsf{T} \nabla_{\mathbf{xx}}^2 \mathcal{L}(\mathbf{x}^*, \boldsymbol{\lambda}^*) \mathbf{d} \geq 0$ for any vector $\mathbf{d} \in \mathbb{R}^n$ such that $(\nabla c_i(\mathbf{x}^*))^\mathsf{T} \mathbf{d} = 0$ for all $i \in 1, \ldots, l$.

Let us consider again the example of the minimization problem (4.23). We previously showed that $\mathbf{x}^* = [-\frac{47}{210}, -\frac{47}{1050}, -\frac{47}{1470}]^\mathsf{T}$ is a regular point and satisfies the first-order necessary conditions for $\boldsymbol{\lambda}^* = [\frac{6}{5}, \frac{1}{7}]^\mathsf{T}$. Let us now examine the second-order necessary conditions. Constraint functions $c_1(\mathbf{x}) := x_1 - 5x_2$ and $c_2(\mathbf{x}) := x_1 - 7x_3$ are C^2 (in fact, they are $C^{+\infty}$). The same holds for the objective function $f(\mathbf{x}) := 3x_1^2 + 6x_2 + x_3$. Let $\mathbf{d}$ be any vector that lies in the tangent space $T(\mathbf{x}^*)$ at $\mathbf{x}^*$:

$$\mathbf{d} \in T(\mathbf{x}^*) := \{\mathbf{d} \in \mathbb{R}^n \mid (\nabla c_i(\mathbf{x}^*))^\mathsf{T} \mathbf{d} = 0 \text{ for all } i \in 1, \ldots, l\}$$

We have that:

$$T(\mathbf{x}^*) := \{\mathbf{d} \in \mathbb{R}^n \mid (\nabla c_i(\mathbf{x}^*))^\mathsf{T} \mathbf{d} = 0 \text{ for all } i \in 1, \ldots, l\}$$

where:

$$T(\mathbf{x}^*) := \begin{cases} (\nabla c_1(\mathbf{x}^*))^\mathsf{T} \mathbf{d} = 0 \Rightarrow [1, -5, 0] \begin{bmatrix} d_1 \\ d_2 \\ d_3 \end{bmatrix} = 0 \Rightarrow d_1 - 5d_2 = 0 \\ (\nabla c_2(\mathbf{x}^*))^\mathsf{T} \mathbf{d} = 0 \Rightarrow [1, 0, -7] \begin{bmatrix} d_1 \\ d_2 \\ d_3 \end{bmatrix} = 0 \Rightarrow d_1 - 7d_3 = 0 \end{cases}$$

From the above system, we obtain $[d_1, d_2, d_3]^\mathsf{T} = [d_1, \frac{d_1}{5}, \frac{d_1}{7}]^\mathsf{T}$ where $d_1 \in \mathbb{R}$. In addition,

$$\nabla_{\mathbf{xx}}^2 \mathcal{L}(\mathbf{x}^*) = \begin{bmatrix} 6 & 0 & 0 \\ 0 & 0 & 0 \\ 0 & 0 & 0 \end{bmatrix}$$

Thus,

$$\mathbf{d}^\mathsf{T} \nabla_{\mathbf{xx}}^2 \mathcal{L}(\mathbf{x}^*, \boldsymbol{\lambda}^*) \mathbf{d} = \begin{bmatrix} d_1 & d_1/5 & d_1/7 \end{bmatrix} \begin{bmatrix} 6 & 0 & 0 \\ 0 & 0 & 0 \\ 0 & 0 & 0 \end{bmatrix} \begin{bmatrix} d_1 \\ d_1/5 \\ d_1/7 \end{bmatrix} = 6d_1^2$$

which is greater than or equal to 0 for any $d_1 \in \mathbb{R}$. We thus conclude that $\mathbf{x}^*$ satisfies the second-order necessary conditions.

Tips

As we will later see, $\mathbf{x}^*$ is a *strict* local minimizer because it satisfies also the *second-order sufficient conditions* requesting that $\mathbf{d}^\mathsf{T} \nabla_{\mathbf{xx}}^2 \mathcal{L}(\mathbf{x}^*, \boldsymbol{\lambda}^*)\mathbf{d} > 0$ for any $\mathbf{d} \in T(\mathbf{x}^*)$ such that $\mathbf{d} \neq \mathbf{0}$. If we were solving an unconstrained minimization problem, this would not have been true because there would exist a $\mathbf{d} \in \mathbb{R}^n$, i.e., $\mathbf{d} = [0, 1, 0]^\mathsf{T}$, for which $\mathbf{d} \neq \mathbf{0}$ and $\mathbf{d}^\mathsf{T} \nabla_{\mathbf{xx}}^2 \mathcal{L}(\mathbf{x}^*, \boldsymbol{\lambda}^*)\mathbf{d} = 6 \cdot 0^2 = 0$.

Finally, observe that the expression $(\nabla c_i(\mathbf{x}^*))^\mathsf{T}\mathbf{d} = 0$ for all $i = 1, \ldots, l$ can be written as $\nabla \mathbf{c}(\mathbf{x}^*)\mathbf{d} = \mathbf{0}$, where $\nabla \mathbf{c}(\mathbf{x}^*)$ is the $l \times n$ Jacobian matrix of vector-valued function $\mathbf{c} : \mathbb{R}^n \to \mathbb{R}^l$ and $\mathbf{0}$ an l-valued column vector of zeros. The tangent space at $\mathbf{x}^*$ is then written as:

$$T(\mathbf{x}^*) := \{\mathbf{d} \in \mathbb{R}^n \mid \nabla \mathbf{c}(\mathbf{x}^*)\mathbf{d} = \mathbf{0}\}$$

To take advantage of this succinctly written form of the tangent space, we will make use of the following definitions.

Definition 4.4 The *null space* Null(A) or *kernel* ker(A) of a linear map represented by matrix $A \in \mathbb{R}^{l \times n}$ is the set of solutions to $\mathbf{Ad} = \mathbf{0}$, where $\mathbf{d} \in \mathbb{R}^n$ and $\mathbf{0}$ an l-valued column vector of zeros. That is, Null(A) := $\{\mathbf{d} \in \mathbb{R}^n \mid \mathbf{Ad} = \mathbf{0}\}$.

From the above definition follows that any $\mathbf{d} \in \mathbb{R}^n$ that lies in the tangent space $T(\mathbf{x}^*)$ is in the null space of the $l \times n$ Jacobian matrix $\nabla \mathbf{c}(\mathbf{x}^*)$. Let us now define the span and the basis of a vector space V.

Definition 4.5 The *span* of vectors $\mathbf{v}_1, \ldots, \mathbf{v}_n$, denoted as span($\mathbf{v}_1, \ldots, \mathbf{v}_n$), is the set vector space V of all linear combinations $a_1\mathbf{v}_1 + \cdots + a_n\mathbf{v}_n$ of vectors $\mathbf{v}_1, \ldots, \mathbf{v}_n$, where $a_1, \ldots, a_n \in \mathbb{R}$.

Definition 4.6 A *basis* $\mathcal{B}$ of a vector space V is a linearly independent subset of V that spans V. That is, $\mathcal{B}$ is a basis of V if vectors $\mathbf{v}_1, \ldots, \mathbf{v}_m$ of $\mathcal{B}$ are linearly independent (linear independence property) and span($\mathbf{v}_1, \ldots, \mathbf{v}_m$) = V (spanning property) ensuring that any vector $\mathbf{v}_i \in V$ is a linear combination of vectors in $\mathcal{B}$.

Given the above, if $\mathcal{B}$ is a basis for the null space of $\nabla \mathbf{c}(\mathbf{x}^*)$, then $\mathbf{d} \in \text{span}(\mathcal{B})$ since Null($\nabla \mathbf{c}(\mathbf{x}^*)$) $\equiv$ span($\mathcal{B}$). Any basis $\mathcal{B}$ for the null space of $\nabla \mathbf{c}(\mathbf{x}^*)$ is called *null basis* and its vectors $\mathbf{v}_1, \ldots, \mathbf{v}_m \in \mathcal{B}$ are called *null vectors* [1]. Although a vector space V might have several bases, the number of vectors in any basis of V is the same and this number is called the *dimension* of V. The dimension of the null space of a matrix $\mathbf{A} \in \mathbb{R}^{l \times n}$, denoted as nullity($\mathbf{A}$), is equal to $n - \text{rank}(\mathbf{A})$, where rank($\mathbf{A}$) is the maximum number of linearly independent column or row vectors in matrix $\mathbf{A}$. Of particular interest are the orthonormal bases, where an orthonormal basis can be derived from an arbitrary basis by imposing the additional requirements that all of its vectors have a norm (length) of 1 and are pairwise orthogonal.

Let us now return to the tangent space $T(\mathbf{x}^*) := \{\mathbf{d} \in \mathbb{R}^n \mid \nabla \mathbf{c}(\mathbf{x}^*)\mathbf{d} = \mathbf{0}\}$. Let $t = \text{nullity}(\nabla \mathbf{c}(\mathbf{x}^*))$. Since $\mathbf{d}$ is any vector in the null space of $\nabla \mathbf{c}(\mathbf{x}^*)$, we have that $\mathbf{d} = \mathbf{Zu}$, where $\mathbf{Z} \in \mathbb{R}^{n \times t}$ is an *orthonormal null basis* of $\nabla \mathbf{c}(\mathbf{x}^*)$ and $\mathbf{u} \in \mathbb{R}^t$. Because $\mathbf{u}$ can be any vector in $\mathbb{R}^t$, instead of requesting that

$$\mathbf{d}^\mathsf{T} \nabla_{xx}^2 \mathcal{L}(\mathbf{x}^*, \boldsymbol{\lambda}^*)\mathbf{d} \geq 0 \text{ for any vector } \mathbf{d} \in \mathbb{R}^n \mid \nabla \mathbf{c}(\mathbf{x}^*)\mathbf{d} = \mathbf{0}$$

we can use the orthonormal basis for the null space to request that:

$$\mathbf{u}^\mathsf{T} \mathbf{Z}^\mathsf{T} \nabla_{xx}^2 \mathcal{L}(\mathbf{x}^*, \boldsymbol{\lambda}^*)\mathbf{Zu} \geq 0 \;\; \forall \mathbf{u} \in \mathbb{R}^t \;\; \Leftrightarrow \;\; \mathbf{Z}^\mathsf{T} \nabla_{xx}^2 \mathcal{L}(\mathbf{x}^*, \boldsymbol{\lambda}^*)\mathbf{Z} \text{ is positive semi-definite}$$

The two expressions presented above are equivalent and matrix $\mathbf{Z}^\mathsf{T} \nabla_{xx}^2 \mathcal{L}(\mathbf{x}^*, \boldsymbol{\lambda}^*)\mathbf{Z}$ is known as the *Projected Hessian* [2] of the Lagrangian function at $\mathbf{x}^*$.

Let us consider the example of the minimization problem in (4.23). We have shown that:

$$\nabla \mathbf{c}(\mathbf{x}^*) = \begin{bmatrix} 1 & -5 & 0 \\ 1 & 0 & -7 \end{bmatrix}$$

The rank of this matrix is $\text{rank}(\nabla \mathbf{c}(\mathbf{x}^*)) = 2$. Thus, $t = n - \text{rank}(\nabla \mathbf{c}(\mathbf{x}^*)) = 1$. This means that any basis for the null space of $\nabla \mathbf{c}(\mathbf{x}^*)$ will consist of 1 vector with $n = 3$ elements. That is, $\mathbf{Z} \in \mathbb{R}^{3 \times 1}$. Let, for brevity reasons, refer to the Jacobian $\nabla \mathbf{c}(\mathbf{x}^*)$ as matrix $\mathbf{A}$. An orthonormal null basis $\mathbf{Z}$ of matrix $\mathbf{A}$ can be computed with the use of the Singular Value Decomposition (SVD) which is a factorization of a real matrix [3]:

$$\mathbf{A} = \mathbf{USV}^\mathsf{T}$$

where $\mathbf{U} \in \mathbb{R}^{l \times l}$ such that $\mathbf{UU}^\mathsf{T} = \mathbf{I}$, $\mathbf{S} \in \mathbb{R}^{l \times n}$ is a rectangular diagonal matrix, and $\mathbf{V} \in \mathbb{R}^{n \times n}$ such that $\mathbf{VV}^\mathsf{T} = \mathbf{I}$. In our case,

$$\mathbf{A} \simeq \begin{bmatrix} 0.04156 & 0.99914 \\ 0.99914 & -0.04156 \end{bmatrix} \begin{bmatrix} 7.07401 & 0 & 0 \\ 0 & 5.09494 & 0 \end{bmatrix} \begin{bmatrix} 0.14712 & -0.02937 & -0.98868 \\ 0.18795 & -0.98052 & 0.05710 \\ 0.97110 & 0.19422 & 0.13873 \end{bmatrix}$$

The last rows of $\mathbf{V}^\mathsf{T}$ with row number $r > t$ provide an orthonormal basis for the null space of $\mathbf{A}$. That is, $\mathbf{Z} = [0.97110, 0.19422, 0.13873]^\mathsf{T}$. This results in the projected Hessian:

$$\mathbf{Z}^\mathsf{T} \nabla_{xx}^2 \mathcal{L}(\mathbf{x}^*, \boldsymbol{\lambda}^*)\mathbf{Z} \simeq 5.658$$

which is positive definite (its eigenvalue is 5.658).

Practitioner's Corner
SONC for Equality Constrained Minimization

We can examine whether a minimization problem satisfies the second-order necessary conditions (SONC) for local optimality in Python 3 using SymPy, which is a Python library for symbolic mathematics. Consider, for example, the minimization problem in (4.23). We first compute the values $\mathbf{x}^*$, $\boldsymbol{\lambda}^*$ as follows.

```python
import sympy as sp
x1, x2, x3, l1, l2 = sp.var('x1,x2,x3,l1,l2',real=True)
f = 3*x1**2 + 6*x2 + x3; c1 = x1-5*x2; c2 = x1-7*x3
c=sp.Matrix([c1,c2]); l=sp.Matrix([l1,l2])
L = f + l.dot(c)
gradL = [sp.diff(L,x) for x in [x1,x2,x3,l1,l2]]
stationary_points = sp.solve(gradL, [x1, x2, x3, l1, l2])
print('stationary_points',stationary_points)
```

resulting in solution $[x_1^*, x_2^*, x_3^*, \lambda_1^*, \lambda_2^*]^\mathsf{T} = [-47/210, -47/1050, -47/1470, 6/5, 1/7]^\mathsf{T}$. We then investigate whether $\mathbf{x}^* = [-47/210, -47/1050, -47/1470]^\mathsf{T}$ is a regular point as follows.

```python
a1, a2 = sp.var('a1,a2', real = True)
a=sp.Matrix([a1,a2])
jac_c = c.jacobian([x1,x2,x3])
values={"x1":-47/210,"x2":-47/1050,"x3":-47/1470}
jac_c=jac_c.subs(values)
LICQ_expression = jac_c.T*a
LICQ_test = sp.solve(LICQ_expression, [a1, a2])
print('LICQ_test',LICQ_test)
```

This results in $a_1 = a_2 = 0$ which means that the LICQ test is satisfied and $\mathbf{x}^*$ is a regular point. Finally, we check if $\mathbf{Z}^\mathsf{T} \nabla_{\mathbf{xx}}^2 \mathcal{L}(\mathbf{x}^*, \boldsymbol{\lambda}^*) \mathbf{Z}$ is positive semi-definite, where $\mathbf{Z}$ is an orthonormal basis for the null space of $\nabla \mathbf{c}(\mathbf{x})^*$ calculated by the scipy.linalg.null_space method that employs SVD factorization.

```python
import scipy.linalg as sc
import numpy as np
jac_c=np.array(jac_c).astype(np.float64)
Z=sc.null_space(jac_c)
Hessian_L=[[L.diff(x).diff(y) for x in [x1,x2,x3]] for
    y in [x1,x2,x3]]
Hessian_L=sp.Matrix(Hessian_L)
values={'x1':-47/210,'x2':-47/1050,'x3':-47/1470,'l1'
    :6/5,'l2':1/7}
Hessian_L=Hessian_L.subs(values)
Projected_Hessian=Z.T*Hessian_L*Z
print('Projected_Hessian',Projected_Hessian)
print('eigenvalues',Projected_Hessian.eigenvects())
```

This results in eigenvalue $2450/433 \simeq 5.658$, which is positive, meaning that the projected Hessian is positive definite.

4.1.3   Necessary Conditions for Opimization Problems with Equality and Inequality Constraints (Karush-Kuhn-Tucker)

We now proceed to the general case of optimization problems with both *equality* and *inequality* constraints. First, we state that an inequality constraint $g_i(\mathbf{x}) \leq 0$ is active at a point $\mathbf{x}^*$ if $g_i(\mathbf{x}^*) = 0$ and inactive if $g_i(\mathbf{x}^*) < 0$. Consider, for instance, the minimization problem:

$$\min_{\mathbf{x} \in \mathbb{R}^n} f(\mathbf{x})$$

$$\text{subject to: } \mathbf{h}(\mathbf{x}) = \mathbf{0} \qquad\qquad (4.26)$$

$$\mathbf{g}(\mathbf{x}) \leq \mathbf{0}$$

where $f : \mathbb{R}^n \to \mathbb{R}$, $\mathbf{h} : \mathbb{R}^n \to \mathbb{R}^l$ and $\mathbf{g} : \mathbb{R}^n \to \mathbb{R}^m$. Note that $\mathbf{h}(\mathbf{x})$ and $\mathbf{g}(\mathbf{x})$ are vector-valued functions. This problem can be equivalently written as:

$$\min_{\mathbf{x} \in \mathbb{R}^n} f(\mathbf{x}) \qquad\qquad (4.27)$$

$$\text{subject to: } h_i(\mathbf{x}) = 0, \qquad\qquad i \in L = \{1 \dots, l\} \qquad (4.28)$$

$$g_i(\mathbf{x}) \leq 0, \qquad\qquad i \in M = \{1 \dots, m\} \qquad (4.29)$$

The *Active Set* at any feasible point $\mathbf{x}^*$ of the optimization problem is defined as follows.

Definition 4.7 *Active Set* at any feasible point $\mathbf{x}^*$ of an optimization problem is the set $\mathcal{A}(\mathbf{x}^*)$ which is the union of the set of all equality constraints L and the subset of inequality constraints $i \in M$ for which $g_i(\mathbf{x}^*) = 0$. That is, $\mathcal{A}(\mathbf{x}^*) = L \cup \{i \in M \mid g_i(\mathbf{x}^*) = 0\}$.

For the minimization problem that contains both equality and inequality constraints, the Lagrangian function $\mathcal{L} : \mathbb{R}^n \times \mathbb{R}^l \times \mathbb{R}^m$ is:

$$\mathcal{L}(\mathbf{x}, \boldsymbol{\lambda}, \boldsymbol{\mu}) := f(\mathbf{x}) + \boldsymbol{\lambda}^\mathsf{T}\mathbf{h}(\mathbf{x}) + \boldsymbol{\mu}^\mathsf{T}\mathbf{g}(\mathbf{x}) = f(\mathbf{x}) + \sum_{i \in L} \lambda_i h_i(\mathbf{x}) + \sum_{i \in M} \mu_i g_i(\mathbf{x})$$

$$(4.30)$$

where the dual variables $\boldsymbol{\lambda} = [\lambda_1, \dots, \lambda_\lambda]^\mathsf{T}$ and $\boldsymbol{\mu} = [\mu_1, \dots, \mu_m]^\mathsf{T}$ are now called Karush-Kuhn-Tucker (KKT) multipliers [4, 5]. If the problem was a maximization problem, then

$$\mathcal{L}(\mathbf{x}, \boldsymbol{\lambda}, \boldsymbol{\mu}) := f(\mathbf{x}) - \boldsymbol{\lambda}^\mathsf{T}\mathbf{h}(\mathbf{x}) - \boldsymbol{\mu}^\mathsf{T}\mathbf{g}(\mathbf{x}) \qquad\qquad (4.31)$$

In the remainder of this section, we will focus on the optimality conditions for minimization problems. For maximization, the only change is that we would need to use the Lagrangian function in (4.31). The dual of the minimization problem in (4.26) is:

$$\max_{\lambda,\mu} \inf_{x} \ \mathcal{L}(\mathbf{x}, \boldsymbol{\lambda}, \boldsymbol{\mu})$$

$$\text{s.t. } \boldsymbol{\mu} \geq \mathbf{0}$$

The first-order *necessary* optimality conditions are provided below.

First-order Necessary conditions of multi-variable minimization problems with equality and inequality constraints

Assume that $f(\mathbf{x})$ and $h_i(\mathbf{x}) \ \forall i \in L$, $g_i(\mathbf{x}) \ \forall i \in M$ are C^1 in an open neighborhood of $\mathbf{x}^*$. If $\mathbf{x}^*$ is a local minimum and $\mathbf{x}^*$ is a *regular point*, then there exist KKT multipliers $\boldsymbol{\lambda}^*, \boldsymbol{\mu}^*$ such that the following conditions (known as KKT conditions) are satisfied:

1. $\nabla_{\mathbf{x}}\mathcal{L}(\mathbf{x}^*, \boldsymbol{\lambda}^*, \boldsymbol{\mu}^*) = \mathbf{0} \Rightarrow \nabla_{\mathbf{x}}f(\mathbf{x}^*) + \sum_{i \in L} \lambda_i^* \nabla_{\mathbf{x}}h_i(\mathbf{x}^*) + \sum_{i \in M} \mu_i^* \nabla_{\mathbf{x}}g_i(\mathbf{x}^*) = \mathbf{0}$
   (stationarity)
2. $h_i(\mathbf{x}^*) = 0 \ \forall i \in L$ (primal feasibility)
3. $g_i(\mathbf{x}^*) \leq 0 \ \forall i \in M$ (primal feasibility)
4. $\mu_i^* \geq 0 \ \forall i \in M$ (dual feasibility)
5. $\mu_i^* g_i(\mathbf{x}^*) = 0 \ \forall i \in M$ (complementarity conditions, known as complementary slackness)

Note that $\mu_i^* g_i(\mathbf{x}^*) = 0 \ \forall i \in M$ can be also written as $\sum_{i \in M} \mu_i^* g_i(\mathbf{x}^*) = 0$. Using vector notation, we can rewrite the KKT conditions as:

1. $\nabla_{\mathbf{x}}f(\mathbf{x}^*) + \boldsymbol{\lambda}^{\mathsf{T}}\nabla_{\mathbf{x}}\mathbf{h}(\mathbf{x}^*) + \boldsymbol{\mu}^{\mathsf{T}}\nabla_{\mathbf{x}}\mathbf{g}(\mathbf{x}^*) = \mathbf{0}$ (stationarity)
2. $\mathbf{h}(\mathbf{x}^*) = \mathbf{0}$ (primal feasibility)
3. $\mathbf{g}(\mathbf{x}^*) \leq \mathbf{0}$ (primal feasibility)
4. $\boldsymbol{\mu}^* \geq \mathbf{0}$ (dual feasibility)
5. $(\boldsymbol{\mu}^*)^{\mathsf{T}}\mathbf{g}(\mathbf{x}^*) = \mathbf{0}$ (complementary slackness)

The complementarity conditions, also known as complementary slackness, imply that either inequality constraint $i \in M$ is active at $\mathbf{x}^*$, thus $g_i(\mathbf{x}^*) = 0$, or $\mu_i^* = 0$, or both. The primal feasibility constraints ensure that $\mathbf{x}^*$ satisfies all constraints of the optimization program. Finally, dual feasibility ensures that the KKT multipliers associated with inequality constraints cannot take negative values. Note that if from the KKT conditions we remove conditions 3-5 that refer to inequality constraints, then we scale back to the first-order necessary conditions for multi-variable optimization problems with equality constraints.

Finally, the second-order necessary conditions of multi-variable minimization problems with equality and inequality constraints are presented below. In the second-order necessary conditions, the Lagrangian function must be locally convex (or locally concave in case of maximization) in the feasible directions.

Second-order Necessary conditions of multi-variable minimization problems with equality and inequality constraints

Assume that $f(\mathbf{x})$ and $h_i(\mathbf{x})\,\forall i \in L$, $g_i(\mathbf{x})\,\forall i \in M$ are C^2 in an open neighborhood of $\mathbf{x}^*$. If $\mathbf{x}^*$ is a *local minimizer* and $\mathbf{x}^*$ is a *regular point*, then there exist KKT multipliers $\boldsymbol{\lambda}^*$, $\boldsymbol{\mu}^*$ such that the KKT conditions are satisfied and $\mathbf{d}^{\mathsf{T}} \nabla_{\mathbf{xx}}^2 \mathcal{L}(\mathbf{x}^*, \boldsymbol{\lambda}^*, \boldsymbol{\mu}^*)\mathbf{d} \geq 0$ for any vector $\mathbf{d}$ such that:

$$\mathbf{d} \in C(\mathbf{x}^*) := \begin{cases} (\nabla h_i(\mathbf{x}^*))^{\mathsf{T}}\mathbf{d} = 0 \text{ for all } i \in L \\ (\nabla g_i(\mathbf{x}^*))^{\mathsf{T}}\mathbf{d} \leq 0 \text{ for all } i \in M \cap \mathcal{A}(\mathbf{x}^*) \text{ with } \mu_i^* = 0 \\ (\nabla g_i(\mathbf{x}^*))^{\mathsf{T}}\mathbf{d} = 0 \text{ for all } i \in M \cap \mathcal{A}(\mathbf{x}^*) \text{ with } \mu_i^* > 0 \end{cases}$$

It is worth noting that $C(\mathbf{x}^*)$ is called (strong) *Critical Cone* [6], and we essentially request that the Hessian of the Lagrangian function is positive semidefinite on the critical cone at $\mathbf{x}^*$. Reckon that we do not request the Hessian of the Lagrangian function to be positive semidefinite in any direction in the neighborhood of $\mathbf{x}^*$ because we solve a constrained optimization problem. In constrained optimization, we are only interested in directions $\mathbf{d} \in C(\mathbf{x}^*)$ [6].

4.2 Second-order Sufficient Conditions for Local Optimality

We now proceed to the sufficient conditions for a solution $\mathbf{x}^*$ to be a strict local minimizer of a multivariable constrained minimization problem. In the case of solving a minimization problem with equality constraints, we request that the Hessian of the Lagrangian function is locally positive definite (locally strictly convex). This results in the following sufficient conditions.

Second-order Sufficient conditions of multi-variable minimization problem with equality constraints

If $f(\mathbf{x})$ and $c_i(\mathbf{x})$ are C^2 in an open neighborhood of $\mathbf{x}^*$ of the minimization problem

$$\min_{\mathbf{x}} f(\mathbf{x}) \quad \text{s.t.} \quad c_i(\mathbf{x}) = 0 \text{ for } i = 1, \ldots, l$$

and:

- $\nabla c_i(\mathbf{x}^*)$ are linearly independent vectors so that $\mathbf{x}^*$ is a regular point
- There is $\boldsymbol{\lambda}^*$ such that $\nabla \mathcal{L}(\mathbf{x}^*, \boldsymbol{\lambda}^*) = \mathbf{0}$
- $\mathbf{d}^{\mathsf{T}} \nabla_{\mathbf{xx}}^2 \mathcal{L}(\mathbf{x}^*, \boldsymbol{\lambda}^*)\mathbf{d} > 0$ for any vector $\mathbf{d} \neq \mathbf{0}$ such that

$$\mathbf{d} \in T(\mathbf{x}^*) := \{(\nabla c_i(\mathbf{x}^*))^\mathsf{T}\mathbf{d} = 0 \text{ for all } i \in 1, \dots, l\}$$

then $\mathbf{x}^*$ is a strict local minimizer of the constrained minimization problem.

This is expanded in the case of minimization problems with equality and inequality constraints, as follows.

Second-order Sufficient conditions of multi-variable minimization problem with equality and inequality constraints

If $f(\mathbf{x})$, $h_i(\mathbf{x})$ and $g_i(\mathbf{x})$ are C^2 in an open neighborhood of $\mathbf{x}^*$ of the minimization problem

$$\min_{\mathbf{x}} f(\mathbf{x}) \quad \text{s.t.} \quad h_i(\mathbf{x}) = 0 \text{ for } i \in L, \quad g_i(\mathbf{x}) \le 0 \text{ for } i \in M$$

and:

- $\mathbf{x}^*$ is a regular point (i.e., LICQ holds at $\mathbf{x}^*$)
- There exist $\boldsymbol{\lambda}^*$, $\boldsymbol{\mu}^*$ that satisfy the KKT conditions at $\mathbf{x}^*$
- $\mathbf{d}^\mathsf{T}\nabla_{\mathbf{xx}}^2 \mathcal{L}(\mathbf{x}^*, \boldsymbol{\lambda}^*, \boldsymbol{\mu}^*)\mathbf{d} > 0$ for any vector $\mathbf{d} \ne \mathbf{0}$ such that

$$\mathbf{d} \in C(\mathbf{x}^*) := \begin{cases} (\nabla h_i(\mathbf{x}^*))^\mathsf{T}\mathbf{d} = 0 \text{ for all } i \in L \\ (\nabla g_i(\mathbf{x}^*))^\mathsf{T}\mathbf{d} \le 0 \text{ for all } i \in M \cap \mathcal{A}(\mathbf{x}^*) \text{ with } \mu_i^* = 0 \\ (\nabla g_i(\mathbf{x}^*))^\mathsf{T}\mathbf{d} = 0 \text{ for all } i \in M \cap \mathcal{A}(\mathbf{x}^*) \text{ with } \mu_i^* > 0 \end{cases}$$

then $\mathbf{x}^*$ is a strict local minimizer of the constrained minimization problem.

Consider, for example, the following constrained minimization problem:

$$\min_{\mathbf{x} \in \mathbb{R}^2} f(\mathbf{x}) := x_1^2 - 4x_1 + 2x_2^2 - 4x_2 \tag{4.32}$$

$$\text{subject to: } g_1(\mathbf{x}) := -x_1 \le 0 \tag{4.33}$$

$$g_2(\mathbf{x}) := x_2 - 2/3 \le 0 \tag{4.34}$$

$$g_3(\mathbf{x}) := x_1 + x_2 - 2 \le 0 \tag{4.35}$$

Its Lagrangian function is:

$$\mathcal{L}(\mathbf{x}, \boldsymbol{\mu}) = x_1^2 - 4x_1 + 2x_2^2 - 4x_2 - \mu_1 x_1 + \mu_2(x_2 - 2/3) + \mu_3(x_1 + x_2 - 2)$$

and the KKT conditions:

$$\nabla_{\mathbf{x}}\mathcal{L}(\mathbf{x},\boldsymbol{\mu}) = 0 \Rightarrow [2x_1 - 4 - \mu_1 + \mu_3, 4x_2 - 4 + \mu_2 + \mu_3]^\mathsf{T} = [0,0]^\mathsf{T} \quad (4.36)$$

$$-x_1 \leq 0 \quad (4.37)$$

$$x_2 - 2/3 \leq 0 \quad (4.38)$$

$$x_1 + x_2 - 2 \leq 0 \quad (4.39)$$

$$\mu_1, \mu_2, \mu_3 \geq 0 \quad (4.40)$$

$$\mu_1(-x_1) = 0 \quad (4.41)$$

$$\mu_2(x_2 - 2/3) = 0 \quad (4.42)$$

$$\mu_3(x_1 + x_2 - 2) = 0 \quad (4.43)$$

Let us solve the system of equations that contain all equality constraints (4.36), (4.41)–(4.43). Solving this system of equations results in solutions:

- Solution a: $[x_1^a, x_2^a, \mu_1^a, \mu_2^a, \mu_3^a] = [0, 2, -8, 0, -4]$
- Solution b: $[x_1^b, x_2^b, \mu_1^b, \mu_2^b, \mu_3^b] = [0, 1, -4, 0, 0]$
- Solution c: $[x_1^c, x_2^c, \mu_1^c, \mu_2^c, \mu_3^c] = [0, 2/3, -4, 4/3, 0]$
- Solution d: $[x_1^d, x_2^d, \mu_1^d, \mu_2^d, \mu_3^d] = [2, 1, 0, 0, 0]$
- Solution e: $[x_1^e, x_2^e, \mu_1^e, \mu_2^e, \mu_3^e] = [1.333, 0.666, 0, 0, 1.333]$
- Solution f: $[x_1^f, x_2^f, \mu_1^f, \mu_2^f, \mu_3^f] = [2, 0.666, 0, 1.333, 0]$

From the above solutions, only solution $\mathbf{x}^e$ satisfies all inequality constraints of the KKT conditions as expressed in constraints (4.36)–(4.43) and it is a *KKT point*. Solution $\mathbf{x}^e$ satisfies the *first-order necessary conditions* if it is also a regular point (i.e., LICQ holds for this solution). For this solution, constraints (4.38) and (4.39) are active, and the expression:

$$a_1 \nabla g_2(\mathbf{x}^e) + a_2 \nabla g_3(\mathbf{x}^e) = a_1 \begin{bmatrix} 0 \\ 1 \end{bmatrix} + a_2 \begin{bmatrix} 1 \\ 1 \end{bmatrix} = \begin{bmatrix} a_2 \\ a_1 + a_2 \end{bmatrix}$$

is equal to $[0, 0]^\mathsf{T}$ if, and only if, $a_1, a_2 = 0$. Thus, the gradients of the active inequality constraints are linearly independent and LICQ holds for $\mathbf{x}^e$. Now that the first-order necessary conditions are satisfied for $\mathbf{x}^e$, we proceed with examining if it is a (strict) local minimizer. To do so, we proceed with the *second-order sufficient conditions* for (strict) local optimality.

For solution $\mathbf{x}^e$, constraints (4.38) and (4.39) are active. In addition, $\mu_2^* = 0$ and $\mu_3^* > 0$ for solution $\mathbf{x}^e$. Thus, the critical cone $C(\mathbf{x}^e)$ for solution $\mathbf{x}^e$ comprises of:

$$C(\mathbf{x}^e) := \begin{cases} (\nabla g_2(\mathbf{x}^e))^\mathsf{T}\mathbf{d} \leq 0 \Rightarrow [0,1]\begin{bmatrix} d_1 \\ d_2 \end{bmatrix} \leq 0 \Rightarrow d_1 \in \mathbb{R}, d_2 \leq 0 \\ \\ (\nabla g_3(\mathbf{x}^e))^\mathsf{T}\mathbf{d} = 0 \Rightarrow [1,1]\begin{bmatrix} d_1 \\ d_2 \end{bmatrix} = 0 \Rightarrow d_1 + d_2 = 0 \end{cases}$$

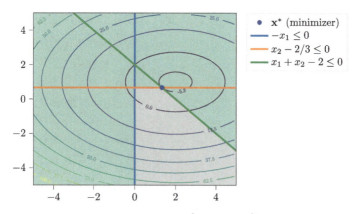

Fig. 4.2 2-dimensional contour plot of $f(\mathbf{x}) := x_1^2 - 4x_1 + 2x_2^2 - 4x_2$ and its strict local minimizer $\mathbf{x}^* = [1.333, 0.666]^\mathsf{T}$ when considering constraints $-x_1 \leq 0$, $x_2 - 2/3 \leq 0$, and $x_1 + x_2 - 2 \leq 0$ forming the gray-colored feasible region

We also have $\nabla_{\mathbf{xx}}^2 \mathcal{L}(\mathbf{x}^e, \boldsymbol{\mu}^e) = \begin{bmatrix} 2 & 0 \\ 0 & 4 \end{bmatrix}$. Thus, $\mathbf{d}^\mathsf{T} \nabla_{\mathbf{xx}}^2 \mathcal{L}(\mathbf{x}^e, \boldsymbol{\mu}^e)\mathbf{d} = 2d_1^2 + 4d_2^2$.

For vector $\mathbf{d} \in C(\mathbf{x}^e)$ such that $\mathbf{d} \neq \mathbf{0}$, we have:

$$\mathbf{d}^\mathsf{T} \nabla_{\mathbf{xx}}^2 \mathcal{L}(\mathbf{x}^e, \boldsymbol{\mu}^e)\mathbf{d} = 2d_1^2 + 4d_2^2 \overset{d_2 = -d_1}{=} 6d_1^2$$

which is always greater than 0 for $\mathbf{d} \neq \mathbf{0}$. Thus, feasible solution $\mathbf{x}^e$ satisfies the *second-order sufficient conditions* and it is a strict local minimizer of the minimization problem in (4.32)–(4.35).

To illustrate that the KKT point $\mathbf{x}^e = [1.333, 0.666]^\mathsf{T}$ is a strict local minimizer, we show the contour plot of $f(\mathbf{x}) := x_1^2 - 4x_1 + 2x_2^2 - 4x_2$ in Fig. 4.2. In Fig. 4.2 we show also the graphs of the inequality constraint functions that result in the feasible region of the grey-colored area. Let denote $\mathbf{x}^e$ as $\mathbf{x}^*$. Notice that $\mathbf{x}^*$ lies on the graphs of $g_2(\mathbf{x}) := x_2 - 2/3$ and $g_3(\mathbf{x}) := x_1 + x_2 - 2$. This was expected because we knew that constraints $g_2(\mathbf{x})$ and $g_3(\mathbf{x})$ were active at $\mathbf{x}^*$. In addition, $\mathbf{x}^*$ does not lie on the graph of $g_1(\mathbf{x}) := -x_1$ because this constraint was inactive. Finally, $\mathbf{x}^*$ intersects the contour plot of f at $f(\mathbf{x}^*) \simeq -5.332$. Notice that there is no other point in the vicinity of $\mathbf{x}^*$ with a better performance when considering the feasible directions in the grey-colored area. As we will later see, because the objective function is convex and the feasible region is a convex set, solution $\mathbf{x}^*$ is also a global minimizer.

Tips

If the problem was an unconstrained optimization one, the global minimimizer would have been point $\hat{\mathbf{x}} = [2, 1]^\mathsf{T}$ with $f(\hat{\mathbf{x}}) = -6$. Notice that in the unconstrained case there are descent directions from $\mathbf{x}^*$ because we are not restricted by the feasible region.

4.3 Global Optimality

We discussed the *first-order necessary conditions, second-order necessary conditions* and the *second-order sufficient conditions for local optimality* when solving continuous constrained optimization problems with equality constraints and inequality constraints.

As in unconstrained optimization, convexity is the key to determine whether a local minimizer is a global minimizer. This is expressed in the following conditions.

Sufficient Condition for Global Optimality in a constrained optimization problem with equality constraints

A solution $\mathbf{x}^*$ is a global minimizer (maximizer) if the objective function f is convex (concave), the equality constraints $c_1, \ldots, c_l$ are affine functions and $\mathbf{x}^*$ satisfies the first-order necessary conditions for local optimality of equality-constrained optimization problems. Namely, $\mathbf{x}^*$ is a *regular point* and $\nabla \mathcal{L}(\mathbf{x}^*, \boldsymbol{\lambda}^*) = \mathbf{0}$ for some Lagrange multiplier $\boldsymbol{\lambda}^*$.

Notice that we request from the equality constraints to be affine functions. The reason is that we want the feasible region to be a convex set, and in case of equality constraints this can be achieved in the case of affine functions [43]. In more detail, any equality constraint $c(\mathbf{x}) = 0$ can be written as two inequality constraints, $c(\mathbf{x}) \leq 0$ and $c(\mathbf{x}) \geq 0$, and if the one is convex, the other will be concave (except for the case where $c(\mathbf{x})$ is an affine—or linear—function).

Sufficient Condition for Global Optimality in a constrained optimization problem with equality and inequality constraints

A solution $\mathbf{x}^*$ is a global minimizer (maximizer) if the objective function is convex (concave), the equality constraints are affine functions, the inequality constraints are continuously differentiable convex functions, and $\mathbf{x}^*$ satisfies the first-order necessary conditions for local optimality of constrained optimization problems. Namely, $\mathbf{x}^*$ is a regular point and there exist $\boldsymbol{\lambda}^*$ and $\boldsymbol{\mu}^*$ such that:

1. $\nabla_{\mathbf{x}} f(\mathbf{x}^*) + \boldsymbol{\lambda}^{\mathsf{T}} \nabla_{\mathbf{x}} \mathbf{h}(\mathbf{x}^*) + \boldsymbol{\mu}^{\mathsf{T}} \nabla_{\mathbf{x}} \mathbf{g}(\mathbf{x}^*) = \mathbf{0}$ \quad (stationarity)
2. $\mathbf{h}(\mathbf{x}^*) = \mathbf{0}$ \quad (primal feasibility)
3. $\mathbf{g}(\mathbf{x}^*) \leq \mathbf{0}$ \quad (primal feasibility)
4. $\boldsymbol{\mu}^* \geq \mathbf{0}$ \quad (dual feasibility)
5. $(\boldsymbol{\mu}^*)^{\mathsf{T}} \mathbf{g}(\mathbf{x}^*) = 0$ \quad (complementary slackness)

As before, if the inequality constraints are convex functions and the equality constraints affine functions, the feasible region will be a convex set as the intersection of convex sets. Minimizing (maximizing) a convex (concave) function over a convex feasible region ensures that a locally optimal solution is a globally optimal one. From the above, follows the corollary:

Corollary 4.1 *A regular point that satisfies the KKT conditions is a global minimizer of a constrained optimization problem if the optimization problem is convex.*

We also express the relation between KKT conditions and Duality in the following theorems.

Theorem 4.3 *For any optimization problem with differentiable objective and constraint functions for which strong duality holds, any pair of primal and dual optimal points $\mathbf{x}^*$, $\boldsymbol{\lambda}^*$, $\boldsymbol{\mu}^*$ must satisfy the KKT conditions.*

Proof Let $\mathbf{x}^*$ and $\boldsymbol{\lambda}^*, \boldsymbol{\mu}^*$ be primal and dual solutions, respectively. Let $f(\mathbf{x})$ be the objective function of the primal minimization problem and $q(\boldsymbol{\lambda}, \boldsymbol{\mu}) = \inf_{\mathbf{x}} \mathcal{L}(\mathbf{x}, \boldsymbol{\lambda}, \boldsymbol{\mu})$ the objective function of the dual maximization problem. We have zero duality gap, $f(\mathbf{x}^*) = q(\boldsymbol{\lambda}^*, \boldsymbol{\mu}^*)$, since strong duality holds. Then,

$$
\begin{aligned}
f(\mathbf{x}^*) &= q(\boldsymbol{\lambda}^*, \boldsymbol{\mu}^*) \\
&= \inf_{\mathbf{x}} \mathcal{L}(\mathbf{x}, \boldsymbol{\lambda}^*, \boldsymbol{\mu}^*) \\
&= \inf_{\mathbf{x}} \left(f(\mathbf{x}) + \sum_{i=1}^{p} \lambda_i^* h_i(\mathbf{x}) + \sum_{i=1}^{m} \mu_i^* g_i(\mathbf{x}) \right) \\
&\leq f(\mathbf{x}^*) + \sum_{i=1}^{p} \lambda_i^* h_i(\mathbf{x}^*) + \sum_{i=1}^{m} \mu_i^* g_i(\mathbf{x}^*) \\
&\leq f(\mathbf{x}^*)
\end{aligned}
$$

where the last inequality holds because $\sum_{i=1}^{m} \mu_i^* g_i(\mathbf{x}^*) \leq 0$. This means that

$$
\sum_{i=1}^{m} \mu_i^* g_i(\mathbf{x}^*) = 0
$$

since $\boldsymbol{\mu} \geq \mathbf{0}$ and $\mathbf{g}(\mathbf{x}^*) \leq \mathbf{0}$ and this gives us the complementary slackness conditions:

$$
\boldsymbol{\mu}^{\mathsf{T}} \mathbf{g}(\mathbf{x}) = 0
$$

which can be also written as:

$$
\mu_i g_i(\mathbf{x}) = 0 \;\; \forall i \in M
$$

In addition, we have that $\mathbf{x}^*$ minimizes $\mathcal{L}(\mathbf{x}, \boldsymbol{\lambda}^*, \boldsymbol{\mu}^*)$ over $\mathbf{x}$ since $f(\mathbf{x}^*)$ must be equal to $\inf_{\mathbf{x}} \mathcal{L}(\mathbf{x}, \boldsymbol{\lambda}^*, \boldsymbol{\mu}^*)$. That is,

$$\nabla_{\mathbf{x}}\mathcal{L}(\mathbf{x}^*, \boldsymbol{\lambda}^*, \boldsymbol{\mu}^*) = \mathbf{0} \Rightarrow \nabla_{\mathbf{x}}f(\mathbf{x}^*) + \sum_{i=1}^{p}\lambda_i^*\nabla_{\mathbf{x}}h_i(\mathbf{x}^*) + \sum_{i=1}^{m}\mu_i^*\nabla_{\mathbf{x}}g_i(\mathbf{x}^*) = \mathbf{0}$$

which gives us the stationarity KKT condition. The remaining KKT conditions related to the primal and dual feasibility hold because $\mathbf{x}^*$ and $\boldsymbol{\lambda}^*$, $\boldsymbol{\mu}^*$ are optimal (and thus feasible) primal and dual solutions. Thus, we have proved that when strong duality holds in a minimization problem, KKT conditions are satisfied. $\square$

Theorem 4.4 *If an optimization problem satisfies the refined Slater's condition, then the KKT conditions provide necessary and sufficient conditions for optimality.*

Proof If an optimization problem satisfies the refined Slater's conditions, then strong duality holds. Thus, a pair $\mathbf{x}^*, \boldsymbol{\lambda}^*, \boldsymbol{\mu}^*$ that satisfies the KKT conditions has zero duality gap, meaning that this is the pair of optimal primal and dual solutions. $\square$

4.4 More Advanced Topics in Linear Algebra

In the remainder of this chapter we will focus on solution methods for constrained optimization problems. In these solution methods we will use matrix multiplications and gradients of expressions with matrices. To be able to follow the underlying theory of the solution methods, we present first some more elaborated topics in linear algebra. We first present again the multiplication properties of matrices.

Matrix multiplication properties

1. Matrix multiplication is associative: $(\mathbf{AB})\mathbf{C} = \mathbf{A}(\mathbf{BC})$
2. Matrix multiplication is distributive: $\mathbf{A}(\mathbf{B} + \mathbf{C}) = \mathbf{AB} + \mathbf{AC}$
3. Matrix multiplication is not commutative: $\mathbf{AB} \neq \mathbf{BA}$ for any $\mathbf{A}$ and $\mathbf{B}$

We continue with the properties of transpose matrices.

Properties of Matrix Transpose

1. Matrix multiplication obeys the transpose property: $(\mathbf{AB})^{\mathsf{T}} = \mathbf{B}^{\mathsf{T}}\mathbf{A}^{\mathsf{T}}$
2. $(\mathbf{AB}\cdots\mathbf{Z})^{\mathsf{T}} = \mathbf{Z}^{\mathsf{T}}\cdots\mathbf{B}^{\mathsf{T}}\mathbf{A}^{\mathsf{T}}$
3. $(\mathbf{A} + \mathbf{B})^{\mathsf{T}} = \mathbf{A}^{\mathsf{T}} + \mathbf{B}^{\mathsf{T}}$
4. If $\mathbf{C} = \mathbf{AB}$, then $\mathbf{C}^{\mathsf{T}} = \mathbf{B}^{\mathsf{T}}\mathbf{A}^{\mathsf{T}}$
5. Transpose has higher precedence than multiplication or addition: $\mathbf{AB}^{\mathsf{T}} = \mathbf{A}(\mathbf{B}^{\mathsf{T}})$

6. Transpose has higher precedence than addition/subtraction: $\mathbf{A} + \mathbf{B}^\mathsf{T} = \mathbf{A} + (\mathbf{B}^\mathsf{T})$
7. $(k\mathbf{A})^\mathsf{T} = k\mathbf{A}^\mathsf{T}$ where k is a scalar
8. $(\mathbf{A}^\mathsf{T})^\mathsf{T} = \mathbf{A}$
9. $\det(\mathbf{A}^\mathsf{T}) = \det(\mathbf{A})$
10. $\mathbf{A}\mathbf{A}^\mathsf{T}$ with $\mathbf{A} \in \mathbb{R}^{m \times n}$ is a square $m \times m$ matrix and $\mathbf{A}^\mathsf{T}\mathbf{A}$ a square $n \times n$ matrix
11. If $\mathbf{x} \in \mathbb{R}^n$ and $\mathbf{y} \in \mathbb{R}^n$, then $\mathbf{x}^\mathsf{T}\mathbf{y} = \mathbf{y}^\mathsf{T}\mathbf{x}$ and the product is a scalar

We continue with the properties of inverse matrices. Reckon that a matrix $\mathbf{A}$ is *singular* or *degenerate* if it is not invertible, i.e., there is no matrix $\mathbf{C}$ such that $\mathbf{A}\mathbf{C} = \mathbf{I}$, where $\mathbf{I}$ is the identity matrix. In reverse, a matrix $\mathbf{A}$ is *nonsingular* or *nondegenerate* (invertible) if there exists inverse matrix $\mathbf{A}^{-1}$ such that $\mathbf{A}^{-1}\mathbf{A} = \mathbf{I}$. Note that only *square* matrices that have the same number of rows and columns can be nonsignular. We can test whether a square matrix is nonsingular by computing its determinant, which should be nonzero. Other tests can emerge from the Invertible Matrix Theorem [7].

Theorem 4.5 *Invertible Matrix theorem: for a square $A \in \mathbb{R}^{n \times n}$ matrix the following statements are equivalent:*

1. *A is invertible.*
2. *The column vectors of A are linearly independent.*
3. *The determinant $\det(A) \neq 0$.*
4. *$Null(A) = \{0\}$.*
5. *The column vectors of A span $\mathbb{R}^n$.*
6. *$Ax = b$ has a unique solution for each b in $\mathbb{R}^n$.*
7. *The only solution to $Ax = 0$ is the trivial solution.*
8. *There is no eigenvalue of A that has a value equal to 0.*

Typical methods of finding the inverse of a matrix are the Gauss-Jordan elimination and the Lower-Upper (LU) decomposition. The main properties of matrix inverse are presented below.

Properties of Matrix Inverse, if such a matrix exists

1. $(k\mathbf{A}^{-1}) = \frac{1}{k}\mathbf{A}^{-1}$ for nonzero scalar k
2. $(\mathbf{A}^{-1})^{-1} = \mathbf{A}$
3. $(\mathbf{A}^\mathsf{T})^{-1} = (\mathbf{A}^{-1})^\mathsf{T}$
4. $(\mathbf{A}^n)^{-1} = (\mathbf{A}^{-1})^n = \mathbf{A}^{-n}$
5. $(\mathbf{A}\mathbf{B})^{-1} = \mathbf{B}^{-1}\mathbf{A}^{-1}$
6. $(\mathbf{A}_1\mathbf{A}_2 \cdots \mathbf{A}_n)^{-1} = \mathbf{A}_n^{-1}\mathbf{A}_{n-1}^{-1} \cdots \mathbf{A}_1^{-1}$
7. If $\mathbf{C} = \mathbf{A}\mathbf{B}$, then $\mathbf{C}^{-1} = \mathbf{B}^{-1}\mathbf{A}^{-1}$
8. $\det(\mathbf{A}^{-1}) = \det(\mathbf{A})^{-1}$

9. If $\mathbf{A} \in \mathbb{R}^{n \times n}$ is nonsingular, then the equation $\mathbf{Ax} = \mathbf{b}$, where variables $\mathbf{x} \in \mathbb{R}^n$ and parameters $\mathbf{b} \in \mathbb{R}^n$ has a unique solution $\mathbf{x} = \mathbf{A}^{-1}\mathbf{b}$

As already discussed, another important property of a matrix $\mathbf{A} \in \mathbb{R}^{m \times n}$ with m rows and n columns is the *rank* of the matrix, written as rank$(\mathbf{A})$. We proceed with the following definitions.

Definition 4.8 The *row rank* of a matrix $\mathbf{A} \in \mathbb{R}^{m \times n}$ is the number of its linearly independent rows. It follows that the row rank is less than or equal to m.

Definition 4.9 The *column rank* of a matrix $\mathbf{A} \in \mathbb{R}^{m \times n}$ is the number of its linearly independent rows. It follows that the column rank is less than or equal to n.

Note that the row rank is equal to the column rank of a matrix and this number is called *rank* of the matrix, as it is defined below.

Definition 4.10 The rank of matrix $\mathbf{A} \in \mathbb{R}^{m \times n}$ is the maximum number of linearly independent column or row vectors in the matrix. By definition, rank$(\mathbf{A}) \leq \min\{m, n\}$.

We call $\mathbf{A} \in \mathbb{R}^{m \times n}$ a *full-rank* matrix if rank$(\mathbf{A}) = \min\{m, n\}$. We also say that a matrix has *full row rank* if rank$(\mathbf{A}) = m$ and *full column rank* if rank$(\mathbf{A}) = n$. Clearly, all row vectors of a *full row rank* matrix are linearly independent and all column vectors of a *full column rank* are linear independent, respectively.

The rank of a matrix can be computed by using the Minor method or the Echelon form method [9].

We also present properties related to matrix differentiation. These properties can be written in a different way based on the selected layout notation. As already discussed, there are two main layout notations. The *numerator* layout notation and the *denominator* layout notation. We present these two types below.

Numerator and Denominator layout notations

Scalar-by-vector derivative (Gradient vector):

$$\nabla_{\mathbf{x}} f(\mathbf{x}) = \frac{\partial f(\mathbf{x})}{\partial \mathbf{x}} = \underbrace{\left[\frac{\partial f(\mathbf{x})}{\partial x_1}, \ldots, \frac{\partial f(\mathbf{x})}{\partial x_n}\right]}_{\text{numerator layout}} \text{ or } \underbrace{\begin{bmatrix} \frac{\partial f(\mathbf{x})}{\partial x_1} \\ \vdots \\ \frac{\partial f(\mathbf{x})}{\partial x_n} \end{bmatrix}}_{\text{denominator layout}}$$

Vector-by-scalar derivative:

$$\frac{\partial \mathbf{f}(x)}{\partial x} = \underbrace{\begin{bmatrix} \frac{\partial f_1(x)}{\partial x} \\ \vdots \\ \frac{\partial f_m(x)}{\partial x} \end{bmatrix}}_{\text{numerator layout}} \quad \text{or} \quad \underbrace{\begin{bmatrix} \frac{\partial f_1(x)}{\partial x}, & \dots, & \frac{\partial f_m(x)}{\partial x} \end{bmatrix}}_{\text{denominator layout}}$$

Vector-by-vector derivative (Jacobian matrix):

$$\frac{\partial \mathbf{f}(\mathbf{x})}{\partial \mathbf{x}} = \underbrace{\begin{bmatrix} \frac{\partial f_1(\mathbf{x})}{\partial x_1} & \dots & \frac{\partial f_1(\mathbf{x})}{\partial x_n} \\ \vdots & \ddots & \vdots \\ \frac{\partial f_m(\mathbf{x})}{\partial x_1} & \dots & \frac{\partial f_m(\mathbf{x})}{\partial x_n} \end{bmatrix}}_{\text{numerator layout } (m \times n)} \quad \text{or} \quad \underbrace{\begin{bmatrix} \frac{\partial f_1(\mathbf{x})}{\partial x_1} & \dots & \frac{\partial f_m(\mathbf{x})}{\partial x_1} \\ \vdots & \ddots & \vdots \\ \frac{\partial f_1(\mathbf{x})}{\partial x_n} & \dots & \frac{\partial f_m(\mathbf{x})}{\partial x_n} \end{bmatrix}}_{\text{denominator layout } (n \times m)}$$

Scalar-by-matrix derivative $\mathbf{X} \in \mathbb{R}^{p \times q}$:

$$\frac{\partial f(\mathbf{X})}{\partial \mathbf{X}} = \underbrace{\begin{bmatrix} \frac{\partial f(\mathbf{X})}{\partial x_{11}} & \dots & \frac{\partial f(\mathbf{X})}{\partial x_{p1}} \\ \vdots & \ddots & \vdots \\ \frac{\partial f(\mathbf{X})}{\partial x_{1q}} & \dots & \frac{\partial f(\mathbf{X})}{\partial x_{pq}} \end{bmatrix}}_{\text{numerator layout } (q \times p)} \quad \text{or} \quad \underbrace{\begin{bmatrix} \frac{\partial f(\mathbf{X})}{\partial x_{11}} & \dots & \frac{\partial f(\mathbf{X})}{\partial x_{1q}} \\ \vdots & \ddots & \vdots \\ \frac{\partial f(\mathbf{X})}{\partial x_{p1}} & \dots & \frac{\partial f(\mathbf{X})}{\partial x_{pq}} \end{bmatrix}}_{\text{denominator layout } (p \times q)}$$

Based on the above definitions, we can proceed to the calculation of derivatives that involve vectors, scalars and matrices. In Table 4.1 we present the main outcomes of derivatives when using the nominator and the denominator layout notation.

We can use the above rules to compute derivatives between scalars, vectors and matrices.

Proposition 4.1 *Consider* $\mathbf{a}, \mathbf{x} \in \mathbb{R}^n$. *Then,* $\frac{\partial \mathbf{a}^\mathsf{T} \mathbf{x}}{\partial \mathbf{x}} = \frac{\partial \mathbf{x}^\mathsf{T} \mathbf{a}}{\partial \mathbf{x}} = \mathbf{a}^\mathsf{T}$ *when using the numerator and* $\mathbf{a}$ *when using the denominator layout notation.*

Proof Consider the numerator layout notation.

$$\frac{\partial \mathbf{a}^\mathsf{T} \mathbf{x}}{\partial \mathbf{x}} = \frac{\partial \sum_{i=1}^n a_i x_i}{\partial \mathbf{x}} = [a_1, a_2, \dots, a_n] = \mathbf{a}^\mathsf{T}$$

In addition,

$$\frac{\partial \mathbf{x}^\mathsf{T} \mathbf{a}}{\partial \mathbf{x}} = \frac{\partial \sum_{i=1}^n x_i a_i}{\partial \mathbf{x}} = [a_1, a_2, \dots, a_n] = \mathbf{a}^\mathsf{T}$$

Table 4.1 Basic derivation rules

Input	Expression	Numerator layout	Denominator layout
$\mathbf{a} \in \mathbb{R}^m, \mathbf{x} \in \mathbb{R}^n$	$\frac{\partial \mathbf{a}}{\partial \mathbf{x}}$	$\underset{m \times n}{\mathbf{0}}$	$\underset{n \times m}{\mathbf{0}}$
$\mathbf{x} \in \mathbb{R}^n$	$\frac{\partial \mathbf{x}}{\partial \mathbf{x}}$	$\underset{n \times n}{\mathbf{I}}$	$\underset{n \times n}{\mathbf{I}}$
$\mathbf{x} \in \mathbb{R}^n, \mathbf{f}(\mathbf{x}): \mathbb{R}^n \to \mathbb{R}^m$	$\frac{\partial a\mathbf{f}(\mathbf{x})}{\partial \mathbf{x}}$	$a\,\underset{m \times n}{\frac{\partial \mathbf{f}(\mathbf{x})}{\partial \mathbf{x}}}$	$a\,\underset{n \times m}{\frac{\partial \mathbf{f}(\mathbf{x})}{\partial \mathbf{x}}}$
$\mathbf{x} \in \mathbb{R}^n, \mathbf{a} \in \mathbb{R}^m, f(\mathbf{x}): \mathbb{R}^n \to \mathbb{R}$	$\frac{\partial f(\mathbf{x})\mathbf{a}}{\partial \mathbf{x}}$	$\underset{m \times 1}{\mathbf{a}}\,\underset{1 \times n}{\frac{\partial f(\mathbf{x})}{\partial \mathbf{x}}}$	$\underset{n \times 1}{\frac{\partial f(\mathbf{x})}{\partial \mathbf{x}}}\,\underset{1 \times m}{\mathbf{a}^\mathsf{T}}$
$\mathbf{a} \in \mathbb{R}^m, \mathbf{X} \in \mathbb{R}^{m \times n}, \mathbf{b} \in \mathbb{R}^n$	$\frac{\partial \mathbf{a}^\mathsf{T}\mathbf{X}\mathbf{b}}{\partial \mathbf{X}}$	$\mathbf{b}\mathbf{a}^\mathsf{T}$	$\mathbf{a}\mathbf{b}^\mathsf{T}$
$\mathbf{a} \in \mathbb{R}^m, \mathbf{X} \in \mathbb{R}^{n \times m}, \mathbf{b} \in \mathbb{R}^n$	$\frac{\partial \mathbf{a}^\mathsf{T}\mathbf{X}^\mathsf{T}\mathbf{b}}{\partial \mathbf{X}}$	$\mathbf{a}\mathbf{b}^\mathsf{T}$	$\mathbf{b}\mathbf{a}^\mathsf{T}$
$\mathbf{x} \in \mathbb{R}^n, \mathbf{A} \in \mathbb{R}^{m \times p}, \mathbf{f}(\mathbf{x}): \mathbb{R}^n \to \mathbb{R}^p$	$\frac{\partial \mathbf{A}f(\mathbf{x})}{\partial \mathbf{x}}$	$\underset{m \times p}{\mathbf{A}}\,\underset{p \times n}{\frac{\partial \mathbf{f}(\mathbf{x})}{\partial \mathbf{x}}}$	$\underset{n \times p}{\frac{\partial \mathbf{f}(\mathbf{x})}{\partial \mathbf{x}}}\,\underset{p \times m}{\mathbf{A}^\mathsf{T}}$
$\mathbf{x} \in \mathbb{R}^n, f(\mathbf{x}): \mathbb{R}^n \to \mathbb{R}, \mathbf{g}(\mathbf{x}): \mathbb{R}^n \to \mathbb{R}^m$	$\frac{\partial f(\mathbf{x})\mathbf{g}(\mathbf{x})}{\partial \mathbf{x}}$	$f(\mathbf{x})\underset{m \times n}{\frac{\partial \mathbf{g}(\mathbf{x})}{\partial \mathbf{x}}} + \underset{m \times 1}{\mathbf{g}(\mathbf{x})}\,\underset{1 \times n}{\frac{\partial f(\mathbf{x})}{\partial \mathbf{x}}}$	$f(\mathbf{x})\underset{n \times m}{\frac{\partial \mathbf{g}(\mathbf{x})}{\partial \mathbf{x}}} + \underset{n \times 1}{\frac{\partial f(\mathbf{x})}{\partial \mathbf{x}}}\,\underset{1 \times m}{\mathbf{g}^\mathsf{T}(\mathbf{x})}$
$\mathbf{x} \in \mathbb{R}^n, \mathbf{f}(\mathbf{x}): \mathbb{R}^n \to \mathbb{R}^m, \mathbf{g}(\mathbf{x}): \mathbb{R}^m \to \mathbb{R}^p$	$\frac{\partial \mathbf{g}(\mathbf{f}(\mathbf{x}))}{\partial \mathbf{x}}$	$\underset{p \times m}{\frac{\partial \mathbf{g}(\mathbf{f}(\mathbf{x}))}{\partial \mathbf{f}(\mathbf{x})}}\,\underset{m \times n}{\frac{\partial \mathbf{f}(\mathbf{x})}{\partial \mathbf{x}}}$	$\underset{n \times m}{\frac{\partial \mathbf{f}(\mathbf{x})}{\partial \mathbf{x}}}\,\underset{m \times p}{\frac{\partial \mathbf{g}(\mathbf{f}(\mathbf{x}))}{\partial \mathbf{f}(\mathbf{x})}}$
$\mathbf{x} \in \mathbb{R}^n, f, g(\mathbf{x}): \mathbb{R}^n \to \mathbb{R}$	$\frac{\partial g(f(\mathbf{x}))}{\partial \mathbf{x}}$	$\underset{1 \times 1}{\frac{\partial g(f(\mathbf{x}))}{\partial f(\mathbf{x})}}\,\underset{1 \times n}{\frac{\partial f(\mathbf{x})}{\partial \mathbf{x}}}$	$\underset{n \times 1}{\frac{\partial f(\mathbf{x})}{\partial \mathbf{x}}}\,\underset{1 \times 1}{\frac{\partial g(f(\mathbf{x}))}{\partial f(\mathbf{x})}}$
$\mathbf{x} \in \mathbb{R}^n, f, g(\mathbf{x}): \mathbb{R}^n \to \mathbb{R}$	$\frac{\partial g(\mathbf{x})f(\mathbf{x})}{\partial \mathbf{x}}$	$f(\mathbf{x})\underset{1 \times n}{\frac{\partial g(\mathbf{x})}{\partial \mathbf{x}}} + g(\mathbf{x})\underset{1 \times n}{\frac{\partial f(\mathbf{x})}{\partial \mathbf{x}}}$	$f(\mathbf{x})\underset{n \times 1}{\frac{\partial g(\mathbf{x})}{\partial \mathbf{x}}} + g(\mathbf{x})\underset{n \times 1}{\frac{\partial f(\mathbf{x})}{\partial \mathbf{x}}}$
$\mathbf{x} \in \mathbb{R}^n, \mathbf{f}(\mathbf{x}): \mathbb{R}^n \to \mathbb{R}^m, \mathbf{g}(\mathbf{x}): \mathbb{R}^m \to \mathbb{R}^p$	$\frac{\partial (\mathbf{f}(\mathbf{x}) \cdot \mathbf{g}(\mathbf{x}))}{\partial \mathbf{x}} = \frac{\partial \mathbf{f}^\mathsf{T}(\mathbf{x})\mathbf{g}(\mathbf{x})}{\partial \mathbf{x}}$	$\underset{1 \times m}{\mathbf{f}^\mathsf{T}(\mathbf{x})}\,\underset{m \times n}{\frac{\partial \mathbf{g}(\mathbf{x})}{\partial \mathbf{x}}} + \underset{1 \times m}{\mathbf{g}^\mathsf{T}(\mathbf{x})}\,\underset{m \times n}{\frac{\partial \mathbf{f}(\mathbf{x})}{\partial \mathbf{x}}}$	$\underset{n \times m}{\frac{\partial \mathbf{g}(\mathbf{x})}{\partial \mathbf{x}}}\,\underset{m \times 1}{\mathbf{f}(\mathbf{x})} + \underset{n \times m}{\frac{\partial \mathbf{f}(\mathbf{x})}{\partial \mathbf{x}}}\,\underset{m \times 1}{\mathbf{g}(\mathbf{x})}$

Consider the denominator layout notation. In this case,

$$\frac{\partial \mathbf{a}^\mathsf{T}\mathbf{x}}{\partial \mathbf{x}} = \frac{\partial \mathbf{x}^\mathsf{T}\mathbf{a}}{\partial \mathbf{x}} = \frac{\partial \sum_{i=1}^{n} a_i x_i}{\partial \mathbf{x}} = \begin{bmatrix} a_1 \\ a_2 \\ \vdots \\ a_n \end{bmatrix} = \mathbf{a}$$

and this completes the proof.

Proposition 4.2 *Consider $A \in \mathbb{R}^{m \times n}$ and $\mathbf{x} \in \mathbb{R}^n$. Then, $\frac{\partial A\mathbf{x}}{\partial \mathbf{x}} = A$ when using the numerator and A^T when using the denominator layout notation.*

Proof $A\mathbf{x}$ is an $m \times 1$ vector. The Jacobian when using the numerator layout notation is:

$$\frac{\partial A\mathbf{x}}{\partial \mathbf{x}} = \begin{bmatrix} \frac{\partial \sum_{k=1}^{n} a_{1k} x_k}{\partial \mathbf{x}} \\ \frac{\partial \sum_{k=1}^{n} a_{2k} x_k}{\partial \mathbf{x}} \\ \vdots \\ \frac{\partial \sum_{k=1}^{n} a_{mk} x_k}{\partial \mathbf{x}} \end{bmatrix} = \begin{bmatrix} a_{11} & a_{12} & \cdots & a_{1n} \\ a_{21} & a_{22} & \cdots & a_{2n} \\ \vdots & \vdots & \ddots & \vdots \\ a_{m1} & a_{m2} & \cdots & a_{mn} \end{bmatrix} = A$$

When using the denominator layout notation:

$$\frac{\partial \mathbf{Ax}}{\partial \mathbf{x}} = \begin{bmatrix} \frac{\partial \sum_{k=1}^{n} a_{1k}x_k}{\partial \mathbf{x}} & \frac{\partial \sum_{k=1}^{n} a_{2k}x_k}{\partial \mathbf{x}} & \cdots & \frac{\partial \sum_{k=1}^{n} a_{mk}x_k}{\partial \mathbf{x}} \end{bmatrix} = \begin{bmatrix} a_{11} & a_{21} & \cdots & a_{m1} \\ a_{12} & a_{22} & \cdots & a_{m2} \\ \vdots & \vdots & \ddots & \vdots \\ a_{1n} & a_{2n} & \cdots & a_{mn} \end{bmatrix} = \mathbf{A}^\mathsf{T}$$

Proposition 4.3 *For $\mathbf{x} \in \mathbb{R}^n$, $\frac{\partial \mathbf{x}\mathbf{x}^\mathsf{T}}{\partial \mathbf{x}} = 2\mathbf{x}^\mathsf{T}$ when using the numerator and $2\mathbf{x}$ when using the denominator layout notation.*

Proof When using the numerator layout notation:

$$\frac{\partial \mathbf{x}\mathbf{x}^\mathsf{T}}{\partial \mathbf{x}} = \frac{\partial \sum_{i=1}^{n} x_i x_i}{\partial \mathbf{x}} = \begin{bmatrix} \frac{\partial \sum_{i=1}^{n} x_i x_i}{\partial x_1}, & \frac{\partial \sum_{i=1}^{n} x_i x_i}{\partial x_2}, & \cdots, & \frac{\partial \sum_{i=1}^{n} x_i x_i}{\partial x_n} \end{bmatrix} = \begin{bmatrix} 2x_1, & 2x_2, & \cdots, & 2x_n \end{bmatrix} = 2\mathbf{x}^\mathsf{T}$$

When using the denominator layout notation:

$$\frac{\partial \mathbf{x}\mathbf{x}^\mathsf{T}}{\partial \mathbf{x}} = \frac{\partial \sum_{i=1}^{n} x_i x_i}{\partial \mathbf{x}} = \begin{bmatrix} \frac{\partial \sum_{i=1}^{n} x_i x_i}{\partial x_1} \\ \frac{\partial \sum_{i=1}^{n} x_i x_i}{\partial x_2} \\ \vdots \\ \frac{\partial \sum_{i=1}^{n} x_i x_i}{\partial x_n} \end{bmatrix} = \begin{bmatrix} 2x_1 \\ 2x_2 \\ \cdots \\ 2x_n \end{bmatrix} = 2\mathbf{x}$$

Proposition 4.4 *Consider $\mathbf{x} \in \mathbb{R}^n$, $\mathbf{A} \in \mathbb{R}^{m \times n}$, $\mathbf{b} \in \mathbb{R}^m$. Then, $\frac{\partial \mathbf{b}^\mathsf{T}\mathbf{Ax}}{\partial \mathbf{x}} = \mathbf{b}^\mathsf{T}\mathbf{A}$ when using the numerator and $\mathbf{A}^\mathsf{T}\mathbf{b}$ when using the denominator layout notation.*

Proof We can write $\mathbf{b}^\mathsf{T}\mathbf{Ax}$ as $\mathbf{b}^\mathsf{T}\mathbf{f}(\mathbf{x})$ where $\mathbf{f}(\mathbf{x}) = \mathbf{Ax}$ is an $m \times 1$ matrix (=vector). With the numerator layout notation:

$$\frac{\partial \mathbf{b}^\mathsf{T}\mathbf{Ax}}{\partial \mathbf{x}} = \frac{\partial \mathbf{b}^\mathsf{T}\mathbf{f}(\mathbf{x})}{\partial \mathbf{x}} = \mathbf{b}^\mathsf{T}\frac{\partial \mathbf{f}(\mathbf{x})}{\partial \mathbf{x}} + \mathbf{f}(\mathbf{x})^\mathsf{T}\frac{\partial \mathbf{b}}{\partial \mathbf{x}} = \mathbf{b}^\mathsf{T}\frac{\partial \mathbf{f}(\mathbf{x})}{\partial \mathbf{x}} = \mathbf{b}^\mathsf{T}\frac{\partial \mathbf{Ax}}{\partial \mathbf{x}} = \mathbf{b}^\mathsf{T}\mathbf{A}$$

With the denominator layout notation:

$$\frac{\partial \mathbf{b}^\mathsf{T}\mathbf{Ax}}{\partial \mathbf{x}} = \frac{\partial \mathbf{b}^\mathsf{T}\mathbf{f}(\mathbf{x})}{\partial \mathbf{x}} = \frac{\partial \mathbf{f}(\mathbf{x})}{\partial \mathbf{x}}\mathbf{b} + \frac{\partial \mathbf{b}}{\partial \mathbf{x}}\mathbf{f}(\mathbf{x}) = \frac{\partial \mathbf{f}(\mathbf{x})}{\partial \mathbf{x}}\mathbf{b} = \frac{\partial \mathbf{Ax}}{\partial \mathbf{x}}\mathbf{b} = \mathbf{A}^\mathsf{T}\mathbf{b}$$

Proposition 4.5 *Consider $\mathbf{x} \in \mathbb{R}^n$, $\mathbf{A} \in \mathbb{R}^{n \times m}$, $\mathbf{b} \in \mathbb{R}^m$. Then, $\frac{\partial \mathbf{x}^\mathsf{T}\mathbf{Ab}}{\partial \mathbf{x}} = \mathbf{b}^\mathsf{T}\mathbf{A}^\mathsf{T}$ when using the numerator and $\mathbf{Ab}$ when using the denominator layout notation.*

Proof We can write $\mathbf{Ab}$ as $\mathbf{c}$ where $\mathbf{c}$ is an n-valued column vector. With the numerator layout notation:

$$\frac{\partial \mathbf{x}^\mathsf{T}\mathbf{c}}{\partial \mathbf{x}} = \mathbf{c}^\mathsf{T} = (\mathbf{Ab})^\mathsf{T} = \mathbf{b}^\mathsf{T}\mathbf{A}^\mathsf{T}$$

With the denominator layout notation:

$$\frac{\partial \mathbf{x}^\mathsf{T}\mathbf{c}}{\partial \mathbf{x}} = \mathbf{c} = \mathbf{Ab}$$

Proposition 4.6 *Consider* $x \in \mathbb{R}^n$, $f(x) : \mathbb{R}^n \to \mathbb{R}^m$, $g(x) : \mathbb{R}^n \to \mathbb{R}^p$, *and* $A \in \mathbb{R}^{m \times p}$. *Then,* $\frac{\partial f^T(x)Ag(x)}{\partial x} = f^T(x)A\frac{\partial g(x)}{\partial x} + g^T(x)A^T\frac{\partial f(x)}{\partial x}$ *when using the numerator and* $\frac{\partial g(x)}{\partial x}A^Tf(x) + \frac{\partial f(x)}{\partial x}Ag(x)$ *when using the denominator layout notation.*

Proof We can write $\mathbf{f}^T(\mathbf{x})A\mathbf{g}(\mathbf{x})$ as $\mathbf{f}^T(\mathbf{x})\mathbf{z}(\mathbf{x})$ where $\mathbf{z}(\mathbf{x}) = A\mathbf{g}(\mathbf{x})$ is an m-valued column vector. From the vector-by-vector dot product when using the numerator layout notation:

$$\frac{\partial \mathbf{f}^T(\mathbf{x})\mathbf{z}(\mathbf{x})}{\partial \mathbf{x}} = \mathbf{f}^T(\mathbf{x})\frac{\partial \mathbf{z}(\mathbf{x})}{\partial \mathbf{x}} + \mathbf{z}^T(\mathbf{x})\frac{\partial \mathbf{f}(\mathbf{x})}{\partial \mathbf{x}} = \mathbf{f}^T(\mathbf{x})\frac{\partial A\mathbf{g}(\mathbf{x})}{\partial \mathbf{x}} + (A\mathbf{g}(\mathbf{x}))^T\frac{\partial \mathbf{f}(\mathbf{x})}{\partial \mathbf{x}} =$$
$$= \mathbf{f}^T(\mathbf{x})A\frac{\partial \mathbf{g}(\mathbf{x})}{\partial \mathbf{x}} + \mathbf{g}^T(\mathbf{x})A^T\frac{\partial \mathbf{f}(\mathbf{x})}{\partial \mathbf{x}}$$

Consider now the denominator layout notation:

$$\frac{\partial \mathbf{f}^T(\mathbf{x})\mathbf{z}(\mathbf{x})}{\partial \mathbf{x}} = \frac{\partial \mathbf{z}(\mathbf{x})}{\partial \mathbf{x}}\mathbf{f}(\mathbf{x}) + \frac{\partial \mathbf{f}(\mathbf{x})}{\partial \mathbf{x}}\mathbf{z}(\mathbf{x}) = \frac{\partial A\mathbf{g}(\mathbf{x})}{\partial \mathbf{x}}\mathbf{f}(\mathbf{x}) + \frac{\partial \mathbf{f}(\mathbf{x})}{\partial \mathbf{x}}(A\mathbf{g}(\mathbf{x})) =$$
$$= \frac{\partial \mathbf{g}(\mathbf{x})}{\partial \mathbf{x}}A^T\mathbf{f}(\mathbf{x}) + \frac{\partial \mathbf{f}(\mathbf{x})}{\partial \mathbf{x}}A\mathbf{g}(\mathbf{x})$$

Proposition 4.7 *Consider* $x \in \mathbb{R}^n$, $A \in \mathbb{R}^{n \times n}$. *Then,* $\frac{\partial x^TAx}{\partial x} = x^T(A + A^T)$ *when using the numerator and* $(A + A^T)x$ *when using the denominator layout notation.*

Proof We can write $A\mathbf{x}$ as $\mathbf{f}(\mathbf{x}) = A\mathbf{x}$, where $\mathbf{f}(\mathbf{x})$ is an n-valued function. With the numerator layout notation:

$$\frac{\partial \mathbf{x}^T\mathbf{f}(\mathbf{x})}{\partial \mathbf{x}} = \mathbf{x}^T\frac{\partial \mathbf{f}(\mathbf{x})}{\partial \mathbf{x}} + \mathbf{f}^T(\mathbf{x})\frac{\partial \mathbf{x}}{\partial \mathbf{x}} = \mathbf{x}^TA\frac{\partial \mathbf{x}}{\partial \mathbf{x}} + (A\mathbf{x})^T\frac{\partial \mathbf{x}}{\partial \mathbf{x}} = \mathbf{x}^TA + \mathbf{x}^TA^T = \mathbf{x}^T(A + A^T)$$

With the denominator layout notation:

$$\frac{\partial \mathbf{x}^T\mathbf{f}(\mathbf{x})}{\partial \mathbf{x}} = \frac{\partial \mathbf{x}}{\partial \mathbf{x}}\mathbf{f}(\mathbf{x}) + \frac{\partial \mathbf{f}(\mathbf{x})}{\partial \mathbf{x}}\mathbf{x} = A\mathbf{x} + \frac{\partial A\mathbf{x}}{\partial \mathbf{x}}\mathbf{x} = A\mathbf{x} + A^T\mathbf{x} = (A + A^T)\mathbf{x}$$

Proposition 4.8 *Consider* $x \in \mathbb{R}^n$, $A \in \mathbb{R}^{n \times n}$. *Then,* $\frac{\partial x^TAx}{\partial x} = 2x^TA$ *when using the numerator and* $2Ax$ *when using the denominator layout notation.*

Proof This is proved trivially by noticing that $A = A^T$ since A is symmetric. $\square$

Proposition 4.9 *Consider* $x \in \mathbb{R}^n$, $a \in \mathbb{R}^m$ *and* $f(x) : \mathbb{R}^n \to \mathbb{R}^m$. *Then,* $\frac{\partial a^Tf(x)}{\partial x} = a^T\frac{\partial f(x)}{\partial x}$ *when using the numerator and* $\frac{\partial f(x)}{\partial x}a$ *when using the denominator layout notation.*

Proof With the numerator layout notation:

$$\frac{\partial \mathbf{a}^T\mathbf{f}(\mathbf{x})}{\partial \mathbf{x}} = \mathbf{a}^T\frac{\partial \mathbf{f}(\mathbf{x})}{\partial \mathbf{x}} + \mathbf{f}(\mathbf{x})^T\frac{\partial \mathbf{a}}{\partial \mathbf{x}} = \mathbf{a}^T\frac{\partial \mathbf{f}(\mathbf{x})}{\partial \mathbf{x}}$$

With the denominator layout notation:

$$\frac{\partial \mathbf{a}^\top \mathbf{f}(\mathbf{x})}{\partial \mathbf{x}} = \frac{\partial \mathbf{f}(\mathbf{x})}{\partial \mathbf{x}} \mathbf{a} + \frac{\partial \mathbf{a}}{\partial \mathbf{x}} \mathbf{f}(\mathbf{x}) = \frac{\partial \mathbf{f}(\mathbf{x})}{\partial \mathbf{x}} \mathbf{a}$$

Proposition 4.10 *Consider $\mathbf{x} \in \mathbb{R}^m$, $A \in \mathbb{R}^{m \times n}$. Then, $\frac{\partial \mathbf{x}^\top A}{\partial \mathbf{x}} = A^\top$ when using the numerator and A when using the denominator layout notation.*

Proof With the numerator layout notation:

$$\frac{\partial \mathbf{x}^\top A}{\partial \mathbf{x}} = \begin{bmatrix} \frac{\partial \sum_{k=1}^m x_k a_{k1}}{\partial \mathbf{x}} \\ \vdots \\ \frac{\partial \sum_{k=1}^m x_k a_{kn}}{\partial \mathbf{x}} \end{bmatrix} = \begin{bmatrix} \frac{\partial \sum_{k=1}^m x_k a_{k1}}{\partial x_1} & \frac{\partial \sum_{k=1}^m x_k a_{k1}}{\partial x_2} & \cdots & \frac{\partial \sum_{k=1}^m x_k a_{k1}}{\partial x_m} \\ \vdots & \vdots & \ddots & \vdots \\ \frac{\partial \sum_{k=1}^m x_k a_{kn}}{\partial x_1} & \frac{\partial \sum_{k=1}^m x_k a_{kn}}{\partial x_2} & \cdots & \frac{\partial \sum_{k=1}^m x_k a_{kn}}{\partial x_m} \end{bmatrix} = A^\top$$

Similarly, $\frac{\partial \mathbf{x}^\top A}{\partial \mathbf{x}} = A$ when using the denominator layout notation. $\square$

In the remainder of this chapter we will present solution methods for continuous constrained optimization problems, after categorizing them as follows:

- Linear problems
- Quadratic problems (which are a sub-category of nonlinear problems)
- Nonlinear problems

Concerning convexity, linear problems are always convex (and concave). Quadratic minimization problems can be convex or non-convex depending on the convexity of their objective function and nonlinear minimization problems can be convex or non-convex depending on the convexity of their objective function and the convexity of their feasible set.

4.5   Linear Programming

One special case of continuous constrained optimization is the *linear programming (LP)* case where the objective and the (in)equality constraints of the mathematical program are linear/affine functions. The feasible set of a linear program is convex (a polyhedron) and because its objective function is both convex and concave, locally optimal solutions of minimization and maximization LPs are also globally optimal ones. Because of this property, an important task of modelers is to investigate whether nonlinear problems can be reformulated as linear ones. We note that LPs have no solutions if their feasible region is empty (*infeasible case*) or the objective function is not bounded from below/above in case of solving a minimization/maximization problem, respectively (*unbounded case*).

Fig. 4.3 Polyhedron $\mathcal{P} = \{\mathbf{x} \in \mathbb{R}^n_{\geq 0} \mid \mathbf{A}\mathbf{x} \leq \mathbf{b}\}$ for the LP in Eqs. (4.44)–(4.47). Note that this polyhedron is bounded and thus it is also a polytope. If we remove constraint $x_2 \geq 0$ the polyhedron becomes unbounded

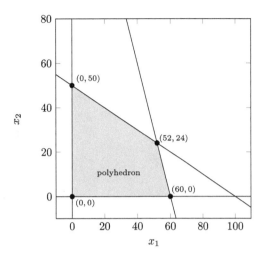

We will first focus on the geometry of LPs. Any LP can be written in the canonical form provided below.

Canonical Form of LPs

$$\max \mathbf{c}^{\mathsf{T}}\mathbf{x} \text{ s.t. } \mathbf{A}\mathbf{x} \leq \mathbf{b} \wedge \mathbf{x} \in \mathbb{R}^n_{\geq 0}$$

where the feasible region $\mathcal{P} = \{\mathbf{x} \in \mathbb{R}^n_{\geq 0} \mid \mathbf{A}\mathbf{x} \leq \mathbf{b}\}$ is a polyhedron. This means that any LP is equivalent to minimizing a linear form $\mathbf{c}^{\mathsf{T}}\mathbf{x}$ over a polyhedral feasible region $\mathcal{P} = \{\mathbf{x} \in \mathbb{R}^n_{\geq 0} \mid \mathbf{A}\mathbf{x} \leq \mathbf{b}\}$. This is a very important property because the maximum (or minimum) of a linear form over a polyhedron is attained at a vertex of the polyhedron, defined below.

Definition 4.11 A point $\mathbf{x}^* \in \mathcal{P}$ is a vertex of polyhedron $\mathcal{P}$ if $\mathbf{x}^*$ is not a convex combination of two other points in $\mathcal{P}$ (geometrically, $\mathbf{x}^*$ is a corner of the polyhedron).

Consider, for example, the LP:

$$\max \ 15x_1 + 12x_2 \tag{4.44}$$
$$\text{s.t. } 3x_1 + x_2 \leq 180 \tag{4.45}$$
$$x_1 + 2x_2 \leq 100 \tag{4.46}$$
$$x_1, x_2 \geq 0 \tag{4.47}$$

This results in the feasible region of Fig. 4.3.

From this graphical representation of the feasible region we can make several observations about our LP. First, our LP is not infeasible because its feasible region

Fig. 4.4 Trimmed lines
$\mathbf{c}^{\mathsf{T}}\mathbf{x} = z$ with increasing
values of z until these lines
leave the feasible region

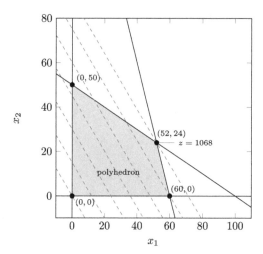

Fig. 4.4 Trimmed lines $\mathbf{c}^{\mathsf{T}}\mathbf{x} = z$ with increasing values of z until these lines leave the feasible region

is not empty. Second, the feasible region is bounded and it is thus a polytope. In addition, it has four vertices, namely $(0, 0)$, $(0, 50)$, $(52, 24)$, $(60, 0)$. We can see that the linear form $\mathbf{c}^{\mathsf{T}}\mathbf{x} = 15x_1 + 12x_2$ attains its maximum value at one of the vertices of the polyhedron by making the following observation. Notice that $\mathbf{c}^{\mathsf{T}}\mathbf{x} = 15x_1 + 12x_2$ is a line that represents the solution of equation $\mathbf{c}^{\mathsf{T}}\mathbf{x} = z$ for some value $z \in \mathbb{R}$. As we change z, we make the line move continuously across $\mathbb{R}^2$. Our objective is to make z as large as possible while maintaining feasibility (see Fig. 4.4). For this, we keep moving the line by increasing the value of z until the line no longer intersects with the feasible region. At this last point before leaving the feasible region, the line will intersect with a vertex of the polyhedron. This is illustrated in Fig. 4.4 where we draw the first dashed line that crosses from $(0, 0)$ by setting $z = 0$ and plotting the line $15x_1 + 12x_2 = z$. We then increase the value of z by 180 and plot the second dashed line $15x_1 + 12x_2 = 180$. This continues until plotting line $15x_1 + 12x_2 = 1068$ with $z = 1068$ which is the last line that intersects the feasible region. Notice that this line passes by vertex $[52, 24]^{\mathsf{T}}$, meaning that this vertex is the optimal solution of the LP.

If one computes the objective function values at the vertices of the polyhedron will have $z = 0$ for $[x_1, x_2]^{\mathsf{T}} = [0, 0]^{\mathsf{T}}$, $z = 600$ for $[x_1, x_2]^{\mathsf{T}} = [600, 0]^{\mathsf{T}}$, $z = 1068$ for $[x_1, x_2]^{\mathsf{T}} = [52, 24]^{\mathsf{T}}$, and $z = 900$ for $[x_1, x_2]^{\mathsf{T}} = [60, 0]^{\mathsf{T}}$. Notice that the maximum and minimum values of z are at vertices $(0, 0)$ and $(52, 24)$, after which the lines $\mathbf{c}^{\mathsf{T}}\mathbf{x}$ leave the feasible region. In fact, if we were solving a minimization instead of a maximization problem then its solution would have been $[x_1, x_2]^{\mathsf{T}} = [0, 0]^{\mathsf{T}}$.

There are two more observations when using this graphical approach to solve an LP. First, if the feasible region is empty (meaning that our problem is infeasible) we cannot obtain any solution. Second, if our LP is unbounded from below, it can still have a solution when solving a maximization problem (in reverse, if it is unbounded from above, it can still have a solution when solving a minimization problem). Consider, for instance, the case where our example LP does not have constraint $x_1 \geq 0$. That would mean that our LP is unbounded from below as presented in Fig. 4.5,

Fig. 4.5 Feasible region of our example LP without $x_1 \geq 0$, which is unbounded from below

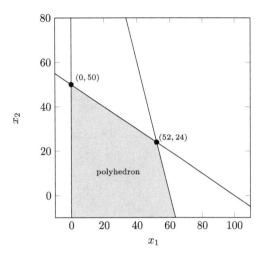

but it still has an optimal solution at $(52, 24)$. If we were solving a minimization problem, however, there would have been no solution because we would have been able to reduce the value of z as much as we want without the line $\mathbf{c}^\mathsf{T}\mathbf{x} = z$ exiting the feasible region.

Although the graphical approach is intuitive when solving problems in $\mathbb{R}^2$, we need more advanced solution methods to solve LPs with many variables. One of the most prevalent solution methods for LPs is the *simplex* algorithm, which requires first to transform an LP to its *standard form*, where:

- The objective function is maximized.
- All constraints are equality constraints.
- All variables are non-negative.

The standard form of an LP is expressed as follows.

! Standard form of a linear program

$$\max_{\mathbf{x}\in\mathbb{R}^n} \mathbf{c}^\mathsf{T}\mathbf{x} \tag{4.48}$$

$$\text{s.t.: } \mathbf{A}\mathbf{x} = \mathbf{b} \tag{4.49}$$

$$\mathbf{x} \geq \mathbf{0} \tag{4.50}$$

where $\mathbf{c}$ and $\mathbf{x}$ are column vectors in $\mathbb{R}^n$, $\mathbf{b}$ is a vector in $\mathbb{R}^l$, and $\mathbf{A}$ is an $l \times n$ matrix, where l is the number of equality constraints. To apply solution algorithms, such as simplex, there is the additional requirement that matrix $\mathbf{A}$ has *full row rank*, meaning that rank$(\mathbf{A}) = l$. If this is not the case, we can remove rows with repetitive equations before applying the solution method. Notice that the formulation can also be applied to minimization problems, as minimization can be transformed to maximization by changing the objective function:

$$\max_{\mathbf{x} \in \mathbb{R}^n} \mathbf{c}^\mathsf{T}\mathbf{x} = -\min_{\mathbf{x} \in \mathbb{R}^n} -\mathbf{c}^\mathsf{T}\mathbf{x}$$

Consider, for example, the minimization problem:

$$\min 5x_1 + 3x_2$$
$$\text{subject to: } 3x_1 + x_2 \le 27$$
$$2x_1 + 3x_2 \le 66$$
$$3x_1 + 2x_2 = 72$$
$$x_2 \ge 0$$
$$x_1 \in \mathbb{R}$$

To transform this LP to its standard form, we first add non-negative *slack* variables z_1 and z_2 that will transform the two inequality constraints into equality constraints:

$$\min 5x_1 + 3x_2$$
$$\text{subject to: } 3x_1 + x_2 + z_1 = 27$$
$$2x_1 + 3x_2 + z_2 = 66$$
$$3x_1 + 2x_2 = 72$$
$$x_2, z_1, z_2 \ge 0$$
$$x_1 \in \mathbb{R}$$

Notice that the two mathematical programs are equivalent because if $3x_1 + x_2 < 27$, then $z_1 = 27 - (3x_1 + x_2)$ resulting in $3x_1 + x_2 + z_1 = 27$ and if $2x_1 + 3x_2 + z_2 < 66$, then $z_2 = 66 - (2x_1 + 3x_2)$ resulting in $2x_1 + 3x_2 + z_2 = 66$. It is also clear that if $3x_1 + x_2 = 27$, then $z_1 = 0$ and if $2x_1 + 3x_2 + z_2 = 66$, then $z_2 = 0$. By introducing non-negative slack variables, all constraints are now equality constraints.

The only remaining part is that variable x_1 can take any value in $\mathbb{R}$ even if all variables should be non-negative. To rectify this, we can replace x_1 by two variables y_1 and y_2 such that $x_1 = y_1 - y_2$ and $y_1, y_2 \ge 0$. Notice that if $y_1, y_2 \ge 0$ and $x_1 = y_1 - y_2$, then $y_1 = \max(0, x_1)$ and $y_2 = \max(0, -x_2)$. Translating also the LP to a maximization problem by multiplying the objective function by -1 results in its standard form:

$$\max \; -5(y_1 - y_2) - 3x_2$$
$$\text{subject to: } 3(y_1 - y_2) + x_2 + z_1 = 27$$
$$2(y_1 - y_2) + 3x_2 + z_2 = 66$$
$$3(y_1 - y_2) + 2x_2 = 72$$
$$x_2, y_1, y_2, z_1, z_2 \geq 0$$

We can now set the vector of variables $\mathbf{x} = [y_1, y_2, x_2, z_1, z_2]^{\mathsf{T}}$, vector $\mathbf{c} = [-5, 5, -3, 0, 0]^{\mathsf{T}}$, vector $\mathbf{b} = [27, 66, 72]^{\mathsf{T}}$ and matrix

$$\mathbf{A} = \begin{bmatrix} 3 & -3 & 1 & 1 & 0 \\ 2 & -2 & 3 & 0 & 1 \\ 3 & -3 & 2 & 0 & 0 \end{bmatrix}$$

in order to cast the LP as:

$$\max_{\mathbf{x}} \mathbf{c}^{\mathsf{T}}\mathbf{x} \quad \text{subject to: } \mathbf{Ax} = \mathbf{b}, \; \mathbf{x} \geq \mathbf{0} \tag{4.51}$$

This formulation is known as the *primal* problem to distinguish it from its *dual* that we will discuss later. Our standard form LP has a convex feasible region, a convex/concave objective function and has equality and inequality constraints. Thus, to find its optimal solution it suffices to find a solution that satisfies the *first-order necessary* conditions for multivariate constrained optimization problems with equality and inequality constraints. These conditions consist of the KKT conditions and the linear independence of the active constraint gradients (LICQ test). The latter is guaranteed because all constraints are linear. Let us examine the KKT conditions. If we have l equality constraints, then we have KKT multipliers $\boldsymbol{\lambda} = [\lambda_1, \dots \lambda_l]^{\mathsf{T}}$ for the equality constraints. If we have n variables, then we also have n inequality constraints that force every variable to be non-negative. That is, we have additional KKT multipliers $\boldsymbol{\mu} = [\mu_1, \dots, \mu_n]^{\mathsf{T}}$ corresponding to constraints $-\mathbf{x} \leq \mathbf{0}$. We can now write the Lagrangian function, where we subtract the constraint functions from the objective function because we solve a maximization problem:

$$\mathcal{L}(\mathbf{x}, \boldsymbol{\lambda}, \boldsymbol{\mu}) := \mathbf{c}^{\mathsf{T}}\mathbf{x} - \boldsymbol{\lambda}^{\mathsf{T}}(\mathbf{Ax} - \mathbf{b}) - \boldsymbol{\mu}^{\mathsf{T}}(-\mathbf{x}) = \mathbf{c}^{\mathsf{T}}\mathbf{x} - \boldsymbol{\lambda}^{\mathsf{T}}(\mathbf{Ax} - \mathbf{b}) + \boldsymbol{\mu}^{\mathsf{T}}\mathbf{x}$$

notice that we have the multiplication of matrices with the following dimensions:

$$\underset{1 \times n}{\mathbf{c}^{\mathsf{T}}} \; \underset{n \times 1}{\mathbf{x}} - \underset{1 \times l}{\boldsymbol{\lambda}^{\mathsf{T}}}(\underset{l \times n}{\mathbf{A}} \; \underset{n \times 1}{\mathbf{x}} - \underset{l \times 1}{\mathbf{b}}) + \underset{1 \times n}{\boldsymbol{\mu}^{\mathsf{T}}}\underset{n \times 1}{\mathbf{x}}$$

The scalar-by-vector derivative (gradient) with respect to $\mathbf{x}$ when using the *denominator* layout notation is:

$$\nabla_\mathbf{x} \mathcal{L}(\mathbf{x}, \lambda, \mu) = \mathbf{0} \Rightarrow \nabla_\mathbf{x} \mathbf{c}^\mathsf{T} \mathbf{x} - \nabla_\mathbf{x}(\lambda^\mathsf{T} \mathbf{A}\mathbf{x} - \mathbf{b}) + \nabla_\mathbf{x} \mu^\mathsf{T} \mathbf{x} = \mathbf{0}$$
$$\Rightarrow \mathbf{c} - \nabla_\mathbf{x} \lambda^\mathsf{T} \mathbf{A}\mathbf{x} + \mu = \mathbf{0} \Rightarrow \mathbf{c} - \mathbf{A}^\mathsf{T}\lambda + \mu = \mathbf{0}$$

The dual problem of the primal LP in standard form is:

DUAL PROBLEM: $\quad\displaystyle \min_{\lambda, \mu} \sup_\mathbf{x} \ (\mathbf{c}^\mathsf{T}\mathbf{x} - \lambda^\mathsf{T}(\mathbf{A}\mathbf{x} - \mathbf{b}) + \mu^\mathsf{T}\mathbf{x})$

$\qquad\qquad\qquad\qquad\quad$ s.t.: $\mu \geq \mathbf{0}$ $\hfill$ (4.52)

where:

$$\mathbf{c}^\mathsf{T}\mathbf{x} - \lambda^\mathsf{T}(\mathbf{A}\mathbf{x} - \mathbf{b}) + \mu^\mathsf{T}\mathbf{x} = \lambda^\mathsf{T}\mathbf{b} + (\mathbf{c}^\mathsf{T} - \lambda^\mathsf{T}\mathbf{A} + \mu^\mathsf{T})\mathbf{x}$$

and thus:

$$\sup_{\mathbf{x} \in \mathbb{R}^n} \ (\mathbf{c}^\mathsf{T}\mathbf{x} - \lambda^\mathsf{T}(\mathbf{A}\mathbf{x} - \mathbf{b}) + \mu^\mathsf{T}\mathbf{x}) = \begin{cases} \lambda^\mathsf{T}\mathbf{b} & \text{if } \mathbf{c}^\mathsf{T} - \lambda^\mathsf{T}\mathbf{A} + \mu^\mathsf{T} = \mathbf{0}^\mathsf{T} \Leftrightarrow \mathbf{c} - \mathbf{A}^\mathsf{T}\lambda + \mu = \mathbf{0} \\ +\infty & \text{otherwise.} \end{cases}$$

In addition, because λ and $\mathbf{b}$ are vectors in $\mathbb{R}^m$, $\lambda^\mathsf{T}\mathbf{b} = \mathbf{b}^\mathsf{T}\lambda$. Thus, to find the *tightest* upper bound of our primal LP we need to solve the dual problem:

$$\min_{\lambda, \mu} \ \mathbf{b}^\mathsf{T}\lambda$$
$$\text{s.t.: } \mathbf{c} - \mathbf{A}^\mathsf{T}\lambda + \mu = \mathbf{0} \qquad (4.53)$$
$$\mu \geq \mathbf{0}$$

The KKT conditions of the primal LP are:

$$\begin{array}{ll} \nabla_\mathbf{x}\mathcal{L}(\mathbf{x}, \lambda, \mu) = \mathbf{0} \Rightarrow \mathbf{c} - \mathbf{A}^\mathsf{T}\lambda + \mu = \mathbf{0} & \text{stationarity} \\ \mathbf{A}\mathbf{x} = \mathbf{b} & \text{primal feasibility} \\ \mathbf{x} \geq \mathbf{0} & \text{primal feasibility} \\ \mu \geq \mathbf{0} & \text{dual feasibility} \\ \mu^\mathsf{T}\mathbf{x} = 0 & \text{complementary slackness} \end{array}$$

Let $(\mathbf{x}^*, \lambda^*, \mu^*)$ be a feasible solution that satisfies the KKT conditions. From stationarity we have that:

$$\mathbf{c} = \mathbf{A}^\mathsf{T}\lambda^* - \mu^*$$

Multiplying both sides by $\mathbf{x}^{*\mathsf{T}}$ we have that:

$$\mathbf{x}^{*\mathsf{T}}\mathbf{c} = \mathbf{x}^{*\mathsf{T}}(\mathbf{A}^\mathsf{T}\lambda^* - \mu^*) = \mathbf{x}^{*\mathsf{T}}\mathbf{A}^\mathsf{T}\lambda^* = (\mathbf{A}\mathbf{x})^\mathsf{T}\lambda^* = \mathbf{b}^\mathsf{T}\lambda^*$$

Because $\mathbf{c}, \mathbf{x}$ are vectors in $\mathbb{R}^n$ we have that $\mathbf{c}^\mathsf{T}\mathbf{x}^* = \mathbf{x}^{*\mathsf{T}}\mathbf{c}$. That is, $\mathbf{c}^\mathsf{T}\mathbf{x}^* = \mathbf{b}^\mathsf{T}\lambda^*$ meaning that if the primal LP has a finite optimal solution, the objective function value of the *primal* is equal to the optimal objective function value of the *dual* and the

Table 4.2 Rules to transform a maximization primal LP to its dual

PRIMAL (maximize)	DUAL (minimize)
ith constraint ≤ 0	λ_i variable ≥ 0
ith constraint ≥ 0	λ_i variable ≤ 0
ith constraint $= 0$	λ_i variable $\in \mathbb{R}$ (unrestricted)
x_j variable ≥ 0	jth constraint $\geq$
x_j variable ≤ 0	jth constraint $\leq$
x_j variable unrestricted	jth constraint $=$

duality gap is zero. This was expected because any feasible linear program satisfies the refined Slater's conditions meaning that *Strong Duality* holds.

Let us now consider the *dual* problem of the primal LP expressed in Eq. (4.53). Observe that μ play the role of slack variables and can be dropped to result in the more compact formulation:

$$\min_{\lambda} \ \mathbf{b}^{\mathsf{T}}\lambda$$
$$\text{s.t.: } \mathbf{c} - \mathbf{A}^{\mathsf{T}}\lambda \leq \mathbf{0} \tag{4.54}$$

To summarize, a primal LP in standard form:

$$\text{PRIMAL LP}: \quad \max_{\mathbf{x} \in \mathbb{R}^n} \mathbf{c}^{\mathsf{T}}\mathbf{x}, \text{ subject to: } \mathbf{A}\mathbf{x} = \mathbf{b}, \ \mathbf{x} \geq \mathbf{0}$$

has the dual:

$$\text{DUAL LP}: \quad \min_{\lambda \in \mathbb{R}^l} \mathbf{b}^{\mathsf{T}}\lambda, \text{ subject to: } \mathbf{A}^{\mathsf{T}}\lambda \geq \mathbf{c}$$

Observe that the primal LP is the dual of the dual LP and the *dual* LP is derived from the *primal* LP in standard form such that:

- Each variable in the primal LP is a constraint in the dual LP;
- Each constraint in the primal LP is a variable in the dual LP;
- The objective direction changes (from maximization to minimization).

If the primal LP is not written in its standard form, we can apply the principles of Table 4.2 to find its dual.

There are two main properties pertaining to the primal and dual LPs (the *weak* duality and the *strong* duality). These are presented below.

Theorem 4.6 *Weak Duality: for any feasible solutions x and y of the primal maximization LP and its dual minimization LP, we have that $c^{\mathsf{T}}x \leq b^{\mathsf{T}}\lambda$.*

Proof Without loss of generality, consider a primal LP in its standard form $\max \mathbf{c}^{\mathsf{T}}\mathbf{x}$ subject to $\mathbf{A}\mathbf{x} = \mathbf{b}, \mathbf{x} \geq \mathbf{0}$. This has the dual $\min \mathbf{b}^{\mathsf{T}}\lambda$ subject to $\mathbf{A}^{\mathsf{T}}\lambda \geq \mathbf{c}$.

Consider $\mathbf{x}$ to be a feasible solution of the primal and λ a feasible solution of the dual. Because $\mathbf{x} \geq \mathbf{0}$, from the inequality constraint of the dual $\mathbf{A}^\mathsf{T}\lambda \geq \mathbf{c}$ follows that $\mathbf{x}^\mathsf{T}\mathbf{A}^\mathsf{T}\lambda \geq \mathbf{x}^\mathsf{T}\mathbf{c}$. In addition, because $\mathbf{c}$ and $\mathbf{x}$ are n-valued column vectors, we have that $\mathbf{c}^\mathsf{T}\mathbf{x} = \mathbf{x}^\mathsf{T}\mathbf{c}$. This results in:

$$\mathbf{x}^\mathsf{T}\mathbf{A}^\mathsf{T}\lambda \geq \mathbf{c}^\mathsf{T}\mathbf{x}$$

Finally, $\mathbf{x}^\mathsf{T}\mathbf{A}^\mathsf{T}\lambda = (\mathbf{A}\mathbf{x})^\mathsf{T}\lambda = \mathbf{b}^\mathsf{T}\lambda$. This results in $\mathbf{c}^\mathsf{T}\mathbf{x} \leq \mathbf{b}^\mathsf{T}\lambda$. $\square$

It follows from the weak duality theorem that the objective function value of the dual LP at any feasible solution is always an upper bound on the objective function value of the primal LP. This has the following implications:

Corollary 4.2 *Unboundedness of Primal LP: if the optimal solution of the primal LP has an objective function value of $c^\mathsf{T}x = +\infty$ (unbounded from above), then the dual problem is infeasible because $c^\mathsf{T}x \leq b^\mathsf{T}\lambda$ cannot be satisfied.*

Corollary 4.3 *Infeasibility of Primal LP: if the optimal solution of the dual LP has an objective function value of $b^\mathsf{T}\lambda = -\infty$ (unbounded from below), then the primal problem is infeasible because $c^\mathsf{T}x \leq b^\mathsf{T}\lambda$ cannot be satisfied.*

Finally, because LPs satisfy the refined Slater's condition, strong duality always holds for LPs—with the exception of infeasible problems. The strong duality theorem states that:

Theorem 4.7 *Strong Duality: if the primal has a finite optimal solution x^* or the dual has a finite optimal solution λ^*, then $c^\mathsf{T}x^* = b^\mathsf{T}\lambda^*$ meaning that their objective function values are equal.*

The proof of the strong duality was already provided when discussing the KKT conditions of the primal LP. From the strong duality theorem follows that if we find feasible solutions $\mathbf{x}^*$ and λ^* such that $\mathbf{c}^\mathsf{T}\mathbf{x}^* = \mathbf{b}^\mathsf{T}\lambda^*$, then these are the optimal solutions for our primal and dual LPs, respectively. That is, if the primal has a finite optimal solution then the dual has also a finite optimal solution, and their objective function values are equal.

It is also worth noting that if our primal LP is a minimization and not a maximization problem, we should make some sign modifications when converting it to its dual. These are presented in Table 4.3.

Note that the *weak duality* theorem when our primal is a minimization problem states that the objective function of the dual LP at any feasible solution is always a lower bound on the objective function of the primal LP, that is $\mathbf{c}^\mathsf{T}\mathbf{x} \geq \mathbf{b}^\mathsf{T}\lambda$.

With this theoretical background, we proceed to the description of the simplex solution method for linear programming.

Table 4.3 Rules to transform a minimization primal LP to its dual.

PRIMAL (minimize)	DUAL (maximize)
ith constraint ≤ 0	λ_i variable ≤ 0
ith constraint ≥ 0	λ variable ≥ 0
ith constraint $= 0$	λ variable $\in \mathbb{R}$ (unrestricted)
x_j variable ≥ 0	jth constraint $\leq$
x_j variable ≤ 0	jth constraint $\geq$
x_j variable unrestricted	jth constraint $=$

4.5.1 Simplex

The simplex method was proposed by Dantzig in 1947 and it is the most commonly used method for linear programming. To apply simplex, the LP must be in its standard form:

$$\max \ \mathbf{c}^\mathsf{T}\mathbf{x}$$
$$\text{s.t. } \mathbf{A}\mathbf{x} = \mathbf{b}$$
$$\mathbf{x} \geq \mathbf{0}$$

and matrix $\mathbf{A} \in \mathbb{R}^{l \times n}$ must have *full row rank*, meaning that $\text{rank}(\mathbf{A}) = l$. Simplex is based on the following observations that were discussed when we introduced the graphical approach of solving LPs:

1. An LP is equivalent to minimizing a linear form $\mathbf{c}^\mathsf{T}\mathbf{x}$ over a polyhedral feasible region $\mathcal{P} = \{\mathbf{x} \in \mathbb{R}^n_+ \mid \mathbf{A}\mathbf{x} \leq \mathbf{b}\}$.
2. The maximum $\mathbf{x}^*$ of a linear form over a polyhedron $\mathcal{P}$ is attained at a vertex of the polyhedron.
3. Ensuring that a vertex of the polyhedron is a local optimum of the linear form requires checking the objective function values of its adjacent vertices and comparing them with each other.
4. Because any LP has a linear objective function which is convex and concave and a convex feasible region, the local optimum is also a global optimum.

Based on these observations, if the feasible region of an LP is not empty, simplex starts from a feasible vertex of the polyhedron and moves to an adjacent vertex with a higher objective function value. If we reach a vertex that does not have adjacent vertices with higher objective function values, simplex terminates and declares this vertex as global maximum [9]. The advantage of simplex is that it only moves from one vertex to a better vertex, thereby (potentially) skipping large numbers of suboptimal vertices.

If the LP is feasible and bounded, each iteration generated by simplex returns a basic feasible solution (bfs) of the primal LP, which is a vertex of the polyhedron. We should state here that even it is called a basic feasible solution, this solution is not optimal - it is just a feasible point. Before introducing the bfs, we first focus on the constraints $\mathbf{Ax} = \mathbf{b}$. Because simplex is applicable to problems with *full row rank* matrix $\mathbf{A} \in \mathbb{R}^{l \times n}$, rank$(\mathbf{A}) = l \le n$. If $\mathbf{a}_1, \ldots, \mathbf{a}_n$ are the n column vectors of $\mathbf{A}$, each linearly independent subset of l vectors from the set $\mathbf{a}_1, \ldots, \mathbf{a}_n$ is a basis $\mathbf{B}$ of the vector space represented by matrix $\mathbf{A}$. Consider, for instance, the two following equality constraints:

$$x_1 + x_2 + x_3 = 40$$
$$2x_1 + x_2 + x_4 = 60$$

resulting in full row rank matrix $\mathbf{A} = \begin{bmatrix} 1 & 1 & 1 & 0 \\ 2 & 1 & 0 & 1 \end{bmatrix}$. The dimensions of this matrix are $l = 2$ and $n = 4$. Let us denote its column vectors as $\mathbf{a}_1, \ldots, \mathbf{a}_4$. Selecting two linearly independent vectors from this set of column vectors, let us say $\mathbf{a}_1, \mathbf{a}_4$, forms a basis of $\mathbf{A}$, denoted as $\mathbf{B} = [\mathbf{a}_1, \mathbf{a}_4]$. $\mathbf{B}$ is a basis of $\mathbf{A}$ because any column vector $\mathbf{a}_1, \mathbf{a}_2, \mathbf{a}_3, \mathbf{a}_4$ in $\mathbf{A}$ can be written as a linear combination of the column vectors in $\mathbf{B}$ (that is, $\mathbf{B}$ spans $\mathbf{A}$). For instance, $\mathbf{a}_2 = \beta_1 \mathbf{a}_1 + \beta_4 \mathbf{a}_4$ for scalars $\beta_1 = 1$ and $\beta_4 = -1$, and $\mathbf{a}_3 = \beta_1 \mathbf{a}_1 + \beta_4 \mathbf{a}_4$ for scalars $\beta_1 = 1$ and $\beta_4 = -2$. Notice that matrix $\mathbf{A}$ might have several bases, but all of them should have the same number of column vectors.

If we define $\mathcal{B}$ as the set of indexes of the column vectors that belong to the basis $\mathbf{B}$ (in the above example, $\mathcal{B} = \{1, 4\}$), then we have the following definition of basic and nonbasic variables for a given basis $\mathbf{B}$ of an LP in standard form.

Definition 4.12 For a given basis $\mathbf{B}$ of an LP in standard form, the l variables $\mathbf{x}_\mathcal{B} = \{x_i \mid i \in \mathcal{B}\}$ corresponding to the indices $i \in \mathcal{B}$ of the column vectors in the basis are called *basic variables*. The remaining $n - l$ variables $\mathbf{x}_\mathcal{N} = \{x_i \mid i \in \mathcal{N}\}$ where $\mathcal{N} = \{1, 2, \ldots, n\} \setminus \mathcal{B}$ are called *nonbasic variables*.

The basic and nonbasic variables of a basis are used to find a basic solution to the LP.

Definition 4.13 *Basic Solution.* The *basic solution* of an LP in standard form for a given basis $\mathbf{B}$ is a solution $\mathbf{x}$ for which:

- The $n - l$ nonbasic variables $\mathbf{x}_\mathcal{N} = \{x_i \mid i \in \mathcal{N}\}$ are set equal to 0.
- The remaining variables $\mathbf{x}_\mathcal{B}$ assume unique values by solving $\mathbf{Ax} = \mathbf{b}$.

Observe that if we set all nonbasic variables to zero, $\mathbf{x}_\mathcal{N} = \mathbf{0}$, then we can obtain the unique values of the basic variables by solving the reduced system:

$$\mathbf{Bx}_\mathcal{B} = \mathbf{b}$$

instead of $\mathbf{Ax} = \mathbf{b}$. The values of $\mathbf{x}_\mathcal{B}$ are unique because, according to the Invertible Matrix Theorem (see 4.5), for symmetric matrix $\mathbf{B}$ with linear independent columns, $\mathbf{Bx}_\mathcal{B} = \mathbf{b}$ has a unique solution for each $\mathbf{b} \in \mathbb{R}^l$. In addition, from the same theorem, matrix **B** is invertible (nonsingular). By multiplying both sides of the equation by the inverse of the nonsingular **B**, we have:

$$\mathbf{x}_\mathcal{B} = \mathbf{B}^{-1}\mathbf{b}$$

By definition, a *basic solution* satisfies only the equality constraints $\mathbf{Ax} = \mathbf{b}$ of the LP, and thus it might be infeasible since $\mathbf{x} \geq 0$ is not guaranteed. This differentiates it from a basic feasible solution, defined below.

Definition 4.14 *Basic Feasible Solution*. The basic solution $\mathbf{x}_\mathcal{B} = \mathbf{B}^{-1}\mathbf{b}$ and $\mathbf{x}_N = \mathbf{0}$ of an LP for a given basis **B** which satisfies $\mathbf{x}_\mathcal{B} \geq \mathbf{0}$ is a *basic feasible solution (bfs)*.

From the definition of basic feasible solutions (bfs) it follows that:

- A bfs has at most l nonzero-valued variables and at least $n - l$ zero-valued variables.
- An LP in standard form has *at most* $\binom{n}{l}$ basic feasible solutions because from n variables a set of l basic variables can be chosen in $\binom{n}{l} = \frac{n!}{(n-l)!l!}$ different ways.

> **Time Complexity**

Because the potential number of basic feasible solutions is

$$\binom{n}{l} = \frac{n!}{(n-l)!l!}$$

and, at the worst-case, simplex might need to visit them all, simplex is not a polynomial time algorithm. For this reason, linear programming was considered as a nondeterministic polynomial time problem until the development of the Ellipsoid method of Khachiyan [10] in 1979 and the *interior point method* of Karmarkar [11] in 1984, which are polynomial-time algorithms.

To find the basic feasible solutions of an LP we do not need any information from the objective function. We only use information from the constraints. Consider, for example, the following LP in its standard form, for which $\mathbf{A} = \begin{bmatrix} 1 & 1 & 1 & 0 \\ 2 & 1 & 0 & 1 \end{bmatrix}$.

$$
\begin{aligned}
\max \ & 15x_1 + 4x_2 \\
\text{s.t.:} \ & x_1 + x_2 + x_3 = 40 \\
& 2x_1 + x_2 + x_4 = 60 \\
& x_1, x_2, x_3, x_4 \geq 0
\end{aligned}
\tag{4.55}
$$

This problem has at most $\frac{4!}{(4-2)!2!} = 6$ bfs. To find all basic feasible solutions, we have to perform the following steps for each basis $\mathbf{B}$ that contains l linearly independent column vectors of $\mathbf{A} \in \mathbb{R}^{l \times n}$:

- Select the $n - l$ nonbasic variables of $\mathbf{B}$ and set them to 0. That is, $\mathbf{x}_N = \mathbf{0}$.
- Find the *basic solution* $\mathbf{x}_\mathcal{B} = \mathbf{B}^{-1}\mathbf{b}$.
- check if $\mathbf{x}_\mathcal{B} \geq \mathbf{0}$, which would mean that the *basic solution* is also a *basic feasible solution*.

Let us start with basis $\mathbf{B}$ of column vectors 1 and 2 of $\mathbf{A}$. That is, $\mathcal{B} = \{1, 2\}$ and $\mathcal{N} = \{3, 4\}$. The basis has $\det(\mathbf{B}) = -1 \neq 0$ and, according to the Invertible Matrix Theorem, its columns are linearly independent. We thus set $x_3 = x_4 = 0$. We can then find the basic solution for this case:

$$\mathbf{x}_\mathcal{B} = \mathbf{B}^{-1}\mathbf{b} = \begin{bmatrix} -1 & 1 \\ 2 & -1 \end{bmatrix}\begin{bmatrix} 40 \\ 60 \end{bmatrix} = \begin{bmatrix} 20 \\ 20 \end{bmatrix}$$

resulting in $x_1 = 20$, $x_2 = 20$. This basic solution is a bfs because $x_1, x_2, x_3, x_4 \geq 0$, thus satisfying the constraints of the LP.

Consider now $\mathcal{B} = \{1, 3\}$. This will give us basis matrix $\mathbf{B} = \begin{bmatrix} 1 & 1 \\ 2 & 0 \end{bmatrix}$ with linearly independent column vectors ($\det(\mathbf{B}) = -2 \neq 0$). Thus,

$$\mathbf{x}_\mathcal{B} = \mathbf{B}^{-1}\mathbf{b} = \begin{bmatrix} 0 & 1/2 \\ 1 & -1/2 \end{bmatrix}\begin{bmatrix} 40 \\ 60 \end{bmatrix} = \begin{bmatrix} 30 \\ 10 \end{bmatrix}$$

This basic solution is also feasible for the original LP because $x_1, x_2, x_3, x_4 \geq 0$. Thus, this basic solution is also a bfs.

For $\mathcal{B} = \{1, 4\}$ we have the system of equations $x_1 = 40$ and $2x_1 + x_4 = 60$ which gives us basic solution $x_1 = 40$, $x_4 = -20$. This basic solution is, however, infeasible for our LP because $x_4 < 0$. Thus, it is not a bfs. For $\mathcal{B} = \{2, 3\}$ we have the system of equations $x_2 + x_3 = 40$ and $x_2 = 60$ which gives us basic solution $x_2 = 60$, $x_3 = -20$. This solution is also infeasible for our LP because $x_3 < 0$. For $\mathcal{B} = \{2, 4\}$ we have $x_2 = 40$ and $x_4 = 20$, which is a bfs. Finally, for $\mathcal{B} = \{3, 4\}$ we have $x_3 = 40$ and $x_4 = 60$. This is a feasible solution because $x_1, x_2, x_3, x_4 \geq 0$.

Summarizing, we found all 4 basic feasible solutions of our LP from a total of 6 basic solutions. In this case it was possible to find all bfs because the size of the problem was small (we could have had at most 6 bfs). In larger problems, however, it will be very difficult to enumerate all possible bfs. Luckily, simplex requires as starting input only one bfs and it will generate from it adjacent basic feasible solutions in each iteration until reaching a bfs that does not have any adjacent bfs with better performance. That is, we are required to find just one bfs before applying simplex. Let us consider again the required properties of a bfs. Namely, a bfs should (i) have at least $n - l$ zero-valued variables, (ii) be a feasible solution to the original LP, (iii) have a basis $l \times l$ matrix $\mathbf{B} = [\mathbf{A}_i]$ for $i \in \mathcal{B}$ with linear independent column vectors. We can take advantage of these requirements to easily find an initial bfs by constructing a standard form LP formulation such that:

- For every equality constraint $i = \{1, \ldots, l\}$ the right-hand side value b_i is non-negative.
- In every equality constraint equation $i \in \{1, \ldots, l\}$ there is at least one variable x_j with coefficient $a_{ij} = 1$, and for this variable x_j its coefficient a_{kj} at any other equality constraint equation $k \in \{1, \ldots, l\} \setminus \{i\}$ is equal to 0.

If we are able to have such a standard LP formulation, we can consider as the *basic variable* of each equality constraint $i \in \{1, \ldots, l\}$ the respective variable x_j for which $a_{ij} = 1$ and $a_{kj} = 0$ at any other equality constraint equation $k \in \{1, \ldots, l\} \setminus \{1\}$. By doing so, our basis is formed by the l basic variables corresponding to the l equality constraints. The resulting basis will be an $l \times l$ matrix with a nonzero determinant because each column will have a different element equal to 1 resulting in no rows/columns being equal to 0 and no rows/columns being identical. According to the Invertible Matrix Theorem, this means that the columns of the basis are linearly independent. All other $n - l$ variables will be set to 0. We can use this basis to compute a basic solution, where each variable $x_j \in \mathcal{B}$ which is a basic variable to some equality constraint i will be equal to $x_j = b_i$ since the other basic variables have a 0 coefficient in this equality constraint and all nonbasic variables are set to 0. Having ensured that $\mathbf{b} \geq \mathbf{0}$ in our problem's formulation, will result in a basic solution $\mathbf{x}_{\mathcal{B}} \geq \mathbf{0}$ which is a bfs.

Notice that our standard form LP in (4.55) meets already these requirements because it has positive right-hand side values in the equality constraints ($b_1 = 40$ and $b_2 = 60$). In addition, there exists variable x_3 in the first equality constraint that has coefficient $a_{13} = 1$ and a coefficient of 0 at any other equality constraint (namely, $a_{23} = 0$ in constraint 2). Furthermore, there exists variable x_4 at the second equality constraint that has coefficient $a_{24} = 1$ and a coefficient of 0 at any other equality constraint (namely, $a_{14} = 0$). That is, x_3 can be the basic variable of the first constraint with value $x_3 = b_1 = 40$ and x_4 the basic variable of the second constraint with value $x_4 = b_2 = 60$. This will result in basis $\mathbf{B}$ with linearly independent columns (its determinant is $\det(\mathbf{B}) = \det\left(\begin{bmatrix} 1 & 0 \\ 0 & 1 \end{bmatrix}\right) = 1 \neq 0$). All other variables are moved to the set of nonbasic variables resulting in basic solution $[x_1, x_2, x_3, x_4]^\mathsf{T} = [0, 0, 40, 60]^\mathsf{T}$, which is a basic feasible solution. We will later see that if it is not easy to formulate our LP in such a way, we can add artificial variables by using the *two-phase simplex* method.

Geometrically, each bfs corresponds to a vertex of the polyhedron of feasible solutions. There is an interesting theorem on bfs stating that for each bfs there exists a cost vector $\mathbf{c} \in \mathbb{R}^n$ such that this bfs is the unique optimal solution of the LP. This is shown below.

Theorem 4.8 *For an LP with feasible region $\{x \in \mathbb{R}_{\geq 0}^n \mid Ax = b\}$ there exists a cost vector $c \in \mathbb{R}^n$ such that a bfs x' is the optimal solution when considering the objective function $c^\mathsf{T}x$.*

Proof It suffices to find a cost vector $\mathbf{c} \in \mathbb{R}^n$ for which $\mathbf{x}'$ is an optimal solution, meaning that $\mathbf{c}^\mathsf{T}\mathbf{x}' \geq \mathbf{c}^\mathsf{T}\mathbf{x}$ for any $\mathbf{x} \geq \mathbf{0} \mid \mathbf{A}\mathbf{x} = \mathbf{b}$ in the case of solving a maximization problem. As a bfs, $\mathbf{x}'$ can be split into $\mathbf{x}'_\mathcal{B}$ and $\mathbf{x}'_\mathcal{N}$, where $\mathbf{x}'_\mathcal{N} = \mathbf{0}$. Let us choose $\mathbf{c}$ such that $c_j = 0$ for all j in $\mathcal{B}$ and $c_j = -1$ for all j in $\mathcal{N}$. Then, $\mathbf{c}^\mathsf{T}\mathbf{x}'$ results in $\mathbf{c}_\mathcal{B}^\mathsf{T}\mathbf{x}' + \mathbf{c}_\mathcal{N}^\mathsf{T}\mathbf{x}' = \mathbf{0}^\mathsf{T}\mathbf{x}'_\mathcal{B} - \mathbf{1}^\mathsf{T}\mathbf{x}'_\mathcal{N} = 0$. For any other $\mathbf{x} \geq \mathbf{0} \mid \mathbf{A}\mathbf{x} = \mathbf{b}$ we have $\mathbf{c}^\mathsf{T}\mathbf{x} \leq 0$ since $\mathbf{x} \geq \mathbf{0}$ and $\mathbf{c} < \mathbf{0}$. Thus, $\mathbf{c}^\mathsf{T}\mathbf{x}' \geq \mathbf{c}^\mathsf{T}\mathbf{x}$ for any $\mathbf{x} \geq \mathbf{0} \mid \mathbf{A}\mathbf{x} = \mathbf{b}$.   □

This is an interesting theorem which states that a bfs might not be an optimal solution of an LP with a cost vector $\mathbf{c}$, but if the objective function of the LP changes, there exists a cost vector that can make this bfs optimal.

Let us now show that a bfs is a vertex of the LP's polyhedron.

Theorem 4.9 *A basic feasible solution of an LP with feasible region* $\mathcal{P} = \{\mathbf{x} \in \mathbb{R}^n_{\geq 0} \mid \mathbf{A}\mathbf{x} = \mathbf{b}\}$ *is a vertex of* $\mathcal{P}$.

Proof Let $\mathbf{x}$ be a basic feasible solution with $\mathcal{B} = \{\mathcal{B}_1, \dots, \mathcal{B}_l\}$ and $\mathcal{N} = \{1, \dots, n\} \setminus \mathcal{B} = \{\mathcal{N}_1, \dots, \mathcal{N}_{n-l}\}$. The basis matrix $\mathbf{B}$ is nonsingular and $\mathbf{x}_\mathcal{N} = \mathbf{0}$. Suppose that the bfs $\mathbf{x}$ is not a vertex of $\mathcal{P}$. This means that there exist two other points $\mathbf{y}$ and $\mathbf{z}$ in $\mathcal{P} \setminus \{\mathbf{x}\}$ for which $\mathbf{x}$ is a convex combination of them. Reckon that the *convex combination* of two points $\mathbf{y}$ and $\mathbf{z}$ is:

$$\beta_1\mathbf{y} + \beta_2\mathbf{z}, \text{ where } \begin{array}{l} \beta_1, \beta_2 \in \mathbb{R} \\ \beta_1 + \beta_2 = 1 \\ \beta_1, \beta_2 \geq 0 \end{array}$$

Then, we can find $\beta_1 \in [0, 1]$ for which $\mathbf{x} = \beta_1\mathbf{y} + (1 - \beta_1)\mathbf{z}$. We now have three cases.

Case 1: for $\beta_1 = 0$ we have $\mathbf{x} = \mathbf{z}$ meaning that $\mathbf{z}$ must be the same point as $\mathbf{x}$.

Case 2: for $\beta_1 = 1$ we have $\mathbf{x} = \mathbf{y}$ meaning that $\mathbf{y}$ must be the same point as $\mathbf{x}$.

Case 3: for $\beta_1 \in (0, 1)$ we have that β_1 and $1 - \beta_1$ are positive. We can write $\mathbf{x} = \beta_1\mathbf{y} + (1 - \beta_1)\mathbf{z}$ as $\begin{bmatrix} \mathbf{x}_\mathcal{B} \\ \mathbf{x}_\mathcal{N} \end{bmatrix} = \beta_1 \begin{bmatrix} \mathbf{y}_\mathcal{B} \\ \mathbf{y}_\mathcal{N} \end{bmatrix} + (1 - \beta_1) \begin{bmatrix} \mathbf{z}_\mathcal{B} \\ \mathbf{z}_\mathcal{N} \end{bmatrix}$ and because $\mathbf{x}_\mathcal{N} = \mathbf{0}$ we must have $y_i = z_i = 0$ for any $i \in \mathcal{N}$ to ensure that $\mathbf{x} = \beta_1\mathbf{y} + (1 - \beta_1)\mathbf{z}$. That would mean $\mathbf{x}_\mathcal{N} = \mathbf{y}_\mathcal{N} = \mathbf{z}_\mathcal{N} = \mathbf{0}$. We also know that all points must satisfy $\mathbf{A}\mathbf{x} = \mathbf{A}\mathbf{y} = \mathbf{A}\mathbf{z} = \mathbf{b}$ to be feasible. Because $\mathbf{y}_\mathcal{N} = \mathbf{z}_\mathcal{N} = \mathbf{0}$ that would mean $\mathbf{B}\mathbf{x}_\mathcal{B} = \mathbf{B}\mathbf{y}_\mathcal{B} = \mathbf{B}\mathbf{z}_\mathcal{B} = \mathbf{b}$. Because $\mathbf{B}$ is nonsingular, $\mathbf{B}^{-1}\mathbf{B}\mathbf{x}_\mathcal{B} = \mathbf{B}^{-1}\mathbf{B}\mathbf{y}_\mathcal{B} = \mathbf{B}^{-1}\mathbf{B}\mathbf{z}_\mathcal{B} = \mathbf{B}^{-1}\mathbf{b}$ resulting in $\mathbf{x}_\mathcal{B} = \mathbf{y}_\mathcal{B} = \mathbf{z}_\mathcal{B}$. That is, $\mathbf{x} = \mathbf{y} = \mathbf{z}$.

Concluding, we showed that there do not exist two feasible points $\mathbf{y}$ and $\mathbf{z}$ in $\mathcal{P} \setminus \{\mathbf{x}\}$ that are a convex combination of $\mathbf{x}$ because at least one of these two points must be the same as $\mathbf{x}$. Thus, we reached a contradiction and $\mathbf{x}$ is a vertex of $\mathcal{P}$.   □

Let us consider again the following LP example:

Fig. 4.6 Feasible
polyhedron of our example's
LP. Notice that it is bounded,
and thus it is also a polytope.
The bfs of the LP in standard
form are the vertices of the
polytope. Namely, (20, 20)
corresponds to the 1st bfs,
(30, 0) to the 2nd, (0, 40) to
the 3rd, and (0, 0) to the 4th

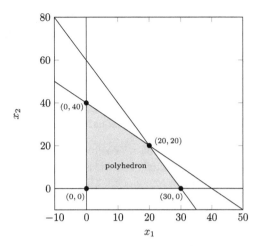

$$\max\ 15x_1 + 4x_2$$
$$\text{s.t.: } x_1 + x_2 + x_3 = 40$$
$$2x_1 + x_2 + x_4 = 60$$
$$x_1, x_2, x_3, x_4 \geq 0$$

that had the following four bfs: $[x_1, x_2, x_3, x_4]^\mathsf{T} = [20, 20, 0, 0]^\mathsf{T}$, $[x_1, x_2, x_3, x_4]^\mathsf{T} = [30, 0, 10, 0]^\mathsf{T}$, $[x_1, x_2, x_3, x_4]^\mathsf{T} = [0, 40, 0, 20]^\mathsf{T}$ and $[x_1, x_2, x_3, x_4]^\mathsf{T} = [0, 0, 40, 60]^\mathsf{T}$. In this LP x_3 and x_4 are slack variables which are used to write the LP in its standard form. Removing them results in the LP:

$$\max\ 15x_1 + 4x_2$$
$$\text{s.t.: } x_1 + x_2 \leq 40$$
$$2x_1 + x_2 \leq 60$$
$$x_1, x_2 \geq 0$$

which can be solved in 2 dimensions with the graphical method and has the poly-hedron $\mathcal{P} = \{\mathbf{x} \in \mathbb{R}^n_{\geq 0} \mid x_1 + x_2 \leq 40 \wedge 2x_1 + x_2 \leq 60\}$ as its feasible region. In Fig. 4.6 we see that this polyhedron has 4 vertices which are the four basic feasible solutions of the standard form LP without the values of the slack variables.

The intuitive interpretation of the values of the slack variables x_3, x_4 is that for the 1st bfs constraints $x_1 + x_2 \leq 40$ and $2x_1 + x_2 \leq 60$ are active, meaning that the left hand side and the right hand side have the same values, for the 2nd bfs constraint $2x_1 + x_2 \leq 60$ is inactive, for the 3rd bfs constraint $x_1 + x_2 \leq 40$ is inactive, and for the 4th bfs both constraints are inactive.

According to Bauer's maximum principle [12], a convex (i.e., linear) objective function defined on a set that is convex and compact attains its maximum at some extreme point of the set (vertex). Thus, if there exists an optimal solution of the LP, then there exists an optimal bfs. Taking advantage of this, simplex iterates among basic feasible solutions until finding the optimal basic feasible solution. We should note here that simplex might fail to find a solution when encountering the same bfs more than once since this might result in a never-ending loop, known as cycling [13]. This can occur when the LP is *degenerate*, where degeneracy is defined below.

Definition 4.15 Degenerate LP: An LP is *degenerate* if it has at least one bfs in which a *basic variable* is equal to 0 [14].

We note that a basic feasible solution that has a basic variable equal to zero is called *degenerate bfs*. Provided that we start from a bfs, simplex might fail if the LP is degenerate or unbounded. Thus, it follows that:

Corollary 4.4 *If an LP is nondegenerate and bounded, then simplex terminates at a basic feasible solution which is optimal.*

Simplex will move from a bfs to an adjacent bfs. The definition of adjacent basic feasible solutions is provided below.

Definition 4.16 For any LP in standard form with l equality constraints, two basic feasible solutions are adjacent if their sets of basic variables have $l - 1$ basic variables in common. That is, the basis **B** of a bfs differs from the basis of its adjacent bfs by exactly one column vector.

Simplex moves from the current bfs to an adjacent bfs at each iteration according to the procedure described in the remainder of this section.

Let us consider the following LP in its standard form.

$$\text{PRIMAL LP}: \quad \max_{\mathbf{x}} \ \sum_{j=1}^{n} c_j x_j$$

$$\text{subject to:} \ \sum_{j=1}^{n} a_{ij} x_j = b_i, \ \forall i \in \{1, \dots, l\}$$

$$x_j \geq 0, \ \forall j \in \{1, \dots, n\}$$

From this LP, we can consider a new variable z as the objective function value:

$$z = c_1 x_1 + c_2 x_2 + \cdots + c_n x_n$$

We then create a tableau where each row $1, \dots, l$ represents an equality constraint and row 0 represents the objective function written in the form

$$z - c_1 x_1 - c_2 x_2 - \cdots - c_n x_n = 0$$

The cells of the tableau have the values of the coefficients of the variables (see Table 4.4 where $b_0 = 0$). Note that at the last column of the tableau we present the

Table 4.4 Simplex Tableau

Row	z	x_1	x_2	...	x_n	rhs
0	1	$-c_1$	$-c_2$	...	$-c_n$	b_0
1	0	a_{11}	a_{12}	...	a_{1n}	b_1
2	0	a_{21}	a_{22}	...	a_{2n}	b_2
⋮	⋮	⋮	⋮	⋮	⋮	⋮
l	0	a_{l1}	a_{l2}	...	a_{ln}	b_l

values of the right-hand side (rhs) of each equality constraint. It is also important to observe that z acts as the *basic variable* of the constraint in row 0 because it has a coefficient of 1 in this row and of 0 at any other row.

Ignoring the column related to z, we can write the simplex tableau more succinctly as: $\begin{bmatrix} -\mathbf{c}^\mathsf{T} & b_0 \\ \mathbf{A} & \mathbf{b} \end{bmatrix}$

As we will later see, the coefficient values at the tableau will change with the simplex iterations following a procedure called *pivoting* to move from a bfs to an adjacent bfs with a better performance.

Let us assume that we have an initial bfs with l basic variables in $\mathcal{B}$ and $n - l$ nonbasic variables in $\mathcal{N}$. Given the values of the variables $\mathbf{x} = [\mathbf{x}_\mathcal{B}, \mathbf{x}_\mathcal{N}]^\mathsf{T}$ for this bfs, we can compute the objective function value z. In the next iteration, simplex checks if the value of z can increase by moving to an adjacent bfs. To find an adjacent bfs, we have to replace a column vector $\mathbf{a}_i$ that belongs to our current basis $\mathbf{B}$ by another column vector $\mathbf{a}_j$, where $j \in \mathcal{N}$. The question is which column vector $\mathbf{a}_j$ to choose to enter our basis. The strategy is to choose the one that has the highest potential to increase the objective function value. If there is a nonbasic variable $x_j \in \mathcal{N}$ that has a negative coefficient $-c_j < 0$ in row 0, then if the current value of $x_j = 0$ increases the objective function z will also increase by $c_j x_j$. If there are more nonbasic variables in row 0 with a negative coefficient $-c_j$, then we can select the nonbasic variable x_q with the most negative coefficient. That is, $-c_q < 0$ and $-c_q \leq -c_j$ for all $j \in \mathcal{N}$.

The nonbasic variable x_q which is selected is called *entering variable* because its column vector will *enter the basis* at this simplex iteration. To complete the construction of the adjacent bfs, we need to find which column vector will leave the basis. This is done by the so-called *ratio test*. When increasing the value of x_q, other basic variables would have to change their values to maintain feasibility. In fact, we can increase the value of x_q as long as no basic variable becomes negative. Because of this, we use the ratio test to compute the maximum value of x_q such that:

$$x_q = \min_{i \in \{1, \dots, l\} \mid a_{iq} > 0} \left\{ \frac{b_i}{a_{iq}} \right\}$$

In more detail, in the ratio test we compute the ratio $\frac{b_i}{a_{iq}}$ for all constraint rows $i = \{1, 2, \dots, l\}$ for which the coefficient of x_q is positive ($a_{iq} > 0$) and we select the minimum value from all these ratio values. If row w is the row with the minimum ratio value, we call this row *privot row* and the procedure *pivoting*. From this row w, we check which basic variable x_p, where $p \in \mathcal{B}$, became equal to 0 when x_q took the value of $\frac{b_w}{a_{wq}}$. This basic variable is the *leaving variable* and its column vector will be removed to form the adjacent bfs.

The new basis $\mathbf{B}$ can be used to update the tableau $\begin{bmatrix} -\mathbf{c}^\mathsf{T} & b_0 \\ \mathbf{A} & \mathbf{b} \end{bmatrix}$ with the use of *elementary row operations* (EROs) that can make the coefficient of x_q take the value of 1 in row w and the value of 0 at any other row. This will make x_q the basic variable of the w-th equality constraint. These operations can be described as follows:

1. Set $\bar{a}_{wq} = a_{wq}$ and for any $j \in \{1, \ldots, n\}$ set $a_{wj} = \frac{a_{wj}}{\bar{a}_{wq}}$. Set also $b_w = \frac{b_w}{\bar{a}_{wq}}$. This will ensure that the coefficient of x_q in row w becomes $a_{wq} = 1$ and all other coefficients values in that row are updated accordingly.
2. For any other row $i = \{1, \ldots, l\} \setminus \{w\}$ for which $a_{iq} \neq 0$ set $\bar{a}_{iq} = a_{iq}$. Then, set $a_{ij} = a_{ij} - a_{wj}\bar{a}_{iq}$ for $j \in \{1, \ldots, n\}$ and $b_i = b_i - b_w\bar{a}_{iq}$. This will ensure that a_{iq} takes the value of 0 at any other row because $a_{iq} = a_{iq} - a_{wq}\bar{a}_{iq} = 0$ and it will update all other coefficient values accordingly.
3. Using the new values of matrix $\mathbf{A}$ and setting $\bar{c}_q = c_q$, update each $-c_j$ in $\mathbf{c}$ by $-c_j = -c_j + a_{wj}\bar{c}_q$ and b_0 by $b_0 = b_0 + b_w\bar{c}_q$.

Note that pivoting steps 1 and 2 can be also performed by replacing matrix $\mathbf{A}$ by $\mathbf{B}^{-1}\mathbf{A}$ and vector $\mathbf{b}$ by $\mathbf{B}^{-1}\mathbf{b}$.

By completing this procedure, we have updated our tableau and we can move to the next iteration. If in the next iteration all updated coefficients $-c_j$ for $j \in \mathcal{N}$ are non-negative, simplex stops because we have found our optimal bfs since increasing the value of any nonbasic variable will not increase the value of z. It is important to note that if for an *entering variable* x_q in one of the simplex iterations all its constraint coefficients a_{iq} in rows $i = 1, \ldots, l$ have nonpositive values ($a_{iq} \leq 0, \forall i \in \{1, \ldots, l\}$), then the LP is *unbounded* and we should terminate simplex before continuing any further.

Algorithm 14 formalizes this process.

It is important to note that although we select *entering variable* x_q that has the most negative coefficient $-c_j$ in row 0 among all nonbasic variables, this is not a strict requirement. We can select any nonbasic variable with a negative coefficient $-c_j$ and this will still allow simplex to converge to the optimal bfs, if such bfs exists. Let us consider the implementation of simplex in the following example (Table 4.5).

$$
\begin{aligned}
\max \; & 70x_1 - 8x_2 + 5x_3 - 4x_4 \\
\text{s.t.: } & 4x_1 + 8x_2 - 5x_3 + x_4 = 6 \\
& 5x_1 + 2x_2 + 3x_3 + x_5 = 4 \\
& 2x_1 + 7x_2 + 9x_3 + x_6 = 7 \\
& x_2 + x_7 = 5 \\
& x_1, x_2, x_3, x_4, x_5, x_6, x_7 \geq 0
\end{aligned}
$$

Algorithm 14 Simplex for maximization LPs in standard form

1: find initial bfs with basis $\mathbf{B}$ resulting in sets $\mathcal{B}$ and $\mathcal{N} = \{1, 2, \ldots, n\} \setminus \mathcal{B}$
2: **repeat**
3:   find *entering variable* x_q that has the most negative coefficient $-c_j$ of all $j \in \mathcal{N}$

4:   **if** all coefficients a_{iq} for $i = 1, \ldots, l$ are nonpositive **then**
5:     **terminate** because the LP is *unbounded*
6:   **end if**
7:   set $x_q = \min_{i \in \{1, \ldots, m\} \mid a_{iq} > 0} \left\{ \frac{b_i}{a_{iq}} \right\}$
8:   set pivot row $w = arg\ \min_{i \in \{1, \ldots, m\} \mid a_{iq} > 0} \left\{ \frac{b_i}{a_{iq}} \right\}$
9:   find $p \in \mathcal{B}$ in pivot row w for which $x_p = 0$ and declare it as the *leaving variable*.
     If there are more than one, the problem is *degenerate*.
10:   add q to $\mathcal{B}$ in the place of p, which is moved to $\mathcal{N}$
11:   set $\bar{a}_{wq} = a_{wq}, \bar{c}_q = c_q$, and $\bar{a}_{iq} = a_{iq}$
12:   for any $j \in \{1, \ldots, n\}$ set $a_{wj} = \frac{a_{wj}}{\bar{a}_{wq}}$. Set also $b_w = \frac{b_w}{\bar{a}_{wq}}$
13:   **for** for any other row $i = \{1, \ldots, l\} \setminus \{w\}$ for which $\bar{a}_{iq} \neq 0$ **do**
14:     set $a_{ij} = a_{ij} - a_{wj}\bar{a}_{iq}$ for any $j \in \{1, \ldots, n\}$. Set also $b_i = b_i - b_w\bar{a}_{iq}$
15:   **end for**
16:   set $-c_j = -c_j + a_{wj}\bar{c}_q$ for any $j \in \{1, \ldots, n\}$ and $b_0 = b_0 + b_w\bar{c}_q$
17: **until** all coefficient values $-c_j$ with $j \in \mathcal{N}$ are nonnegative

---

Table 4.5 Simplex Tableau

Row	z	x_1	x_2	x_3	x_4	x_5	x_6	x_7	Rhs	Basic variable
0	1	−70	8	−5	4	0	0	0	$b_0 = 0$	z
1	0	4	8	−5	1	0	0	0	$b_1 = 6$	x_4
2	0	5	2	3	0	1	0	0	$b_2 = 4$	x_5
3	0	2	7	9	0	0	1	0	$b_3 = 7$	x_6
4	0	0	1	0	0	0	0	1	$b_4 = 5$	x_7

Notice that x_4 is a basic variable for constraint 1, x_5 for constraint 2, x_6 for constraint 3, and x_7 for constraint 4 because they have a coefficient equal to 1 at these equality constraints and 0 at any other. In addition, the rhs of every equality constraint is positive. Thus, we can easily obtain the bfs with $\mathcal{B} = \{4, 5, 6, 7\}$ with $x_4 = 6, x_5 = 4, x_6 = 7, x_7 = 5$ and $\mathcal{N} = \{1, 2, 3\}$. From the problem formulation, we have the tableau of Table 4.5.

Reckon that z is considered to be a basic variable in row 0 because its coefficient is 1 in that row and 0 at any other row. From the nonbasic variables $\mathcal{N} = \{x_1, x_2, x_3\}$ the one with the most negative coefficient in row 0 is x_1 because $-c_1 = -70$. Thus, x_1 is selected as the entering variable ($q = 1$). According to the ratio test, the value of x_1 will increase from 0 to:

Table 4.6 Simplex Tableau after iteration 1

Row	z	x_1	x_2	x_3	x_4	x_5	x_6	x_7	Rhs	Basic variable
0	1	0	36	37	4	14	0	0	$b_0 = 56$	z
1	0	0	6.4	−7.4	1	−0.8	0	0	$b_1 = 2.8$	x_4
2	0	1	0.4	0.6	0	0.2	0	0	$b_2 = 0.8$	x_1
3	0	0	6.2	7.8	0	−0.4	1	0	$b_3 = 5.4$	x_6
4	0	0	1	0	0	0	0	1	$b_4 = 5$	x_7

$$x_1 = \min_{i \in \{1,2,3\}} \left\{ \frac{b_i}{a_{i1}} \right\} = \min \left\{ \frac{6}{4}, \frac{4}{5}, \frac{7}{2} \right\} = \frac{4}{5}$$

In this ratio test we did not include row 4 because a_{41} is not greater than 0 in row 4. It follows that $w = 2$ is the pivot row and x_1 will replace x_5, meaning that $q = 1$ will enter $\mathcal{B}$ and $p = 5$ will leave $\mathcal{B}$. This results in the new basis:

$$\mathbf{B} = \begin{bmatrix} 1 & 4 & 0 & 0 \\ 0 & 5 & 0 & 0 \\ 0 & 2 & 1 & 0 \\ 0 & 0 & 0 & 1 \end{bmatrix}$$

for which:

$$\mathbf{A} = \mathbf{B}^{-1}\mathbf{A} = \begin{bmatrix} 0 & 6.4 & -7.4 & 1 & -0.8 & 0 & 0 \\ 1 & 0.4 & 0.6 & 0 & 0.2 & 0 & 0 \\ 0 & 6.2 & 7.8 & 0 & -0.4 & 1 & 0 \\ 0 & 1 & 0 & 0 & 0 & 0 & 1 \end{bmatrix}$$

and

$$\mathbf{b} = \mathbf{B}^{-1}\mathbf{b} = \begin{bmatrix} 1 & -0.8 & 0 & 0 \\ 0 & 0.2 & 0 & 0 \\ 0 & -0.4 & 1 & 0 \\ 0 & 0 & 0 & 1 \end{bmatrix} \begin{bmatrix} 6 \\ 4 \\ 7 \\ 5 \end{bmatrix} = \begin{bmatrix} 2.8 \\ 0.8 \\ 5.4 \\ 5 \end{bmatrix}$$

Notice that the updated coefficient values of tables $\mathbf{A}$ and $\mathbf{b}$ could have been also computed by performing the EROs described in lines 11–15 of Algorithm 14. In addition, we set $\bar{c}_q = c_q = 70$ and we update $-c_j = -c_j + a_{wj}\bar{c}_q$ for any $j = \{1, \ldots, n\}$ and $b_0 = b_0 + b_w\bar{c}_q$. This results in the updated tableau of Table 4.6.

Observe that coefficients $-c_j$ of all nonbasic variables in row 0 are non-negative. Namely, $-c_5 = 14$, $-c_2 = 36$, and $-c_3 = 37$. That is, we cannot increase the value of z by adding one of the nonbasic variables to our basis and our bfs is the optimal bfs. The objective function value of the optimal bfs is:

$$z + 36x_2 + 37x_3 + 4x_4 + 14x_5 = 56 \Rightarrow z = 56 - 4 \cdot 2.8 = 44.8$$

and simplex has terminated successfully with solution $\mathbf{x}^* = [0.8, 0, 0, 2.8, 0, 5.4, 5]^\mathsf{T}$.

Consider now the case that our LP is exactly the same, but all constraint coefficients of x_1 have a negative sign:

$$\max\ 70x_1 - 8x_2 + 5x_3 - 4x_4$$
$$\text{s.t.:}\ -4x_1 + 8x_2 - 5x_3 + x_4 = 6$$
$$-5x_1 + 2x_2 + 3x_3 + x_5 = 4$$
$$-2x_1 + 7x_2 + 9x_3 + x_6 = 7$$
$$x_2 + x_7 = 5$$
$$x_1, x_2, x_3, x_4, x_5, x_6, x_7 \geq 0$$

In this case, x_1 will also be selected as the entering variable but all its coefficients at the constraint rows 1–4 are non-positive. This means that simplex will stop without returning a solution and declare that the LP is unbounded.

4.5.2 Two-phase Simplex

When implementing simplex we assumed that we start from a given basic feasible solution and we move to adjacent basic feasible solutions. For many problems, we can easily find an initial bfs by inspection. This, however, is not always the case and finding an initial bfs might be a complex problem by itself.

As its name suggests, the two-phase simplex method splits the problem into two:

1. The first problem is the problem of finding a bfs (known as Phase I LP).
2. The second problem is the problem of using the bfs to find the optimal bfs or declare that the LP is unbounded (known as Phase II LP).

The method presented in the previous section tackles the second part of the problem. Let us now see how one can find an initial bfs if we do not have a variable in every constraint row with a coefficient equal to 1 in that row and 0 at any other row. The main method to find an initial bfs is to relax the LP with the introduction of *artificial variables* so that finding a bfs becomes trivial. We can add at most one artificial variable to each row, and this artificial variable is restricted to be nonnegative $a_i \geq 0$, where i is the number of the row (i.e., $1, 2, \ldots, l$). The objective is to have one variable at each row which has a coefficient equal to 1 at this row and 0 at any other row.

Adding artificial variables can help finding an initial bfs, but can also result in a problem that is not equivalent to the original LP. For this reason, we need to ensure that the optimal solution of the new problem with the added artificial variables results in all artificial variables being equal to 0. To do so, we change the objective function of the reformulated problem, which will now maximize the sum of all artificial variables. This new problem is called *Phase I LP* to distinguish it from the *Phase II LP* described in the previous section.

When using simplex, we assumed that there exists an initial bfs. This implied that the LP had a feasible solution. Thus, simplex was able to indicate whether a problem is unbounded or it has an optimal solution. When using the two-phase simplex method to find an initial bfs, we might realize that this is not possible because the original LP is infeasible. In fact, if by solving the reformulated Phase I LP we see that at least one artificial variable a_i has a positive value $a_i > 0$ in the optimal solution, then the original LP is infeasible.

The first question is in which rows to add artificial variables. After bringing the original LP to its standard form, we also multiply equality constraints with a negative rhs value b_i by -1 to have positive right-hand sides. Then, for each equality constraint that:

- Does not have a slack variable
- Or the sign of its slack variable is opposite to the sign of the right-hand side

we add an artificial variable. Let us consider, for example, the following LP:

$$
\begin{aligned}
\min\ & 6x_1 + 3x_2 \\
\text{s.t.}\ & x_1 + x_2 \geq 1 \\
& -2x_1 + x_2 \leq -1 \\
& -3x_2 \geq -2 \\
& x_1, x_2 \geq 0
\end{aligned}
\tag{4.56}
$$

We first multiply the constraints of this problem that have a negative right-hand-side value by -1:

$$
\begin{aligned}
\min\ & 6x_1 + 3x_2 \\
\text{s.t.}\ & x_1 + x_2 \geq 1 \\
& 2x_1 - x_2 \geq 1 \\
& 3x_2 \leq 2 \\
& x_1, x_2 \geq 0
\end{aligned}
\tag{4.57}
$$

Then, we turn the LP to its standard form by introducing slack variables $x_3, x_4, x_5 \geq 0$:

Original LP in standard form:

$$
\begin{aligned}
\max\ & -6x_1 - 3x_2 \\
\text{s.t.}\ & x_1 + x_2 - x_3 = 1 \\
& 2x_1 - x_2 - x_4 = 1 \\
& 3x_2 + x_5 = 2 \\
& x_1, x_2, x_3, x_4, x_5 \geq 0
\end{aligned}
\tag{4.58}
$$

This problem is our original LP in its standard form. Notice that it is not trivial to find a bfs for this problem. For instance, there is no variable in constraint $x_1 + x_2 -$

$x_3 = 1$ with a coefficient equal to 1 that has a 0 coefficient at any other constraint. For this reason, we develop the *Phase-I LP* by first adding artificial variables $a_1 \geq 0$ and $a_2 \geq 0$ to the first two constraints because the signs of their non-negative slack variables are opposite to the signs of the right-hand sides. We also change the objective function and we request to maximize the sum of the artificial variables multiplied by -1 resulting in the Phase I LP:

Phase I LP:

$$\max \ - a_1 - a_2 \tag{4.59}$$
$$\text{s.t. } x_1 + x_2 - x_3 + a_1 = 1 \tag{4.60}$$
$$2x_1 - x_2 - x_4 + a_2 = 1 \tag{4.61}$$
$$3x_2 + x_5 = 2 \tag{4.62}$$
$$x_1, x_2, x_3, x_4, x_5, a_1, a_2 \geq 0 \tag{4.63}$$

Then, we solve the Phase I LP with simplex. In our Phase I LP variable a_1 is the basic variable of the first constraint because its coefficient at any other constraint is 0. Thus $a_1 = b_1 = 1$. Variable a_2 is the basic variable of constraint 2, and thus $a_2 = b_2 = 1$, and x_5 is the basic variable of constraint 3 resulting in $x_5 = b_3 = 2$. In our example, the initial basis of the Phase I LP corresponds to the basic variables $\{a_1, a_2, x_5\}$ with $a_1 = 1, a_2 = 1, x_5 = 2$. The nonbasic variables $\{x_1, x_2, x_3, x_4\}$ are equal to 0. In addition, we would not like to express our objective function in terms of a_1, a_2 because they are basic variables. Using the equality constraint equations where artificial variables a_1, a_2 appear, we can rewrite $-a_1 - a_2$ as:

$$-a_1 - a_2 = x_1 + x_2 - x_3 - 1 + 2x_1 - x_2 - x_4 - 1 = 3x_1 - x_3 - x_4 - 2$$

and express the objective function as:

$$z - 3x_1 + x_3 + x_4 = -2$$

This will result in the tableau of Table 4.7.
Solving the Phase I LP with simplex will result in one of the following cases:

Table 4.7 Phase I LP Simplex Tableau

Row	z	x_1	x_2	x_3	x_4	x_5	a_1	a_2	Rhs	Basic variable
0	1	−3	0	1	1	0	0	0	−2	z
1	0	1	1	−1	0	0	1	0	1	a_1
2	0	2	−1	0	−1	0	0	1	1	a_2
3	0	0	3	0	0	1	0	0	2	x_5

Table 4.8 Optimal tableau of Phase I LP

Row	z	x_1	x_2	x_3	x_4	x_5	a_1	a_2	Rhs
0	1	0	0	0	0	0	1	1	0
1	0	0	1	$-2/3$	1/3	0	2/3	$-1/3$	1/3
2	0	1	0	$-1/3$	$-1/3$	0	1/3	1/3	2/3
3	0	0	0	2	-1	1	-2	1	1

1. The objective function value of the optimal solution of the Phase I LP is positive instead of 0, meaning that the original LP is *infeasible*.
2. The objective function value of the optimal solution of the Phase I LP is equal to 0. This case results in two subcases:

   - The basis of the optimal bfs of the Phase I LP does not include any column vectors corresponding to the artificial variables (all artificial variables are nonbasic). Then, we can drop all columns in the optimal Phase I tableau that correspond to the artificial variables. We can then combine the original objective function with the constraints from the optimal Phase I LP tableau to form the Phase II LP and solve the Phase II LP (its optimal solution is the optimal solution of the original LP).
   - The basis of the optimal bfs of the Phase I LP includes at least one artificial variable. Then we can drop from the Phase I optimal tableau all nonbasic artificial variables and any variable from the original problem that has a negative coefficient in row 0 to form our Phase II LP.

For our example Phase I LP in Eqs. (4.59)–(4.63) the optimal bfs is presented in the tableau of Table 4.8.

The tableau of Table 4.8 has basic variables $x_1 = 2/3$, $x_2 = 1/3$, $x_5 = 1$ and nonbasic variables $x_3, x_4, a_1, a_2 = 0$. Note that the objective function value of the optimal bfs is equal to 0:

$$z + a_1 + a_2 = 0 \Rightarrow z = 0$$

and the artificial variables a_1, a_2 are nonbasic variables. Thus, the original LP is feasible and we proceed to solve the Phase II LP after dropping the two columns of our artificial variables and using the objective function of the original LP. The Phase II LP is:

$$
\begin{aligned}
\text{Phase II LP:} \\
\max \quad & -6x_1 - 3x_2 \\
\text{s.t.} \quad & x_2 - 2/3x_3 + 1/3x_4 = 1/3 \\
& x_1 - 1/3x_3 - 1/3x_4 = 2/3 \\
& 2x_3 - x_4 + x_5 = 1 \\
& x_1, x_2, x_3, x_4, x_5 \geq 0
\end{aligned}
\tag{4.64}
$$

Table 4.9 Initial tableau of Phase II LP

Row	z	x_1	x_2	x_3	x_4	x_5	Rhs
0	1	0	0	4	1	0	−5
1	0	0	1	−2/3	1/3	0	1/3
2	0	1	0	−1/3	−1/3	0	2/3
3	0	0	0	2	−1	1	1

The objective function of the Phase II LP can be written as:

$$z + 6x_1 + 3x_2 = 0$$

Because x_1 and x_2 are basic variables of the optimal Phase I LP bfs, we need to replace them in the objective function by nonbasic variables. We have that $x_2 = 1/3 + 2/3x_3 - 1/3x_4$ and $x_1 = 2/3 + 1/3x_3 + 1/3x_4$, resulting in objective function:

$$z + 4x_3 + x_4 = -5$$

The initial tableau of our Phase II LP is presented in Table 4.9.

Notice that all nonbasic variables x_3, x_4 in row 0 of the tableau have non-negative coefficients. Thus, the initial bfs is the optimal bfs for the Phase II LP, and, consequently, it is also the optimal bfs of the original LP. This means that our optimal solution is $x_1 = 2/3$, $x_2 = 1/3$, $x_3 = 0$, $x_4 = 0$, $x_5 = 1$ with objective function score $z = -5$. Note that because we turned our minimization LP in (4.56) to a maximization one, we should multiply z by -1 to compute the optimal function score of our original problem, which is $-z = 5$.

The two-phase simplex method requires several problem reformulations. For instance, we should decide in which rows to introduce artificial variables and how to formulate the objective functions of the Phase I and Phase II LPs to include nonbasic variables. When we formulate an original LP, however, we would like an optimization solver to perform these internal reformulation steps. These steps can be automated with the cost of adding more artificial variables than needed. For instance, if we provide to an optimization solver our original LP formulated in standard form with non-negative right-hand side values, we would offer the following input:

$$\text{Original LP:}$$
$$\max \ \mathbf{c}^{\mathsf{T}}\mathbf{x}$$
$$\text{s.t. } \mathbf{Ax} = \mathbf{b} \tag{4.65}$$
$$\mathbf{x} \geq \mathbf{0}$$

where $\mathbf{c}$ is an n-valued column vector with the coefficients of the variables in the objective function, $\mathbf{x}$ an n-valued column vector including all variables, $\mathbf{A}$ the $l \times n$-valued *full row rank* matrix, and $\mathbf{b} \geq \mathbf{0}$ an l-valued column vector with the right-hand side values of each equality constraint. Using this as input, we can generate the Phase I LP in an automated way by adding one artificial variable at each equality constraint:

Original LP with artificial variables:

$$\max \ \mathbf{c}^{\mathsf{T}} \mathbf{x}$$
$$\text{s.t. } \mathbf{Ax} + \mathbf{Ia} = \mathbf{b} \qquad\qquad (4.66)$$
$$\mathbf{x} \geq \mathbf{0}$$
$$\mathbf{a} \geq \mathbf{0}$$

where $\mathbf{a}$ is an l-valued column vector and $\mathbf{I}$ an $l \times l$-valued identity matrix. In this way, we can use $\mathbf{a}$ as our basic and $\mathbf{x}$ as nonbasic variables. This will give us a basic feasible solution because if we set $a_i = b_i$ for any $i \in \{1, \ldots, l\}$ we have that:

• $\mathbf{a}$ is a feasible solution to our problem because $\mathbf{a} \geq \mathbf{0}$.
• The basis $\mathbf{B} = \mathbf{I}$ is a symmetric, $l \times l$ matrix with nonzero determinant (that is, its column vectors are linearly independent).

The Phase I LP is formed by changing the objective function to:

Phase I LP:

$$\max \ -\mathbf{1}^{\mathsf{T}} \mathbf{a}$$
$$\text{s.t. } \mathbf{Ax} + \mathbf{Ia} = \mathbf{b} \qquad\qquad (4.67)$$
$$\mathbf{x} \geq \mathbf{0}$$
$$\mathbf{a} \geq \mathbf{0}$$

where $\mathbf{1}$ is an l-valued vector of ones. To introduce the nonbasic variables $\mathbf{x}$ in the objective function, we replace each artificial variable a_i by:

$$a_i = b_i - \alpha_{i1} x_1 - \alpha_{i2} x_2 - \cdots - \alpha_{in} x_n \ \ \forall i \in \{1, \ldots, l\}$$

where $\alpha_{ij} \in \mathbf{A}$ are the coefficients of $\mathbf{x}$ at the equality constraints. We then solve the Phase I LP with simplex. If the optimal bfs of the Phase I LP:

• Has a non-zero objective function value, then we terminate declaring the original LP as infeasible.
• Has an objective function equal to 0 and all artificial variables are not basic, we proceed to the Phase II LP.
• Has an objective function equal to 0 and some artificial variables are basic, we continue with more pivots until all artificial variables become nonbasic and then we proceed to the Phase II LP. This is required to automate the transition from the Phase I to the Phase II LP.

Finally, we move to solve the Phase II LP with simplex after we remove all columns related to artificial variables and use the original objective function in row 0.

Practitioner's Corner
Simplex in Python 3

To implement simplex, one can use the `scipy.optimize.linprog` in Python 3. This library accepts LPs that might not be in standard form. All equality constraint coefficients are stored at A_eq and b_eq. All inequality constraint coefficients are stored at A_ub and b_ub. If the LP is in standard form, A_ub and b_ub are not needed since $\mathbf{A}$ =A_eq and $\mathbf{b}$ =b_eq. If not, all inequality constraints should be transformed to ($\leq$) first. Finally, it solves minimization problems (that is, we should change the signs of the objective function coefficients $\mathbf{c}$ if our LP is a maximization one). An example of solving the LP in (4.56) is provided below.

```
from scipy.optimize import linprog,
    linprog_verbose_callback
c = [6, 3]
A = [[-1, -1], [-2, 1], [0,3]]
b = [-1, -1, 2]
x1_bounds = (0, None)
x2_bounds = (0, None)
res = linprog(c, A_ub=A, b_ub=b, method='simplex',
    bounds=[x1_bounds, x2_bounds], callback=
    linprog_verbose_callback)
```

resulting in solution $\mathbf{x}^* = [0.6667, 0.3333]^\mathsf{T}$ with objective function score $\mathbf{c}^\mathsf{T}\mathbf{x}^* = 5$ after 3 iterations.

4.5.3 Revised Simplex

When using simplex, the entire tableau is updated at each iteration even if we use a small part of it. The revised simplex method uses the same steps, but it tries to reduce the computational costs. Let us consider an LP transformed to its standard form:

$$\max \ z = \mathbf{c}^\mathsf{T}\mathbf{x}$$
$$\text{s.t. } \mathbf{Ax} = \mathbf{b}$$
$$\mathbf{x} \geq \mathbf{0}$$

Reckon that the necessary KKT conditions are also sufficient optimality conditions for linear programming problems. The Lagrangian function is:

$$\mathcal{L}(\mathbf{x}, \lambda, \mu) = \mathbf{c}^\mathsf{T}\mathbf{x} - \lambda^\mathsf{T}(\mathbf{Ax} - \mathbf{b}) - \mu^\mathsf{T}(-\mathbf{x})$$

and the KKT conditions are:

$$
\begin{aligned}
&\nabla_{\mathbf{x}}\mathcal{L}(\mathbf{x},\lambda,\mu)=\mathbf{0}\Rightarrow \mathbf{c}-\mathbf{A}^{\mathsf{T}}\lambda+\mu=\mathbf{0} && \text{stationarity}\\
&\mathbf{A}\mathbf{x}=\mathbf{b} && \text{primal feasibility}\\
&\mathbf{x}\geq\mathbf{0} && \text{primal feasibility}\\
&\mu\geq\mathbf{0} && \text{dual feasibility}\\
&\mu^{\mathsf{T}}\mathbf{x}=0 && \text{complementary slackness}
\end{aligned}
$$

where $\lambda=[\lambda_1,\ldots\lambda_l]^{\mathsf{T}}$ are the KKT multipliers for the equality constraints and $\mu=[\mu_1,\ldots\mu_n]^{\mathsf{T}}$ the KKT multipliers for the inequality constraints. Let us consider that we start from a basic feasible solution with indexes $\mathcal{B}=\{\mathcal{B}_1,\ldots,\mathcal{B}_l\}$ indicating the l linearly independent columns of $\mathbf{A}$ in our basis $\mathbf{B}$, and $\mathcal{N}=\{\mathcal{N}_1,\ldots,\mathcal{N}_{n-l}\}=\{1,2,\ldots,n\}\setminus\mathcal{B}$. We will also use matrix $\mathbf{N}$ to represent all column vectors with indexes in $\mathcal{N}$. Then,

$$
\mathbf{x}=\begin{bmatrix}\mathbf{x}_{\mathcal{B}}\\\mathbf{x}_{\mathcal{N}}\end{bmatrix}=\begin{bmatrix}\mathbf{B}^{-1}\mathbf{b}\\\mathbf{0}\end{bmatrix}
$$

where $\mathbf{x}_{\mathcal{B}}=[x_{\mathcal{B}_1},\ldots,x_{\mathcal{B}_l}]^{\mathsf{T}}\geq\mathbf{0}$ because we start from a bfs. Note that any bfs satisfies the primal feasibility conditions. In addition, we have:

$$
\mathbf{c}=\begin{bmatrix}\mathbf{c}_{\mathcal{B}}\\\mathbf{c}_{\mathcal{N}}\end{bmatrix}
$$

and

$$
\mu=\begin{bmatrix}\mu_{\mathcal{B}}\\\mu_{\mathcal{N}}\end{bmatrix}
$$

To satisfy the complementary slackness condition we need:

$$
\mu^{\mathsf{T}}\mathbf{x}=0\Rightarrow \mu_{\mathcal{B}}^{\mathsf{T}}\mathbf{x}_{\mathcal{B}}+\mu_{\mathcal{N}}^{\mathsf{T}}\mathbf{x}_{\mathcal{N}}=0\Rightarrow \mu_{\mathcal{B}}^{\mathsf{T}}\mathbf{x}_{\mathcal{B}}=0
$$

That is, we can choose $\mu_{\mathcal{B}}=\mathbf{0}$ to satisfy these conditions. To compute λ and $\mu_{\mathcal{N}}$ let us write $\mathbf{c}-\mathbf{A}^{\mathsf{T}}\lambda+\mu=\mathbf{0}$ as:

$$
\mathbf{A}^{\mathsf{T}}\lambda-\begin{bmatrix}\mu_{\mathcal{B}}\\\mu_{\mathcal{N}}\end{bmatrix}=\begin{bmatrix}\mathbf{c}_{\mathcal{B}}\\\mathbf{c}_{\mathcal{N}}\end{bmatrix}
$$

This will yield:

$$
\mathbf{B}^{\mathsf{T}}\lambda-\mu_{\mathcal{B}}=\mathbf{c}_{\mathcal{B}}\Rightarrow \lambda=(\mathbf{B}^{\mathsf{T}})^{-1}\mathbf{c}_{\mathcal{B}}
$$

and

$$
\mathbf{N}^{\mathsf{T}}\lambda-\mu_{\mathcal{N}}=\mathbf{c}_{\mathcal{N}}\Rightarrow \mu_{\mathcal{N}}=-\mathbf{c}_{\mathcal{N}}+\mathbf{N}^{\mathsf{T}}\lambda
$$

We thus need $\boldsymbol{\mu}_N = -\mathbf{c}_N + \mathbf{N}^\mathsf{T}\boldsymbol{\lambda}$ to be greater than or equal to zero for all $i \in N$ to satisfy the dual feasibility conditions. If this is the case, all KKT conditions are satisfied and the current bfs is our optimal solution. If, however, $\boldsymbol{\mu}_N = [\mu_{N_1}, \ldots, \mu_{N_{n-l}}]^\mathsf{T}$ has some variable with $\mu_{N_j} < 0$, then our current bfs is not an optimal solution and N_j will enter $\mathcal{B}$. To avoid cycling, one can implement Bland's rule [15] and choose the lowest-numbered (i.e., leftmost) nonbasic column with a negative (reduced) cost. The step of calculating $\boldsymbol{\mu}_N$ is called *price-out* or *pricing*. For brevity, let us name $q = N_j$. We then extract the l-valued column vector $\mathbf{a}_q$ from matrix $\mathbf{A}$ that corresponds to column q of the n columns of the matrix. We can then compute the l-valued column $\mathbf{d}$ as:

$$\mathbf{d} = \underbrace{\mathbf{B}^{-1}}_{l \times l} \underbrace{\mathbf{a}_q}_{l \times 1}$$

If $d_i \leq 0$ for any $i \in \{1, \ldots, l\}$ we can terminate the algorithm because the LP is unbounded. If, however, this is not the case, we can compute the value of x_q with the following ratio test:

$$x_q = \min_{i \in \{1,\ldots,l\} \mid d_i > 0} \left\{ \frac{x_{\mathcal{B}_i}}{d_i} \right\}$$

To find the leaving variable p from the basis, we perform pivoting by finding pivot row $w \in \{1, \ldots, l\}$ such that:

$$w = \mathrm{argmin}_{i \in \{1,\ldots,l\} \mid d_i > 0} \left\{ \frac{x_{\mathcal{B}_i}}{d_i} \right\}$$

The leaving variable from the basis is then $p = \mathcal{B}_w$. The values of all basic variables are updated as:

$$\mathbf{x}_{\mathcal{B}} = \mathbf{x}_{\mathcal{B}} - x_q \mathbf{B}^{-1} \mathbf{a}_q$$

Finally, we add q to the place of $\mathcal{B}_w$, and we move p to N. When doing so, we set $x_{\mathcal{B}_w} = x_q$. This procedure repeats itself until $\boldsymbol{\mu}_N \geq \mathbf{0}$ because then we meet the KKT conditions and we have found the optimal bfs. The Revised Simplex algorithm is provided in Algorithm 15. Note that the revised simplex algorithm requires an initial bfs, which can be obtained by the Phase I LP of the two-phase simplex method.

Let us consider again the example:

$$
\begin{aligned}
\max \quad & 70x_1 - 8x_2 + 5x_3 - 4x_4 \\
\text{s.t.:} \quad & 4x_1 + 8x_2 - 5x_3 + x_4 = 6 \\
& 5x_1 + 2x_2 + 3x_3 + x_5 = 4 \\
& 2x_1 + 7x_2 + 9x_3 + x_6 = 7 \\
& x_2 + x_7 = 5 \\
& x_1, x_2, x_3, x_4, x_5, x_6, x_7 \geq 0
\end{aligned}
\tag{4.68}
$$

Algorithm 15 Revised Simplex for maximization LPs in standard form

---
1: given $\mathbf{A}, \mathbf{b}, \mathbf{c}$ and initial bfs with $\mathcal{B} = \{\mathcal{B}_1, \ldots, \mathcal{B}_l\}$ and $\mathcal{N} = \{\mathcal{N}_1, \ldots, \mathcal{N}_{n-l}\}$
2: generate $\mathbf{B}, \mathbf{N}, \mathbf{c}_\mathcal{B}, \mathbf{c}_\mathcal{N}$
3: compute $\mathbf{x}_\mathcal{B} = \mathbf{B}^{-1}\mathbf{b}$ and set $\mathbf{x}_\mathcal{N} = \mathbf{0}$
4: compute $\lambda = (\mathbf{B}^\mathsf{T})^{-1}\mathbf{c}_\mathcal{B}$
5: compute $\mu_\mathcal{N} = -\mathbf{c}_\mathcal{N} + \mathbf{N}^\mathsf{T}\lambda$ (pricing step)
6: **if** $\mu_{\mathcal{N}_j} \geq 0$ for any $\mathcal{N}_j \in \mathcal{N}$ **then**
7:      terminate and return solution $\mathbf{x}_\mathcal{B}$
8: **end if**
9: **repeat**
10:     find the first nonbasic variable $\mathcal{N}_j \in \mathcal{N}$ for which $\mu_{\mathcal{N}_j} \leq 0$ and set $q = \mathcal{N}_j$ as
        the index of the *entering variable* to the basis
11:     set $\mathbf{a}_q$ as the l-valued column vector of matrix $\mathbf{A}$ that corresponds to q
12:     compute $\mathbf{d} = \mathbf{B}^{-1}\mathbf{a}_q$
13:     **if** $d_i \leq 0$ for any $i \in \{1, \ldots, l\}$ **then**
14:         **terminate** because the LP is *unbounded*
15:     **end if**
16:     initialize $x_q = +\infty$
17:     **for** $i \in \{1, \ldots, l\}$ **do**
18:         **if** $d_i > 0$ and $x_q \geq \frac{x_{\mathcal{B}_i}}{d_i}$ **then**
19:             set $x_q = \frac{x_{\mathcal{B}_i}}{d_i}$
20:             set *leaving variable* index $p = \mathcal{B}_i$ and pivot row $w = i$
21:         **end if**
22:     **end for**
23:     compute $\mathbf{x}_\mathcal{B} = \mathbf{x}_\mathcal{B} - x_q\mathbf{B}^{-1}\mathbf{a}_q$
24:     replace p by q in $\mathcal{B}$ and q by p in $\mathcal{N}$ by setting $\mathcal{B}_w = q$ and $\mathcal{N}_j = p$
25:     update the matrices $\mathbf{B}, \mathbf{N}$ and vectors $\mathbf{c}_\mathcal{B}, \mathbf{c}_\mathcal{N}$
26:     set $\mathbf{x}_\mathcal{N} = \mathbf{0}$ and $x_{\mathcal{B}_w} = x_q$
27:     compute $\lambda = (\mathbf{B}^\mathsf{T})^{-1}\mathbf{c}_\mathcal{B}$
28:     compute $\mu_\mathcal{N} = -\mathbf{c}_\mathcal{N} + \mathbf{N}^\mathsf{T}\lambda$ (pricing step)
29: **until** $\mu_{\mathcal{N}_j} \geq \mathbf{0}$ for any $\mathcal{N}_j \in \mathcal{N}$

---

with $\mathcal{B} = \{\mathcal{B}_1, \mathcal{B}_2, \mathcal{B}_3, \mathcal{B}_4\} = \{4, 5, 6, 7\}$ and $\mathcal{N} = \{\mathcal{N}_1, \mathcal{N}_2, \mathcal{N}_3\} = \{1, 2, 3\}$. This
gives us:

$$
\mathbf{A} = \begin{bmatrix} 4 & 8 & -5 & 1 & 0 & 0 & 0 \\ 5 & 2 & 3 & 0 & 1 & 0 & 0 \\ 2 & 7 & 9 & 0 & 0 & 1 & 0 \\ 0 & 1 & 0 & 0 & 0 & 0 & 1 \end{bmatrix} \quad \mathbf{B} = \begin{bmatrix} 1 & 0 & 0 & 0 \\ 0 & 1 & 0 & 0 \\ 0 & 0 & 1 & 0 \\ 0 & 0 & 0 & 1 \end{bmatrix} \quad \mathbf{N} = \begin{bmatrix} 4 & 8 & -5 \\ 5 & 2 & 3 \\ 2 & 7 & 9 \\ 0 & 1 & 0 \end{bmatrix}
$$

$$\mathbf{b} = [6, 4, 7, 5]^\mathsf{T}$$
$$\mathbf{c} = [70, -8, 5, -4, 0, 0, 0]^\mathsf{T}$$
$$\mathbf{c}_\mathcal{B} = [-4, 0, 0, 0]^\mathsf{T}$$
$$\mathbf{c}_\mathcal{N} = [70, -8, 5]^\mathsf{T}$$
$$\mathbf{x}_\mathcal{B} = [6, 4, 7, 5]^\mathsf{T}$$
$$\boldsymbol{\lambda} = (\mathbf{B}^\mathsf{T})^{-1}\mathbf{c}_\mathcal{B} = [-4, 0, 0, 0]^\mathsf{T}$$
$$\boldsymbol{\mu}_\mathcal{N} = -\mathbf{c}_\mathcal{N} + \mathbf{N}^\mathsf{T}\boldsymbol{\lambda} = [-86, -24, 15,]^\mathsf{T}$$

Given that $\mu_{\mathcal{N}_1} = -86$ and $\mu_{\mathcal{N}_2} = -24$ are negative, we have to proceed to the next iteration. Using Bland's rule, we choose the first listed nonbasic variable $\mathcal{N}_j$ with a negative $\mu_{\mathcal{N}_j}$ score as our entering variable. That is, $q = \mathcal{N}_1 = 1$. This results in $\mathbf{a}_q = [4, 5, 2, 0]^\mathsf{T}$ and $\mathbf{d} = \mathbf{B}^{-1}\mathbf{a}_q = [4, 5, 2, 0]^\mathsf{T}$. We also have that:

$$x_q = \min_{i \in \{1,\ldots,l\} \mid d_i > 0} \left\{ \frac{x_{\mathcal{B}_i}}{d_i} \right\} = 0.8$$

with pivot row:

$$w = \operatorname{argmin}_{i \in \{1,\ldots,l\} \mid d_i > 0} \left\{ \frac{x_{\mathcal{B}_i}}{d_i} \right\} = 2$$

and leaving variable $p = \mathcal{B}_w = 5$. This results in:

$$\mathbf{x}_\mathcal{B} = \mathbf{x}_\mathcal{B} - x_q \mathbf{B}^{-1}\mathbf{a}_q = [2.8, 0, 5.4, 5]^\mathsf{T}$$

We then set $\mathcal{B}_w = q$ and $\mathcal{N}_j = p$ resulting in $\mathcal{B} = \{4, 1, 6, 7\}$ and $\mathcal{N} = \{5, 2, 3\}$. We thus update:

$$\mathbf{B} = \begin{bmatrix} 1 & 4 & 0 & 0 \\ 0 & 5 & 0 & 0 \\ 0 & 2 & 1 & 0 \\ 0 & 0 & 0 & 1 \end{bmatrix} \text{ and } \mathbf{N} = \begin{bmatrix} 0 & 8 & -5 \\ 1 & 2 & 3 \\ 0 & 7 & 9 \\ 0 & 1 & 0 \end{bmatrix}$$

$$\mathbf{c}_\mathcal{B} = [-4, 70, 0, 0]^\mathsf{T} \text{ and } \mathbf{c}_\mathcal{N} = [0, -8, 5]^\mathsf{T}$$

In addition, $x_{\mathcal{B}_w}$ is replaced by x_q giving us:

$$\mathbf{x}_\mathcal{B} = [2.8, 0.8, 5.4, 5]^\mathsf{T}$$

and we also have $\mathbf{x}_\mathcal{N} = [0, 0, 0]^\mathsf{T}$. We then compute:

$$\lambda = (\mathbf{B}^\mathsf{T})^{-1}\mathbf{c}_{\mathcal{B}} = [-4, 17.2, 0, 0]^\mathsf{T}$$
$$\mu_\mathcal{N} = -\mathbf{c}_\mathcal{N} + \mathbf{N}^\mathsf{T}\lambda = [17.2, 10.4, 66.6]^\mathsf{T}$$

Note that $\mu_\mathcal{N} \geq \mathbf{0}$, and thus our algorithm terminates with optimal solution $\mathbf{x}_\mathcal{B} = [2.8, 0.8, 5.4, 5]^\mathsf{T}$ that has an objective function score $\mathbf{c}^\mathsf{T}\mathbf{x} = \mathbf{c}_\mathcal{B}^\mathsf{T}\mathbf{x}_\mathcal{B} = 44.8$.

Practitioner's Corner
Revised Simplex in Python 3

To implement revised simplex, one can use the `scipy.optimize.linprog` in Python 3. An example of solving the LP in (4.68) is provided below.

```
from scipy.optimize import linprog,
    linprog_verbose_callback
c=[-70,8,-5,4,0,0,0]
A=[[4,8,-5,1,0,0,0], [5,2,3,0,1,0,0],
    [2,7,9,0,0,1,0],[0,1,0,0,0,0,1]]
b=[6,4,7,5]
x_bounds = [(0, None) for i in range(0,len(c))]
print(x_bounds)
res = linprog(c, A_eq=A, b_eq=b, method='revised
    simplex', bounds=x_bounds, callback=
    linprog_verbose_callback)
```

resulting in solution $\mathbf{x}^* = [0.8, 0, 0, 2.8, 0, 5.4, 5]^\mathsf{T}$ with objective function score $\mathbf{c}^\mathsf{T}\mathbf{x}^* = -44.8$ which, because (4.68) was a maximization problem, should be turned to 44.8.

4.5.4 Dual Simplex

Reckon that the strong duality theorem states that if the primal LP has a finite optimal solution, then the objective function value of this solution is the same as the objective function value of the solution of the dual LP. One of the practical implications of this is that we can solve the dual LP instead of the primal LP in case this provides a computational benefit (i.e., if the number of constraints of the primal LP is much greater than the number of variables). Let us consider our primal LP in standard form:

$$\max_{\mathbf{x}} \ \mathbf{c}^\mathsf{T}\mathbf{x}$$
$$\text{s.t. } \mathbf{A}\mathbf{x} = \mathbf{b}$$
$$\mathbf{x} \geq \mathbf{0}$$

which has the dual (see Eq. (4.53)):

$$\min_{\lambda,\mu} \mathbf{b}^\mathsf{T}\lambda$$

$$\text{s.t. } \mathbf{A}^\mathsf{T}\lambda - \mu = \mathbf{c}$$

$$\mu \geq \mathbf{0}$$

From the KKT conditions we have:

$$\nabla_{\mathbf{x}}\mathcal{L}(\mathbf{x}, \lambda, \mu) = \mathbf{0} \Rightarrow \mathbf{c} - \mathbf{A}^\mathsf{T}\lambda + \mu = \mathbf{0} \quad \text{stationarity}$$
$$\mathbf{Ax} = \mathbf{b} \quad \text{primal feasibility}$$
$$\mathbf{x} \geq \mathbf{0} \quad \text{primal feasibility}$$
$$\mu \geq \mathbf{0} \quad \text{dual feasibility}$$
$$\mu^\mathsf{T}\mathbf{x} = 0 \quad \text{complementary slackness}$$

The *primal* revised simplex algorithm started from a feasible solution $\mathbf{x}$ that satisfied the primal feasibility constraints, namely $\mathbf{Ax} = \mathbf{b}$ and $\mathbf{x} \geq \mathbf{0}$. After computing the value of μ by enforcing the stationary and complementary slackness constraints, this initial feasible solution might not have satisfied the dual feasibility $\mu \geq \mathbf{0}$. Thus, the algorithm continued until finding the optimal solution $\mathbf{x}$ for which $\mu \geq \mathbf{0}$ while maintaining primal feasibility at each step. This optimal solution (if existed) was optimal for both the primal and the dual LP.

There is also another alternative approach, called *dual simplex*, where we start from a point (λ, μ) which is feasible for the dual LP, meaning that $\mu \geq \mathbf{0}$ and $\mathbf{A}^\mathsf{T}\lambda - \mu - \mathbf{c} = \mathbf{0}$. We can select our basis such that $\mu_{\mathcal{B}} = \mathbf{0}$ and $\mu_{\mathcal{N}} \geq \mathbf{0}$. Note that this point satisfies the stationary KKT condition, but it is not necessarily primal feasible. Indeed, this (λ, μ) point has a corresponding primal point $\mathbf{x}$ for which $\mathbf{x}_{\mathcal{N}} = \mathbf{0}$ that satisfies $\mu^\mathsf{T}\mathbf{x} = 0$ and $\mathbf{Ax} = \mathbf{b}$, but it is not necessarily primal feasible because $\mathbf{x} \geq \mathbf{0}$ might not hold for some $\mathbf{x}_{\mathcal{B}}$.

As it might have been expected, in dual simplex we will make changes between $\mathcal{B}$ and $\mathcal{N}$ until reaching a feasible primal point $\mathbf{x} \geq \mathbf{0}$ signifying optimality since the triplet $(\mathbf{x}, \lambda, \mu)$ would satisfy the KKT conditions. We can compute the primal and dual variables at each iteration as follows:

$$\mathbf{x}_{\mathcal{N}} = \mathbf{0}$$
$$\mathbf{x}_{\mathcal{B}} = \mathbf{B}^{-1}\mathbf{b}$$
$$\lambda = (\mathbf{B}^\mathsf{T})^{-1}\mathbf{c}_{\mathcal{B}}$$
$$\mu_{\mathcal{B}} = \mathbf{0}$$
$$\mu_{\mathcal{N}} = \mathbf{N}^\mathsf{T}\lambda - \mathbf{c}_{\mathcal{N}} \geq \mathbf{0}$$

where the algorithm terminates if $\mathbf{x}_{\mathcal{B}} \geq \mathbf{0}$. At each dual simplex iteration we select as a leaving variable with index $q \in \mathcal{B}$ a variable for which $x_q < 0$. The leaving variable is identified from a ratio test which strives to ensure that μ will still be greater than or equal to $\mathbf{0}$.

4.5.5 Karmarkar's Interior Point Method

Although simplex is very efficient in solving practical problems, its nonpolynomial time complexity, which is related to the potential number of bfs, motivated many researchers to develop linear programming algorithms with polynomial time complexity. The first algorithm was the Ellipsoid method of Khachiyan [10]. However, despite its theoretical improvement in computational complexity, in practical problems simplex is faster. On the other side, Karmarkar's interior point method [11] has both guaranteed polynomial time complexity and it performs well in practical problems. In addition, its computational complexity, $O(n^{3.5}L^2)$, where n is the number of variables and L the number of bits used to represent numbers, is lower than the computational complexity $O(n^6 L^2)$ of the Ellipsoid method. Unlike simplex, where bfs are extreme points on the boundary of the feasible region, Karmarkar's projective algorithm moves through the interior of the feasible polyhedron, improving the approximation of the optimal solution with every iteration (see Fig. 4.7).

As described in [16], the basic strategy of the Karmarkar's projective algorithm is to take an interior point, transform the space so as to place the point near the center of the polytope, and then move it in the direction of steepest descent, but not all the way to the boundary of the feasible region so that the point remains interior.

To apply Karmarkar's projective algorithm:

1. We must be able to write the original LP to its *homogeneous form*,
2. At the center of the "simplex" $\{x_1, x_2, \ldots, x_n\} = \{\frac{1}{n}, \ldots, \frac{1}{n}\}$, the problem must have a feasible solution,
3. The optimal value of the objective function of the problem should be 0.

Fig. 4.7 Difference in optimum search path between Simplex and Karmarkar's Algorithm. Simplex moves from an initial bfs (vertex) to an adjacent vertex at each iteration. Karmarkar's algorithm moves from a feasible point in the interior of the feasible region to other feasible points until finding the optimal point close to the boundary of the feasible region

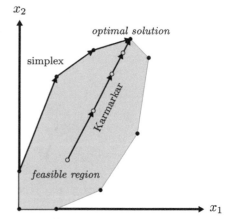

4.5.5.1 Solving Homogeneous Form LPs with an Optimal Objective Function Value Equal to 0

The homogeneous form of an LP is provided below.

Homogeneous Form of an LP

$$\min \ \mathbf{c}^{\mathsf{T}}\mathbf{x} \tag{4.69}$$

$$\text{s.t. } \mathbf{A}\mathbf{x} = \mathbf{0} \tag{4.70}$$

$$\mathbf{e}^{\mathsf{T}}\mathbf{x} = 1 \tag{4.71}$$

$$\mathbf{x} \geq \mathbf{0} \tag{4.72}$$

where $\mathbf{A}$ is an $l \times n$ matrix and $\mathbf{e}$ a vector column of ones.

An example problem in this form provided in [16] is:

$$\min \ 2x_1 + x_2 + x_3 \tag{4.73}$$

$$\text{s.t. } 2x_1 + x_2 - 3x_3 = 0 \tag{4.74}$$

$$x_1 + x_2 + x_3 = 1 \tag{4.75}$$

$$x_1, x_2, x_3 \geq 0 \tag{4.76}$$

Note that the initial interior point $\{x_1, x_2, x_3\} = \{\frac{1}{3}, \frac{1}{3}, \frac{1}{3}\}$ is indeed a feasible solution of the homogeneous form of the LP. Before proceeding, we provide the definition of *simplex* in Karmarkar's method. This term does not have the meaning of the simplex algorithm described in the previous sections.

Definition 4.17 The set of points satisfying the normalization constraint $\mathbf{e}^{\mathsf{T}}\mathbf{x} = 1$ and the non-negative constraints $\mathbf{x} \geq \mathbf{0}$ is called *simplex*.

Starting from an interior point $\mathbf{x}_k$ with $k = 0$, Karmarkar's algorithm:

step 1: Transforms the problem and its interior point $\mathbf{x}_k$ via a *projective transformation* to another point $\mathbf{y}_k$ so as to put the current point $\mathbf{x}_k$ at the center of the simplex

step 2: Computes a projected *steepest descent* direction $\mathbf{d}_k$

step 3: Takes a step a_k along this direction: $\mathbf{y}_{k+1} = \mathbf{y}_k + a_k \mathbf{d}_k$

step 4: Maps the new point $\mathbf{y}_{k+1}$ back to the original space $\mathbf{x}_{k+1}$

Step 1 transforms the problem so as to put the current point at the center of the simplex because this will lead to making a considerable step towards the optimal solution. To transform a feasible point $\mathbf{x}_k$ to the center of the transformed unit simplex, we use transformation T such that:

$$\mathbf{y} = T(\mathbf{x}) = \frac{[Diag(\mathbf{x}_k)]^{-1}\mathbf{x}}{\mathbf{e}^\mathsf{T}[Diag(\mathbf{x}_k)]^{-1}\mathbf{x}}$$

where $Diag(\mathbf{x}_k)$ is a diagonal $n \times n$ matrix constructed from vector $\mathbf{x}_k$ such that:

$$Diag(\mathbf{x}_k) = \begin{bmatrix} x_1 & 0 & \cdots & 0 \\ 0 & x_2 & \cdots & 0 \\ \vdots & \vdots & \ddots & \vdots \\ 0 & 0 & \cdots & x_n \end{bmatrix}$$

Note that this transformation transforms the current point $\mathbf{x}_k$ to the point $\frac{\mathbf{e}}{n} = [\frac{1}{n}, \ldots, \frac{1}{n}]^\mathsf{T}$ which is the center point of the constraints $\mathbf{e}^\mathsf{T}\mathbf{x} = 1, \mathbf{x} \geq \mathbf{0}$ (center of the simplex).

Step 2 computes a projected *steepest descent* direction $\mathbf{d}_k$ for the transformed problem which is provided below:

Transformed Problem: $\min_{\mathbf{y}} \dfrac{[Diag(\mathbf{x}_k)\mathbf{c}]^\mathsf{T}\mathbf{y}}{\mathbf{e}^\mathsf{T} Diag(\mathbf{x}_k)\mathbf{y}}$

$$\text{s.t. } \mathbf{A}Diag(\mathbf{x}_k)\mathbf{y} = \mathbf{0}$$
$$\mathbf{e}^\mathsf{T}\mathbf{y} = 1$$
$$\mathbf{y} \geq \mathbf{0}$$

This is a linear fractional programming problem, but since we seek a solution with objective function equal to zero and the denominator is positive and bounded, we can remove the denominator of the objective function and solve the problem:

$$\min_{\mathbf{y}} [Diag(\mathbf{x}_k)\mathbf{c}]^\mathsf{T}\mathbf{y}$$
$$\text{s.t. } \mathbf{A}Diag(\mathbf{x}_k)\mathbf{y} = 0$$
$$\mathbf{e}^\mathsf{T}\mathbf{y} = 1$$
$$\mathbf{y} \geq \mathbf{0}$$

Using steepest descent, the objective function of the transformed problem drops in the direction of the negative gradient, which at $\frac{\mathbf{e}}{n}$ is parallel to $-Diag(\mathbf{x}_k)\mathbf{c}$. The constraint matrix of the transformed problem is:

$$\mathbf{B} = \begin{bmatrix} \mathbf{A}Diag(\mathbf{x}_k) \\ \mathbf{e}^\mathsf{T} \end{bmatrix}$$

and the corresponding orthogonal projection matrix is:

$$\mathbf{P} = \mathbf{I} - \mathbf{B}^\mathsf{T}(\mathbf{B}\mathbf{B}^\mathsf{T})^{-1}\mathbf{B}$$

The projected steepest-descent direction is:

$$\mathbf{d}_k = -\frac{\mathbf{c}_p}{\|\mathbf{c}_p\|_2}$$

where

$$\mathbf{c}_p = \mathbf{P}Diag(\mathbf{x}_k)\mathbf{c}$$

Step 3 takes a step with length a_k along the projected steepest descent direction:

$$\mathbf{y}_{k+1} = \mathbf{y}_k + a_k\mathbf{d}_k$$

The step length a_k should move in the direction of the steepest descent, but not so far as to leave the feasible set. Karmarkar proposed to avoid infeasibility by inscribing a circle of radius $r = [n(n-1)]^{-1/2}$ with its center at $\frac{\mathbf{r}}{n}$. Since the circle contains only feasible points, we can move across a distance $a_k = \theta r$, where $0 < \theta \leq 1$ and it is typically selected to be $\theta = \frac{n-1}{3n}$ for $n \geq 2$ or $\theta = 1/3$.

Step 4 finally maps $\mathbf{y}_{k+1}$ back to the x-space using transformation T^{-1}:

$$\mathbf{x}_{k+1} = \frac{[Diag(\mathbf{x}_k)]\mathbf{y}_{k+1}}{\mathbf{e}^\mathsf{T}[Diag(\mathbf{x}_k)]\mathbf{y}_{k+1}}$$

and checks if the objective function value $\mathbf{c}^\mathsf{T}\mathbf{x}_{k+1}$ is close to 0. If not, the previous steps are repeated. The algorithm is presented in Algorithm 16.

To provide an example, let us consider the following LP in the homogeneous form with an optimal objective function value equal to 0:

$$\begin{aligned}
\min\ & 2x_2 - x_3 \\
\text{s.t. } & x_1 - 2x_2 + x_3 = 0 \\
& x_1 + x_2 + x_3 = 1 \\
& x_1, x_2, x_3 \geq 0
\end{aligned}$$

In step 1 we set $k = 0$ and we start with an interior point $\mathbf{x}_k = \frac{\mathbf{e}}{n} = [1/3, 1/3, 1/3]^\mathsf{T}$ of the feasible polytope. We then have to put point $\mathbf{x}_k$ to the center of simplex using the transformation:

$$\mathbf{y}_k = T(\mathbf{x}_k) = \frac{[Diag(\mathbf{x}_k)]^{-1}\mathbf{x}_k}{\mathbf{e}^\mathsf{T}[Diag(\mathbf{x}_k)]^{-1}\mathbf{x}_k}$$

The numerator yields:

$$\begin{bmatrix} 3 & 0 & 0 \\ 0 & 3 & 0 \\ 0 & 0 & 3 \end{bmatrix} \begin{bmatrix} 1/3 \\ 1/3 \\ 1/3 \end{bmatrix} = \begin{bmatrix} 1 \\ 1 \\ 1 \end{bmatrix}$$

Algorithm 16 Karmarkar's algorithm for LPs in homogeneous form with $z = 0$ at the optimal solution

---

1: set solution performance tolerance $\tau \in \mathbb{R}_{\geq 0}$ with $\tau \approx 0$ and iteration $k = 0$
2: start with an interior point $\mathbf{x}_k$ of the feasible polyhedron
3: set $z = \mathbf{c}^\mathsf{T}\mathbf{x}_k$
4: **repeat**
5:     transform $\mathbf{x}_k$ to $\mathbf{y}_k$ at the center of the simplex by $\mathbf{y}_k = \frac{[Diag(\mathbf{x}_k)]^{-1}\mathbf{x}_k}{\mathbf{e}^\mathsf{T}[Diag(\mathbf{x}_k)]^{-1}\mathbf{x}_k}$
6:     set $\mathbf{B} = \begin{bmatrix} \mathbf{A}Diag(\mathbf{x}_k) \\ \mathbf{e}^\mathsf{T} \end{bmatrix}$
7:     set $\mathbf{P} = \mathbf{I} - \mathbf{B}^\mathsf{T}(\mathbf{B}\mathbf{B}^\mathsf{T})^{-1}\mathbf{B}$
8:     **if** $\mathbf{P}Diag(\mathbf{x}_k)\mathbf{c} = \mathbf{0}$ **then**
9:         **terminate** because any feasible solution becomes an optimal solution.
10:    **else**
11:       Compute $\mathbf{c}_p = \mathbf{P}Diag(\mathbf{x}_k)\mathbf{c}$
12:       Compute projected steepest-descent direction: $\mathbf{d}_k = -\frac{\mathbf{c}_p}{\|\mathbf{c}_p\|_2}$
13:    **end if**
14:    set $a_k = \frac{\theta}{[n(n-2)]^{1/2}}$ to maintain feasibility
15:    take step $\mathbf{y}_{k+1} = \mathbf{y}_k + a_k\mathbf{d}_k$
16:    map $\mathbf{y}_{k+1}$ back to the x-space with $\mathbf{x}_{k+1} = \frac{[Diag(\mathbf{x}_k)]\mathbf{y}_{k+1}}{\mathbf{e}^\mathsf{T}[Diag(\mathbf{x}_k)]\mathbf{y}_{k+1}}$
17:    set $k \leftarrow k + 1$
18: **until** $|z| = |\mathbf{c}^\mathsf{T}\mathbf{x}_k| \leq \tau$

---

and the denominator yields:

$$[1, 1, 1] \begin{bmatrix} 3 & 0 & 0 \\ 0 & 3 & 0 \\ 0 & 0 & 3 \end{bmatrix} \begin{bmatrix} 1/3 \\ 1/3 \\ 1/3 \end{bmatrix} = 3 \cdot \frac{1}{3} + 3 \cdot \frac{1}{3} + 3 \cdot \frac{1}{3} = 3$$

Thus,

$$\mathbf{y}_k = \begin{bmatrix} 1/3 \\ 1/3 \\ 1/3 \end{bmatrix}$$

Note that $\mathbf{y}_k = \mathbf{x}_k$ because the selected interior point $\mathbf{x}_k$ was already the center of the simplex.

We now proceed to step 2 by computing:

$$\mathbf{A}Diag(\mathbf{x}_k) = \begin{bmatrix} 1 & -2 & 1 \end{bmatrix} \begin{bmatrix} 1/3 & 0 & 0 \\ 0 & 1/3 & 0 \\ 0 & 0 & 1/3 \end{bmatrix} = [1/3, -2/3, 1/3]$$

which results in:

$$\mathbf{B} = \begin{bmatrix} \mathbf{A}Diag(\mathbf{x}_k) \\ \mathbf{e}^\mathsf{T} \end{bmatrix} = \begin{bmatrix} 1/3 & -2/3 & 1/3 \\ 1 & 1 & 1 \end{bmatrix}$$

We also have:

$$\mathbf{B}\mathbf{B}^\mathsf{T} = \begin{bmatrix} 2/3 & 0 \\ 0 & 3 \end{bmatrix}$$

$$(\mathbf{B}\mathbf{B}^\mathsf{T})^{-1} = \begin{bmatrix} 1.5 & 0 \\ 0 & 1/3 \end{bmatrix}$$

$$\mathbf{B}^\mathsf{T}(\mathbf{B}\mathbf{B}^\mathsf{T})^{-1}\mathbf{B} = \begin{bmatrix} 0.5 & 0 & 0.5 \\ 0 & 1 & 0 \\ 0.5 & 0 & 0.5 \end{bmatrix}$$

$$\mathbf{P} = \mathbf{I} - \mathbf{B}^\mathsf{T}(\mathbf{B}\mathbf{B}^\mathsf{T})^{-1}\mathbf{B} = \begin{bmatrix} 0.5 & 0 & -0.5 \\ 0 & 0 & 0 \\ -0.5 & 0 & 0.5 \end{bmatrix}$$

$$\mathbf{c}_p = \mathbf{P}Diag(\mathbf{x}_k)\mathbf{c} = \begin{bmatrix} 1/6 \\ 0 \\ -1/6 \end{bmatrix}$$

$$\|\mathbf{c}_p\|_2 = \sqrt{(1/6)^2 + 0^2 + (-1/6)^2} \simeq 0.2357$$

For $a_k = \frac{n-1}{3n}r = 2/3\frac{1}{\sqrt{n(n-1)}} = 0.09072$ we have:

$$\mathbf{y}_{k+1} = \mathbf{y}_k - a_k \frac{\mathbf{c}_p}{\|\mathbf{c}_p\|_2} = \begin{bmatrix} 1/3 \\ 1/3 \\ 1/3 \end{bmatrix} - \frac{0.09072}{0.2357}\begin{bmatrix} 1/6 \\ 0 \\ -1/6 \end{bmatrix} = \begin{bmatrix} 0.26918 \\ 0.33333 \\ 0.39748 \end{bmatrix}$$

Mapping $\mathbf{y}_{k+1}$ back to the x-space we have:

$$\mathbf{x}_{k+1} = \frac{[Diag(\mathbf{x}_k)]\mathbf{y}_{k+1}}{\mathbf{e}^\mathsf{T}[Diag(\mathbf{x}_k)]\mathbf{y}_{k+1}} = \begin{bmatrix} 0.26918 \\ 0.33333 \\ 0.39748 \end{bmatrix}$$

We can now set $k \leftarrow k + 1$ and check the updated value of the objective function:

$$|z| = |\mathbf{c}^\mathsf{T}\mathbf{x}_k| = \left| [0, 2, -1]\begin{bmatrix} 0.26918 \\ 0.33333 \\ 0.39748 \end{bmatrix} \right| = 0.26918$$

The objective function value is close to 0 for the new $\mathbf{x}_k$, but not sufficiently close. Let us say that we impose a tolerance of $\tau = 0.01$ allowing the algorithm to stop when $|z| \leq \tau$. This will result in the next iteration ($k = 1$), where:

$$\mathbf{x}_k = \begin{bmatrix} 0.26918 \\ 0.33333 \\ 0.39748 \end{bmatrix}$$

$$\mathbf{y}_k = T(\mathbf{x}_k) = \frac{[Diag(\mathbf{x}_k)]^{-1}\mathbf{x}_k}{\mathbf{e}^{\mathsf{T}}[Diag(\mathbf{x}_k)]^{-1}\mathbf{x}_k} = \begin{bmatrix} 1/3 \\ 1/3 \\ 1/3 \end{bmatrix}$$

$$\mathbf{A}Diag(\mathbf{x}_k) = \begin{bmatrix} 1 & -2 & 1 \end{bmatrix} \begin{bmatrix} 0.26918 & 0 & 0 \\ 0 & 0.33333 & 0 \\ 0 & 0 & 0.39748 \end{bmatrix} = [0.26918, -0.66667, 0.39748]$$

$$\mathbf{B} = \begin{bmatrix} \mathbf{A}Diag(\mathbf{x}_k) \\ \mathbf{e}^{\mathsf{T}} \end{bmatrix} = \begin{bmatrix} 0.26918 & -0.66667 & 0.39748 \\ 1 & 1 & 1 \end{bmatrix}$$

$$\mathbf{B}\mathbf{B}^{\mathsf{T}} = \begin{bmatrix} 0.6749 & 0 \\ 0 & 3 \end{bmatrix}$$

$$(\mathbf{B}\mathbf{B}^{\mathsf{T}})^{-1} = \begin{bmatrix} 1.4817 & 0 \\ 0 & 0.3333 \end{bmatrix}$$

$$\mathbf{P} = \mathbf{I} - \mathbf{B}^{\mathsf{T}}(\mathbf{B}\mathbf{B}^{\mathsf{T}})^{-1}\mathbf{B} = \begin{bmatrix} 0.5593 & -0.0674 & -0.4919 \\ -0.0674 & 0.0081 & 0.0593 \\ -0.4919 & 0.0593 & 0.4326 \end{bmatrix}$$

$$\mathbf{c}_p = \mathbf{P}Diag(\mathbf{x}_k)\mathbf{c} = \begin{bmatrix} 0.1505 \\ -0.0181 \\ -0.1324 \end{bmatrix}$$

$$\|\mathbf{c}_p\|_2 = 0.2013$$

$$a_{k+1} = \frac{n-1}{3n}r = 2/3\frac{1}{\sqrt{n(n-1)}} = 0.09072$$

$$\mathbf{y}_{k+1} = \mathbf{y}_k - a_{k+1}\frac{\mathbf{c}_p}{\|\mathbf{c}_p\|_2} = \begin{bmatrix} 1/3 \\ 1/3 \\ 1/3 \end{bmatrix} - \frac{0.09072}{0.2013}\begin{bmatrix} 0.1505 \\ -0.0181 \\ -0.1324 \end{bmatrix} = \begin{bmatrix} 0.26548 \\ 0.34151 \\ 0.393 \end{bmatrix}$$

Mapping $\mathbf{y}_k$ back to the x-space we have:

$$\mathbf{x}_{k+1} = \frac{[Diag(\mathbf{x}_k)]\mathbf{y}_{k+1}}{\mathbf{e}^{\mathsf{T}}[Diag(\mathbf{x}_k)]\mathbf{y}_{k+1}} = \begin{bmatrix} 0.20925 \\ 0.33333 \\ 0.45741 \end{bmatrix}$$

We can now set $k \leftarrow k + 1$ and check the updated value of the objective function:

$$|z| = |\mathbf{c}^{\mathsf{T}}\mathbf{x}_k| = \left| [0, 2, -1] \begin{bmatrix} 0.20925 \\ 0.33333 \\ 0.45741 \end{bmatrix} \right| = 0.20925$$

Thus, $|z|$ is still greater than τ and we should proceed to the next iteration. In Table 4.10 we present all iterations with the $\mathbf{y}_k$ and $\mathbf{x}_k$ solutions until convergence. It is important to note that using Karmarkar's method does not ensure that the objective function value $z = \mathbf{c}^{\mathsf{T}}\mathbf{x}$ will decrease at each iteration. Indeed, $\mathbf{c}^{\mathsf{T}}\mathbf{x}_{k+1}$ might be

Table 4.10 Karmarkar's algorithm iterations until $|z^{k+1}| \leq \tau$, for $\tau = 0.01$.

Iteration	$\mathbf{x}_k$	$\mathbf{y}_{k+1}$	$\mathbf{x}_{k+1}$	$|z^{k+1}|$	$f(\mathbf{x}_{k+1})$
0	$[1/3, 1/3, 1/3]^{\mathsf{T}}$	$[0.269, 0.333, 0.397]^{\mathsf{T}}$	$[0.269, 0.333, 0.397]^{\mathsf{T}}$	0.26918	−0.60351
1	$[0.269, 0.333, 0.397]^{\mathsf{T}}$	$[0.265, 0.342, 0.393]^{\mathsf{T}}$	$[0.209, 0.333, 0.457]^{\mathsf{T}}$	0.20925	−1.24758
2	$[0.209, 0.333, 0.457]^{\mathsf{T}}$	$[0.263, 0.349, 0.388]^{\mathsf{T}}$	$[0.158, 0.333, 0.509]^{\mathsf{T}}$	0.15763	−1.92102
3	$[0.158, 0.333, 0.509]^{\mathsf{T}}$	$[0.261, 0.355, 0.384]^{\mathsf{T}}$	$[0.116, 0.333, 0.551]^{\mathsf{T}}$	0.11601	−2.61288
4	$[0.116, 0.333, 0.551]^{\mathsf{T}}$	$[0.260, 0.359, 0.380]^{\mathsf{T}}$	$[0.084, 0.333, 0.583]^{\mathsf{T}}$	0.08400	−3.31514
5	$[0.084, 0.333, 0.583]^{\mathsf{T}}$	$[0.260, 0.363, 0.377]^{\mathsf{T}}$	$[0.060, 0.333, 0.607]^{\mathsf{T}}$	0.06016	−4.02292
6	$[0.060, 0.333, 0.607]^{\mathsf{T}}$	$[0.260, 0.365, 0.375]^{\mathsf{T}}$	$[0.043, 0.333, 0.624]^{\mathsf{T}}$	0.04277	−4.73350
7	$[0.043, 0.333, 0.624]^{\mathsf{T}}$	$[0.259, 0.367, 0.374]^{\mathsf{T}}$	$[0.030, 0.333, 0.636]^{\mathsf{T}}$	0.03025	−5.44548
8	$[0.030, 0.333, 0.636]^{\mathsf{T}}$	$[0.259, 0.368, 0.373]^{\mathsf{T}}$	$[0.021, 0.333, 0.645]^{\mathsf{T}}$	0.02134	−6.15816
9	$[0.021, 0.333, 0.645]^{\mathsf{T}}$	$[0.260, 0.369, 0.372]^{\mathsf{T}}$	$[0.015, 0.333, 0.652]^{\mathsf{T}}$	0.01501	−6.87118
10	$[0.015, 0.333, 0.652]^{\mathsf{T}}$	$[0.260, 0.369, 0.371]^{\mathsf{T}}$	$[0.011, 0.333, 0.656]^{\mathsf{T}}$	0.01054	−7.58437
11	$[0.011, 0.333, 0.656]^{\mathsf{T}}$	$[0.259, 0.369, 0.371]^{\mathsf{T}}$	$[0.007, 0.333, 0.659]^{\mathsf{T}}$	0.00740	−8.29764

greater than $\mathbf{c}^{\mathsf{T}}\mathbf{x}^k$ for some k. For this reason, the convergence progress $\mathbf{c}^{\mathsf{T}}\mathbf{x}^k \to 0$ is measured with the *potential function*:

$$f(\mathbf{x}_k) := n \ln \mathbf{c}^{\mathsf{T}}\mathbf{x}^k - \sum_{j=1}^{n} \ln x_j^k = \sum_{j=1}^{n} \ln \frac{\mathbf{c}^{\mathsf{T}}\mathbf{x}^k}{x_j^k}$$

which is proven to always reduce from iteration k to $k + 1$ by some constant value, δ (see [17]). Note that the *potential function* is used to monitor the convergence progress because, unlike the *objective function*, it will decrease in value from iteration to iteration. In addition, as $f(\mathbf{x}_k)$ reduces from iteration to iteration, the objective function $z_k \to 0$. This is illustrated in Table 4.10. For brevity, in this table we do not present the $\mathbf{y}_k$ values because they are always equal to $[1/3, 1/3, 1/3]^{\mathsf{T}}$.

The optimal solution $\mathbf{x}^* = [0.007, 0.333, 0.659]^{\mathsf{T}}$ satisfies all constraints and results in an objective function score of $z^* = 0.0074 \simeq 0$. Notice also that the potential function reduces steadily from iteration to iteration by, approximately, $\delta = 0.7$. This can be used as a termination criterion with a typical termination condition being $f(\mathbf{x}_k) \leq -14n$ (see [18]). This is the key to prove the polynomial time complexity of the algorithm since the stable decrease of the potential function will require $O(n)$ iterations until convergence, where the cost of each iteration is $O(n^{2.5}L)$.

4.5.5.2 Solving Homogeneous Form LPs with an Unknown Optimal Objective Function Value

Applying Karmarkar's method requires writing an LP to its homogeneous form and the objective function to be equal to 0 at the optimal solution. Although satisfying the first requirement is possible for LP problems, the second is difficult to ensure because we typically do not know the objective function value z^* of the optimal solution when starting to solve the problem. If z^* was known, we could have replaced the objective function $\mathbf{c}^\mathsf{T}\mathbf{x}$ by $(\mathbf{c} - z^*\mathbf{e})^\mathsf{T}\mathbf{x}$ since $\mathbf{e}^\mathsf{T}\mathbf{x} = 1$ for any feasible solution. This means that we could have used Karmarkar's algorithm with $\mathbf{c}$ being replaced by $(\mathbf{c} - z^*\mathbf{e})$. In practice, however, z^* is not known. To address this, we can replace z^* by a lower bound u_k which updates itself from iteration to iteration. By doing so, $\mathbf{c}$ in our objective function can be replaced by $\mathbf{c}(u_k) = \mathbf{c} - u_k\mathbf{e}$. To ensure that $u_k \to z^*$ as the algorithm progresses, we can identify a dual feasible solution $(\mathbf{v}_k, u_k)$ to the primal LP. In more detail, the primal LP at iteration k:

$$\text{PRIMAL LP:} \quad \min \ (\mathbf{c} - u_k\mathbf{e})^\mathsf{T}\mathbf{x}_k$$
$$\text{s.t. } \mathbf{Ax}_k = \mathbf{0}$$
$$\mathbf{e}^\mathsf{T}\mathbf{x}_k = 1$$
$$\mathbf{x}_k \geq \mathbf{0}$$

has a dual where constraints $\mathbf{Ax}_k = \mathbf{0}$ are replaced by dual variables $\mathbf{v}_k$ and constraint $\mathbf{e}^\mathsf{T}\mathbf{x}_k = 1$ by dual variable u. Note that for an $l \times n$ matrix $\mathbf{A}$, we have an l-valued vector $\mathbf{v}_k$ of dual variables. Because the constraints of our primal are equality constraints, variables $\mathbf{v}_k, u$ can take any value in $\mathbb{R}$. Let $\mathbf{b} = [0, \ldots, 0]^\mathsf{T}$ symbolize the rhs values of the primal LP constraints $\mathbf{Ax}_k = \mathbf{0}$ and $\bar{b} = 1$ the rhs value of constraint $\mathbf{e}^\mathsf{T}\mathbf{x}_k = 1$. Then, the objective function of the dual LP becomes $\max \mathbf{b}^\mathsf{T}\mathbf{v}_k + \bar{b}u = u$. In addition, because the primary variables $\mathbf{x}_k \geq \mathbf{0}$ and the primal LP is a minimization problem, all constraints of the dual problem will be inequalities with a ($\leq$) sign. The constraints of the dual LP will thus be $\mathbf{A}^\mathsf{T}\mathbf{v}_k + \mathbf{e}u \leq \mathbf{c} - u_k\mathbf{e}$ resulting in:

$$\text{DUAL LP:} \quad \max \ u$$
$$\text{s.t. } \mathbf{A}^\mathsf{T}\mathbf{v}_k + \mathbf{e}u \leq \mathbf{c} - u_k\mathbf{e}$$

Because the primal LP is a minimization problem, from the weak duality theorem follows that a feasible solution u of the dual LP is a lower bound of a feasible solution $\mathbf{x}_k$ of the primal LP: $u \leq (\mathbf{c} - u_k\mathbf{e})^\mathsf{T}\mathbf{x}_k$. Now, from the strong duality theorem we know that $u = (\mathbf{c} - u_k\mathbf{e})^\mathsf{T}\mathbf{x}_k$ at the optimal solution. This means that $u = z^*$. We will use this property to solve LPs with unknown optimal objective function values z^*.

Note that for any $\mathbf{v}_k$, $(\mathbf{v}_k, u)$ is a feasible solution of the dual if $u = \min_{j=1,\ldots,n}\{(\mathbf{c} - u_k\mathbf{e}) - \mathbf{A}^\mathsf{T}\mathbf{v}_k\}_j$. Todd and Burrell [19] proposed to use feasible dual solution $(\mathbf{v}_k, u)$ computed as:

$$\mathbf{v}_k = (\mathbf{AA}^\mathsf{T})^{-1}\mathbf{A}(\mathbf{c} - u_k\mathbf{e})$$
$$u = \min_{j=1,\ldots,n} \{(\mathbf{c} - u_k\mathbf{e}) - \mathbf{A}^\mathsf{T}\mathbf{v}_k\}_j$$

where if $u > u_k$ the new lower bound has increased and we set $u_{k+1} = u$ in the next iteration. Otherwise, $u_{k+1} = u_k$ because the lower bound has not improved.

Because the objective function of the primal LP changes at each iteration, we make the following revisions in Karmarkar's algorithm according to [19]:

Step 0: We set $k = 0$ and $\mathbf{v}_k = (\mathbf{AA}^\mathsf{T})^{-1}\mathbf{Ac}$ and $u_k = \min_{j=1,\dots,n}\{\mathbf{c} - \mathbf{A}^\mathsf{T}\mathbf{v}_k\}_j$.

Step 1: With a projective transformation we obtain $\mathbf{y}_k$ from the interior point $\mathbf{x}_k$ to the polytope.

Step 2: We update the dual variables in the transformed space. Consider $\mathbf{X}_k = Diag(\mathbf{x}_k)$. Then,

$$\mathbf{v}_{k+1} = (\mathbf{AX}_k\mathbf{X}_k\mathbf{A}^\mathsf{T})^{-1}\mathbf{AX}_k\mathbf{X}_k(\mathbf{c} - u_k\mathbf{e})$$
$$u = \min_{j=1,\dots,n}\{\mathbf{c} - \mathbf{A}^\mathsf{T}\mathbf{v}_k\}_j$$

and we set $u_{k+1} = u$ if $u > u_k$. Otherwise, $u_{k+1} = u_k$. After updating u_{k+1} we also update

$$\mathbf{v}_{k+1} = (\mathbf{AX}_k\mathbf{X}_k\mathbf{A}^\mathsf{T})^{-1}\mathbf{AX}_k\mathbf{X}_k(\mathbf{c} - u_{k+1}\mathbf{e})$$

We then compute the steepest descent direction $\mathbf{d}_k$ by calculating:

$$\mathbf{B} = \begin{bmatrix} \mathbf{A}Diag(\mathbf{x}_k) \\ \mathbf{e}^\mathsf{T} \end{bmatrix}$$
$$\mathbf{P} = I - \mathbf{B}^\mathsf{T}(\mathbf{BB}^\mathsf{T})^{-1}\mathbf{B}$$
$$\mathbf{c}_p = \mathbf{P}Diag(\mathbf{x}_k)(\mathbf{c} - u_{k+1}\mathbf{e})$$
$$\mathbf{d}_k = -\frac{\mathbf{c}_p}{\|\mathbf{c}_p\|_2}$$

Steps 3–4: in these step we compute $\mathbf{y}_{k+1} = \mathbf{y}_k + a_k\mathbf{d}_k$ and map $\mathbf{y}_{k+1}$ back to the x-space:

$$\mathbf{x}_{k+1} = \frac{[Diag(\mathbf{x}_k)]\mathbf{y}_{k+1}}{\mathbf{e}^\mathsf{T}[Diag(\mathbf{x}_k)]\mathbf{y}_{k+1}}$$

Now, if $(\mathbf{c} - u_{k+1}\mathbf{e})^\mathsf{T}\mathbf{x}_{k+1}$ is close to 0, we terminate the algorithm. Otherwise, we set $k = k + 1$ and we repeat steps 1–4.

Let us implement these steps in the following example provided by [16]:

$$\min \ 2x_1 + x_2 + x_3$$
$$\text{s.t. } 2x_1 + x_2 - 3x_3 = 0$$
$$x_1 + x_2 + x_3 = 1$$
$$x_1, x_2, x_3 \geq 0$$

Note that the initial interior point $\{x_1, x_2, x_3\} = \{\frac{1}{3}, \frac{1}{3}, \frac{1}{3}\}$ is indeed a feasible solution of the homogeneous form of the LP, but the objective function at the optimal solution is not zero (in fact, as we will later see, it is $z^* = 1$). We initially have $\mathbf{c} = [2, 1, 1]^\mathsf{T}$ and $\mathbf{A} = [2, 1, -3]$. The problem has only one equality constraint in $\mathbf{A}$. Thus, $l = 1$ and the dual variables are v^k, u_k. At step 0 we set $k = 0$ and we have:

$$\mathbf{v}_k = (\mathbf{A}\mathbf{A}^\mathsf{T})^{-1}\mathbf{A}\mathbf{c} \simeq 0.1429$$
$$u_k = \min_{j=1,\dots,n} \{\mathbf{c} - \mathbf{A}^\mathsf{T}\mathbf{v}_k\}_j \simeq 0.8571$$

Proceeding to step 1, starting with feasible solution $\mathbf{x}_k = [\frac{1}{3}, \frac{1}{3}, \frac{1}{3}]^\mathsf{T}$ we have the projective transformation to $\mathbf{y}_k = [\frac{1}{3}, \frac{1}{3}, \frac{1}{3}]^\mathsf{T}$. In step 2 we update the dual variables in the transformed space as:

$$\mathbf{v}_{k+1} = (\mathbf{A}\mathbf{X}_k\mathbf{X}_k\mathbf{A}^\mathsf{T})^{-1}\mathbf{A}\mathbf{X}_k\mathbf{X}_k(\mathbf{c} - u_k\mathbf{e}) \simeq 0.1429$$
$$u = \min_{j=1,\dots,n} \{\mathbf{c} - \mathbf{A}^\mathsf{T}\mathbf{v}_k\}_j \simeq 0.8571$$

Note that $u \leq u_k$, thus $u_{k+1} = u_k = 0.8571$ and

$$\mathbf{v}_{k+1} = (\mathbf{A}\mathbf{X}_k\mathbf{X}_k\mathbf{A}^\mathsf{T})^{-1}\mathbf{A}\mathbf{X}_k\mathbf{X}_k(\mathbf{c} - u_{k+1}\mathbf{e}) \simeq 0.1429$$

We now compute the steepest descent direction $\mathbf{d}_k$ by calculating:

$$\mathbf{B} = \begin{bmatrix} \mathbf{A}Diag(\mathbf{x}_k) \\ \mathbf{e}^\mathsf{T} \end{bmatrix} = \begin{bmatrix} 2/3 & 1/3 & -1 \\ 1 & 1 & 1 \end{bmatrix}$$

$$\mathbf{P} = \mathbf{I} - \mathbf{B}^\mathsf{T}(\mathbf{B}\mathbf{B}^\mathsf{T})^{-1}\mathbf{B} \simeq \begin{bmatrix} 0.381 & -0.476 & 0.095 \\ -0.476 & 0.595 & -0.119 \\ 0.095 & -0.119 & 0.024 \end{bmatrix}$$

$$\mathbf{c}_p = \mathbf{P}Diag(\mathbf{x}_k)(\mathbf{c} - u_{k+1}\mathbf{e}) \simeq \begin{bmatrix} 0.1270 \\ -0.1587 \\ 0.0317 \end{bmatrix}$$

$$\mathbf{d}_k = -\frac{\mathbf{c}_p}{\|\mathbf{c}_p\|_2} \simeq \begin{bmatrix} -0.6172 \\ 0.7715 \\ -0.1543 \end{bmatrix}$$

We now proceed to steps 3-4. Consider $\theta = \frac{n-1}{3n}$, $r = [n(n-1)]^{-\frac{1}{2}}$ and $a_k = \theta r \simeq 0.09072$. We have:

$$\mathbf{y}_{k+1} = \mathbf{y}_k + a_k\mathbf{d}_k = \begin{bmatrix} 1/3 \\ 1/3 \\ 1/3 \end{bmatrix} + 0.09072 \begin{bmatrix} -0.6172 \\ 0.7715 \\ -0.1543 \end{bmatrix} \simeq \begin{bmatrix} 0.27734 \\ 0.40333 \\ 0.31933 \end{bmatrix}$$

Mapping $\mathbf{y}_{k+1}$ back to the x-space we have:

$$\mathbf{x}_{k+1} = \frac{[Diag(\mathbf{x}_k)]\mathbf{y}_{k+1}}{\mathbf{e}^{\mathsf{T}}[Diag(\mathbf{x}_k)]\mathbf{y}_{k+1}} = \begin{bmatrix} 0.27734 \\ 0.40333 \\ 0.31933 \end{bmatrix}$$

After this iteration $(\mathbf{c} - u_{k+1}\mathbf{e})^{\mathsf{T}}\mathbf{x}_{k+1} \simeq 0.4202$ which is still far from 0. Note also that the actual objective function $\mathbf{c}^{\mathsf{T}}\mathbf{x}_{k+1} \simeq 1.2773$. We will later see that when $(\mathbf{c} - u_{k+1}\mathbf{e})^{\mathsf{T}}\mathbf{x}_{k+1} \rightarrow 0, \mathbf{c}^{\mathsf{T}}\mathbf{x}_{k+1} \rightarrow 1$ indicating that optimal objective function value of the problem is $z^* = \mathbf{c}^{\mathsf{T}}\mathbf{x}^* = 1$ and the optimal dual solution $u^* = z^* = 1$.

Let us now set $k = k + 1$ and proceed to the next iteration where we will repeat steps 1–4.

Proceeding to step 1, starting with $\mathbf{x}_k = [0.27734, 0.40333, 0.31933]^{\mathsf{T}}$ we have the projective transformation to $\mathbf{y}_k = [\frac{1}{3}, \frac{1}{3}, \frac{1}{3}]^{\mathsf{T}}$. In step 2 we update the dual variables in the transformed space as:

$$\mathbf{v}_{k+1} = (\mathbf{A}\mathbf{X}_k\mathbf{X}_k\mathbf{A}^{\mathsf{T}})^{-1}\mathbf{A}\mathbf{X}_k\mathbf{X}_k(\mathbf{c} - u_k\mathbf{e}) \simeq 0.1119$$
$$u = \min_{j=1,\ldots,n} \{\mathbf{c} - \mathbf{A}^{\mathsf{T}}\mathbf{v}_k\}_j \simeq 0.8881$$

Note that $u > u^k$, thus $u_{k+1} = u = 0.8881$ and

$$\mathbf{v}_{k+1} = (\mathbf{A}\mathbf{X}_k\mathbf{X}_k\mathbf{A}^{\mathsf{T}})^{-1}\mathbf{A}\mathbf{X}_k\mathbf{X}_k(\mathbf{c} - u_{k+1}\mathbf{e}) \simeq 0.1117$$

We now compute the steepest descent direction $\mathbf{d}_k$ by calculating:

$$\mathbf{B} = \begin{bmatrix} \mathbf{A}Diag(\mathbf{x}_k) \\ \mathbf{e}^{\mathsf{T}} \end{bmatrix} \simeq \begin{bmatrix} 0.5547 & 0.4033 & -0.9580 \\ 1 & 1 & 1 \end{bmatrix}$$

$$\mathbf{P} = \mathbf{I} - \mathbf{B}^{\mathsf{T}}(\mathbf{B}\mathbf{B}^{\mathsf{T}})^{-1}\mathbf{B} \simeq \begin{bmatrix} 0.445 & -0.494 & 0.049 \\ -0.494 & 0.549 & -0.055 \\ 0.049 & -0.0550.0055 \end{bmatrix}$$

$$\mathbf{c}_p = \mathbf{P}Diag(\mathbf{x}_k)(\mathbf{c} - u_{k+1}\mathbf{e}) \simeq \begin{bmatrix} 0.1167 \\ -0.1297 \\ 0.0130 \end{bmatrix}$$

$$\mathbf{d}_k = -\frac{\mathbf{c}_p}{\|\mathbf{c}_p\|_2} \simeq \begin{bmatrix} -0.6671 \\ 0.7413 \\ -0.0742 \end{bmatrix}$$

We now proceed to steps 3-4 where we have $a_k = \theta r \simeq 0.09072$ and:

$$\mathbf{y}_{k+1} = \mathbf{y}_k + a_k\mathbf{d}_k = \begin{bmatrix} 1/3 \\ 1/3 \\ 1/3 \end{bmatrix} + 0.09072 \begin{bmatrix} -0.6671 \\ 0.7413 \\ -0.0742 \end{bmatrix} \simeq \begin{bmatrix} 0.27281 \\ 0.40058 \\ 0.32660 \end{bmatrix}$$

Mapping $\mathbf{y}_{k+1}$ back to the x-space we have:

$$\mathbf{x}_{k+1} = \frac{[Diag(\mathbf{x}_k)]\mathbf{y}_{k+1}}{\mathbf{e}^{\mathsf{T}}[Diag(\mathbf{x}_k)]\mathbf{y}_{k+1}} = \begin{bmatrix} 0.22154 \\ 0.47307 \\ 0.30531 \end{bmatrix}$$

After this iteration $(\mathbf{c} - u_{k+1}\mathbf{e})^{\mathsf{T}}\mathbf{x}_{k+1} \simeq 0.3335$ and $\mathbf{c}^{\mathsf{T}}\mathbf{x}_{k+1} \simeq 1.2215$. Proceeding to the next iterations, we have the results of Table 4.11 when considering convergence tolerance $\tau = 0.0001$ and termination criterion $|(\mathbf{c} - u_{k+1}\mathbf{e})^{\mathsf{T}}\mathbf{x}_{k+1}| \leq \tau$.

From Table 4.11 we can validate some previously made statements. First, the value of the dual variable $u_{k+1} = 1$ at the optimal solution is almost equal to the value of $z^* = \mathbf{c}^{\mathsf{T}}\mathbf{x}_{k+1} = 1.0001$. This was expected from the strong duality theorem. In addition, u_{k+1} is a lower bound of $\mathbf{c}^{\mathsf{T}}\mathbf{x}_{k+1}$ at each iteration. Further, the optimal solution of the primal $\mathbf{x}_k = [0.0001, 0.7499, 0.25]^{\mathsf{T}}$ is very close to the globally optimal solution, which is $[0, 0.75, 0.25]^{\mathsf{T}}$. The reason of their small difference is that we allow a convergence tolerance of 0.0001 for the algorithm's termination.

4.5.5.3   Converting an LP to its Homogeneous Form

Closing the presentation of Karmarkar's method, although it might seem unlikely, we can write any LP to its homogeneous form. This, however, might require increasing the number of variables and constraints of the initial problem. Tomlin [20] proposed a conversion method that turns an LP to its homogeneous form and returns an initial feasible solution. We previously showed that any LP can be written in its standard form:

$$\max \ \mathbf{c}^{\mathsf{T}}\mathbf{x}$$
$$\text{s.t. } \mathbf{Ax} = \mathbf{b}$$
$$\mathbf{x} \geq \mathbf{0}$$

To turn it to its homogeneous form, we start first by changing its objective function to min $\tilde{\mathbf{c}}^{\mathsf{T}}\mathbf{x}$ where $\tilde{\mathbf{c}} = -\mathbf{c}$. If the LP in its standard form has l equality constraints and n variables, we introduce variables $\mathbf{y} = [y_1, \ldots, y_n]^{\mathsf{T}}$ and we rescale the problem by replacing $\mathbf{x}$ by $\mathbf{y} = \mathbf{x}/\sigma$, where σ is chosen in order to ensure that any feasible solution $\mathbf{x}$ satisfies $\sum_{j=1}^{n} x_j < \sigma$ (note that such σ exists if the original problem is not unbounded). After rescaling, the original LP becomes:

$$\min \ \sigma\tilde{\mathbf{c}}^{\mathsf{T}}\mathbf{y}$$
$$\text{s.t. } \mathbf{Ay} = \frac{\mathbf{b}}{\sigma}$$
$$\mathbf{y} \geq \mathbf{0}$$

Table 4.11 Karmarkar's algorithm iterations for a non-zero objective function value at the optimal solution

k	$\mathbf{x}^k$	v^{k+1}	u_{k+1}	$\mathbf{y}_{k+1}$	$\mathbf{x}_{k+1}$	$(\mathbf{c}-u_{k+1}\mathbf{e})^T\mathbf{x}_{k+1}$	$\mathbf{c}^T\mathbf{x}_{k+1}$
0	$[1/3, 1/3, 1/3]^T$	0.1429	0.8571	$[0.2773, 0.4033, 0.3193]^T$	$[0.2773, 0.4033, 0.3193]^T$	0.4202	1.2773
1	$[0.2773, 0.4033, 0.3193]^T$	0.1117	0.8881	$[0.2728, 0.4006, 0.3266]^T$	$[0.2215, 0.4731, 0.3053]^T$	0.3335	1.2215
2	$[0.2215, 0.4731, 0.3053]^T$	0.0807	0.9183	$[0.2685, 0.3968, 0.3347]^T$	$[0.1702, 0.5372, 0.2926]^T$	0.2519	1.1702
3	$[0.1702, 0.5372, 0.2926]^T$	0.0536	0.9444	$[0.2650, 0.3922, 0.3428]^T$	$[0.1267, 0.5917, 0.2817]^T$	0.1822	1.1267
4	$[0.1267, 0.5917, 0.2817]^T$	0.0330	0.9645	$[0.2625, 0.3875, 0.3500]^T$	$[0.0921, 0.6350, 0.2730]^T$	0.1276	1.0921
5	$[0.0921, 0.6349, 0.273]^T$	0.0191	0.9784	$[0.261, 0.3833, 0.3558]^T$	$[0.0659, 0.6676, 0.2665]^T$	0.0875	1.0659
6	$[0.0659, 0.6676, 0.2665]^T$	0.0107	0.9874	$[0.2601, 0.3798, 0.3601]^T$	$[0.0468, 0.6915, 0.2617]^T$	0.0594	1.0468
7	$[0.0468, 0.6915, 0.2617]^T$	0.0058	0.9929	$[0.2597, 0.3771, 0.3632]^T$	$[0.033, 0.7087, 0.2583]^T$	0.0401	1.0330
8	$[0.033, 0.7087, 0.2583]^T$	0.0031	0.9961	$[0.2595, 0.3752, 0.3653]^T$	$[0.0232, 0.721, 0.2558]^T$	0.0272	1.0232
9	$[0.0232, 0.721, 0.2558]^T$	0.0016	0.9979	$[0.2594, 0.3738, 0.3669]^T$	$[0.0163, 0.7296, 0.2541]^T$	0.0184	1.0163
10	$[0.0163, 0.7296, 0.2541]^T$	0.0008	0.9989	$[0.2593, 0.3728, 0.3679]^T$	$[0.0114, 0.7357, 0.2529]^T$	0.0126	1.0114
11	$[0.0114, 0.7357, 0.2529]^T$	0.0004	0.9994	$[0.2593, 0.3721, 0.3687]^T$	$[0.008, 0.74, 0.252]^T$	0.0086	1.0080
12	$[0.008, 0.74, 0.252]^T$	0.0002	0.9997	$[0.2593, 0.3716, 0.3692]^T$	$[0.0056, 0.743, 0.2514]^T$	0.0059	1.0056
13	$[0.0056, 0.743, 0.2514]^T$	0.0001	0.9998	$[0.2593, 0.3712, 0.3695]^T$	$[0.0039, 0.7451, 0.251]^T$	0.0041	1.0039
14	$[0.0039, 0.7451, 0.251]^T$	0.0001	0.9999	$[0.2593, 0.371, 0.3698]^T$	$[0.0028, 0.7466, 0.2507]^T$	0.0028	1.0028
15	$[0.0028, 0.7466, 0.2507]^T$	0.0000	1.0000	$[0.2593, 0.3708, 0.37]^T$	$[0.0019, 0.7476, 0.2505]^T$	0.0020	1.0019
16	$[0.0019, 0.7476, 0.2505]^T$	0.0000	1.0000	$[0.2593, 0.3707, 0.3701]^T$	$[0.0014, 0.7483, 0.2503]^T$	0.0014	1.0014
17	$[0.0014, 0.7483, 0.2503]^T$	0.0000	1.0000	$[0.2593, 0.3706, 0.3702]^T$	$[0.0009, 0.7488, 0.2502]^T$	0.0010	1.0009
18	$[0.0009, 0.7488, 0.2502]^T$	0.0000	1.0000	$[0.2593, 0.3705, 0.3702]^T$	$[0.0007, 0.7492, 0.2502]^T$	0.0007	1.0007
19	$[0.0007, 0.7492, 0.2502]^T$	0.0000	1.0000	$[0.2593, 0.3705, 0.3703]^T$	$[0.0005, 0.7494, 0.2501]^T$	0.0005	1.0005
20	$[0.0005, 0.7494, 0.2501]^T$	0.0000	1.0000	$[0.2593, 0.3704, 0.3703]^T$	$[0.0003, 0.7496, 0.2501]^T$	0.0003	1.0003
21	$[0.0003, 0.7496, 0.2501]^T$	0.0000	1.0000	$[0.2593, 0.3704, 0.3703]^T$	$[0.0002, 0.7497, 0.2501]^T$	0.0002	1.0002
22	$[0.0002, 0.7497, 0.2501]^T$	0.0000	1.0000	$[0.2593, 0.3704, 0.3703]^T$	$[0.0002, 0.7498, 0.25]^T$	0.0002	1.0002
23	$[0.0002, 0.7498, 0.25]^T$	0.0000	1.0000	$[0.2593, 0.3704, 0.3703]^T$	$[0.0001, 0.7499, 0.25]^T$	0.0001	1.0001
24	$[0.0001, 0.7499, 0.25]^T$	0.0000	1.0000	$[0.2593, 0.3704, 0.3704]^T$	$[0.0001, 0.7499, 0.25]^T$	0.0001	1.0001

Instead of solving this problem, we introduce slack variables $y_{n+1}, y_{n+2}, y_{n+3}$ and parameter M that has a very large positive value. We then solve the related problem which is in a homogeneous form and the center of its simplex $[\mathbf{y}, y_{n+1}, y_{n+2}, y_{n+3}]^\mathsf{T} = [1/(n+3), \ldots, 1/(n+3)]^\mathsf{T}$ is a feasible solution:

$$\min \; [\sigma\tilde{\mathbf{c}}^\mathsf{T} \; 0 \; 0 \; M] \begin{bmatrix} \mathbf{y} \\ y_{n+1} \\ y_{n+2} \\ y_{n+3} \end{bmatrix}$$

$$\text{s.t.} \; \begin{bmatrix} \mathbf{A} & 0 & -(n+3)\frac{\mathbf{b}}{\sigma} & (n+3)\frac{\mathbf{b}}{\sigma} - \mathbf{Ae} \\ -\mathbf{e}^\mathsf{T} & -1 & n+2 & -1 \end{bmatrix} \begin{bmatrix} \mathbf{y} \\ y_{n+1} \\ y_{n+2} \\ y_{n+3} \end{bmatrix} = \begin{bmatrix} 0 \\ 0 \end{bmatrix}$$

$$\mathbf{e}^\mathsf{T}\mathbf{y} + y_{n+1} + y_{n+2} + y_{n+3} = 1$$

$$y_1, \ldots, y_n, y_{n+1}, y_{n+2}, y_{n+3} \geq 0$$

We first show that the center of the simplex $[\mathbf{y}, y_{n+1}, y_{n+2}, y_{n+3}]^\mathsf{T} = [1/(n+3), \ldots, 1/(n+3)]^\mathsf{T}$ is always a feasible solution of our reformulated problem. Let us replace the values of the variables by $1/(n+3)$ in our equality constraints. Note that if every variable takes the value of $1/(n+3)$ we can write $\mathbf{y}$ as $\frac{\mathbf{e}}{n+3}$. Then we have:

$$\mathbf{A}\frac{\mathbf{e}}{n+3} - (n+3)\frac{\mathbf{b}}{\sigma}\frac{1}{n+3} + [(n+3)\frac{\mathbf{b}}{\sigma} - \mathbf{Ae}]\frac{1}{n+3} =$$

$$\mathbf{A}\frac{\mathbf{e}}{n+3} - \frac{\mathbf{b}}{\sigma} + \frac{\mathbf{b}}{\sigma} - \mathbf{A}\frac{\mathbf{e}}{n+3} = 0$$

satisfying the first set of equality constraints. In addition, $-\mathbf{e}^\mathsf{T}\mathbf{y} - y_{n+1} + (n+2)y_{n+2} - y_{n+3}$ becomes:

$$-\mathbf{e}^\mathsf{T}\frac{\mathbf{e}}{n+3} - \frac{1}{n+3} + \frac{n+2}{n+3} - \frac{1}{n+3} = -\frac{n}{n+3} - \frac{2}{n+3} + \frac{n+2}{n+3} = 0$$

satisfying the second set of equality constraints. Finally, $\mathbf{e}^\mathsf{T}\mathbf{y} + y_{n+1} + y_{n+2} + y_{n+3}$ becomes:

$$\mathbf{e}^\mathsf{T}\frac{\mathbf{e}}{n+3} + \frac{1}{n+3} + \frac{1}{n+3} + \frac{1}{n+3} = \frac{n+3}{n+3} = 1$$

satisfying the last set of equality constraints. Thus, the center of the simplex of the reformulated program is a feasible solution. It now remains to show that our reformulated, homogeneous LP has the same optimal solution as the min $\sigma\tilde{\mathbf{c}}^\mathsf{T}\mathbf{y}$ s.t. $\mathbf{Ay} = \frac{\mathbf{b}}{\sigma}$, $\mathbf{y} \geq \mathbf{0}$ if the latter is feasible. We write again the equations related to the equality constraints of our homogeneous LP:

$$\mathbf{Ay} - (n + 3)\frac{\mathbf{b}}{\sigma}y_{n+2} + [(n + 3)\frac{\mathbf{b}}{\sigma} - \mathbf{Ae}]y_{n+3} = 0 \tag{4.77}$$

$$-\mathbf{e}^{\mathsf{T}}\mathbf{y} - y_{n+1} + (n + 2)y_{n+2} - y_{n+3} = 0 \tag{4.78}$$

$$\mathbf{e}^{\mathsf{T}}\mathbf{y} + y_{n+1} + y_{n+2} + y_{n+3} = 1 \tag{4.79}$$

Notice that any feasible solution of our homogeneous LP has to set the values of the slack variables $y_{n+1}, y_{n+2}, y_{n+3}$ such that all its constraints are satisfied. By adding equation (4.79) to equation (4.78) we have:

$$(n + 2)y_{n+2} + y_{n+2} = 1 \Rightarrow y_{n+2} = \frac{1}{n + 3}$$

Thus, y_{n+2} is a slack variable that should always be equal to $\frac{1}{n+3}$. Replacing the value of y_{n+2} in (4.79) we have:

$$y_{n+1} = 1 - \mathbf{e}^{\mathsf{T}}\mathbf{y} - \frac{1}{n + 3} - y_{n+3}$$

This means that slack variable y_{n+1} absorbs the difference between 1 and $\mathbf{e}^{\mathsf{T}}\mathbf{y} + y_{n+2} + y_{n+3}$ so that equation (4.79) can be satisfied for any $\mathbf{y}$ which is a feasible solution of min $\sigma\tilde{\mathbf{c}}^{\mathsf{T}}\mathbf{y}$ s.t. $\mathbf{Ay} = \frac{\mathbf{b}}{\sigma}$, $\mathbf{y} \geq \mathbf{0}$. Thus, it only remains to show that our homogeneous LP forces $\mathbf{y}$ to satisfy $\mathbf{Ay} = \frac{\mathbf{b}}{\sigma}$ at the optimal solution. The objective function of our homogeneous LP can be written as:

$$\text{min }\sigma\tilde{\mathbf{c}}^{\mathsf{T}}\mathbf{y} + My_{n+3}$$

We now have two cases:

- If $y_{n+3} = 0$ at the optimal solution of the homogeneous LP, then the homogeneous LP and min $\sigma\tilde{\mathbf{c}}^{\mathsf{T}}\mathbf{y}$ s.t. $\mathbf{Ay} = \frac{\mathbf{b}}{\sigma}$, $\mathbf{y} \geq \mathbf{0}$ have the same optimal solution y^* resulting in solution $\mathbf{x}^* = \sigma y^*$ of the original LP.
- If $y_{n+3} > 0$ at the optimal solution of the homogeneous LP, then min $\sigma\tilde{\mathbf{c}}^{\mathsf{T}}\mathbf{y}$ s.t. $\mathbf{Ay} = \frac{\mathbf{b}}{\sigma}$, $\mathbf{y} \geq \mathbf{0}$ and, subsequently, the original LP are infeasible.

In more detail, observe that if $y_{n+3} = 0$ at the optimal solution $\mathbf{y}^*$ of the homogeneous LP, then:

$$\sigma\tilde{\mathbf{c}}^{\mathsf{T}}\mathbf{y}^* + My_{n+3} = \sigma\tilde{\mathbf{c}}^{\mathsf{T}}\mathbf{y}^*$$

In addition, $\mathbf{y}^*$ in our homogeneous LP satisfies:

$$\mathbf{Ay}^* - (n + 3)\frac{\mathbf{b}}{\sigma}y_{n+2} + ((n + 3)\frac{\mathbf{b}}{\sigma} - \mathbf{Ae})y_{n+3} = 0$$

which, for $y_{n+2} = \frac{1}{n+3}$ and $y_{n+3} = 0$, gives us:

$$\mathbf{Ay}^* = \frac{\mathbf{b}}{\sigma}$$

meaning that $\mathbf{y}^*$ is also the optimal solution of min $\sigma\tilde{\mathbf{c}}^\mathsf{T}\mathbf{y}$ s.t. $\mathbf{Ay} = \frac{\mathbf{b}}{\sigma}$, $\mathbf{y} \geq \mathbf{0}$. Consider the application of this approach to the following example of an LP in standard form.

$$\text{Standard Form LP:} \quad \max \; -x_1 - 2x_2$$
$$\text{s.t. } x_1 + x_2 - x_3 = 2$$
$$3x_1 - x_2 = 0 \tag{4.80}$$
$$x_1, x_2, x_3 \geq 0$$

The standard form LP can be transformed to its homogeneous form by introducing slack variables y_4, y_5, y_6 since $n = 3$. In addition, we replace $\mathbf{x}$ by $\mathbf{y}$ such that $\mathbf{y} = \frac{\mathbf{x}}{\sigma}$ where σ ensures that any feasible solution $\mathbf{x}$ satisfies $\sum_{j=1}^{n} x_j \leq \sigma$. The homogeneous LP is:

Homogeneous LP:

$$\min \; \sigma y_1 + 2\sigma y_2 + M y_6$$

$$\text{s.t. } \begin{bmatrix} 1 & 1 & -1 \\ 3 & -1 & 0 \end{bmatrix} \begin{bmatrix} y_1 \\ y_2 \\ y_3 \end{bmatrix} - \frac{6y_5}{\sigma} \begin{bmatrix} 2 \\ 0 \end{bmatrix} + 6y_6 \left(\frac{1}{\sigma} \begin{bmatrix} 2 \\ 0 \end{bmatrix} - \begin{bmatrix} 1 & 1 & -1 \\ 3 & -1 & 0 \end{bmatrix} \begin{bmatrix} 1 \\ 1 \\ 1 \end{bmatrix} \right) = 0$$

$$- y_1 - y_2 - y_3 - y_4 + 5y_5 - y_6 = 0$$
$$y_1 + y_2 + y_3 + y_4 + y_5 + y_6 = 0$$
$$y_1, y_2, y_3, y_4, y_5, y_6 \geq 0$$

For $\sigma = 10$ and a large positive $M = 10000$ we can solve the homogeneous form LP with Karmarkar's algorithm for problems with unknown optimal objective function using $\mathbf{c} = [10, 20, 0, 0, 0, 10000]^\mathsf{T}$ and

$$\mathbf{A} = \begin{bmatrix} 1 & 1 & -1 & 0 & -1.2 & 0.2 \\ 3 & -1 & 0 & 0 & 0 & -2 \\ -1 & -1 & -1 & -1 & 5 & -1 \end{bmatrix}$$

For a tolerance of $\tau = 0.0001$ we have solution $[y_1, y_2, y_3, y_4, y_5, y_6]^\mathsf{T} \simeq [0.05, 0.15, 0, 0.633, 0.166, 0]^\mathsf{T}$. The optimal objective function score $z^* = 10y_1 + 20y_2 = 3.5$. Because $y_6 = 0$, the original LP is feasible and its solution is $\mathbf{x} = \sigma[y_1^*, y_2^*, y_3^*]^\mathsf{T} = [0.5, 1.5, 0]^\mathsf{T}$ with objective function value -3.5.

Fig. 4.8 Convergence of simplex and Karmarkar's method when solving the problem in (4.44)–(4.47)

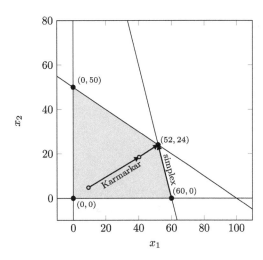

Practitioner's Corner
Interior Point for LP in Python 3

To implement the interior point method for linear programming, one can use the `scipy.optimize.linprog` in Python 3. This library uses the MOSEK interior point algorithm of [21]. An example of solving the LP in (4.80) is provided below.

```
from scipy.optimize import linprog,
    linprog_verbose_callback
c=[1,2,0]
A=[[1,1,-1],[3,-1,0]]
b=[2,0]
x_bounds = [(0, None) for i in range(0,len(c))]
res = linprog(c, A_eq=A, b_eq=b, method='interior-
    point', bounds=x_bounds, callback=
    linprog_verbose_callback,options={'tol':1e-8})
print(res)
```

resulting in solution $\mathbf{x}^* = [0.5, 1.5, 0.0]^{\mathsf{T}}$ with objective function score 3.5 which, because (4.80) was a maximization problem, should be turned to -3.5.

To highlight the difference between the simplex solution method and the interior point solution method, let us consider the maximization problem expressed in (4.44)–(4.47). Solving both problems with `linprog` using simplex and interior-point as solution methods will result in the convergence to the optimal point $[52, 24]^{\mathsf{T}}$ after 1 and 2 iterations, respectively (see Fig. 4.8).

4.6 Quadratic Programming

Quadratic programming (QP) is a special case of nonlinear programming. A quadratic program has a quadratic objective function $f : \mathbb{R}^n \to \mathbb{R}$ and affine equality/inequality constraints. A quadratic program can be written in the following form.

Quadratic Program

$$\min_{\mathbf{x} \in \mathbb{R}^n} \mathbf{c}^{\mathsf{T}}\mathbf{x} + \frac{1}{2}\mathbf{x}^{\mathsf{T}}\mathbf{Q}\mathbf{x} \tag{4.81}$$

$$\text{s.t. } \mathbf{A}\mathbf{x} \leq \mathbf{b} \tag{4.82}$$

$$\mathbf{E}\mathbf{x} = \mathbf{d} \tag{4.83}$$

where $\mathbf{A}$ is an $m \times n$ matrix, where m is the number of inequality constraints and $\mathbf{E}$ is an $l \times n$ matrix with $l \leq n$, where l is the number of equality constraints. In addition, $\mathbf{Q}$ is an $n \times n$ symmetric matrix, $\mathbf{c}$ a column vector with n values, and $\mathbf{x} \in \mathbb{R}^n$. Note that $\mathbf{Q}$ is the Hessian matrix of f:

$$\mathbf{Q} = \begin{bmatrix} \frac{\partial^2 f(\mathbf{x})}{\partial x_1^2} & \frac{\partial^2 f(\mathbf{x})}{\partial x_1 x_2} & \cdots & \frac{\partial^2 f(\mathbf{x})}{\partial x_1 x_n} \\ \frac{\partial^2 f(\mathbf{x})}{\partial x_2 x_1} & \frac{\partial^2 f(\mathbf{x})}{\partial x_2^2} & \cdots & \frac{\partial^2 f(\mathbf{x})}{\partial x_2 x_n} \\ \vdots & \vdots & \ddots & \vdots \\ \frac{\partial^2 f(\mathbf{x})}{\partial x_n x_1} & \frac{\partial^2 f(\mathbf{x})}{\partial x_n x_2} & \cdots & \frac{\partial^2 f(\mathbf{x})}{\partial x_n^2} \end{bmatrix}$$

where $f(\mathbf{x})$ is the objective function of the original problem. In addition, any constants contained in the objective function $f(\mathbf{x})$ are left out of the general formulation in (4.81)–(4.83) because they do not affect the solution search.

The feasible region of a QP is a convex set because it consists of affine equality and inequality constraint functions. It thus suffices to explore the convexity of the objective function $\mathbf{c}^{\mathsf{T}}\mathbf{x} + \frac{1}{2}\mathbf{x}^{\mathsf{T}}\mathbf{Q}\mathbf{x}$. Consequently, if $\mathbf{Q}$ is positive semidefinite the QP is *convex*, whereas if it is positive definite the QP is *strictly convex*. If $\mathbf{Q}$ is indefinite, then the QP is not convex and it can have several local minima that are not necessarily global minima resulting in an NP-Hard problem [22, 23].

4.6.1 Equality-Constrained Convex Quadratic Programs

We start from the simpler case of a convex QP with equality constraints:

$$\min \mathbf{c}^{\mathsf{T}}\mathbf{x} + \frac{1}{2}\mathbf{x}^{\mathsf{T}}\mathbf{Q}\mathbf{x} \tag{4.84}$$

$$\text{s.t. } \mathbf{A}\mathbf{x} = \mathbf{b} \tag{4.85}$$

where $\mathbf{x} \in \mathbb{R}^n$, $\mathbf{c} \in \mathbb{R}^n$, $\mathbf{A}$ an $l \times n$ matrix with $l \leq n$ and *full row rank l*, and $\mathbf{b} \in \mathbb{R}^l$. Because we have only equality constraints and the program is convex, the first-order necessary conditions of multi-variable minimization problems with equality constraints are also sufficient for finding a globally optimal solution. The Lagrangian function of our program is:

$$\mathcal{L}(\mathbf{x}, \boldsymbol{\lambda}) = \mathbf{c}^\mathsf{T}\mathbf{x} + \frac{1}{2}\mathbf{x}^\mathsf{T}\mathbf{Q}\mathbf{x} + \boldsymbol{\lambda}^\mathsf{T}(\mathbf{A}\mathbf{x} - \mathbf{b})$$

and we seek $\mathbf{x}^*$ and Lagrangian multipliers $\boldsymbol{\lambda}^* \in \mathbb{R}^l$ such that $\nabla\mathcal{L}(\mathbf{x}^*, \boldsymbol{\lambda}^*) = \mathbf{0}$. Using the denominator layout notation:

$$\nabla_\mathbf{x}\mathcal{L}(\mathbf{x}, \boldsymbol{\lambda}) = \mathbf{0} \Rightarrow \nabla_\mathbf{x}\mathbf{c}^\mathsf{T}\mathbf{x} + \frac{1}{2}\nabla_\mathbf{x}\mathbf{x}^\mathsf{T}\mathbf{Q}\mathbf{x} + \nabla_\mathbf{x}\boldsymbol{\lambda}^\mathsf{T}(\mathbf{A}\mathbf{x} - \mathbf{b}) = \mathbf{0}$$

$$\Rightarrow \mathbf{c} + \frac{1}{2}(\mathbf{Q} + \mathbf{Q}^\mathsf{T})\mathbf{x} + \mathbf{A}^\mathsf{T}\boldsymbol{\lambda} = \mathbf{0}$$

Because $\mathbf{Q}$ is symmetric, $\mathbf{Q}^\mathsf{T} = \mathbf{Q}$. Thus,

$$\nabla_\mathbf{x}\mathcal{L}(\mathbf{x}, \boldsymbol{\lambda}) = \mathbf{0} \Rightarrow \mathbf{Q}\mathbf{x} + \mathbf{A}^\mathsf{T}\boldsymbol{\lambda} = -\mathbf{c} \tag{4.86}$$

We also have:

$$\nabla_\boldsymbol{\lambda}\mathcal{L}(\mathbf{x}, \boldsymbol{\lambda}) = \mathbf{0} \Rightarrow \nabla_\boldsymbol{\lambda}\boldsymbol{\lambda}^\mathsf{T}\mathbf{A}\mathbf{x} - \nabla_\boldsymbol{\lambda}\boldsymbol{\lambda}^\mathsf{T}\mathbf{b} = \mathbf{0} \Rightarrow \mathbf{A}\mathbf{x} - \mathbf{b} = \mathbf{0} \tag{4.87}$$

Thus, $\nabla\mathcal{L}(\mathbf{x}, \boldsymbol{\lambda}) = \mathbf{0}$ results in the linear system:

$$\begin{bmatrix} \mathbf{Q} & \mathbf{A}^\mathsf{T} \\ \mathbf{A} & \mathbf{0} \end{bmatrix} \begin{bmatrix} \mathbf{x} \\ \boldsymbol{\lambda} \end{bmatrix} = \begin{bmatrix} -\mathbf{c} \\ \mathbf{b} \end{bmatrix}$$

where matrix $\begin{bmatrix} \mathbf{Q} & \mathbf{A}^\mathsf{T} \\ \mathbf{A} & \mathbf{0} \end{bmatrix}$ is called the KKT matrix. Solving this linear system will result in the optimal values of $\mathbf{x}$, $\boldsymbol{\lambda}$. If the KKT matrix is nonsingular, then we can inverse it to compute the unique solution:

$$\begin{bmatrix} \mathbf{x} \\ \boldsymbol{\lambda} \end{bmatrix} = \begin{bmatrix} \mathbf{Q} & \mathbf{A}^\mathsf{T} \\ \mathbf{A} & \mathbf{0} \end{bmatrix}^{-1} \begin{bmatrix} -\mathbf{c} \\ \mathbf{b} \end{bmatrix}$$

More solution methods that relax the nonsingularity requirement are provided below.

4.6.1.1 Range-space Methods for Equality QPs

One set of methods for solving equality QPs is the *range-space methods*. Let us write again the system of (4.86)–(4.87) as:

$$Qx + A^\mathsf{T}\lambda = -c \tag{4.88}$$

$$Ax = b \tag{4.89}$$

Solving Eq. (4.88) for x we get:

$$x = Q^{-1}(-A^\mathsf{T}\lambda - c)$$

Substituting x to Eq. (4.89) we get:

$$AQ^{-1}(c + A^\mathsf{T}\lambda) = -b \Rightarrow AQ^{-1}c + AQ^{-1}A^\mathsf{T}\lambda = -b$$

from which we can obtain λ:

$$\lambda = (AQ^{-1}A^\mathsf{T})^{-1}(-AQ^{-1}c - b)$$

This can be written in a more compact form by using the Schur complement defined below.

Definition 4.18 The *Schur complement* of the block A of a matrix $\begin{bmatrix} A & B \\ C & D \end{bmatrix}$ is $D - CA^{-1}B$.

From this definition, we have that the Schur complement S of block Q of the KKT matrix $\begin{bmatrix} Q & A^\mathsf{T} \\ A & 0 \end{bmatrix}$ is $S = -AQ^{-1}A^\mathsf{T}$. Thus, λ can be written as:

$$\lambda = S^{-1}(AQ^{-1}c + b)$$

The approach is summarized as follows:

Algorithm 17 Solving KKT system by block elimination

1: given KKT matrix with nonsingular $Q > 0$
2: calculate the Schur complement $S = -AQ^{-1}A^\mathsf{T}$
3: determine λ by solving $S\lambda = (AQ^{-1}c + b)$
4: determine x by solving $Qx = -A^\mathsf{T}\lambda - c$

Consider, for example, the constrained QP:

$$\min \; x_1^2 - 4x_1 + 2x_2^2 - 4x_2 + x_3^2 + 1 \tag{4.90}$$

$$\text{s.t. } x_1 + 4x_2 = 3 \tag{4.91}$$

$$2x_1 + 3x_2 + x_3 = 5 \tag{4.92}$$

This problem can be written as:

$$\min \ [-4, -4, 0] \begin{bmatrix} x_1 \\ x_2 \\ x_3 \end{bmatrix} + \frac{1}{2}[x_1, x_2, x_3] \begin{bmatrix} 2\ 0\ 0 \\ 0\ 4\ 0 \\ 0\ 0\ 2 \end{bmatrix} \begin{bmatrix} x_1 \\ x_2 \\ x_3 \end{bmatrix} \tag{4.93}$$

$$\text{s.t.} \ \begin{bmatrix} 1\ 4\ 0 \\ 2\ 3\ 1 \end{bmatrix} \begin{bmatrix} x_1 \\ x_2 \\ x_3 \end{bmatrix} = \begin{bmatrix} 3 \\ 5 \end{bmatrix} \tag{4.94}$$

Note that we drop the constant term $+1$ from the objective function because it does not affect the solution of the problem. We have that:

$$\mathbf{Q} = \begin{bmatrix} 2\ 0\ 0 \\ 0\ 4\ 0 \\ 0\ 0\ 2 \end{bmatrix} \quad \mathbf{c} = \begin{bmatrix} -4 \\ -4 \\ 0 \end{bmatrix} \quad \mathbf{A} = \begin{bmatrix} 1\ 4\ 0 \\ 2\ 3\ 1 \end{bmatrix}$$

and $\mathbf{b} = [3, 5]^\mathsf{T}$. The Schur complement of block $\mathbf{Q}$ of the KKT matrix is:

$$\mathbf{S} = -\mathbf{A}\mathbf{Q}^{-1}\mathbf{A}^\mathsf{T} = \begin{bmatrix} -4.5\ \ -4 \\ -4\ \ -4.75 \end{bmatrix}$$

This results in solution:

$$\boldsymbol{\lambda}^* = \mathbf{S}^{-1}(\mathbf{A}\mathbf{Q}^{-1}\mathbf{c} + \mathbf{b}) \simeq [1.1628, -0.5581]^\mathsf{T}$$
$$\mathbf{x}^* = \mathbf{Q}^{-1}(-\mathbf{A}^\mathsf{T}\boldsymbol{\lambda} - \mathbf{c}) = [85/43, 11/43, 12/43]^\mathsf{T}$$

with objective function value -4.814. Because the objective function of the original problem included also the $+1$ constant, the objective function value of the optimal solution to the original problem is -3.814.

Practitioner's Corner
Range-space Method for equality QP in Python 3

Using `numpy.linalg`, we can write a function of the range-space method in Python 3, as follows. We can also call it to solve our example resulting in $\mathbf{x}^* = [1.976744, 0.255814, 0.279070]^\mathsf{T}$ and $\boldsymbol{\lambda}^* = [1.1627907, -0.55813953]^\mathsf{T}$.

```python
def rsm_equality_QP(Q,A,b,c):
    import numpy as np
    from numpy.linalg import solve
    A=np.array(A);Q=np.array(Q);c=np.array(c);b=np.
    array(b)
    Q_inverse=np.linalg.inv(Q)
    Schur_complement = -np.dot(np.dot(A,Q_inverse),A.T
    )
    l=solve(Schur_complement,np.dot(np.dot(A,Q_inverse
    ),c)+b)
    x=solve(Q, -np.dot(A.T, l) - c)
    return(x,l)
```

```
Q=[[2,0,0],[0,4,0],[0,0,2]]
A=[[1,4,0],[2,3,1]]
b=[3,5];c=[-4,-4,0]
print(rsm_equality_QP(Q,A,b,c))
```

4.6.1.2   Null-space Methods for Equality QPs

Convex QPs can also be solved with the more general *null-space* method that does not require the nonsigularity of $\mathbf{Q}$. The null-space method eliminates the affine equality constraints to solve an unconstrained minimization problem. In the null-space method, we introduce basis $\mathbf{Z} \in \mathbb{R}^{n \times (n-l)}$ composed of $n - l$ linearly independent column vectors of the nullspace of $\mathbf{A}$, resulting in $\mathbf{AZ} = \mathbf{0}$. We also introduce vector $\mathbf{y} \in \mathbb{R}^n$ which is any feasible solution of:

$$\mathbf{Ay} = \mathbf{b}$$

Then, we replace $\mathbf{x}$ by:

$$\mathbf{x} = \mathbf{y} + \mathbf{Zz}$$

where $\mathbf{z} \in \mathbb{R}^{n-l}$ is the new set of decision variables. Note that because $\mathbf{AZ} = \mathbf{0}$ the equality constraints of the original QP are satisfied for any $\mathbf{z}$ since $\mathbf{Ax} = \mathbf{b} \Rightarrow \mathbf{A}(\mathbf{y} + \mathbf{Zz}) = \mathbf{b} \Rightarrow \mathbf{Ay} + \mathbf{AZz} = \mathbf{b} \Rightarrow \mathbf{Ay} = \mathbf{b}$ and $\mathbf{y}$ is a feasible solution. Thus, given a feasible solution $\mathbf{y}$ and a basis $\mathbf{Z}$ of null($\mathbf{A}$) we can rewrite our equality-constrained QP as an unconstrained minimization problem with only $n - l$ variables:

$$\min_{\mathbf{z} \in \mathbb{R}^{n-l}} \frac{1}{2}(\mathbf{y} + \mathbf{Zz})^\mathsf{T} \mathbf{Q}(\mathbf{y} + \mathbf{Zz}) + \mathbf{c}^\mathsf{T}(\mathbf{y} + \mathbf{Zz})$$

which can be solved with trust-region/line search solution methods for unconstrained optimization or, directly, by setting:

$$\nabla_\mathbf{z} \left(\frac{1}{2}(\mathbf{y} + \mathbf{Zz})^\mathsf{T} \mathbf{Q}(\mathbf{y} + \mathbf{Zz}) + \mathbf{c}^\mathsf{T}(\mathbf{y} + \mathbf{Zz}) \right) = \mathbf{0}$$

in order to find $\mathbf{z}$. Let us set $\mathbf{f}(\mathbf{z}) = \mathbf{y} + \mathbf{Zz}$ to find $\mathbf{z}$ directly. Using the denominator layout notation, we have:

$$\frac{1}{2}\frac{\partial \mathbf{f}^\mathsf{T}(\mathbf{z})\mathbf{Q}\mathbf{f}(\mathbf{z})}{\partial \mathbf{z}} + \frac{\partial \mathbf{c}^\mathsf{T}\mathbf{f}(\mathbf{z})}{\partial \mathbf{z}} = \frac{1}{2}\left(\frac{\partial \mathbf{f}(\mathbf{z})}{\partial \mathbf{z}}\mathbf{Q}\mathbf{f}(\mathbf{z}) + \frac{\partial \mathbf{f}(\mathbf{z})}{\partial \mathbf{z}}\mathbf{Q}^\mathsf{T}\mathbf{f}(\mathbf{z}) \right) + \frac{\partial \mathbf{f}(\mathbf{z})}{\partial \mathbf{z}}\mathbf{c} = \frac{\partial \mathbf{f}(\mathbf{z})}{\partial \mathbf{z}}\mathbf{Q}\mathbf{f}(\mathbf{z}) + \frac{\partial \mathbf{f}(\mathbf{z})}{\partial \mathbf{z}}\mathbf{c}$$

$$= \frac{\partial (\mathbf{y} + \mathbf{Zz})}{\partial \mathbf{z}}\mathbf{Q}(\mathbf{y} + \mathbf{Zz}) + \frac{\partial (\mathbf{y} + \mathbf{Zz})}{\partial \mathbf{z}}\mathbf{c} = \mathbf{Z}^\mathsf{T}\mathbf{Q}(\mathbf{y} + \mathbf{Zz}) + \mathbf{Z}^\mathsf{T}\mathbf{c}$$

Thus, setting $\mathbf{Z}^\mathsf{T}\mathbf{Q}(\mathbf{y} + \mathbf{Zz}) + \mathbf{Z}^\mathsf{T}\mathbf{c} = \mathbf{0}$ results in:

$$\mathbf{Z}^\mathsf{T}\mathbf{Q}\mathbf{Z}\mathbf{z} = -\mathbf{Z}^\mathsf{T}\mathbf{Q}\mathbf{y} - \mathbf{Z}^\mathsf{T}\mathbf{c} \Rightarrow \mathbf{z} = (\mathbf{Z}^\mathsf{T}\mathbf{Q}\mathbf{Z})^{-1}(-\mathbf{Z}^\mathsf{T}\mathbf{Q}\mathbf{y} - \mathbf{Z}^\mathsf{T}\mathbf{c})$$

After finding solution $\mathbf{z}^* = (\mathbf{Z}^\mathsf{T}\mathbf{Q}\mathbf{Z})^{-1}(-\mathbf{Z}^\mathsf{T}\mathbf{Q}\mathbf{y} - \mathbf{Z}^\mathsf{T}\mathbf{c})$ of the unconstrained min-imization problem, we obtain $\mathbf{x}^* = \mathbf{y} + \mathbf{Z}\mathbf{z}^*$ and $\lambda^* = -(\mathbf{A}\mathbf{A}^\mathsf{T})^{-1}\mathbf{A}\nabla\mathbf{f}(\mathbf{x}^*)$.

Consider the application of the nullspace approach in our previous example. A feasible solution can be easily obtained by solving a *Phase-I LP* since we do not have to consider the objective function of the QP and its constraints are affine/linear functions. Using this approach, we compute feasible solution $\mathbf{y} = [2.2, 0.2, 0]^\mathsf{T}$. We also have:

$$\mathbf{Z} = \text{Basis of null}(\mathbf{A}) \simeq \begin{bmatrix} -0.61721 \\ 0.15430 \\ 0.77152 \end{bmatrix}$$

resulting in:

$$\mathbf{z}^* = (\mathbf{Z}^\mathsf{T}\mathbf{Q}\mathbf{Z})^{-1}(-\mathbf{Z}^\mathsf{T}\mathbf{c} - \mathbf{Z}^\mathsf{T}\mathbf{Q}\mathbf{y}) = 0.36171576$$

Thus, $\mathbf{x}^* = \mathbf{y} + \mathbf{Z}\mathbf{z}^* = [2.2, 0.2, 0]^\mathsf{T} + [-0.61721, 0.15430, 0.77152]^\mathsf{T}0.36172 = [85/43, 11/43, 12/43]^\mathsf{T}$.

4.6.2 Convex Quadratic Programs with Equality and Inequality Constraints

Inequality-constrained quadratic programs are QPs that contain inequality constraints and (possibly) equality constraints. *Active set* methods have been used for solving inequality-constrained QPs since the 1970s, while *Interior Point* methods have been also very common, especially after the 1990s, because they work well for large-scale convex inequality QPs.

We can cast an inequality constraint QP as:

$$\min_{\mathbf{x}\in\mathbb{R}^n} \mathbf{c}^\mathsf{T}\mathbf{x} + \frac{1}{2}\mathbf{x}^\mathsf{T}\mathbf{Q}\mathbf{x} \tag{4.95}$$

$$\text{s.t. } \mathbf{A}\mathbf{x} \le \mathbf{b} \tag{4.96}$$

$$\mathbf{E}\mathbf{x} = \mathbf{d} \tag{4.97}$$

where $\mathbf{A} \in \mathbb{R}^{m \times n}$ is the set of inequality constraints and $\mathbf{E} \in \mathbb{R}^{l \times n}$ is the set of equality constraints with $l \le n$. This QP is convex if $\mathbf{Q} \in \mathbb{R}^{n \times n}$ is positive semidefinite. In that case, strong duality holds because the problem satisfies the refined Slater's conditions since all inequality constraints are affine functions.

Let, for convenience, call the set of equality constraints L. That is, $L = \{1, 2, \ldots, l\}$. In addition, let $M = \{1, \ldots, m\}$. The Lagrangian of our inequality

QP is:

$$\mathcal{L}(\mathbf{x}, \lambda, \mu) = \mathbf{c}^\mathsf{T}\mathbf{x} + \frac{1}{2}\mathbf{x}^\mathsf{T}\mathbf{Q}\mathbf{x} + \lambda^\mathsf{T}(\mathbf{E}\mathbf{x} - \mathbf{d}) + \mu^\mathsf{T}(\mathbf{A}\mathbf{x} - \mathbf{b})$$

where $\lambda \in \mathbb{R}^l$ and $\mu \in \mathbb{R}^m_+$ are KKT multipliers. The dual of this problem is:

$$\max_{\lambda, \mu} \inf_{\mathbf{x}} \mathcal{L}(\mathbf{x}, \lambda, \mu)$$

$$\text{s.t. } \mu \geq \mathbf{0}$$

In addition, with the use of the denominator layout notation, we have:

$$\nabla_\mathbf{x}\mathcal{L}(\mathbf{x}, \lambda, \mu) = \mathbf{0} \Rightarrow \mathbf{c} + \frac{1}{2}(\mathbf{Q} + \mathbf{Q}^\mathsf{T})\mathbf{x} + \mathbf{E}^\mathsf{T}\lambda + \mathbf{A}^\mathsf{T}\mu = \mathbf{0} \xRightarrow{\mathbf{Q}^\mathsf{T}=\mathbf{Q}} \mathbf{Q}\mathbf{x} + \mathbf{E}^\mathsf{T}\lambda + \mathbf{A}^\mathsf{T}\mu = -\mathbf{c}$$

resulting in the following first-order necessary KKT conditions, which are also sufficient for a positive semidefinite $\mathbf{Q}$:

1. $\mathbf{Q}\mathbf{x} + \mathbf{E}^\mathsf{T}\lambda + \mathbf{A}^\mathsf{T}\mu = -\mathbf{c}$ (stationarity)
2. $\mathbf{E}\mathbf{x} = \mathbf{d}$ (primal feasibility)
3. $\mathbf{A}\mathbf{x} \leq \mathbf{b}$ (primal feasibility)
4. $\mu \geq \mathbf{0}$ (dual feasibility)
5. $\mu^\mathsf{T}\mathbf{x} = 0$ (complementary slackness)

A prominent solution method for inequality QPs is the Active Set method. The active set method starts by finding a feasible point during an initial phase and then finds iterates that remain feasible while steadily decreasing the objective function by solving a sequence of equality-constrained QPs. Unlike the simplex method in linear programming where the optimal solution was at the vertices of the feasible polyhedron, neither the iterates nor the solution need to be vertices of the feasible set.

Reckon from definition 4.7 that an *active set* at any feasible point $\mathbf{x}^*$ of an optimization problem is the set $\mathcal{A}(\mathbf{x}^*)$ which is the union of the set of all equality constraints L and the subset of inequality constraints $i \in M$ for which $c_i(\mathbf{x}^*) = 0$. That is, $\mathcal{A}(\mathbf{x}^*) = L \cup \{i \in M \mid \sum_{j=1}^n a_{ij}x_j^* = b_i\}$. Observe that if an *oracle* tells us that the optimal solution $\mathbf{x}^*$ of the inequality QP has active set $\mathcal{A}(\mathbf{x}^*)$, then we can solve the inequality QP as an equality-constrained QP by considering only the constraints in the active set since all other constraints will be inactive. This is the essence of the *active set* method which is based on the following steps:

1. Start from a feasible solution $\mathbf{x}_k$ of the inequality QP where $k = 0$ and find its active set $\mathcal{A}(\mathbf{x}_k)$.
2. Find the next iterate $\mathbf{x}_{k+1} = \mathbf{x}_k + \alpha_k\mathbf{p}_k$ where $\mathbf{p}_k$ is the step direction and α_k the step length.
3. Terminate when finding an optimal $\mathbf{x}_{k+1}$ that satisfies all constraints.

The question is how to determine the step direction and the step length at each iteration. Let us consider that we seek to determine the step direction $\mathbf{p}_k$ that will move us from $\mathbf{x}_k$ to $\mathbf{x}_{k+1}$. This can be achieved by setting $\mathbf{x} = \mathbf{x}_k + \mathbf{p}$ and solving the equality QP with respect to $\mathbf{p}$ that considers only the active constraints at $\mathbf{x}_k$:

$$
\begin{aligned}
\min_{\mathbf{p}} \quad & \frac{1}{2}(\mathbf{x}_k + \mathbf{p})^\mathsf{T} \mathbf{Q}(\mathbf{x}_k + \mathbf{p}) + \mathbf{c}^\mathsf{T}(\mathbf{x}_k + \mathbf{p}) \\
\text{s.t.} \quad & \mathbf{e}_i^\mathsf{T}(\mathbf{x}_k + \mathbf{p}) = d_i \quad \forall i \in L \\
& \mathbf{a}_i^\mathsf{T}(\mathbf{x}_k + \mathbf{p}) = b_i \quad \forall i \in M \cap \mathcal{A}(\mathbf{x}_k)
\end{aligned}
\tag{4.98}
$$

where $\mathbf{e}_i^\mathsf{T}$ is the ith row of matrix $\mathbf{E}$ that contains equality constraint $i \in L$ and $\mathbf{a}_i^\mathsf{T}$ is the ith row of matrix $\mathbf{A}$ that contains the active inequality constraint $i \in M \cap \mathcal{A}(\mathbf{x}_k)$. In essence, solving (4.98) completely disregards the inequality constraints that are not in the active set of solution guess $\mathbf{x}_k$. We will later see that after solving this problem, we should check whether these inequality constraints are satisfied in order to decide about the feasibility of the provided solution. For now, we can rewrite the constraints in the active set of $\mathbf{x}_k$ as follows:

$$
\mathbf{e}_i^\mathsf{T}(\mathbf{x}_k + \mathbf{p}) = d_i \Rightarrow \mathbf{e}_i^\mathsf{T}\mathbf{p} = d_i - \mathbf{e}_i^\mathsf{T}\mathbf{x}_k \xrightarrow{\mathbf{x}_k \text{ is feasible}} \mathbf{e}_i^\mathsf{T}\mathbf{p} = 0 \quad \forall i \in L
$$

$$
\mathbf{a}_i^\mathsf{T}(\mathbf{x}_k + \mathbf{p}) = b_i \Rightarrow \mathbf{a}_i^\mathsf{T}\mathbf{p} = b_i - \mathbf{a}_i^\mathsf{T}\mathbf{x}_k \xrightarrow{\mathbf{x}_k \text{ is feasible}} \mathbf{a}_i^\mathsf{T}\mathbf{p} = 0 \quad \forall i \in M \cap \mathcal{A}(\mathbf{x}_k)
$$

Let $\tilde{\mathbf{A}}$ be the submatrix of $\mathbf{A}$ which contains all columns of matrix $\mathbf{A}$ and the active rows $i \in M \cap \mathcal{A}(\mathbf{x}_k)$ of matrix $\mathbf{A}$. Then, we can rewrite the minimization problem (4.98) of finding the step direction $\mathbf{p}_k$ at iterate k as:

$$
\begin{aligned}
\min_{\mathbf{p}} \quad & \frac{1}{2}(\mathbf{x}_k + \mathbf{p})^\mathsf{T} \mathbf{Q}(\mathbf{x}_k + \mathbf{p}) + \mathbf{c}^\mathsf{T}(\mathbf{x}_k + \mathbf{p}) \\
\text{s.t.} \quad & \mathbf{E}\mathbf{p} = \mathbf{0} \\
& \tilde{\mathbf{A}}\mathbf{p} = \mathbf{0}
\end{aligned}
\tag{4.99}
$$

with Lagrangian function:

$$
\mathcal{L}(\mathbf{p}, \boldsymbol{\lambda}, \boldsymbol{\mu}) = \frac{1}{2}(\mathbf{x}_k + \mathbf{p})^\mathsf{T} \mathbf{Q}(\mathbf{x}_k + \mathbf{p}) + \mathbf{c}^\mathsf{T}(\mathbf{x}_k + \mathbf{p}) + \boldsymbol{\lambda}^\mathsf{T}\mathbf{E}\mathbf{p} + \boldsymbol{\mu}^\mathsf{T}\tilde{\mathbf{A}}\mathbf{p}
$$

where $\boldsymbol{\lambda}$ and $\boldsymbol{\mu}$ are Lagrange multipliers. We thus have, based on the denominator layout notation,

$$
\nabla_{\mathbf{p}}\mathcal{L}(\mathbf{p}, \boldsymbol{\lambda}, \boldsymbol{\mu}) = \mathbf{0} \Rightarrow \frac{1}{2}\left(\frac{\partial(\mathbf{x}_k + \mathbf{p})}{\partial \mathbf{p}}\mathbf{Q}(\mathbf{x}_k + \mathbf{p}) + \frac{\partial(\mathbf{x}_k + \mathbf{p})}{\partial \mathbf{p}}\mathbf{Q}^\mathsf{T}(\mathbf{x}_k + \mathbf{p}) \right) + \frac{\partial(\mathbf{x}_k + \mathbf{p})}{\partial \mathbf{p}}\mathbf{c} + \mathbf{E}^\mathsf{T}\boldsymbol{\lambda} + \tilde{\mathbf{A}}^\mathsf{T}\boldsymbol{\mu} = \mathbf{0}
$$

$$
\xrightarrow{\mathbf{Q}=\mathbf{Q}^\mathsf{T}} \mathbf{Q}(\mathbf{x}_k + \mathbf{p}) + \mathbf{c} + \mathbf{E}^\mathsf{T}\boldsymbol{\lambda} + \tilde{\mathbf{A}}^\mathsf{T}\boldsymbol{\mu} = \mathbf{0}
$$

$$
\Rightarrow \mathbf{Q}\mathbf{p} + \mathbf{E}^\mathsf{T}\boldsymbol{\lambda} + \tilde{\mathbf{A}}^\mathsf{T}\boldsymbol{\mu} = -(\mathbf{Q}\mathbf{x}_k + \mathbf{c})
$$

and the optimal solution should satisfy:

$$\mathbf{Q}\mathbf{p} + \mathbf{E}^\mathsf{T}\boldsymbol{\lambda} + \tilde{\mathbf{A}}^\mathsf{T}\boldsymbol{\mu} = -(\mathbf{Q}\mathbf{x}_k + \mathbf{c})$$
$$\mathbf{E}\mathbf{p} = \mathbf{0}$$
$$\tilde{\mathbf{A}}\mathbf{p} = \mathbf{0}$$

Setting $\mathbf{g}_k = \mathbf{Q}\mathbf{x}_k + \mathbf{c}$ for brevity, we have the system:

$$\begin{bmatrix} \mathbf{Q} & \mathbf{E}^\mathsf{T} & \tilde{\mathbf{A}}^\mathsf{T} \\ \mathbf{E} & \mathbf{0} & \mathbf{0} \\ \tilde{\mathbf{A}} & \mathbf{0} & \mathbf{0} \end{bmatrix} \begin{bmatrix} \mathbf{p} \\ \boldsymbol{\lambda} \\ \boldsymbol{\mu} \end{bmatrix} = \begin{bmatrix} -\mathbf{g}_k \\ \mathbf{0} \\ \mathbf{0} \end{bmatrix}$$

which we can solve to obtain $\mathbf{p}_k, \boldsymbol{\lambda}_k, \boldsymbol{\mu}_k$. Instead of solving this system, an alternative way to find $\mathbf{p}_k, \boldsymbol{\lambda}_k, \boldsymbol{\mu}_k$ is to apply an equality QP solution method for the equality QP in (4.99). To do so, we should cast (4.99) as:

$$\min_{\mathbf{p}} \ \mathbf{g}_k^\mathsf{T}\mathbf{p} + \frac{1}{2}\mathbf{p}^\mathsf{T}\mathbf{Q}\mathbf{p} \tag{4.100}$$
$$\text{s.t. } \mathbf{A}'\mathbf{p} = \mathbf{0}$$

where $\mathbf{A}' = \begin{bmatrix} \mathbf{E} \\ \tilde{\mathbf{A}} \end{bmatrix}$ and $\mathbf{g}_k^\mathsf{T} = (\mathbf{Q}\mathbf{x}_k + \mathbf{c})^\mathsf{T}$.

Given the solution $\mathbf{p}_k, \boldsymbol{\lambda}_k, \boldsymbol{\mu}_k$ of the equality-constrained QP at the k-th iteration, we update $\mathbf{x}_{k+1} = \mathbf{x}_k + \alpha_k \mathbf{p}_k$ where $\alpha_k \in [0, 1]$ according to the following cases.

Case 1: suppose that the optimal step direction $\mathbf{p}_k$ is equal to $\mathbf{0}$. This means that $\mathbf{x}_k$ is not only a feasible solution to the original problem, but it is also an optimal solution for the equality constrained QP with active set $\mathcal{A}(\mathbf{x}_k)$. If the KKT multipliers of the inequality constraints in the active set are nonnegative, $\boldsymbol{\mu}_k \geq \mathbf{0}$, then we can terminate the active set algorithm because we have reached an optimal solution with $\mathbf{x}_{k+1} = \mathbf{x}_k$. If, however, some components of $\boldsymbol{\mu}_k$ are negative, we remove the most negative component from the active set $\mathcal{A}(\mathbf{x}_k)$ and we solve again the problem to obtain a new $\mathbf{p}_k, \boldsymbol{\lambda}_k, \boldsymbol{\mu}_k$ triplet.

Case 2: suppose that the optimal $\mathbf{p}_k \neq \mathbf{0}$. We then need to decide how far we will move in the direction of $\mathbf{p}_k$ to maintain feasibility while reducing the objective function value. For this, we choose step length $\alpha_k \in [0, 1]$ to be the largest value for which all constraints are satisfied for $\mathbf{x}_{k+1} = \mathbf{x}_k + \alpha_k \mathbf{p}_k$.

Note that for any $i \in \mathcal{A}(\mathbf{x}_k)$ we can maintain feasibility regardless of the step length α_k because $\mathbf{a}_i^\mathsf{T}(\mathbf{x}_k + \alpha_k \mathbf{p}_k) = b_i$ for any α_k. This is evident since: (i) $\mathbf{a}_i^\mathsf{T}\mathbf{x}_k = b_i$ because $\mathbf{x}_k$ is a feasible solution for the active constraints, and (ii) $\mathbf{a}_i^\mathsf{T}\mathbf{p}_k = 0$ as a solution of (4.100). This means that we only need our step length α_k to satisfy the inequality constraints that are not active at $\mathbf{x}_k$, and thus were completely disregarded when we obtained solution $\mathbf{p}_k$. Notice that if $\mathbf{a}_i^\mathsf{T}\mathbf{p}_k \leq 0$ for some inequality constraint $i \notin \mathcal{A}(\mathbf{x}_k)$, then this inequality constraint will be satisfied regardless the step length $\alpha_k \in [0, 1]$ because $\mathbf{a}_i^\mathsf{T}\mathbf{x}_k \leq b_i$ since $\mathbf{x}_k$ is feasible for the original problem and $\mathbf{a}_i^\mathsf{T}\mathbf{x}_k + \alpha_k \mathbf{a}_i^\mathsf{T}\mathbf{p}_k \leq \mathbf{a}_i^\mathsf{T}\mathbf{x}_k$. If, however, $\mathbf{a}_i^\mathsf{T}\mathbf{p}_k > 0$ we cannot guarantee that $\mathbf{a}_i^\mathsf{T}\mathbf{x}_k + \alpha_k \mathbf{a}_i^\mathsf{T}\mathbf{p}_k \leq b_i$. To ensure this, we should restrict our step length α_k so that:

$$\mathbf{a}_i^{\mathsf{T}} \mathbf{x}_{k+1} \leq b_i \; \forall i \in M \setminus \mathcal{A}(\mathbf{x}_k) \text{ and } \mathbf{a}_i^{\mathsf{T}} \mathbf{p}_k > 0$$

Replacing $\mathbf{x}_{k+1}$ by $\mathbf{x}_k + \alpha_k \mathbf{p}_k$ we have:

$$\mathbf{a}_i^{\mathsf{T}} \mathbf{x}_k + \alpha_k \mathbf{a}_i^{\mathsf{T}} \mathbf{p}_k \leq b_i \; \forall i \in M \setminus \mathcal{A}(\mathbf{x}_k) \text{ and } \mathbf{a}_i^{\mathsf{T}} \mathbf{p}_k > 0$$

This gives us:

$$\alpha_k \leq \frac{b_i - \mathbf{a}_i^{\mathsf{T}} \mathbf{x}_k}{\mathbf{a}_i^{\mathsf{T}} \mathbf{p}_k} \; \forall i \in M \setminus \mathcal{A}(\mathbf{x}_k) \text{ and } \mathbf{a}_i^{\mathsf{T}} \mathbf{p}_k > 0$$

We can thus select the largest possible step length α_k that will allow us to maintain feasibility:

$$\alpha_k = \min_{i \notin \mathcal{A}(\mathbf{x}_k) \wedge \mathbf{a}_i^{\mathsf{T}} \mathbf{p}_k > 0} \frac{b_i - \mathbf{a}_i^{\mathsf{T}} \mathbf{x}_k}{\mathbf{a}_i^{\mathsf{T}} \mathbf{p}_k}$$

Now, we have three sub-cases:

1. If $a_k = 0$ this means that we have a constraint i which was active at $\mathbf{x}_k$ but we did not include it in the active set $\mathbf{A}(\mathbf{x}_k)$. Thus, we need to add it and solve again the problem of determining $\mathbf{p}_k, \lambda_k, \mu_k$.
2. If $a_k \geq 1$, no new constraints become active if we take step $a_k = 1$, which is the allowed maximum step in $a_k \in [0, 1]$. In such case, we set $\mathbf{x}_{k+1} = \mathbf{x}_k + \mathbf{p}_k$ and proceed to solve again the problem of determining $\mathbf{p}_k, \lambda_k, \mu_k$ without altering the active set.
3. Finally, if $0 < a_k < 1$ we have some inequality constraint $i \in M \setminus \mathcal{A}(\mathbf{x}_k)$ outside of the current active set that becomes active for this value of a_k. In this case, we set $\mathbf{x}_{k+1} = \mathbf{x}_k + \alpha_k \mathbf{p}_k$ and add that *blocking constraint*, i, to the active set before proceeding to solve again the equality-constrained QP.

This process is summarized in Algorithm 18.

Note that to apply the active set method we need to find first an initial feasible solution $\mathbf{x}_{k=0}$. Because the constraints are affine functions, this can be easily achieved with the use of the Phase I LP method that was proposed in the *two-phase simplex* method.

Let us consider the example inequality QP:

$$\min_{\mathbf{x} \in \mathbb{R}^2} x_1^2 - x_1 x_2 + x_2^2 - 4x_1 - 7x_2$$

$$\text{s.t.} \; -x_1 + 2x_2 \leq 2$$

$$x_1 + 3x_2 \leq 4 \tag{4.101}$$

$$x_1 - 2x_2 \leq 3$$

$$-x_1 \leq 0$$

Algorithm 18 Active Set method for inequality QP

1: set $k = 0$ and find feasible solution $\mathbf{x}_k$
2: find the active set $\mathcal{A}(\mathbf{x}_k)$ of the feasible solution $\mathbf{x}_k$
3: set $\mathbf{g}_k = \mathbf{Q}\mathbf{x}_k + \mathbf{c}$
4: solve the equality-constrained QP:

$$\min_{\mathbf{p}} \ \mathbf{g}_k^\mathsf{T}\mathbf{p} + \frac{1}{2}\mathbf{p}^\mathsf{T}\mathbf{Q}\mathbf{p}$$
$$\text{s.t. } \mathbf{e}_i^\mathsf{T}\mathbf{p} = 0 \quad \forall i \in L$$
$$\mathbf{a}_i^\mathsf{T}\mathbf{p} = 0 \quad \forall i \in M \cap \mathcal{A}(\mathbf{x}_k)$$

5: **if** $\mathbf{p}_k = \mathbf{0}$ and $\boldsymbol{\mu}_k \geq \mathbf{0}$ **then**
6:     **terminate** because solution $\mathbf{x}_k$ is optimal.
7: **end if**
8: **if** $\mathbf{p}_k = \mathbf{0}$, but some components of $\boldsymbol{\mu}_k$ are negative **then**
9:     Remove the most negative component from the active set $\mathcal{A}(\mathbf{x}_k)$ and solve again the equality-constrained QP.
10: **end if**
11: **if** $\mathbf{p}_k \neq \mathbf{0}$ compute $\alpha_k = \min_{i \notin \mathcal{A}(\mathbf{x}_k) \wedge \mathbf{a}_i^\mathsf{T}\mathbf{p}_k > 0} \frac{b_i - \mathbf{a}_i^\mathsf{T}\mathbf{x}_k}{\mathbf{a}_i^\mathsf{T}\mathbf{p}_k}$ **then**
12:     **if** $\alpha_k \geq 1$ **then**
13:         Set $\mathbf{x}_{k+1} = \mathbf{x}_k + \mathbf{p}_k$, $\mathcal{A}(\mathbf{x}_{k+1}) = \mathcal{A}(\mathbf{x}_k)$, and solve again the equality-constrained QP.
14:     **end if**
15:     **if** $\alpha_k = 0$ **then**
16:         Select inequality constraint $i \in M \setminus \mathcal{A}(\mathbf{x}_k)$ for which $\mathbf{a}_i^\mathsf{T}\mathbf{p}_k > 0$ and $b_i = \mathbf{a}_i^\mathsf{T}\mathbf{x}_k$
17:         Set $\mathbf{x}_{k+1} = \mathbf{x}_k$ and $\mathcal{A}(\mathbf{x}_{k+1}) = \mathcal{A}(\mathbf{x}_k) \cup \{i\}$ and solve again the equality-constrained QP.
18:     **end if**
19:     **if** $0 < \alpha_k < 1$ **then**
20:         Select $i \in M \setminus \mathcal{A}(\mathbf{x}_k)$ with $\mathbf{a}_i^\mathsf{T}\mathbf{p}_k > 0$ for which $\alpha_k = \frac{b_i - \mathbf{a}_i^\mathsf{T}\mathbf{x}_k}{\mathbf{a}_i^\mathsf{T}\mathbf{p}_k}$.
21:         Set $\mathbf{x}_{k+1} = \mathbf{x}_k + \alpha_k\mathbf{p}_k$ and $\mathcal{A}(\mathbf{x}_{k+1}) = \mathcal{A}(\mathbf{x}_k) + \{i\}$ and solve again the equality-constrained QP.
22:     **end if**
23: **end if**

---

with no equality constraints $L = \emptyset$ and inequality constraints $M = \{1, 2, 3, 4\}$. We can easily obtain an initial feasible solution $\mathbf{x}_k = [1, 0]^\mathsf{T}$ resulting in active set $\mathcal{A}(\mathbf{x}_k) = \emptyset$. We also have the Hessian of the objective function $\mathbf{Q} = \begin{bmatrix} 2 & -1 \\ -1 & 2 \end{bmatrix}$ and

$$\mathbf{A} = \begin{bmatrix} -1 & 2 \\ 1 & 3 \\ 1 & -2 \\ -1 & 0 \end{bmatrix}, \tilde{\mathbf{A}} = \emptyset, \mathbf{E} = \emptyset, \mathbf{b} = [2, 4, 3, 0]^\mathsf{T}, \mathbf{d} = \emptyset \text{ and } \mathbf{c} = [-4, -7]^\mathsf{T}. \text{ Thus,}$$

$$\begin{bmatrix} \mathbf{Q} & \mathbf{E}^\mathsf{T} & \tilde{\mathbf{A}}^\mathsf{T} \\ \mathbf{E} & 0 & 0 \\ \tilde{\mathbf{A}} & 0 & 0 \end{bmatrix} \begin{bmatrix} \mathbf{p} \\ \lambda \\ \mu \end{bmatrix} = \begin{bmatrix} -(\mathbf{Q}\mathbf{x}_k + \mathbf{c}) \\ 0 \\ 0 \end{bmatrix}$$

results in:

$$\begin{bmatrix} 2 & -1 \\ -1 & 2 \end{bmatrix} \mathbf{p} = \begin{bmatrix} 2 \\ 8 \end{bmatrix}$$

which has solution $\mathbf{p}_k = [4, 6]^\mathsf{T}$. Because $\mathbf{p}_k \neq \mathbf{0}$ we compute

$$\alpha_k = \min_{i \notin \mathcal{A}(\mathbf{x}_k) \wedge \mathbf{a}_i^\mathsf{T} \mathbf{p}_k > 0} \frac{b_i - \mathbf{a}_i^\mathsf{T} \mathbf{x}_k}{\mathbf{a}_i^\mathsf{T} \mathbf{p}_k}$$

from which we get $\alpha_k \simeq 0.13636$ where constraint 2 is the blocking constraint. We thus set $\mathbf{A}(\mathbf{x}_{k+1}) = \{2\}$ and we move to $\mathbf{x}_{k+1} = \mathbf{x}_k + \alpha_k \mathbf{p}_k \simeq [1.54545, 0.81818]^\mathsf{T}$. We then set $k = k + 1$ and we move to the next iteration where we have the system:

$$\begin{bmatrix} 2 & -1 & 1 \\ -1 & 2 & 3 \\ 1 & 3 & 0 \end{bmatrix} \begin{bmatrix} \mathbf{p} \\ \mu \end{bmatrix} = \begin{bmatrix} 1.7273 \\ 6.9091 \\ 0 \end{bmatrix}$$

which has solution $\mathbf{p}_k \simeq [-0.1993, 0.0664]^\mathsf{T}$, $\mu_k \simeq 2.1923$. Since $\mathbf{p}_k \neq \mathbf{0}$ we compute

$$\alpha_k = \min_{i \notin \mathcal{A}(\mathbf{x}_k) \wedge \mathbf{a}_i^\mathsf{T} \mathbf{p}_k > 0} \frac{b_i - \mathbf{a}_i^\mathsf{T} \mathbf{x}_k}{\mathbf{a}_i^\mathsf{T} \mathbf{p}_k}$$

from which we get $\alpha_k \geq 1$ and thus we make the full step $\alpha_k = 1$ resulting in $\mathbf{x}_{k+1} = \mathbf{x}_k + \mathbf{p}_k = [1.3462, 0.8846]^\mathsf{T}$. We then set $k = k + 1$ and we move to the next iteration having the same active set. In the next iteration, we have the system:

$$\begin{bmatrix} 2 & -1 & 1 \\ -1 & 2 & 3 \\ 1 & 3 & 0 \end{bmatrix} \begin{bmatrix} \mathbf{p} \\ \mu \end{bmatrix} = \begin{bmatrix} 2.1923 \\ 6.5769 \\ 0 \end{bmatrix}$$

which has solution $\mathbf{p}_k = [0, 0]^\mathsf{T}$, $\mu_k \simeq 2.1923$ which meets the optimality criteria, and thus our optimal point is $\mathbf{x}^* = [1.3462, 0.8846]^\mathsf{T}$. The iterative solutions of the

Fig. 4.9 2-dimensional contour plot of $f(\mathbf{x}) := x_1^2 - x_1 x_2 + x_2^2 - 4x_1 - 7x_2$. The values of the objective function for specific combinations of x_1, x_2 appear in the contours, whereas the feasible region is presented in gray color. We move from $\mathbf{x}_0 = [1, 0]^\mathsf{T}$ with $f(\mathbf{x}_0) = -3$ to $\mathbf{x}_1$ and finally to $\mathbf{x}_2 = [1.3462, 0.8846]^\mathsf{T}$ with $f(\mathbf{x}_2) = -10.1731$

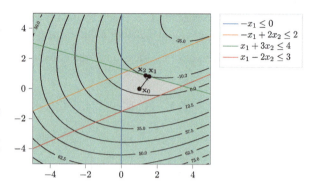

active set method are presented in Fig. 4.9. Starting from feasible solution $[1, 0]$ without active constraints, we convergence to solution $[1.3462, 0.8846]^\mathsf{T}$ for which constraint 2 is active. Notice that if the problem was unconstrained, the objective function value could have been reduced even further.

Closing the presentation of the active set method, we should note that it is not necessary to use all active constraints at each iteration k. Some approaches use a working set $\mathcal{W}(\mathbf{x}_k)$, which is a subset of the active set $\mathcal{A}(\mathbf{x}_k)$. Doing so will not affect the final outcome of the problem, but it might modify some steps of the solution search resulting in a slower or faster convergence.

4.7 Sequential Quadratic Programming

Until now, we presented solution methods for linear and quadratic optimization problems. We now turn our attention to general nonlinear optimization problems (NLPs). One effective approach to solve NLPs is the sequential quadratic programming (SQP) approach that can be used both in line search and trust-region frameworks, and is appropriate for large-scale problems. SQP is an iterative method which solves an inequality-constrained quadratic program at each iteration. SQP can be viewed as an extension of Newton's method for unconstrained optimization to constrained optimization problems. Let us consider the general formulation of a constrained nonlinear program:

$$\min_{\mathbf{x} \in \mathbb{R}^n} \ f(\mathbf{x}) \tag{4.102}$$

$$\text{s.t. } \mathbf{h}(\mathbf{x}) = \mathbf{0} \tag{4.103}$$

$$\mathbf{g}(\mathbf{x}) \leq \mathbf{0} \tag{4.104}$$

where $f : \mathbb{R}^n \to \mathbb{R}, \mathbf{h} : \mathbb{R}^n \to \mathbb{R}^l$ and $\mathbf{g} : \mathbb{R}^n \to \mathbb{R}^m$ meaning that functions $\mathbf{h}(\mathbf{x}) = [h_1(\mathbf{x}), \ldots, h_l(\mathbf{x})]^\mathsf{T}$ are all the equality constraint functions and $\mathbf{g}(\mathbf{x}) = [g_1(\mathbf{x}), \ldots, g_m(\mathbf{x})]^\mathsf{T}$ are all the inequality constraint functions of the mathematical program.

An NLP contains as special cases the linear (LP) and quadratic programming (QP) problems that we have already discussed. For instance, the general NLP formulation can represent an LP if $\mathbf{h}$ and $\mathbf{g}$ are affine functions and f is a linear function. Similarly, if $\mathbf{h}$ and $\mathbf{g}$ are affine functions and f is a quadratic function, then the NLP is reduced to a QP.

The SQP approach is an iterative approach for solving NLPs. It approximates the NLP for a given iterate $\mathbf{x}_k$ with the use of an inequality QP subproblem, which is solved to approximate the next iterate $\mathbf{x}_{k+1}$. The SQP objective is that the sequence $(\mathbf{x}_k)$ will converge to a local optimum of the NLP for some k. Every minimization NLP has the Lagrangian function:

$$\mathcal{L}(\mathbf{x}, \boldsymbol{\lambda}, \boldsymbol{\mu}) = f(\mathbf{x}) + \boldsymbol{\lambda}^\mathsf{T}\mathbf{h}(\mathbf{x}) + \boldsymbol{\mu}^\mathsf{T}\mathbf{g}(\mathbf{x})$$

where $\boldsymbol{\lambda} \in \mathbb{R}^l$ and $\boldsymbol{\mu} \in \mathbb{R}^m$ are the KKT multipliers related to the equality and inequality constraints, respectively. Let $L = \{1, \ldots, l\}$ be the set of equality and $M = \{1, \ldots, m\}$ the set of inequality constraints. We can now state that an NLP solution is a *strict local minimizer* if it satisfies the second-order sufficient conditions (SOSC):

- $\mathbf{x}^*$ is a regular point (i.e., LICQ holds at $\mathbf{x}^*$)
- There exist $\boldsymbol{\lambda}^*, \boldsymbol{\mu}^*$ that satisfy the KKT conditions at $\mathbf{x}^*$
- $\mathbf{d}^\mathsf{T}\nabla_{\mathbf{xx}}^2\mathcal{L}(\mathbf{x}^*, \boldsymbol{\lambda}^*, \boldsymbol{\mu}^*)\mathbf{d} > 0$ for any vector $\mathbf{d} \neq \mathbf{0}$ such that

$$\mathbf{d} \in C := \begin{cases} (\nabla h_i(\mathbf{x}^*))^\mathsf{T}\mathbf{d} = 0 \text{ for all } i \in L \\ (\nabla g_i(\mathbf{x}^*))^\mathsf{T}\mathbf{d} \leq 0 \text{ for all } i \in M \cap \mathcal{A}(\mathbf{x}^*) \text{ with } \mu_i^* = 0 \\ (\nabla g_i(\mathbf{x}^*))^\mathsf{T}\mathbf{d} = 0 \text{ for all } i \in M \cap \mathcal{A}(\mathbf{x}^*) \text{ with } \mu_i^* > 0 \end{cases}$$

where the KKT conditions are:

$$\begin{aligned} \nabla_{\mathbf{x}}\mathcal{L}(\mathbf{x}^*, \boldsymbol{\lambda}^*, \boldsymbol{\mu}^*) &= \mathbf{0} \text{ (stationarity)} \\ \mathbf{h}(\mathbf{x}^*) &= \mathbf{0} \quad \text{(primal feasibility)} \\ \mathbf{g}(\mathbf{x}^*) &\leq \mathbf{0} \quad \text{(primal feasibility)} \\ \boldsymbol{\mu}^* &\geq \mathbf{0} \quad \text{(dual feasibility)} \\ (\boldsymbol{\mu}^*)^\mathsf{T}\mathbf{g}(\mathbf{x}^*) &= 0 \quad \text{(complementary slackness)} \end{aligned}$$

Starting from a feasible solution $\mathbf{x}_k$ with $k = 0$, SQP finds the next iterate $\mathbf{x}_{k+1} = \mathbf{x}_k + a_k\mathbf{p}_k$, where a_k is the step length and $\mathbf{p}_k$ the step direction. Instead of solving a nonlinear problem to find the step direction $\mathbf{p}_k$, SQP approximates the nonlinear objective function $f(\mathbf{x})$ by a second-order Taylor series:

$$f(\mathbf{x}_k + \mathbf{p}) \approx f(\mathbf{x}_k) + \mathbf{p}^\mathsf{T}\nabla f(\mathbf{x}_k) + \frac{1}{2}\mathbf{p}^\mathsf{T}\nabla_{\mathbf{xx}}^2\mathcal{L}(\mathbf{x}_k, \boldsymbol{\lambda}_k, \boldsymbol{\mu}_k)\mathbf{p}$$

The Hessian matrix of the Lagrangian, $\nabla_{\mathbf{xx}}^2 \mathcal{L}(\mathbf{x}_k, \lambda_k, \mu_k)$, is used instead of $\nabla_{\mathbf{xx}}^2 f(\mathbf{x}_k)$ because of the potential presence of nonlinear constraints that might result in unbounded inequality QPs [24, 25].

Until now, SQP has strong similarities to Newton's method that was applied to unconstrained, nonlinear optimization problems. The additional aspect is the presence of constraint functions, which might be nonlinear. In SQP constraint functions are also approximated by local affine functions (note that we use the denominator layout notation):

$$\mathbf{h}(\mathbf{x}) \approx \underbrace{\mathbf{h}(\mathbf{x}_k)}_{l \times 1} + \underbrace{(\nabla \mathbf{h}(\mathbf{x}_k))^\mathsf{T}}_{l \times n} \underbrace{\mathbf{p}}_{n \times 1}$$

$$\mathbf{g}(\mathbf{x}) \approx \underbrace{\mathbf{g}(\mathbf{x}_k)}_{m \times 1} + \underbrace{(\nabla \mathbf{g}(\mathbf{x}_k))^\mathsf{T}}_{m \times n} \underbrace{\mathbf{p}}_{n \times 1}$$

Thus, at the k-th iteration we have to solve the following inequality QP to derive the step direction $\mathbf{p}_k$:

$$\min_{\mathbf{p} \in \mathbb{R}^n} \mathbf{p}^\mathsf{T} \nabla f(\mathbf{x}_k) + \frac{1}{2}\mathbf{p}^\mathsf{T} \nabla_{\mathbf{xx}}^2 \mathcal{L}(\mathbf{x}_k, \lambda_k, \mu_k)\mathbf{p} \tag{4.105}$$

$$\text{s.t. } \mathbf{h}(\mathbf{x}_k) + (\nabla \mathbf{h}(\mathbf{x}_k))^\mathsf{T}\mathbf{p} = \mathbf{0} \tag{4.106}$$

$$\mathbf{g}(\mathbf{x}_k) + (\nabla \mathbf{g}(\mathbf{x}_k))^\mathsf{T}\mathbf{p} \le \mathbf{0} \tag{4.107}$$

If we assume a step length equal to 1 (like in *pure* Newton's method), with step direction $\mathbf{p}_k$ we proceed to the next iteration with $\mathbf{x}_{k+1} = \mathbf{x}_k + \mathbf{p}_k$ with $\lambda_{k+1} = \lambda^*$ and $\mu_{k+1} = \mu^*$. Note that in the inequality QP of (4.105)–(4.107) we dropped term $f(\mathbf{x}_k)$ from the objective function because it has a fixed value that does not affect the solution of the problem.

One of the main issues with the QP in (4.105)–(4.107) is that the approximate linearization of the nonlinear constraints might lead to inconsistencies, meaning that the QP at the k-th iteration has no feasible solution even if the original NLP is a feasible problem. For instance, if a problem has constraints $x \le 1$ and $-x^2 \le -5$, a linearization at point $x_k = 1$ will lead to inequalities $p \le 0$ and $4 - 2p \le 0 \Rightarrow p \ge 2$ which cannot be satisfied for any $p \in \mathbb{R}$. Powell [26] proposed to overcome this difficulty by adding an extra variable δ resulting in an $(n + 1)$-dimensional problem with consistent constraints (see [27]):

$$\min_{\mathbf{p} \in \mathbb{R}^n, \delta \in \mathbb{R}} \mathbf{p}^\mathsf{T} \nabla f(\mathbf{x}_k) + \frac{1}{2}\mathbf{p}^\mathsf{T} \nabla_{\mathbf{xx}}^2 \mathcal{L}(\mathbf{x}_k, \lambda_k, \mu_k)\mathbf{p} + \frac{1}{2}\psi_k \delta^2$$

$$\begin{aligned} \text{s.t.} \quad & (\nabla \mathbf{h}(\mathbf{x}_k))^\mathsf{T}\mathbf{p} + (1 - \delta)\mathbf{h}(\mathbf{x}_k) = \mathbf{0} \\ & (\nabla g_i(\mathbf{x}_k))^\mathsf{T}\mathbf{p} + (1 - \delta)g_i(\mathbf{x}_k) \le 0 && \forall i \in M_1 \\ & (\nabla g_i(\mathbf{x}_k))^\mathsf{T}\mathbf{p} + g_i(\mathbf{x}_k) \le 0 && \forall i \in M_2 \\ & 0 \le \delta \le 1 \end{aligned} \tag{4.108}$$

where set $M_2 = M \setminus \{M_1\}$ and set M_1:

$$M_1 := \{i \in M \ : \ g_i(\mathbf{x}_k) > 0\}$$

In addition, $\psi_k > 0$ is a penalty that will force δ to take the minimum possible value [28]. For instance, the constraints $x \leq 1$ and $-x^2 \leq -5$ of the previous problem will be linearized at point $x_k = 1$ as $p \leq 0$ and $4(1 - \delta) - 2p \leq 0 \Rightarrow p \geq 2(1 - \delta)$ which can be satisfied for some $\delta \in [0, 1]$, i.e., $\delta = 0$.

One of the computational intensive issues of SQP is the computation of the Hessian $\mathbf{H}_k = \nabla^2_{\mathbf{xx}} \mathcal{L}(\mathbf{x}_k, \lambda_k, \mu_k)$ at each iteration. As in unconstrained optimization, we can use quasi-Newton inspired approaches to approximate the Hessian. We can, for instance, use the BFGS formula (3.12):

$$\mathbf{H}_{k+1} = \mathbf{H}_k + \frac{\mathbf{y}_k \mathbf{y}_k^{\mathsf{T}}}{\mathbf{y}_k^{\mathsf{T}} \mathbf{d}_k} - \frac{\mathbf{H}_k \mathbf{d}_k \mathbf{d}_k^{\mathsf{T}} \mathbf{H}_k}{\mathbf{d}_k \mathbf{H}_k \mathbf{d}_k} \quad \text{(BFGS)}$$

where $\mathbf{y}_k$ is the difference in gradients of the Lagrangian function:

$$\mathbf{y}_k = \nabla_{\mathbf{x}} \mathcal{L}(\mathbf{x}_{k+1}, \lambda_{k+1}, \mu_{k+1}) - \nabla_{\mathbf{x}} \mathcal{L}(\mathbf{x}_k, \lambda_{k+1}, \mu_{k+1})$$

and $\mathbf{d}_k = \mathbf{x}_{k+1} - \mathbf{x}_k$. Note that if $\mathbf{H}_k$ is positive definite and $\mathbf{y}_k^{\mathsf{T}} \mathbf{d}_k > 0$, then $\mathbf{H}_{k+1}$ is also positive definite. This, however, cannot be always guaranteed. For this reason, one can apply the Powell-SQP update [29] by replacing $\mathbf{y}_k$ by:

$$\hat{\mathbf{y}}_k = \theta_k \mathbf{y}_k + (1 - \theta_k) \mathbf{H}_k \mathbf{d}_k$$

where $\theta_k = 1$ if $\mathbf{y}_k^{\mathsf{T}} \mathbf{d}_k > 0$. Otherwise, θ_k can take some value in $[0, 1)$, which can be as close to 1 as possible to limit the distortion of the elements of $\mathbf{y}_k$, such that $\hat{\mathbf{y}}_k^{\mathsf{T}} \mathbf{d}_k$ is sufficiently positive, i.e., $\hat{\mathbf{y}}_k^{\mathsf{T}} \mathbf{d}_k \geq s \mathbf{d}_k^{\mathsf{T}} \mathbf{H}_k \mathbf{d}_k$ for some $s \in (0, 1)$, where $s = 0.1$ is a common choice. The choice of θ_k ensures that $\mathbf{H}_{k+1}$ is positive definite because, in the most extreme case, θ_k can be set equal to 0 resulting in

$$\mathbf{H}_{k+1} = \mathbf{H}_k + \frac{\mathbf{H}_k \mathbf{d}_k \mathbf{d}_k^{\mathsf{T}} \mathbf{H}_k}{\mathbf{d}_k^{\mathsf{T}} \mathbf{H}_k \mathbf{d}_k} - \frac{\mathbf{H}_k \mathbf{d}_k \mathbf{d}_k^{\mathsf{T}} \mathbf{H}_k}{\mathbf{d}_k^{\mathsf{T}} \mathbf{H}_k \mathbf{d}_k} = \mathbf{H}_k,$$

which is positive definite. Powell [26, 30] proposed that:

$$\theta_k = \begin{cases} 1 & \text{if } \mathbf{d}_k^{\mathsf{T}} \mathbf{y}_k \geq 0.2 \mathbf{d}_k^{\mathsf{T}} \mathbf{H}_k \mathbf{d}_k \\ \frac{0.8 \mathbf{d}_k^{\mathsf{T}} \mathbf{H}_k \mathbf{d}_k}{\mathbf{d}_k^{\mathsf{T}} \mathbf{H}_k \mathbf{d}_k - \mathbf{d}_k^{\mathsf{T}} \mathbf{y}_k} & \text{otherwise.} \end{cases}$$

which guarantees the condition $\mathbf{d}_k^{\mathsf{T}} \hat{\mathbf{y}}_k \geq 0.2 \mathbf{d}_k^{\mathsf{T}} \mathbf{H}_k \mathbf{d}_k$. Powell's generalization of BFGS holds $\mathbf{H}_{k+1}$ positive definite within the linear manifold defined by the active constraints at $\mathbf{x}_{k+1}$ [25].

If we compute directly the Hessian $\mathbf{H}_k = \nabla^2_{\mathbf{xx}}\mathcal{L}(\mathbf{x}_k, \boldsymbol{\lambda}_k, \boldsymbol{\mu}_k)$ at each iteration without using a quasi-Newton approach, such as BFGS, it will be prudent to ensure that the Hessian is positive definite to warrant that the QP can be solved to global optimality if it has a feasible solution. This can be achieved by using *Hessian Modification* via setting $\mathbf{H}_k = \nabla^2_{\mathbf{xx}}\mathcal{L}(\mathbf{x}_k, \boldsymbol{\lambda}_k, \boldsymbol{\mu}_k) + \theta\mathbf{I}$ for some $\theta \geq 0$ such that $\mathbf{H}_k$ is positive definite (refer to Algorithm 11 for an example of Hessian Modification with Cholesky decomposition).

Finally, we note that SQP may need a line search strategy to determine the step length a_k in order to achieve convergence from any starting point (known as *global* instead of *local* convergence), especially if the assumptions on the Hessian of the Lagrangian or the Jacobian of the constraints do not hold at a particular iterate k. To do so, we use a step length a_k resulting in:

$$\mathbf{x}_{k+1} = \mathbf{x}_k + a_k\mathbf{p}_k$$
$$\boldsymbol{\lambda}_{k+1} = \boldsymbol{\lambda}_k + a_k(\boldsymbol{\lambda}^* - \boldsymbol{\lambda}_k)$$
$$\boldsymbol{\mu}_{k+1} = \boldsymbol{\mu}_k + a_k(\boldsymbol{\mu}^* - \boldsymbol{\mu}_k)$$

where $\boldsymbol{\lambda}^*$ and $\boldsymbol{\mu}^*$ are the optimal solutions of the QP in (4.105)–(4.107).

In unconstrained minimization, a_k is typically determined so that the objective function reduces its value in each iteration (reckon the examples of exact and inexact line search). In constrained optimization, deciding the step length a_k is not as easy because we aim at reducing the objective function value, but, at the same time, we want to maintain feasibility or—at least—move towards a feasible solution. Often these two aims are in conflict. To rectify this, SQP uses a *merit* function to decide about a_k. The merit function ensures that, under certain assumptions, we will converge to a (potential) solution even if the initial solution guess is not close to a solution [31]. A merit function is chosen in such a way that the solutions of the NLP are also unconstrained minimizers of the merit function. If ϕ is the selected merit function, the step length a_k at the k-th iteration is determined such that:

$$\phi(\mathbf{x}_k + a_k\mathbf{p}_k) < \phi(\mathbf{x}_k)$$

in order to ensure a sufficient decrease of the merit function. This can be done with backtracking line search, i.e., starting with $a_k = 1$ and decreasing its value until the aforementioned constraint is satisfied. The most popular merit functions are l_p-norms and augmented Lagrangians. The l_1 merit function was proposed by Han in 1997 [32] who employed it to prove a global convergence theorem in the convex case. The l_1 merit function is:

$$\phi_\rho(\mathbf{x}) := f(\mathbf{x}) + \sum_{i \in L} \rho_i |h_i(\mathbf{x})| + \sum_{i \in M} \rho_i \max\{0, g_i(\mathbf{x})\}$$

where $\rho > 0$ are the penalty parameters of the l_1 merit function. Using the l_1 merit function in an exact line search, we seek to find a_k by minimizing the following

1-dimensional problem (notice that $\mathbf{x}_k$ and $\mathbf{p}_k$ are fixed-value parameters):

$$\min_a \phi_\rho(\mathbf{x}_k + a\mathbf{p}_k)$$

where:

$$\phi_\rho(\mathbf{x}_k + a\mathbf{p}_k) := f(\mathbf{x}_k + a\mathbf{p}_k) + \sum_{i \in L} \rho_i |h_i(\mathbf{x}_k + a\mathbf{p}_k)| + \sum_{i \in M} \rho_i \max\{0, g_i(\mathbf{x}_k + a\mathbf{p}_k)\}$$

Han [32] proved that minimizing the 1-dimensional function $\phi_\rho(\mathbf{x}_k + a\mathbf{p}_k)$ leads to a step size a_k guaranteeing a move to a KKT point for values of the penalty parameters ρ greater than some lower bound if the eigenvalues of $\mathbf{H}_k$ are bounded above and are bounded away from zero [30]. In particular, each penalty parameter ρ_i should be greater than or equal to $|\lambda_i|$ if $i \in L$ and $|\mu_i|$ if $i \in M$. Since an upper bound to the multipliers is not known in practice, Powell [26] proposed to update the penalty parameters ρ according to:

$$\rho_{k,i} = \max_i\{\lambda_{ki}, \frac{\rho_{k-1,i} + \lambda_{ki}}{2}\} \quad \forall i \in L$$

$$\rho_{k,i} = \max_i\{\mu_{ki}, \frac{\rho_{k-1,i} + \mu_{ki}}{2}\} \quad \forall i \in M$$

where $\rho_{k-1,i}$ was the penalty parameter value at the previous iteration $\rho_{k-1} = [\rho_{k-1,1}, \dots, \rho_{k-1,l+m}]^\mathsf{T}$, $\boldsymbol{\lambda}_k = [\lambda_{k1}, \dots, \lambda_{kl}]^\mathsf{T}$, and $\boldsymbol{\mu}_k = [\mu_{k1}, \dots, \mu_{km}]^\mathsf{T}$. When $k = 0$, we can initialize the values of ρ as:

$$\rho_{ki} = \frac{\|\nabla f(\mathbf{x}_k)\|_2}{\|\nabla h_i(\mathbf{x}_k)\|_2} \quad \forall i \in L \qquad \rho_{ki} = \frac{\|\nabla f(\mathbf{x}_k)\|_2}{\|\nabla g_i(\mathbf{x}_k)\|_2} \quad \forall i \in M$$

The main issue with the l_1 merit function is that it is not differentiable, and it is thus not easy to optimize. To overcome this, we can perform an inexact line search instead of an exact one by merely seeking to find a_k such that:

$$\phi(\mathbf{x}_k + a_k\mathbf{p}_k) < \phi(\mathbf{x}_k)$$

We can make this more explicit by requesting to have a *sufficient decrease*:

$$\phi(\mathbf{x}_k + a_k\mathbf{p}_k) \le \phi(\mathbf{x}_k) + \eta a_k \mathrm{D}_{\mathbf{p}_k}(\phi(\mathbf{x}_k))$$

for some $\eta \in (0, 1)$, where $\mathrm{D}_{\mathbf{p}_k}(\phi(\mathbf{x}_k))$ is the directional derivative of ϕ in the direction $\mathbf{p}_k$. If we use slack variables $\boldsymbol{\mu} \ge \mathbf{0}$ for the $\mathbf{g}(\mathbf{x}) \le \mathbf{0}$ inequality constraints, then $\mathbf{g}(\mathbf{x}) + \boldsymbol{\mu} = \mathbf{0}$ and we can use the new equality constraint functions $\mathbf{c}(\mathbf{x}) = \mathbf{h}(\mathbf{x}) \cup (\mathbf{g}(\mathbf{x}) + \boldsymbol{\mu})$ to compute the directional derivative:

$$\mathrm{D}_{\mathbf{p}_k}(\phi(\mathbf{x}_k)) = \nabla f(\mathbf{x}_k)^\mathsf{T}\mathbf{p}_k - \tau\|\mathbf{c}(\mathbf{x}_k)\|$$

This requirement is analogous to the Armijo condition for unconstrained optimization if $\mathbf{p}_k$ is a descent direction, i.e., $\mathrm{D}_{\mathbf{p}_k}(\phi(\mathbf{x}_k)) < 0$, which holds if we choose a sufficiently large τ.

A basic framework for the SQP is provided below.

Algorithm 19 Basic SQP framework

1: set $k = 0$ and choose $\mathbf{x}_k$, $\boldsymbol{\lambda}_k$, $\boldsymbol{\mu}_k$ and merit function $\phi(\mathbf{x})$
2: **repeat**
3:     Compute $\nabla^2_{\mathbf{xx}}\mathcal{L}(\mathbf{x}_k, \boldsymbol{\lambda}_k, \boldsymbol{\mu}_k)$ and turn it to positive definite with Hessian modification, or compute a positive definite $\mathbf{H}_k$ with a quasi-Newton approach (i.e., BFGS)
4:     solve the consistent inequality QP in (4.108) to obtain $\mathbf{p}^*$, $\boldsymbol{\lambda}^*$, $\boldsymbol{\mu}^*$
5:     set $\mathbf{p}_k = \mathbf{p}^*$
6:     choose step length a_k with *backtracking* so that $\phi(\mathbf{x}_k + a_k\mathbf{p}_k) < \phi(\mathbf{x}_k)$
7:     set:

$$\mathbf{x}_{k+1} = \mathbf{x}_k + a_k\mathbf{p}_k$$
$$\boldsymbol{\lambda}_{k+1} = \boldsymbol{\lambda}_k + a_k(\boldsymbol{\lambda}^* - \boldsymbol{\lambda}_k)$$
$$\boldsymbol{\mu}_{k+1} = \boldsymbol{\mu}_k + a_k(\boldsymbol{\mu}^* - \boldsymbol{\mu}_k)$$

8:     compute $\nabla^2_{\mathbf{xx}}\mathcal{L}(\mathbf{x}_{k+1}, \boldsymbol{\lambda}_{k+1}, \boldsymbol{\mu}_{k+1})$
9:     set $k = k + 1$
10: **until** a convergence test is satisfied

SQP can be viewed as a general framework with different versions. So far, we have discussed the version of solving an inequality QP sub-problem at each iteration. There are other important implementation versions. One is the Sequential linear-quadratic programming (SLQP) where a linear program is solved first to determine the active set, and then we solve an equality-constrained QP on the equality constraints of the active set to compute the step direction. The linear program uses a first-order Taylor series approximation of the objective function and it considers the same linearizations of the constraints:

$$\min_{\mathbf{p}\in\mathbb{R}^n} (\nabla f(\mathbf{x}_k))^\mathsf{T}\mathbf{p}$$
$$\text{s.t. } \mathbf{h}(\mathbf{x}_k) + (\nabla\mathbf{h}(\mathbf{x}_k))^\mathsf{T}\mathbf{p} = \mathbf{0} \tag{4.109}$$
$$\mathbf{g}(\mathbf{x}_k) + (\nabla\mathbf{g}(\mathbf{x}_k))^\mathsf{T}\mathbf{p} \le \mathbf{0}$$

The solution of this linear program, $\mathbf{p}^*$, will have a set $\mathcal{A}(\mathbf{p}^*)$ of active constraints:

$$\mathcal{A}(\mathbf{p}^*) := \{i \in L : h_i(\mathbf{x}_k) + (\nabla h_i(\mathbf{x}_k))^\mathsf{T}\mathbf{p}^* = 0\} \cup \{i \in M : h_i(\mathbf{x}_k) + (\nabla g_i(\mathbf{x}_k))^\mathsf{T}\mathbf{p}^* = 0\}$$

Then, the step direction is computed at the second sub-problem, which is the equality-constrained QP:

$$\min_{\mathbf{p}\in\mathbb{R}^n} \mathbf{p}^\mathsf{T}\nabla f(\mathbf{x}_k) + \frac{1}{2}\mathbf{p}^\mathsf{T}\nabla_{\mathbf{xx}}^2\mathcal{L}(\mathbf{x}_k, \boldsymbol{\lambda}_k, \boldsymbol{\mu}_k)\mathbf{p}$$
$$\text{s.t. } h_i(\mathbf{x}_k) + (\nabla h_i(\mathbf{x}_k))^\mathsf{T}\mathbf{p} = 0 \quad \forall i \in L \cap \mathcal{A}(\mathbf{p}^*) \qquad (4.110)$$
$$g_i(\mathbf{x}_k) + (\nabla g_i(\mathbf{x}_k))^\mathsf{T}\mathbf{p} = 0 \quad \forall i \in M \cap \mathcal{A}(\mathbf{p}^*)$$

resulting in the step direction $\mathbf{p}_k$.

Another approach proposed by Schittkowski [28] and developed as an algorithmic package [33] is the Sequential Least Squares quadratic programming (SLSQP). SLSQP is similar to the SQP method that solves an inequality QP at each iteration, but replaces the quadratic objective function with a linear least squares function using LDL^T decomposition:

$$\nabla_{\mathbf{xx}}^2\mathcal{L}(\mathbf{x}_k, \boldsymbol{\lambda}_k, \boldsymbol{\mu}_k) = \mathbf{L}_k\mathbf{D}_k\mathbf{L}_k^\mathsf{T}$$

where $\mathbf{L}$ is a lower triangular matrix and $\mathbf{D}$ is a diagonal matrix with $\mathbf{p}$ at its diagonal. The resulting sub-problem at each iteration k becomes a least squares one:

$$\min_{\mathbf{p}\in\mathbb{R}^n} \|\mathbf{D}_k^{1/2}\mathbf{L}_k^\mathsf{T}\mathbf{p} + \mathbf{D}_k^{-1/2}\mathbf{L}_k^{-1}\nabla f(\mathbf{x}_k)\|$$
$$\text{s.t. } \mathbf{h}(\mathbf{x}_k) + (\nabla \mathbf{h}(\mathbf{x}_k))^\mathsf{T}\mathbf{p} = \mathbf{0} \qquad (4.111)$$
$$\mathbf{g}(\mathbf{x}_k) + (\nabla \mathbf{g}(\mathbf{x}_k))^\mathsf{T}\mathbf{p} \leq \mathbf{0}$$

Matrix $\mathbf{D}^{1/2} = \text{diag}(\sqrt{p_1}, \ldots, \sqrt{p_n})$ is well defined when guaranteeing that the Hessian approximation $\mathbf{H}_k \approx \nabla_{\mathbf{xx}}^2\mathcal{L}(\mathbf{x}_k, \boldsymbol{\lambda}_k, \boldsymbol{\mu}_k)$ is always positive definite. To account for the inconsistencies of the constraint linearizations, the problem can be reformulated as the $(n + 1)$-dimensional problem:

$$\min_{\mathbf{p}\in\mathbb{R}^n, \delta\in\mathbb{R}} \|\mathbf{D}_k^{1/2}\mathbf{L}_k^\mathsf{T}\mathbf{p} + \mathbf{D}_k^{-1/2}\mathbf{L}_k^{-1}\nabla f(\mathbf{x}_k) + \psi_k\delta\|$$
$$\text{s.t. } (\nabla \mathbf{h}(\mathbf{x}_k))^\mathsf{T}\mathbf{p} + (1 - \delta)\mathbf{h}(\mathbf{x}_k) = \mathbf{0}$$
$$(\nabla g_i(\mathbf{x}_k))^\mathsf{T}\mathbf{p} + (1 - \delta)g_i(\mathbf{x}_k) \leq 0 \qquad \forall i \in M_1 \qquad (4.112)$$
$$(\nabla g_i(\mathbf{x}_k))^\mathsf{T}\mathbf{p} + g_i(\mathbf{x}_k) \leq 0 \qquad \forall i \in M_2$$
$$0 \leq \delta \leq 1$$

where set $M_2 = M \setminus \{M_1\}$ and $M_1 := \{i \in M : g_i(\mathbf{x}_k) > 0\}$. An implementation of the SLSQP method is provided below and the results are presented in Table 4.12.

Table 4.12 SLSQP iterations.

k	$\mathbf{x}_k$	$f(\mathbf{x}_k)$	$\phi(\mathbf{x}_k)$
0	$[0.5000, 0]^\mathsf{T}$	0.125000011	1.25E+00
1	$[0.3810, -0.1587]^\mathsf{T}$	0.00993413	2.24E-01
2	$[0.2132, -0.0560]^\mathsf{T}$	−0.004784168	9.33E-02
3	$[0.2201, -0.0722]^\mathsf{T}$	−0.005481559	6.87E-03
4	$[0.2216, -0.0739]^\mathsf{T}$	−0.00548682	5.66E-04
5	$[0.2220, -0.0740]^\mathsf{T}$	−0.005486945	3.13E-04
6	$[0.2222, -0.0741]^\mathsf{T}$	−0.005486968	1.08E-05

Practitioner's Corner
SLSQP in Python 3

Method SLSQP in library `scipy.optimize.minimize` applies the SLSQP method in Python 3. An example of the application of this method to the constrained nonlinear problem:

$$\min \ x_1^3 + 2x_1x_2 + 3x_2^2$$
$$\text{s.t. } 1 - x_1 - 2x_2 \geq 0$$
$$1 - x_1^2 - x_2 \geq 0$$
$$1 - x_1^2 + x_2 \geq 0$$

is presented below.

```
import numpy as np
from scipy.optimize import minimize
def f(x):
    return x[0]**3+2*x[1]*x[0]+3*x[1]**2

ineq_cons = {'type': 'ineq',
    'fun' : lambda x: np.array([1 - x[0] - 2*x[1],
                                1 - x[0]**2 - x[1],
                                1 - x[0]**2 + x[1]])}
x0 = np.array([0.5, 0])
res = minimize(f, x0, method='SLSQP',
                constraints=ineq_cons, options={'ftol':
    1e-9, 'disp': True,'iprint': 10})
print(res.x)
```

which will result in solution $\mathbf{x}^* = [0.22222311, -0.07407612]^\mathsf{T}$. The detailed iterations are presented in Table 4.12.

4.8 Interior Point Method in Nonlinear Programming

Interior-point methods (IPMs), which are also known as barrier methods, are a class of algorithms that encode the feasible set using a barrier. Barrier methods were developed in the 1960s, but they were of little use until 1990 and the success of Karmarkar's interior point algorithm for linear programming. In this section, we will discuss interior point methods for nonlinear programming. In the early stages, interior point methods were solving problems transformed to the form:

$$\min_{\mathbf{x}} \ f(\mathbf{x})$$

$$\text{s.t. } \mathbf{g}(\mathbf{x}) \geq \mathbf{0}$$

and they were using the *Barrier Function*:

$$B(\mathbf{x}, \mu) := f(\mathbf{x}) - \mu \sum_{i=1}^{m} \ln g_i(\mathbf{x})$$

to approximate the original problem by the minimization problem:

$$\min_{\mathbf{x}, \mu} \ B(\mathbf{x}, \mu)$$

for some parameter $\mu > 0$, known as *barrier parameter*. The name *interior point methods* was based on the fact that the barrier forces $\mathbf{x}$ to be feasible with respect to the inequality constraints $\mathbf{g}(\mathbf{x}) \geq \mathbf{0}$. More recent interior point methods reformulate any original problem with equality and inequality constraints:

$$\min_{\mathbf{x}} \ f(\mathbf{x}) \tag{4.113}$$

$$\text{s.t. } \mathbf{h}(\mathbf{x}) = \mathbf{0} \tag{4.114}$$

$$\mathbf{g}(\mathbf{x}) \leq \mathbf{0} \tag{4.115}$$

To the following, equivalent, standard form formulation:

Standard form of nonlinear program used in Interior Point Methods

$$\min_{\mathbf{x}, \mathbf{s}} \ f(\mathbf{x}) \tag{4.116}$$

$$\text{s.t. } \mathbf{h}(\mathbf{x}) = \mathbf{0} \tag{4.117}$$

$$\mathbf{g}(\mathbf{x}) + \mathbf{s} = \mathbf{0} \tag{4.118}$$

$$\mathbf{s} \geq \mathbf{0} \tag{4.119}$$

where $\mathbf{h}(\mathbf{x}) = [h_1(\mathbf{x}), h_2(\mathbf{x}), \ldots, h_l(\mathbf{x})]^\mathsf{T}$ is an l-valued vector function $\mathbb{R}^n \to \mathbb{R}^l$ of equality constraints, $\mathbf{g}(\mathbf{x}) = [g_1(\mathbf{x}), g_2(\mathbf{x}), \ldots, g_m(\mathbf{x})]^\mathsf{T}$ is an m-valued vector function $\mathbb{R}^n \to \mathbb{R}^m$, and $\mathbf{s}$ an m-valued vector of slack variables, where each one of them is associated with an inequality constraint.

The barrier function associated to the NLP in standard form presented in Eqs. (4.116)–(4.119) is:

$$B(\mathbf{x}, \mathbf{s}, \mu) := f(\mathbf{x}) - \mu \sum_{i=1}^{m} \ln(s_i)$$

for some $\mu > 0$ and the problem in Eqs. (4.116)–(4.119) can be approximated by:

$$\min_{\mathbf{x}, \mathbf{s}, \mu} \; f(\mathbf{x}) - \mu \sum_{i=1}^{m} \ln(s_i) \tag{4.120}$$

$$\text{s.t. } \mathbf{h}(\mathbf{x}) = \mathbf{0} \tag{4.121}$$

$$\mathbf{g}(\mathbf{x}) + \mathbf{s} = \mathbf{0} \tag{4.122}$$

Notice that we no longer need the inequality constraints $\mathbf{s} \geq \mathbf{0}$ because $\lim_{s_i \to 0^+} \ln(s_i) = -\infty$ resulting in $B(\mathbf{x}, \mathbf{s}, \mu) \to +\infty$ when a variable s_i approaches 0. The original problem in Eqs. (4.116)–(4.119) and the problem in Eqs. (4.120)–(4.122) are not equivalent; however, as the barrier parameter μ approaches 0, the minimum value of $B(\mathbf{x}, \mathbf{s}, \mu)$ should converge to the minimum value of the objective function of the original problem. To understand this better, consider the following minimization problem in standard form:

$$\min_{x} \; (x + 1)^2 \quad \text{subject to: } x \geq 0$$

that can be approximated by:

$$\min_{x} \; (x + 1)^2 - \mu \ln x$$

In Fig. 4.10 we present the objective functions of the original problem and the approximated problem with different barrier parameter values $\mu = 10$, $\mu = 5$, $\mu = 0.5$. As you can see, as μ gets closer to 0, its objective function value is closer to the optimal objective function value for the original problem.

Fig. 4.10 Plot of $(x + 1)^2$ with optimal solution $x^* = 0$ when considering constraint $x \geq 0$ and plots of the barrier functions $(x + 1)^2 - \mu \ln x$ for $\mu = 10$, $\mu = 5$, and $\mu = 0.5$. Notice that as $\mu \to 0$ the optimal solution of $(x + 1)^2 - \mu \ln x$ gets closer to x^*.

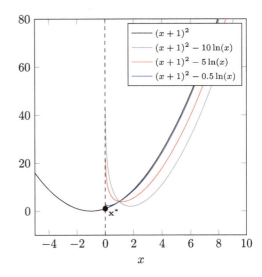

4.8.1 Step Direction

To solve problems in higher dimensions, let us consider the KKT conditions of the barrier problem in Eqs. (4.120)–(4.122). The Lagrangian of this problem is:

$$\mathcal{L}(\mathbf{x}, \mathbf{s}, \boldsymbol{\lambda}, \mathbf{z}) = f(\mathbf{x}) - \mu \sum_{i=1}^{m} \ln(s_i) + \boldsymbol{\lambda}^{\mathsf{T}} \mathbf{h}(\mathbf{x}) + \mathbf{z}^{\mathsf{T}} (\mathbf{g}(\mathbf{x}) + \mathbf{s})$$

where $\boldsymbol{\lambda} \in \mathbb{R}^l$ and $\mathbf{z} \in \mathbb{R}^m$ are Lagrange multipliers corresponding to the equality constraints. There are no dual variables related to inequality constraints, thus our first-order necessary conditions, using the denominator layout notation, are:

$$\nabla \mathcal{L}(\mathbf{x}, \mathbf{s}, \boldsymbol{\lambda}, \mathbf{z}) = \mathbf{0} \Rightarrow [\nabla_{\mathbf{x}} \mathcal{L}(\mathbf{x}, \mathbf{s}, \boldsymbol{\lambda}, \mathbf{z}), \nabla_{\mathbf{s}} \mathcal{L}(\mathbf{x}, \mathbf{s}, \boldsymbol{\lambda}, \mathbf{z}), \nabla_{\boldsymbol{\lambda}} \mathcal{L}(\mathbf{x}, \mathbf{s}, \boldsymbol{\lambda}, \mathbf{z}), \nabla_{\mathbf{z}} \mathcal{L}(\mathbf{x}, \mathbf{s}, \boldsymbol{\lambda}, \mathbf{z})]^{\mathsf{T}} = \mathbf{0}$$

We have:

$$\nabla_{\mathbf{x}} \mathcal{L}(\mathbf{x}, \mathbf{s}, \boldsymbol{\lambda}, \mathbf{z}) = \mathbf{0} \Rightarrow \underbrace{\nabla_{\mathbf{x}} f(\mathbf{x})}_{n \times 1} + \underbrace{\nabla_{\mathbf{x}} \mathbf{h}(\mathbf{x})}_{n \times l} \underbrace{\boldsymbol{\lambda}}_{l \times 1} + \underbrace{\nabla_{\mathbf{x}} \mathbf{g}(\mathbf{x})}_{n \times m} \underbrace{\mathbf{z}}_{m \times 1} = \underbrace{\mathbf{0}}_{n \times 1} \qquad (4.123)$$

$$\nabla_{\mathbf{s}} \mathcal{L}(\mathbf{x}, \mathbf{s}, \boldsymbol{\lambda}, \mathbf{z}) = \mathbf{0} \Rightarrow -\mu \begin{bmatrix} \frac{\partial \sum_{i=1}^{m} \ln s_i}{\partial s_1} \\ \frac{\partial \sum_{i=1}^{m} \ln s_i}{\partial s_2} \\ \vdots \\ \frac{\partial \sum_{i=1}^{m} \ln s_i}{\partial s_m} \end{bmatrix} + \frac{\partial \mathbf{z}^{\mathsf{T}} \mathbf{s}}{\partial \mathbf{s}} = \mathbf{0} \Rightarrow -\mu \underbrace{\mathbf{s}^{-1}}_{m \times 1} + \underbrace{\mathbf{z}}_{m \times 1} = \underbrace{\mathbf{0}}_{m \times 1}$$

$$(4.124)$$

$$\nabla_\lambda \mathcal{L}(\mathbf{x}, \mathbf{s}, \lambda, z) = 0 \Rightarrow \underbrace{\mathbf{h}(\mathbf{x})}_{l \times 1} = \underbrace{\mathbf{0}}_{l \times 1} \tag{4.125}$$

$$\nabla_z \mathcal{L}(\mathbf{x}, \mathbf{s}, \lambda, z) = 0 \Rightarrow \underbrace{\mathbf{g}(\mathbf{x})}_{m \times 1} + \underbrace{\mathbf{s}}_{m \times 1} = \underbrace{\mathbf{0}}_{m \times 1} \tag{4.126}$$

Notice that if we add $m \times m$ diagonal matrix $\mathbf{S}$ for which $S_{ii} = s_i$ and an m-valued column vector $\mathbf{e} = [1, 1, \ldots, 1]^\mathsf{T}$, this allows us to rewrite Eq. (4.124) as:

$$-\mu \underbrace{\mathbf{S}^{-1}}_{m \times m} \underbrace{\mathbf{e}}_{m \times 1} + \underbrace{z}_{m \times 1} = \underbrace{\mathbf{0}}_{m \times 1} \tag{4.127}$$

Because the diagonal elements of $\mathbf{S}^{-1}$ are forced to be positive, we can multiply this equation by $\mathbf{S}$ resulting in [34]:

$$-\mu \underbrace{\mathbf{e}}_{m \times 1} + \underbrace{\mathbf{S}}_{m \times m} \underbrace{z}_{m \times 1} = \underbrace{\mathbf{0}}_{m \times 1} \tag{4.128}$$

Equations (4.123), (4.125), (4.126), and (4.128) result in a system with $n + 2m + l$ equations. Let us apply Newton's method to this system. Reckon that Newton's method tries to approximate a vector-valued function $\mathbf{q}(\mathbf{x})$ near the current iterate $\mathbf{x}_k$ by a function $\mathbf{q}(\mathbf{x})$ for which the system of equations $\mathbf{q}(\mathbf{x})$ is easy to solve, and then uses this solution as the next iterate $\mathbf{x}_{k+1}$. A suitable choice of $\mathbf{q}(\mathbf{x})$ is the linear approximation of $\mathbf{q}(\mathbf{x})$ at $\mathbf{x}_k$, whose graph is the tangent space to $\mathbf{q}(\mathbf{x})$ at $\mathbf{x}_k$:

$$\mathbf{q}(\mathbf{x}) \approx \mathbf{q}(\mathbf{x}_k) + (\nabla_\mathbf{x}\mathbf{q}(\mathbf{x}_k))(\mathbf{x} - \mathbf{x}_k)$$

where $(\mathbf{x} - \mathbf{x}_k)$ is the step direction, written also as $\mathbf{p} = (\mathbf{x} - \mathbf{x}_k)$. In our case, we have variables $\mathbf{x}, \mathbf{s}, \lambda, z$, thus the step direction is $[\mathbf{p}_\mathbf{x}, \mathbf{p}_\mathbf{s}, \mathbf{p}_\lambda, \mathbf{p}_z]^\mathsf{T}$. In addition, in our case $\mathbf{q}(\mathbf{x})$ is:

$$\mathbf{q}(\mathbf{x}, \mathbf{s}, \lambda, z) = \begin{bmatrix} \nabla_\mathbf{x} f(\mathbf{x}) + \nabla_\mathbf{x}\mathbf{h}(\mathbf{x})\lambda + \nabla_\mathbf{x}\mathbf{g}(\mathbf{x})z \\ -\mu\mathbf{e} + \mathbf{S}z \\ \mathbf{h}(\mathbf{x}) \\ \mathbf{g}(\mathbf{x}) + \mathbf{s} \end{bmatrix}$$

and the Jacobian, when using the denominator layout notation, is:

$$\nabla\mathbf{q}(\mathbf{x}, \mathbf{s}, \lambda, z) = \begin{bmatrix} \frac{\partial \nabla_\mathbf{x} f(\mathbf{x}) + \nabla_\mathbf{x}\mathbf{h}(\mathbf{x})\lambda + \nabla_\mathbf{x}\mathbf{g}(\mathbf{x})z}{\partial \mathbf{x}} & \frac{\partial -\mu\mathbf{e} + \mathbf{S}z}{\partial \mathbf{x}} & \frac{\partial \mathbf{h}(\mathbf{x})}{\partial \mathbf{x}} & \frac{\partial \mathbf{g}(\mathbf{x}) + \mathbf{s}}{\partial \mathbf{x}} \\ \frac{\partial \nabla_\mathbf{x} f(\mathbf{x}) + \nabla_\mathbf{x}\mathbf{h}(\mathbf{x})\lambda + \nabla_\mathbf{x}\mathbf{g}(\mathbf{x})z}{\partial \mathbf{s}} & \frac{\partial -\mu\mathbf{e} + \mathbf{S}z}{\partial \mathbf{s}} & \frac{\partial \mathbf{h}(\mathbf{x})}{\partial \mathbf{s}} & \frac{\partial \mathbf{g}(\mathbf{x}) + \mathbf{s}}{\partial \mathbf{s}} \\ \frac{\partial \nabla_\mathbf{x} f(\mathbf{x}) + \nabla_\mathbf{x}\mathbf{h}(\mathbf{x})\lambda + \nabla_\mathbf{x}\mathbf{g}(\mathbf{x})z}{\partial \lambda} & \frac{\partial -\mu\mathbf{e} + \mathbf{S}z}{\partial \lambda} & \frac{\partial \mathbf{h}(\mathbf{x})}{\partial \lambda} & \frac{\partial \mathbf{g}(\mathbf{x}) + \mathbf{s}}{\partial \lambda} \\ \frac{\partial \nabla_\mathbf{x} f(\mathbf{x}) + \nabla_\mathbf{x}\mathbf{h}(\mathbf{x})\lambda + \nabla_\mathbf{x}\mathbf{g}(\mathbf{x})z}{\partial z} & \frac{\partial -\mu\mathbf{e} + \mathbf{S}z}{\partial z} & \frac{\partial \mathbf{h}(\mathbf{x})}{\partial z} & \frac{\partial \mathbf{g}(\mathbf{x}) + \mathbf{s}}{\partial z} \end{bmatrix}$$

which results in:

$$\nabla \mathbf{q}(\mathbf{x}, \mathbf{s}, \boldsymbol{\lambda}, z) = \begin{bmatrix} \underbrace{\nabla_{\mathbf{xx}} \mathcal{L}(\mathbf{x}, \mathbf{s}, \boldsymbol{\lambda}, z)}_{n \times n} & \underbrace{\mathbf{0}}_{n \times m} & \underbrace{\nabla_{\mathbf{x}} \mathbf{h}(\mathbf{x})}_{n \times l} & \underbrace{\nabla_{\mathbf{x}} \mathbf{g}(\mathbf{x})}_{n \times m} \\ \underbrace{\mathbf{0}}_{m \times n} & \underbrace{\mathbf{Z}}_{m \times m} & \underbrace{\mathbf{0}}_{m \times l} & \underbrace{\mathbf{S}}_{m \times m} \\ \underbrace{(\nabla_{\mathbf{x}} \mathbf{h}(\mathbf{x}))^{\mathsf{T}}}_{l \times n} & \mathbf{0} & \mathbf{0} & \mathbf{0} \\ \underbrace{(\nabla_{\mathbf{x}} \mathbf{g}(\mathbf{x}))^{\mathsf{T}}}_{m \times n} & \underbrace{\mathbf{I}}_{m \times m} & \mathbf{0} & \mathbf{0} \end{bmatrix}$$

where $\mathbf{Z}$ is the diagonal matrix of z with $Z_{ii} = z_i$ and $\mathbf{I}$ the identity matrix.

Applying Newton's method to the nonlinear system in the variables $\mathbf{x}, \mathbf{s}, \boldsymbol{\lambda}, z$ results in:

$$\nabla \mathbf{q}(\mathbf{x}, \mathbf{s}, \boldsymbol{\lambda}, z)[\mathbf{p_x}, \mathbf{p_s}, \mathbf{p_\lambda}, \mathbf{p_z}]^{\mathsf{T}} = -\mathbf{q}(\mathbf{x}, \mathbf{s}, \boldsymbol{\lambda}, z) \Rightarrow$$

$$\begin{bmatrix} \nabla_{\mathbf{xx}} \mathcal{L}(\mathbf{x}, \mathbf{s}, \boldsymbol{\lambda}, z) & 0 & \nabla_{\mathbf{x}} \mathbf{h}(\mathbf{x}) & \nabla_{\mathbf{x}} \mathbf{g}(\mathbf{x}) \\ 0 & \mathbf{Z} & 0 & \mathbf{S} \\ (\nabla_{\mathbf{x}} \mathbf{h}(\mathbf{x}))^{\mathsf{T}} & 0 & 0 & 0 \\ (\nabla_{\mathbf{x}} \mathbf{g}(\mathbf{x}))^{\mathsf{T}} & \mathbf{I} & 0 & 0 \end{bmatrix} \begin{bmatrix} \mathbf{p_x} \\ \mathbf{p_s} \\ \mathbf{p_\lambda} \\ \mathbf{p_z} \end{bmatrix} = - \begin{bmatrix} \nabla_{\mathbf{x}} f(\mathbf{x}) + \nabla_{\mathbf{x}} \mathbf{h}(\mathbf{x})\boldsymbol{\lambda} + \nabla_{\mathbf{x}} \mathbf{g}(\mathbf{x})z \\ -\mu \mathbf{e} + \mathbf{S}z \\ \mathbf{h}(\mathbf{x}) \\ \mathbf{g}(\mathbf{x}) + \mathbf{s} \end{bmatrix}$$

This formulation is known as the *primal-dual* system that gives a primal-dual search direction. Note that all values in this system refer to the k-th iteration. For instance, $\mathbf{s}$ corresponds to $(\mathbf{s})_k$, $\mathbf{x}$ to $(\mathbf{x})_k$, and so forth. We can turn $\nabla \mathbf{q}(\mathbf{x}, \mathbf{s}, \boldsymbol{\lambda}, z)$ into a symmetric matrix to reduce the computation costs by multiplying the second row by $\mathbf{S}^{-1}$ resulting in:

$$\begin{bmatrix} \nabla_{\mathbf{xx}} \mathcal{L}(\mathbf{x}, \mathbf{s}, \boldsymbol{\lambda}, z) & 0 & \nabla_{\mathbf{x}} \mathbf{h}(\mathbf{x}) & \nabla_{\mathbf{x}} \mathbf{g}(\mathbf{x}) \\ 0 & \mathbf{S}^{-1}\mathbf{Z} & 0 & \mathbf{I} \\ (\nabla_{\mathbf{x}} \mathbf{h}(\mathbf{x}))^{\mathsf{T}} & 0 & 0 & 0 \\ (\nabla_{\mathbf{x}} \mathbf{g}(\mathbf{x}))^{\mathsf{T}} & \mathbf{I} & 0 & 0 \end{bmatrix} \begin{bmatrix} \mathbf{p_x} \\ \mathbf{p_s} \\ \mathbf{p_\lambda} \\ \mathbf{p_z} \end{bmatrix} = - \begin{bmatrix} \nabla_{\mathbf{x}} f(\mathbf{x}) + \nabla_{\mathbf{x}} \mathbf{h}(\mathbf{x})\boldsymbol{\lambda} + \nabla_{\mathbf{x}} \mathbf{g}(\mathbf{x})z \\ -\mu \mathbf{S}^{-1} \mathbf{e} + z \\ \mathbf{h}(\mathbf{x}) \\ \mathbf{g}(\mathbf{x}) + \mathbf{s} \end{bmatrix}$$

This formulation allows to use a symmetric linear equations solver to obtain the step direction $\mathbf{p_x}, \mathbf{p_s}, \mathbf{p_\lambda}, \mathbf{p_z}$ at iteration k.

4.8.2 Step Length

To enforce $\mathbf{s} \geq \mathbf{0}$ and $z \geq \mathbf{0}$ we need to choose an appropriate step length. We can enforce this by using the *fraction-to-boundary-rule* to select initial step lengths:

$$\begin{aligned} a_{primal} &= \max\{a \in (0, 1] : \mathbf{s}_k + a\mathbf{p_s} \geq (1 - \tau)\mathbf{s}_k\} \\ a_{dual} &= \max\{a \in (0, 1] : z_k + a\mathbf{p_z} \geq (1 - \tau)z_k\} \end{aligned} \tag{4.129}$$

where $\tau \in (0, 1)$, e.g., $\tau = 0.995$ [35], is the *fraction-to-the-boundary parameter* [36]. By restricting the primal and dual step length with this fraction-to-the-boundary rule, we guarantee positivity of **s** and z at each iterate.

In addition to the above, we want our step lengths to ensure that we are making progress from iteration to iteration. For this, we can employ backtracking line search using a merit function, such as the l_1 or l_2 merit function, i.e.,

$$\phi_\rho(\mathbf{x}, \mathbf{s}) = f(\mathbf{x}) - \mu \sum_{i=1}^{m} \log s_i + \rho \|\mathbf{h}(\mathbf{x})\|_2 + \rho \|\mathbf{g}(\mathbf{x}) + \mathbf{s}\|_2$$

where $\rho > 0$. With backtracking line search, we further reduce the step lengths in successive trial steps:

$$a_{primal} = c_1 a_{primal} \quad \text{where } c_1 \in (0, 1)$$
$$a_{dual} = c_2 a_{dual} \quad \text{where } c_2 \in (0, 1)$$

until achieving a sufficient decrease of the merit function. Then, we can use the step lengths to update the values of our variables and proceed to the next iteration:

$$\mathbf{x}_{k+1} = \mathbf{x}_k + a_{primal}\mathbf{p_x}$$
$$\mathbf{s}_{k+1} = \mathbf{s}_k + a_{primal}\mathbf{p_s}$$
$$\boldsymbol{\lambda}_{k+1} = \boldsymbol{\lambda}_k + a_{dual}\mathbf{p_\lambda}$$
$$z_{k+1} = z_k + a_{dual}\mathbf{p_z}$$

4.8.3 Termination Criterion

There are many variations for the stopping criteria of the interior point methods. One of them is that we have found a solution $\mathbf{x}^*, \mathbf{s}^*, \boldsymbol{\lambda}^*, z^*$ so that:

$$\nabla \mathcal{L}(\mathbf{x}^*, \mathbf{s}^*, \boldsymbol{\lambda}^*, z^*) \approx \mathbf{0}$$

Given a tolerance ϵ, we can terminate the algorithm when the four following conditions hold for some $\mathbf{x}^*, \mathbf{s}^*, \boldsymbol{\lambda}^*, z^*$ [37]:

$$\|\nabla_{\mathbf{x}} f(\mathbf{x}) + \nabla_{\mathbf{x}}\mathbf{h}(\mathbf{x})\boldsymbol{\lambda} + \nabla_{\mathbf{x}}\mathbf{g}(\mathbf{x})z\|_\infty \leq \epsilon$$

$$\| - \mu\mathbf{e} + \mathbf{S}z\|_\infty \leq \epsilon$$

$$\|\mathbf{h}(\mathbf{x})\|_\infty \leq \epsilon$$

$$\|\mathbf{g}(\mathbf{x}) + \mathbf{s}\|_\infty \le \epsilon$$

where $\|\mathbf{x}\|_\infty = \max\{|x_1|, |x_2|, \ldots, |x_n|\}$. We can, equivalently, write this as:

$$E(\mathbf{x}, \mathbf{s}, \boldsymbol{\lambda}, \mathbf{z}, \mu = 0) \le \epsilon$$

where:

$$E(\mathbf{x}, \mathbf{s}, \boldsymbol{\lambda}, \mathbf{z}, \mu) = \max\{\|\nabla_\mathbf{x} f(\mathbf{x}) + (\nabla_\mathbf{x}\mathbf{h}(\mathbf{x}))^\mathsf{T}\boldsymbol{\lambda} + (\nabla_\mathbf{x}\mathbf{g}(\mathbf{x}))^\mathsf{T}\mathbf{z}\|_\infty, \| - \mu\mathbf{e} + \mathbf{S}\mathbf{z}\|_\infty, \|\mathbf{h}(\mathbf{x})\|_\infty, \|\mathbf{g}(\mathbf{x}) + \mathbf{s}\|_\infty\}$$

4.8.4 Updating the Barrier Parameter

As already discussed, the approximation of the original problem by using a barrier function is accurate when μ gets close to 0. One approach, known as the *monotone* approach, reduces the value of μ by a specific ratio only when the optimality conditions of the barrier problem are satisfied with some accuracy.

The monotone approach, also known as the Fiacco-McCormick approach [38], fixes the barrier parameter by giving it an initial value μ_k until:

$$\max\{\|\nabla_\mathbf{x} f(\mathbf{x}) + (\nabla_\mathbf{x}\mathbf{h}(\mathbf{x}))^\mathsf{T}\boldsymbol{\lambda} + (\nabla_\mathbf{x}\mathbf{g}(\mathbf{x}))^\mathsf{T}\mathbf{z}\|_\infty, \| - \mu\mathbf{e} + \mathbf{S}\mathbf{z}\|_\infty, \|\mathbf{h}(\mathbf{x})\|_2, \|\mathbf{g}(\mathbf{x}) + \mathbf{s}\|_\infty\} \le \mu_k$$

Then, the barrier parameter is decreased by the rule $\mu_{k+1} = \sigma_k\mu_k$ with $\sigma_k \in (0, 1)$. A typical value of σ_k used in solvers is 1/5. In essence, with the monotone approach we require the approximate solution of the barrier problem defined for μ_k to satisfy the tolerance μ_k before the algorithm continues to the solution of the next barrier problem with a smaller μ_{k+1}.

Summarizing the provided information, a basic interior-point framework is provided below.

4.8.5 Dealing with Nonconvexity and Practical Implementation

When solving the primal dual system at each Newton iteration, $\nabla_{\mathbf{xx}}\mathcal{L}(\mathbf{x}, \mathbf{s}, \boldsymbol{\lambda}, \mathbf{z})$ might be indefinite. If this is the case, i.e., if we identify this with Cholesky factorization, we can replace $\nabla_{\mathbf{xx}}\mathcal{L}(\mathbf{x}, \mathbf{s}, \boldsymbol{\lambda}, \mathbf{z})$ by $\nabla_{\mathbf{xx}}\mathcal{L}(\mathbf{x}, \mathbf{s}, \boldsymbol{\lambda}, \mathbf{z}) + \theta\mathbf{I}$. We can start with a small value of $\theta > 0$ and increase it iteratively until Cholesky's decomposition is successful.

Interior point approaches offer a framework for solving constrained nonlinear optimization problems. Slight modifications can lead to different versions of these methods, in the sense that there is not a single interior point algorithm that supersedes all others in all cases. In terms of practical implementation, a specific interior point algorithm for large-scale constrained nonlinear problems is provided by the open source Ipopt (Interior Point Optimizer). Ipopt offers a specific interior point imple-

---

Algorithm 20 Basic Interior-Point framework

1: set $k = 0$ and choose $\mathbf{x}_k$, $\mathbf{s}_k \geq 0$, $\boldsymbol{\lambda}_k$ and $z_k \geq 0$
2: select initial barrier parameter $\mu_k > 0$, parameters $\sigma, \tau, c_1, c_2 \in (0, 1)$, and solution tolerance ϵ
3: select merit function ϕ
4: **repeat**
5:   **repeat**
6:     solve the symmetric primal-dual system to obtain the search direction $\mathbf{p} = [\mathbf{p_x}, \mathbf{p_s}, \mathbf{p_\lambda}, \mathbf{p_z}]^\mathsf{T}$
7:     compute $a_{primal} = \max\{a \in (0, 1] : \mathbf{s}_k + a\mathbf{p_s} \geq (1 - \tau)\mathbf{s}_k\}$
8:     compute $a_{dual} = \max\{a \in (0, 1] : z_k + a\mathbf{p_z} \geq (1 - \tau)z_k\}$
9:     update the values $\mathbf{x}_{k+1} = \mathbf{x}_k + a_{primal}\mathbf{p_x}$ and $\mathbf{s}_{k+1} = \mathbf{s}_k + a_{primal}\mathbf{p_s}$
10:    update the values $\boldsymbol{\lambda}_{k+1} = \boldsymbol{\lambda}_k + a_{dual}\mathbf{p_\lambda}$ and $z_{k+1} = z_k + a_{dual}\mathbf{p_z}$
11:    **if** the merit function is not sufficiently decreased with the new variables **then**
12:       With backtracking line search, steadily reduce the step lengths in successive trial steps: $a_{primal} = c_1 a_{primal}$ and $a_{dual} = c_2 a_{dual}$
13:    **end if**
14:    considering the new step lengths, update the values $\mathbf{x}_{k+1} = \mathbf{x}_k + a_{primal}\mathbf{p_x}$, $\mathbf{s}_{k+1} = \mathbf{s}_k + a_{primal}\mathbf{p_s}$, $\boldsymbol{\lambda}_{k+1} = \boldsymbol{\lambda}_k + a_{dual}\mathbf{p_\lambda}$ and $z_{k+1} = z_k + a_{dual}\mathbf{p_z}$
15:    set $\mu_{k+1} = \mu_k$
16:    set $k = k + 1$
17: **until** $E(\mathbf{x}, \mathbf{s}, \boldsymbol{\lambda}, z, \mu_k) \leq \mu_k$
18: set $\mu_{k+1} = \sigma\mu_k$
19: **until** $E(\mathbf{x}, \mathbf{s}, \boldsymbol{\lambda}, z, 0) \leq \epsilon$

---

mentation, described in Wächter and Biegler [39]. It can be used to solve constrained nonlinear programming problems with the only requirement that the objective and constraint functions should be twice continuously differentiable (convexity is not requested).

Let us consider the example:

$$\min\ x_1^3 + 2x_1x_2 + 3x_2^2$$
$$\text{s.t. } 1 - x_1 - 2x_2 \geq 0$$
$$1 - x_1^2 - x_2 \geq 0$$
$$1 - x_1^2 + x_2 \geq 0$$

From the above, we have $\mathbf{g}(\mathbf{x}) := -[1 - x_1 - 2x_2, 1 - x_1^2 - x_2, 1 - x_1^2 + x_2]^\mathsf{T}$. Using slack variables $s_1, s_2, s_3 \geq 0$ for the inequality constraints, we have:

$$\mathcal{L}(\mathbf{x}, \mathbf{s}, z) = f(\mathbf{x}) - \mu \sum_{i=1}^{m} \ln(s_i) + z^\mathsf{T}(\mathbf{g}(\mathbf{x}) + \mathbf{s})$$

For initial solution guess $\mathbf{x}_k = [0.5, 0]^\mathsf{T}$, this problem can be solved with Ipopt as follows.

Practitioner's Corner

Interior Point for constrained nonlinear programming in Python 3

We can use the `cyipopt` library which offers a wrapper for the Ipopt optimization package of [39] in Python 3. The application of the interior point solver to the previous example is presented below.

```python
from cyipopt import minimize_ipopt
import numpy as np
def obj(x):
    return x[0]**3+2*x[1]*x[0]+3*x[1]**2
def g1(x):
    return (1 - x[0] - 2*x[1])
def g2(x):
    return (1 - x[0]**2 - x[1])
def g3(x):
    return (1 - x[0]**2 + x[1])
cons = [
    {'type': 'ineq', 'fun': g1},{'type': 'ineq', 'fun'
    : g2},{'type': 'ineq', 'fun': g3}
]
x0 = np.array([0.5, 0])
res = minimize_ipopt(obj, x0, constraints=cons,
    options={'disp':12})
print(res)
```

This will return solution $\mathbf{x}^* = [0.22222223, -0.07407408]^\mathsf{T}$ after 10 iterations.

4.9 Penalty and Augmented Lagrangian Methods

4.9.1 Penalty Methods

Penalty methods replace a constrained optimization problem by a series of single-function *unconstrained* optimization problems which consider the effects of the constraint violations and (ideally) converge to the solution of the original problem. The single function of each unconstrained minimization problem consists of:

- The *objective function* of the constrained optimization problem
- The *penalty function* that adds one additional term for each constraint which, when solving a minimization problem, is positive when the constraint is violated to indicate the constraint violation penalty, and zero otherwise.

The additional penalty terms associated with the constraints are multiplied by a positive coefficient which typically takes a high value to penalize constraint violations and prioritize them over the objective function.

Let us consider a minimization problem with equality constraints:

$$\min_{\mathbf{x}} \ f(\mathbf{x})$$

$$\text{s.t. } \mathbf{c}(\mathbf{x}) = \mathbf{0}$$

with $\mathbf{c} : \mathbb{R}^n \to \mathbb{R}^l$. This problem can be translated to a single-function problem with no constraints by using function:

$$Q(\mathbf{x}, \mu) = f(\mathbf{x}) + \mu \sum_{i=1}^{l} |c_i(\mathbf{x})|$$

where μ is the *penalty parameter*. Notice that for a large positive value of μ a minimization solver will try to set each $c_i(\mathbf{x})$ to 0 in order to avoid the penalty from the constraint violations.

Another typical penalty method that penalizes progressively the violation of constraints (the more a constraint is violated, the higher the inflicted penalty) is the *quadratic penalty method* with single function:

$$Q(\mathbf{x}, \mu) = f(\mathbf{x}) + \frac{\mu}{2} \sum_{i=1}^{l} c_i^2(\mathbf{x})$$

Typically, we can start at iteration $k = 0$ with a low value μ_k to compute $\mathbf{x}_k$ by solving:

$$\min_{\mathbf{x}} Q(\mathbf{x}, \mu_k)$$

For a small value of μ_k, solution $\mathbf{x}_k$ might not be feasible to the original problem because some constraints are violated. This is why the penalty methods are called *exterior point* methods since we are initially allowed to violate constraints (as opposed to interior point methods that maintain feasibility at each step). Our original constrained optimization problem is approximated by a series of $\min_{\mathbf{x}} Q(\mathbf{x}, \mu_k)$ problems, where:

- There is an increasing sequence $\{\mu_k\}$ where $\mu_{k+1} > \mu_k$
- The solution of $\mathbf{x}_k = \arg\min_{\mathbf{x}} Q(\mathbf{x}, \mu_k)$ at iteration k is used as initial solution guess at the next iteration

If successful, solutions $\mathbf{x}_k$ will stabilize after a number of iterations, indicating that we have eventually converged to the solution of the original constrained problem. The basic algorithmic framework of a penalty method is provided below.

In the following theorem, we show that convergence is guaranteed for some $\mu_k \to +\infty$ when solving convex problems.

Algorithm 21 Basic framework of Penalty Methods

1: choose $\mu_0 > 0$ and initial starting point $\mathbf{x}_0^s$ that might be infeasible to the constrained optimization problem
2: set $k = 0$
3: **repeat**
4:    solve min $Q(\mathbf{x}, \mu_k)$ to find $\mathbf{x}_k$ with an optimization method for unconstrained optimization starting from $\mathbf{x}_k^s$
5:    choose $\mu_{k+1} > \mu_k$
6:    set new starting point $\mathbf{x}_{k+1}^s = \mathbf{x}_k$
7:    set $k = k + 1$
8: **until** convergence is achieved

Theorem 4.10 *Suppose that $Q(\mathbf{x}, \mu_k)$ is convex and each solution $\mathbf{x}_k$ is a global optimum of $Q(\cdot, \mu_k)$. Then, a limit point $\bar{\mathbf{x}}$ of the sequence $\{\mathbf{x}_k\}$ is a globally optimal solution of the constrained optimization problem.*

Proof Let $\mathbf{x}^*$ be a global minimizer of min $f(\mathbf{x})$ s.t. $\mathbf{c}(\mathbf{x}) = \mathbf{0}$ for $\mathbf{x} \in \mathbb{R}^n$ and $\mathbf{c} : \mathbb{R}^n \to \mathbb{R}^l$. Then,

$$f(\mathbf{x}^*) \leq f(\mathbf{x}) \;\; \forall \mathbf{x} : \mathbf{c}(\mathbf{x}) = \mathbf{0}$$

In addition, since $\mathbf{x}_k$ is the global minimizer of $Q(\cdot, \mu_k)$ for each k, we have:

$$Q(\mathbf{x}_k, \mu_k) \leq Q(\mathbf{x}^*, \mu_k) \Rightarrow$$

$$f(\mathbf{x}_k) + \frac{\mu_k}{2} \sum_{i=1}^{l} c_i^2(\mathbf{x}_k) \leq f(\mathbf{x}^*) + \frac{\mu_k}{2} \sum_{i=1}^{l} c_i^2(\mathbf{x}^*) \xrightarrow{\mathbf{c}(\mathbf{x}^*) = \mathbf{0}}$$

$$f(\mathbf{x}_k) + \frac{\mu_k}{2} \sum_{i=1}^{l} c_i^2(\mathbf{x}_k) \leq f(\mathbf{x}^*) \Rightarrow$$

$$\sum_{i=1}^{l} c_i^2(\mathbf{x}_k) \leq \frac{2}{\mu_k}(f(\mathbf{x}^*) - f(\mathbf{x}_k))$$

Suppose now that $\bar{\mathbf{x}}$ is a limit point of the sequence $\{\mathbf{x}_k\}$, meaning that $\bar{\mathbf{x}} = \mathbf{x}_k$ for infinitely many values of sequence $\{\mathbf{x}_k\}$. That is, $\mathbf{x}_k = \bar{\mathbf{x}}$ for $k \to +\infty$ which also means that $\mu_k \to +\infty$. When $\mu_k \to +\infty$ we have:

$$\sum_{i=1}^{l} c_i^2(\mathbf{x}_k) \leq \frac{2}{\mu_k}(f(\mathbf{x}^*) - f(\mathbf{x}_k)) \Rightarrow$$

$$\sum_{i=1}^{l} c_i^2(\bar{\mathbf{x}}) \leq \frac{2}{\mu_k}(f(\mathbf{x}^*) - f(\mathbf{x}_k)) \xrightarrow{\mu_k \to +\infty} \sum_{i=1}^{l} c_i^2(\bar{\mathbf{x}}) \leq 0 \Rightarrow \mathbf{c}(\mathbf{x}) = \mathbf{0}$$

and thus $\bar{\mathbf{x}}$ is feasible. In addition, for $\mu_k \to +\infty$:

$$f(\mathbf{x}_k) + \frac{\mu_k}{2} \sum_{i=1}^{l} c_i^2(\mathbf{x}_k) \le f(\mathbf{x}^*) \xRightarrow{c(\bar{\mathbf{x}})=0} f(\bar{\mathbf{x}}) \le f(\mathbf{x}^*)$$

meaning that $\bar{\mathbf{x}}$ is a globally optimal solution of the original constrained optimization problem.

Consider, for example, the nonlinear Rosenbrock's function in two dimensions under a single equality constraint:

$$\min\ 100(x_2 - x_1^2)^2 + (1 - x_1)^2 \qquad\qquad (4.130)$$
$$\text{s.t. } x_1 + x_2 = 3 \qquad\qquad (4.131)$$

The quadratic penalty method results in the formation of the single function:

$$Q(\mathbf{x}, \mu_k) = 100(x_2 - x_1^2)^2 + (1 - x_1)^2 + \mu_k(x_1 + x_2 - 3)^2$$

The gradient of Q at each iteration k is:

$$\begin{bmatrix} \frac{\partial 100(x_2-x_1^2)^2+(1-x_1)^2+\mu_k(x_1+x_2-3)^2}{\partial x_1} \\ \frac{\partial 100(x_2-x_1^2)^2+(1-x_1)^2+\mu_k(x_1+x_2-3)^2}{\partial x_2} \end{bmatrix} = \begin{bmatrix} -400x_1(-x_1^2 + x_2) + 2x_1 - 2 + 2\mu_k x_1 \\ -200x_1^2 + 200x_2 + 2\mu_k x_2 \end{bmatrix}$$

and its Hessian matrix:

$$\nabla_{\mathbf{x}}^2 Q(\mathbf{x}, \mu_k) = \begin{bmatrix} 1200x_1^2 - 400x_2 + 2 + 2\mu_k & -400x_1 \\ -400x_1 & 200 + 2\mu_k \end{bmatrix}$$

Using Newton line search with Hessian modification (Algorithm 11) with starting $\mu_k = 2$ and initial solution guess $\mathbf{x}_k^s = [x_1, x_2]^\mathsf{T} = [5, 5]^\mathsf{T}$, the solution of the unconstrained problem $\min_{\mathbf{x}} Q(\mathbf{x}, \mu_k)$ is $\mathbf{x}_k = [1.28056896, 1.64064472]^\mathsf{T}$. We proceed to the next iteration by updating $\mu_k = 20$ and using $\mathbf{x}_k^s = [1.28056896, 1.64064472]^\mathsf{T}$ as initial solution guess to obtain new solution $\mathbf{x}_k = [1.30022991, 1.69143166]^\mathsf{T}$. In the next iteration, for $\mu_k = 200$ we obtain $\mathbf{x}_k = [1.30231036, 1.69685096]^\mathsf{T}$. Proceeding with four more iterations results in a sufficient convergence, as presented in Table 4.13.

We should note here that the selection of μ_0, the initial solution guess $\mathbf{x}_0^s$, and the updating of penalty parameter μ_k are very important for the successive application of penalty methods due to the danger of ill-conditioning. In general:

- μ_0 should not be too large to avoid ill-conditioning.
- μ_k should not be increased too fast because too much ill-conditioning might be forced upon the unconstrained optimization too early.

Table 4.13 Quadratic Penalty Method iterations with Newton's line search with Armijo backtracking when minimizing Rosenbrock's function.

k	$\mathbf{x}_k^s$	μ_k	$\mathbf{x}_k$	$f(\mathbf{x}_k)$	$Q(\mathbf{x}_k, \mu_k)$
0	[5,5]	2	[1.28057,1.64064]	0.0787810	0.0849883
1	[1.28057,1.64064]	$2 \cdot 10$	[1.30023,1.69143]	0.0902075	0.0909028
2	[1.30023,1.69143]	$2 \cdot 10^2$	[1.30231,1.69685]	0.0914619	0.0915322
3	[1.30231,1.69685]	$2 \cdot 10^3$	[1.30252,1.69740]	0.0915885	0.0915956
4	[1.30252,1.69740]	$2 \cdot 10^4$	[1.30254,1.69745]	0.0916012	0.0916019
5	[1.30254,1.69745]	$2 \cdot 10^5$	[1.30254,1.69746]	0.0916025	0.0916025
6	[1.30254,1.69746]	$2 \cdot 10^6$	[1.30254,1.69746]	0.0916026	0.0916026

- μ_k should not be increased too slowly because it might require solving too many unconstrained optimization sub-problems resulting in poor convergence.

Typically, if Newton or quasi-Newton methods are used to solve the unconstrained optimization sub-problems, μ_0 should receive a moderate value and each update of μ_k can happen so that $\mu_{k+1} = \beta \mu_k$ for some $\beta \in [1.5, 10]$.

Consider now the general nonlinear optimization case with equality and inequality constraints:

$$\min_{\mathbf{x}} \ f(\mathbf{x})$$

$$\text{s.t. } \mathbf{h}(\mathbf{x}) = \mathbf{0}$$

$$\mathbf{g}(\mathbf{x}) \leq \mathbf{0}$$

with $\mathbf{h} : \mathbb{R}^n \to \mathbb{R}^l$ and $\mathbf{g} : \mathbb{R}^n \to \mathbb{R}^m$. For this, we can construct the quadratic penalty function:

$$Q(\mathbf{x}, \mu) = f(\mathbf{x}) + \frac{\mu}{2} \sum_{i=1}^{l} h_i^2(\mathbf{x}) + \frac{\mu}{2} \sum_{i=1}^{m} \max\{0, g_i(\mathbf{x})\}^2$$

Note that $\max\{0, g_i(\mathbf{x})\}$ adds a positive penalty if, and only if, $g_i(\mathbf{x}) > 0$ for a point $\mathbf{x}$. We can solve this problem using unconstrained optimization solution methods. Notice, however, that $Q(\mathbf{x}, \mu)$ is not twice continuously differentiable when considering inequality constraints.

4.9.2 Augmented Lagrangian Methods

Augmented Lagrangian is related to the Quadratic Penalty Method. It introduces Lagrange multiplier estimates into the function to be minimized, thus reducing the possibility of ill-conditioning. The resulting function is known as the *augmented Lagrangian function*. Its advantage compared to penalty methods is that it generally preserves smoothness. In essence, the difference of Augmented Lagrangian methods is that they add yet another term to the penalty function, designed to mimic a Lagrange multiplier.

The method, originally known as the method of multipliers, was presented in the 1970s as an alternative to penalty methods [40, 41]. When solving a constrained optimization problem with equality constraints, the augmented Lagrangian method uses the following unconstrained objective function at each iteration k:

$$A(\mathbf{x}, \boldsymbol{\lambda}, \mu_k) = \underbrace{f(\mathbf{x}) + \boldsymbol{\lambda}^\mathsf{T}\mathbf{c}(\mathbf{x})}_{\text{Lagrangian function}} + \underbrace{\frac{\mu_k}{2}\|\mathbf{c}(\mathbf{x})\|^2}_{\text{augmentation}} \tag{4.132}$$

$$= f(\mathbf{x}) + \sum_{i=1}^{l} \lambda_i c_i(\mathbf{x}) + \frac{\mu_k}{2}\sum_{i=1}^{l} c_i^2(\mathbf{x}) \tag{4.133}$$

where the additional term $\sum_{i=1}^{l} \lambda_i c_i(\mathbf{x})$ mimics the Lagrange multipliers and $\boldsymbol{\lambda}$ is updated at each iteration as:

$$\boldsymbol{\lambda}_{k+1} = \boldsymbol{\lambda}_k + \mu_k \mathbf{c}(\mathbf{x}_k)$$

The variable $\boldsymbol{\lambda}$ is an estimate of the Lagrange multiplier which has an improved accuracy at every step. Unlike the penalty method, it is not necessary to increase μ_k up to $+\infty$ in order to solve the original constrained problem. Instead, because of the presence of the Lagrange multiplier term, μ_k can stay much smaller, thus avoiding ill-conditioning. At each iteration of the augmented Lagrangian method, we consider μ_k and $\boldsymbol{\lambda}_k$ as fixed and we solve the following unconstrained minimization problem with respect to $\mathbf{x}$ to derive $\mathbf{x}_k$:

$$\min_{\mathbf{x}} f(\mathbf{x}) + \boldsymbol{\lambda}_k^\mathsf{T}\mathbf{c}(\mathbf{x}) + \frac{\mu_k}{2}\sum_{i=1}^{l} c_i^2(\mathbf{x}) \tag{4.134}$$

The general algorithmic framework of the augmented Lagrangian method in the case of equality constraints is provided below.

Consider, for example, the nonlinear Rosenbrock's function in two dimensions under a single equality constraint:

$$\min\ 100(x_2 - x_1^2)^2 + (1 - x_1)^2 \tag{4.135}$$
$$\text{s.t. } x_1 + x_2 = 3 \tag{4.136}$$

The augmented Lagrangian function at iteration k is:

Algorithm 22 Basic framework of Augmented Lagrangian method for equality constraints

---
1: choose $\mu_0 > 0$ and initial starting points $\mathbf{x}_0^s$, λ_0 that might be infeasible to the constrained optimization problem
2: set $k = 0$
3: **repeat**
4:   starting from $\mathbf{x}_k^s$, solve (4.134) with an unconstrained optimization solver to find $\mathbf{x}_k$
5:   set $\lambda_{k+1} = \lambda_k + \mu_k \mathbf{c}(\mathbf{x}_k)$
6:   choose $\mu_{k+1} > \mu_k$
7:   set new starting point $\mathbf{x}_{k+1}^s = \mathbf{x}_k$
8:   set $k = k + 1$
9: **until** convergence is achieved

Table 4.14 Augmented Lagrangian Method's iterations.

k	$\mathbf{x}_k$	μ_k	λ_k	$f(\mathbf{x}_k)$	$A(\mathbf{x}_k, \lambda_k, \mu_k)$
0	$[5, 5]^\mathsf{T}$	2	2	–	–
1	$[1, 1]^\mathsf{T}$	$2 \cdot 10$	0	-1	-1
2	$[1.30023, 1.69143]^\mathsf{T}$	$2 \cdot 10^2$	-0.166773	0.089512138	0.090902811
3	$[1.30254, 1.69745]^\mathsf{T}$	$2 \cdot 10^3$	-0.167842	0.091601715	0.091602611
4	$[1.30254, 1.69746]^\mathsf{T}$	$2 \cdot 10^4$	-0.167842	0.091602614	0.091602614

$$A(\mathbf{x}, \lambda_k, \mu_k) = 100(x_2 - x_1^2)^2 + (1 - x_1)^2 + \frac{\mu_k}{2}(x_1 + x_2 - 3)^2 + \lambda_k(x_1 + x_2 - 3)$$

We have only one constraint, and thus only one Lagrange multiplier. Using Newton line search with Hessian modification (Algorithm 11) with starting $\mu_k = 2$, $\lambda_k = 2$ and initial solution guess $\mathbf{x}_k^s = [x_1, x_2]^\mathsf{T} = [5, 5]^\mathsf{T}$, we have solution $\mathbf{x}_k = [1, 1]^\mathsf{T}$ resulting in $\lambda_{k+1} = \lambda_k + \mu_k(x_1 + x_2 - 3) = 2 + 2(1 + 1 - 3) = 0$. We proceed to the next iteration $k = k + 1$ by setting $\mu_k = 20$, $\lambda_k = 0$ and using $\mathbf{x}_k^s = [1, 1]^\mathsf{T}$ as initial solution guess to obtain new solution $\mathbf{x}_k = [1.30023, 1.69143]^\mathsf{T}$. Proceeding with two more iterations results in a sufficient convergence, as presented in Table 4.14.

Notice that the advantage of the augmented Lagrangian method is that it does not necessarily require $\mu_k \to +\infty$ to converge, thus avoiding ill-conditioning of the Hessian of the augmented Lagrangian function (something that cannot be guaranteed when using penalty methods). This is proved in the following theorem which shows that if we have knowledge of the optimal Lagrange multiplier $\boldsymbol{\lambda}^*$, the solution $\mathbf{x}^*$ of the original equality-constrained optimization problem is a strict minimizer of $A(\mathbf{x}, \boldsymbol{\lambda}^*, \mu_k)$ for a sufficiently large μ_k, but not necessarily $\mu_k \to +\infty$. This is very important because it indicates that we can find $\mathbf{x}^*$ even for small values of μ_k, thus avoiding ill-conditioning that occurs when $\mu_k \to +\infty$.

Theorem 4.11 *Let x^* and $\boldsymbol{\lambda}^*$ satisfy the second-order sufficient conditions for local optimality of the original equality-constrained optimization problem. If μ_k is larger than a threshold value (but not necessarily $\mu_k \to +\infty$), then x^* is a strict local minimizer of the augmented Lagrangian $A(x, \mu_k, \boldsymbol{\lambda}^*)$ corresponding to $\boldsymbol{\lambda}^*$.*

Proof Because $\mathbf{x}^*$ and $\boldsymbol{\lambda}^*$ satisfy the second-order sufficient conditions, we have that for the original equality constrained problem $\min_{\mathbf{x}} f(\mathbf{x})$ s.t. $\mathbf{c}(\mathbf{x}) = \mathbf{0}$:

- The LICQ is satisfied, meaning that $\nabla c_i(\mathbf{x}^*)$ are linearly independent vectors making $\mathbf{x}^*$ a regular point
- $\nabla \mathcal{L}(\mathbf{x}^*, \boldsymbol{\lambda}^*) = \mathbf{0}$
- $\mathbf{d}^{\mathsf{T}} \nabla^2_{\mathbf{xx}} \mathcal{L}(\mathbf{x}^*, \boldsymbol{\lambda}^*) \mathbf{d} > 0$ $\forall \mathbf{d} \neq \mathbf{0}$ such that $\nabla c_i(\mathbf{x}^*) \mathbf{d} = 0$

where, using the denominator layout notation,

$$\nabla \mathcal{L}(\mathbf{x}^*, \boldsymbol{\lambda}^*) = \begin{cases} \nabla_{\mathbf{x}} \mathcal{L}(\mathbf{x}^*, \boldsymbol{\lambda}^*) = \nabla f(\mathbf{x}^*) + \nabla \mathbf{c}(\mathbf{x}^*) \boldsymbol{\lambda}^* \\ \nabla_{\boldsymbol{\lambda}} \mathcal{L}(\mathbf{x}^*, \boldsymbol{\lambda}^*) = \mathbf{c}(\mathbf{x}^*) \end{cases}$$

Now, the augmented Lagrangian function $A(\mathbf{x}, \boldsymbol{\lambda}, \mu) = f(\mathbf{x}) + \boldsymbol{\lambda}^{\mathsf{T}} \mathbf{c}(\mathbf{x}) + \frac{\mu}{2} \sum_{i \in I}^{l} c_i^2(\mathbf{x})$ has strict local minimizer $\mathbf{x}^*, \boldsymbol{\lambda}^*$ if, and only if,

$$\nabla_{\mathbf{x}} A(\mathbf{x}^*, \boldsymbol{\lambda}^*, \mu) = \mathbf{0} \text{ and } \nabla^2_{\mathbf{xx}} A(\mathbf{x}^*, \boldsymbol{\lambda}^*, \mu) \text{ is positive definite}$$

in order to satisfy the second-order sufficient conditions of *strict* local optimality for unconstrained minimization.

The gradient of the augmented Lagrangian at $\mathbf{x}^*, \boldsymbol{\lambda}^*$ is:

$$\nabla_{\mathbf{x}} A(\mathbf{x}^*, \boldsymbol{\lambda}^*, \mu) = \nabla \mathcal{L}(\mathbf{x}^*, \boldsymbol{\lambda}^*) + \underbrace{\nabla \mathbf{c}(\mathbf{x}^*)}_{n \times m} \mu \underbrace{\mathbf{c}(\mathbf{x}^*)}_{m \times 1}$$

which is equal to $\mathbf{0}$ independently of the value of μ because $\mathcal{L}(\mathbf{x}^*, \boldsymbol{\lambda}^*) = 0$ and $\mathbf{c}(\mathbf{x}^*) = \mathbf{0}$. It now remains to prove that the Hessian of the augmented Lagrangian function is positive definite. The Hessian matrix is:

$$\nabla^2_{\mathbf{xx}} A(\mathbf{x}^*, \boldsymbol{\lambda}^*, \mu) = \nabla^2_{\mathbf{xx}} \mathcal{L}(\mathbf{x}^*, \boldsymbol{\lambda}^*) + \mu \underbrace{\nabla \mathbf{c}(\mathbf{x}^*)}_{n \times m} \underbrace{\nabla \mathbf{c}^{\mathsf{T}}(\mathbf{x}^*)}_{m \times n}$$

To prove that this matrix is positive definite, we use the following lemma: for two symmetric matrices $\nabla^2_{\mathbf{xx}} \mathcal{L}(\mathbf{x}^*, \boldsymbol{\lambda}^*)$ and $\underbrace{\nabla \mathbf{c}(\mathbf{x}^*) \nabla \mathbf{c}^{\mathsf{T}}(\mathbf{x}^*)}_{n \times n}$ such that $\mathbf{d}^{\mathsf{T}} \nabla^2_{\mathbf{xx}} \mathcal{L}(\mathbf{x}^*, \boldsymbol{\lambda}^*) \mathbf{d} > 0$ for all $\mathbf{d} \neq \mathbf{0}$ with $\mathbf{d}^{\mathsf{T}} \nabla \mathbf{c}(\mathbf{x}^*) \nabla \mathbf{c}^{\mathsf{T}}(\mathbf{x}^*) \mathbf{d} = 0$, there exists a positive scalar $\bar{\mu}$ such that $\nabla^2_{\mathbf{xx}} \mathcal{L}(\mathbf{x}^*, \boldsymbol{\lambda}^*) + \bar{\mu} \nabla \mathbf{c}(\mathbf{x}^*) \nabla \mathbf{c}^{\mathsf{T}}(\mathbf{x}^*)$ is positive definite.

That is, for a sufficiently large $\mu \geq \bar{\mu}$ solution $\mathbf{x}^*, \boldsymbol{\lambda}^*$ is a strict local minimizer of the augmented Lagrangian. Note that $\nabla \mathbf{c}(\mathbf{x}^*) \nabla \mathbf{c}^{\mathsf{T}}(\mathbf{x}^*)$ is symmetric as the product of a matrix with its transpose.

Let us now consider the general case of constrained optimization problems with equality and inequality constraints:

$$\min_{\mathbf{x}} \ f(\mathbf{x})$$
$$\text{s.t. } \mathbf{h}(\mathbf{x}) = \mathbf{0}$$
$$\mathbf{g}(\mathbf{x}) \leq \mathbf{0}$$

with $\mathbf{h} : \mathbb{R}^n \to \mathbb{R}^l$ and $\mathbf{g} : \mathbb{R}^n \to \mathbb{R}^m$. For this, we can construct the augmented Lagrangian:

$$A(\mathbf{x}, \boldsymbol{\lambda}, \boldsymbol{\nu}, \mu) = f(\mathbf{x}) + \boldsymbol{\lambda}^\mathsf{T} \mathbf{h}(\mathbf{x}) + \frac{\mu}{2} \sum_{i=1}^{l} h_i^2(\mathbf{x}) + \boldsymbol{\nu}^\mathsf{T} \mathbf{g}(\mathbf{x}) + \frac{\mu}{2} \sum_{i=1}^{m} \max\{0, g_i(\mathbf{x})\}^2$$

The term $\max\{0, g_i(\mathbf{x})\}$ adds a positive penalty if, and only if, $g_i(\mathbf{x}) > 0$ for a point $\mathbf{x}$. Notice that $A(\mathbf{x}, \mu)$ is not twice continuously differentiable when considering inequality constraints. A common approach to deal with this issue, used by augmented Lagrangian solvers like LANCELOT [42], is to turn the problem into a boundary-constrained problem with the introduction of slack variables $\mathbf{s} \in \mathbb{R}^m$ that result in the reformulation:

$$\min_{\mathbf{x},\mathbf{s}} \ f(\mathbf{x})$$
$$\text{s.t. } \mathbf{h}(\mathbf{x}) = \mathbf{0}$$
$$\mathbf{g}(\mathbf{x}) + \mathbf{s} = \mathbf{0}$$
$$\mathbf{s} \geq \mathbf{0}$$

Let us do some rearrangements. First, gather all $\mathbf{x} \in \mathbb{R}^n$, $\mathbf{s} \in \mathbb{R}^m$ variables into vector $\bar{\mathbf{x}} \in \mathbb{R}^r$ where $r = n + m$. Let also name all $l + m$ equality constraints as $\mathbf{c}(\bar{\mathbf{x}}) = \mathbf{0}$ and add boundary conditions for our new set of variables $\bar{\mathbf{x}}$ resulting in:

$$\min_{\bar{\mathbf{x}}} \ f(\bar{\mathbf{x}})$$
$$\text{s.t. } \mathbf{c}(\bar{\mathbf{x}}) = \mathbf{0}$$
$$l_i \leq \bar{x}_i \leq u_i \ \forall i \in \{1, \dots, r\}$$

where l_i and u_i are the lower and upper bounds associated to each variable $\bar{x}_i$. Notice that if $\bar{x}_i$ is referring to an $\mathbf{x}$ variable, then $l_i = -\infty$ and $u_i = +\infty$, whereas if $\bar{x}_i$ is referring to an $\mathbf{s}$ variable, then $l_i = 0$ and $u_i = +\infty$. This will yield the augmented Lagrangian:

$$A(\bar{\mathbf{x}}, \boldsymbol{\lambda}, \mu) = f(\bar{\mathbf{x}}) + \boldsymbol{\lambda}^\mathsf{T} \mathbf{c}(\bar{\mathbf{x}}) + \frac{\mu}{2} \sum_{i=1}^{l+m} c_i^2(\bar{\mathbf{x}})$$

where $\lambda \in \mathbb{R}^r$. Using this bound-constrained approach, at each iteration k we have to solve the bound-constrained minimization problem:

$$\min_{\mathbf{x}} A(\bar{\mathbf{x}}, \lambda_k, \mu_k) := f(\bar{\mathbf{x}}) + \lambda_k^{\mathsf{T}} \mathbf{c}(\bar{\mathbf{x}}) + \frac{\mu_k}{2} \sum_{i=1}^{l+m} c_i^2(\bar{\mathbf{x}}) \qquad (4.137)$$

$$\text{s.t. } l_i \leq \bar{x}_i \leq u_i \ \forall i \in \{1, \ldots, r\} \qquad (4.138)$$

in order to receive a new point $\bar{\mathbf{x}}_k$ and update $\lambda_{k+1} = \lambda_k + \mu_k \mathbf{c}(\bar{\mathbf{x}}_k)$ before proceeding to the next iteration. Notice that we no longer have an unconstrained optimization problem, but a bound-constrained one. This problem can be solved with the nonlinear gradient projection method described in the following steps.

Step 1: Start with iteration $s = 0$ and solution guess $\bar{\mathbf{x}}_s$
Step 2: At the current iterate s form the quadratic model, following the denominator layout notation:

$$q_s(\bar{\mathbf{x}}) = A(\bar{\mathbf{x}}_s, \lambda_k, \mu_k) + \underbrace{(\nabla_{\bar{\mathbf{x}}} A(\bar{\mathbf{x}}_s, \lambda_k, \mu_k))^{\mathsf{T}}}_{1 \times r}(\mathbf{x} - \mathbf{x}_s) + \frac{1}{2}(\bar{\mathbf{x}} - \bar{\mathbf{x}}_s)^{\mathsf{T}} \mathbf{B}_s (\bar{\mathbf{x}} - \bar{\mathbf{x}}_s)$$

where $\mathbf{B}_s$ is a positive definite approximation to the Hessian $\nabla_{\bar{\mathbf{x}}\bar{\mathbf{x}}}^2 A(\bar{\mathbf{x}}_s, \lambda_k, \mu_k)$.
Step 3: Use the gradient projection method or the active set method for quadratic programming to find an approximate solution $\hat{\mathbf{x}}$ of the subproblem:

$$\min_{\bar{\mathbf{x}}} q_s(\bar{\mathbf{x}})$$

$$\text{s.t. } l_i \leq \bar{x}_i \leq u_i \ \forall i \in \{1, \ldots, r\}$$

Step 4: Set step direction $\mathbf{p}_s = \hat{\mathbf{x}} - \bar{\mathbf{x}}_s$ and step length $a_s \in (0, 1]$ after applying backtracking line search to satisfy:

$$A(\bar{\mathbf{x}}_s + a_s \mathbf{p}_s, \lambda_k, \mu_k) \leq A(\bar{\mathbf{x}}_s, \lambda_k, \mu_k) + \eta a_s (\nabla_{\bar{\mathbf{x}}} A(\bar{\mathbf{x}}_s, \lambda_k, \mu_k))^{\mathsf{T}} \mathbf{p}_s$$

for some $\eta \in (0, 1)$.
Step 5: update $\bar{\mathbf{x}}_{s+1} = \bar{\mathbf{x}}_s + a_s \mathbf{p}_s$ and proceed to the next iteration until converging to solution $\bar{\mathbf{x}}_k$ of Eqs. (4.137)–(4.138).

After finding $\bar{\mathbf{x}}_k$, we update the values of λ_k and μ_k and we proceed to the next iteration of the Augmented Lagrangian method presented in Algorithm 22.

Exercises

4.1 Duality
Write the dual of:

$$\max_{\mathbf{x}\in\mathbb{R}^n} f(\mathbf{x})$$
$$\text{s.t. } \mathbf{h}(\mathbf{x}) = \mathbf{0}$$
$$\mathbf{g}(\mathbf{x}) \geq \mathbf{0}$$

4.2 Local Optimality
Using the Projected Hessian, prove that the second-order necessary conditions for local optimality of the following problem are satisfied for some $\mathbf{x}^*, \boldsymbol{\lambda}^*$:

$$\min_{\mathbf{x}\in\mathbb{R}^3} 2x_1^2 + 6x_2 + x_3$$
$$\text{s.t. } x_1 = 5x_2$$
$$x_1 = 9x_3$$

4.3 Strict Local Optimality
Prove that the following minimization problem has a strict local minimizer:

$$\min_{\mathbf{x}\in\mathbb{R}^2} x_1^4 - 4x_1 + 2x_2^2 - 4x_2$$
$$\text{s.t. } -x_1 \leq 0$$
$$x_2 - 3 \leq 0$$
$$x_1 + x_2 - 2 \leq 0$$

4.4 Basic Feasible Solutions

(a) Describe the difference between a basic solution and a basic feasible solution in a linear program.

(b) Find all basic solutions and basic feasible solutions of:

$$\max_{\mathbf{x}\in\mathbb{R}^4} 35x_1 + 3x_2$$
$$\text{s.t. } 3x_1 + x_2 + 2x_3 = 40$$
$$2x_1 + x_2 + x_4 = 60$$
$$x_1, x_2, x_3, x_4 \geq 0$$

Justify which is the optimal solution to this problem.

4.5 Two-phase Simplex
Write a two-phase Simplex algorithm and solve the linear program in (4.58) in a programming language of your choice.

4.6 Karmarkar's method

Solve the following linear program with Karmarkar's method after applying Tomlin's conversion to derive its homogeneous form.

$$\max \ -6x_1 - 3x_2$$
$$\text{s.t. } x_1 + x_2 - x_3 = 1$$
$$2x_1 - x_2 - x_4 = 1$$
$$3x_2 + x_5 = 2$$
$$x_1, x_2, x_3, x_4, x_5 \geq 0$$

4.7 Equality QP

Apply the null-space method to solve the following quadratic program after using a Phase I LP to find an initial feasible solution:

$$\min_{\mathbf{x} \in \mathbb{R}^3} \ x_1^2 - 4x_1 + 2x_2^2 + x_3^2$$
$$\text{s.t. } x_1 + 2x_2 = 5$$
$$2x_1 + 2x_2 + x_3 = 5$$

4.8 Inequality QP

Solve the following inequality QP by using Phase I LP to obtain an initial feasible solution and the active set method.

$$\min_{\mathbf{x} \in \mathbb{R}^2} \ x_1^2 - x_1 x_2 + x_2^2 - 7x_1 - 3x_2$$
$$\text{s.t. } -x_1 + 2x_2 \leq 4$$
$$x_1 + 3x_2 \leq 7$$
$$x_1 - 2x_2 \leq 5$$
$$-x_1 \leq 0$$

4.9 SQP

Solve the previous inequality QP with sequential quadratic programming.

4.10 Interior Point Method

Solve the following nonlinear program with the interior point method using initial solution guess $\mathbf{x}_0 = [0.5, 0]^\mathsf{T}$:

$$\min_{\mathbf{x} \in \mathbb{R}^2} \ x_1^3 + 2x_1 x_2$$
$$\text{s.t. } 1 - x_1 - 2x_2 \geq 0$$
$$1 - x_1^2 - x_2 \geq 0$$

Examine whether the updated solutions $\mathbf{x}_k$ at each iterate k of the interior point method remain within the feasible region.

4.11 Penalty and Augmented Lagrangian Methods

Solve the following mathematical program

$$\min_{\mathbf{x} \in \mathbb{R}^2} 100(x_2 - x_1^2)^2 + (1 - x_1)^2 \qquad \text{subject to: } x_1 + x_2 = 16$$

with:

(a) The Quadratic Penalty method with initial solution guess $\mathbf{x}_0 = [12, 2]^{\mathsf{T}}$. Start with $\mu_k = 0.2$ and update it as $\mu_{k+1} \leftarrow 10\mu_k$ at each iteration. Solve the resulting unconstrained nonlinear problem for each μ_k value with Newton's line search with Hessian modification with precision error $\tau = 0.0001$, a value of $\beta = 0.075$ in Cholesky's decomposition and backtracking line search with Armijo's condition with $a = 1$, $\rho = 0.99$, and $c = 10^{-4}$.

(b) The Augmented Lagrangian method with initial solution guess $\mathbf{x}_0 = [12, 2]^{\mathsf{T}}$ and $\lambda_0 = 2$. Use the same settings as in (a) to solve the unconstrained nonlinear problem at each iteration with Newton's line search.

References

1. T.F. Coleman, A. Pothen, SIAM J. Algebraic Discrete Methods **8**(4), 544 (1987)
2. J. Nocedal, M.L. Overton, SIAM J. Numer. Anal. **22**(5), 821 (1985)
3. G.W. Stewart, SIAM Rev. **35**(4), 551 (1993)
4. W. Karush, M.Sc. Dissertation. Department of Mathematics. (University of Chicago, 1939)
5. H. Kuhn, A.W. Tucker, in *Proceedings of 2nd Berkeley Symposium on Mathematical Statistics and Probability* (1951), pp. 481–492
6. R. Andreani, R. Behling, G. Haeser, P.J. Silva, Optim. Methods Softw. **32**(1), 22 (2017)
7. M. Wawro, Int. J. Res. Undergrad. Math. Educ. **1**(3), 315 (2015)
8. R. Johnston, G. Barton, M. Brisk, Int. J. Control **40**(2), 257 (1984)
9. G.B. Dantzig, Oper. Res. **50**(1), 42 (2002)
10. L.G. Khachiyan, in *Doklady Akademii Nauk*, vol. 244 (Russian Academy of Sciences, 1979), pp. 1093–1096
11. N. Karmarkar, in *Proceedings of the Sixteenth Annual ACM Symposium on Theory of Computing* (ACM, 1984), pp. 302–311
12. H. Bauer, Archiv der Mathematik **11**(1), 200 (1960)
13. T.C. Kotiah, N. Slater, Oper. Res. **21**(2), 597 (1973)
14. I. Elhallaoui, A. Metrane, G. Desaulniers, F. Soumis, INFORMS J. Comput. **23**(4), 569 (2011)
15. R.G. Bland, Math. Oper. Res. **2**(2), 103 (1977)
16. J. Hooker, Interfaces **16**(4), 75 (1986)
17. M.J. Todd, *Anticipated Behavior of Karmarkar's Algorithm* (Cornell University Operations Research and Industrial Engineering, Tech. rep., 1989)
18. M. Powell, *The Complexity of Karmarkar's Algorithm for Linear Programming* (University of Cambridge, Department of Applied Mathematics and Theoretical, 1991)
19. M.J. Todd, B.P. Burrell, Algorithmica **1**(1), 409 (1986)
20. J. Tomlin, in *Computation Mathematical Programming* (Springer, 1987), pp. 175–191
21. E.D. Andersen, K.D. Andersen, in *High Performance Optimization* (Springer, 2000), pp. 197–232
22. S. Sahni, SIAM J. Comput. **3**(4), 262 (1974)

23. P.M. Pardalos, S.A. Vavasis, J. Global Optim. **1**(1), 15 (1991)
24. R.B. Wilson, Ph.D. Dissertation, Graduate School of Bussiness Administration (1963)
25. D. Kraft, et al., (1988)
26. M.J. Powell, in *Numerical Analysis* (Springer, 1978), pp. 144–157
27. K. Schittkowski, Optim. Lett. **5**(2), 283 (2011)
28. K. Schittkowski, Numer. Math. **38**(1), 115 (1981)
29. P. Boggs, J. Tolle,
30. M.J. Powell, Math. Program. **14**(1), 224 (1978)
31. P.T. Boggs, J.W. Tolle, Acta Numerica **4**, 1 (1995)
32. S.P. Han, J. Optim. Theory Appl. **22**(3), 297 (1977)
33. K. Schittkowski, Ann. Oper. Res. **5**(2), 485 (1986)
34. R.J. Vanderbei, D.F. Shanno, Comput. Optim. Appl. **13**(1), 231 (1999)
35. J. Nocedal, A. Wächter, R.A. Waltz, SIAM J. Optim. **19**(4), 1674 (2009)
36. M. Schmidt, EURO J. Comput. Optim. **3**(4), 309 (2015)
37. R.H. Byrd, M.E. Hribar, J. Nocedal, SIAM J. Optim. **9**(4), 877 (1999)
38. A.V. Fiacco, G.P. McCormick, *Nonlinear Programming: Sequential Unconstrained Minimization Techniques* (SIAM, 1990)
39. A. Wächter, L.T. Biegler, Math. Program. **106**(1), 25 (2006)
40. M.R. Hestenes, J. Optim. Theory Appl. **4**(5), 303 (1969)
41. M.J. Powell, Optimization. pp. 283–298 (1969)
42. A.R. Conn, G. Gould, P.L. Toint, *LANCELOT: A Fortran Package for Large-Scale Nonlinear Optimization (Release A)*, vol. 17 (Springer Science & Business Media, 2013)
43. S. Boyd, S.P. Boyd, L. Vandenberghe, *Convex Optimization* (Cambridge University Press, 2004)

Chapter 5
Discrete Optimization

Abstract When solving optimization problems we typically have to make some discrete decisions because of natural constraints that may restrict decision variables to take discrete values. This chapter expands the theory of continuous optimization to discrete problems where (at least some) of the decision variables are restricted to take discrete values. Specific emphasis is placed on exact methods (simple enumeration, branch and bound, branch and cut) and approaches that can guarantee the computation of solutions with a provable convergence tolerance.

5.1 Introduction

Discrete optimization refers to a class of problems which have at least one variable that is restricted to take values from a set with finite elements. We can generally express a discrete optimization problem as follows.

$$\min_{\mathbf{x}} \ f(\mathbf{x}) \tag{5.1}$$

$$\text{s.t. } \mathbf{g}(\mathbf{x}) \leq \mathbf{0} \tag{5.2}$$

$$\mathbf{x} \in \mathbb{R}^{n-r} \times \mathcal{X}^r \tag{5.3}$$

where n is the total number of variables, $f : \mathbb{R}^{n-r} \times \mathcal{X}^r \to \mathbb{R}$, $\mathcal{X}$ is a set with finite elements, $r \geq 1$ is the number of variables that are forced to take values from a discrete set $\mathcal{X}$, and $\mathbf{g}(\mathbf{x}) \leq \mathbf{0}$ are all other constraints of the problem expressed in the form of inequalities. If set $\mathcal{X}$ is the set of integer numbers, then $\mathcal{X}^r$ can be replaced by $\mathbb{Z}^r$ and the problem is a mixed-integer problem. If $\mathcal{X}$ is the binary set, then $\mathcal{X}^r$ can be replaced by $\{0, 1\}^r$. We might even have different finite sets $\mathcal{X}_1, \ldots, \mathcal{X}_r$ corresponding to each discrete variable. For brevity reasons, it is also very common to declare all sets with finite elements as $\mathcal{X}$ without making any distinction among

Fig. 5.1 Example of the discrete optimization problem $\min 5x^2$ with $x \in \{-400, -350, \ldots, +350, +400\}$

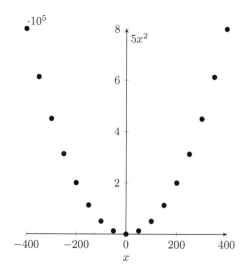

the different discrete variables. I.e., even if we have r discrete variables, we state that they all belong to a set $\mathcal{X}$ with finite elements.

Discrete optimization problems result in discontinuities and they are not convex (see example in Fig. 5.1). This limits the ability to find a globally optimal solution within a reasonable time. In more detail, Karp proved that even the binary (0–1) integer programming problem is NP-complete. In this chapter, we will present methods for solving discrete optimization problems to optimality. Note, though, that these methods are not polynomial-time methods and do not guarantee computational tractability for large problem instances. In addition to the naive approach of simple enumeration, we will also focus on *branch and bound* for general discrete optimization problems and *branch and cut/cut and branch* for discrete linear problems. This chapter will also present discrete optimization formulations for common combinatorial and set ordering problems.

5.2   Branch and Bound

One of the most popular approaches for solving discrete optimization problems by iteratively solving continuously relaxed subproblems is the method of Branch and Bound (B&B). Although we present B&B in the context of discrete optimization, we must state that it has much broader applications because it is a general-purpose search strategy for global optimization. Consider a non-convex minimization problem where all equality constraints are turned into inequalities:

$$\min \ f(\mathbf{x}) \tag{5.4}$$

$$\text{s.t. } \mathbf{g}(\mathbf{x}) \le \mathbf{0} \tag{5.5}$$

$$\mathbf{x} \in \mathcal{X} \tag{5.6}$$

where $f : \mathbb{R}^n \to \mathbb{R}$ is possibly non-convex and the feasible region is also possibly non-convex. Let us name $\mathcal{F}$ the feasible region of this minimization problem and suppose that we can find a lower bound L and an upper bound U of the problem in (5.4)–(5.6), so that $L \le \min_{\mathbf{x} \in \mathcal{F}} f(\mathbf{x}) \le U$. Because of non-convexity, it is very hard to compute the global minimizer of $\min_{\mathbf{x} \in \mathcal{F}} f(\mathbf{x})$. If, however, there is an easy way to compute the lower and upper bounds L, U and we show that $L - U \le \epsilon$ where $\epsilon \ge 0$ is a small solution tolerance, then we have found a solution which is very close to the global minimizer of the non-convex problem. This is easier understood in the extreme case where $\epsilon = 0$ and $L = \min_{\mathbf{x} \in \mathcal{F}} f(\mathbf{x}) = U$.

Although it is very hard to find the global minimizer of a non-convex problem, we can find a local minimizer by starting from an initial solution guess. By definition, this local minimizer will be an *upper bound* to the global minimizer $\mathbf{x}^* = \operatorname*{argmin}_{\mathbf{x} \in \mathcal{F}} f(\mathbf{x})$. We call this local minimizer $\mathbf{x}^l$ and we have that $U = f(\mathbf{x}^l)$. Note that $f(\mathbf{x}^l) \ge f(\mathbf{x}^*)$.

To find a lower bound L, we can *relax* the non-convex problem into an approximate convex problem:

$$\min \ \hat{f}(\mathbf{x}) \tag{5.7}$$

$$\text{s.t. } \hat{\mathbf{g}}(\mathbf{x}) \le \mathbf{0} \tag{5.8}$$

$$\mathbf{x} \in \mathbf{conv} \ \mathcal{X} \tag{5.9}$$

where the relaxed problem is convex, $\hat{f}(\mathbf{x}) \le f(\mathbf{x})$ and $\hat{\mathbf{g}}(\mathbf{x}) \le \mathbf{g}(\mathbf{x})$ for all $\mathbf{x} \in \mathbb{R}^n \mid \mathcal{X} \subseteq \mathbf{conv} \ \mathcal{X}$. By doing so, we find a globally optimal solution of the relaxed convex problem which is a *lower bound* to the global minimum of (5.4)–(5.6) since this solution is (most probably) infeasible for the original non-convex problem. That is, if $\mathbf{x}^p$ is the globally optimal solution of the relaxed problem then $L = f(\mathbf{x}^p) \le f(\mathbf{x}^*)$.

When computing U and L it can be the case that $U - L > \epsilon$ and we have to improve the bounds. B&B is a strategy to do so by performing the following steps:

1. Partition the feasible set $\mathcal{F}$ into subsets $\mathcal{F} = \mathcal{F}_1 \cup \mathcal{F}_2 \cup \cdots \cup \mathcal{F}_N$
2. Compute lower and upper bounds $L^{\mathcal{F}_i}$ and $U^{\mathcal{F}_i}$ for each subset
3. Compute the global lower and upper bounds $\min_{i \in \{1,...,N\}} L^{\mathcal{F}_i} \le f(\mathbf{x}^*) \le \min_{i \in \{1,...,N\}} U^{\mathcal{F}_i}$
4. If $\min_{i \in \{1,...,N\}} U^{\mathcal{F}_i} - \min_{i \in \{1,...,N\}} L^{\mathcal{F}_i} \le \epsilon$ terminate
5. Else, refine the partition step and repeat

Notice that because of the convergence criterion, B&B provides a certificate of ϵ-suboptimality. This is very important because it offers a guarantee of reaching a globally optimal solution with some accuracy ϵ in the case of non-convex optimization. As N increases, the difference $\min_{i \in \{1,...,N\}} U^{\mathcal{F}_i} - \min_{i \in \{1,...,N\}} L^{\mathcal{F}_i}$ becomes smaller and smaller. However, increasing the number of partitions, increases also the number of subproblems that need to be solved to obtain their lower and upper bounds. To limit this, given a partition $\mathcal{F} = \mathcal{F}_1 \cup \mathcal{F}_2 \cup \cdots \cup \mathcal{F}_N$, branch-and-bound typically picks region $\mathcal{F}_j$ such that $\mathrm{L}^{\mathcal{F}_j} = \min_{i \in \{1,\dots,N\}} \mathrm{L}^{\mathcal{F}_i}$ and splits it into two halves in order to concentrate on the most promising area. Once selected, subset $\mathcal{F}_j$ cannot be visited again in future because it is replaced by its two halves. This process is called *branching*. At each iteration we can also eliminate all subsets $\mathcal{F}_j$ for which $\mathrm{L}^{F_j} > \min_{i \in \{1,\dots,N\}} \mathrm{U}^{\mathcal{F}_i}$ because the global minimum cannot be found in them. This process is called *pruning* and helps to explore intelligently the solution space.

The B&B guarantee of ϵ-suboptimality when solving nonconvex problems is also applicable when dealing with discrete optimization problems, because they are nonconvex due to the discontinuities in their feasible region. In discrete optimization, some or all of the variables should be drawn from sets with *finitely* many elements. This is achieved by adding the so-called *integrality constraints* to the optimization problem. Integrality constraints have the form $x_i \in \mathcal{P}$, where x_i is a variable and $\mathcal{P}$ a set with finitely many elements. Notably, if $\mathcal{P}$ is a set of integers, the discrete optimization problem is referred to as an *integer program*. Reckon the two sub-categories of integer programs:

- Pure integer optimization problems: problems where the variables can take values from integer or binary sets (the latter are also called binary optimization problems).
- Mixed-integer optimization problems: problems where we have a nonempty subset of integer variables and a subset of real-valued (continuous) variables [1].

Another important category is the combinatorial optimization problems, which are discrete optimization problems with a combinatorial origin, meaning that the problem is one of graph theory, or arranging objects in a particular way. A combinatorial optimization problem can be seen as an integer optimization or a mixed-integer optimization problem, depending on its nature.

To explain in more detail the convergence of B&B, we present again the definition of tightness for discrete optimization problems.

Definition 5.1 *Tightness* of a discrete optimization problem refers to the distance between the discrete solution and the associated solution of the continuous relaxation. This distance indicates the search space that the optimization solver requires to explore in order to compute the optimal discrete solution.

Let us consider the general case of a discrete optimization problem written as a mixed-integer program:

$$\min_{\mathbf{x}, \nu} \; f(\mathbf{x}, \nu) \tag{5.10}$$

$$\text{s.t.} \; \mathbf{h}(\mathbf{x}, \nu) = \mathbf{0} \tag{5.11}$$

$$\mathbf{g}(\mathbf{x}, \nu) \leq \mathbf{0} \tag{5.12}$$

$$\mathbf{x} \in \mathbb{R}^n, \nu \in \mathbb{Z}^r \tag{5.13}$$

Note that ν is a column vector of all integer variables and $\mathbf{x}$ a column vector of all continuous variables. This problem is much more complex compared to its continuous counterpart because it is nonconvex due to the discontinuities caused by the integrality constraints of ν. That is, even if the continuous version of the problem (in which ν are continuous variables) is convex, the mixed-integer version is nonconvex.

Because the set of feasible solutions of the discrete optimization problem is not convex, examining the immediate neighborhood of a proposed solution can prove the existence only of a local optimum. To solve this problem to global optimality, one can use a naive method where we split the problem into many subproblems where each one assumes different values of ν. The k-th such subproblem would be:

$$\min_{\mathbf{x}} \; f(\mathbf{x}, \nu_k) \tag{5.14}$$

$$\text{s.t. } \mathbf{h}(\mathbf{x}, \nu_k) = \mathbf{0} \tag{5.15}$$

$$\mathbf{g}(\mathbf{x}, \nu_k) \leq \mathbf{0} \tag{5.16}$$

$$\mathbf{x} \in \mathbb{R}^n \tag{5.17}$$

where ν_k is fixed. Notice that if these sub-problems are convex, we can compute the globally optimal solution for each one of them. If we create as many sub-problems as the possible values of ν, a process called *simple enumeration*, *brute force* or *exhaustive search*, then the best-performing solution from all these subproblems is the globally optimal solution of the original discrete optimization problem. This is, however, extremely costly because of the sheer volume of potential sub-problems. Even in the most favorable case where every discrete variable $\nu_i \in \nu$ is binary and can take only two values (i.e., 0 or 1), this will require solving 2^r sub-problems. This demonstrates that the number of generated sub-problems grows exponentially with the number of discrete variables and we can only solve small-sized problems.

B&B is a more clever search technique that searches the solution space more efficiently, oftentimes requiring to solve much fewer sub-problems than the brute force method. We should note, though, that there is no theoretical guarantee for that and the running time of B&B is still exponential at the worst-case scenario. Although B&B is not a polynomial-time algorithm, it can perform well in practical applications bringing significant benefits compared to naive methods.

Starting from a discrete optimization problem, B&B should compute an upper bound U and a lower bound L. An upper bound can be initially set to $+\infty$ or another value $U \leq +\infty$ that can be attained by solving the original problem with a (meta)heuristic which can provide a feasible solution within a short time. To compute a lower bound, we can relax the discrete optimization problem by dropping its integrality constraints and solving its continuous relaxation (see the following definition).

Definition 5.2 A *continuous relaxation* of a discrete optimization problem is a new problem obtained by dropping all integrality constraints.

For instance, the continuous relaxation of our example problem in (5.10)–(5.13) is:

$$\min_{\mathbf{x},\nu}\ f(\mathbf{x},\nu) \tag{5.18}$$

$$\text{s.t.}\ \ \mathbf{h}(\mathbf{x},\nu) = \mathbf{0} \tag{5.19}$$

$$\mathbf{g}(\mathbf{x},\nu) \le \mathbf{0} \tag{5.20}$$

$$\mathbf{x} \in \mathbb{R}^n, \nu \in \mathbb{R}^r \tag{5.21}$$

Consider that this relaxation is convex (if not, it can be turned into convex by using approximations). Then, solving this problem to global optimality will provide a lower bound L to our discrete optimization problem. Having computed L and U, we can check whether $U - L \le \epsilon$. If not, we proceed with the partition of the feasible set (*branching*).

5.2.1 Mixed-integer Problems with Binary Variables

To better understand the B&B process, consider the case of a mixed-integer problem with binary variables:

$$\min_{\mathbf{x},\nu}\ f(\mathbf{x}) \tag{5.22}$$

$$\text{s.t.}\ \ \mathbf{h}(\mathbf{x},\nu) = \mathbf{0} \tag{5.23}$$

$$\mathbf{g}(\mathbf{x},\nu) \le \mathbf{0} \tag{5.24}$$

$$\mathbf{x} \in \mathbb{R}^n, \nu \in \{0, 1\}^r \tag{5.25}$$

where the scalar-valued f and the vector-valued $\mathbf{h}$, $\mathbf{g}$ functions are convex. Although we do not know its value yet, let us denote the globally optimal solution of this discrete optimization problem as p^*. As previously discussed, solving this discrete optimization problem with brute force requires solving 2^r convex optimization sub-problems. It is important to understand that using B&B to solve this problem might still require solving 2^r convex optimization sub-problems at the worst-case scenario, that is, degenerate to an exhaustive search. In practice though, B&B is considerably more efficient. To apply B&B, we introduce the following continuous relaxation of the discrete problem. Note that we assume this continuous relaxation to be a convex program:

$$\min_{\mathbf{x}, \nu} \; f(\mathbf{x}) \tag{5.26}$$

$$\text{s.t.} \; \mathbf{h}(\mathbf{x}, \nu) = \mathbf{0} \tag{5.27}$$

$$\mathbf{g}(\mathbf{x}, \nu) \le \mathbf{0} \tag{5.28}$$

$$\mathbf{x} \in \mathbb{R}^n \tag{5.29}$$

$$0 \le \nu_i \le 1 \quad \forall i \in \{1, \dots, r\} \tag{5.30}$$

As a continuous convex problem, this continuous relaxation is easily solved to obtain L^0—which is a lower bound of p^*. If $\mathrm{L}^0 = +\infty$, then the original discrete problem is infeasible. To derive an upper bound, we also need to compute a feasible solution for the discrete problem in (5.22)–(5.25). We can also start with an upper bound $\mathrm{U}^0 = +\infty$ if obtaining a feasible solution for the discrete problem is not easy. Observe that if $\mathrm{U}^0 - \mathrm{L}^0 \le \epsilon$ we can terminate the algorithm because we have a guaranteed ϵ-suboptimality. If not, we are going to partition (branch). We can pick any binary variable $\nu_i \in \nu$ and form two subproblems: one for $\nu_i = 0$ and one for $\nu_i = 1$. Namely,

Subproblem 1:

$$\min_{\mathbf{x}, \nu} \; f(\mathbf{x})$$

$$\text{s.t.} \; \mathbf{h}(\mathbf{x}, \nu) = \mathbf{0}$$

$$\mathbf{g}(\mathbf{x}, \nu) \le \mathbf{0}$$

$$\mathbf{x} \in \mathbb{R}^n, \nu \in \{0, 1\}^r$$

$$\nu_i = 0$$

Subproblem 2:

$$\min_{\mathbf{x}, \nu} \; f(\mathbf{x})$$

$$\text{s.t.} \; \mathbf{h}(\mathbf{x}, \nu) = \mathbf{0}$$

$$\mathbf{g}(\mathbf{x}, \nu) \le \mathbf{0}$$

$$\mathbf{x} \in \mathbb{R}^n, \nu \in \{0, 1\}^r$$

$$\nu_i = 1$$

These are sub-problems of the original and solving them to global optimality will give us the global optimum of the original, which is either the globally optimal solution of subproblem 1 or 2. It is very important to note that the generated subproblems should be constructed in such a way that the globally optimal solution of the original problem can be computed by one of them. To provide a counterexample, if subproblem 1 had two variables ν_i and ν_j fixed to 0 and subproblem 2 had two variables $\nu_i = \nu_j = 1$ where $i \ne j$, these subproblems are not valid because the globally optimal solution of the original might have been part of a subproblem where $\nu_i = 0$ and $\nu_j = 1$.

Let us return to our two formulated subproblems. Notice that these subproblems have $r - 1$ binary variables. We can solve the continuous (and convex) relaxations of these two sub-problems to obtain lower bounds L^1 and L^2, respectively. By definition of these sub-problems, $\min\{\mathrm{L}^1, \mathrm{L}^2\} \ge \mathrm{L}^0$ meaning that we have found a tighter lower bound of p^*. If, by luck, the solution of any of the two subproblems provides discrete values to ν, then this solution is feasible for the original problem and provides a tighter upper bound. If not, the upper bounds of the two subproblems are $\mathrm{U}^1, \mathrm{U}^2 = +\infty$. We then have $\min\{\mathrm{L}^1, \mathrm{L}^2\} \le p^* \le \min\{\mathrm{U}^0, \mathrm{U}^1, \mathrm{U}^2\}$. Note that we do not consider L^0 anymore because we have branched out from it.

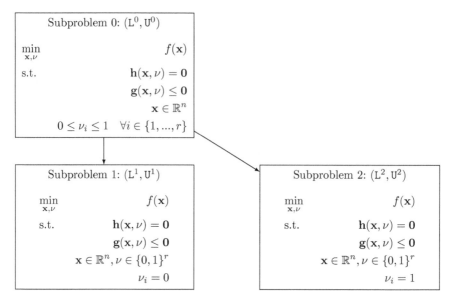

Fig. 5.2   Boolean rooted tree created by the B&B method after 1 iteration

In this process we have created a boolean rooted tree, with the root being the original problem and the leaves (nodes) being the subproblems (see Fig. 5.2).

As previously noted, we have already branched out of the root, so we will not branch from the root again. To continue branching, one can pick the most promising node (leaf), i.e., the sub-problem with the smallest lower bound value. Let us say that this is sub-problem 1 with $\nu_i = 0$. If so, we branch out by generating subproblems 3 and 4 such that another variable $\nu_j \in \boldsymbol{\nu}$ is equal to 0 and 1, respectively:

Subproblem 3:

$$\min_{\mathbf{x}, \boldsymbol{\nu}} f(\mathbf{x})$$

$$\text{s.t.}\ \mathbf{h}(\mathbf{x}, \boldsymbol{\nu}) = \mathbf{0}$$

$$\mathbf{g}(\mathbf{x}, \boldsymbol{\nu}) \le \mathbf{0}$$

$$\mathbf{x} \in \mathbb{R}^n, \boldsymbol{\nu} \in \{0, 1\}^r$$

$$\nu_i = 0$$

$$\nu_j = 0$$

Subproblem 4:

$$\min_{\mathbf{x}, \boldsymbol{\nu}} f(\mathbf{x})$$

$$\text{s.t.}\ \mathbf{h}(\mathbf{x}, \boldsymbol{\nu}) = \mathbf{0}$$

$$\mathbf{g}(\mathbf{x}, \boldsymbol{\nu}) \le \mathbf{0}$$

$$\mathbf{x} \in \mathbb{R}^n, \boldsymbol{\nu} \in \{0, 1\}^r$$

$$\nu_i = 0$$

$$\nu_j = 1$$

Note that one of the sub-problems 2, 3, and 4 contains the globally optimal solution to the original problem. Proceeding like this, we compute the lower bounds of sub-problems 3 and 4 (L^3, L^4) by solving their continuously relaxed versions. Their upper bounds can be set to $+\infty$ or to the value of the lower bound if, by luck, the solution of the continuous relaxed sub-problem is a feasible solution for the original problem.

The B&B search continues in a similar fashion in the next iterations, performing one branching per iteration. If, at some point, we find a solution to some continuously relaxed sub-problem that is also a feasible solution to the original problem we have found an upper bound with a value less than $+\infty$. If this is the case, we can remove (prune) all leaves (nodes) that have a lower bound value greater than this upper bound because the sub-problems represented by these nodes do not contain the globally optimal solution of the original problem. The reason for this is that the upper bound value refers to the value of a feasible solution for the original discrete problem and if a sub-problem has a higher lower bound value than that, then it cannot contain the globally optimal solution of the original problem.

Combining these two techniques (branching and pruning), the algorithm terminates when $U - L \leq \epsilon$, where U is the smallest upper bound value of all leaf nodes and L the smallest lower bound value of all leaf nodes from which we have not branched out yet (known as *solution candidate* nodes). Notice that if we have k iterations, we would have $2k$ generated leaf nodes (sub-problems). In addition, if L_k and U_k is the best lower and upper bound at each iteration k, $U_k - L_k \leq U_{k-1} - L_{k-1}$. This observation ensures convergence.

We can now formalize further the B&B steps in case of mixed-integer problems, where all integer variables are binary in Algorithm 23.

Algorithm 23 B&B framework for mixed-integer minimization problems with binary and continuous variables

---

1: solve the convex continuous relaxation of the original problem to compute L^0. If this solution is feasible for the original problem, then $U^0 = L^0$. Otherwise, set $U^0 = +\infty$ or compute another feasible solution with a heuristic to attain U^0
2: Set $k = 0$ and $L_k = L^0$, $U_k = U^0$, and select $\epsilon > 0$
3: initialize *candidate list* $\mathcal{P} = \{0\}$ which contains all leaf nodes from which we have not performed branching
4: **repeat**
5:    set $k = k + 1$
6:    from the leaf node in $\mathcal{P}$ with the smallest lower bound value, generate two sub-problems by fixing the values of one of the binary variables to 0 and 1, respectively (branching)
7:    for the two generated sub-problems, compute their lower and upper bounds
8:    add these two sub-problems to $\mathcal{P}$ and remove the parent leaf node of these two sub-problems from $\mathcal{P}$
9:    set L_k as the smallest lower bound from the lower bounds of all sub-problems in $\mathcal{P}$. That is, $L_k = \min_{i \in \mathcal{P}} L^i$
10:   set U_k as the smallest upper bound of all sub-problems
11:   remove all leaf nodes (sub-problems) from $\mathcal{P}$ for which their lower bound values are greater than U_k (pruning)
12: **until** $U_k - L_k \leq \epsilon$

It is important to note that finding an initial feasible solution U^0 which is close to the optimal one is crucial to the computational time of the B&B method. Finding a low value for the upper bound U^0 will result in pruning early by eliminating large parts of the solution space associated with sub-problems that have a lower bound with higher value than U^0. This can be achieved by using the so-called *primal heuristics*, which can be heuristics that round the values of integer variables after solving a continuous problem or metaheuristics (i.e., Simulated Annealing, Tabu search, Genetic Algorithms).

Let us consider, for example, the minimization problem:

$$
\begin{aligned}
\min \ & 4x_1^2 - x_1x_2 + x_2^2 + 8x_3^2 + 8x_4^2 \\
\text{s.t. } & x_1 + x_2 + x_3 + x_4 = 6 \\
& x_1, x_2 \in \mathbb{R} \\
& x_3, x_4 \in \{0, 1\}
\end{aligned}
\tag{5.31}
$$

We can easily obtain a feasible solution for this problem $\mathbf{x} = [1, 1, 2, 2]^\mathsf{T}$ with objective function value $U^0 = 68$. Its continuous relaxation with $0 \le x_3 \le 1$ and $0 \le x_4 \le 1$ has solution $\mathbf{x} = [1.2973, 3.89189, 0.405405, 0.405405]^\mathsf{T}$ with objective function value $L^0 = 19.4595$. We set $U_0 = U^0$ and $L_0 = L^0$ and we proceed to iteration $k = 1$. From set $\mathcal{P} = \{0\}$ we branch out from node 0 by creating two subproblems, where $x_3 = 0$ in subproblem 1 and $x_3 = 1$ in subproblem 2. Subproblem 1 has solution $\mathbf{x} = [1.3913, 4.17391, 0, 0.434783]^\mathsf{T}$ with $L^1 = 20.8696$. This solution is not a feasible solution to the original problem, so U_0 is not updated. Subproblem 2 has solution $\mathbf{x} = [1.15942, 3.47826, 1, 0.362319]^\mathsf{T}$ with $L^2 = 22.4928$. Similarly to subproblem 1, this solution is not feasible for the original problem and U_0 is not updated.

This results in the updated set $\mathcal{P} = \{1, 2\}$, where subproblem 1 has the smallest lower bound. In the next iteration, $k = 2$, we branch out from subproblem 1 creating subproblem 3 with $x_4 = 0$ and subproblem 4 with $x_4 = 1$. Subproblem 3 has solution $\mathbf{x} = [1.5, 4.5, 0, 0]^\mathsf{T}$ with $L^3 = 22.5$. This solution is feasible to the original problem, thus $U^3 = 22.5$. Because $U^3 < U_0$ the upper bound is now U^3. Subproblem 4 has solution $\mathbf{x} = [1.25, 3.75, 0, 1]^\mathsf{T}$ with $L^4 = 23.625$. This solution is also feasible for the original problem, thus $U^4 = 23.625$. Because $U^4 > U^3$ the upper bound after the end of iteration 2 is $U_{k=2} = U^3 = 22.5$. From the resulting set of solution candidate subproblems $\mathcal{P} = \{2, 3, 4\}$ we have that $L^4 > U_2$. Thus, we perform pruning by removing subproblem 4 resulting in set $\mathcal{P} = \{2, 3\}$.

In the next iteration, $k = 3$, we branch out from node 2 resulting in subproblem 5 where $x_4 = 0$ and subproblem 6 where $x_4 = 1$. Subproblem 5 has solution $\mathbf{x} = [1.25, 3.75, 1, 0]^\mathsf{T}$ with $L^5 = 23.625$. This solution is removed (the node is pruned) because $L^5 > U_2$. Subproblem 6 has solution $\mathbf{x} = [1, 3, 1, 1]^\mathsf{T}$ with $L^6 = 26$. This solution is also excluded because $L^6 > U_2$. Thus, $\mathcal{P} = \{3\}$ and the algorithm terminates because $U_2 - L^3 = 0$, resulting in solution $\mathbf{x}^* = [1.5, 4.5, 0, 0]^\mathsf{T}$.

5.2.2 Problems with Continuous and Discrete Variables

B&B can be applied to more general discrete problems where the variables can take values from discrete sets and are not necessarily binary. This is the general case that can cover all discrete optimization problems. It was proposed by Land and Doig in 1960 [2] and is one of the most commonly used tools for solving NP-hard optimization problems in the areas of mixed- or pure-integer linear programming, mixed- or pure-integer quadratic and nonlinear convex programming, and combinatorial optimization.

Land and Doig [2] introduced B&B as an algorithm for solving integer linear programs because many nonlinear optimization problems can be reformulated into linear ones with additional constraints that force some or all of the variables must take only integral values. This is very common in public transport optimization problems, making B&B a very useful algorithm.

Let us consider the general case of an optimization problem with continuous and discrete variables:

$$\min_{\mathbf{x}, \nu} \ f(\mathbf{x}) \tag{5.32}$$

$$\text{s.t. } \mathbf{h}(\mathbf{x}, \nu) = \mathbf{0} \tag{5.33}$$

$$\mathbf{g}(\mathbf{x}, \nu) \leq \mathbf{0} \tag{5.34}$$

$$\mathbf{x} \in \mathbb{R}^n, \nu \in \mathcal{X}^r \tag{5.35}$$

where $\mathcal{X}$ is a set of discrete values ordered from the lowest to the highest value. Let also assume, without loss of generality, that set $\mathcal{X}$ is the same for all discrete variables $\nu_1, \ldots, \nu_r$. Let $\mathcal{X}_{\min}$ and $\mathcal{X}_{\max}$ be the lower and higher values in that set, respectively. In that case, the original problem can be split into two sub-problems by selecting a discrete variable ν_i and requesting from it to be within set $\mathcal{X}_1$ or $\mathcal{X}_2$. The immediate question is how to select these subsets $\mathcal{X}_1$ and $\mathcal{X}_2$. The main requirement is that $\mathcal{X}_1$ and $\mathcal{X}_2$ should include all points in $\mathcal{X}$ ensuring that the globally optimal solution of one of the generated subproblems is equal to the globally optimal solution of the original problem. The other requirement is that subsets $\mathcal{X}_1$ and $\mathcal{X}_2$ should be disjoint. That is:

$$\mathcal{X}_1 \cup \mathcal{X}_2 = \mathcal{X} \text{ and } \mathcal{X}_1 \cap \mathcal{X}_2 = \emptyset$$

To speed-up the search, we can make an intelligent selection of sets $\mathcal{X}_1$ and $\mathcal{X}_2$ by exploiting the solution of the continuous relaxation of (5.32)–(5.35). Let us say that this solution resulted in a value ν_i^* for variable ν_i. This value will, most probably, be a real number that does not belong to the discrete set $\mathcal{X}$ since it is a solution of the continuously relaxed problem. We can find two discrete values $\mathcal{X}_{\text{lower}}^i$ and $\mathcal{X}_{\text{upper}}^i$ from set $\mathcal{X}$ such that $\mathcal{X}_{\text{lower}}^i \leq \nu_i^* \leq \mathcal{X}_{\text{upper}}^i$ so that $\mathcal{X}_{\text{lower}}^i = \sup_{i \in \mathcal{X} : i \leq \nu_i^*} \mathcal{X}$ and $\mathcal{X}_{\text{upper}}^i = \inf_{i \in \mathcal{X} : i \geq \nu_i^*} \mathcal{X}$. To put it simply, $\mathcal{X}_{\text{lower}}^i$ is the highest value of set $\mathcal{X}$ which is less than or equal to ν_i^* and $\mathcal{X}_{\text{upper}}^i$ the lowest value of set $\mathcal{X}$ which is greater than or equal to ν_i^*. Then, set $\mathcal{X}_1$ will include $\{\mathcal{X}_{\min}, \ldots, \mathcal{X}_{\text{lower}}^i\}$, and $\mathcal{X}_2$ will include $\{\mathcal{X}_{\text{upper}}^i, \ldots, \mathcal{X}_{\max}\}$. Note that $\mathcal{X}_1 \cup \mathcal{X}_2 = \mathcal{X}$ and $\mathcal{X}_1 \cap \mathcal{X}_2 = \emptyset$.

By splitting the discrete set $\mathcal{X}$ this way, one of the two generated sub-problems will include the globally optimal solution of the original problem. The next steps are similar to the case of binary problems, with the only difference that subsets $\mathcal{X}_1$ and $\mathcal{X}_2$ can be further split to create more sub-problems.

Consider, for instance, that the continuous relaxation of the original problem has solution ν^* providing a lower bound value L^0. Let us now branch out from the root of the tree using variable ν_i to generate two sub-problems. The *continuous relaxation* of the sub-problem considering variable ν_i as part of set $\mathcal{X}_1$ is:

$$\min_{\mathbf{x}, \nu} \; f(\mathbf{x}) \tag{5.36}$$

$$\text{s.t.} \;\; \mathbf{h}(\mathbf{x}, \nu) = \mathbf{0} \tag{5.37}$$

$$\mathbf{g}(\mathbf{x}, \nu) \le \mathbf{0} \tag{5.38}$$

$$\mathbf{x} \in \mathbb{R}^n, \nu \in \mathbb{R}^r \tag{5.39}$$

$$\mathcal{X}_{\min} \le \nu_i \le \mathcal{X}_{\text{lower}}^i \tag{5.40}$$

where $\mathcal{X}_{\text{lower}}^i = \sup_{i \in \mathcal{X}\, :\, i \le \nu_i^*} \mathcal{X}$. The second sub-problem will have $\mathcal{X}_{\text{upper}}^i \le \nu_i \le \mathcal{X}_{\max}$, where $\mathcal{X}_{\text{upper}}^i = \inf_{i \in \mathcal{X}\, :\, i \ge \nu_i^*} \mathcal{X}$.

To demonstrate the steps of the algorithm, let us consider the mixed-integer linear programming problem where variables x_1, x_2 are integer and x_3 continuous:

$$\min \; -6x_1 - 5x_2 + 2x_3 \tag{5.41}$$

$$\text{s.t.} \; 3x_1 + 1.5x_2 + x_3 \le 11 \tag{5.42}$$

$$-x_1 + 2x_2 \le 5 \tag{5.43}$$

$$x_1, x_2, x_3 \ge 0 \tag{5.44}$$

$$x_1, x_2 \in \mathbb{Z}, x_3 \in \mathbb{R} \tag{5.45}$$

A trivial upper bound for this problem can be obtained by setting $x_1 = x_2 = x_3 = 0$ resulting in a feasible solution with objective function score $\text{U}^0 = 0$. This is a tighter upper bound compared to considering $\text{U}^0 = +\infty$. Let us now compute a lower bound. The continuous relaxation of this discrete problem allows x_1, x_2 to take any value in $\mathbb{R}$. Solving this problem with a linear programming solver results in solution $[x_1^*, x_2^*, x_3^*]^{\mathsf{T}} = [1.93333, 3.46667, 0]^{\mathsf{T}}$ with lower bound value $\text{L}^0 = -6 \cdot 1.93333 - 5 \cdot 3.46667 + 2 \cdot 0 = -28.9333$. Let us branch from the root of our tree by taking variable x_1 and considering sets $\mathcal{X}_1 = \{0, 1\}$ and $\mathcal{X}_2$, where $\mathcal{X}_2$ is the set of all integers with a value greater than or equal to 2.

We then have the two following sub-problems at iteration $k = 1$:

Subproblem 1:

$$\min \quad -6x_1 - 5x_2 + 2x_3$$
$$\text{s.t. } 3x_1 + 1.5x_2 + x_3 \le 11$$
$$- x_1 + 2x_2 \le 5$$
$$x_1, x_2, x_3 \ge 0$$
$$x_1 \le 1$$
$$x_1, x_2, x_3 \in \mathbb{R}$$

Subproblem 2:

$$\min \quad -6x_1 - 5x_2 + 2x_3$$
$$\text{s.t. } 3x_1 + 1.5x_2 + x_3 \le 11$$
$$- x_1 + 2x_2 \le 5$$
$$x_1, x_2, x_3 \ge 0$$
$$x_1 \ge 2$$
$$x_1, x_2, x_3 \in \mathbb{R}$$

By solving the continuous relaxations of subproblems 1 and 2, the lower bound of subproblem 1 is $L^1 = -21$ with solution $[x_1^*, x_2^*, x_3^*]^\mathsf{T} = [1, 3, 0]^\mathsf{T}$, and the lower bound of subproblem 2 is $L^2 = -28.6667$ with solution $[x_1^*, x_2^*, x_3^*]^\mathsf{T} = [2, 3.333, 0]^\mathsf{T}$. We also update the set of leaf nodes (subproblems) without branches at the current iteration. That is $\mathcal{P} = \{1, 2\}$, where $\mathcal{P}$ is the candidate list. Because subproblem 2 has the smallest lower bound of all subproblems in $\mathcal{P}$, we branch from this leaf node in the next iteration. It is important to note that the solution of subproblem 1 has integer values for the integer variables x_1 and x_2. That is, it is a feasible solution for the original problem and we can update the upper bound value of our problem to $U^1 = -21$ because it is smaller than U^0. Any future subproblem with a lower bound value greater than U^1, will be pruned from the set of solution candidates.

Subproblem 2 has the smallest lower bound from the leaf nodes in $\mathcal{P}$ and it has solution $[x_1^*, x_2^*, x_3^*]^\mathsf{T} = [2, 3.333, 0]^\mathsf{T}$ where x_1^* is integer. We can branch out from this node considering variable x_2^* and creating the two following subproblems.

Subproblem 3:

$$\min \quad -6x_1 - 5x_2 + 2x_3$$
$$\text{s.t. } 3x_1 + 1.5x_2 + x_3 \le 11$$
$$- x_1 + 2x_2 \le 5$$
$$x_1, x_2, x_3 \ge 0$$
$$x_1 \ge 2$$
$$x_2 \le 3$$
$$x_1, x_2, x_3 \in \mathbb{R}$$

Subproblem 4:

$$\min \quad -6x_1 - 5x_2 + 2x_3$$
$$\text{s.t. } 3x_1 + 1.5x_2 + x_3 \le 11$$
$$- x_1 + 2x_2 \le 5$$
$$x_1, x_2, x_3 \ge 0$$
$$x_1 \ge 2$$
$$x_2 \ge 4$$
$$x_1, x_2, x_3 \in \mathbb{R}$$

By solving the continuous relaxations of subproblems 3 and 4, the lower bound of subproblem 3 is $L^3 = -28$ with solution $[x_1^*, x_2^*, x_3^*]^\mathsf{T} = [2.16667, 3, 0]^\mathsf{T}$, and the lower bound of subproblem 4 does not exist because this subproblem is infeasible due to constraint $3x_1 + 1.5x_2 + x_3 \le 11$ which is violated for $x_1 \ge 2$, $x_3 \ge 0$, and $x_2 \ge 4$. We update the set of leaf nodes (subproblems) without branches at the current iteration as $\mathcal{P} = \{1, 3\}$. Note that subproblem 4 is excluded from further consideration. Because subproblem 3 has the smallest lower bound of all subproblems in $\mathcal{P}$, we branch from this leaf node in the next iteration generating the following two subproblems for $x_1 \le 2$ and $x_1 \ge 3$:

Subproblem 6:

Subproblem 5:

$$\min \ -6x_1 - 5x_2 + 2x_3$$
$$\text{s.t. } 3x_1 + 1.5x_2 + x_3 \le 11$$
$$-x_1 + 2x_2 \le 5$$
$$x_1, x_2, x_3 \ge 0$$
$$2 \le x_1 \le 2$$
$$x_2 \le 3$$
$$x_1, x_2, x_3 \in \mathbb{R}$$

$$\min \ -6x_1 - 5x_2 + 2x_3$$
$$\text{s.t. } 3x_1 + 1.5x_2 + x_3 \le 11$$
$$-x_1 + 2x_2 \le 5$$
$$x_1, x_2, x_3 \ge 0$$
$$x_1 \ge 2$$
$$x_1 \ge 3$$
$$x_2 \le 3$$
$$x_1, x_2, x_3 \in \mathbb{R}$$

Subproblem 5 has lower bound $L^5 = -27$ with solution $[x_1^*, x_2^*, x_3^*]^\mathsf{T} = [2, 3, 0]^\mathsf{T}$, and subproblem 6 has lower bound $L^6 = -24.6667$ with solution $[x_1^*, x_2^*, x_3^*]^\mathsf{T} = [3, 1.3333, 0]^\mathsf{T}$. Set $\mathcal{P}$ is updated as $\mathcal{P} = \{1, 5, 6\}$. Importantly, the solution of subproblem 5 has integer values for x_1 and x_2, meaning that it is a feasible solution for the original problem. Its objective function value is lower than the current upper bound of $U^1 = -21$. Thus, we update U which now is equal to $U^5 = -27$. Because subproblems 1 and 6 in $\mathcal{P}$ have a lower bound greater than U^5, they are pruned (removed from $\mathcal{P}$) and B&B terminates with globally optimal solution $[x_1^*, x_2^*, x_3^*]^\mathsf{T} = [2, 3, 0]^\mathsf{T}$. The process is summarized in Fig. 5.3. Notice that B&B required to solve 6 subproblems. Solving this problem with brute-force would require the solution of $|\mathbb{Z}_+|^3$ subproblems, where $|\mathbb{Z}_+|$ is the number of all positive integer numbers. This makes evident the practical importance of B&B.

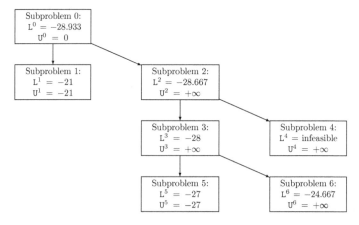

Fig. 5.3 Boolean rooted tree created by the B&B method until convergence. If the subproblem of a leaf results in an infeasible solution for the original discrete problem, its upper bound value is set to $+\infty$. The algorithm terminates when solving subproblem 5 because $U^5 = \min\{L^1, L^5, L^6\}$, where 1, 5 and 6 are the leaf nodes in $\mathcal{P}$ from which we have not branched out yet

---

Practitioner's Corner
Branch and Bound for convex nonlinear programs in Python 3

Programming a B&B algorithm can be a tedious task because we need to automate the process of generating subproblems. To automate this process, one can use a Python package, like GEKKO, which applies the B&B steps and is coupled with open source solvers for linear, quadratic, nonlinear, and mixed-integer programming (LP, QP, NLP, MILP, MINLP). Below follows an example of using GEKKO to solve the mixed-integer quadratic program presented in (5.31).

```python
from gekko import GEKKO
m = GEKKO() # Initialize gekko
m.options.SOLVER=1 # Non-Dynamic problem
# Initialize variables
x1 = m.Var()
x2 = m.Var()
x3 = m.Var(value=0,lb=0,ub=1,integer=True)
x4 = m.Var(value=0,lb=0,ub=1,integer=True)
# Equations
m.Equation(x1+x2+x3+x4==6)
m.Obj(4*x1**2-x1*x2+x2**2+8*x3**2+8*x4**2) # Objective
m.solve(disp=False) # Solve
print('x1: ' + str(x1.value), 'x2: ' + str(x2.value), 'x3: ' + str(x3
    .value), 'x4: ' + str(x4.value))
print('Objective: ' + str(m.options.objfcnval))
```

Implementing this code returns solution $\mathbf{x}^* = [1.5, 4.5, 0, 0]^\mathsf{T}$ with performance 22.5.

5.3 Branch and Bound Decisions

5.3.1 Leaf Node Selection

One of the key B&B decisions is the selection of the leaf node at each iteration from which we will be branching out. So far, we have discussed the approach of selecting the leaf node i from the candidate list $\mathcal{P}$ that has the smallest lower bound: $\mathrm{L}^i = \min_{j \in \mathcal{P}} \mathrm{L}^j$. This approach is called *Best First* approach and it seeks to minimize the size of the rooted tree. The *Best First* approach guarantees that we will examine only critical nodes, where the critical nodes are defined below.

Definition 5.3 *Critical node* is every *candidate node* that its lower bound is smaller than the upper bound at the current iteration, which represents the best currently known feasible solution to the original discrete problem.

The best first approach minimizes the size of the search tree. It is generally very common, but it has the following drawbacks:

- It typically does not find feasible solutions of the original discrete problem too quickly, i.e., feasible solutions are more likely to be found deep in the rooted tree.
- The continuous relaxations of the discrete program may change quite drastically by moving from one leaf node to another leaf node from iteration to iteration.
- Memory usage is high because we need to store the candidate list and the tree can grow broad.

An alternative approach is the *Depth First* method. In this approach, we always choose the deepest node to process next. By doing so, we branch aggressively until we prune. Consider, for instance, original problem (0) and the generated subproblems (1) and (2). We will branch from leaf node (2) to create subproblems (3) and (4) and then we will branch from leaf node (4) selecting every time the deepest node until finding a feasible solution for the original problem. By doing so, the lower bound values of the leaf nodes are not taken into consideration in the selection of the next leaf node. When we find a feasible solution, we go back up in our search tree (i.e., to leaf node (1)) and conduct the same steps. The advantages of this approach are:

- We do not need to keep a long list $\mathcal{P}$ of candidate leaf nodes, thus saving memory.
- Evaluating leaf nodes is easier because the sub-problems do not change much from iteration to iteration.
- Feasible solutions are found quickly.

The disadvantage of the *Depth First* method is that we may need to process too many non-critical nodes.

5.3.2 Selection of Branching Variable

So far we have not discussed in detail which variable to choose when branching from a leaf node. We used to say that if ν is the column vector of discrete variables, we select variable ν_i to branch and we used the following branching rules:

- If ν_i is a binary variable, the first branch (subproblem) will have $\nu_i = 0$ and the other branch $\nu_i = 1$.
- If ν_i is a discrete/integer variable, then $\nu_i \in \{X_{\min}, \ldots, X_{\text{lower}}\}$ on the one branch, and $\nu_i \in \{X_{\text{upper}}, \ldots, X_{\max}\}$ on the other branch where X_{lower} is the largest value in the discrete set X which is lower than or equal to ν_i^*, and X_{lower} is the smallest value in the discrete set X which is higher than or equal to ν_i^*.

Reckon that ν_i^* is the optimal value of variable ν_i^* when solving the continuously relaxed problem of the leaf node from which we are currently branching. So far, we have only considered one requirement when selecting variable $\nu_i \in \nu$ based on which we will create the two new subproblems. The only requirement was that value ν_i^* does not belong already to the discrete set X. If it does, we have to find another variable. Obviously, if all variables $\nu_i^* \in \nu^*$ belong to the discrete set X then we have found a feasible solution to the original problem and we will not branch from this leaf node.

Let us take the case of being at a leaf node with multiple variables that can be selected for branching. The key question is, which variable to branch on? One approach, that we have used until now, is to arbitrarily use one variable. If we want, however, to choose the branching that probably reduces the total number of generated subproblems we would need to make an intelligent selection of the branching variable. An obvious idea is to branch on the variable that will cause the lower bounds to increase the most. This will lead to more pruning. To do so, we need to quickly estimate the lower bounds that would result from branching on a variable. This can be done by:

- *Strong branching*, where we solve the continuous relaxation of each subproblem for each potential branching variable.
- Using *pseudo-costs*, where, mimicking gradients, we estimate the per-unit change in the objective function from enforcing the value of each variable to be rounded up or down.
- *Most infeasible branching*, where we select the most infeasible variable (i.e., the variable that has value ν_i^* closer to 0.5 when solving a binary problem or $z + 0.5$ when solving an integer problem, where z is the closest integer number to ν_i^*).

Strong branching is obviously computationally expensive. On the other hand, most infeasible branching and pseudo-cost branching are easy-to-use strategies. Most infeasible branching is easy to implement because we simply need to check the values of ν^*. Pseudo-cost branching will require from us to know the change of the objective function when changing each variable in order to choose the variable that is predicted to have the most change on the objective function. Because it might be expensive to produce such an estimation, we can initially assume that all variables have the same effect on the objective function change and, as we branch for a variable, keep track of the change in the objective function so that we have an estimate for the future.

5.4 Branch and Cut for Integer Linear Programming

In the special case of integer linear programming, we can speed up the computation of a globally optimal solution with the use of cutting planes. The idea is to try to improve the bounds so that more leaf nodes of the rooted tree can be pruned (fathomed) as early as possible by adding constraints that eliminate fractional—but not integer—solutions of the relaxed integer linear program.

If we include additional, well-selected inequality constraints to the original problem, then every extreme point of the continuous relaxation of the integer linear program can be an extreme point for the original problem as well. These additional inequality constraints are known as *cutting planes*. In essence, Branch and Cut involves implementing a B&B algorithm and using cutting planes to tighten the linear programming relaxations. If cuts are only used to tighten the initial linear program relaxation, the solution strategy is called Cut and Branch.

5.4.1 Intuition of Cutting Planes

To understand branch and cut, it is important to understand the *cutting-plane* methods, which are a variety of optimization methods that iteratively refine a feasible set by means of linear inequalities (known as *cuts*). Let us consider the following integer linear program.

$$\max \; z := 3x_1 + 2x_2 \tag{5.46}$$

$$\text{s.t. } 4x_1 + 2x_2 \le 15 \tag{5.47}$$

$$x_1 + 2x_2 \le 8 \tag{5.48}$$

$$x_1 + x_2 \le 5 \tag{5.49}$$

$$x_1, x_2 \in \mathbb{Z}_+ \tag{5.50}$$

Using the graphical method, the optimal value of the linear relaxation of this problem is provided in Fig. 5.4. This optimal value is $[x_1^*, x_2^*]^\mathsf{T} = [2.5, 2.5]^\mathsf{T}$ and it is infeasible for the original problem, the feasible values of which are presented with black dots in Fig. 5.4.

Notice that we can add *valid inequalities*, which we can call *cutting planes* or *cuts* to improve the upper bound produced from the relaxation, which is $U = 3 \cdot 2.5 + 2 \cdot 2.5 = 12.5$. Because we solve a maximization problem, we have to change the terminology used so far regarding lower and upper bounds. Now, the solution of a relaxed problem at a leaf node offers an upper bound to the solution of the discrete

Fig. 5.4 Trimmed lines $\mathbf{c}^\mathsf{T}\mathbf{x} = z$ with increasing values of z represent the objective function values until leaving the feasible region of the relaxed linear problem. Black dots represent the feasible solutions of the integer linear problem. The remaining lines present the inequality constraints. Solution $[2.5, 2.5]^\mathsf{T}$ is the optimal solution of the continuously relaxed problem, which is infeasible for the integer linear problem

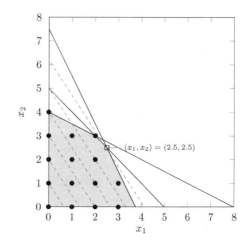

problem represented at the leaf node and the lower bound can be initially set to $-\infty$. To proceed further, we provide the definition of a valid inequality.

Definition 5.4 A *valid inequality* of a mixed-integer linear problem with feasible region $X = \{\mathbf{x} \in \mathbb{R}^n : \mathbf{Ax} \leq \mathbf{b}, \; x_j \in \mathbb{Z}, \; \forall j \in J\}$, where J is the subset of integer variables, is an inequality $\boldsymbol{\pi}^\mathsf{T}\mathbf{x} \leq \pi_0$ for which $\boldsymbol{\pi}^\mathsf{T}\mathbf{x} \leq \pi_0$ for all $\mathbf{x} \in X$, where $\boldsymbol{\pi} \in \mathbb{R}^n$ and $\pi_0 \in \mathbb{R}$. That is, it is an inequality constraint that does not eliminate any feasible solutions to the mixed-integer linear problem.

In our maximization problem, $x_1 \leq 2$ is not a valid inequality because it excludes points $[x_1, x_2]^\mathsf{T} = [3, 0]^\mathsf{T}$ and $[x_1, x_2]^\mathsf{T} = [3, 1]^\mathsf{T}$ which are part of X. In contrast, $x_2 \leq 6$ is a valid inequality because this inequality is satisfied for all $\mathbf{x} \in X$. Notice though that if we add this inequality to our relaxed linear program, the solution of the problem will not change. That is, this cut is not beneficial. We can think of other cuts that remove fractional solutions without eliminating any feasible integer solutions. The goal is to cut off part of the feasible region of the relaxed problem without excluding any feasible solutions to the discrete problem. In particular, if we add cuts so that we can limit ourselves to the convex hull of X, then solving the relaxed linear program with these cuts will result in the solution of the mixed-integer linear program. This is the crux of cutting planes in mixed-integer linear programming. Consider, for instance, the cut $2x_1 + x_2 \leq 7$. Adding this cut to the continuous relaxation of the original problem will result in the relaxed problem:

$$\text{Relaxed problem with cut :} \tag{5.51}$$

$$\max \; z := 3x_1 + 2x_2 \tag{5.52}$$

$$\text{s.t. } 4x_1 + 2x_2 \leq 15 \tag{5.53}$$

$$x_1 + 2x_2 \leq 8 \tag{5.54}$$

$$x_1 + x_2 \leq 5 \tag{5.55}$$

$$2x_1 + x_2 \leq 7 \tag{5.56}$$

$$x_1, x_2 \in \mathbb{R}_+ \tag{5.57}$$

which, when solved, returns the globally optimal solution $\mathbf{x}^* = [2, 3]^\mathsf{T}$ of the mixed-integer problem (see Fig. 5.5).

Notice that in Fig. 5.5 we have not added all valid inequalities needed to create the convex hull of X. Despite this, our added valid inequality (cut) is enough to lead us to the computation of the globally optimal solution of the mixed-integer problem. This is because, in practice, we do not need to add all valid inequalities that will return the convex hull of X. Adding the required inequality constraints around the optimal points is enough. Of course, if we do not have any information about the optimal points, we can try to generate all valid inequalities that will result in obtaining the convex hull of X since this guarantees that the solution of the relaxation of the problem is the same as the solution of its original, mixed-integer version. Trying to find the convex hull in large-scale problems is very difficult both because this will

Fig. 5.5 Black dots
represent the feasible
solutions of the integer linear
problem. The added valid
inequality represents the cut

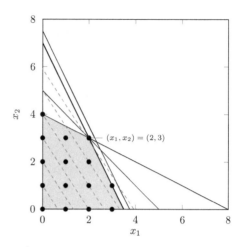

result in too many constraints and because these constraints are hard to find. For this
reason, cutting plane algorithms try to find useful inequality constraints and update
them by solving repetitive relaxed linear programs.

A general cutting plain algorithmic framework for integer linear problems is
provided below.

Algorithm 24 Cutting Plane framework for ILPs

1: Given $\max_{\mathbf{x}} \mathbf{c}^\mathsf{T}\mathbf{x}$ s.t. $\mathbf{A}\mathbf{x} \le \mathbf{b}$, $\mathbf{x} \in \mathbb{Z}^n$ with feasible region $\mathcal{P}_I = \{\mathbf{x} \in \mathbb{Z}^n \mid \mathbf{A}\mathbf{x} \le \mathbf{b}\}$

2: Solve LP relaxation $\max_{\mathbf{x}} \mathbf{c}^\mathsf{T}\mathbf{x}$ s.t. $\mathbf{A}\mathbf{x} \le \mathbf{b}$, $\mathbf{x} \in \mathbb{R}^n$ with feasible region $\mathcal{P} = \{\mathbf{x} \in \mathbb{R}^n \mid \mathbf{A}\mathbf{x} \le \mathbf{b}\}$ to obtain $\mathbf{x}^*$

3: **repeat**

4:    find valid inequality $\boldsymbol{\pi}^\mathsf{T}\mathbf{x} \le \pi_0$ for which $\boldsymbol{\pi}^\mathsf{T}\mathbf{x} \le \pi_0$ for all $\mathbf{x} \in \mathcal{P}_I$ and $\boldsymbol{\pi}^\mathsf{T}\mathbf{x}^* > \pi_0$ separating $\mathbf{x}^*$ from $\mathcal{P}_I$

5:    add valid inequality to $\mathcal{P}$ making $\mathbf{x}^*$ infeasible

6:    solve LP relaxation: $\mathbf{x}^* = \mathrm{argmax}\ \mathbf{c}^\mathsf{T}\mathbf{x}$ s.t. $\mathbf{x} \in \mathcal{P}$

7: **until** solution $\mathbf{x}^*$ is a vector of integer values

---

In practice, there are many methods to generate valid inequalities (cutting planes).
Typical methods are the gomory cuts, cover cuts, and mixed-integer rounding cuts.

5.4.2 Gomory Cutting Planes

The main objective of cutting planes is to cut off parts of the feasible region of the linear relaxation of a mixed-integer linear program without removing any feasible solutions. Gomory proposed the first finite cutting plane algorithm for pure ILPs (1958) [3] and extended it for mixed-integer LPs in 1960. We consider first the case of a pure ILP:

$$\max_{\mathbf{x}} \ \mathbf{c}^\mathsf{T}\mathbf{x}$$

$$\text{s.t. } \mathbf{A}\mathbf{x} \le \mathbf{b}$$

$$\mathbf{x} \in \mathbb{Z}^n$$

with m constraints and n variables. Its feasible region is $\mathcal{P}_I = \{\mathbf{x} \in \mathbb{Z}^n \mid \mathbf{A}\mathbf{x} \le \mathbf{b}\}$ and the feasible region of its LP relaxation is $\mathcal{P} = \{\mathbf{x} \in \mathbb{R}^n \mid \mathbf{A}\mathbf{x} \le \mathbf{b}\}$. As previously discussed, if we knew the convex hull $\mathcal{P}_c = \mathbf{conv}\{\mathbf{x} \in \mathbb{R}^n \mid \mathbf{A}\mathbf{x} \le \mathbf{b}, \ \mathbf{x} \text{ integer}\}$ we could have solved an LP considering $\mathcal{P}_c$ to obtain the globally optimal solution of the pure ILP.

Gomory cut for ILPs

Consider $\max_{\mathbf{x} \in \mathcal{P}_I} \mathbf{c}^\mathsf{T}\mathbf{x}$ where $\mathcal{P}_I = \{\mathbf{x} \in \mathbb{Z}^n \mid \mathbf{A}\mathbf{x} \le \mathbf{b}\}$. Let vector $\boldsymbol{\lambda} \in [0, 1]^m$ such that:

- $\underbrace{\boldsymbol{\lambda}^\mathsf{T}}_{1 \times m} \underbrace{\mathbf{A}}_{m \times n} \mathbf{x} \le \boldsymbol{\lambda}^\mathsf{T}\mathbf{b}$ holds for all $\mathbf{x} \in \mathcal{P}_I$
- $\boldsymbol{\lambda}^\mathsf{T}\mathbf{A}$ is an integer n-valued vector

  Then, $\boldsymbol{\lambda}^\mathsf{T}\mathbf{A} \le \lfloor \boldsymbol{\lambda}^\mathsf{T}\mathbf{b} \rfloor$ is a Gomory cut, known also as Chvátal-Gomory cut.

If we use simplex to solve the linear programming relaxation, gomory cuts are used to the equations provided by the optimal tableau. Consider an integer linear program in standard form:

$$\max_{\mathbf{x}} \ \mathbf{c}^\mathsf{T}\mathbf{x}$$

$$\text{s.t. } \mathbf{A}\mathbf{x} = \mathbf{b}$$

$$\mathbf{x} \ge \mathbf{0}$$

$$\mathbf{x} \in \mathbb{Z}^n$$

where $\mathbf{A}$ is an $m \times n$ matrix and some variables $\mathbf{x}$ might be slack variables. Note that we request all variables to be integers, even the slack variables. To ensure that, if we have inequality constraints that need to be transformed to equality constraints with the use of slack variables, we need to change the constraint coefficients $\mathbf{A}$ and the right-hand-side coefficients $\mathbf{b}$ first, so that they take integer values. For

Table 5.1 Simplex Tableau

Row	z	x_1	x_2	$\ldots$	x_n	Rhs
0	1	$-c_1$	$-c_2$	$\ldots$	$-c_n$	b_0
1	0	a_{11}	a_{12}	$\ldots$	a_{1n}	b_1
2	0	a_{21}	a_{22}	$\ldots$	a_{2n}	b_2
$\vdots$	$\vdots$	$\vdots$	$\vdots$	$\vdots$	$\vdots$	$\vdots$
m	0	a_{m1}	a_{m2}	$\ldots$	a_{mn}	b_m

instance, if one of the constraints of the problem is $0.4x_1 + 0.2x_2 \leq 3.6$ we convert it first to $4x_1 + 2x_2 \leq 36$ and then we add integer slack variable $x_3 \geq 0$ so that $4x_1 + 2x_2 + x_3 = 36$. This results in an ILP with $\mathbf{A} \in \mathbb{Z}^{m \times n}$ and $\mathbf{b} \in \mathbb{Z}^m$. Notice that this transformation is possible for any problem with coefficient values that are rational numbers ($\mathbb{Q}$), meaning that they can be expressed as a fraction of two integers.

If we drop the integrality constraints, then we can solve the relaxed (continuous) linear program with simplex to obtain a basic feasible solution. This basic feasible solution (if the problem is not infeasible or unbounded) is a vertex of the polyhedron consisting of all feasible points. As we previously discussed, if the variable values of that vertex are not integer, then we need to add cutting planes. Geometrically, a useful cutting plane is a hyperplane with the vertex on the one side and all feasible points of the integer linear problem on the other.

We will now discuss the Gomory strategy of generating cutting planes. We first solve the linear relaxation of the ILP by starting from the simplex tableau presented in Table 5.1.

Solving this problem with simplex, the coefficient in each row of the table will change values until the updated coefficient values $-\bar{c}_j$ for all nonbasic variables are nonnegative, indicating that we have found an optimal solution:

$$\mathbf{x}^* = \arg\max \mathbf{c}^\mathsf{T}\mathbf{x} \text{ s.t. } \mathbf{x} \in \mathcal{P}$$

where $\mathcal{P} = \{\mathbf{x} \in \mathbb{R}_{\geq 0}^n \mid \mathbf{A}\mathbf{x} = \mathbf{b}\}$.

When finding this solution, each one of the $1, \ldots, m$ rows of the tableau will have a basic variable with a coefficient equal to 1. At the next step, we can pick any row of the updated tableau that has a fractional right-hand side value. That is, we pick a row p for which $\bar{b}_p$ is noninteger. The selection can be arbitrary, although, to speed up the process, the row for which the fractional part of the coefficient's value is closer to 1/2 is typically selected.

Let us say that we select row p with basic variable x_p and fractional rhs value $\bar{b}_p$. Then, from this row we have the equality constraint:

$$\sum_{j=1}^n \bar{a}_{pj} x_j = \bar{b}_p$$

where $\bar{a}_{pj}$ and $\bar{b}_p$ are the updated values of the coefficients in the tableau. We then split every coefficient $\bar{a}_{pj}$ in two parts:

- The part that is rounded down to the closest integer, $\lfloor \bar{a}_{pj} \rfloor$
- The remaining fractional part, $\bar{a}_{pj} - \lfloor \bar{a}_{pj} \rfloor$

We do the same for coefficients $\bar{b}_p$. We can then rewrite the row as:

$$\sum_{j=1}^{n} \left(\lfloor \bar{a}_{pj} \rfloor + (\bar{a}_{pj} - \lfloor \bar{a}_{pj} \rfloor) \right) x_j = \lfloor \bar{b}_p \rfloor + (\bar{b}_p - \lfloor \bar{b}_p \rfloor)$$

We also rewrite the equation so that the integer parts are on the left hand side and the fractional parts are on the right hand side:

$$\sum_{j=1}^{n} \lfloor \bar{a}_{pj} \rfloor x_j - \lfloor \bar{b}_p \rfloor = (\bar{b}_p - \lfloor \bar{b}_p \rfloor) - \sum_{i=1}^{n} \left(\bar{a}_{pj} - \lfloor \bar{a}_{pj} \rfloor \right) x_j \qquad (5.58)$$

where the inequality:

$$\sum_{j=1}^{n} \lfloor \bar{a}_{pj} \rfloor x_j - \lfloor \bar{b}_p \rfloor \leq 0 \qquad (5.59)$$

or, equivalently,

$$(\bar{b}_p - \lfloor \bar{b}_p \rfloor) - \sum_{i=1}^{n} \left(\bar{a}_{pj} - \lfloor \bar{a}_{pj} \rfloor \right) x_j \leq 0 \qquad (5.60)$$

is the *Gomory cut*. We will show why this cut is feasible for the integer linear program. Consider again equation (5.58) rewritten as:

$$\underbrace{\sum_{j=1}^{n} \lfloor \bar{a}_{pj} \rfloor x_j - \lfloor \bar{b}_p \rfloor}_{\text{integer}} + \underbrace{\sum_{i=1}^{n} \left(\bar{a}_{pj} - \lfloor \bar{a}_{pj} \rfloor \right) x_j}_{\text{nonnegative}} = \underbrace{(\bar{b}_p - \lfloor \bar{b}_p \rfloor)}_{\text{some value in } (0,1)} \qquad (5.61)$$

Notice that for any feasible solution of the ILP, $\lfloor \bar{a}_{pj} \rfloor x_j - \lfloor \bar{b}_p \rfloor$ is integer. In addition, any $x_j \geq 0$ to satisfy constraint $\mathbf{x} \geq 0$. Thus, $\sum_{i=1}^{n} \left(\bar{a}_{pj} - \lfloor \bar{a}_{pj} \rfloor \right) x_j$ is nonnegative. Finally, $\bar{b}_p$ is fractional, and thus $0 < \bar{b}_p - \lfloor \bar{b}_p \rfloor < 1$. We can exclude the nonnegative term from this equality constraint and get the following inequality constraint which is satisfied for any feasible solution of the ILP:

$$\underbrace{\sum_{j=1}^{n} \lfloor \bar{a}_{pj} \rfloor x_j - \lfloor \bar{b}_p \rfloor}_{\text{integer}} \leq \underbrace{(\bar{b}_p - \lfloor \bar{b}_p \rfloor)}_{\text{some value in } (0,1)} \qquad (5.62)$$

Now that we established that the inequality constraint (5.62) holds for any feasible solution of the integer linear program, we can further tighten this inequality by requesting from the fractional right-hand-side to be also integer since the left-hand-side is always integer. If we round its value down, then the right-hand side becomes 0 yielding:

$$\sum_{j=1}^{n} \lfloor \bar{a}_{pj} \rfloor x_j - \lfloor \bar{b}_p \rfloor \leq 0 \Leftrightarrow \sum_{j=1}^{n} \lfloor \bar{a}_{pj} \rfloor x_j \leq \lfloor \bar{b}_p \rfloor \tag{5.63}$$

which is the Gomory cut, and it is satisfied for any feasible solution of the original ILP.

To summarize, using either Eqs. (5.59) or (5.60) the Gomory cut at row p is:

$$\sum_{j=1}^{n} \lfloor \bar{a}_{pj} \rfloor x_j \leq \lfloor \bar{b}_p \rfloor$$

or, equivalently,

$$\sum_{i=1}^{n} \left(\bar{a}_{pj} - \lfloor \bar{a}_{pj} \rfloor \right) x_j \geq \bar{b}_p - \lfloor \bar{b}_p \rfloor$$

Note that both valid inequalities are equivalent, but the second one is more practical when using simplex because it is easily incorporated into the tableau. To add it to the tableau and solve the next linear programming relaxation that considers this valid inequality, we have to turn it into an equality first. This is done by adding one more integer slack variable $x_{n+1} \geq 0$ such that:

$$-x_{n+1} + \sum_{i=1}^{n} \left(\bar{a}_{pj} - \lfloor \bar{a}_{pj} \rfloor \right) x_j = \bar{b}_p - \lfloor \bar{b}_p \rfloor \Rightarrow$$

$$x_{n+1} - \sum_{i=1}^{n} \left(\bar{a}_{pj} - \lfloor \bar{a}_{pj} \rfloor \right) x_j = -(\bar{b}_p - \lfloor \bar{b}_p \rfloor) \tag{5.64}$$

This indicates that with any additional cut, we add one more row and one more slack variable in the simplex tableau. Note that the added slack variable is the basic variable of the added row. The algorithmic framework of using simplex and Gomory cuts to solve an ILP is provided in Algorithm 25.

Algorithm 25 Simplex with Gomory cuts for ILPs

1: Given $\max_{\mathbf{x}} \mathbf{c}^\mathsf{T}\mathbf{x}$ s.t. $\mathbf{A}\mathbf{x} = \mathbf{b}$, $\mathbf{x} \in \mathbb{Z}^n_{\geq 0}$ with feasible region $\mathcal{P}_I = \{\mathbf{x} \in \mathbb{Z}^n_{\geq 0} \mid \mathbf{A}\mathbf{x} = \mathbf{b}\}$ and coefficients $\mathbf{A} \in \mathbb{Z}^{m \times n}$, $\mathbf{b} \in \mathbb{Z}^m$

2: Use simplex to solve LP relaxation $\max_{\mathbf{x}} \mathbf{c}^\mathsf{T}\mathbf{x}$ s.t. $\mathbf{x} \in \mathcal{P}$ with feasible region $\mathcal{P} = \{\mathbf{x} \in \mathbb{R}^n_{\geq 0} \mid \mathbf{A}\mathbf{x} \leq \mathbf{b}\}$ and obtain $\mathbf{x}^*$

3: Obtain updated simplex tableau corresponding to solution $\mathbf{x}^*$

4: **repeat**

5:     select a row p from the updated tableau for which $\bar{b}_p$ is fractional

6:     generate integer slack variable x_{n+1}

7:     add $x_{n+1} \geq 0$ to $\mathcal{P}$

8:     add Gomory cut $x_{n+1} - \sum_{i=1}^n \left(\bar{a}_{pj} - \lfloor \bar{a}_{pj} \rfloor \right) x_j = -(\bar{b}_p - \lfloor \bar{b}_p \rfloor)$ to $\mathcal{P}$

9:     update indexes $n \leftarrow n + 1$, $m \leftarrow m + 1$

10:    solve LP relaxation: $\mathbf{x}^* = \operatorname{argmax} \mathbf{c}^\mathsf{T}\mathbf{x}$ s.t. $\mathbf{x} \in \mathcal{P}$ with simplex

11: **until** all $\bar{b}_j \in \bar{\mathbf{b}}$ values in the simplex tableau are integer

---

Let us now consider the more general case of a mixed-integer linear program. In 1960, Gomory [4] extended his approach to mixed-integer linear programs (MILPs), inventing the *mixed-integer cuts*, known today as GMI cuts. To apply simplex, the MILP is written as:

$$\max_{\mathbf{x}} \mathbf{c}^\mathsf{T}\mathbf{x}$$

$$\text{s.t. } \mathbf{A}\mathbf{x} = \mathbf{b}$$

$$\mathbf{x} \geq \mathbf{0}$$

$$\mathbf{x} \in \mathbb{R}^n$$

$$x_j \in \mathbb{Z} \ \forall j \in J$$

where set $J \subseteq \{1, 2, \ldots, n\}$ is the set of integer variables. Alternatively, $\mathbf{x}$ can be symbolized as a vector $\mathbf{x} \in \mathbb{Z}^{|J|} \times \mathbb{R}^{n-|J|}$. The basic idea is to find cutting planes using coefficients for integer variables. In the mixed-integer case, we perform the following steps.

In the first step, we solve the linear relaxation of the problem by removing the integrality constraints $x_j \in \mathbb{Z} \ \forall j \in J$. Starting from a simplex tableau, we apply simplex in the relaxed linear program resulting in an optimal solution $\mathbf{x}^*$ and a revised tableau, as presented in Table 5.2.

In step 2, we select row $p \in \{1, \ldots, m\}$ from the revised simplex tableau for which:

- The right-hand side $\bar{b}_p$ is noninteger
- The basic variable x_p of this row belongs to the set J of integer variables

In step 3, from the selected row $\sum_{j=1}^n \bar{a}_{pj} x_j = \bar{b}_p$ we write $\bar{b}_p = \lfloor \bar{b}_p \rfloor + f_0$ where $0 < f_0 < 1$ and $\bar{a}_{pj} = \lfloor \bar{a}_{pj} \rfloor + f_j$ where $0 \leq f_j < 1$.

Table 5.2 Revised Simplex Tableau corresponding to the optimal solution $\mathbf{x}^*$ of the relaxed mixed-integer linear program

Row	z	x_1	x_2	$\ldots$	x_n	Rhs
0	1	$-\bar{c}_1$	$-\bar{c}_2$	$\ldots$	$-\bar{c}_n$	$\bar{b}_0$
1	0	$\bar{a}_{11}$	$\bar{a}_{12}$	$\ldots$	$\bar{a}_{1n}$	$\bar{b}_1$
2	0	$\bar{a}_{21}$	$\bar{a}_{22}$	$\ldots$	$\bar{a}_{2n}$	$\bar{b}_2$
$\vdots$	$\vdots$	$\vdots$	$\vdots$	$\vdots$	$\vdots$	$\vdots$
m	0	$\bar{a}_{m1}$	$\bar{a}_{m2}$	$\ldots$	$\bar{a}_{mn}$	$\bar{b}_m$

Step 4 concerns the addition of the mixed-integer gomory cut provided below. Let $N = \{1, \ldots, n\}$ be the set of all variables, from which the variables in J are integer and the variables in $N - J$ can take any real value. Then,

Mixed-integer Gomory cut (GMI)

$$\sum_{\substack{j \in J \\ f_j \leq f_0}} \frac{f_j}{f_0} x_j + \sum_{\substack{j \in J \\ f_j > f_0}} \frac{1 - f_j}{1 - f_0} x_j + \sum_{\substack{j \in N - J \\ \bar{a}_{pj} \geq 0}} \frac{\bar{a}_{pj}}{f_0} x_j - \sum_{\substack{j \in N - J \\ \bar{a}_{pj} < 0}} \frac{\bar{a}_{pj}}{1 - f_0} x_j \geq 1$$

Finally, in step 5 we add the mixed-integer gomory cut and solve the linear relaxation until all right-hand side values of the simplex tableau are integer for the rows that have basic variables which should take integer values. Below we show that GMI is a cut of the original MILP.

Theorem 5.1 *GMI is a cut, meaning that if $\mathbf{x}$ is a feasible point of the original MILP it satisfies also the GMI inequality constraint.*

Proof Notice that the feasible region of the MILP concerning the selected row p is $P_I := \{\mathbf{x} \in \mathbb{Z}_+^{|J|} \times \mathbb{R}_+^{n-|J|} : \sum_{j=1}^{n} \bar{a}_{pj} x_j = b_p\}$. We can write the equality constraint of this row:

$$\sum_{j=1}^{n} \bar{a}_{pj} x_j = \bar{b}_p$$

as:

$$\sum_{\substack{j \in J \\ f_j \leq f_0}} f_j x_j + \sum_{\substack{j \in J \\ f_j > f_0}} (f_j - 1) x_j + \sum_{j \in N - J} \bar{a}_{pj} x_j = f_0 + k$$

for some integer k. There are two options: $k \geq 0$ or $k \leq -1$. For $k \geq 0$ we get:

$$\sum_{\substack{j \in J \\ f_j \leq f_0}} \frac{f_j}{f_0} x_j - \sum_{\substack{j \in J \\ f_j > f_0}} \frac{1 - f_j}{f_0} x_j + \sum_{j \in N-J} \frac{\bar{a}_{pj}}{f_0} x_j \geq 1$$

and for $k \leq -1$:

$$- \sum_{\substack{j \in J \\ f_j \leq f_0}} \frac{f_j}{1 - f_0} x_j + \sum_{\substack{j \in J \\ f_j > f_0}} \frac{1 - f_j}{1 - f_0} x_j - \sum_{j \in N-J} \frac{\bar{a}_{pj}}{1 - f_0} x_j \geq 1$$

To get a valid inequality that is valid for both $k \geq 0$ and $k \leq -1$ we take the maximum coefficient on the left hand side:

$$\sum_{\substack{j \in J \\ f_j \leq f_0}} \frac{f_j}{f_0} x_j + \sum_{\substack{j \in J \\ f_j > f_0}} \frac{1 - f_j}{1 - f_0} x_j + \sum_{\substack{j \in N-J \\ \bar{a}_{pj} \geq 0}} \frac{\bar{a}_{pj}}{f_0} x_j - \sum_{\substack{j \in N-J \\ \bar{a}_{pj} < 0}} \frac{\bar{a}_{pj}}{1 - f_0} x_j \geq 1$$

Thus, we proved that the GMI cut is a valid inequality of P_I. $\square$

5.4.3 Branch & Cut and Cut & Branch

Although cutting plane methods were developed in the 1960s, they were considered impractical due to numerical instability. In addition, the number of cuts to be added, though finite, is usually quite large. Because of the above, cutting planes were important from a theoretical perspective, rather than a practical one. The rejuvenation of cutting plane methods started in 1990s, when their successful combination with branch-and-bound resulted in effective branch-and-cut and cut-and-branch algorithms.

In essence, *branch and cut* is used to describe an algorithmic framework that implements a branch and bound algorithm and uses cutting planes to tighten the linear programming relaxations. If cuts are only used to tighten the initial LP relaxation, then the approach is called *cut and branch*.

An important term in branch and cut is the *local cut*. This term is used for the cut generated for a specific leaf node of the branch and bound tree, since this cut will be satisfied by all solutions fulfilling the side constraints from the currently considered branch and bound subtree. The branch and cut approach is provided in the following algorithmic framework, which adapts the branch and bound algorithm to the case of discrete linear programs. Algorithm 26 refers to a mixed-integer linear program with r integer variables and $n - r$ continuous variables.

Algorithm 26 Branch and Cut framework for mixed-integer linear problems

---

1: given $\max_{\mathbf{x}} \mathbf{c}^\mathsf{T}\mathbf{x}$ s.t. $\mathbf{x} \in \mathcal{P}_I$ with feasible region $\mathcal{P}_I := \{\mathbf{x} \in \mathbb{Z}_{\geq 0}^r \times \mathbb{R}_{\geq 0}^{n-r} \mid \mathbf{A}\mathbf{x} = \mathbf{b}\}$

2: solve its continuous relaxation with feasible region $\mathcal{P} := \{\mathbf{x} \in \mathbb{R}_{\geq 0}^n \mid \mathbf{A}\mathbf{x} = \mathbf{b}\}$ to compute upper bound U^0.

3: if desired, search for cutting planes and solve again the continuous relaxation of the problem to tighten U^0.

4: If this solution is feasible for the original problem, then set lower bound $\mathsf{L}^0 = \mathsf{U}^0$. Otherwise, set $\mathsf{L}^0 = -\infty$ or compute another feasible solution with a heuristic to attain L^0.

5: Set $k = 0$ and $\mathsf{L}_k = \mathsf{L}^0$, $\mathsf{U}_k = \mathsf{U}^0$.

6: initialize *candidate list* $\mathcal{P} = \{0\}$ which contains all active leaf nodes from which we have not performed branching.

7: **repeat**

8:     from the leaf node in $\mathcal{P}$ with the smallest upper bound value, generate two sub-problems k' and k'' (branching).

9:     for the two generated sub-problems, compute their upper bounds $\mathsf{U}^{k'}, \mathsf{U}^{k''}$ by solving their continuous relaxations.

10:     if desired, search for cutting planes when solving the continuous relaxations of these sub-problems to tighten their upper bounds $\mathsf{U}^{k'}, \mathsf{U}^{k''}$.

11:     add these two sub-problems to $\mathcal{P}$ and remove the parent leaf node of these two sub-problems from $\mathcal{P}$.

12:     set U^k as the smallest upper bound from the upper bounds of all sub-problems in $\mathcal{P}$.

13:     if solution $\mathbf{x}_{k'}$ belongs to $\mathcal{P}_I$ and $\mathsf{U}^{k'} > \mathsf{L}_k$, set $\mathsf{L}_k \leftarrow \mathsf{U}^{k'}$ (do the same for $\mathbf{x}_{k''}$).

14:     remove all leaf nodes (sub-problems) from $\mathcal{P}$ for which their upper bound values are lower than L_k (pruning).

15:     set $k = k + 1$

16: **until** candidate list $\mathcal{P} = \emptyset$ or $\mathsf{U}_k - \mathsf{L}_k \leq \epsilon$ for some tolerance $\epsilon > 0$

Note that the main change of this branch and cut algorithm compared to branch and bound is line 10 which introduces local cuts. If we use cut and branch, we have to remove line 10 from Algorithm 26 and implement only the global cuts produced in line 3.

Let us consider the example of the compact formulation of the Traveling Salesman Problem with MTZ sub-tour elimination constraints:

$$\min \sum_{i \in N} \sum_{j \in N \setminus \{i\}} c_{ij} x_{ij}$$

$$\text{subject to:} \quad \sum_{j \in N \setminus \{i\}} x_{ij} = 1 \qquad\qquad \forall i \in N$$

$$\sum_{i \in N \setminus \{j\}} x_{ij} = 1 \qquad\qquad \forall j \in N \qquad\qquad (5.65)$$

$$(n-1)x_{ij} + u_i - u_j \le n - 2 \quad \forall i, j \in \{2, 3, \dots, n\} \mid i \ne j$$

$$1 \le u_i \le n - 1 \qquad\qquad \forall i \in \{2, 3, \dots, n\}$$

$$u_i \in \mathbb{Z} \qquad\qquad\qquad \forall i \in \{2, 3, \dots, n\}$$

$$x_{ij} \in \{0, 1\} \qquad\qquad \forall i \in N, j \in N$$

This is an integer linear program and can be solved with branch and cut, where we use branch and bound and cutting planes (for instance, Gomory cuts). Let us consider that we have $n = 16$ cities and the cost c_{ij} when traveling from i to j is the same as the cost when traveling from j to i (symmetric TSP). To automate the implementation of branch and bound with cuts, one can use a solver for integer linear problems. Below we provide an implementation example when we only allow Gomory cuts.

Practitioner's Corner
Branch and Cut in Python 3 for ILPs

One can use PuLP, an open-source linear programming (LP) package which uses Python syntax and interacts with many industry-standard solvers to solve integer and mixed-integer linear programs with the use of branch and cut. Apart from commercial solvers (CPLEX, Gurobi), PuLP can also use the open source COIN-OR Branch-and-Cut solver (CBC) solver. For cut generators, CBC relies on the COIN-OR Cut Generation Library (CGL). Cut generation callbacks (CGC) are called at each node of the B&B tree where a fractional solution is found. Below follows an example of modeling the TSP of (5.65) in PuLP and solving it with the CBC solver. For demonstration purposes, we do not use primal heuristics to provide better bounds and we only allow the application of Gomory cuts. The symmetric cost matrix $\mathbf{C} = \{c_{ij}\}$ is provided as input.

```python
from itertools import product
import pulp
from pulp import *
# travel costs in an upper triangular matrix
dists = [[83, 81, 113, 52, 42, 73, 44, 23, 91, 105, 90, 124, 57,
    42, 26],
        [161, 160, 39, 89, 151, 110, 90, 99, 177, 143, 193, 100,
    101, 105],
        [90, 125, 82, 13, 57, 71, 123, 38, 72, 59, 82, 86, 52],
        [123, 77, 81, 71, 91, 72, 64, 24, 62, 63, 51, 56],
        [51, 114, 72, 54, 69, 139, 105, 155, 62, 101, 81],
        [70, 25, 22, 52, 90, 56, 105, 16, 39, 42],
        [45, 61, 111, 36, 61, 57, 70, 52, 61],
        [23, 71, 67, 48, 85, 29, 31, 49],
```

```
                 [74, 89, 69, 107, 36, 52, 47],
                 [117, 65, 125, 43, 41, 50],
                 [54, 22, 84, 71, 29],
                 [60, 44, 61, 53],
                 [97, 52, 101],
                 [81, 72],
                 [69],
                 []]
# number of nodes and list of vertices
n = len(dists); N = set(range(1,n+1)); V = set(range(0,n))
c = [[0 if i == j
      else dists[i][j-i-1] if j > i
      else dists[j][i-j-1]
      for j in V] for i in V]
model = LpProblem("Traveling_Salesman_Problem", LpMinimize)
# variables
x = LpVariable.dicts("x", (N, N), None,None, LpBinary)
u = LpVariable.dicts("u", N, None,None, LpInteger)
# objective function: minimize the total cost
model += (sum(c[i-1][j-1]*x[i][j] for i in N for j in N))
# constraints:
for i in N: model += sum(x[i][j] for j in N - {i}) == 1
for i in N: model += sum(x[j][i] for j in N - {i}) == 1
for (i, j) in product(N - {1}, N - {1}):
    if i != j: model += (n-1)*x[i][j]+u[i]-u[j] <= n-2
for i in N-{1}: model += 1 <= u[i]; model += u[i] >= n-1
# optimizing
solver = pulp.PULP_CBC_CMD(mip=True, msg=True, timeLimit=None,
    warmStart=False, path=None, mip_start=False, timeMode='elapsed',
    options=["passCuts=0","preprocess=off","presolve=off","cuts=off",
    "gomoryCuts=on","heuristics=off","greedyHeuristic=off"])
model.solve(solver)
print ("Status:", LpStatus[model.status])
for v in model.variables():
    if v.varValue>0: print (v.name, "=", v.varValue)
print ("Optimal Solution = ", value(model.objective))
```

Implementing this code will solve the TSP using the branch and cut approach of the CBC solver resulting in an optimal solution with a total travel cost of 602. The detailed output of the solver is provided below, together with its explanation.

```
Problem MODEL has 272 rows, 255 columns and 1140 elements
```

This informs us that the problem has 272 constraints (rows) and 255 variables (columns).

```
Continuous objective value is 433.267 - 0.00 seconds
```

The solver found a lower bound $L^0 = 433.267$ after solving a continuous relaxation of the integer LP in 0 seconds.

```
Cbc0010I After 0 nodes, 1 on tree, 1e+50 best solution, best possible
    433.26667 (0.02 seconds)
```

The solver informs us that it has processed 0 nodes (we are still at the root node) and there is 1 node at the B&B tree (the root). The best possible solution is 433.267, which is the current lower bound L_k of our problem. The upper bound (best solution) is set to the very large value of 1e+50, meaning that $U_k = +\infty$.

```
Cbc0016I Integer solution of 650 found by strong branching after 283
    iterations and 8 nodes (0.21 seconds)
Cbc0016I Integer solution of 612 found by strong branching after 524
    iterations and 15 nodes (0.28 seconds)
Cbc0004I Integer solution of 610 found after 1162 iterations and 30
    nodes (0.41 seconds)
Cbc0016I Integer solution of 607 found by strong branching after 1775
    iterations and 44 nodes (0.50 seconds)
Cbc0004I Integer solution of 602 found after 1890 iterations and 48
    nodes (0.51 seconds)
```

The solver found the first integer solution after exploring 8 nodes of the B&B tree, and this updated our upper bound to $U = 650$. The upper bound was later updated 4 more times and it was 602 after the exploration of 48 nodes of the B&B tree.

```
Cbc0001I Search completed - best objective 602, took 2355 iterations
    and 53 nodes (0.53 seconds)
```

The solver informs us that the B&B search is completed after searching 53 nodes and the optimal solution is 602.

```
Gomory was tried 88 times and created 376 cuts of which 0 were active
    after adding rounds of cuts (0.040 seconds)
```

The solves informs us that Gomory cuts were applied 376 times, and 0 of them were active.

```
Result - Optimal solution found
Objective value:            602.00000000
Enumerated nodes:           53
Total iterations:           2355
Time (CPU seconds):         0.53
Time (Wallclock seconds):   0.53
```

This message informs us that an optimal solution is found after exploring 53 nodes in 0.53 seconds.

The provided example did not use an initial feasible solution and it assumed $U^0 = +\infty$ because we were solving a minimization problem. Implementing primal heuristics can help to find feasible solutions and reduce the explored nodes of the B&B tree since we will perform more aggressive pruning. The same applies to the use of more cutting plane methods (probing cuts, knapsack cuts, clique cuts). In the following example, we solve the same problem by allowing primal heuristics and more cut options to demonstrate this effect. The only difference is that we replace the options of the solver as follows:

```
solver = pulp.PULP_CBC_CMD(mip=True, msg=True, timeLimit=None,
    warmStart=False, path=None, mip_start=False, timeMode='elapsed',
    options=["passCuts=100","preprocess=on","presolve=off","cuts=on",
    "gomoryCuts=on","heuristics=on","greedyHeuristic=on","
    passFeasibilityPump=100"])
```

This will result in the following detailed output, which is followed by its explanation.

```
Continuous objective value is 433.267 - 0.00 seconds
```

Solving the continuous version of the TSP, the optimal solution is 433.269. Thus, $L^0 = 433.269$.

```
Cbc0038I After 0.46 seconds - Feasibility pump exiting with objective
    of 882 - took 0.45 seconds
```

The solver applied the feasibility pump heuristic of Fischetti et al. [5] and found a feasible solution with an improved upper bound $U^0 = 882$. This provides us with a much better upper bound than $U^0 = +\infty$ that we had when we did not use a primal heuristic.

```
Cbc0013I At root node, 18 cuts changed objective from 433.26667 to
    598 in 10 passes
Cbc0010I After 0 nodes, 1 on tree, 882 best solution, best possible
    598 (0.63 seconds)
```

The above message informs us that the solver, while still at the root node, implements cuts that increase the lower bound from $L^0 = 433.26667$ to $L^0 = 598$. Knowing that the best possible solution has a performance of 598 tightens the solution space. In fact, while still at the root of the B&B tree, we have $L_k = 598$ and $U_k = 882$ which is a much tighter gap compared to the $[L_k, U_k] = [433.26667, 1e+50]$ gap of the previous implementation. Then, we have:

```
Cbc0016I Integer solution of 615 found by strong branching after 921
    iterations and 13 nodes (0.91 seconds)
Cbc0016I Integer solution of 606 found by strong branching after 1262
    iterations and 21 nodes (0.96 seconds)
Cbc0004I Integer solution of 603 found after 1379 iterations and 22
    nodes (0.96 seconds)
Cbc0016I Integer solution of 602 found by strong branching after 1411
    iterations and 23 nodes (0.97 seconds)
Cbc0001I Search completed - best objective 602, took 1740 iterations
    and 29 nodes (0.98 seconds)
```

indicating that a better feasible solution is found after exploring 13 nodes of the B&B tree and the optimal solution is found after exploring 29 nodes. Note that without using primal heuristics and numerous cuts, we required 53 node explorations to solve the same problem.

Let us now consider an alternative mixed-integer formulation of the TSP problem:

$$\min \sum_{i \in N} \sum_{j \in N \setminus \{i\}} c_{ij} x_{ij}$$

$$\text{subject to:} \quad \sum_{j \in N \setminus \{i\}} x_{ij} = 1 \qquad \forall i \in N$$

$$\sum_{i \in N \setminus \{j\}} x_{ij} = 1 \qquad \forall j \in N \qquad (5.66)$$

$$y_i - (n+2)x_{ij} \geq y_j - (n+1) \quad \forall i, j \in \{2, 3, \ldots, n\} \mid i \neq j$$

$$y_i \in \mathbb{R} \qquad \forall i \in \{2, 3, \ldots, n\}$$

$$x_{ij} \in \{0, 1\} \qquad \forall i \in N, j \in N$$

where $y_i - (n+2)x_{ij} \geq y_j - (n+1)$ $\forall i, j \in \{2, 3, \ldots, n\} \mid i \neq j$ is the newly defined subtour elimination constraint that uses vector **y** of real values. This problem can be solved as follows and results in the same solution with total cost 602 after exploring 40 leaf nodes.

Practitioner's Corner
Branch and Cut in Python 3 for MILPs

```python
from itertools import product
import pulp
from pulp import *
# travel costs in an upper triangular matrix
dists = [[83, 81, 113, 52, 42, 73, 44, 23, 91, 105, 90, 124, 57,
    42, 26],
        [161, 160, 39, 89, 151, 110, 90, 99, 177, 143, 193, 100,
    101, 105],
        [90, 125, 82, 13, 57, 71, 123, 38, 72, 59, 82, 86, 52],
        [123, 77, 81, 71, 91, 72, 64, 24, 62, 63, 51, 56],
        [51, 114, 72, 54, 69, 139, 105, 155, 62, 101, 81],
        [70, 25, 22, 52, 90, 56, 105, 16, 39, 42],
        [45, 61, 111, 36, 61, 57, 70, 52, 61],
        [23, 71, 67, 48, 85, 29, 31, 49],
        [74, 89, 69, 107, 36, 52, 47],
        [117, 65, 125, 43, 41, 50],
        [54, 22, 84, 71, 29],
        [60, 44, 61, 53],
        [97, 52, 101],
        [81, 72],
        [69],
        []]
# number of nodes and list of vertices
n = len(dists); N = set(range(1,n+1)); V = set(range(0,n))
c = [[0 if i == j
    else dists[i][j-i-1] if j > i
    else dists[j][i-j-1]
    for j in V] for i in V]
model = LpProblem("Traveling_Salesman_Problem", LpMinimize)
# variables
```

```
x = LpVariable.dicts("x", (N, N), None,None, LpBinary)
y = LpVariable.dicts("u", N, None,None, LpContinuous)
# objective function: minimize the total cost
model += (sum(c[i-1][j-1]*x[i][j] for i in N for j in N))
# constraints:
for i in N: model += sum(x[i][j] for j in N - {i}) == 1
for i in N: model += sum(x[j][i] for j in N - {i}) == 1
for (i, j) in product(N - {1}, N - {1}):
    if i != j:
        model += y[i] - (n + 2) * x[i][j] >= y[j] - (n+1)
# optimizing
solver = pulp.PULP_CBC_CMD(mip=True, msg=True, timeLimit=None,
    warmStart=False, path=None, mip_start=False, timeMode='elapsed',
    options=["passCuts=100","preprocess=on","presolve=off","cuts=on",
    "gomoryCuts=on","heuristics=on","greedyHeuristic=on","
    passFeasibilityPump=100"])
model.solve(solver)
print ("Status:", LpStatus[model.status])
for v in model.variables():
    if v.varValue>0: print (v.name, "=", v.varValue)
print ("Optimal Solution = ", value(model.objective))
```

5.5 Integer and Mixed-integer Formulations of Common Combinatorial Problems

Closing this chapter, we present integer and mixed-integer formulations of common combinatorial optimization problems, such as optimization versions of some of the NP-complete decision problems of Karp [6].

Because combinatorial problems are of graph nature, it is useful to use a standardized terminology based on the following definitions. We note that from now on we refer to graphs where no edge starts and ends at the same vertex.

Definition 5.5 *Undirected Graph* is a graph $G = (V, E)$ with a set V of vertices and a set E of edges. Each edge $e \in E$ connects two vertices in the graph in both directions (i.e., (i, j) and (j, i) with $i \in V$, $j \in V$) and vertices i and j are called incident to edge e.

Definition 5.6 *Directed Graph* or *Digraph* is a graph $G = (V, A)$ with a set V of vertices and a set A of directed edges (also called arcs). Each arc $a \in A$ connects two vertices $i \in V$ and $j \in V$ in the graph with direction from i to j.

An example of an undirected and a directed graph is presented in Fig. 5.6.

We can present the edges and arcs of an undirected and a directed graph using sets. For instance, $A = \{(a, b), (b, c), (c, f), (a, d), (d, e), (e, f)\}$ for the graph in Fig. 5.6. Notice that an edge can be expressed as two antiparallel (two-way) arcs and an undirected graph can be seen as a directed graph with:

Fig. 5.6 Example of undirected graph $G = (V, E)$ and digraph $G = (V, A)$

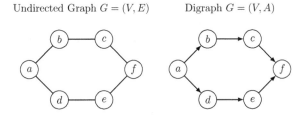

$$E = \{(a, b), (b, a), (b, c), (c, b), (c, f), (f, c), (a, d), (d, a), (d, e), (e, d), (e, f), (f, e)\}$$

We can also present the edges (or arcs) of an undirected (or directed) graph with $|V|$ vertices using an *adjacency matrix* $\mathbf{A} \in \{0, 1\}^{|V| \times |V|}$ where $a_{ij} = 1$ if $(i, j) \in E$ (or $(i, j) \in A$) and $a_{ij} = 0$ otherwise. Notice that $a_{ij} = 1$ if, and only if, i and j are adjacent vertices in graph G. The adjacency matrix for the undirected graph in Fig. 5.6 is:

$$
\mathbf{A} = \begin{array}{c}
\quad \begin{array}{cccccc} a & b & c & d & e & f \end{array} \\
\begin{array}{c} a \\ b \\ c \\ d \\ e \\ f \end{array}
\left[\begin{array}{cccccc}
0 & 1 & 0 & 1 & 0 & 0 \\
1 & 0 & 1 & 0 & 0 & 0 \\
0 & 1 & 0 & 0 & 0 & 1 \\
1 & 0 & 0 & 0 & 1 & 0 \\
0 & 0 & 0 & 1 & 0 & 1 \\
0 & 0 & 1 & 0 & 1 & 0
\end{array} \right]
\end{array}
$$

The adjacency matrix of the digraph in Fig. 5.6 is:

$$
\mathbf{A} = \begin{array}{c}
\quad \begin{array}{cccccc} a & b & c & d & e & f \end{array} \\
\begin{array}{c} a \\ b \\ c \\ d \\ e \\ f \end{array}
\left[\begin{array}{cccccc}
0 & 1 & 0 & 1 & 0 & 0 \\
0 & 0 & 1 & 0 & 0 & 0 \\
0 & 0 & 0 & 0 & 0 & 1 \\
0 & 0 & 0 & 0 & 1 & 0 \\
0 & 0 & 0 & 0 & 0 & 1 \\
0 & 0 & 0 & 0 & 0 & 0
\end{array} \right]
\end{array}
$$

Observe that the adjacency matrix of an undirected graph is *symmetric* because $a_{ij} = a_{ji}$ for any $i \in V$ and $j \in V$.

We also present the definitions of the *complete graph, simple graph, walk, trail, path, cycle,* and *circuit.*

Definition 5.7 *Complete graph* is a graph G if, and only if, for every $i \in V$ there exists some $j \in V$ such that i and j are adjacent. In the case of digraphs, we request that $(i, j) \in A$.

Definition 5.8 *Simple graph* is an undirected graph $G = (V, E)$ that does not contain any loops or multiple (parallel) edges between two vertices. That is, any vertex i is *not adjacent to itself* and it if it is adjacent to another vertex j, then there exists *only one edge* connecting i and j.

Definition 5.9 *Walk* in a graph $G = (V, E)$ is a sequence of vertices with the property that each vertex in the sequence is adjacent to the vertex next to it. In a finite walk, we have a vertex sequence $\{v_1, v_2, \ldots, v_n\}$ where v_i and v_{i+1} are incident to an edge $e_i \in E$. It follows that a walk has also an edge sequence $\{e_1, e_2, \ldots, e_{n-1}\}$. In a walk, any vertex in the vertex sequence and any edge in the edge sequence can be repeated (visited more than once).

Definition 5.10 *Trail* in a graph $G = (V, E)$ is a sequence of vertices with the property that each vertex in the sequence is adjacent to the vertex next to it and all edges in the edge sequence are distinct (visited only once). In a trail, any vertex in the vertex sequence can be repeated (visited more than once) but this does not apply to the edges in the edge sequence.

Definition 5.11 *Path* or *simple path* in a graph $G = (V, E)$ is a sequence of vertices with the property that each vertex in the sequence is adjacent to the vertex next to it and:

- Each vertex in the vertex sequence is distinct.
- Each edge in the edge sequence is distinct.

That is, a simple path is a *trail* with distinct vertices in the vertex sequence.

Definition 5.12 *Cycle* in a graph $G = (V, E)$ is a sequence of vertices with the property that each vertex in the sequence is adjacent to the vertex next to it and all edges in the edge sequence are distinct (visited only once). In addition, all vertices in the vertex sequence are distinct except for the first and the last vertices which are the same. That is, a cycle is a *trail* in which only the first and last vertices in the vertex sequence are equal.

Definition 5.13 *Circuit* in a graph $G = (V, E)$ is a sequence of vertices with the property that each vertex in the sequence is adjacent to the vertex next to it and all edges in the edge sequence are distinct (visited only once). In addition, the first and the last vertices in the vertex sequence are the same. That is, a circuit is a *trail* in which the first and last vertices in the vertex sequence are equal.

To provide an example, a walk in Fig. 5.7 is the vertex sequence $\{a, b, a\}$, a trail is the vertex sequence $\{a, b, e, c, b\}$, a simple path is the vertex sequence $\{a, b, c\}$, a cycle is the vertex sequence $\{a, b, c, f, e, d, a\}$, and a circuit is the vertex sequence $\{a, b, e, c, b, d, a\}$. Note that any cycle is also a circuit, but the reverse is not always true. This is why a cycle is also referred to as *simple circuit*.

With this terminology, we proceed to the formulation of the following common combinatorial optimization problems.

Fig. 5.7 Example of
undirected graph
$G = (V, E)$

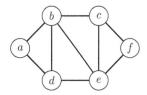

Definition 5.14 *Maximum Weight Clique Problem:* consider a weighted undirected graph $G = (V, E)$ where each vertex $i \in V$ has a positive weight c_i. Find a maximum weight clique in a weighted graph $G = (V, E)$, where a clique is a subset of vertices, all adjacent to each other.

Let n be the size of set V. Consider binary variables $\mathbf{x} \in \{0, 1\}^n$ where $x_i = 1$ if vertex $i \in V$ is part of the maximum weight clique and $x_i = 0$ otherwise. To symbolize the edges in the undirected graph, it suffices to represent the edge in one direction. For instance, every edge that connects two vertices i and j can be expressed as edge $(i, j) \in E$ implying that it connects vertices i and j in both directions. Then, the maximum weight clique formulation can be expressed as follows.

Maximum Weight Clique formulation

$$\max_{\mathbf{x}} \ \mathbf{c}^\mathsf{T}\mathbf{x} \tag{5.67}$$

$$\text{s.t.} \ x_i + x_j \le 1 \qquad\qquad \forall (i, j) \notin E \tag{5.68}$$

$$\mathbf{x} \in \{0, 1\}^n \tag{5.69}$$

The solution of this problem cannot contain any non-adjacent vertices because constraints $x_i + x_j \le 1$ for any $(i, j) \notin E$ force non-adjacent i and j to not be both part of the maximum weight clique.

The *maximum clique* problem that seeks to find the maximum possible number of vertices in V, all adjacent to each other, can be expressed by replacing the objective function $\mathbf{c}^\mathsf{T}\mathbf{x}$ with $\sum_{i=1}^{n} x_i$.

Definition 5.15 *Minimum Weight Set Cover Problem:* consider a set S with m elements and a collection V of n sub-sets of S whose union is equal to S. Find the smallest weight sub-collection of V whose union equals to S given that any sub-set i of V has positive weight c_i.

Consider binary variables $\mathbf{x} \in \{0, 1\}^n$ where $x_i = 1$ if sub-set $i \in V$ is part of the minimum weight set cover and $x_i = 0$ otherwise. Consider also parameter matrix $\mathbf{A} \in \{0, 1\}^{m \times n}$ where $a_{ei} = 1$ if element e of set S is part of sub-set i and $a_{ei} = 0$ otherwise. Then, the minimum weight set cover formulation can be expressed as follows.

Minimum Weight Set Cover formulation

$$\min_{\mathbf{x}} \ \mathbf{c}^\mathsf{T}\mathbf{x} \tag{5.70}$$

$$\text{s.t.} \ \sum_{i \in V} a_{ei} x_i \geq 1 \qquad\qquad \forall e \in S \tag{5.71}$$

$$\mathbf{x} \in \{0, 1\}^n \tag{5.72}$$

Constraints $\sum_{i \in V} a_{ei} x_i \geq 1$ for any $e \in S$ ensure that each element e of set S will be covered, e.g., be a member of, at least one sub-set $i \in V$. They are equivalent to constraints $\sum_{i \in V: e \in i} x_i \geq 1$ $\forall e \in S$ that do not use indicator parameters a_{ei}. These constraints can be compactly written as $\mathbf{Ax} \geq \mathbf{1}$.

The *minimum set cover* problem is expressed by replacing the objective function $\mathbf{c}^\mathsf{T}\mathbf{x}$ with $\sum_{i=1}^{n} x_i$.

Definition 5.16 *Minimum Weight Set Partitioning Problem:* consider a set S with m elements and a collection V of n sub-sets of S whose union is equal to S. Each sub-set i of V has positive weight c_i. Find the smallest weight sub-collection P of V whose union equals to S and $i \cap j = \emptyset$ for any $i \in P$ and $j \in P : j \neq i$.

Consider binary variables $\mathbf{x} \in \{0, 1\}^n$ where $x_i = 1$ if sub-set $i \in V$ is part of the minimum weight set partition and $x_i = 0$ otherwise. Consider also parameter matrix $\mathbf{A} \in \{0, 1\}^{m \times n}$ where $a_{ei} = 1$ if element e of set S is part of sub-set i and $a_{ei} = 0$ otherwise. Then, the minimum weight set partitioning expressed as follows.

Minimum Weight Set Partitioning formulation

$$\min_{\mathbf{x}} \ \mathbf{c}^\mathsf{T}\mathbf{x} \tag{5.73}$$

$$\text{s.t.} \ \sum_{i \in V} a_{ei} x_i = 1 \qquad\qquad \forall e \in S \tag{5.74}$$

$$\mathbf{x} \in \{0, 1\}^n \tag{5.75}$$

Constraints (5.74) ensure that each element e of set S will be covered by exactly one sub-set $i \in V$. Because of this, it is evident that the union of all selected sub-sets i of the problem's solution will be equal to set S. In addition, sub-sets $i \in V$ contain only elements of S because, by definition, the union of all sub-sets in collection V is S. Thus, constraints (5.74) also ensure that the selected sub-sets $i, j, \ldots$ of the optimal solution will satisfy $i \cap j = \emptyset$ for $i \neq j$. Constraints (5.74) can be compactly written as $\mathbf{Ax} = \mathbf{1}$.

The *minimum set partitioning* problem is expressed by replacing the objective function $\mathbf{c}^\mathsf{T}\mathbf{x}$ with $\sum_{i=1}^n x_i$.

Definition 5.17 *Maximum Weight Set Packing Problem:* consider a set S with m elements and a collection V of n sub-sets of S. Each sub-set i of V has positive weight c_i. Find the largest weight sub-collection P of V where each element e of S must appear in at most one sub-set of collection P [7].

Consider binary variables $\mathbf{x} \in \{0, 1\}^n$ where $x_i = 1$ if sub-set $i \in V$ is part of the minimum weight set packing and $x_i = 0$ otherwise. Consider also parameter matrix $\mathbf{A} \in \{0, 1\}^{m \times n}$ where $a_{ei} = 1$ if element e of set S is part of sub-set i and $a_{ei} = 0$ otherwise. Then, the maximum weight set packing formulation can be expressed as follows.

Maximum Weight Set Packing formulation

$$\max_{\mathbf{x}} \ \mathbf{c}^\mathsf{T}\mathbf{x} \tag{5.76}$$

$$\text{s.t.} \ \sum_{i \in V} a_{ei} x_i \leq 1 \qquad\qquad \forall e \in S \tag{5.77}$$

$$\mathbf{x} \in \{0, 1\}^n \tag{5.78}$$

Constraints (5.77) ensure that each element e of set S will appear in at most one sub-set $i \in V$. Because each element e of set S will appear in at most one sub-set $i \in V$, the sub-sets of the optimal solution are disjoint.

Definition 5.18 *Minimum Vertex Coloring Problem:* consider an undirected graph $G = (V, E)$. Consider also a set M with m available colors, where m is at least equal to the number of vertices, $|V|$, in G. Assign a color to each vertex $i \in V$ in such a way that colors on adjacent vertices are different and the number of colors used is minimized [8].

Consider binary variables $\mathbf{y} \in \{0, 1\}^m$ where $y_j = 1$ if color $j \in M$ is used in the optimal solution and $y_j = 0$ if not. Consider also binary variables $\mathbf{x} \in \{0, 1\}^{n \times m}$ where $x_{ij} = 1$ if vertex $i \in V$ is assigned to color $j \in M$ and $x_{ij} = 0$ otherwise. Let us express every edge that connects two vertices i and k in set E as (i, k), implying that it also connects the opposite direction (k, i). Then, the minimum vertex coloring problem formulation can be expressed as follows.

Minimum Vertex Coloring formulation

$$\min_{\mathbf{x},\mathbf{y}} \sum_{j \in M} y_j \tag{5.79}$$

$$\text{s.t.} \sum_{j \in M} x_{ij} = 1 \qquad\qquad \forall i \in V \tag{5.80}$$

$$x_{ij} + x_{kj} \leq y_j \qquad\qquad \forall (i,k) \in E, j \in M \tag{5.81}$$

$$x_{ij} \in \{0, 1\} \qquad\qquad \forall i \in V, j \in M \tag{5.82}$$

$$y_j \in \{0, 1\} \qquad\qquad \forall j \in M \tag{5.83}$$

The objective function minimizes the number of required colors. Constraints (5.80) ensure that each vertex $i \in V$ is colored by exactly one color. Constraints (5.81) ensure that at most one of a pair of adjacent vertices receives the same color. They also ensure that if a vertex i is assigned to color j ($x_{ij} = 1$), then $y_j = 1$ indicating that color j is used in the graph.

If every color $j \in M$ has an associated cost (weight) c_j, the *minimum weight vertex coloring* problem can be expressed by replacing the objective function with $\mathbf{c}^\mathsf{T}\mathbf{y}$.

Definition 5.19 *Minimum Weight Vertex Cover Problem:* consider an undirected graph $G = (V, E)$. Each vertex $i \in V$ has a positive weight c_i. Find the minimum weight vertex cover $V' \subseteq V$ such that every edge has at least one endpoint in vertex cover V'.

Consider n to be the size of set V and binary variables $\mathbf{x} \in \{0, 1\}^n$ where $x_i = 1$ if vertex $i \in V$ is in the vertex cover and $x_i = 0$ if not. Let us express every edge that connects two vertices i and j in set E as (i, j), implying that it also connects the opposite direction (j, i). Then, the minimum weight vertex cover problem formulation can be expressed as follows.

Minimum Weight Vertex Cover formulation

$$\min_{\mathbf{x}} \mathbf{c}^\mathsf{T}\mathbf{x} \tag{5.84}$$

$$\text{s.t.} \; x_i + x_j \geq 1 \qquad\qquad \forall (i, j) \in E \tag{5.85}$$

$$x_i \in \{0, 1\} \qquad\qquad \forall i \in V \tag{5.86}$$

Constraints (5.85) ensure that at least one vertex from any adjacent pair of vertices $(i, j) \in E$ in graph G is covered.

Fig. 5.8 Example of
undirected graph
$G = (V, E)$ where set V
with 14 vertices is
partitioned into disjoint and
independent sets A and B
and we have a perfect
matching M represented by
all edges in solid lines. Edges
represented by trimmed lines
are part of set $E \supset M$

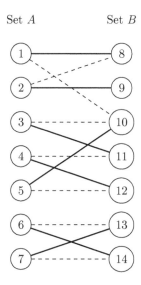

Definition 5.20 *Maximum Weight Bipartite Matching Problem:* consider a bipartite
graph $G = (V, E)$ where vertices V are partitioned into two disjoint and independent
sets (A and B) and edges E connect the vertices between these two partitioned sets.
Each edge has a positive weight c_{ij}. Find the maximum weight bipartite matching.

From the problem definition it follows that if two vertices belong to the same
partitioned set (i.e., $i \in A$ and $j \in A$), then $(i, j) \notin E$ for G to be bipartite, where
an edge (i, j) also connects the opposite direction (j, i). A matching $M \subseteq E$ is a
subset of edges such that every vertex of V is incident to at most one edge of M,
where any edge that connects i and j is said to be incident to i and j. In case any
vertex in V is incident to exactly one edge in M, then M is a *perfect matching* and
its cardinality is equal to $|A| = |B|$. An example of perfect matching is presented in
Fig. 5.8.

If $|A| \neq |B|$ we can add fictional vertices to the set with the less elements in order
to create a *balanced* bipartite graph with $|A| = |B|$. We can also add fictional edges
with a cost of minus infinity to make the bipartite graph *complete* by having $|A| \times |B|$
edges. Consider binary variables $\mathbf{x} \in \{0, 1\}^{|E|}$ where $x_{ij} = 1$ if edge $(i, j) \in E$ is
part of the perfect bipartite matching and $x_{ij} = 0$ if not. Then, the maximum weight
perfect bipartite matching problem formulation can be expressed as follows.

Maximum Weight Perfect Bipartite Matching formulation

$$\max_{\mathbf{x}} \sum_{(i,j)\in E} c_{ij}x_{ij} \qquad (5.87)$$

$$\text{s.t.} \sum_{j:(i,j)\in E} x_{ij} = 1 \qquad \forall i \in A \qquad (5.88)$$

$$\sum_{i:(i,j)\in E} x_{ij} = 1 \qquad\qquad \forall j \in B \qquad\qquad (5.89)$$

$$x_{ij} \in \{0, 1\} \qquad\qquad \forall (i, j) \in E \qquad\qquad (5.90)$$

Constraints (5.88) and (5.89) ensure that each vertex from set A is matched with exactly one vertex from set B, and vice versa. In more detail, $\sum_{j:(i,j)\in E} x_{ij} = 1$ $\forall i \in A$ ensure that for any $i \in A$ there is exactly one j in $(i, j) \in E$ such that $x_{ij} = 1$. Note that this j belongs to B because $i \in A$ and set E does not contain edges between vertices of the same partitioned set. Similarly, $\sum_{i:(i,j)\in E} x_{ij} = 1$ $\forall j \in B$ ensure that for any $j \in B$ there is exactly one $i \in A$ such that $x_{ij} = 1$.

This formulation is also known as the *Assignment Problem* where we have $|A|$ resources and $|B|$ service providers with $|A| = |B|$ and we need to assign exactly one resource to exactly one service provider and exactly one service provider to exactly one resource. Notice that we can rewrite the formulation of the maximum weight perfect bipartite matching problem as:

$$\max_{\mathbf{x}} \sum_{(i,j)\in E} c_{ij} x_{ij} \qquad\qquad\qquad\qquad (5.91)$$

$$\text{s.t.} \sum_{j:(i,j)\in E} x_{ij} = 1 \qquad\qquad \forall i \in A \qquad\qquad (5.92)$$

$$\sum_{i:(i,j)\in E} x_{ij} = 1 \qquad\qquad \forall j \in B \qquad\qquad (5.93)$$

$$0 \leq x_{ij} \leq 1 \qquad\qquad \forall (i, j) \in E \qquad\qquad (5.94)$$

$$x_{ij} \in \mathbb{Z} \qquad\qquad \forall (i, j) \in E \qquad\qquad (5.95)$$

and we can even drop the requirement that $x_{ij} \leq 1$ from constraints (5.94) since constraints (5.92) and (5.92) do not allow x_{ij} to take a value greater than 1. Importantly, we can obtain the same solution by solving the following linear program in equations (5.96)–(5.99) which drops the integrality constraints $x_{ij} \in \mathbb{Z}$ because this linear program always has an optimal solution where the variables take integer values since its constraint matrix is *totally unimodular*.

$$\max_{\mathbf{x} \in \mathbb{R}^{|E|}} \sum_{(i,j)\in E} c_{ij} x_{ij} \qquad\qquad\qquad\qquad (5.96)$$

$$\text{s.t.} \sum_{j:(i,j)\in E} x_{ij} = 1 \qquad\qquad \forall i \in A \qquad\qquad (5.97)$$

$$\sum_{i:(i,j)\in E} x_{ij} = 1 \qquad\qquad \forall j \in B \qquad\qquad (5.98)$$

$$x_{ij} \geq 0 \qquad\qquad \forall (i, j) \in E \qquad\qquad (5.99)$$

Definition 5.21 *Minimum Weight Hamiltonian Circuit Problem:* consider an undirected graph $G = (V, E)$. There is a positive weight c_{ij} associated with every edge $(i, j) \in E$. Find the minimum weight Hamiltonian circuit, where a Hamiltonian circuit is a circuit that starts and ends at the same vertex and visits every other vertex in V exactly once.

The minimum weight Hamiltonian circuit problem has the mixed-integer formulation of the Traveling Salesman Problem (see (5.66)). Notice that because the Hamiltonian circuit has to visit every vertex exactly once, it is actually a *cycle* since there will be no repeated vertices in the vertex sequence, except for the first vertex in the sequence which is the same as the last one.

Definition 5.22 *Minimum Weight Multi-commodity Flow Problem:* consider a graph $G = (V, E)$ where each edge $(i, j) \in E$ that connects both (i, j) and (j, i) has capacity u_{ij} and a positive weight c_{ij}. There is a set of K commodities where s_k and t_k is the source and sink vertex of commodity $k \in K$ and d_k its demand. Find the minimum weight flow for all commodities subject to edge capacity constraints.

Consider binary variables $x_{k,ij} \in \{0, 1\}$ indicating the flow of commodity $k \in K$ through edge $(i, j) \in E$, where $x_{k,ij} = 0$ if edge (i, j) is not part of the path of commodity k. Then, the minimum weight multi-commodity flow problem can be expressed as follows.

Minimum Weight Multi-commodity Flow formulation

$$\min_{\mathbf{x}} \quad \sum_{(i,j) \in E} c_{ij} \sum_{k \in K} x_{k,ij} \tag{5.100}$$

$$\text{s.t.} \sum_{k \in K} x_{k,ij} d_k \leq u_{ij} \qquad\qquad \forall (i, j) \in E \tag{5.101}$$

$$\sum_{j:(i,j) \in E} x_{k,ij} - \sum_{j:(i,j) \in E} x_{k,ji} = 0 \qquad \forall k \in K, \forall i \in V \setminus \{s_k, t_k\} \tag{5.102}$$

$$\sum_{j:(s_k,j) \in E} x_{k,s_k j} - \sum_{j:(s_k,j) \in E} x_{k,js_k} = 1 \quad \forall k \in K \tag{5.103}$$

$$\sum_{i:(i,t_k) \in E} x_{k,it_k} - \sum_{i:(i,t_k) \in E} x_{k,t_k i} = 1 \qquad \forall k \in K \tag{5.104}$$

$$\mathbf{x} \in \{0, 1\}^{|K| \times |E|} \tag{5.105}$$

The objective function minimizes the weighted flow of all commodities. Constraints (5.101) ensure that the capacity of any edge $(i, j) \in E$ is not violated. Constraints (5.102) ensure that the incoming flow of a commodity k at any transit vertex $i \in V \setminus \{s_k, t_k\}$ is equal to the outgoing flow of that commodity, ensuring the flow conservation. Constraints (5.103) enforce that commodity k should exit the source vertex by using an edge but not return to it. Constraints (5.104) enforce that commodity k should return to the sink vertex and not use an outgoing edge after that.

It is worth noting that this formulation assumes that the flow $x_{k,ij}$ is a binary variable, meaning that commodity k will be fully transported through an edge (i, j) or it will not use edge (i, j) at all. That is, commodity k will be transported through a single path. If we allow a fraction of commodity k to be transported through a path and another fraction through another path, we have to use continuous variables $x_{k,ij}$ describing the flow of k through (i, j), where $x_{k,ij} \in \mathbb{R}$ and $0 \le x_{k,ij} \le 1$. This will result in a linear programming formulation of the minimum weight multi-commodity flow problem.

Exercises

5.1 Branch and Bound and Convexity
Can we solve a discrete optimization problem with a convex continuous relaxation to global optimality with branch and bound? Justify your answer.

5.2 Branch and Bound
Write a branch and bound algorithm and solve the following mixed-integer program.

$$\max \ 3x_1 + 20x_2^2 + 5x_3$$
$$\text{s.t. } 4x_1 + 2x_2 \le 22$$
$$x_1 + 2x_2 + x_3 \le 9$$
$$x_1 + x_2 \le 5$$
$$x_1, x_2 \in \mathbb{Z}_+, x_3 \in \mathbb{R}$$

5.3 Valid Inequalities
Propose a valid inequality for the following integer linear program.

$$\max \ 2x_1 + 3.5x_2 + 3.5x_3$$
$$\text{s.t. } 4.5x_1 - 7.5x_2 + 2.2x_3 + x_4 = 22$$
$$x_1 + 4.5x_2 + x_3 = 9$$
$$2.5x_1 + x_2 + x_5 = 12$$
$$x_1, x_2, x_3, x_4, x_5 \in \mathbb{Z}_{\ge 0}$$

5.4 Maximum value Knapsack problem
Consider a set of objects with weights $\mathbf{w} = [w_1, \ldots, w_n]^\mathsf{T}$ and values $\mathbf{p} = [p_1, \ldots, p_n]^\mathsf{T}$. One wants to select a subset with maximum profit such that the summation of the weights of the selected items is less or equal to the knapsack's capacity c. Formulate this problem as an integer linear program. Then, solve

this problem with branch and bound using the following parameter values: $c = 47$, $\mathbf{p} = [10, 13, 18, 31, 7, 15]^\intercal$, $\mathbf{w} = [11, 15, 20, 35, 10, 33]^\intercal$.

5.5 Maximum Independent Set

Consider an undirected graph $G = (V, E)$. An independent set S in G is a set of vertices, no two of which are adjacent (if $i \in S$ and $j \in S$, then $(i, j) \notin E$). Formulate the maximum independent set as an integer linear program, which seeks to find the independent set of largest possible size in G. Describe the steps of solving this problem with branch and bound.

5.6 Hithcock's Transportation Problem

Suppose there are 3 factories that supply a product to 4 cities. Due to freight rates, the cost of supplying a unit of the product to a particular city varies according to which factory supplies it, and it also varies from city to city. These costs are:

$$
\mathbf{c} = \begin{bmatrix} 12 & 11 & 7 & 6 \\ 5 & 2 & 9 & 3 \\ 7 & 5 & 6 & 9 \end{bmatrix}
$$

In addition, the total products shipped from factory 1 are 11, from factory 2 are 6, and from factory 3 are 5. The total products that must be supplied to every city 1,2,3,4 are 6,8,4, and 4, respectively. Consider that the amount of product x_{ij} transported from a factory i to a city j is a natural number. Formulate this problem as an integer program and solve it with branch and bound having as objective the minimization of shipping costs.

5.7 Path and Hamiltonian Path

Consider the undirected graph in the following figure. Provide the vertex sequence of a simple path and a Hamiltonian path in this figure. Provide also the vertex sequence of a cycle and a Hamiltonian cycle (Fig. 5.9).

Fig. 5.9 Example of undirected graph $G = (V, E)$

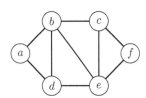

References

1. L.A. Wolsey, *Wiley Encyclopedia of Computer Science and Engineering* (2007), pp. 1–10
2. A.H. Land, A.G. Doig, in *50 Years of Integer Programming 1958–2008* (Springer, 2010), pp. 105–132
3. R. Gomory, in *Bulletin of the American Mathematical Society* (1958)
4. R. Gomory, *An Algorithm for the Mixed Integer Problem* (Tech. rep, RAND CORP SANTA MONICA CA, 1960)
5. M. Fischetti, F. Glover, A. Lodi, Math. Program. **104**(1), 91 (2005)
6. R.M. Karp, in *Complexity of Computer Computations* (Springer, 1972), pp. 85–103
7. X. Delorme, X. Gandibleux, J. Rodriguez, Eur. J. Oper. Res. **153**(3), 564 (2004)
8. E. Malaguti, M. Monaci, P. Toth, Dis. Optim. **8**(2), 174 (2011)

Part II
Solution Approximation with Artificial Intelligence: The Case of Metaheuristics

As of now, we have mostly explored numerical (iterative) optimization approaches to solve continuous and discrete optimization problems. These approaches are deterministic in the sense that they will always produce the same output (solution) when provided with the same input and provide, subject to certain conditions, guarantees regarding local or global optimality. In many complex mathematical formulations, however, (i.e., non-convex problems) it is very time-consuming, if not practically impossible, to converge to an optimal solution. In such situations, it is often the case that we need to compute a solution within a reasonable time and we are willing to compromise with attaining a suboptimal solution. This is typically achieved by using artificial intelligence approaches expressed in the form of metaheuristics that involve stochastic processes when searching the solution space.

Metaheuristics are master processes that produce solutions following stochastic steps during their search. This allows them to search the solution space faster, but there is no guarantee of optimality when applying such methods given their stochastic search nature. The use of artificial intelligence methods is very common when solving NP-Hard optimization problems that cannot be solved with numerical optimization methods for large-scale instances. Many public transport problems, such as the vehicle scheduling problem, the vehicle routing problem, or the dial-a-ride problem belong to this category. This part of the textbook will present metaheuristic approaches for continuous and discrete optimization problems. It will also present the basics of multi-objective optimization theory and metaheuristics tailored to multi-objective optimization problems which are more complex than single-objective ones. After completing this part of the textbook, the reader is expected to be able to:

- Develop metaheuristic solution methods to solve optimization problems of continuous and discrete nature.
- Familiarize himself/herself with open-source packages that implement metaheuristic solution methods and be able to apply them to practical problems.
- Formulate mathematical optimization problems with multiple objectives and solve them using appropriate solution methods.

- Understand the advantages and disadvantages of metaheuristics and be able to identify which metaheuristics are best suited for specific problem formulations.

Chapter 6
Metaheuristics

Abstract Non-convex problems are notoriously hard to solve. Many of these problems are NP-Hard and cannot be solved to global optimality within a reasonable time. This chapter presents metaheuristic approaches that return suboptimal solutions to such problems. It also clarifies the differences between heuristics and metaheuristics. Metaheuristics can be used to provide a fast, suboptimal solution to a complex problem. These solutions can be adopted by the decision maker or they can be used as an initial solution guess when applying branch and bound. Classic approaches for discrete and continuous optimization problems are described, including genetic algorithms, simulated annealing, tabu search, differential evolution, and particle swarm optimization.

6.1 Heuristics vs Metaheuristics

To better understand meta-heuristics, we first provide a brief discussion on heuristics. This will allow understanding the differences between heuristics and meta-heuristics which—oftentimes—are not easily distinguishable. In literature, there are many definitions of heuristics and meta-heuristics that may make the two hard to distinguish. A practical and useful distinction—although not always valid—is that heuristics, similarly to exact optimization approaches, return always the same solution when applied to the same problem. That is, if we have a specific problem P with specific input I and we apply a specific heuristic H, we will get the same solution every time we apply it. This does not necessarily mean that the computed solution is the globally optimal solution of the optimization problem: it merely means that the solution-finding process is deterministic, and thus stable. Many heuristics, such as Dijkstra or best-first search, follow a deterministic search process, as defined below.

Definition 6.1 A deterministic search process (algorithm) is a process that, given a particular input, will always produce the same output, with the underlying search always passing through the same sequence of states.

In addition, heuristics are constructed for specific problems, exploiting their special nature. In contrast, meta-heuristics are problem-independent and do not exploit specific problem attributes. Furthermore, meta-heuristics involve stochastic implementation processes and if a meta-heuristic M is applied to problem P with input I multiple times, it might generate different solutions [1].

There are two important observations here. First, although the name *heuristic* may suggest otherwise, heuristic approaches might solve some specific problems to global optimality and offer guarantees for it since their solutions are not changing between repeated runs in case of following deterministic search processes. On the other side, meta-heuristics cannot guarantee (theoretical) convergence to a globally optimal solution even for specific problems because their solutions are not stable (they might vary from one algorithmic run to another). Given the above, one might wonder why using a meta-heuristic if its search process is not stable. The main reason for this is that the involvement of stochastic processes in the implementation of meta-heuristics allows them to avoid getting trapped to local optima and find solutions which are closer to the global optimum—something very important when solving non-convex problems. When dealing with such problems, many heuristics are trapped to local optima and cannot escape from them due to their deterministic search rules. This is why many heuristics are used as sub-routines in metaheuristic algorithms.

More formal definitions of heuristics and meta-heuristics follow below. Note that there are numerous definitions of heuristics and metaheuristics. The ones provided are just a reference that tries to underline their differences.

Definition 6.2 "*Heuristics* are criteria, methods, or principles for deciding which among several alternative courses of action promises to be the most effective in order to achieve some goal. They represent compromises between two requirements: the need to make such criteria simple and, at the same time, the desire to see them discriminate correctly between good and bad choices. A heuristic may be a rule of thumb that is used to guide one's action [2]."

Definition 6.3 "A *meta-heuristic* is an iterative master process that guides and modifies the operations of subordinate heuristics to efficiently produce high-quality solutions. It may manipulate a complete (or incomplete) single solution or a collection of solutions at each iteration. The subordinate heuristics may be high (or low) level procedures, or a simple local search, or just a construction method [3]."

In the next section of this chapter, we will present commonly-used meta-heuristics in optimization problems. Many of them use heuristics as sub-routines. To provide a practical understanding of heuristics, we present here the *hill-climbing* heuristic which is one of the most general-purpose heuristics. Let us consider a maximization problem:

$$\max_{\mathbf{x} \in X} f(\mathbf{x})$$

where X is the feasible region and $\mathbf{x}$ an n-valued column vector. Given a starting point $\mathbf{x}$, hill climbing attempts to maximize (or minimize) a target function $f(\mathbf{x})$ (i.e., an objective function of an optimization problem) by creating neighbor solutions via adjusting the value of variables in **x**. The term *neighbor solutions* is used to describe solutions that are very close to the current solution (e.g., a small number of variables are mildly modified). If **x** is the current solution, an example neighborhood $N(\mathbf{x})$ can be all feasible solutions $\mathbf{x}'$ contained in the n-dimensional closed unit ball centered at **x**:

$$N(\mathbf{x}) := \{\mathbf{x}' \in X \; : \; \|\mathbf{x}' - \mathbf{x}\|_2 \le 1\}$$

The right-hand-side of the inequality can be replaced by a larger or lower value if we want to search in a larger or a smaller neighborhood.

After the neighbor solutions of the current solution are generated, hill climbing follows a *greedy approach* by selecting the neighbor which is a feasible solution of the optimization problem and has the best performance compared to all other feasible neighbors and the current solution. The best-performing neighbor replaces the current solution and this process continues until reaching an iteration where we cannot find a neighbor that improves the current solution.

From the aforementioned description, it is clear that hill-climbing follows a deterministic set of rules based on a greedy search and, when provided with the same input, it will return the same output. However, hill-climbing will terminate once it finds a solution with inferior neighbors, even if this solution is very far from the global optimum.

Let us consider, for instance, the application of hill-climbing to the following discrete optimization problem:

$$\min_{\mathbf{x}\in\mathbb{Z}^2} -10x_1 + x_2^2 \quad \text{s.t.} \quad \begin{aligned} x_1 + x_2 &= 6 \\ 2 \le x_1 &\le 5 \\ 2 \le x_2 &\le 7 \end{aligned}$$

We can construct a hill-climbing method that starts from an initial solution guess $[x_1, x_2]^{\mathsf{T}} = [2, 4]^{\mathsf{T}}$ with performance $-10 \cdot 2 + 4^2 = -20 + 16 = -4$ and searches the performance of its neighbors. Let us define the neighbors of $\mathbf{x} = [x_1, x_2]^{\mathsf{T}}$ as the set of points which have a -1 or $+1$ value for each variable of the current solution. This will give us neighbor points (1,3), (1,4), (1,5), (2,3), (2,4), (2,5), (3,3), (3,4), (3,5). From all these neighbors, neighbor 5 is the same as the current solution and neighbors 1–4, 6, 8, 9 are infeasible. The 7th neighbor is a feasible point and has performance $-10 \cdot 3 + 3^2 = -30 + 9 = -21$ which is better than the performance of the current solution. Replacing the current solution by $[3, 3]^{\mathsf{T}}$ we proceed to the next iteration with neighbors (2,2), (2,3), (2,4), (3,2), (3,3), (3,4), (4,2), (4,3), (4,4). Neighbors 1, 2, 4, 6, 8, 9 are infeasible and neighbor 5 is the current solution. The remaining neighbors 3 and 7 have performances -4 and -36, respectively. Thus, the best-performing neighbor is $[4, 2]^{\mathsf{T}}$ and it updates the current solution. In the next iteration the neighbors of the current solution are (3, 1), (3, 2), (3, 3), (4, 1), (4, 2), (4, 3), (5, 1), (5, 2), (5, 3). Neighbors 1–4, 6–9 are infeasible and neighbor 5 is the same as the current solution. Thus, hill-climbing terminates with solution $[x_1, x_2]^{\mathsf{T}} = [4, 2]^{\mathsf{T}}$.

Notice that we do not know whether this solution is a globally optimal one. Notice also that if we run this algorithm repeatedly, we will always get solution $[x_1, x_2]^\mathsf{T} = [4, 2]^\mathsf{T}$ when the input is the same (that is, the starting point is $[2, 4]^\mathsf{T}$). The latter is not always guaranteed when applying meta-heuristics.

In the remainder of this chapter, we will focus on meta-heuristics. We will start with metaheuristics that are mostly applied to discrete optimization problems (genetic algorithms, simulated annealing, tabu search) and then move to metaheuristics that are mostly applied to continuous optimization problems (differential evolution, particle swarm optimization).

6.2   Genetic Algorithms

Genetic algorithms (GAs) are metaheuristics developed by John Holland in the 1960s and 1970s [4] that are inspired by Charles Darwin's theory of *natural evolution*. They reflect the process of *natural selection*: the fittest individuals are selected for reproduction to produce the next generation. A GA requires a general representation of the search space and a *fitness function* to evaluate the performance of different points of the search space.

One can think of the fitness function as the objective function of the optimization problem in unconstrained optimization. A GA starts from an initial population of individuals that are points of the solution space. These individuals are sometimes called candidate solutions or phenotypes. Because individuals in a population act like independent agents, the population (or any subgroup) can explore the search space in many directions simultaneously. This enables parallelism, one of the key advantages of a GA. GA is an iterative process which performs the following steps leading to the population's evolution:

- *Selection*: Selection of individuals that will be reproduced based on their fitness.
- *Crossover*: The process of mating two individuals.
- *Mutation*: The process of mutating an individual.

We should note here that GAs are search strategies and there is no standardized approach of performing the aforementioned steps. In fact, encoding the individuals of the population, defining a selection, crossover and mutation strategy, and defining the algorithm's termination criteria are important decisions that can considerably affect the performance of the algorithm. For this reason, GAs need careful tailoring to the optimization problem that they are applied to and there can be several trials with different encoding schemes, parameter values and termination criteria to develop a GA that performs well for a specific problem. This process is known as sensitivity analysis.

Because GAs require considerable tailoring to the problem they try to solve, there is not a single GA version that overperforms all others at all times. Instead, GAs should be seen as abstract solution strategies. In this section we will discuss the main elements of a GA, which need to be tweaked when solving different problems.

Before proceeding further, we report the main disadvantages of GAs:

- The repeated evaluation of the fitness function, which can be computationally expensive.
- The required tuning of a GA to a specific problem that involves several tests with different encoding strategies, parameter values and termination criteria.
- It is not easy to define a termination criterion that provides some assurance of finding a solution close to the globally optimal one.

A general GA framework is provided below. Note that this framework is abstract and needs tailoring and tuning in order to be problem-specific (Algorithm 27).

Algorithm 27 Basic GA framework

1: Decide about the encoding mechanism of chromosomes and the fitness function.
2: Generate an initial random population of $\{1, 2, ..., |P|\}$ individuals.
3: **repeat**
4:    Select the $k \leq |P|$ fittest individuals for reproduction
5:    Crossover pairs of the selected individuals to generate offsprings
6:    Perform mutation to population's individuals
7: **until** a termination criterion is satisfied

6.2.1 Encoding

Let us discuss in more detail the main elements of a genetic algorithm. We first have to encode our population by using information from the optimization problem. Every individual of the population is characterized by the selected values of a set of decision variables, known as *genes*. Genes are joined into a *string* to form a *chromosome*. A chromosome encapsulates the values of all decision variables of an individual and can be seen as a potential solution of the optimization problem (that is, a point in the search space).

An example of the encoding of a chromosome as a bit string is presented in Fig. 6.1. Each chromosome is a population member (individual). Because every chromosome can be seen as a string that resembles a row vector, the entire population can be seen as a matrix. If the size of the population is $m = |P|$ and each chromosome has n genes, then the population is an $m \times n$ matrix.

6.2.1.1 Encoding for Binary Problems

It is important to note that a chromosome, which refers to a candidate solution of an optimization problem, can be encoded in different ways. The simplest encoding of

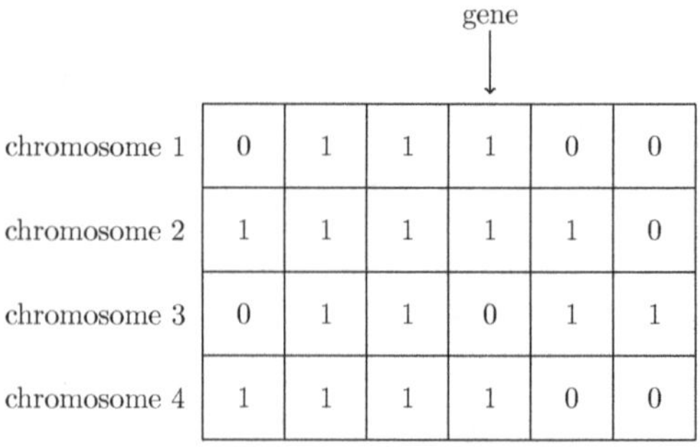

Fig. 6.1 Example of the encoding of a population with $m = 4$ members as a bit string

gene

chromosome 1	0	1	1	1	0	0
chromosome 2	1	1	1	1	1	0
chromosome 3	0	1	1	0	1	1
chromosome 4	1	1	1	1	0	0

a chromosome as a bit-string occurs when we solve a binary optimization problem. Consider :

$$\max_{\mathbf{x} \in \{0,1\}^n} f(\mathbf{x}) \quad \text{s.t. } \mathbf{h}(\mathbf{x}) = \mathbf{0}, \ \mathbf{g}(\mathbf{x}) \leq \mathbf{0}$$

This optimization problem has n binary decision variables and each chromosome can be encoded as a set of n genes, each one taking the value of 0 or 1. Let us say that our initial population P has $|P|$ individuals. Each chromosome $\mathbf{y} \in \{1, ..., |P|\}$ will have n binary values, where $\mathbf{y} \in \{0, 1\}^n$. This is the most trivial encoding representation. If the number of binary decision variables is $n = 6$ and the number of population members is $m = 4$, this encoding will resemble Fig. 6.1.

6.2.1.2 Encoding for Integer Problems

In integer optimization problems, a chromosome can be represented by genes that take integer values (not necessarily 0–1 values). Let us consider, for instance, the following integer optimization problem:

$$\max_{\mathbf{x} \in \mathbb{Z}^n} f(\mathbf{x}) \quad \text{s.t. } \mathbf{h}(\mathbf{x}) = \mathbf{0}, \ \mathbf{g}(\mathbf{x}) \leq \mathbf{0}$$

In this case, when generating the chromosome of an individual we can have n genes, where each gene receives an integer value. An example for a problem with 4 population members and $n = 6$ is provided in Fig. 6.2. Notice that if we allow any gene to take any arbitrary integer value we cannot ensure that the generated chromosome will be a feasible solution to the optimization problem. In fact, chromosomes might be infeasible and improve their performance in terms of feasibility in subsequent population evolutions. This applies in all cases (i.e., when solving binary, integer or continuous problems).

If we want to use bit-strings to encode integer numbers, this is also possible. Let us consider, for instance, that each variable $x_i \in \mathbf{x}$ is an integer that can take a value in the range 0 to 850. Then, x_i can be represented by a $\lceil \log_2 850 \rceil = 10$-bit string. For instance, 850 can be encoded as a bit-string 1101010010 and decoded as:

Fig. 6.2 Example of the encoding of a population with $m = 4$ members and $n = 6$ integer decision variables

gene

chromosome 1	109	71	15	252	91	26
chromosome 2	17	21	91	10	11	27
chromosome 3	22	49	58	31	49	52
chromosome 4	81	93	17	101	108	51

$$1 \cdot 2^9 + 1 \cdot 2^8 + 0 \cdot 2^7 + 1 \cdot 2^6 + 0 \cdot 2^5 + 1 \cdot 2^4 + 0 \cdot 2^3 + 0 \cdot 2^2 + 1 \cdot 2^1 + 0 \cdot 2^0 = 850$$

If integers are allowed to take negative values, i.e., $-850 \le x_i \le 850$, then we can add one more bit at the left of the bit-string denoting the sign. For instance, the 11-bit string 01101010010 denotes 850 and the 11-bit string 11101010010 denotes -850, where the left-most bit is 0 when we have a plus sign and 1 when we have a minus sign. Notice that if we have n variables, we need a bit-string with $11n$ bits to represent a chromosome (1 byte = 8 bits). In contrast, in binary optimization problems each chromosome was represented by n bits.

6.2.1.3 Encoding for Continuous Problems

For real-valued continuous optimization problems, one can use a floating-point chromosome representation. A floating-point number is represented with a fixed number of significant digits (the *significand*) and scaled using an exponent in some fixed base (base of 2 or 10). For instance, 2.3454 can be written as:

$$2.3454 = \underbrace{23454}_{\text{significand}} \cdot \underbrace{10^{-4}}_{\text{exponent}}$$

With this representation, each chromosome is represented by a floating-point vector instead of a bit-string. This has significant consequences, however, since the crossover and mutation steps are based on bit-strings and should be modified to account for floating points. For example, the mutation operation can choose a floating-point number within a particular range to change the value of a gene instead of flipping a bit [5].

It is worth noting that we can use a bit-string representation even for floating points. Using the IEEE 754-1985 standard, we can represent a floating-point number with single precision in 32 bits, where the left-most bit is the sign of the number, bits 2–9 are the exponent bits (with bias $2^7 - 1 = 127$), and bits 10–32 are the fraction.

If we denote the sign bit with s, the exponent with e and the fraction with f, the value of a represented number is $(-1)^s \cdot 1.f \cdot 2^{e-127}$. Consider, for example, the 32-bit string 00111110 00100000 00000000 00000000. We have:

- sign $(-1)^s = (-1)^0 = +1$
- $e = 0 \cdot 2^7 + 1 \cdot 2^6 + 1 \cdot 2^5 + 1 \cdot 2^4 + 1 \cdot 2^3 + 1 \cdot 2^2 + 0 \cdot 2^1 + 0 \cdot 2^0 = 124$
- $2^{e-127} = 2^{-3} = 0.125$
- $1.f = 1 + \sum_{i=1}^{23} b_{9+i} 2^{-i} = 1 + 1 \cdot 2^{-2} = 1.25$

Thus, this bit-string is decoded as $(-1)^s \cdot 1.f \cdot 2^{e-127} = +1 \cdot 1.25 \cdot 0.125 = 0.15625$. Notice that with a 32-bit encoding, a chromosome with n variables will be a $32n$ bit-string. If we use a double precision 64-bit encoding, then the chromosome will be a $64n$ bit-string.

6.2.2 Fitness Function

The next step is to create the fitness function. In an unconstrained maximization problem,

$$\max f(\mathbf{x})$$

the fitness function can be the objective function. We can use the objective function as the fitness function even if the problem has integrality constraints (i.e., it is a discrete optimization problem).

Let us now consider the general case of a constrained optimization problem where all equality constraints are turned into inequalities:

$$\max y(\mathbf{x}) \text{ s.t. } \mathbf{g}(\mathbf{x}) \leq \mathbf{0}$$

In constrained optimization problems, we have more options. For instance, one option is to set the fitness function as:

$$f(\mathbf{x}) = \begin{cases} y(\mathbf{x}) & \text{if } \mathbf{x} \text{ is feasible.} \\ -\infty & \text{otherwise.} \end{cases}$$

By using this fitness function, every chromosome $\mathbf{x}$ for which $\mathbf{g}(\mathbf{x}) > \mathbf{0}$ will have a fitness of $-\infty$, practically deeming infeasible chromosomes as unfit options. The problem with this fitness function is that all infeasible chromosomes are treated equally, although some infeasible chromosomes might be closer to a feasible solution than others. If we wish to reward chromosomes that are closer to a feasible solution, then we can use a quadratic penalty function that penalizes infeasibilities:

$$f(\mathbf{x}) = \begin{cases} y(\mathbf{x}) & \text{if } \mathbf{x} \text{ is feasible.} \\ v - \mu \sum_{i=1}^{m} \max\{0, g_i(\mathbf{x})\}^2 & \text{otherwise.} \end{cases}$$

where v is a constant which ensures that an infeasible solution has lower fitness value than any feasible solution and m is the number of constraints. For instance, if all feasible solutions belong to set $\mathcal{F}$, then v takes a value such that $v < \inf \mathcal{F}$. In addition, μ is a positive number that penalizes progressively infeasible solutions by reducing their fitness values in proportion to the degree of constraint violation. Note that if $g_i(\mathbf{x}) \leq 0$, then there is no added penalty to the fitness function because $\max\{0, g_i(\mathbf{x})\} = 0$, whereas if $g_i(\mathbf{x}) > 0$ this constraint violation reduces the value of the fitness function by $-\mu g_i(\mathbf{x})^2$.

There are different versions of penalty functions that can be employed, including adaptive penalty functions [6].

6.2.3 Selection for Reproduction

After the encoding phase and the generation of a number of individuals equal to the required population size, the next step is to select individuals for reproduction. A typical strategy is to select the *fittest individuals* and let them reproduce to pass their genes to the next generation. In the context of an optimization problem, the fittest individuals are the individuals with the best fitness function scores. Some of the approaches to select the fittest individuals are:

- Truncation selection: taking the best half, third or another proportion of the individuals.
- Roulette-wheel selection: individuals are chosen based on a probability proportional to their fitness.
- Tournament selection: running "tournaments" among a few individuals chosen at random from the population. The winner of each tournament (the one with the best fitness) is selected for crossover.
- Elitism selection: a small portion (i.e., 1%) of the best individuals from the population is carried over to the next generation without any changes.

One can choose truncation, roulette-wheel, or tournament selection to select a portion of the population's individuals that will be later used as parents for performing crossovers and generating offsprings. The portion of the population's individuals that will be selected for generating offsprings is a parameter of the GA (i.e., we can set it to 30%, 50% or any other percentage). Note that the elitism selection can be used in combination with one of the above because it does not refer to the selected individuals for generating offsprings, but to the best individuals from this generation that will be passed on to the next generation without changes. GAs that use elitism selection are called *elitist* GAs because they have an important difference compared to *standard* GAs: their best solution cannot deteriorate from generation to generation.

Although truncation and tournament selection are self-explanatory, roulette-wheel selection requires more details, especially in the case of individuals with negative fitness values. When using the *roulette wheel* selection in a maximization problem, each individual $i \in P$ has probability to be selected for reproduction:

$$p_i = \frac{f_i}{\sum\limits_{j \in P} f_j},$$

where:

- f_i is the fitness score of individual i
- P the set of all population members

Because some individuals (chromosomes) might have negative fitness values, the selection probability formula of individual i can be adjusted as:

$$p_i = \frac{f_i + |\inf\limits_{k \in P} f_k|}{\sum\limits_{j \in P} (f_j + |\inf_{k \in P} f_k|)}$$

This process is called *normalization* and removes all negative numbers by adding the absolute term of the most negative fitness value to the roulette wheel formula. Note that this will give a selection probability of 0 to the individual with the worst performance.

6.2.4   Crossover

After selecting the fittest individuals for reproduction, we proceed to the crossover step (also known as *recombination*). In this step, a GA stochastically generates new solutions (*offsprings*) from two existing individuals (*parents*) that are chosen for reproduction. This mimics the crossover that happens during reproduction in biology. Crossover is an abstract term, and there are several different types of crossover strategies. Which crossover strategy is better, depends on the characteristics of the optimization problem.

Two common crossover strategies, the *single-point* crossover and the *uniform* crossover, are presented in Figs. 6.3 and 6.4. As can be seen in these figures, the single-point crossover selects randomly a specific point and alters the genes of each chromosome after that point by replacing them with the genes of the other chromosome. In contrast, in uniform crossover each offspring is generated by two parents by choosing any of its genes from either parent with equal probability. Other strategies that generalize the single-point crossover by adding more crossover points are the *two-point* crossover and the *k-point* crossover, where k is an integer less than n.

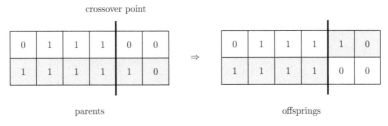

Fig. 6.3 Example of single-point crossover

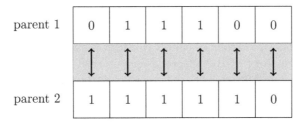

Fig. 6.4 Example of uniform crossover: each gene of the generated offspring can be chosen by either parent with equal probability

Because of the randomness of the crossover process, a generated offpsring might have inferior performance compared to its parents. When using the *plus-selection* mechanism from evolution strategies, a generated offspring is kept only if its fitness is not inferior to the fitness of its parents. If not, it is discarded and we perform another crossover.

Except for the crossover type, another important GA parameter is the *crossover probability*, which determines the chance of an existing solution (parent) to pass its genes to new trial solutions (offsprings). A randomly generated floating-point value is compared to the crossover probability, and if it is less than the probability, crossover is performed; otherwise, the offsprings are identical to the parents. If the crossover probability is 0, we have no mating (we simply skip the crossover step) [7].

6.2.5 Mutation

After crossover, the next step is the mutation of individuals. Mutation is applied to maintain genetic diversity from one generation of a population to the next. Mutation is used to explore other parts of the solution space and avoid local optima traps. Typically, the value of each gene of a chromosome can change based on a very small mutation probability (Fig. 6.5). If we use a 0–1 encoding, a mutation scheme is the bit-flip where a gene with a value of 0 is turned to 1, and vice versa. Other mutation schemes are swap mutation or scramble mutation.

Fig. 6.5 Illustration of
mutation process

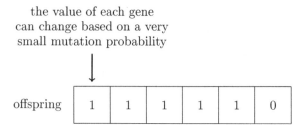

the value of each gene
can change based on a very
small mutation probability

offspring

1	1	1	1	1	0

The mutation probability is typically very small in order to avoid widespread changes. A commonly used probability for the mutation of every bit is $\frac{1}{n}$, where n is the number of bits of a chromosome [8]. That is, if we have a mutation probability of 0.01 and a bit-string of 100 bits, on average we will be changing the value of 1 bit. In dynamic implementations, one can increase the mutation probability if the best-found solution is not improving from generation to generation.

6.2.6 Population Evolution

GAs perform the steps of selecting the fittest individuals for reproduction, crossover, and mutation in order to create the population of the next generation (population evolution). Each generation can be seen as an iteration of the GA which includes the selection, crossover, and mutation steps. Typically, if our initial population has m individuals, we would like to pass m individuals to the next generation.

As previously discussed, in elitist GAs a small portion of the best individuals can be passed on to the next generation without performing any changes to them. The total number of allowed population generations can be one of the termination criteria of a GA, as described below.

6.2.7 Termination Criteria

Because GAs are metaheuristics and cannot guarantee the computation of a globally optimal solution, a key issue is when to terminate them. In practice, one might select one or more criteria to trigger the termination of a GA:

- Termination after reaching a pre-defined number of population evolutions;
- Termination if the best-found solution has reached a plateau such that successive generations no longer produce better results;
- Termination when the pre-defined maximum computation time is reached;
- Combinations of the above.

Note that the above-mentioned termination criteria do not guarantee optimality. Summarizing, typical parameters of a GA include:

- Population size.
- Elitism ratio.
- Parent selection ratio (and selection type).
- Crossover probability (and crossover type).
- Mutation probability (and mutation type).
- Maximum number of population evolutions.
- Maximum number of iterations without improvement.

Because a GA algorithm involves stochastic processes, implementing a GA with the same parameter values to the same problem might result in different results. In addition, its parameters need to be tuned to improve its performance for a particular problem and we should perform a sensitivity analysis to explore the sensitivity of the produced solutions to the changes in the parameter values.

Let us consider an example with a GA implementation at a binary constrained optimization problem: the 0–1 knapsack problem where we have n items with a cost value $\mathbf{c} = [c_1, ..., c_n]^\mathsf{T}$ and a weight $\mathbf{a} = [a_1, ..., a_n]^\mathsf{T}$. The objective is to select the most valued items that have a total weight less than the weight of the knapsack, b. Introducing binary decision variables $\mathbf{x} \in \{0, 1\}^n$, we can formulate this problem as:

$$\max_{\mathbf{x}} \ \mathbf{c}^\mathsf{T}\mathbf{x}$$
$$\text{s.t. } \mathbf{a}^\mathsf{T}\mathbf{x} \le b \qquad\qquad (6.1)$$
$$\mathbf{x} \in \{0, 1\}^n$$

As a binary problem, this is a hard-to-solve optimization problem (it is actually NP-Hard). To apply a GA, we first need to encode the chromosomes. An intuitive encoding, given that we solve a binary problem, is to have n genes at any chromosome $\mathbf{y}$ where each gene can take a 0 or 1 value. In this n-valued bit-string, the i-th gene corresponds to variable x_i and it can take a value of 0 or 1. If our population is of size m, then we have a matrix $m \times n$ representing all individuals in the population. The next step is to create the fitness function. Since we solve a constrained optimization problem, one option for our problem is to set the fitness function as:

$$f(\mathbf{y}) = \begin{cases} \mathbf{c}^\mathsf{T}\mathbf{y} & \text{if } \mathbf{y} \text{ is feasible } (\mathbf{a}^\mathsf{T}\mathbf{y} \le b). \\ \nu - \mu \max\{0, (\mathbf{a}^\mathsf{T}\mathbf{y} - b)\}^2 & \text{otherwise.} \end{cases}$$

with $\mu = 10^2$. Because $\inf_{\mathbf{x} \in \{0,1\}^n} \mathbf{c}^\mathsf{T}\mathbf{x} = 0$ we can select $\nu = -1$. Note that by using this fitness function, every chromosome $\mathbf{y}$ for which $\mathbf{a}^\mathsf{T}\mathbf{y} > b$ will reduce the fitness function value by $\mu(\mathbf{a}^\mathsf{T}\mathbf{y} - b)^2$.

Let us consider that $n = 6$ with weights $\mathbf{a} = [3, 14, 10, 9, 8, 6]^\mathsf{T}, b = 30$, and cost values $\mathbf{c} = [213, 144, 125, 189, 208, 306]^\mathsf{T}$. Let us also select a population size of $m = 8$. The termination criterion is to reach 50 population evolutions. As previously discussed, the implementation of the GA is stochastic, meaning that every time we apply the GA algorithm to the same problem we might obtain a different solution. In fact, GAs, as other metaheuristics, are run several times for the same problem to increase the chances of computing a better solution which, hopefully, is close to the globally optimal solution of the problem.

Solving the knapsack problem, we initialize a population with randomly selected individuals that can be presented as an $m \times n$ matrix:

$$
\mathbf{P} = \begin{bmatrix}
1 & 1 & 1 & 1 & 1 & 1 \\
1 & 1 & 0 & 1 & 1 & 0 \\
1 & 0 & 1 & 1 & 0 & 0 \\
1 & 0 & 0 & 0 & 0 & 0 \\
1 & 1 & 0 & 1 & 1 & 0 \\
1 & 0 & 1 & 1 & 1 & 1 \\
0 & 0 & 1 & 1 & 0 & 1 \\
0 & 0 & 1 & 0 & 0 & 0
\end{bmatrix}
$$

Every row of this matrix is a chromosome with $n = 6$ genes representing an individual. The fitness of each individual when using the fitness function is:

$$[f_1, ..., f_6]^\mathsf{T} = [-40001, -1601, 527, 213, -1601, -3601, 620, 125]^\mathsf{T}$$

Using a truncation selection and a parent selection ratio of 50%, we select the 50% of individuals with the highest fitness values to perform crossover (3rd, 4th, 7th and 8th). We also use a crossover probability of 50% and an elitism ratio of 0.125, meaning that the individual with the best performance out of the 8 population members is passed to the next generation without any changes (in this case, the 7th individual). In the crossover phase, we generate four offsprings by pairing individuals 3 and 8, and 4 and 8 resulting in:

- Offspring 1: 101 000
- Offspring 2: 001 100
- Offspring 3: 10 1000
- Offspring 4: 00 0000

Note that crossover is a stochastic process and we can randomly pair different individuals based on the 50% crossover probability. That is, the generated offsprings can be completely different if we run this process again. The crossover point is randomly selected and an offspring is accepted if it is not the same as another already existing population member. In this implementation, we do not use plus-selection that discards offsprings which do not perform better than their parents. Allowing the generated offsprings to replace individuals 1, 2, 5 and 6 updates the population to:

$$\mathbf{P} = \begin{bmatrix} 1\,0\,1\,0\,0\,0 \\ 0\,0\,1\,1\,0\,0 \\ 1\,0\,1\,1\,0\,0 \\ 1\,0\,0\,0\,0\,0 \\ 1\,0\,1\,0\,0\,0 \\ 0\,0\,0\,0\,0\,0 \\ 0\,0\,1\,1\,0\,1 \\ 0\,0\,1\,0\,0\,0 \end{bmatrix}$$

Performing a mutation to each individual—except the 7th individual which should remain unchanged due to elitism—can result in the next generation:

$$\mathbf{P} = \begin{bmatrix} 1\,0\,1\,0\,0\,0 \\ 1\,0\,1\,1\,0\,0 \\ 1\,0\,1\,1\,1\,0 \\ 1\,0\,0\,0\,0\,0 \\ 1\,0\,1\,0\,1\,0 \\ 0\,0\,0\,1\,0\,0 \\ 0\,0\,1\,1\,0\,1 \\ 0\,0\,1\,0\,0\,1 \end{bmatrix}$$

with fitness values:

$$[f_1, ..., f_6]^\mathsf{T} = [338, 527, 735, 213, 546, 189, 620, 431]^\mathsf{T}$$

Notice that our new elite member is the 3rd individual and all population members of the new generation are feasible solutions. We can repeat these steps 49 more times to generate new population generations and then stop the GA since its termination criterion is selected to be reaching 50 population evolutions.

Practitioner's Corner
Genetic Algorithms in Python 3

There are many libraries that implement different versions of GAs. A comprehensive evolutionary computation framework is Deap, which includes a set of evolutionary algorithms. A simpler, GA-specific library, is geneticalgorithm. Its implementation to the Knapsack problem of our previous example is presented below. This library solves minimization problems, thus the fitness function is multiplied by -1. In the following code, mu refers to $\mu = 10^2$ and nu to $\nu = -1$.

```python
import numpy as np
from geneticalgorithm import geneticalgorithm as ga
def fitness_eval(X):
    a = np.array([3, 14, 10, 9, 8, 6])
    c = np.array([213,144,125,189,208,306])
    b=30; mu=1e+2; nu=-1
```

```
    if np.dot(a.transpose(), X) <= b:
        outcome = np.dot(c.transpose(), X)
    else:
        outcome = nu - mu * max(0, np.dot(a.transpose(), X) - b) ** 2
    return -outcome

algorithm_param = {'max_num_iteration': 50,\
                   'population_size':8,\
                   'mutation_probability':0.1,\
                   'elit_ratio': 0.125,\
                   'crossover_probability': 0.5,\
                   'parents_portion': 0.5,\
                   'crossover_type':'one_point',\
                   'max_iteration_without_improv':None}
model=ga(function=fitness_eval,dimension=6,variable_type='bool',
    algorithm_parameters=algorithm_param)
model.run()
```

Implementing this code can return different solutions because it contains stochastic processes. For instance, running this code one time returned solution $\mathbf{x} = [1, 0, 0, 1, 1, 1]^\mathsf{T}$ with fitness score 916.0. Running it again returned solution $\mathbf{x} = [1, 1, 0, 0, 0, 1]^\mathsf{T}$ with fitness score 663.0. It is clear that the first solution is better than the second. In practice, we can run the GA several times and keep the solution with the best performance.

6.3 Simulated Annealing

Simulated annealing is another metaheuristic. The name of the algorithm comes from *annealing* in metallurgy, a technique involving heating and controlled cooling of a material to increase the size of its crystals and reduce their defects. As with GAs, it is often used when the search space is discrete.

The idea behind simulated annealing is that we do not always want to find new solutions which improve the current status: sometimes we also need to *explore* to avoid local optima. E.g., in annealing repeated heating and cooling strengthens the blade. The name simulated annealing was proposed in 1983 in the work of Kirkpatrick et al. [9] who applied it to the traveling salesman problem. However, similar techniques were proposed in the 1980s. As an algorithm, it has a strong similarity to the classic Metropolis algorithm [10].

The goal of simulated annealing is to bring the system from an arbitrary initial state to the state with the *minimum* or *maximum possible energy* by optimizing an *energy* function $E(\mathbf{s})$, where $\mathbf{s}$ is the current state of our system. The state of our system can be seen as a potential solution of an optimization problem which is updated from iteration to iteration.

Each iteration t of simulated annealing consists of:

- The update of the *temperature* parameter value, T_t
- The selection of a random *neighbor solution* $\mathbf{y}$ of the current state $\mathbf{s}_t$
- The decision about *replacing the current solution* $\mathbf{s}_t$ with the neighbor solution $\mathbf{y}$ based on a probability $P(\mathbf{y}, \mathbf{s}_t, T_t)$ that depends on the performance of solutions $\mathbf{y}, \mathbf{s}_t$ and the current temperature T_t

Simulated annealing is based on the *improve* and *explore* operations related to the updating of the current solution. The current solution is replaced when the neighbor solution is better (*improve*). There are cases, however, where we allow the current solution to be replaced even if the neighbor solution is not better, based on a small probability (*explore*).

6.3.1 Temperature Update

An important part of simulated annealing is the update of temperature T_t from iteration to iteration because this affects the probability of exploring and replacing the current solution with a neighbor solution which might have inferior performance. Initially, the temperature value should be a *high value* (or infinity), and then it should be decreased at each step. The temperature T_t should *reduce gradually* from iteration to iteration and $T_t \to 0$ at the last iterations. This would allow the following:

- At the initial stages of the algorithm, we provide more freedom to a neighbor solution to replace the current solution, even if the neighbor solution has an inferior performance (thus, supporting aggressive exploration).
- At the last iterations of the algorithm, we accept a neighbor solution only if it is better than the current solution (we restrain further exploration).

We can view this process as follows: the process consists of first *melting* the system being optimized at a high effective temperature, then lowering the temperature by slow stages until the system *freezes* and no further changes occur [11]. An example of updating the temperature is the so-called *fast cooling*, calculated as follows:

$$T_t = T_0/(t + 1) \tag{6.2}$$

where T_0 is the initial temperature (a very large positive number). We add 1 in the denominator to avoid diving by 0 if the initial iteration is $t = 0$. An extension of the fast cooling is the *linear multiplicative* cooling:

$$T_t = T_0/(at + 1) \tag{6.3}$$

where $a > 0$ is a constant.

Another common cooling scheme for updating the temperature is the *exponential cooling scheme* defined as:

$$T_t = T_0 \cdot a^t \tag{6.4}$$

where a is a constant in the range 0.9–0.99 [12]. Finally, another cooling scheme is the *logarithmical multiplicative cooling* based on the asymptotical convergence condition of simulated annealing [13], but incorporating a factor $a > 1$ that speeds up cooling:

$$T_t = T_0/(1 + a \log(t)) \tag{6.5}$$

6.3.2 Replacing Probability

When solving a minimization problem with objective (energy) function E, the probability of replacing the current solution $\mathbf{s}_t$ by a selected neighbor solution $\mathbf{y}$ is based on the Boltzmann distribution when $\mathbf{y}$ has inferior performance [9]:

$$P(\mathbf{y}, \mathbf{s}_t, T_t) = \begin{cases} 1 & \text{if } E(\mathbf{y}) \le E(\mathbf{s}_t) \\ \exp \frac{-(E(\mathbf{y}) - E(\mathbf{s}_t))}{T_t} & \text{otherwise} \end{cases}$$

That is, the current solution $\mathbf{s}_t$ is always replaced by its neighbor if its neighbor has superior performance ($E(\mathbf{y}) \le E(\mathbf{s}_t)$ when solving a minimization problem) and can be replaced with a probability $\exp \frac{-(E(\mathbf{y}) - E(\mathbf{s}_t))}{T_t}$ if the neighbor solution has inferior performance ($E(\mathbf{y}) > E(\mathbf{s}_t)$).

From the above probability, one can notice that:

- For large T_t values, $\frac{-(E(\mathbf{y}) - E(\mathbf{s}_t))}{T_t} \to 0$ and $\exp \frac{-(E(\mathbf{y}) - E(\mathbf{s}_t))}{T_t} \to 1$. Thus, we tend to accept the neighbor solution even if it does not offer an improvement.
- For small T_t values approaching 0, $\frac{-(E(\mathbf{y}) - E(\mathbf{s}_t))}{T_t} \to -\infty$ and $\exp \frac{-(E(\mathbf{y}) - E(\mathbf{s}_t))}{T_t} \to 0$. Thus, we tend to reject the neighbor solution if it does not offer an improvement.

The basic simulated annealing framework is provided in Algorithm 28. Similarly to other meta-heuristics, there are different implementations of simulated annealing. Its main parameters that require tuning are the initial value of the temperature and the mechanism of reducing the temperature at each iteration. The selection mechanism of a neighbor solution at each iteration is also something that can change in different simulation annealing implementations. Regarding the termination criteria of simulated annealing, iterations can be repeated until the system reaches a state that is stable (the solution does not change after repeated iterations), until a maximum number of iterations has been reached, or until a given computation budget has been exhausted.

Algorithm 28 Basic Simulated Annealing framework

1: Given a minimization problem min $f(\mathbf{x})$ s.t. $\mathbf{h}(\mathbf{x}) = \mathbf{0}$, $\mathbf{g}(\mathbf{x}) \leq \mathbf{0}$
2: Decide the formulation of the energy function $E(\mathbf{s})$, where $\mathbf{s}$ plays the role of $\mathbf{x}$.
3: Set $t = 1$ and choose initial solution guess $\mathbf{s}_t$
4: Set T_t equal to a very large positive number
5: **repeat**
6:    Choose neighbor solution $\mathbf{y}$
7:    **if** $E(\mathbf{y}) \leq E(\mathbf{s}_t)$ **then**
8:       set $\mathbf{s}_t = \mathbf{y}$
9:    **else**
10:      with probability $\exp \frac{-(E(\mathbf{y})-E(\mathbf{s}_t))}{T_t}$, set $\mathbf{s}_t = \mathbf{y}$
11:   **end if**
12:   set $t = t + 1$
13:   reduce the value of T_t while ensuring that $T_t > 0$
14: **until** a termination criterion is satisfied

Note that line 10 of the algorithm can be implemented in a computer program by randomly selecting a number k from a continuous uniform probability distribution $U(0, 1)$ and setting $\mathbf{s}_t = \mathbf{y}$ if $k \leq \exp \frac{-(E(\mathbf{y})-E(\mathbf{s}_t))}{T_t}$.

One open issue is the selection of neighbor $\mathbf{y}$ of a current solution $\mathbf{s}_t$. For this, we can perform a *greedy search* in the neighborhood of $\mathbf{s}_t$ and select the best-performing $\mathbf{y}$. This is very similar to an iteration of the hill-climbing heuristic, but, unlike hill-climbing, simulated annealing can escape from local optima when the temperature is not close to 0. Defining the neighborhood of the greedy search is based on the properties of the problem. A general selection mechanism is to select all points $\mathbf{x}$ in the neighborhood $\mathcal{N}(\mathbf{s}_t)$ of $\mathbf{s}_t$ which can be a closed ball with radius $r > 0$:

$$\mathcal{N}(\mathbf{s}_t) := \{\mathbf{x} \in \mathcal{X} \ : \ \|\mathbf{x} - \mathbf{s}_t\|_2 \leq r\}$$

where $\mathcal{X}$ is the feasible region of the problem. Notice that in the case of greedy search, $\mathbf{y}$ is the neighbor for which $E(\mathbf{y}) \leq E(\mathbf{x}) \ \forall \mathbf{x} \in \mathcal{N}(\mathbf{s}_t) \setminus \{\mathbf{s}_t\}$.

Let us consider the knapsack problem of Eq. (6.1) with $b = 30$, item weights $\mathbf{a} = [3, 14, 10, 9, 8, 6]^\mathsf{T}$, and item values $\mathbf{c} = [213, 144, 125, 189, 208, 306]^\mathsf{T}$. An example of solving this knapsack problem with initial temperature $50 \cdot 10^3$ and an exponential cooling scheme is provided in Fig. 6.6 where we perform two runs of the metaheuristic allowing 100 iterations. Notice that the steps taken in the exploration steps are quite different between the two runs due to the stochastic search process. In the first run that appears in the left-hand-side of the figure, the energy function was reduced further resulting in a value of -916. In the second run, the energy function reduced to -727. Notice that at the early stages of the algorithm it is allowed to accept new solutions, even if they are not improving the objective function. As the algorithm progresses, this becomes harder due to the change in the temperature parameter.

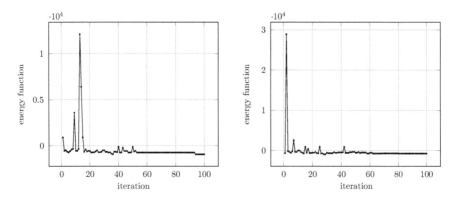

Fig. 6.6 Example of two runs of the simulated annealing algorithm to the same problem with the same input. The y-axis presents the change in the energy function from iteration to iteration

Practitioner's Corner
Simulated Annealing in Python 3

A simple implementation of Simulated Annealing for the knapsack problem in (6.1) is provided below. We select a maximum of 100 iterations as termination criterion.

```python
import numpy as np
import math, random
# define objective function
def energy_fun(x):
    a, c = np.array([3, 14, 10, 9, 8, 6]), np.array
    ([213,144,125,189,208,306])
    b=30; mu=1e+2; nu=-1
    if np.dot(a.transpose(), x) <= b:
        outcome = np.dot(c.transpose(), x)
    else:
        outcome = nu - mu * max(0, np.dot(a.transpose(), x) - b) ** 2
    return -outcome
x_start = np.array([0, 0, 0, 0, 0, 0]) # Start location
T0, Tt = 50e+3, 50e+3 # Initial temperature
n=100; t=1 # Number of iterations
while(t!=n):
    print('current temperature: ', Tt)
    t = t+1
    y=np.copy(x_start)
    for i in range(0,len(x_start)):
        y[i] = random.randint(0, 1)
    if energy_fun(y) <= energy_fun(x_start):
        x_start = np.copy(y)
    else:
        prob = math.exp(-(energy_fun(y)-energy_fun(x_start))/Tt)
        choice = random.random() # Draw a number from U(0,1)
        if choice <= prob:
            x_start = np.copy(y)
```

```
    # Apply linear multiplicative cooling
    a = 0.9; Tt = T0*(a**t)
print('Solution:', x_start, 'Energy function:', energy_fun(x_start))
```

6.4 Tabu Search

Tabu search is a metaheuristic search method employing local search. Tabu search was proposed by Glover in 1986 followed by two subsequent papers in 1989 and 1990 [14–16]. Tabu search is used for discrete optimization problems of combinatorial nature. It mainly strives to avoid getting trapped to local optima when implementing a local search by:

- Accepting worsening moves if we are at a local optimum and no improving moves are available.
- Marking the previously visited solutions and using problem-specific sets of rules to mark solutions as *tabu* (forbidden) so that they are not explored repeatedly.

Let us consider a discrete optimization problem:

$$(P) \qquad \min_{\mathbf{x} \in \mathcal{X}} \ f(\mathbf{x})$$

where $\mathcal{X}$ is the discrete set of feasible solutions and $\mathbf{x}$ an n-valued vector.

Given a trial solution $\mathbf{x} \in \mathcal{X}$, we can define a subset $\mathcal{Y}(\mathbf{x}) \subseteq \mathcal{X}$ of the neighborhood of $\mathbf{x}$. We can also define as a *move*, denoted as s, a specific mapping that maps trial solution $\mathbf{x}$ to $\mathbf{x}' \in \mathcal{Y}(\mathbf{x})$. That is, $s(\mathbf{x}) = \mathbf{x}'$ for some $\mathbf{x}' \in \mathcal{Y}(\mathbf{x})$. For a given trial solution $\mathbf{x} \in \mathcal{X}$ there can be several moves s that map $\mathbf{x}$ to another feasible solution $\mathbf{x}' \in \mathcal{Y}(\mathbf{x})$. Thus, associated with trial solution $\mathbf{x} \in \mathcal{X}$ is the set of mappings:

$$S(\mathbf{x}) = \{s \in \mathcal{S} \ : \ s(\mathbf{x}) \in \mathcal{Y}(\mathbf{x})\}$$

This set of mappings is the set of allowed moves from trial solution $\mathbf{x}$ and it can be viewed as a set of neighborhood functions that map trial solution $\mathbf{x}$ to any other neighborhood solution $\mathbf{x}' \in \mathcal{Y}(\mathbf{x})$. This set of moves uniquely defines the neighborhood of $\mathbf{x}$.

To provide an example, we can express the hill-climbing heuristic, that was presented at the beginning of this chapter, with the use of moves as follows in Algorithm 29.

Algorithm 29 Hill Climbing heuristic

1: Given a minimization problem $\min_{\mathbf{x} \in \mathcal{X}} f(\mathbf{x})$

2: Select an initial solution $\mathbf{x}$

3: **repeat**

4:    Greedy search: select move $s \in \mathcal{S}(\mathbf{x})$ such that $f(s(\mathbf{x})) \leq f(s'(\mathbf{x}))$ for all other moves $s' \in \mathcal{S}(\mathbf{x})$

5:    **if** $f(s(\mathbf{x})) < f(\mathbf{x})$ **then**

6:        Set $\mathbf{x} := s(\mathbf{x})$

7:    **else**

8:        $\mathbf{x}$ is a local optimum

9:    **end if**

10: **until** $\mathbf{x}$ is a local optimum

Because the hill climbing heuristic is trapped in a local optimum, tabu search is a meta-heuristic that guides this heuristic to continue its exploration without being confined by an absence of improving moves, and without falling back into a local optimum from which it previously emerged. The former can be achieved by allowing to select solution $s(\mathbf{x})$ that has worst performance than $\mathbf{x}$. Doing so, however, might result in cycling, causing the process to move from $s(\mathbf{x})$ to the better-performing $\mathbf{x}$ in the next iterations leading to returning repeatedly to the same solution. This is why tabu search is also using a memory structure that forbids certain moves that would return to a recently visited solution [17].

Two key elements of tabu search are the constraining of the search by classifying certain of its moves as forbidden (i.e., tabu), and the freeing of the search by a short-term memory function that provides strategic forgetting [15]. To implement the above, a subset $\mathcal{T} \subseteq \mathcal{S}$ is created whose elements are called *tabu moves*. To determine the elements of $\mathcal{T}$, we utilize historical information from the search process, extending up to t iterations in the past. To be a member of $\mathcal{T}$, pre-determined *tabu conditions* must be satisfied. The tabu search framework provided by [15] is presented in Algorithm 30.

Note that each iteration moves from the current solution $\mathbf{x}$ to $s_k(\mathbf{x})$ that yields the greatest improvement in the neighborhood of $\mathbf{x}$. That is, $f(s_k(\mathbf{x})) \leq f(s'(\mathbf{x}))$ for all $s' \in \mathcal{S} - \mathcal{T}$ which considers the restrictions of tabu moves. We can view $s_k(\mathbf{x})$ as the best solution in the neighborhood that is not tabu. If many solutions are considered tabu during a particular iteration, the selected $s_k(\mathbf{x})$ might not perform better than $\mathbf{x}$. Despite this, we will replace $\mathbf{x}$ by $s_k(\mathbf{x})$ even if $f(s_k(\mathbf{x})) > f(\mathbf{x})$.

Concerning the best solution currently found, $\mathbf{x}^*$, this solution is only updated when $f(s_k(\mathbf{x})) < f(\mathbf{x}^*)$. Another important point is the update mechanism of the list of tabu moves $\mathcal{T}$ at each iteration. There are generally three different memory structures that contribute to the updating of the list:

Algorithm 30 Tabu Search meta-heuristic

1: Given a minimization problem $\min_{\mathbf{x} \in \mathcal{X}} f(\mathbf{x})$
2: Select an initial solution $\mathbf{x}$ and set the best solution currently found $\mathbf{x}^* := \mathbf{x}$
3: Set $k = 0$ and $\mathcal{T} = \emptyset$
4: **repeat**
5:   Set $k = k + 1$
6:   Select $s_k \in \mathcal{S}(\mathbf{x}) - \mathcal{T}$ such that $f(s_k(\mathbf{x})) \leq f(s'(\mathbf{x}))$ for all $s' \in \mathcal{S} - \mathcal{T}$
7:   **if** $f(s_k(\mathbf{x})) < f(\mathbf{x}^*)$ **then**
8:     Set $\mathbf{x}^* := s_k(\mathbf{x})$
9:   **end if**
10:   update $\mathcal{T}$
11: **until** $\mathcal{S}(\mathbf{x}) - \mathcal{T}$ is empty or the maximum number of allowed iterations is reached.

- Short-term: keeps the list of solutions recently considered. If a potential solution appears on the tabu list, it cannot be revisited until it reaches an expiration point (reaching t iterations).
- Intermediate-term: *Intensification rules* intended to focus the search on a small, promising area of the search space.
- Long-term: *Diversification rules* that drive the search into new regions when the search is stuck in local optima.

If we want to apply the short-term memory structure, an easy approach is to add the best-performing neighbor $s_k(\mathbf{x})$ of the current iteration to the tabu moves $\mathcal{T}$ at line 11 of the algorithm. That is, line 11 can be replaced by $\mathcal{T} \leftarrow \mathcal{T} \cup \{s_k(\mathbf{x})\}$. If we allow this move after t more iterations, then we should count the additional iterations from now on and remove it from $\mathcal{T}$ when t more iterations are performed. By updating the list of tabu moves, it may happen that in the next iteration we will have to select an inferior solution because better solutions are tabu.

Apart from short-term memory structures that are focused on avoiding cycling, tabu search can also use memory structures to take advantage of a learning process: having visited several solutions, one can observe whether good solutions visited so far have some commonalities and generate an intensification scheme for the search based on these commonalities. The intensification scheme can restrict moves by adding them to $\mathcal{T}$ or relax moves by removing them from $\mathcal{T}$ to favor solutions with properties that occurred often in good solutions.

6.5 Differential Evolution

Differential Evolution (DE) was introduced by Storn and Price in 1996 [18]. Despite its name, Differential Evolution does not require the optimization problem to be differentiable, as is required by most exact optimization methods. In fact, the objective function might have any form (it might even be discontinuous). DE approaches the objective function as a black box, which is only used to evaluate the performance of a solution candidate.

Let $f : \mathbb{R}^n \to \mathbb{R}$ be the possibly non-differentiable objective (fitness) function that needs to be minimized. In DE, every candidate solution is a vector $\mathbf{m}$ with n real numbers. The goal is to find a solution $\mathbf{m}$ that minimizes f by combining the positions of existing candidate solutions. Note that, similarly to other metaheuristics, global optimality cannot be guaranteed. DE uses a number of parameter values for tuning its application. First is the *population size*, which is commonly denoted as P and it should be greater than or equal to 4 ($P \geq 4$) to implement mutation strategies. Second is the *mutation constant*, also known as *differential weight*. In literature it is denoted as F and it is allowed to take values in the range $0 \leq F \leq 2$, with 0.8 being a common value. Another DE parameter is the *crossover probability*, also known as *recombination constant*. In literature it is denoted as CR and it is allowed to take values in the range $0 \leq CR \leq 1$, with $CR = 0.9$ being a typical value. Finally, we have to select a mutation strategy and termination criteria (i.e., maximum number of iterations or solution improvement tolerance).

Using these parameters, we present the main steps of the Differential Evolution algorithm, which is typically applied to constrained, continuous optimization problems:

Step 1: Initialize each population member $\mathbf{m} \in \{1, 2, ..., P\}$ by selecting real number values for the elements of the respective vector $\mathbf{m} = [m_1, m_2, ..., m_n]^\mathsf{T}$

Step 2: For each population member (agent) $\mathbf{m}$ do:

- Select three random agents $\mathbf{x}', \mathbf{x}'', \mathbf{x}'''$ from the population $1, 2, ..., P$, where $\mathbf{x}' \neq \mathbf{x}'' \neq \mathbf{x}''' \neq \mathbf{m}$
- Compute a *donor vector* $\mathbf{v}$ such that $v_i = x_i' + F(x_i'' - x_i''') \ \forall i \in \{1, 2, ..., n\}$
- Randomly select a position $ps \in \{1, 2, ..., n\}$
- Create a *trial vector* $\mathbf{y}$, where $y_i = \begin{cases} v_i \text{ if } r_i < CR \text{ where } r_i \sim U(0, 1) \text{ or } i = ps \\ m_i \text{ otherwise} \end{cases}$
- If $f(\mathbf{y}) \leq f(\mathbf{m})$ in a minimization problem, replace $\mathbf{m}$ by $\mathbf{y}$

Step 3: Perform step 2 repeatedly until triggering a termination criterion.

In the algorithmic framework, we computed a donor vector $\mathbf{v}$ by using the mutation strategy $v_i = x_i' + F(x_i'' - x_i''')$ for each $i \in \{1, 2, ..., n\}$. This mutation strategy is known as rand1bin and it is one of the most common mutation strategies. There are, however, other mutation strategies. Table 6.1 summarizes the most common mutation strategies used to compute each value v_i of the donor vector $\mathbf{v}$.

Let us consider an example of using differential evolution to minimize Ackley's function in two dimensions in its typical domain $[-32.768, 32.768]$ subject to nonlinear constraint $x_1 + x_2^2 \geq 4.1$. Ackley's function is a non-convex function with many local optima. We use mutation constant $F = 0.8$, population size $P = 100$, recombination constant $CR = 0.9$, mutation strategy rand2bin, and when the constraint is

Table 6.1 Common mutation strategies of differential evolution

Mutation strategy	Donor vector
rand1bin	$v_i = x'_i + F(x''_i - x'''_i)$
best1bin	$v_i = x^{best}_i + F(x'_i - x''_i)$
rand2bin	$v_i = x'_i + F(x''_i - x'''_i) + F(x''''_i - x'''''_i)$
best2bin	$v_i = x^{best}_i + F(x'_i - x''_i) + F(x'''_i - x''''_i)$

violated we increase the score of the evaluation function to $+\infty$. An example of this implementation is presented below.

Practitioner's Corner
Differential Evolution in Python 3

An implementation of Differential Evolution is provided in Python, using the `scipy.optimize` library. The method `scipy.optimize.differential_evolution` has the following default options: strategy='best1bin', maxiter=1000, popsize=15, tol=0.01, mutation=(0.5, 1), recombination=0.7, seed=None, init= 'latinhypercube', atol=0, updating='immediate', x0=None.

```python
from scipy.optimize import differential_evolution
import numpy as np
def ackley(x):
    arg1 = -0.2 * np.sqrt(0.5 * (x[0] ** 2 + x[1] ** 2))
    arg2 = 0.5 * (np.cos(2. * np.pi * x[0]) + np.cos(2. * np.pi * x
    [1]))
    if x[0]+x[1]**2 >= 4.1:
        return -20. * np.exp(arg1) - np.exp(arg2) + 20. + np.e
    else:
        return np.infty #some high value
F = 0.8 # Mutation constant
P = 100 # Population size
CR = 0.9 # Recombination constant
bounds = [(-32.768, 32.768), (-32.768, 32.768)]
result = differential_evolution(ackley, bounds, mutation=F, maxiter
    =2000, popsize=P, recombination=CR, polish=False, strategy='
    rand2bin')
print(result.x, result.fun)
```

6.6 Particle Swarm Optimization

Particle Swarm Optimization (PSO) was introduced by Kennedy and Eberhart in 1995 [19]. PSO is a metaheuristic and it is typically applied for constrained continuous optimization. It does not require to have a differentiable objective function and it does not guarantee optimality. In PSO, each population member, i, is called *particle*. A population member is a candidate solution and the entire population is called *swarm*. The number of particles in the swarm is typically denoted as S.

In PSO, the population members (particles) are moved around in the search space according to specific rules [20]. The movements of the particles are guided by their own *best-known position*, $\mathbf{p}^i$, in the search-space as well as the entire *swarm's* best-known position, $\mathbf{g}$.

Let us consider a minimization problem of function $f : \mathbb{R}^n \to \mathbb{R}$, which is not necessarily differentiable. Each candidate solution (particle) i in the swarm of size S has position $\mathbf{x}^i = [x_1^i, x_2^i, ..., x_S^i]^\mathsf{T}$ and velocity $\mathbf{v}^i = (v_1^i, v_2^i, ..., v_S^i)$. In the initialization phase of PSO:

- For each population member $i \in \{1, 2, ..., S\}$, we initialize its position $\mathbf{x}^i = [x_1^i, x_2^i, ..., x_S^i]^\mathsf{T}$ and its velocity $\mathbf{v}^i = [v_1^i, v_2^i, ..., v_S^i]^\mathsf{T}$. The position is a *candidate problem solution*.
- We find the *swarm's* best-known position $\mathbf{g}$, where $f(\mathbf{g}) \leq f(\mathbf{x}^i)$, $\forall i \in \{1, 2, ..., S\}$ when solving a minimization problem.
- For each population member $i \in \{1, 2, ..., S\}$, we initialize the personal best-known position at iteration $t = 1$ as $\mathbf{p}^{i,t} \leftarrow \mathbf{x}^i$.

The personal best-known position and the global best-known position after a number of iterations t are defined as:

$$\mathbf{p}^{i,t} := \arg\min f(\mathbf{x}^{i,k}) \quad \forall i \in \{1, ..., S\} \text{ where } k = 1, ..., t$$

$$\mathbf{g}^t := \arg\min f(\mathbf{x}^{i,k}) \quad \text{where } k = 1, ..., t \text{ and } i = 1, ..., S$$

PSO performs searching via a swarm of particles which update their position from iteration to iteration. Each particle moves in the direction of its previously personal best-known position and the global best-known position. To do so, the position of each particle is updated as:

$$\mathbf{x}^{i,t+1} := \mathbf{x}^{i,t} + \mathbf{v}^{i,t+1} \quad \forall i \in \{1, ..., S\} \tag{6.6}$$

where the velocity is updated as:

$$\mathbf{v}^{i,t+1} := w\mathbf{v}^{i,t} + c_1 r_1 (\mathbf{p}^{i,t} - \mathbf{x}^{i,t}) + c_2 r_2 (\mathbf{g}^t - \mathbf{x}^{i,t}) \quad \forall i \in \{1, ..., S\} \tag{6.7}$$

where w is a real-valued parameter of PSO known as *inertia weight* that balances the global and local exploration, $r_1, r_2 \sim U(0, 1)$ are uniformly distributed random variables in [0,1], and c_1, c_2 are positive PSO parameters known as *acceleration coefficients*. This is the main step that is performed at each iteration to update the positions of the particles. From Eq. (6.6) it follows that if the new position $\mathbf{x}^{i,t+1}$ of particle i is such that $f(\mathbf{x}^{i,t+1}) \leq f(\mathbf{p}^{i,t})$, then $\mathbf{p}^{i,t+1} = \mathbf{x}^{i,t+1}$. The first part of

Eq. (6.7) is the *inertia* of the particle that is dictated by its previous velocity. The second part is known as the *cognitive component* and encourages the particle to move towards its own best-known position. The third part is known as the *cooperation component* and represents the collaborative effect of the particles to find the globally optimal solution [21].

The main implementation of the standard PSO algorithm based on [20] is presented in Algorithm 31.

Algorithm 31 Particle Swarm Optimization meta-heuristic

1: Given a minimization problem $\min_{x \in \mathcal{X}} f(\mathbf{x})$
2: Select PSO parameter values w, c_1, c_2 and set iteration $t = 1$.
3: **for** each particle $i = 1, ..., S$ **do**
4:    initialize particle's position $\mathbf{x}^{i,t} \sim U(LB, UB)$, where LB and UB represent the lower and upper bounds of the search space.
5:    initialize particle's velocity $\mathbf{v}^{i,t} \sim U(-|UB - LB|, |UB - LB|)$.
6:    initialize $\mathbf{p}^i = \mathbf{x}^{i,t}$
7: **end for**
8: initialize $\mathbf{g} = \text{argmin } f(\mathbf{x}^{i,t})$ where $i = 1, ..., S$.
9: **repeat**
10:    **for** each particle $i = 1, ..., S$ **do**
11:       pick uniformly distributed random numbers $r_1, r_2 \sim U(0, 1)$.
12:       set $\mathbf{v}^{i,t+1} = w\mathbf{v}^{i,t} + c_1 r_1 (\mathbf{p}^{i,t} - \mathbf{x}^{i,t}) + c_2 r_2 (\mathbf{g}^t - \mathbf{x}^{i,t})$
13:       set $\mathbf{x}^{i,t+1} = \mathbf{x}^{i,t} + \mathbf{v}^{i,t+1}$
14:       **if** $f(\mathbf{x}^{i,t+1}) \leq f(\mathbf{p}^i)$ **then**
15:          update particle's best-known position: $\mathbf{p}^i = \mathbf{x}^{i,t+1}$
16:          **if** $f(\mathbf{p}^i) \leq f(\mathbf{g})$ **then**
17:             update swarm's best-known position: $\mathbf{g} = \mathbf{p}^i$
18:          **end if**
19:       **end if**
20:    **end for**
21:    set $t = t + 1$
22: **until** a termination criterion is met.

Let us consider the implementation of PSO to minimize Ackley's function in two dimensions in domain $[-5, 5]$ subject to nonlinear constraint $x_1 + x_2^2 \geq 4.1$. We use a population size of $S = 30$ particles in the swarm. The termination criterion is set to be 200 iterations. We also set $c_1 = 2$ and $c_2 = 2$. Solving this problem will result in different solutions every time because of the stochastic processes in the PSO implementation steps. One of these solutions is $\mathbf{x}^* = [0.01472226, -2.02120759]^{\mathsf{T}}$ with objective function score 4.990287. In Fig. 6.7 we show how the particles of the PSO start from widespread positions during the initialization stage and converge towards solution $\mathbf{x}^*$ after 50 iterations. From 50 to 200 iterations the improvement is marginal, so we could have stopped the algorithm earlier.

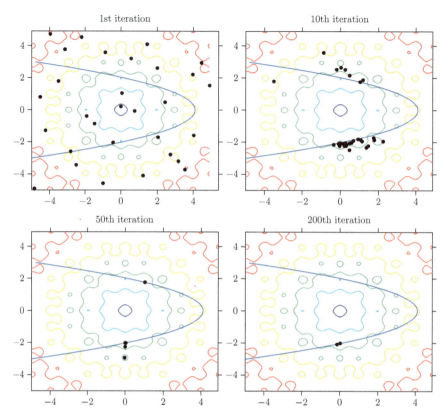

Fig. 6.7 Position of PSO particles (black dots) at the 1st, 10th, 50th, and 200th iteration with respect to Ackley's function presented in a contour plot. The blue parabola expresses constraint $x_1 + x_2^2 \geq 4.1$

Practitioner's Corner
PSO in Python 3

An implementation of PSO is provided in Python, using the `pymoo` library. The method `pymoo.algorithms.soo.nonconvex.pso.PSO` has the following default options: pop_size=25, sampling=LHS(), w=0.9, c1=2.0, c2=2.0, adaptive=True, initial_velocity='random', max_velocity_rate=0.20, pertube_best=True.

Below follows an example of using the PSO method to minimize Ackley's function in two dimensions in domain $[-5, 5]$ subject to nonlinear constraint $x_1 + x_2^2 \geq 4.1$. We use a population size of $S = 30$ particles in the swarm. The termination criterion is set to be 200 iterations.

```
from pymoo.algorithms.soo.nonconvex.pso import PSO
import numpy as np
from pymoo.optimize import minimize
```

```
from pymoo.problems.functional import FunctionalProblem

def ackley(x):
    arg1 = -0.2 * np.sqrt(0.5 * (x[0] ** 2 + x[1] ** 2))
    arg2 = 0.5 * (np.cos(2. * np.pi * x[0]) + np.cos(2. * np.pi * x
    [1]))
    if x[0]+x[1]**2 >= 4.1:
        return -20. * np.exp(arg1) - np.exp(arg2) + 20. + np.e
    else:
        return np.infty #some high value

n_var=2
problem = FunctionalProblem(n_var,ackley,xl=np.array([-5,-5]),xu=np.
    array([5,5]))
algorithm = PSO(pop_size=30)
res = minimize(problem,algorithm,('n_gen', 200),seed=1,verbose=True)

print("Best solution found: \nX = %s\nF = %s" % (res.X, res.F))
```

Exercises

6.1 Differential Evolution
Write your own computer program and solve the following mathematical program
with differential evolution.

$$\min_{x \in \mathbb{R}^2} x_1^2 - x_1 x_2 + x_2^2 - 7x_1 - 3x_2$$

$$\text{s.t. } -x_1 + 2x_2 \le 4$$

$$x_1 + 3x_2 \le 7$$

$$x_1 - 2x_2 \le 5$$

$$-x_1 \le 0$$

6.2 Particle Swarm Optimization
Write your own computer program following Algorithm 31 and solve the following
mathematical program with particle swarm optimization.

$$\min_{x \in \mathbb{R}^2} x_1^2 + 3x_2$$

$$\text{s.t. } -x_1 + 2x_2 \le 4$$

$$x_1 + 3x_2 \le 7$$

$$x_1 - 2x_2 \le 5$$

$$-x_1 \le 0$$

Solve the same problem with the active set method and the compare the results. Justify whether you have found a globally optimal solution when using the active set method.

6.3 Genetic Algorithms

Consider a set of objects with weights $\mathbf{w} = [w_1, ..., w_n]^\mathsf{T}$ and values $\mathbf{c} = [c_1, ..., c_n]^\mathsf{T}$. One wants to select a subset with maximum profit such that the summation of the weights of the selected items is less or equal to the knapsack capacity d. Formulate this problem as an integer linear program. Then, solve this problem with a Genetic Algorithm using the following parameter values: $d = 47$, $\mathbf{c} = [10, 13, 18, 31, 7, 15]^\mathsf{T}$, $\mathbf{w} = [11, 15, 20, 35, 10, 33]^\mathsf{T}$. Solve again this problem with branch and cut and compare the performances of your solutions.

6.4 Tabu Search

Solve the knapsack problem of the previous example with the Tabu search metaheuristic and with the Hill Climbing heuristic. Describe the differences between the two in terms of solution stability.

6.5 Metaheuristics

Justify why it is useful to run a metaheuristic algorithm several times to solve the same problem and why we need to test the parameter values of a metaheuristic when solving a problem.

References

1. Z. Li, I. Kucukkoc, J.M. Nilakantan, Comput. Oper. Res. **84**, 146 (2017)
2. J. Pearl (1984)
3. S. Voß, S. Martello, I.H. Osman, C. Roucairol, *Meta-heuristics: Advances and Trends in Local Search Paradigms for Optimization* (Springer Science & Business Media, 2012)
4. J.H. Holland, *Adaptation in Natural and Artificial Systems* (University of Michigan Press, Ann Arbor, Michigan, 1975)
5. L. Budin, M. Golub, A. Budin, Sign **1**(11), 52 (2010)
6. P. Nanakorn, K. Meesomklin, Comput. Struct. **79**(29–30), 2527 (2001)
7. T.W. Rondeau, C.W. Bostian, in *Cognitive Radio Technology* (Elsevier, 2006), pp. 219–268
8. M.J. Kochenderfer, T.A. Wheeler, *Algorithms for Optimization* (MIT Press, 2019)
9. S. Kirkpatrick, C.D. Gelatt, M.P. Vecchi, Science **220**(4598), 671 (1983)
10. N. Metropolis, A.W. Rosenbluth, M.N. Rosenbluth, A.H. Teller, E. Teller, J. Chem. Phys. **21**(6), 1087 (1953)
11. M. Richey, Am. Math. Mon. **117**(5), 383 (2010)
12. S. Sieniutycz, J. Jeowski, *Energy Optimization in Process Systems and Fuel Cells* (2013), pp. 1–43
13. E.H. Aarts, J.H. Korst, Eur. J. Oper. Res. **39**(1), 79 (1989)
14. F. Glover, Comput. Oper. Res. **13**(5), 533 (1986)
15. F. Glover, ORSA J. Comput. **1**(3), 190 (1989)
16. F. Glover, ORSA J. Comput. **2**(1), 4 (1990)
17. F. Glover, E. Taillard, Ann. Oper. Res. **41**(1), 1 (1993)
18. R. Storn, K. Price, J. Global Optim. **11**(4), 341 (1997)

19. J. Kennedy, R. Eberhart, in *Proceedings of ICNN'95-International Conference on Neural Networks*, vol. 4 (IEEE, 1995), pp. 1942–1948
20. Y. Zhang, S. Wang, G. Ji, Math. Probl. Eng. **2015** (2015)
21. Y.D. Zhang, S. Wang, Z. Dong, Prog. Electromag. Res. **144**, 171 (2014)

Chapter 7
Multi-objective Optimization

Abstract Up to now, we focused on single-objective optimization problems that have a scalar-valued objective function. Many problems, however, have more than one objective and there is a need to find a trade-off between the different objectives. This chapter focuses on multi-objective optimization problems and the theory behind them. It presents traditional approaches, such as the weighted sum, the ϵ-constraint, and the lexicographic method. It also presents multi-objective evolutionary algorithms, which are metaheuristics (i.e. NSGA-II, MOEA/D).

7.1 Multi-objective Optimization

We have already discussed the case of single-objective optimization problems that seek to minimize or maximize a single function $f : \mathbb{R}^n \to \mathbb{R}$ subject to equality and inequality constraints. In multi-objective optimization problems (MOOPs) we have more than one objective function. In addition, the objective functions need to be optimized simultaneously. This can be expressed as:

$$\min_{\mathbf{x} \in \mathcal{X}} \mathbf{f}(\mathbf{x})$$

where $\mathbf{x}$ is an n-valued vector, $\mathcal{X}$ is the feasible region—known as *decision space*, and $\mathbf{f} : \mathcal{X} \to \mathbb{R}^m$ where $\mathbf{f}(\mathbf{x}) = [f_1(\mathbf{x}), ..., f_m(\mathbf{x})]^{\mathsf{T}}$ is a vector-valued function where $\mathbb{R}^m$ is known as the *objective space*. That is, the vector valued function $\mathbf{f}$ maps the decision space $\mathcal{X}$ to the objective space $\mathbb{R}^m$. In literature, multi-objective optimization problems with many functions f_i are typically called *many-objective optimization problems* [1].

Because we have multiple objectives, it becomes apparent that we can have trade-offs between two or more conflicting objectives when trying to find a solution. In a *nontrivial* multi-objective optimization problem, there is no single solution $\mathbf{x}$ that simultaneously optimizes each objective because some of the objective functions are conflicting. The answer to the optimization problem is a set of solutions that define the best trade-offs between competing objectives. In essence, the concept of

K. Gkiotsalitis, *Public Transport Optimization*,
https://doi.org/10.1007/978-3-031-12444-0_7

optimality is replaced with that of Pareto optimality or efficiency, named after the economist Vilfredo Pareto.

In single-objective optimization problems, the superiority of a solution over all other solutions is determined by comparing their objective function values. In multi-objective optimization problems, the goodness of a solution is determined by the concept of dominance and there might be many equally good solutions for a specific problem. A solution of a multi-objective optimization is called nondominated, Pareto optimal, Pareto efficient or noninferior if we cannot improve the value of any objective function without degrading the value(s) of some of the other objective functions.

Pareto Dominance

Let $\mathbf{x}_1 \in \mathcal{X}$ be a feasible solution of a multi-objective minimization problem. Then, $\mathbf{x}_1$ is said to *dominate* another feasible solution $\mathbf{x}_2$ (symbolically, $\mathbf{x}_1 \prec \mathbf{x}_2$) if, and only if:

- $f_i(\mathbf{x}_1) \leq f_i(\mathbf{x}_2) \ \forall i \in \{1, ..., m\}$
- $f_i(\mathbf{x}_1) < f_i(\mathbf{x}_2) \ \text{for at least one } i \in \{1, ..., m\}$

That is, solution $\mathbf{x}_1$ is no worse than $\mathbf{x}_2$ in all objectives $\mathbf{f}$ and it is strictly better than $\mathbf{x}_2$ in at least one objective.

A design is considered Pareto optimal if there does not exist any other design which improves the value of any of its objective criteria without deteriorating at least one other criterion [2]. Let us consider again that $\mathbf{x}_1$ dominates another feasible solution $\mathbf{x}_2$ when $f_i(\mathbf{x}_1) \leq f_i(\mathbf{x}_2) \ \forall i \in \{1, ..., m\}$ and $f_i(\mathbf{x}_1) < f_i(\mathbf{x}_2)$ for at least one $i \in \{1, ..., m\}$. A solution $\mathbf{x}^*$ is called *Pareto Optimal* (or, equivalently, *non-dominated*) and $\mathbf{f}(\mathbf{x}^*)$ a *Pareto Optimal Outcome* if there is no other feasible solution $\mathbf{x}' \in \mathcal{X}$ that dominates solution $\mathbf{x}^*$. That is, $\nexists \mathbf{x}' \in \mathcal{X}$ such that $f_i(\mathbf{x}') \leq f_i(\mathbf{x}^*) \ \forall i \in \{1, ..., m\}$ and $f_i(\mathbf{x}') < f_i(\mathbf{x}^*)$ for at least one $i \in \{1, ..., m\}$. We now proceed to the definitions of the Pareto optimal set and the Pareto front.

Definition 7.1 *Pareto optimal set* or *set of efficient solutions* is called the set of all Pareto Optimal solutions in the decision space $\mathcal{X}_E := \{\mathbf{x}_i \in \mathcal{X} \ : \ \mathbf{x} \nprec \mathbf{x}_i \ \forall \mathbf{x} \in \mathcal{X}\}$.

Definition 7.2 *Pareto front* or *Pareto frontier* is called the set of all Pareto Optimal outcomes $F := \{\mathbf{f}(\mathbf{x}_i) \ : \ \mathbf{x}_i \in \mathcal{X}_E\}$.

The main aim of solving a multi-objective optimization problem is to find or approximate all or a representative set of Pareto optimal (and thus non-dominated) solutions. Figure 7.1 illustrates the difference between the Pareto optimal set and the Pareto front for the case of two decision variables and function $\mathbf{f} : \mathcal{X} \to \mathbb{R}^2$.

According to Hwang and Masud [3], the methods for solving MOOPs can be classified into three broad categories considering the point in time the decision maker provides additional preference information:

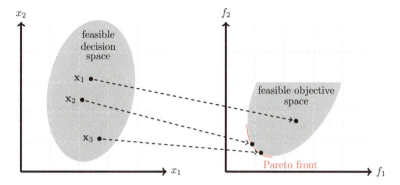

Fig. 7.1 Example of Pareto Front and mapping of points from the decision space to the objective space for a minimization MOOP. Solutions $\mathbf{x}_2$ and $\mathbf{x}_3$ are Pareto optimal (non-dominated). Solution $\mathbf{x}_1$ is dominated. For instance, $\mathbf{x}_2$ dominates $\mathbf{x}_1$ because $f_1(\mathbf{x}_2) < f_1(\mathbf{x}_1)$ and $f_2(\mathbf{x}_2) < f_2(\mathbf{x}_1)$

- a priori methods
- interactive (progressive) methods
- a posteriori methods

In a priori methods, preference information is first asked from the decision maker. Then, a solution best satisfying the decision-maker's preferences is found. This information might include setting goals or weights to the objective functions *prior* to the optimization phase. This implies that a total order is defined between the different objectives. The disadvantage of a priori methods is that it is very difficult for the decision maker to know beforehand and to be able to *accurately quantify* his/her preferences. The main advantage is that a priori information, if successfully quantified, can reduce the complexity of the MOOP (i.e., this information can be used to solve a single-objective optimization problem instead of a MOOP).

In the interactive methods, there are phases of information interchange between the decision maker and the results of calculations. The decision maker progressively drives the search toward the most preferred solution by using input from intermediate calculations. The drawback is that he/she never sees the whole picture (the Pareto set) or an approximation of it [4].

A posteriori methods aim at producing all the Pareto optimal solutions or a sufficient representation of them. After a representative set of Pareto optimal solutions is found, the decision maker must choose one of these non-dominated solutions. By selecting one of the non-dominated solutions, the decision-maker states his/her preferences a posteriori after being informed about the trade-offs among the non-dominated solutions. The main advantage of a posteriori methods is that the decision maker has no involvement in the process of generating the Pareto optimal solutions (or an adequate representation of them). The disadvantage is the increased computational complexity because all Pareto optimal solutions (or an adequate representation of them) must be computed.

7.2 Classic a Priori MOOP Methods

7.2.1 *Weighted Sum Method*

Weighted sum is arguably the simplest method for solving MOOPs. It is oftentimes seen as an a priori method because it requires receiving input from the decision maker regarding the importance of each objective function. Let us consider again the general representation of a MOOP:

$$\min_{\mathbf{x} \in \mathcal{X}} \mathbf{f}(\mathbf{x})$$

where $\mathbf{x}$ is an n-valued vector, $\mathcal{X}$ is the decision space, and $\mathbf{f} : \mathcal{X} \to \mathbb{R}^m$ where $\mathbf{f}(\mathbf{x}) = [f_1(\mathbf{x}), ..., f_m(\mathbf{x})]^\mathsf{T}$.

When using the weighted-sum method, this MOOP is approximated by a single-objective optimization problem expressed as:

$$\min_{\mathbf{x} \in \mathcal{X}} \sum_{i=1}^{m} w_i f_i(\mathbf{x}) \quad \text{(Weighted sum formulation with linear scalarization)}$$

where $w_i \in \mathbb{R}$ is the weight factor associated with each objective function $f_i \in \mathbf{f}$. The values of $\mathbf{w} = [w_1, ..., w_m]^\mathsf{T}$ should be determined before (a priori) solving this single-objective optimization problem based on the decision-maker's preferences. By doing so, we require a priori information about the importance of each objective function. The main advantage of this method is its simplicity because we can solve the MOOP by using any solution method used for single-objective optimization problems. In addition, it guarantees that its solution (if it exists) is on the Pareto optimal set of the MOOP.

Theorem 7.1 *The solution of a weighted sum problem is on the Pareto optimal set of the original minimization MOOP, no matter which weights in $\mathbf{w} \in \mathbb{R}_{\geq 0}^m$ are chosen.*

Proof Let $\mathbf{x}^*$ be the solution of the single-objective weighted sum problem for some $\mathbf{w} \in \mathbb{R}_{\geq 0}^m$. Suppose that this solution is dominated (and thus, not in the Pareto optimal set) of the MOOP. Then, there exists a feasible vector $\mathbf{k} \in \mathcal{X}$ with $f_i(\mathbf{k}) \leq f_i(\mathbf{x}^*)$ for all $i \in \{1, ..., m\}$ and $f_j(\mathbf{k}) < f_j(\mathbf{x}^*)$ for some $j \in \{1, ..., m\}$. Hence, it must also hold that:

$$\sum_{i=1}^{m} w_i f_i(\mathbf{k}) < \sum_{i=1}^{m} w_i f_i(\mathbf{x}^*)$$

which contradicts the assumption that $\mathbf{x}^*$ is a minimal. □

Closing, we should note that the weighted sum method has some considerable disadvantages:

- It is difficult to receive correct information from decision makers to set the values of the weight factors.
- Even if information from decision makers is received, it is difficult to accurately represent the decision maker's preferences with weights.
- If the Pareto front is not convex, there can be points on the Pareto front which are not the solutions of any weighted sum problem.

7.2.2 Lexicographic Method

The Lexicographic method is another a priori method which requires knowing the preferences of the decision maker before solving a MOOP. With the lexicographic method, preferences are imposed by ordering the objective functions according to their importance, rather than by assigning weights [5]. Let us consider again the general representation of a MOOP:

$$\min_{\mathbf{x} \in \mathcal{X}} \mathbf{f}(\mathbf{x})$$

where $\mathbf{x}$ is an n-valued vector, $\mathcal{X}$ is the decision space, and $\mathbf{f} : \mathcal{X} \to \mathbb{R}^m$ where $\mathbf{f}(\mathbf{x}) = [f_1(\mathbf{x}), ..., f_m(\mathbf{x})]^\mathsf{T}$.

The lexicographic method prioritizes the objective functions and orders them from the most to least important. Let, without loss of generality, f_1 be the most important objective function and f_m the least important, where f_i is more important than f_{i+1} for any $i \in \{1, ..., m-1\}$. Then, the lexicographic method returns a single solution by solving the following series of single-optimization problems for $i = 2, ..., m$:

$$\min_{\mathbf{x} \in \mathcal{X}} f_i(\mathbf{x})$$
$$\text{s.t.} \quad f_j(\mathbf{x}) \leq f_j(\mathbf{x}_j^*); \quad j = \{1, ..., i-1\}$$
$$\mathbf{x} \in \mathcal{X}$$

where i is the function's position in the preference sequence, and $f_j(\mathbf{x}_j^*)$ is the minimum value of the jth objective function found at the jth optimization problem. To solve this sequence of single-objective optimization problems, we need to compute first the value of $f_1(\mathbf{x}_1^*)$ which is obtained by solving the single-objective optimization problem:

$$\min_{\mathbf{x} \in \mathcal{X}} f_1(\mathbf{x})$$

Note that each new problem adds one constraint as i goes from 1 to m. If, for example, we have a problem with 3 objective functions ranked from the most important to the least important, then the solution is obtained through the following steps:

$$\mathbf{x}_1^* = \underset{\mathbf{x}\in\mathcal{X}}{\text{argmin}}\ f_1(\mathbf{x})$$

$$\mathbf{x}_2^* = \underset{\mathbf{x}\in\mathcal{X}}{\text{argmin}}\ \left\{f_2(\mathbf{x})\ \text{s.t.:}\ f_1(\mathbf{x}) \le f_1(\mathbf{x}_1^*)\right\}$$

$$\mathbf{x}_3^* = \underset{\mathbf{x}\in\mathcal{X}}{\text{argmin}}\ \left\{f_3(\mathbf{x})\ \text{s.t.:}\ f_1(\mathbf{x}) \le f_1(\mathbf{x}_1^*)\ \text{and}\ f_2(\mathbf{x}) \le f_2(\mathbf{x}_2^*)\right\}$$

The algorithm can be terminated once a unique optimum is determined. This is indicated when two consecutive optimization problems yield the same solution point. To make this more concrete, consider the example of [6] with:

$$f_1(\mathbf{x}) = 4x_1 + x_2$$
$$f_2(\mathbf{x}) = 2x_1 + x_2$$
$$f_3(\mathbf{x}) = x_1 + x_2$$

and feasible region:

$$\mathcal{X} = \left\{\mathbf{x} \in \mathbb{R}^2 : \begin{array}{l} 7 - x_1 - 5x_2 \le 0 \\ 10 - 4x_1 - x_2 \le 0 \\ -7x_1 + 6x_2 - 9 \le 0 \\ -x_1 + 6x_2 - 24 \le 0 \end{array}\right\}$$

We first have:

$$\mathbf{x}_1^* = \underset{\mathbf{x}\in\mathcal{X}}{\text{argmin}}\ f_1(\mathbf{x}) = [1.64516, 3.41935]^{\mathsf{T}}$$

with $f_1(\mathbf{x}_1^*) = 10$. We then proceed with solving the next problem:

$$\mathbf{x}_2^* = \underset{\mathbf{x}\in\mathcal{X}}{\text{argmin}}\ \{f_2(\mathbf{x})\ \text{s.t.:}\ f_1(\mathbf{x}) \le 10\} = [2.26316, 0.947368]^{\mathsf{T}}$$

with $f_2(\mathbf{x}_2^*) = 5.47368$. Note that $\mathbf{x}_1^* \ne \mathbf{x}_2^*$, so we proceed with solving the next problem:

$$\mathbf{x}_3^* = \underset{\mathbf{x}\in\mathcal{X}}{\text{argmin}}\ \{f_3(\mathbf{x})\ \text{s.t.:}\ f_1(\mathbf{x}) \le 10\ \text{and}\ f_2(\mathbf{x}) \le 5.47368\} = [2.26316, 0.947368]^{\mathsf{T}}$$

with $f_3(\mathbf{x}_3^*) = 3.21$. Note that the solution of the second problem, $\mathbf{x}_2^*$, and the solution of the third problem, $\mathbf{x}_3^*$, are identical. Thus, we can terminate the algorithm. The advantages of the lexicographic method are as follows:

- It does not require that the objective functions be normalized.
- It always provides a Pareto optimal solution.

The disadvantages are:

- It can require the solution of many single-objective problems to obtain just one solution point.
- It requires additional constraints to be imposed.
- It is more effective when used with a global optimization engine, which can be expensive.

7.3   A Posteriori MOOP Methods: Mathematical Programming and Metaheuristics

A posteriori (also called *generation*) methods give the whole picture (i.e. the Pareto front or an adequate representation of it) to the decision maker; however, they are more computationally intensive. The decision maker is not involved in the problem-solving phase, but he/she is asked to select a Pareto efficient solution once a representation of the Pareto front is computed and the trade-offs between the non-dominated solutions have been made explicit.

7.3.1   Weighted Sum Method

We have already presented the *a priori* weighted sum method which approximates a MOOP by the single-optimization problem:

$$\min_{\mathbf{x} \in X} \sum_{i=1}^{m} w_i f_i(\mathbf{x})$$

where $\mathbf{w} \in \mathbb{R}^m$ is the weight factor associated with each objective function $f_i \in \mathbf{f}$. In a priori weighted sum methods, the values of the weights $\mathbf{w} \in \mathbb{R}^m$ are predefined (fixed) based on the explicit preferences of the decision maker. In practice, though, it is very hard to quantify the abstract preferences of the decision maker to tangible weight factor values. Contrary to the above, *a posteriori* weighted sum methods try to derive an adequate representation of the Pareto front by using different weights w_i and solving the single-objective optimization problem repeatedly with the hope to compute a representative set of the Pareto front. A main problem with the weighted sum method is that, even if we try numerous combinations of different weights, we might not be able to compute some of the points in the Pareto front if it is not convex.

A second issue is that there might be too many combinations of different weight factor values resulting in solving too many single-objective optimization problems.

7.3.2 ε-Constraint Method

The ϵ-constraint method can provide a representative subset of the Pareto front, which, in most cases, is adequate. With this approach we turn a MOOP into a constrained, single-objective optimization problem. Depending on its structure, it can be categorized as an a priori or an a posteriori method. Let us consider again the general representation of a MOOP:

$$\min_{\mathbf{x} \in \mathcal{X}} \mathbf{f}(\mathbf{x})$$

Using the ϵ-constraint method, we optimize one of the objective functions (let us, without loss of generality, consider $f_1(\mathbf{x})$), and treat the other $m - 1$ objective functions as constraints using constants $\epsilon_2, ..., \epsilon_m$:

$$
\begin{aligned}
\min_{\mathbf{x} \in \mathcal{X}} \ & f_1(\mathbf{x}) \\
\text{s.t. } & f_2(\mathbf{x}) \le \epsilon_2 \\
& f_3(\mathbf{x}) \le \epsilon_3 \\
& \cdots \\
& f_m(\mathbf{x}) \le \epsilon_m
\end{aligned}
\tag{7.1}
$$

If the right-hand-side values ($\epsilon_2, ..., \epsilon_m$) are a priori defined with input from the decision maker, then we have an *a priori* ϵ-constraint method. If, however, we vary the values of the constants and solve (7.1) repeatedly, we can compute a (possibly adequate) subset of the Pareto front without using any information from the decision maker. In the latter case, the ϵ-constraint is an a posteriori (generator) method.

The advantages of the ϵ-constraint method compared to the weighted sum when used as a generator method are explained by Mavrotas [4]:

- For linear problems, the weighted sum method is applied to the original feasible region and results in a corner solution (extreme solution), thus generating only efficient extreme solutions. On the contrary, the ϵ-constraint method alters the original feasible region and is able to produce non-extreme efficient solutions. As a consequence, with the weighting method we may spend a lot of runs that result in the same efficient extreme solution. In contrast, with the ϵ-constraint we can exploit almost every run to produce a different efficient solution, thus obtaining a more rich representation of the efficient set.

- The weighted sum method cannot produce unsupported efficient solutions in multi-objective integer and mixed-integer programming problems, while the ϵ-constraint method does not suffer from this pitfall [7].
- In the weighted sum method the scaling of the objective functions has strong influence in the obtained results. Therefore, we need to scale the objective functions to a common scale before forming the weighted sum. In the ϵ-constrained method, this is not necessary.
- The ϵ-constraint method can control the number of the generated efficient solutions by properly adjusting the number of grid points in each one of the objective function ranges. This is not so easy with the weighted sum method.

The ϵ-constraint method has also some disadvantages. The main one is that the values of the $\epsilon \in \mathbb{R}^{m-1}$ vector should be chosen carefully so that each ϵ_i can take values within the minimum or maximum values of the respective objective function, f_i. This implies that to apply the ϵ-constraint method we need to know the ranges of the objective functions. However, these ranges are sometimes difficult to obtain.

7.3.3 Multi-objective Evolutionary Algorithms (MOEAs)

Multi-objective evolutionary algorithms (MOEAs) are metaheuristics which gradually approach sets of Pareto optimal solutions. The first MOEAs were developed in the 1990s [8, 9] and they were documented in detail in the book of Deb [10]. Two common features of MOEAs are: (i) assigning fitness to population members based on non-dominated sorting, and (ii) preserving diversity among solutions of the same non-dominated front. In contrast to the classic approaches presented in the previous sections that try to solve multiple single-objective optimization problems to approximate the Pareto Front, MOEAs typically try find an approximation of the Pareto Front in a single simulation run. In the remainder of this chapter we present examples of common MOEAs, which are broadly split into three categories:

- Pareto-based MOEAs, such as Non-dominated Sorting Genetic Algorithm II (NSGA-II) [11], strength-Pareto EA 2 (SPEA2).
- Indicator-based MOEAs, such as SMS-EMOA [12].
- Decomposition-based MOEAs, such as the Multi-objective Evolutionary Algorithm based on Decomposition (MOEA/D) [13], NSGA-III.

7.3.3.1 NSGA-II

Presented in Deb et al. [11], NSGA-II is a fast, elitist multi-objective optimization genetic algorithm. NSGA-II differs from GAs for single-objective optimization because it uses a different mechanism to rank individuals. First, it performs non-dominated sorting. Second, individuals which share the same rank after performing non-dominated sorting are ranked according to the crowding distance criterion. The crowding distance is the Manhattan distance in the objective space (defined as the sum of the absolute differences between two vectors) and the extreme points are assigned a crowding distance of infinity because it is desired to keep them at every generation.

Let us start with the non-dominated sorting step. Given a population $P \subseteq \mathcal{X}$ of size $|P|$ and a MOOP with m objective functions, to identify the non-dominated front we need to compare every solution with every other solution. This requires $m|P|$ comparisons for each solution, resulting in a total computational complexity of $O(m|P|^2)$ to find the non-dominated front of population P, denoted as F_1. To sort all population members, we temporarily discount all solutions in the first non-dominated front and we compute the next (second) non-dominated front F_2. The task of finding the second non-dominated front also requires $O(m|P|^2)$. We continue this approach with the third front, fourth front, and so on until all population members are sorted. In the worst case where we have $|P|$ fronts because each front has only one solution, the computational complexity of non-dominated sorting which ranks all population members of P is $O(m|P|^3)$.

To reduce the computational complexity of non-dominated sorting, [11] proposed to find for each individual $p \in P$ its domination count n_p which is equal to the number of solutions $p' \in P$ which dominate p, and the set S_p of solutions $p' \in P$ which p dominates ($S_p = \{p' \in P : p \prec p'\}$). The complexity of this is $O(m|P|^2)$. By doing so, all solutions p in the first non-dominated front that refer to population P will have $n_p = 0$ and rank $p_{rank} = 1$ meaning that they have the highest hierarchy. For each $p \in P$ such that $n_p = 0$ we visit each member p' of set S_p and reduce its domination count by one. In doing so, if for any member p' the domination count $n_{p'}$ becomes 0, we put it in a separate list Q and these members belong to the second non-dominated front. The above procedure is continued with each member of Q and the third front is identified. This process continues until all fronts are identified with a reduced complexity of $O(m|P|^2)$ instead of $O(m|P|^3)$. This fast non-dominated sort routine for a population P is presented in Algorithm 32 where the $O(m|P|^2)$ complexity becomes evident by the number of nested for loops. By the end of this algorithm, we have found the rank of each population member $p \in P$.

Algorithm 32 Fast-non-dominated-sort (P)

1: **for** each $p \in P$ **do**
2:      set $S_p = \emptyset, n_p = 0$
3:      **for** each $p' \in P$ **do**
4:          **if** $p \prec p'$ **then**
5:              $S_p = S_p \cup \{p'\}$
6:          **end if**
7:          **if** $p' \prec p$ **then**
8:              $n_p = n_p + 1$
9:          **end if**
10:     **end for**
11:     **if** $n_p = 0$ **then**
12:         set $p_{rank} = 1$
13:         set $F_1 = F_1 \cup \{p\}$ to add p to the 1st set of non-dominated solutions
14:     **end if**
15: **end for**
16: set $i = 1$ to initialize the front counter
17: **repeat**
18:     set $Q = \emptyset$
19:     **for** each $p \in F_i$ **do**
20:         **for** each $p' \in S_p$ **do**
21:             $n_{p'} = n_{p'} - 1$
22:             **if** $n_{p'} = 0$ **then**
23:                 $p'_{rank} = i + 1$
24:                 $Q = Q \cup \{p'\}$
25:             **end if**
26:         **end for**
27:     **end for**
28:     set $i = i + 1$ and $F_i = Q$
29: **until** $F_i = \emptyset$

After completing the non-dominated sorting, individuals who have the same rank (and thus belong to the same non-dominated set I) are subsequently ranked according to the crowding distance criterion to maintain diversity among population members. The crowding-distance computation requires sorting the population according to each objective function. Thus, for each objective function f_k with $k = 1, ..., m$, the boundary solutions that have the smallest and largest function values are assigned an infinite distance. Every other intermediate solution for objective k is assigned a distance value equal to the absolute normalized difference in the function values of two adjacent solutions. This calculation is continued with other objective functions and the overall crowding-distance value is calculated as the sum of individual distance values corresponding to each objective. The process of computing the crowding distances of non-dominated solutions in I is presented in Algorithm 33.

Algorithm 33 Crowding-distance-assignment (I)

1: set $l = |I|$ indicating the size of the non-dominated set I
2: **for** each $i \in I$ **do**
3:　set distance $I_d[i] = 0$
4: **end for**
5: **for** each objective $k = 1, ..., m$ **do**
6:　set $I = \text{sort}(I, k)$
7:　set $I_d[1] = I_d[l] = +\infty$
8:　**for** $i = 2, ..., l - 1$ **do**
9:　　set $I_d[i] = I_d[i] + \frac{f_k(I[i+1]) - f_k(I[i-1])}{f_k^{\max} - f_k^{\min}}$
10:　**end for**
11: **end for**

After computing the rank p_{rank} and the crowding distance p_d of each population member $p \in P$, NSGA-II uses a crowded-comparison operator ($\prec_n$) to guide the selection process toward a uniformly spread-out Pareto front. The partial order ($\prec_n$), indicating the better fitness of a population member p compared to another population member q, is defined as:

$$p \prec_n q \quad \text{if:} \qquad \begin{array}{c} p_{rank} < q_{rank} \\ \text{or} \\ p_{rank} = q_{rank} \text{ and } p_d > q_d \end{array}$$

Simply put, between two solutions with differing ranks, we prefer the solution with the lower (better) rank. If both solutions belong to the same front, then we prefer the solution that is located in a less crowded region. This fitness evaluation of population members is the key difference between NSGA-II and GAs for the single-objective optimization process. To discuss the entire process, NSGA-II starts from an initial population P_0 which is sorted based on non-domination. Then, NSGA-II uses a binary tournament selection where the fitness of each population member is based on the crowded-comparison operator ($\prec_n$). Then, it performs recombination (crossover), and mutation to create a new offspring population O_0 of size $|P|$. Because NSGA-II uses elitism, at the t-th population generation of NSGA-II are performed the following steps:

1. a combined population $R_t = P_t \cup O_t$ is formed, with size $2|P|$.
2. population R_t is sorted according to non-domination. Since all previous (P_t) and current (O_t) population members are included in R_t, elitism is ensured.
3. the new population generation P_{t+1} has $|P|$ members and is produced by R_t as follows. First, if the best non-dominated solutions F_1 of population R_t are less than $|P|$, then all of them are included in P_{t+1}. If, however, the size of F_1 is greater than $|P|$, then we use the crowded-comparison operator to choose the $|P|$ best solutions in F_1. In case that the size of F_1 is less than $|P|$, we do the same for the second set of non-dominated solutions F_2 and the process can continue for the 3rd set, the 4th set and so on until population P_{t+1} has $|P|$ members.

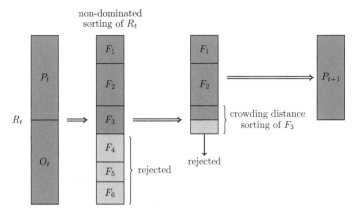

Fig. 7.2 Example of using population P_t and its offsprings O_t to generate population P_{t+1} with NSGA-II

It is important to note that in the population generation P_{t+1} we prioritize population members based on their rank and we apply the crowded-comparison operator only for the non-dominated solutions of a set F_i for which we cannot add all its members to P_{t+1} because this will lead to P_{t+1} having a population size of more than $|P|$. An example of the population generation process is shown in Fig. 7.2.

Let us consider the implementation of NSGA-II to the multi-objective optimization benchmark instance called ZDT1:

$$
\begin{aligned}
&\min \ f_1(\mathbf{x}) \\
&\min \ f_2(\mathbf{x}) \\
&\text{s.t. } f_1(\mathbf{x}) = x_1 \\
&\qquad g(\mathbf{x}) = 1 + \frac{9 \sum_{i=2}^{30} x_i}{(30-1)} \\
&\qquad f_2(\mathbf{x}) = g(\mathbf{x})(1 - \sqrt{x_1/g(\mathbf{x})}) \\
&\qquad 0 \le x_i \le 1 \ \forall i \in \{1, ..., 30\}
\end{aligned}
\tag{7.2}
$$

In Fig. 7.3 we present the results using 80 population members and a termination criterion of 200 population generations. Figure 7.3 presents the Pareto frontier produced by NSGA-II. We should note here that NSGA-II is a metaheuristic which, similarly to GAs, involves stochastic processes in the generation of populations. This means that running the algorithm again might yield different results and another set of Pareto optimal outcomes.

Fig. 7.3 Approximated
Pareto Front when solving
ZDT1 with NSGA-II

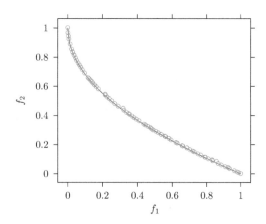

An implementation of NSGA-II as described in [11] is provided in the pymoo
library in Python. The method `pymoo.algorithms.moo.nsga2.NSGA2` is
implemented in the following example to solve the ZDT1 problem instance expressed
in Eq. (7.2).

```
from pymoo.algorithms.moo.nsga2 import NSGA2
from pymoo.factory import get_problem
from pymoo.optimize import minimize
from pymoo.visualization.scatter import Scatter

problem = get_problem("zdt1")
algorithm = NSGA2(pop_size=80)
res = minimize(problem,algorithm,('n_gen', 200),seed=1,verbose=
    True)

plot = Scatter()
plot.add(problem.pareto_front(), plot_type="line", color="black",
    alpha=0.7)
plot.add(res.F, facecolor="none", edgecolor="red")
plot.show()
```

7.3.3.2 MOEA/D

We have already presented traditional MOOP approaches that use scalarizations to
decompose a MOOP into single-objective sub-problems. MOEA/D follows a similar strategy, but incorporates the decomposition into an evolutionary algorithm. Proposed by Zhang and Li in 2007 [13], MOEA/D decomposes a MOOP into a number of
scalar optimization sub-problems and optimizes them simultaneously. Because each

sub-problem is optimized by only using information from its neighboring sub-problems, MOEA/D has lower computational complexity at each generation than NSGA-II. MOEA/D belongs to the broader category of decomposition-based MOEAs. Note that even the early proposed Vector Evaluated Genetic Algorithm (VEGA) of Schaffer in 1985 [14] can be considered as a rudimentary decomposition-based approach.

MOEA/D works with scalarizations, such as the linear weighting scalarizations that we discussed in the weighted sum method. Zhang and Li [13] proposed two other alternatives to linear scalarizations, namely the Chebychev scalarizations and the Boundary Intersection approach. Let us consider the case of Chebychev scalarizations, where the Chebychev scalarization problem is defined below.

Definition 7.3 *Chebychev Scalarization Problem.* Given a minimization MOOP $\min_{\mathbf{x} \in \mathcal{X}} \mathbf{f}(\mathbf{x})$ with $\mathbf{f} : \mathcal{X} \to \mathbb{R}^m$ where $\mathbf{x}$ is an n-valued decision vector, the Chebychev Scalarization Problem is:

$$\min \max_{i=1,\ldots,m} \lambda_i |f_i(\mathbf{x}) - z_i^*|$$

$$\text{s.t. } \mathbf{x} \in \mathcal{X}$$

where $\boldsymbol{\lambda} \in \mathbb{R}_{>0}^m$ is a weight vector and $\mathbf{z}^*$ is a reference objective vector, i.e., the ideal (utopian) objective vector defined as $z_i^* := \inf_{\mathbf{x} \in \mathcal{X}} f_i(\mathbf{x})$ with $i = 1, \ldots, m$.

Given population P_t of the t-th generation, MOEA/D evolves this population using weight vectors $\boldsymbol{\lambda}^p = [\lambda_1^p, \ldots, \lambda_m^p]^\top$ associated to every population member $p \in P_t$. The weight vectors are evenly distributed in the search space. The p-th sub-problem is defined by the Chebychev scalarization function:

$$g(\mathbf{x}|\boldsymbol{\lambda}^p, \mathbf{z}^*) := \max_{i=1,\ldots,m} \lambda_i^p |f_i(\mathbf{x}) - z_i^*|$$

In MOEA/D, a neighborhood of weight vector $\boldsymbol{\lambda}^p$ is defined as a set B_p of its k closest weight vectors in $\{\boldsymbol{\lambda}^1, \ldots, \boldsymbol{\lambda}^{|P|}\}$, where $|P|$ is the population size and k a MOEA/D parameter. The neighborhood of the p-th sup-problem consists of all the sub-problems with weight vectors in the neighborhood of $\boldsymbol{\lambda}^p$. At each generation t, MOEA/D with Chebychev scalarization maintains:

- a population of $|P|$ points $\mathbf{x}^1, \ldots, \mathbf{x}^{|P|} \in \mathcal{X}$, where $\mathbf{x}^p$ is the current solution of the p-th sub-problem.
- $FV^1, \ldots, FV^{|P|}$, where $FV^p = \mathbf{f}(\mathbf{x}^p)$ for each $p = 1, \ldots, |P|$.
- $\mathbf{z}^* = [\mathbf{z}_1^*, \ldots, \mathbf{z}_m^*]^\top$, where $\mathbf{z}_i^*$ is the best value found so far for objective f_i.
- an external population EP which is used to store non-dominated solutions found during the search.

The MOEA/D method is presented in Algorithm 34.

Algorithm 34 MOEA/D with Chebychev scalarization [13]

1: set $EP = \emptyset$, $|P|$, k and initialize $\mathbf{z}^*$
2: initialize $\boldsymbol{\lambda}^1, ..., \boldsymbol{\lambda}^{|P|}$ and generate initial population $P_0 = \{\mathbf{x}^1, ..., \mathbf{x}^{|P|}\}$
3: set $FV^p = \mathbf{f}(\mathbf{x}^p)$ for each population member
4: compute the Euclidean distances between any two weight vectors and store the k closest vectors to each weight vector $\boldsymbol{\lambda}^p$ in set B_p
5: **repeat**
6: 　**for** each population member $p = 1, ..., |P|$ **do**
7: 　　reproduction: randomly select two indexes r, s from B_p and generate new solution $\mathbf{y}$ from $\mathbf{x}^r$ and $\mathbf{x}^s$ by a problem specific recombination operator.
8: 　　local improvement: apply a problem-specific improvement heuristic on $\mathbf{y}$ to produce $\mathbf{y}'$ based on the scalarized objective function $g(\mathbf{y}|\boldsymbol{\lambda}^p, \mathbf{z}^*)$.
9: 　　**for** each $i = 1, ..., m$ **do**
10: 　　　**if** $z_i^* < f_i(\mathbf{y}')$ **then**
11: 　　　　set $z_i^* = f_i(\mathbf{y}')$
12: 　　　**end if**
13: 　　**end for**
14: 　　**for** each index $j \in B_p$ **do**
15: 　　　**if** $g(\mathbf{y}'|\boldsymbol{\lambda}^j, \mathbf{z}^*) \leq g(\mathbf{x}^j|\boldsymbol{\lambda}^j, \mathbf{z}^*)$ **then**
16: 　　　　set $\mathbf{x}^j = \mathbf{y}'$ and $FV^j = \mathbf{f}(\mathbf{y}')$
17: 　　　**end if**
18: 　　**end for**
19: 　　remove from EP all vectors dominated by $\mathbf{y}'$ and add $\mathbf{y}'$ to EP if no vectors in EP dominate $\mathbf{y}'$.
20: 　**end for**
21: **until** a termination criterion is met.

Practitioner's Corner
MOEA/D in Python 3

An implementation of MOEA/D as described in [13] is provided in the pymoo library in Python. The method pymoo.algorithms.moo.moead.MOEAD is implemented below using 80 population members for solving the ZDT1 MOOP. The recombination probability is set to 0.7 and the number of considered neighbors k of a weight vector to 15. The termination criterion is set to be the creation of 200 population generations.

```
from pymoo.algorithms.moo.moead import MOEAD
from pymoo.factory import get_problem, get_reference_directions
from pymoo.optimize import minimize
from pymoo.visualization.scatter import Scatter
problem = get_problem("zdt1")
ref_dirs = get_reference_directions("das-dennis", 2, n_partitions=80)
algorithm = MOEAD(n_neighbors=15, ref_dirs=ref_dirs,
    prob_neighbor_mating=0.7)
```

Fig. 7.4 Approximated Pareto Front when solving ZDT1 with MOEA/D

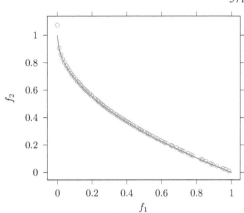

```
res = minimize(problem,algorithm,('n_gen', 200),seed=1,verbose=True)
plot = Scatter()
plot.add(problem.pareto_front(), plot_type="line", color="black",
    alpha=0.7)
plot.add(res.F, facecolor="none", edgecolor="red")
plot.show()
```

Solving this problem will result in the Pareto frontier of Fig. 7.4. We should note here that MOEA/D is a metaheuristic and running the algorithm again might yield different results.

Exercises

7.1 Dominance
Consider the multi-objective optimization problem:

$$\min_{x \in \mathbb{R}} \ \mathbf{f}(x)$$

with $\mathbf{f}(x) = [x^2 - x, x]^{\mathsf{T}}$. Does solution $x_1 = 0$ dominate solution $x_2 = 2$? Justify your answer.

7.2 Lexicographic method
Solve the following MOOP with the lexicographic method.

$$\min_{x \in \mathbb{R}^2} \ \mathbf{f}(\mathbf{x})$$
$$\text{s.t. } x_1 + 2x_2 \le 15$$
$$x_1 + x_2 \ge 0$$

where $f_1(\mathbf{x}) = 8x_1 + x_2$, $f_2(\mathbf{x}) = 4x_1 + x_2$, and $f_3(\mathbf{x}) = x_1 + x_2$. Consider that f_1 has the highest and f_3 the lowest importance.

7.3 Weighted Sum

Consider $\mathbf{f}(\mathbf{x}) = [f_1(\mathbf{x}), f_2(\mathbf{x})]^\mathsf{T} = [x_1^2 + x_2^2, 12x_1 + 6x_2^2]^\mathsf{T}$. Solve the MOOP

$$\min_{x \in \mathbb{R}^2} \mathbf{f}(\mathbf{x})$$
$$\text{s.t. } x_1 + 2x_2 \leq 15$$
$$x_1 + x_2 \geq 0$$

with the weighted sum method by considering five scenarios with weights $(w_1, w_2) = (1, 2)$, $(w_1, w_2) = (1, 3)$, $(w_1, w_2) = (2, 1)$, $(w_1, w_2) = (3, 1)$, and $(w_1, w_2) = (4, 2)$. Present the optimal values of the objective functions f_1 and f_2 for each pair (w_1, w_2) in a 2-dimensional plot. Justify why these values belong to the Pareto Front.

7.4 ϵ-constraint method

Solve the following MOOP with the ϵ-constraint method.

$$\min_{x \in \mathbb{R}^2} \mathbf{f}(\mathbf{x})$$
$$\text{s.t. } x_1 + x_2 \geq 10$$

where $\mathbf{f}(\mathbf{x}) = [x_1^2 + 2x_2, x_1 + 3x_2]^\mathsf{T}$.

7.5 NSGA-II and MOEA/D

Solve the following mathematical program with NSGA-II and MOEA/D

$$\min f_1(\mathbf{x})$$
$$\min f_2(\mathbf{x})$$
$$\text{s.t. } f_1(\mathbf{x}) = x_1$$
$$g(\mathbf{x}) = 1 + \frac{9\sum_{i=2}^{30} x_i}{(30 - 1)}$$
$$f_2(\mathbf{x}) = g(\mathbf{x})(1 - \sqrt{x_1/g(\mathbf{x})})$$
$$0 \leq x_i \leq 1 \ \forall i \in \{1, ..., 30\}$$

7.6 Weighted sum and ϵ-constraint

Describe the differences between the weighted sum and the ϵ-constraint method.

References

1. P.J. Fleming, R.C. Purshouse, R.J. Lygoe, in *International Conference on Evolutionary Multi-criterion Optimization* (Springer, 2005), pp. 14–32
2. S. Brisset, F. Gillon, *Eco-Friendly Innovation in Electricity Transmission and Distribution Networks* (2015), pp. 83–97
3. C.L. Hwang, A.S.M. Masud, *Multiple Objective Decision Making Methods and Applications: a State-of-the-Art Survey*, vol. 164 (Springer Science & Business Media, 2012)
4. G. Mavrotas, Appl. Math. Comput. **213**(2), 455 (2009)
5. J.S. Arora, Introduction to optimum design 657–679 (2012)
6. K.H. Chang, Design theory and methods using CAD/CAE 325–406 (2015)
7. R.E. Steuer, L.R. Gardiner, J. Gray, J. Multi-Criteria Decis. Anal. **5**(3), 195 (1996)
8. F. Kursawe, in *International Conference on Parallel Problem Solving From Nature* (Springer, 1990), pp. 193–197
9. C.M. Fonseca, P.J. Fleming, et al., in *Icga*, vol. 93 (Citeseer, 1993), pp. 416–423
10. K. Deb, *Multi-Objective Optimization Using Evolutionary Algorithms*, vol. 16 (Wiley, 2001)
11. K. Deb, A. Pratap, S. Agarwal, T. Meyarivan, IEEE Trans. Evol. Comput. **6**(2), 182 (2002)
12. M. Emmerich, N. Beume, B. Naujoks, in *International Conference on Evolutionary Multi-Criterion Optimization* (Springer, 2005), pp. 62–76
13. Q. Zhang, H. Li, IEEE Trans. Evol. Comput. **11**(6), 712 (2007)
14. J.D. Schaffer, in *Proceedings of the First International Conference on Genetic Algorithms and Their Applications, 1985* (Lawrence Erlbaum Associates. Inc., Publishers, 1985)

Part III
Public Transport Optimization: From Network Design to Operations

In this part of the textbook, we focus explicitly on public transport optimization problems for designing, planning, and operating more efficient services. Readers with a strong background in operations research can study this part of the textbook independently. Because new public transport problems can emerge every day due to the advancements in technology, regulations, electrification, vehicle automation, and shared mobility, this part of the textbook presents classic public transport problems from the areas of strategic planning, tactical planning, and operational control that can be easily extended to address future needs. Given that the problem descriptions of public transport problems are always evolving, it is important for the reader to be equipped with the required background in operations research so that he/she can follow future developments and design appropriate optimization models. After completing this part of the textbook, the reader is expected to be able to do the following:

1. Formulate, program, and solve classic public transport problems at the strategic planning phase, including the optimal selection of stations/stops, the planning of lines and their routes, and transit network design and frequency settings problems.
2. Formulate, program and solve classic public transport problems at the tactical planning phase, including the optimal selection of service frequencies, the selection of dispatching times of service trips (timetables), the synchronization of timetables at service lines with transfer stops, the design of vehicle schedules, and the design of crew schedules.
3. Formulate, program, and solve classic public transport problems at the operational control stage, including optimizing the skipping of stops, the rescheduling of vehicles, and the holding of vehicles at stops to avoid bunching.
4. Formulate, program, and solve classic problems related to shared mobility and on-demand services, including the traveling salesman problem, the vehicle routing problem from one or more depots, and the dial-a-ride problem which can be extended to carpooling applications.

In practice, there is no textbook that can cover the vast field of all potential public transport problem descriptions. At the time of this writing, many more public transport problem descriptions emerge on a daily basis. The reader, however, will be well-equipped to address these new problems by being aware of the classic problem descriptions/formulations and being able to create new ones. This will allow the reader to adapt the formulations of classic problems to the future needs of public transport service providers by using modeling techniques taught in this textbook.

To provide an idea of the broad spectrum of problem descriptions and mathematical formulations, selected subjects are accompanied with further reading notes which present relevant literature review studies that include links to additional, more recent, public transport problem descriptions.

Chapter 8
Strategic Planning of Public Transport Services

Abstract Optimizing the operations of public transport systems involves many levels, starting from strategic planning and moving to tactical planning and near real-time (operational) control. In a general planning process, we move from the Optimal Stop Location problem to Transit Network Design (TND), Frequency Setting (FS), Transit Network Timetabling (TNT), Vehicle Scheduling (VSP), and Crew Scheduling (CSP). This chapter covers the strategic public transport planning level, including aspects such as the estimation of trip distribution, the optimal stop location problem, routing problems, and the line planning problem that result in the design of the stations and the line routes of a transit network. The decisions about the locations of the stations and the routes of fixed line services are strategic because, once made, it is not easy to amend them within a short time.

8.1 Trip Distribution

To design and operate a public transit service the tasks that have to be performed are:

1. choose the locations of public transport stations/stops;
2. choose a set of routes (lines);
3. allocate service frequencies to the lines;
4. compile detailed timetables for the lines;
5. schedule public transit vehicles to operate the timetables;
6. schedule crews (drivers) to operate the vehicles;
7. schedule the rosters of the crews.

In reality, all these problems interact. This requires the use of a global optimization model that can capture these interactions. Due to the complexity of each one of these problems, however, in practice they are decomposed and solved in sequence [1].

Before designing a public transport system, one should take into consideration the trip generation, trip distribution and mode choice stages considered in traditional transportation planning to estimate the passenger demand and the origin-destination pairs of potential public transport trips. To plan a public transport network at the strategic level, we need first to have a spatio-temporal estimation of the passenger demand. There are several trip generation/trip attraction models based on regression analysis that can provide an estimate of the number of generated and attracted trips in different areas (known as *zones*). Defining zones and establishing the centroids of the zones is a key problem of trip generation/attraction models that provide an estimate of the generated/attracted trips from/to these zones. To design a public transport network, we first need to understand the distribution of these trips in the network and express them in origin-destination matrices that provide the number of trips from a zone to another. In the remainder of this section, we provide a brief overview of common trip distribution approaches that are used to estimate the distribution of trips in a transport network.

8.1.1 Entropy Maximization

Trips can be distributed to origin-destination pairs with the use of the entropy maximization theory proposed by Wilson in 1969 [2]. Consider, for example, three zones of a city for which we know the generated, $\mathbf{d} = [d_1, d_2, d_3]^\mathsf{T}$, and attracted, $\mathbf{o} = [o_1, o_2, o_3]^\mathsf{T}$ trips. In a trip distribution problem, we seek to estimate how many trips generated from zone $i \in I = \{1, 2, 3\}$ are attracted by zone $j \in I$. This results in the origin-destination matrix $\mathbf{T} = \{t_{ij}\}$ expressed below.

The entropy maximization model for trip distribution is provided below.

Sets: The only required set is the set of origins I which is the same as the set of destinations.

Parameters: The values of the generated and attracted trips from/to each zone are parameter values. In more detail, in Table 8.1 the values of the generated $\mathbf{d}$ and attracted $\mathbf{o}$ trips are already known from the trip generation stage. We also know the total number of trips $z = \sum_{i \in I} d_i = \sum_{i \in I} o_i$.

Variables: We seek to estimate the origin-destination matrix $\mathbf{T} = \{t_{ij}\}$. That is, determine every value t_{ij}, where t_{ij} is a positive continuous variable representing the number of trips from i to j. Following the thermodynamic concept of entropy

Table 8.1 Example of origin-destination matrix $\mathbf{T}$

Origin	Destination			
	1	2	3	o_i
1	t_{11}	t_{12}	t_{13}	$o_1 = \sum_{j \in I} t_{1j}$
2	t_{21}	t_{22}	t_{23}	$o_2 = \sum_{j \in I} t_{2j}$
3	t_{31}	t_{32}	t_{33}	$o_3 = \sum_{j \in I} t_{3j}$
d_i	$d_1 = \sum_{i \in I} t_{i1}$	$d_2 = \sum_{i \in I} t_{i2}$	$d_3 = \sum_{i \in I} t_{i3}$	

as a maximum disorder, the entropy-maximizing procedure seeks the most likely configuration of elements within a constrained situation. This is formally expressed by the following mathematical program that estimates the distribution of trips.

Trip distribution with Entropy maximization

$$\max_{\mathbf{T}\in\mathbb{R}_{\geq0}^{|I|\times|I|}} \frac{z!}{\prod_{(i,j):i\in I,j\in I} t_{ij}!} \tag{8.1}$$

$$\text{s.t.:} \sum_{j\in I} t_{ij} = o_i, \ \forall i \in I \tag{8.2}$$

$$\sum_{i\in I} t_{ij} = d_j, \ \forall j \in I \tag{8.3}$$

This maximization problem has a convex feasible region, but a fractional objective function. To simplify this, instead of maximizing the objective function $\dfrac{z!}{\prod_{(i,j):i\in I,j\in I} t_{ij}!}$ one can maximize:

$$\ln \frac{z!}{\prod_{(i,j):i\in I,j\in I} t_{ij}!}$$

Using Stirling's approximation for $\ln t_{ij}! = t_{ij} \ln t_{ij} - t_{ij}$, this results in:

$$\max_{\mathbf{T}} \ln z! - \sum_{i\in I}\sum_{j\in I}(t_{ij} \ln t_{ij} - t_{ij})$$

where $\ln z!$ is a constant and can be removed from the optimization problem. Thus, we have the reformulated problem:

$$\max_{\mathbf{T}\in\mathbb{R}_{\geq0}^{|I|\times|I|}} -\sum_{i\in I}\sum_{j\in I}(t_{ij} \ln t_{ij} - t_{ij}) \tag{8.4}$$

$$\text{s.t.} \sum_{j\in I} t_{ij} = o_i, \ \forall i \in I \tag{8.5}$$

$$\sum_{i\in I} t_{ij} = d_j, \ \forall j \in I \tag{8.6}$$

This maximization problem has *Lagrangian* function:

$$\mathcal{L}(\mathbf{T},\lambda',\lambda'') = -\sum_{i\in I}\sum_{j\in I}(t_{ij} \ln t_{ij} - t_{ij}) - \sum_{i\in I}\lambda_i'\left(\sum_{j\in I} t_{ij} - o_i\right) - \sum_{j\in I}\lambda_j''\left(\sum_{i\in I} t_{ij} - d_j\right)$$

where $\boldsymbol{\lambda}' = [\lambda_1', \lambda_2', \ldots \lambda_{|I|}']^\mathsf{T}$ and $\boldsymbol{\lambda}'' = [\lambda_1'', \lambda_2'', \ldots \lambda_{|I|}'']^\mathsf{T}$ are the Lagrange multipliers.

8.1.2 Growth Factor Method

Growth factor is another method to distribute generated and attracted trips and develop origin-destination matrices. The growth factor method can be *uniform*, *singly-constrained*, or *doubly-constrained*. In the uniform case, we assume that we know the base-year origin-destination matrix, $\mathbf{T}' = \{t_{ij}'\}$, and the growth rate τ for the trips of the whole study area until a future point in time. Then, the estimated number of trips t_{ij} from i to j at that future point in time is:

$$t_{ij} = \tau t_{ij}' \quad \forall i \in I, j \in I \tag{8.7}$$

where $t_{ij}' \in \mathbf{T}'$ is the number of trips from i to j at the base year. The main limitation of the growth factor method is that one should know:

- the base-year trip distribution, $\mathbf{T}'$. That is, we cannot apply the method in public transport networks that are not developed yet.
- the expected trip growth rate τ.

One more limitation of the *uniform* growth factor method is the underlying assumption that all origin-destination trips will increase by the same proportion τ. This, of course, is rarely the case because some areas might develop more than others in the future. To rectify this, one can change the growth term τ by introducing *singly-constrained* and *doubly-constrained* growth factor models.

In singly-constrained growth factor models, the available information includes:

- an origin-specific growth factor τ_i for every origin $i \in I$
- or a destination-specific growth factor τ_j for every destination $j \in I$

It is important to highlight that we either have an origin-specific growth factor or a destination-specific growth factor—not both. If we know the origin-specific growth factor, then the trip distribution formula becomes:

$$t_{ij} = \tau_i t_{ij}' \quad \forall i \in I, j \in I \tag{8.8}$$

Analogously, if we know the destination-specific growth factor, the trip distribution formula becomes:

$$t_{ij} = \tau_j t_{ij}' \quad \forall i \in I, j \in I \tag{8.9}$$

Consider, for instance, the base-year trip distribution matrix of Table 8.2 for a public transport network with 4 stops.

Table 8.2 Base-year trip distribution matrix, $\mathbf{T}' = \{t'_{ij}\}$

Origins	Destinations				$o_i = \sum_{j \in I} t_{ij}$	Future trip generations
	$j = 1$	$j = 2$	$j = 3$	$j = 4$		
$i = 1$	5	50	100	200	$o_1 = 355$	$o_1 = 400$
$i = 2$	50	5	100	300	$o_2 = 455$	$o_2 = 460$
$i = 3$	50	100	5	100	$o_3 = 255$	$o_3 = 400$
$i = 4$	100	200	250	20	$o_4 = 570$	$o_4 = 702$
$d_j = \sum_{i \in I} t_{ij}$	205	355	455	620	$T = 1635$	

Table 8.3 Estimated trip distribution matrix, $\mathbf{T} = \{t_{ij}\}$, with the singly-constrained growth factor method

Origins	Destinations				$o_i = \sum_{j \in I} t_{ij}$
	$j = 1$	$j = 2$	$j = 3$	$j = 4$	
$i = 1$	5.6	56.3	112.7	225.4	$o_1 = 400$
$i = 2$	50.5	5.1	101.1	303.3	$o_2 = 460$
$i = 3$	78.4	156.9	7.8	156.9	$o_3 = 400$
$i = 4$	123.2	246.3	307.3	24.6	$o_4 = 702$
$d_j = \sum_{i \in I} t_{ij}$	257.7	464.4	529.5	701.2	$T = 1962$

Considering the estimated future trip generations presented in the last column of the table, the origin-specific growth factors are $\tau_1 = 400/355 \simeq 1.13$, $\tau_2 = 1.01$, $\tau_3 = 1.57$ and $\tau_4 = 1.23$. This results in the future trip distribution matrix of Table 8.3.

Consider now the *doubly-constrained growth-factor* model where our available information includes:

- Origin-specific growth factors τ_i, $\forall i \in I$
- Destination-specific growth factors γ_j, $\forall i \in I$

If one applies the average growth factor $0.5(\tau_i + \gamma_j)$ in the doubly-constrained growth-factor trip distribution formula:

$$t_{ij} = 0.5(\tau_i + \gamma_j)t'_{ij} \quad \forall i \in I, \ j \in I \tag{8.10}$$

then, $\sum_{j \in I} t_{ij} \neq o_i$ and $\sum_{i \in I} t_{ij} \neq d_j$. That is, the generated/attracted trips from/to each zone do not match with the results from Eq. (8.10). To rectify this, the iterative algorithm of Furness [3] is applied. Let a_i and b_j be new variables indicating the origin-specific and destination-specific growth factors. Then, the algorithm of Furness performs the steps presented in Algorithm 35 until convergence.

Algorithm 35 Furness Iterative Algorithm for the Doubly-constrained Growth factor method

1: set convergence precision error ε, iteration $k = 0$, and $b_j = 1 \ \forall j \in I$
2: **repeat**
3:    set $a_i = \sum\limits_{j \in I} \tau_i t'_{ij} / \sum\limits_{j \in I} b_j t'_{ij} \ \forall i \in I$
4:    with the latest a_i values, find b_j values $b_j = \sum\limits_{i \in I} \gamma_j t'_{ij} / \sum\limits_{i \in I} a_i t'_{ij} \ \forall j \in I$
5: **until** the changes in the values of a_i, b_j from iteration to iteration are less than or equal to ε

Let us consider the following example of Table 8.4.

Applying the iterative algorithm of Furness, the trip distribution matrix, t_{ij}, is presented in Table 8.5. At the last iteration, the origin-specific and destination-specific growth factors have values $a_1 = 1.066$, $a_2 = 0.918$, $a_3 = 1.57$, $a_4 = 1.36$ and $b_1 = 0.978$, $b_2 = 0.821$, $b_3 = 0.915$, $b_4 = 1.192$, respectively.

Table 8.4 Base-year trip distribution matrix, $\mathbf{T}'$, and future trip generations/attractions per zone

Origins	Destinations				$o_i = \sum\limits_{j \in I} t_{ij}$	Future o_i
	$j = 1$	$j = 2$	$j = 3$	$j = 4$		
$i = 1$	5	50	100	200	$o_1 = 355$	$o_1 = 400$
$i = 2$	50	5	100	300	$o_2 = 455$	$o_2 = 460$
$i = 3$	50	100	5	100	$o_3 = 255$	$o_3 = 400$
$i = 4$	100	200	250	20	$o_4 = 570$	$o_4 = 702$
$d_j = \sum\limits_{i \in I} t_{ij}$	205	355	455	620	$T = 1635$	
Future d_j	260	400	500	802		$T = 1962$

Table 8.5 Estimated trip distribution matrix, $\mathbf{T} = \{t_{ij}\}$, with the doubly-constrained growth factor method

Origins	Destinations				$\sum\limits_{j \in I} t_{ij}$	Future o_i
	$j = 1$	$j = 2$	$j = 3$	$j = 4$		
$i = 1$	5.21	43.77	97.54	254.10	400.6	$o_1 = 400$
$i = 2$	44.90	3.77	84.01	328.30	460.9	$o_2 = 460$
$i = 3$	76.80	128.97	7.18	187.17	400.1	$o_3 = 400$
$i = 4$	133.09	223.49	311.27	32.44	700.3	$o_4 = 702$
$\sum\limits_{i \in I} t_{ij}$	260	400	500	802.01	$T \simeq 1962$	
Future d_j	260	400	500	802		$T = 1962$

8.1.3 Gravity Method

The gravity trip distribution model is an analogy to Newton's gravitation law:

$$t_{ij} = a \frac{o_i d_j}{l_{ij}^2} \tag{8.11}$$

where:

a is the gravitational constant

l_{ij} is the distance (cost/disutility) of traveling from zone i to j

o_i, d_j are the generated and attracted trips from the respective zones

The simplest form of the gravity model is the *uniform* gravity model that uses a distance decay function $f(c_{ij})$ to estimate the trips from i to j:

$$t_{ij} := a o_i d_j f(c_{ij}) \tag{8.12}$$

where the distance decay function $f(c_{ij})$ typically takes one of the following forms:

$$
\begin{aligned}
f(c_{ij}) &= \exp(-\beta c_{ij}) && \text{(exponential function)} \\
f(c_{ij}) &= c_{ij}^{-n} && \text{(power function)} \\
f(c_{ij}) &= c_{ij}^{n} \exp(-\beta c_{ij}) && \text{(combined function)}
\end{aligned}
$$

The *singly-constrained* Gravity Model alters the *uniform* Gravity Model by considering origin or destination-specific balancing factors. For an origin-constrained model:

$$t_{ij} = a_i o_i d_j f(c_{ij}) \tag{8.13}$$

where $a_i = 1 / \sum_{j \in I} d_j f(c_{ij})$.

For a destination-constrained model:

$$t_{ij} = o_i b_j d_j f(c_{ij}) \tag{8.14}$$

where $b_j = 1 / \sum_{i \in I} o_i f(c_{ij})$.

Lastly, the *doubly-constrained* Gravity Model estimates the number of trips from any $i \in I$ to any $j \in I$ as:

$$t_{ij} = a_i o_i b_j d_j f(c_{ij}) \tag{8.15}$$

where:

- $a_i = 1 / \sum_{j \in I} b_j d_j f(c_{ij})$
- $b_j = 1 / \sum_{i \in I} a_i o_i f(c_{ij})$

The balancing factors a_i, b_j are interdependent. Hence, t_{ij} should be determined with an iterative algorithm, such as the Furness algorithm presented in Algorithm 36.

Algorithm 36 Furness Iterative Algorithm for the Doubly-constrained Gravity method

1: set convergence precision error ε, iteration $k = 0$, and $b_j = 1 \ \forall j \in I$
2: **repeat**
3:     set $a_i = 1/\sum_{j \in I} b_j d_j f(c_{ij}) \ \forall i \in I$
4:     set $b_j = 1/\sum_{i \in I} a_i o_i f(c_{ij}) \ \forall j \in I$
5: **until** the changes in the values of a_i, b_j from iteration to iteration are less than or equal to ε

Consider, for example, the cost matrix of Table 8.6 when traveling between two zones.

For a distance decay function $f(c_{ij}) = \exp(-\beta c_{ij})$ with $\beta = 0.1$, the trip distribution matrix after 4 iterations is presented in Table 8.7.

Table 8.6 Cost matrix, c_{ij}

Origins	$j = 1$	$j = 2$	$j = 3$	$j = 4$	Target o_i
			Destinations		
$i = 1$	3	11	18	22	400
$i = 2$	12	3	12	19	460
$i = 3$	15.5	13	5	7	400
$i = 4$	24	18	8	5	702
Target d_j	260	400	500	802	$T = 1962$

Table 8.7 Computed Trip distribution, **T**, after 4 iterations

Origins	$j = 1$	$j = 2$	$j = 3$	$j = 4$	Target o_i
			Destinations		
$i = 1$	157.01	100.33	66.10	76.41	$399.9 \simeq 400$
$i = 2$	57.48	201.07	108.46	92.88	$459.9 \simeq 460$
$i = 3$	25.27	46.15	136.26	192.38	$400.06 \simeq 400$
$i = 4$	20.24	52.45	189.17	440.33	$702.2 \simeq 702$
Target d_j	260	400	500	802	$T = 1962$

8.2 Design of Stops

8.2.1 Location Set Covering Problem

An important strategic planning task is the decision about the locations of stops in a public transport network. One of the most common approaches of determining the locations of public transport stops is solving the Location Set Covering Problem (LSCP) which was originally proposed by Toregas et al. in 1971 [4]. The aim of LSCP is to find the smallest number of facilities and their locations so that each demand is covered by at least one facility. This is adapted in the context of public transport, where a facility refers to a public transport stop and the demand refers to the passenger demand at specific areas in the city (see Murray [5]).

LSCP can be formulated as a binary linear programming problem. It assumes that we have a number of potential locations $N = \{1, \ldots, j, \ldots, n\}$ where we can place a public transport stop. There is also a number of areas $M = \{1, \ldots, i, \ldots, m\}$ with passenger demand. Each passenger demand area i refers to a vertex (discrete point) of a continuous terrain. For instance, if we have passenger demand from a number of houses that are too close to each other, we can aggregate it and use a representative vertex (centroid) i to represent this demand. Because LSCP is a set covering problem, we also need a problem-specific definition of cover, as presented below.

Definition 8.1 *Cover of LSCP:* A demand at discrete point i is covered by a facility (public transport stop) j if j is within a maximum distance l of i.

Let us consider, for instance, the example of Fig. 8.1. Demand points 3 and 4 are covered by both stops. In contrast, demand points 1 and 2 are only covered by stop 1 and demand points 5, 6 and 7 by stop 2.

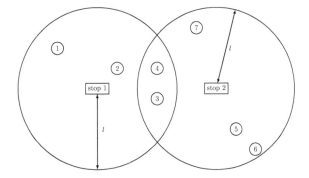

Fig. 8.1 Example of 7 demand points and their cover by 2 stops when considering a maximum distance l

Table 8.8 Nomenclature of LSCP

Sets	
N	set of possible public transport stop locations
M	set of passenger demand points
Parameters	
$\mathbf{A} = \{a_{ij}\}$	$\lvert M \rvert \times \lvert N \rvert$ matrix, where $a_{ij} = 1$ if stop $j \in N$ can cover demand $i \in M$, and $a_{ij} = 0$ otherwise.
Decision Variables	
$\mathbf{x}$	$\mathbf{x} = [x_1, \ldots, x_j, \ldots, x_{\lvert N \rvert}]^{\mathsf{T}}$, where $x_j = 1$ if location j is selected to be a public transport stop and $x_j = 0$ otherwise.

Using this definition of cover, LSCP is formally expressed as follows.

Location Set Covering Problem (LSCP)—Toregas et al. [4]
Find the smallest number of facilities (public transit stops) that are needed such that the demand at each location $i \in M$ is covered.

To formulate this problem description into a binary linear program, the required sets, parameters and variables are presented in Table 8.8.
With this notation, the LSCP is formulated as follows.

Location Set Covering Problem (LSCP)

$$\min_{\mathbf{x}} \sum_{j \in N} x_j$$

$$\text{s.t.:} \sum_{j \in N} a_{ij} x_j \geq 1 \quad \forall i \in M \tag{8.16}$$

$$x_j \in \{0, 1\} \qquad \forall j \in N$$

Note that constraints $\sum_{j \in N} a_{ij} x_j \geq 1$ for all $i \in M$ ensure that each demand location $i \in M$ is covered because they require that $x_j = 1$ for at least one stop $j \in N$ for which $a_{ij} = 1$. That is, these inequality constraints ensure that each demand point is covered by at least one public transport stop. The LSCP formulation is a special case of the *minimum weight set covering problem* which has the same structure but its objective function is $\sum_{j \in N} c_j x_j$ for some weight-related parameter vector $\mathbf{c}$.

A required pre-processing step when applying LSCP is to determine whether a stop j can cover a demand i based on a maximum distance threshold, $l \in \mathbb{R}_{\geq 0}$. To identify that in a real-life road network, we should pre-solve $\lvert N \rvert \times \lvert M \rvert$ shortest path problems to find the shortest distance between any possible pair (i, j) and check if this distance is less than or equal to l. If this is the case, then we can set $a_{ij} = 1$. In essence, it is equivalent to the formulation of the *minimum set covering* problem.

We should also note that there are several reduction rules for the set covering problem that can be also applied to the LSCP to reduce its number of variables and constraints. Some of them proposed by Toregas and ReVelle [6] are the following:

- if there exists at least one demand point i where $a_{ij} = 1$ for a stop location j and $a_{ik} = 0$ for any other stop location $k \in M \setminus \{i\}$, then stop location j is an *essential* public transport stop and we can set $x_j = 1$ or, equivalently, remove row i and column j from the coverage matrix of the linear program.
- if stop location j has $a_{ij} \geq a_{ik}$ for any demand point $i \in M$, then stop location j *dominates* stop location k and we can remove stop location k resulting in a problem without variable x_k.
- if demand point i has $a_{ij} \leq a_{sj}$ for any public transport stop location j, then i dominates demand point s, meaning that if demand i is covered, demand s will be covered as well. Thus, demand point s can be removed from the problem.

Let us consider, for instance, a network with 9 passenger demand nodes and 3 potential public transport stop locations. The shortest path distance between every demand point $i \in \{1, \ldots, 9\}$ and every potential stop location $j = \{1, 2, 3\}$ is presented in Table 8.9.

If we set maximum distance $l = 300$ meters to consider a demand point covered, then we can produce another matrix with the parameter values of a_{ij} (see Table 8.10).

We can observe that demand points $i = 1, 2, 3, 4$ can be covered only by stop $j = 1$. Thus, $j = 1$ is an essential public transport stop and should be selected. Demand points 5 and 6 can also be covered only by stop $j = 2$, which is also an essential stop. Because the size of the problem is small, we can solve it by inspection declaring solution $\mathbf{x} = [1, 1, 0]^\mathsf{T}$. An example of solving this problem with branch-and-cut using a MILP solver is presented below. Using branch-and-cut is useful when solving larger problem instances with more potential stop locations and demand

Table 8.9 Example shortest distance matrix in meters

d_{ij}	$j = 1$	$j = 2$	$j = 3$
$i = 1$	100	600	400
$i = 2$	150	800	800
$i = 3$	200	800	1100
$i = 4$	200	400	600
$i = 5$	400	200	800
$i = 6$	800	200	600
$i = 7$	800	250	200
$i = 8$	600	200	200
$i = 9$	200	800	200

Table 8.10 Values of the cover indicator parameters

a_{ij}	$j = 1$	$j = 2$	$j = 3$
$i = 1$	1	0	0
$i = 2$	1	0	0
$i = 3$	1	0	0
$i = 4$	1	0	0
$i = 5$	0	1	0
$i = 6$	0	1	0
$i = 7$	0	1	1
$i = 8$	0	1	1
$i = 9$	1	0	1

points. Note that LSCP is NP-Hard because the *set cover* problem is one of Karp's NP-complete decision problems.

Practitioner's Corner
Solving LSCP in Python 3

One can use PuLP to program the LSCP model in Python and solve it with solvers that can apply branch-and-cut (CPLEX, Gurobi, CBC). Below follows an example of modeling the aforementioned problem in PuLP and solving it with the CBC solver. For demonstration purposes, we do not use primal heuristics to provide better bounds.

```python
import pulp
from pulp import *
# cover indicator parameters
A = [[1,0,0], [1,0,0], [1,0,0], [1,0,0], [0,1,0], [0,1,0], [0,1,1],
    [0,1,1], [1,0,1]]
# number of possible stop locations and demand points
n, m = 3, 9
N = set(range(0,n)); M = set(range(0,m))
model = LpProblem("LSCP", LpMinimize)
# variables
x = LpVariable.dicts("x", N, None, None, LpBinary)
# objective function
model += (sum(x[j] for j in N))
# constraints:
for i in M: model += sum(A[i][j]*x[j] for j in N) >= 1
# optimizing
solver = pulp.PULP_CBC_CMD(mip=True, msg=True, timeLimit=None,
    warmStart=False, path=None, mip_start=False, timeMode='elapsed',
    options=["passCuts=0","preprocess=off","presolve=off","cuts=on","
    heuristics=off","greedyHeuristic=off"])
model.solve(solver)
print ("Status:", LpStatus[model.status])
for v in model.variables():
```

```
    if v.varValue>0: print (v.name, "=", v.varValue)
print ("Optimal Solution = ", value(model.objective))
```

8.2.2 The Maximal Covering Location Problem

The Maximal Covering Location Problem (MCLP) is another problem formulation
for determining the locations of stops which relaxes the LSCP's requirement of serving each demand point. Presented in 1974 by Church and ReVelle [7], this problem
considers a fixed number of p public transport stops and seeks to find the optimal
public transport stop locations to maximize the served demand. MCLP is not con-
cerned with covering all passenger demand, since this might be impossible given the
restriction of placing exactly p stops. That is, it does not ensure that any demand
point i will have at least a public transport stop j within a maximum distance l.
Instead, it focuses on finding the best locations for the p stops such that they can
serve as much passenger demand as possible.

This problem is particularly relevant when the decision maker (public transport
authority) does not have the resources to build a public transport network to cover
all passenger demand points and needs to cover the passenger demand in the most
efficient way when placing exactly p stops. Stops p should be placed such that we
attain the most possible passengers that can be covered (i.e., be within a distance l of
the locations of the selected stops). To avoid leaving areas with low demand without
public transport stop near them, there might be an additional requirement that all
demand points will at least have a public transport stop within a maximum distance
r, where $r > l$. This maximum distance r is typically much larger than l to ensure
feasibility ($r \gg l$). The MCLP problem can be expressed as follows.

> **Maximal Covering Location Problem (MCLP)—Church and ReVelle** [7]
> Locate a fixed number of facilities (public transport stops) in order to maximize
> the population (passenger demand) covered within a service distance l, while
> maintaining mandatory coverage within a distance of r ($r > l$).

In addition to the notation introduced for the LSCP problem, to formulate MCLP
we need to add the following parameters:

- $\beta_{ij} \in \{0, 1\}$, where $\beta_{ij} = 1$ if stop $j \in N$ is within distance r of demand node
  $i \in M$
- $\mathbf{c} = [c_1, \ldots, c_{|M|}]^\mathsf{T}$, where $c_i \in \mathbb{Z}_{\geq 0}$ is the passenger demand at demand node
  $i \in M$

We also need to add binary decision variables $\mathbf{y} = [y_1, \ldots, y_{|M|}]^{\mathsf{T}}$, where $y_i = 1$ if demand point $i \in M$ is covered by a public transport stop. Then, the problem can be formulated as follows.

Maximum Covering Location Problem (MCLP)

$$\max_{\mathbf{x},\mathbf{y}} \sum_{i \in M} c_i y_i$$

$$\text{s.t.:} \sum_{j \in N} \beta_{ij} x_j \geq 1 \quad \forall i \in M$$

$$\sum_{j \in N} a_{ij} x_j \geq y_i \quad \forall i \in M \tag{8.17}$$

$$\sum_{j \in N} x_j = p$$

$$x_j \in \{0, 1\} \quad \forall j \in N$$

$$y_i \in \{0, 1\} \quad \forall i \in M$$

In the MCLP formulation, constraints $\sum_{j \in N} \beta_{ij} x_j \geq 1$ ensure that there will be at least one public transport stop within a distance r of any demand point $i \in M$. Constraint $\sum_{j \in N} x_j = p$ forces us to select exactly p public transport stops. Constraints $\sum_{j \in N} a_{ij} x_j \geq y_i$ ensure that if a demand point $i \in M$ is covered ($y_i = 1$), then there should be at least one selected public transport stop j ($x_j = 1$) within distance l. Finally, the objective function $\mathbf{c}^{\mathsf{T}} \mathbf{y} = \sum_{i \in M} c_i y_i$ strives to maximize the served passenger demand.

MCLP is a binary linear programming problem. It can be solved to global optimality with branch-and-cut. To improve the computation time of the problem, one can apply a primal heuristic or a metaheuristic to obtain a sub-optimal solution and use it as a warm start solution. A typical greedy heuristic is the Greedy Adding algorithm which starts with an empty solution set and adds a public transport stop to it at each iteration as follows:

- first, the public transport stop which serves the most demand is added.
- second, the public transport stop which serves the most passenger demand, which is not yet covered by the first public transport stop, is added.

This process continues until adding p stops to the set. The Greedy Adding algorithm is applicable when the MCLP formulation does not consider constraints $\sum_{j \in N} \beta_{ij} x_j \geq 1$ related to the low demand areas that should be within a distance r of a stop. Otherwise, its produced solution might not be feasible.

Consider the following example with parameter values $p = 2$, $\mathbf{c} = [15, 21, 33, 52, 68, 71, 89, 32, 61]^{\mathsf{T}}$ and cover indicators a_{ij}, β_{ij} presented in Tables 8.11 and 8.12, respectively.

Table 8.11 Values of the cover indicator parameters for maximum distance l

a_{ij}	$j=1$	$j=2$	$j=3$	$j=4$	$j=5$
$i=1$	1	0	0	1	0
$i=2$	1	0	0	1	1
$i=3$	1	0	0	1	0
$i=4$	1	0	0	0	0
$i=5$	0	0	0	1	0
$i=6$	0	0	1	0	0
$i=7$	0	1	0	0	0
$i=8$	0	1	1	0	1
$i=9$	1	0	1	0	1

Table 8.12 Values of the cover indicator parameters for maximum distance $r > l$

β_{ij}	$j=1$	$j=2$	$j=3$	$j=4$	$j=5$
$i=1$	1	1	1	1	0
$i=2$	1	0	1	1	1
$i=3$	1	1	0	1	0
$i=4$	1	1	0	1	0
$i=5$	1	1	1	1	1
$i=6$	1	1	1	0	1
$i=7$	1	1	1	1	1
$i=8$	1	1	1	0	1
$i=9$	1	0	1	1	1

Solving this problem will result in serving demand points 1,2,3,4 and 7,8,9 by placing two public transport stops at locations 1 and 2. That is, $\mathbf{x}^* = [1, 1, 0, 0, 0]^\mathsf{T}$ and $\mathbf{y}^* = [1, 1, 1, 1, 0, 0, 1, 1, 1]^\mathsf{T}$.

Practitioner's Corner
Solving MCLP in Python 3

One can use PuLP to program the MCLP model in Python and solve it with solvers that can apply branch-and-cut (CPLEX, Gurobi, CBC). Below follows an example of modeling the aforementioned problem in PuLP and solving it with the CBC solver. For demonstration purposes, we do not use primal heuristics to provide better bounds.

```
import pulp
from pulp import *
# cover indicator parameters
A = [[1,0,0,1,0], [1,0,0,1,1], [1,0,0,1,0], [1,0,0,0,0], [0,0,1,1,0],
     [0,0,1,0,0], [0,1,0,0,0], [0,1,1,0,1], [1,0,1,0,1]]
```

```
B = [[1,1,1,1,0], [1,0,1,1,1], [1,1,0,1,0], [1,1,0,1,0], [1,1,1,1,1],
     [1,1,1,0,1], [1,1,1,1,1], [1,1,1,0,1], [1,0,1,1,1]]
# number of possible stop locations and demand points
n, m = 5, 9
p = 2
N = set(range(0,n)); M = set(range(0,m))
# passenger demand at each demand point
c = [15,21,33,52,68,71,89,32,61]
model = LpProblem("MCLP", LpMaximize)
# variables
x = LpVariable.dicts("x", N, None, None, LpBinary)
y = LpVariable.dicts("y", M, None, None, LpBinary)
# objective function
model += (sum(c[i]*y[i] for i in M))
# constraints:
for i in M: model += sum(A[i][j]*x[j] for j in N) >= y[i]
for i in M: model += sum(B[i][j]*x[j] for j in N) >= 1
model += sum(x[j] for j in N) == p
# optimizing
solver = pulp.PULP_CBC_CMD(mip=True, msg=True, timeLimit=None,
    warmStart=False, path=None, mip_start=False, timeMode='elapsed',
    options=["passCuts=0","preprocess=off","presolve=off","cuts=on","
    heuristics=off","greedyHeuristic=off"])
model.solve(solver)
print ("Status:", LpStatus[model.status])
for v in model.variables():
    if v.varValue>0: print (v.name, "=", v.varValue)
print ("Optimal Solution = ", value(model.objective))
```

8.2.3 The m-center Problem

The m-Center Problem (mCP) strives to locate a given number of facilities (public transport stops) anywhere along a road network so as to minimize the maximum distance between these facilities and fixed demand points assigned to them [8]. Minieka [9] proposed a method for solving the m-center problem by solving a finite series of minimum set covering problems. The key difference between the m-center problem and the LSCP, MCLP formulations is that now the possible public transport stop locations are not pre-defined. Instead, they can be any location in the continuous plane of the road network (i.e., any point in an edge (road section) that connects two vertices (demand points)).

m-center Problem (Minieka [9], 1970)
Select a set U of no more than m points (centers) from a road network with
infinite points so as to minimize the maximum distance between the fixed
demand points and their nearest m points.

To formulate the m-center problem, we present the formulation of Minieka [9].
Consider an undirected graph $G = (X, \Gamma)$ where X is the *finite* set of vertices (passenger demand points) and Γ the *finite* set of undirected edges (road sections). Let
P denote the *infinite* set of all points in all edges of G and Q the *infinite* set of all
points on the edges, including the vertices: $Q = P \cup X$.

Concerning the parameters of the problem, we first have that we can select up
to m points (public transport stops). Every edge $e \in \Gamma$ has an associated positive
real number $t_e \in \mathbb{R}_{>0}$ which can be regarded as the length of the edge. The m-center
problem on the plane considers that each edge $e \in \Gamma$ has an infinite set of points where
we can potentially place a public transport stop. A point c in edge e is specified by its
distance from the endpoints of edge e. Let also $d(i, j)$ be a function that determines
the shortest distance from $i \in X$ to $j \in Q$.

Mathematical Description: The m-center problem requests to find a set $U^* \subseteq Q$
with $|U^*| \leq m$ such that for every set $U \subseteq Q$ with $|U| \leq m$:

$$\max_{i \in X} f(i, U) \geq \max_{i \in X} f(i, U^*)$$

where

$$f(i, U) := \min_{j \in U} d(i, j)$$

The members (public transport stops) of U^* are called m-centers of G and set U^*
minimizes the maximum distance $\max_{i \in X} f(i, U^*)$ between a demand vertex i of G
and its nearest m-center.

To solve this problem, we can first reduce the infinite set of points P at every
edge as follows. Let $d(i, j)$ be the shortest path distance between vertex $i \in X$ and
point j in edge e. On edge e, $d_{i,j}$ is a continuous piecewise linear function. Let, for
example, e_s be the start point of edge e and e_t the endpoint. Then, j is situated λt_e
units from e_s, where $0 \leq \lambda \leq 1$, and $(1 - \lambda)t_e$ units from e_t. If $d(i, e_s)$ is the shortest
distance between i and e_s and $d(i, e_t)$ the shortest distance between i and e_t, then
the piecewise linear function $d(i, j)$ is:

$$d(i, j) := \min\{d(i, e_s) + \lambda t_e; d(i, e_t) + (1 - \lambda)t_e\}$$

Let $P' \subseteq P$ be the set of all points in edges of G such that $j \in P'$ if, and only
if, for some $i \in X$ and $y \in X$ where $i \neq y$, j is the unique point in its edge such
that $d(i, j) = d(y, j)$. Let also $Q' = P' \cup X$. Note that $|P'| < +\infty$ and $|Q'| < +\infty$

meaning that they are both finite sets. Minieka [9] proved that there exists an m-center contained in $Q' \subseteq Q$. That is, we no longer need to consider an infinite number of points as potential public transport stop locations, but we can concentrate on the finite set of points Q' and solve the following problem.

Find a set $U^* \subseteq Q'$ with $|U^*| \leq m$ such that for every set $U \subseteq Q'$ with $|U| \leq m$:

$$\max_{i \in X} \min_{j \in U} d(i, j) \geq \max_{i \in X} \min_{j \in U^*} d(i, j)$$

Notice that because X is a finite set and U is also a finite set as a subset of the finite set Q', we can now pre-compute the shortest path distances and store them in a $|X| \times |Q'|$ matrix $\mathbf{D}$ where d_{ij} is the shortest path distance from $i \in X$ to $j \in Q'$.

If we select m arbitrary points from Q' we have an arbitrary set $U_0 = \{u_1, \ldots, u_m\}$ with performance:

$$\tau_{U_0} := \max_{i \in X} \min_{j \in U_0} d_{i,j}$$

Using this new notation, U_0 is a globally optimal m-center if, and only if,

$$\tau_{U_0} \leq \tau_U \text{ for any set } U \subseteq Q' \text{ with } |U| \leq m$$

To validate whether an arbitrary set $U_0 \subseteq Q'$ is indeed a globally optimal solution, consider parameter matrix $\mathbf{A} \in \{0, 1\}^{|X| \times |Q'|}$ where $a_{ij} = 0$ if $d_{ij} \geq \tau_{U_0}$ and $a_{ij} = 1$ if $d_{ij} < \tau_{U_0}$. Using matrix $\mathbf{A}$ we solve the minimal set covering problem:

$$\min_{\mathbf{x}} \sum_{j \in Q'} x_j$$

$$\text{s.t. } \mathbf{Ax} \geq \mathbf{1} \tag{8.18}$$

$$\mathbf{x} \in \{0, 1\}^{|Q'|}$$

If the solution $\mathbf{x}^*$ of this problem has more than m values equal to 1 or does not exist, set U_0 is indeed a globally optimal solution. If not, then the columns of matrix $\mathbf{D}$ corresponding to the components of $\mathbf{x}^*$ that are equal to 1 represent another set U_1 which performs better than U_0: $\tau_{U_1} < \tau_{U_0}$. This minimal set covering problem is solved again using an updated parameter matrix $\mathbf{A}$ corresponding to set U_1 and its τ_{U_1} value.

The process continues by solving subsequent minimal set covering problems until finding a set U^* for which the solution of its corresponding minimal set covering problem has more than m components with a value equal to 1 or does not exist. Then, U^* is declared as a globally optimal solution. The number of solved minimal set covering problems by following this approach is *finite* because at each iteration the number of 0 entries at matrix $\mathbf{A}$ increases by at least one. However, this is an NP-Hard problem because we have to solve a finite number of NP-Hard minimum set covering problems.

8.3 Shortest Paths

After determining the locations of public transport stations, a typical problem at the strategic level of public transit is the generation of lines that minimize operating costs and passenger travel times. Given an OD-matrix, passengers also need to be assigned to lines considering the minimization of their travel times. To achieve this, we typically have to solve multiple shortest path problems. In a shortest path problem, we seek to find the shortest path from an origin vertex $a \in V$ to a destination vertex $f \in V$ given a network $G = (V, E)$ with V vertices and E edges. Occasionally, we also need to solve the k-shortest paths problem which seeks to find the first k shortest paths between two vertices of the network. In the remainder of this section, we present common algorithms for the solution of these problems.

8.3.1 Dijkstra's Algorithm

One of the most common algorithms for the shortest path problem is Dijkstra's heuristic which produces a globally optimal solution for networks with non-negative edge weights. The heuristic was conceived by Dijkstra at the Mathematical Center in Amsterdam in 1956 and presented at [10] three years later. Considering a network in the form of an undirected or directed graph $G = (V, E)$ where V are the vertices (intersections) and E the edges (road sections/track segments), the time complexity of Dijkstra's algorithm is $O(|V|^2)$ when implemented using an array data structure. Notice that this time is polynomial, indicating the computational efficiency of the algorithm. The definition of the single-pair shortest path problem (SPSPP) is provided below.

Single-Pair Shortest Path Problem
Consider an undirected or directed graph $G = (V, E)$ where vertices V are intersections and edges E are road/track segments. Find the shortest path and the associated shortest distance from an origin vertex $a \in V$ to a destination vertex $f \in V$ in graph G.

To apply Dijkstra's algorithm, the graph must have non-negative edge weights that connect each pair of nodes. Dijkstra's algorithm finds the shortest path and the associated shortest distance from an origin vertex $a \in V$ to a destination vertex $f \in V$ in a graph with non-negative edge weights $\mathbf{W}$ that connect each pair of nodes. Non-negative edge weights are required to avoid cycling. Dijkstra's algorithm uses a set V of all vertices in the network and a set B with all unvisited vertices in the network. It also uses an array $\mathbf{d} = [d_1, \ldots, d_{|V|}]^\mathsf{T}$ where d_i is the currently known minimum distance from the origin vertex a to vertex $i \in V$. In addition, it uses an array $\mathbf{p} = [p_1, \ldots, p_{|V|}]^\mathsf{T}$ where p_i indicates the previous vertex of vertex $i \in V$ at the shortest path from a to i. The non-negative weights $w_{ij} \in \mathbf{W}$ represent the length (or travel time) that is needed to traverse the directed edge $(i, j) \in E$. If vertices i and j are not directly connected by a directed edge, we can set $w_{ij} = +\infty$. If graph G is undirected, we also have $w_{ij} = w_{ji}$ for any $i \in V$, $j \in V$.

At the initialization stage of the algorithm, set B contains all vertices in V because they are all unvisited. In addition $p_i =$ undefined for all $i \in V$ because we do not know yet the previous vertex of any vertex $i \in V$ at the shortest path from a to i. Similarly, d_i is set to $+\infty$ for any $i \in V \setminus \{a\}$ since the minimum distance from the origin vertex i to $i \in V \setminus \{a\}$ is unknown. The only known minimum distance is the one from vertex a to itself, which is equal to 0. Thus, $d_a = 0$.

After the initialization, Dijkstra's algorithm iterates by performing the following steps until the destination vertex f is removed from set B:

1. sets *current* vertex $i \in B : d_i \le d_j$, $\forall j \in B$ (initially, $i = a$).
2. removes vertex i from the set of unvisited vertices B.
3. for all remaining unvisited vertices $j \in B$ checks if $d_i + w_{ij} \le d_j$ and if this is true sets $d_j = d_i + w_{ij}$ and $p_j = i$.

This process is summarized in Algorithm 37.

Algorithm 37 Dijkstra's algorithm for the Shortest Path Problem

---

1: **for** $i \in V$ **do**
2:    set $d_i = +\infty$, $p_i =$ undefined
3: **end for**
4: set $d_a = 0$ and $B \equiv V$
5: **repeat**
6:    set tentatively i as an arbitrary vertex in B
7:    **for** each $j \in B$ **do**
8:       if $d_j \le d_i$, set $i = j$
9:    **end for**
10:   set $B = B - \{i\}$
11:   **for** each $j \in B$ **do**
12:      **if** $d_i + w_{ij} \le d_j$ **then**
13:         set $d_j = d_i + w_{ij}$
14:         set $p_j = i$
15:      **end if**
16:   **end for**
17: **until** destination vertex f is not in B

Notice that the algorithm has a main and two nested *for* loops (lines 7–9 and 11–16). The main loop is the repeat loop which terminates when f is not in B and requires performing at most $|V|$ iterations. Each of the nested *for* loops searches all $j \in B$ at every step, which requires performing at most $5|V|$ operations. Thus, Dijkstra's algorithm has a time complexity of $O(|V|^2)$.

Fig. 8.2 Example network
$G = (V, A)$ with
non-negative weights for the
directed edges (arcs)

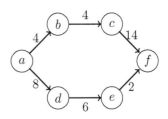

The output of the algorithm is the shortest path distance d_f when traveling from a to f. We also know the previous vertex of f that is in the shortest path from a to f because this information is stored at p_f. We can use backtracking to find all vertices that belong to the shortest path from a to f. To do so, we should implement the following algorithm after the termination of Dijkstra's algorithm in order to derive the shortest path S.

Algorithm 38 Find the Shortest Path with Backtracking

1: initialize the shortest path $S = \emptyset$
2: set $i = p_f$ and $S = \{i\}$
3: **repeat**
4:     replace i by p_i
5:     set $S = S \cup \{i\}$
6: **until** $i = a$
7: reverse the order of the vertices in S

Let us consider the example of directed graph G of Fig. 8.2 with non-negative arc weights. The weight values are presented in the middle of the directed edges (i.e., $w_{ab} = 4$). This network has vertices $V = \{a, b, c, d, e, f\}$. If there is no directed edge connecting two vertices in V, we can set its weight to $+\infty$. For instance, $w_{ac} = +\infty$.

An illustration of Dijkstra's steps is provided in Fig. 8.3 when searching for the shortest path from origin a to destination f. Note that we require 6 iterations to remove vertex f from the set of unvisited vertices B. When doing so, the shortest path distance from a to f is $d_f = 16$ and the previous vertex of f in the shortest path from a to f is $p_f = e$. Using backtracking, we can find $p_e = d$, and $p_d = a$. That is, the shortest path is $S = \{a, d, e, f\}$. Because the directed edge weights are non-negative, Dijkstra's algorithm does not produce cycles meaning that the shortest path from a to f does not include the same vertex twice.

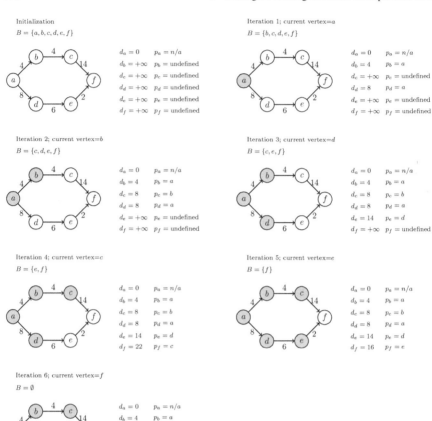

Fig. 8.3 Dijkstra's iterations to find the shortest path from a to f in network $G = (V, E)$ with non-negative directed edge weights

Practitioner's Corner
Dijkstra's algorithm in Python 3

Below we present an implementation of Dijkstra's algorithm for solving the shortest path problem of Fig. 8.2 in Python 3.

```python
def Dijkstra(origin,destination,V,w):
    p={i:"undefined" for i in V}; B=[i for i in V]
    d = {i: float("inf") for i in V}; d[origin] = 0
    while destination in B:
        i=B[0] #temporary initialization of i
        for j in B:
            if d[j]<=d[i]: i=j
        B.remove(i)
```

```
        for j in B:
            if d[i]+w[i,j]<=d[j]:
                d[j]=d[i]+w[i,j]; p[j]=i
    return(d,p)
def Shortest_Path(origin,destination,d,p):
    s_path=[destination];i=destination
    if d[destination]!=float("inf"):
        while i!=origin: #backtracking step
            i=p[i]; s_path.append(i)
        return (list(reversed(s_path)), d[destination])
    else:
        return('shortest path does not exist',d[destination])

V=('a','b','c','d','e','f')
w={(i,j):float("inf")for i in V for j in V}
w['a','b']=4; w['b','c']=4; w['c','f']=14
w['a','d']=8; w['d','e']=6; w['e','f']=2
[d,p]=Dijkstra('a','f',V,w)
print(Shortest_Path('a','f',d,p))
```

8.3.2 Dijkstra's Algorithm with Binary Heap

We showed that the computational complexity of Dijkstra's algorithm using an array data structure is $O(|V|^2)$. This can be further reduced to $O(|E|\log_2|V|)$ by using a binary min-heap data structure introduced by Williams in 1964 [11]. This data structure is defined below.

Definition 8.2 *Binary min-heap* is a data structure that takes the form of a complete binary tree where all levels are completely filled except possibly the last level, which has all keys as left as possible (filled from left to right). As a min-heap, the key at each leaf node is less than or equal to the keys in its leaf node children (min-heap property).

The binary min-heap takes the form of an array and represents a complete binary tree with the aforementioned properties. For instance, the binary min-heap of the complete binary tree in Fig. 8.4 that obeys the properties of the binary min-heap is $\mathbf{d} = [d_1, d_2, d_3, d_4, d_5, d_6]^\mathsf{T} = [1, 3, 6, 5, 10, 8]^\mathsf{T}$. For each key $i = 1, \ldots, 6$ the array representation $\mathbf{d}$ of the binary min-heap has a value d_i. In the case of the shortest path problem, each d_i value refers to the shortest path distance from the origin vertex to key i. To map the keys to our transport network graph, we would need another list $K = \{k_1, \ldots, k_6\}$ where k_i maps the i-th key of the binary min-heap to vertex k_i of the transport network.

Fig. 8.4 Complete binary
tree that obeys the binary
min-heap properties. The
binary min-heap is the array
$\mathbf{d} = [1, 3, 6, 5, 10, 8]^{\mathsf{T}}$

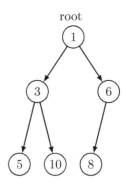

The binary min-heap has the following properties:

- the root element is always stored at d_1 and has the minimum value.
- a leaf i has children leaves $2i$ and $2i + 1$.
- if we are in a leaf i, then $\lfloor \frac{i}{2} \rfloor$ is the key of the parent leaf and $d_{\lfloor \frac{i}{2} \rfloor}$ is its value.

Because of the structure of the binary min-heap, finding the minimum key in the heap has a complexity of $\Theta(1)$ since it is always stored in d_1. That is, the *find-min* complexity is $\Theta(1)$. The complexity of extracting/deleting the minimum key (*delete-min*) is $O(\log_2 |V|)$, where $|V|$ is the total number of elements in the array. This is because we have to perform the following steps to restore the properties of the heap after a deletion:

- replace the root of the heap with the last element on the last level.
- compare the new root with its children:

  - if they are in the correct order, stop.
  - if not, swap the element with its smaller child and return to the previous step.

The aforementioned process restores the binary min-heap property and is known as *downheapify, heapify* or *bubbledown*. Let us consider that we want to extract the root from the previous example. Then, we would need to perform the steps of Fig. 8.5 to restore the binary min-heap property.

To perform the delete-min operation we need to compare the value of the new root ($d_1 = 8$) with its two children and do this at each step until there are no children with a lower key value. The binary tree has $\lceil \log |V| \rceil$ levels, where $|V|$ is the number of leaf nodes. In our example, $\lceil \log_2(6) \rceil \simeq 3$. The first level has one leaf (the root), the second has (at most) two leaf nodes, the third (at most) four leaf nodes, and so forth. Thus, to perform the *delete-min* operation we need to compare the value of a specific leaf node with the values of its two children at each level. This will require, at the worst-case scenario, $2\lceil \log_2 |V| \rceil$ operations resulting in a complexity of $O(\log_2 |V|)$. We can express this in an algorithmic form as in Algorithm 39.

Algorithm 39 Delete-min with downheapify()

1: set $k_1 = k_{|V|}$, $d_1 = d_{|V|}$ and $i = 1$
2: remove $k_{|V|}$ and $d_{|V|}$ from K and $\mathbf{d}$, respectively.
3: **repeat**
4:   set $current\_i = i$ and $no\_children = 0$
5:   set $no\_children = 2$ if $2i \leq |K|$ and $2i + 1 \leq |K|$
6:   set $no\_children = 1$ if $2i \leq |K|$ and $2i + 1 > |K|$
7:   **if** $no\_children = 2$ **then**
8:     **if** $d_{2i} < d_{2i+1}$ and $d_{2i} < d_i$ **then**
9:       swap k_i with k_{2i} and d_i with d_{2i}
10:       set $i = 2i$
11:     **else if** $d_{2i+1} \leq d_{2i}$ and $d_{2i+1} < d_i$ **then**
12:       swap k_i with k_{2i+1} and d_i with d_{2i+1}
13:       set $i = 2i + 1$
14:     **end if**
15:   **end if**
16:   **if** $no\_children = 1$ **then**
17:     **if** $d_{2i} < d_i$ **then**
18:       swap k_i with k_{2i} and d_i with d_{2i}
19:       set $i = 2i$
20:     **end if**
21:   **end if**
22: **until** $i = current\_i$

Another required operation is the *decrease-key* where a leaf node is re-arranged in the complete binary tree when its value is reduced. Let us say that leaf node 5 in Fig. 8.4 with value $d_5 = 10$ has its value updated to $d_5 = 2$. Then, we have to perform the following steps to restore the binary min-heap property:

- compare the value of the updated leaf node with the value of its parent leaf node and swap its position with the position of its parent if its value is lower.
- if the new position of the leaf node is correct, stop. Otherwise, swap it with its current parent and repeat this step until the leaf node is in its correct position.

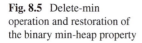

Fig. 8.5 Delete-min operation and restoration of the binary min-heap property

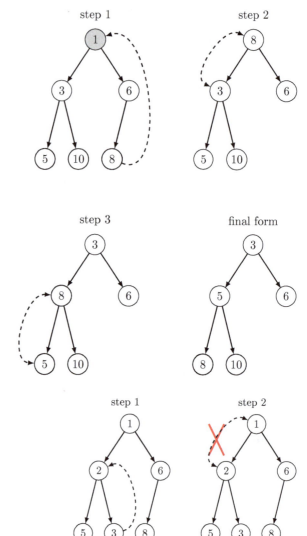

Fig. 8.6 Decrease-key operation and restoration of the binary min-heap property

The aforementioned process can restore the binary min-heap property every time the value of a leaf node is updated and it is known as *upheap* or *upheapify*. We can see the effect of this process in Fig. 8.6 where it becomes evident that its computational complexity is $O(\log_2 |V|)$.

The *decrease-key* operation is briefly presented in an algorithmic form in Algorithm 40.

Let us now consider the implementation of Dijkstra's algorithm with a binary min-heap. Let $K = \{k_1, \ldots, k_{|V|}\}$ be the mapping of each key $1, \ldots, |V|$ of the binary min-heap to each vertex $k_i \in V$ of the shortest path problem. If we consider the

---

Algorithm 40 Decrease-key with upheapify(i)

1: **while** $d_i < d_{\lfloor \frac{i}{2} \rfloor}$ **do**
2:   swap $k_{\lfloor \frac{i}{2} \rfloor}$ with k_i and $d_{\lfloor \frac{i}{2} \rfloor}$ with d_i
3:   set $i = \lfloor \frac{i}{2} \rfloor$
4:   if $i = 1$, **terminate**.
5: **end while**

---

example in Fig. 8.2, we can have the initial mapping $k_1 = a, k_2 = b, k_3 = c, k_4 = d, k_5 = e, k_6 = f$. The mapping of vertices to keys can be random, with only exception the assignment of the origin vertex a which should be the root of the binary min-heap ($k_1 = a$). We can then initialize the shortest path values **d**, where $d_1 = 0$ and $d_i = +\infty$ for all $i \in \{2, 3, \ldots, |V|\}$. Let also create an adjacency set A_{k_i} for each vertex $k_i \in V$ that contains the destination vertices of all outgoing edges of vertex $k_i \in V$.

We will now update the keys and the key values at each iteration of Dijkstra's algorithm until our root becomes the destination vertex f. We will terminate the algorithm if f is our root because then we know that we cannot reduce its shortest path distance any further. At each iteration we start from the root (initially, the origin vertex $k_1 = a$) and we set $r = k_1$, where r is the root. Following that, for every outgoing edge with destination vertex $k_i \in A_r \cap K$ we check if $d_1 + w_{r,k_i} \leq d_i$. If so, we set $d_i = d_1 + w_{r,k_i}$ and we perform *upheapify(i)*. When this process is finished, we perform *delete-min* to extract the root and we update the values of K and **d**. In the next iteration of Dijkstra's algorithm, we perform exactly the same steps and this process continues until our root r is the destination vertex f. The process is summarized in Algorithm 41.

Notice that in the repeat loop we perform *upheapify* with computational complexity $O(\log_2 |V|)$ each time an outgoing edge from the current vertex can result in an improved shortest distance for the edge's destination vertex. Because each vertex can be the root of the binary min-heap only at one iteration of the algorithm, the maximum number of possible explored edges is equal to the total number of edges in our network, $|E|$. Thus, the computational complexity is $O(|E| \log_2 |V|)$.

The binary min-heap has stable performance in practical applications, however, there are also other data structures that can reduce further the computational costs. For instance, implementing the Fibonacci Heap (see Fredman & Tarjan [12], 1984) reduces the time complexity to $O(|E| + |V| \log_2 |V|)$.

Algorithm 41 Dijkstra's algorithm with Binary min-heap

1: Consider a network $G = (V, E)$, an origin a, a destination f, edge distances $\mathbf{W}$, and an adjacency set A_v for each $v \in V$.
2: set $k_1 = a$, $d_1 = 0$ and $r = k_1$.
3: **for** $i \in \{1, \ldots, |V|\}$ **do**
4:    **if** $v_i \neq a$ **then**
5:       set $k_i = v_i$, $d_i = +\infty$, $p_{v_i} =$ undefined.
6:    **end if**
7: **end for**
8: **repeat**
9:    **for** for every outgoing edge with destination vertex $k_i \in A_r \cap K$ **do**
10:       **if** $d_1 + w_{r,k_i} \leq d_i$ **then**
11:          set $d_i = d_1 + w_{r,k_i}$
12:          set $p_{k_i} = r$
13:          perform decrease-key with *upheapify(i)*
14:       **end if**
15:    **end for**
16:    perform delete-min with *downheapify()*
17:    set $r = k_1$
18: **until** k_1 is the destination vertex f

Practitioner's Corner

Dijkstra's algorithm with binary min-heap in Python 3

Below we present an implementation of Dijkstra's algorithm with binary min-heap data structure for solving the shortest path problem of Fig. 8.2 in Python 3. To condense the input, the origin vertex of the shortest path is ordered first and the destination vertex is ordered last in the set of vertices V. We also use a swap function to swap the values of leaf nodes in the binary tree.

```
def swap(K,d,ind1,ind2):
    temp_d = d[ind1]; temp_K = K[ind1]
    d[ind1] = d[ind2]; K[ind1] = K[ind2]
    d[ind2] = temp_d; K[ind2] = temp_K

def upheapify(K,d,i):
    while i // 2 > 0:
        if d[i] < d[i // 2]:
            tmp_d = d[i // 2]; tmp_K = K[i // 2]
            d[i // 2] = d[i]; d[i] = tmp_d
            K[i // 2] = K[i]; K[i] = tmp_K
        i = i // 2
    return(K,d)

def downheapify(K,d):
    length = len(K)
    K[1]=K[length]; d[1]=d[length]
    K.pop(length); d.pop(length)
    i=1; current_i=0
```

```
        while i!=current_i:
            current_i=i; no_children=0
            if 2*i<=len(K) and 2*i+1<=len(K): no_children=2
            if 2*i<=len(K) and 2*i+1>len(K): no_children=1
            if no_children==2:
                if d[2*i]<d[2*i+1] and d[2*i]<d[i]:
                    swap(K, d, i, 2*i)
                    i=2*i
                elif d[2*i+1]<=d[2*i] and d[2*i+1]<d[i]:
                    swap(K, d, i, 2*i+1)
                    i=2*i+1
            if no_children==1:
                if d[2*i]<d[i]:
                    swap(K, d, i, 2*i)
                    i=2*i
    return(K,d)

def Dijkstra_BinaryHeap(V,w,A):
    destination=V[len(V)]
    p = {V[i]: 'undefined' for i in V}
    d = {i: float('inf') for i in range(1,len(V)+1)}; d[1]=0
    K = {i:V[i] for i in range(1,len(V)+1)}; r = K[1]
    while destination!=K[1]:
        edge_destinations = [i for i in A[r][0] if i in K.values()]
        for ki in range(0,len(edge_destinations)):
            k=edge_destinations[ki]
            K_keys = list(K.keys());K_values = list(K.values())
            i=K_keys[K_values.index(k)]
            if d[1]+w[r,k]<=d[i]:
                d[i]=d[1]+w[r,k]
                p[k]=r
                upheapify(K,d,i)
        [K, d] = downheapify(K, d)
        r=K[1]
    return(p,d[1])

V={1:'a',2:'b',3:'c',4:'d',5:'e',6:'f'}
w={('a','b'):4,('b','c'):4,('c','f'):14,('a','d'):8,('d','e'):6,('e','f'):2}
A = {'a': [('d','b')],
     'b': [('c')],
     'c': [('f')],
     'd': [('e')],
     'e': [('f')]} #adjacency sets
print(Dijkstra_BinaryHeap(V,w,A))
```

Implementing this algorithm will result in the following evolution of the binary min-heap at each iteration, as presented in Fig. 8.7. Upon termination, the tree root k_1 is the destination vertex f with shortest path distance $d_1 = 16$.

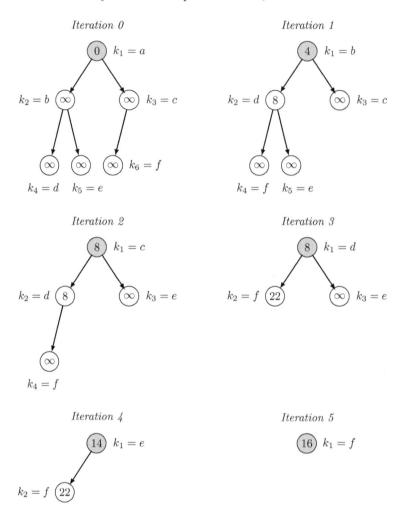

Fig. 8.7 Binary min-heap after each iteration of Dijkstra's algorithm

8.3.3   Shortest Path Problem as an Integer Linear Program

Until now, we have used the polynomial-time heuristic algorithm of Dijkstra that returns the globally optimal solution in networks with nonnegative edge weights. Although Dijkstra's algorithm is computationally efficient, in some cases the problem requirements might not allow us to apply it. For instance, when there are constraints in the formulation (i.e., we might seek to find a path which is not only the shortest, but also consumes less than a pre-determined amount of energy). In these more general cases, we can formulate the shortest path problem as an integer linear program.

Consider a directed graph $G = (V, A)$ with source vertex a, target vertex f, and non-negative cost w_{ij} for each directed edge (arc) $(i, j) \in A$. We can express the shortest path problem as an integer linear program by introducing binary variables x_{ij} for each $(i, j) \in A$, where $x_{ij} = 1$ if arc (i, j) is part of the shortest path from a to f and $x_{ij} = 0$ otherwise. Since our shortest path should minimize the cost (i.e., distance) between a and f, an obvious objective function is:

$$\sum_{(i,j)\in A} w_{ij} x_{ij}$$

This objective function attains its lowest possible value of zero for solution $x_{ij} = 0 \ \forall (i, j) \in A$. Clearly, this solution is not acceptable because it does not represent a path from a to f. For this reason, we need to add constraints that will enforce us to find a solution that represents a path from a to f. Starting from a, the shortest path should utilize an outgoing arc from a to move to the next vertex. This can be enforced by using constraint:

$$\sum_{(i,j)\in A:i=a} x_{ij} - \sum_{(i,j)\in A:j=a} x_{ij} = 1$$

This constraint ensures that the outgoing arcs from a that belong to the shortest path are one more than the incoming ones. In reality, we would like our shortest path to have at most one outgoing arc from every vertex. This can be achieved by setting:

$$\sum_{(i,j)\in A:i=k} x_{ij} \leq 1 \quad \forall k \in V$$

which does not allow our shortest path to have more than one outgoing arc from any vertex, including the origin vertex a. This implicitly enforces to have a single outgoing arc from a in the shortest path.

Let us now consider the destination vertex f of our shortest path problem. We would like our shortest path to include an arc that has f as an outgoing vertex. In addition, we do not want to have an outgoing arc from f in our shortest path since f is our last vertex. This can be enforced by adding the following inequality constraint:

$$\sum_{(i,j)\in A:i=f} x_{ij} - \sum_{(i,j)\in A:j=f} x_{ij} = -1$$

Let us now consider any other vertex $i \in V$ that is not the origin a nor the destination f of the shortest path. There are exactly two possibilities for any such vertex:

- it is not part of the shortest path from a to f and there is no incoming or outgoing arc to/from it that belongs to the shortest path.
- it is part of the shortest path from a to f and there is exactly one incoming and exactly one outgoing arc to/from it that belongs to the shortest path.

Both cases are enforced by the following inequality constraints:

$$\sum_{(i,j)\in A:i=k} x_{ij} - \sum_{(i,j)\in A:j=k} x_{ij} = 0 \quad \forall k \in V \setminus \{a, f\}$$

With this, we derive the following ILP formulation.

ILP formulation for the Shortest Path Problem

$$\min_{\mathbf{x}} \sum_{(i,j)\in A} w_{ij} x_{ij} \tag{8.19}$$

$$\text{s.t.} \sum_{(i,j)\in A:i=a} x_{ij} - \sum_{(i,j)\in A:j=a} x_{ij} = 1 \tag{8.20}$$

$$\sum_{(i,j)\in A:i=f} x_{ij} - \sum_{(i,j)\in A:j=f} x_{ij} = -1 \tag{8.21}$$

$$\sum_{(i,j)\in A:i=k} x_{ij} - \sum_{(i,j)\in A:j=k} x_{ij} = 0 \qquad \forall k \in V \setminus \{a, f\} \tag{8.22}$$

$$\sum_{(i,j)\in A:i=k} x_{ij} \le 1 \qquad \forall k \in V \tag{8.23}$$

$$x_{ij} \in \{0, 1\} \qquad \forall (i, j) \in A \tag{8.24}$$

As a 0–1 integer linear program, it can be solved with branch-and-cut and a solution method for linear programming that can solve the continuous relaxations.

Practitioner's Corner
Solving the Shortest Path ILP Python 3 for ILPs

Below we formulate the shortest path ILP in PuLP and we solve it with the open source COIN-OR Branch-and-Cut solver (CBC) solver. We do not use primal heuristics to provide better bounds and we allow cutting planes resulting in a branch and cut implementation.

```
import pulp
from pulp import *
V={'a','b','c','d','e','f'}
A={'ab','bc','cf','ad','de','ef'}
w={'ab':4,'bc':4,'cf':14,'ad':8,'de':6,'ef':2}
origin='a'; destination='f'
model = LpProblem("SPP_ILP", LpMinimize)
# variables
x = LpVariable.dicts("x", A, None, None, LpBinary)
# objective function
model += (sum(w[i]*x[i] for i in A))
# constraints:
model += sum(x[i] for i in A if i[0]==origin) - sum(x[i] for i in A
    if i[1]==origin) == 1
model += sum(x[i] for i in A if i[0]==destination) - sum(x[i] for i
    in A if i[1]==destination) == -1
for k in V:
    model += sum(x[i] for i in A if i[0]==k) <= 1
    if k!=origin and k!=destination:
        model += sum(x[i] for i in A if i[0]==k) - sum(x[i] for i in
    A if i[1]==k) == 0
# optimizing
solver = pulp.PULP_CBC_CMD(mip=True, msg=True, timeLimit=None,
    warmStart=False, path=None, mip_start=False, timeMode='elapsed',
    options=["passCuts=0","preprocess=off","presolve=off","cuts=on","
    heuristics=off","greedyHeuristic=off"])
model.solve(solver)
print ("Status:", LpStatus[model.status])
for v in model.variables():
    if v.varValue>0: print (v.name, "=", v.varValue)
print ("Optimal Solution = ", value(model.objective))
```

Implementing this code will solve the shortest path problem using the branch and cut approach of the CBC solver resulting in an optimal solution with a shortest path cost of 16. The detailed output of the solver is provided below, together with its explanation.

```
Problem MODEL has 12 rows, 6 columns
```

This informs us that the problem has 12 constraints (rows) and 6 variables (columns).

```
Continuous objective value is 16 - 0.00 seconds
```

The solver found a lower bound $L^0 = 16$ after solving a continuous relaxation of the integer LP in 0 seconds.

```
Cbc0004I Integer solution of 16 found after 0 iterations and 0 nodes (0.00
    seconds)
```

The solver found the first integer solution after exploring 0 nodes of the B&B tree, and this updates our upper bound to $U = 16$.

```
Cbc0001I Search completed - best objective 16, took 0 iterations and 0 nodes
    (0.00 seconds)
```

The solver informs us that the B&B search is completed after searching 0 nodes and the optimal solution is 16.

```
Status: Optimal
x_ad = 1.0
x_de = 1.0
x_ef = 1.0
Optimal Solution =   16.0
```

This message informs us that an optimal solution is found and the shortest path is $a \rightarrow d \rightarrow e \rightarrow f$.

We note that, as with Dijkstra's algorithm, the integer linear program for the singe-pair shortest path problem can be used also in the case of undirected graphs $G = (V, E)$.

8.3.4 All-pairs Shortest Path Problem

The all-pairs shortest path problem finds all shortest paths from any origin to any destination vertex in an undirected or directed graph $G = (V, E)$. This is a generalization of the shortest path problem because we are not explicitly focusing on a specific origin-destination pair. One of the most used algorithms for the all-pairs shortest path problem that does not contain negative cycles is the algorithm of Floyd-Warshall [13] presented in 1962. The definition of the problem is presented below.

> **All-pairs shortest path problem of Floyd-Warshall** [13]
> Find the lengths (summed weights) of shortest paths between all pairs of nodes in a network with non-negative cycles.

The time complexity of the algorithm is $O(|V|^3)$, where $|V|$ is the number of vertices in the graph. The algorithm improves incrementally the estimate on the shortest path between any two vertices, until the estimate is optimal (see Algorithm 42). Each directed edge $(i, j) \in E$ where i and j are vertices in V has a non-negative cost w_{ij} (i.e., distance), thus avoiding cycles. If two vertices are not directly connected by a directed edge, we can set $w_{ij} = +\infty$.

Fig. 8.8 Non-negative edge
weights of the undirected
graph $G = (V, E)$

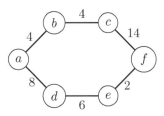

Algorithm 42 Floyd-Warshall algorithm for the All-pairs shortest path problem
---
1: Consider transport network $G = (V, E)$
2: Introduce cost w_{ij} as the nonnegative cost of traversing any directed edge (i, j) $\in E$
3: Initialize $d_{ij} = +\infty$ as the minimum distance between each pair of vertices $(i, j) \mid i \neq j$ where $i \in V$ and $j \in V$;
4: Initialize $d_{ii} = 0$ for all $i \in V$
5: **for** each directed edge $(i, j) \in E$ **do**
6:   set $d_{ij} = w_{ij}$
7: **end for**
8: **for** each vertex $k \in V$ **do**
9:   **for** each vertex $i \in V$ **do**
10:    **for** each vertex $j \in V$ **do**
11:     **if** $d_{ij} > d_{ik} + d_{kj}$ **then**
12:      set $d_{ij} = d_{ik} + d_{kj}$
13:     **end if**
14:    **end for**
15:   **end for**
16: **end for**

Let us consider the example network of the undirected graph presented in Fig. 8.8. Because this graph is undirected, $w_{ij} = w_{ji}$ for all $i \in V$, $j \in V$.

We can find the shortest paths between all pairs of this network as follows.

Practitioner's Corner

Floyd-Warshall algorithm for the all-pairs shortest path problem in Python 3

Below we present an implementation of the Floyd-Warshall algorithm for solving the all-pairs shortest path problem in the network of Fig. 8.2.

```
def Floyd_Warshall(V,W):
    d=W
    for k in V:
        for i in V:
            for j in V:
                if d[i,j] > d[i,k] + d[k,j]:
                    d[i,j] = d[i,k] + d[k,j]
    return(d)
```

```
V=('a','b','c','d','e','f')
W={(i,j):float('inf') for i in V for j in V}
for i in V: W[i,i]=0
W['a','b']=4; W['b','a']=4; W['b','c']=4; W['c','b']=4
W['c','f']=14; W['f','c']=14; W['a','d']=8; W['d','a']=8
W['d','e']=6; W['e','d']=6; W['e','f']=2; W['f','e']=2
print(Floyd_Warshall(V,W))
```

Implementing this algorithm will result in the shortest path distances between any pair of vertices. If there is no shortest path between two vertices because there is no path that connects them in the graph, then this distance will be $+\infty$. The shortest path distances of pairs that have a feasible path are $(a, b) : 4, (a, c) : 8, (a, d) : 8, (a, e) :$ 14, $(a, f) : 16, (b, a) : 4, (b, c) : 4, (b, d) : 12, (b, e) : 18, (b, f) : 18, (c, a) : 8,$ $(c, b) : 4, (c, d) : 16, (c, e) : 16, (c, f) : 14, (d, a) : 8, (d, b) : 12, (d, c) : 16, (d,$ $e) : 6, (d, f) : 8, (e, a) : 14, (e, b) : 18, (e, c) : 16, (e, d) : 6, (e, f) : 2, (f, a) : 16,$ $(f, b) : 18, (f, c) : 14, (f, d) : 8, (f, e) : 2.$

8.3.5   K-shortest Paths Problem

Let us consider again the single-pair shortest path problem. When solving this problem, it is often useful to compute not only the shortest path from the origin to the destination vertex, but also the 2nd, 3rd,..., K-th shortest paths. For instance, if the shortest path is crowded, travelers might use other alternatives. An approach to solve the K-shortest paths problem was proposed by Yen in 1971 [14]. It computes the K shortest loopless paths for a graph with non-negative edge weights and its time complexity is $O(K|V|^3)$ if we use the basic version of Dijkstra's algorithm with the array data structure to generate the K shortest paths.

Let us consider an undirected or directed graph $G = (V, E)$ with nonnegative directed edge weights w_{ij} for any $(i, j) \in E$. We are asked to find the first K shortest paths from an origin vertex a to a destination vertex f. Yen's algorithm uses a container $A = A^1, A^2, \ldots, A^K$ of shortest paths. Initially, $A = \emptyset$. It also uses a container B of the k-th shortest path, where $k \in \{1, \ldots, K\}$. Initially, $B = \emptyset$. Then, it performs the following steps described in Algorithm 43.

From Algorithm 43 we can get a better understanding of the $O(K|V|^3)$ time complexity of Yen's algorithm. Notice that in the worst-case scenario we will have to solve K times (repeat loop starting in line 4) $|V|$ shortest path problems (for loop starting in line 5). Using Dijkstra with an array data structure that has time complexity $O(|V|^2)$ will require a time complexity of $O(K|V|^3)$ to solve all shortest path problems in Yen's algorithm.

Let us consider the implementation of Yen's algorithm in the example network of the digraph $G = (V, A)$ of Fig. 8.9 to find the $K = 4$ shortest paths from a to f.

Algorithm 43 Yen's algorithm for the K-shortest paths problem

---

1: consider an undirected or directed graph $G = (V, E)$ and a request to find the K
   shortest paths from an origin a to a destination f
2: apply a shortest path algorithm (i.e., Dijkstra) to find the shortest path from a to
   f and store it as A^1.
3: set $k = 2$
4: **repeat**
5: **for** each vertex i in the shortest path A^{k-1}, except f **do**
6:    declare path R^k_{ai} that comprises of the i-th first vertices of path A^{k-1} as the
      root path.
7:    for every previous path in A that has the same root path with R^k_{ai}, set the edge
      weight between vertex i and the next vertex in its shortest path to $+\infty$.
8:    compute the shortest path V^k_{if} from vertex i to destination f considering the
      new edge weight values.
9:    derive path $A^k_i = R^k_{ai} \cup V^k_{if}$ and store it to container B.
10:   restore the edge weights that were assigned to $+\infty$ to their previous values.
11: **end for**
12: select path A^k_{i*} that has the lowest cost from all paths in B and add it to container
    A as path A^k.
13: Empty container B and set $k = k + 1$.
14: **until** there are no other feasible paths or $k > K$

Fig. 8.9 Example network
G with non-negative directed
edge weights

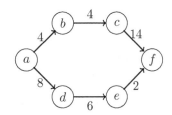

Using Dijkstra's algorithm, the shortest path from a to f is $A^1 = \{a, d, e, f\}$ with
a cost of 16 units. To find the second shortest path, A^2, we have:

- for $i = a$ the root path is $R^2_{aa} = \{a\}$, w_{ad} is set to $+\infty$ because it belongs to
  another shortest path in A, and $V^2_{af} = \{a, b, c, f\}$. Hence, $A^2_a = \{a, b, c, f\}$ and
  A^2_a is added to B.
- for $i = d$ the root path is $R^2_{ad} = \{a, d\}$ and w_{de} is set to $+\infty$ because $\{a, d, e\}$
  belongs to the shortest path A^1. Because of that, there is no shortest path from d
  to f.
- for $i = e$ the root path is $R^2_{ae} = \{a, d, e\}$ and w_{ef} is set to $+\infty$ because $\{a, d, e, f\}$
  belongs to the shortest path A^1. Because of that, there is no shortest path from e
  to f.

Hence, container B includes only path A_a^2 and we thus set $A^2 = \{a, b, c, f\}$. Before moving to the next iteration, we empty B. We also have $A = \{A^1, A^2\}$ where $A^1 = \{a, d, e, f\}$ and $A^2 = \{a, b, c, f\}$.

To find the third shortest path A^3, we have:

- for $i = a$ both w_{ab} and w_{ad} are set equal to $+\infty$ because they belong to past shortest paths in A. Hence, there is no shortest path from a to f.
- for $i = b$ we have root path $R_{ab}^3 = \{a, b\}$ and $w_{bc} = +\infty$ because $\{a, b, c\}$ is part of path A^2. Hence, there is no shortest path from b to f.
- for $i = c$ we have root path $R_{ac}^3 = \{a, b, c\}$ and $w_{c,f} = +\infty$ because $\{a, b, c, f\}$ is part of A^2. Hence, there is no shortest path from c to f.

Notice that B is empty and there are no other feasible paths. Thus, the algorithm terminates with $A = \{A^1, A^2\}$.

8.4 Line Planning

8.4.1 General Outline

As previously discussed, shortest path algorithms are useful when determining the routes of public transport lines. The Line Planning Problem (LPP) is part of the Transit Network Design Problem (TNDP). Depending on the definition of TNDP, oftentimes the LPP and TNDP refer to the same problem. For instance, a common TNDP definition is the problem of defining the set of lines of a transit network given the underlying infrastructure of the network, including the transit stops, and a passenger demand matrix. This definition resembles the LPP. If vehicle sizes are also considered, then the problem becomes the Transit Network Design and Frequency Setting Problem (TNDFSP) because it tries to determine the transit lines and set their operational frequencies at the same time [15]. The LPP typically appears in this form, where determining the lines and their frequencies is seen as a joint task. Based on the latter definition, one can approach the LPP as a TNDFSP where the infrastructure (stops and road/track sections) is already provided and we seek to find the number, the routes of lines, and their frequencies.

There are several line planning approaches, but one abstract formulation that covers bus and rail lines and provides a high-level framework for line planning is the one provided by Schöbel [16] in 2012. Before proceeding to the formulation, we present the following definitions.

Definition 8.3 *Line Plan.* A line plan consists of the number of lines and their routes.

Definition 8.4 *Line concept.* A line concept is a line plan together with the frequencies of the lines.

Using the above definitions, the line planning problem in public transport is the problem of finding a line concept that has some desirable properties, i.e., it ensures that public transport is convenient for passengers and the operational costs are reasonable. Let us consider a transit network $G = (V, E)$ where V is the set of public transport stops and E is the set of direct connections between stops (edges). The transit network is the underlying network of roads or tracks, which is a fixed input to our line planning problem. Given a line network, a *line l* is a path in network G. The *frequency f_l* of a line l indicates how often a service is offered along line l within a time period I (i.e., 1 hour, 1 day), and it is expressed in the form of trips per time period.

Using this notation, a *line concept* can be denoted as $(L, \mathbf{f})$, where L is the set of lines and $f_l \in \mathbf{f}$ the frequency of each line $l \in L$. When trying to determine the optimal line concept, there are three different ways of modeling:

- consider the minimization of operational costs and the maximization of the passengers' convenience as two different (mostly conflicting) objectives, resulting in a multi-objective optimization formulation.
- consider the objective of maximizing the passengers' convenience subject to cost-related constraints for the transit operator (*passenger-oriented* modeling).
- consider the objective of minimizing the transit operator's costs subject to passenger convenience-related constraints (*cost-oriented* modeling).

If we assume that he have homogeneous vehicles at each line and the cost of operating a vehicle of line l is c_l, then the service cost of a line concept $(L, \mathbf{f})$ is:

$$c(L, \mathbf{f}) := \sum_{l \in L} c_l f_l \qquad (8.25)$$

where f_l refers to the frequency of line $l \in L$ over a planning period I. The assigned cost of operating a vehicle of line l, c_l, depends on the vehicle, the round-trip travel time of the line, the cost of every minute of driving, and potentially other factors. Although we assume that the cost depends on the line, it might also depend on other characteristics, such as the vehicle and crew schedules. However, these are not taken into consideration at the line planning stage. Instead, they are considered in detail at a subsequent stage after determining the timetables. In a cost-oriented formulation, we seek to minimize (8.25) subject to passenger convenience-related constraints. In a passenger-oriented formulation, the cost of the line concept is treated as a constraint which should not exceed a pre-determined budget $b > 0$. That is,

$$c(L, \mathbf{f}) := \sum_{l \in L} c_l f_l \leq b$$

To determine the convenience of passengers, the main input is the passenger demand expressed in the form of a $|V| \times |V|$ origin-destination matrix $\mathbf{W}$, where each $w_{uv} \in \mathbf{W}$ refers to the passenger demand from vertex (stop) $u \in V$ to vertex (stop) $v \in V$ during the planning period I. If $w_{uv} > 0$, then (u, v) is an OD-pair. If $w_{uv} = 0$, then there is no demand between these two vertices. Given the OD-matrix, we can define q_e as the passenger load expressed in passengers traveling along edge $e \in E$ in planning period I. To calculate q_e we must use the OD-matrix **W** and perform a passenger assignment to distribute the passengers on paths in the public transportation network before the lines are known. This passenger assignment is independent of the selected line concept and it can be based on assigning every passenger to his/her shortest path or using equilibrium models [17–19]. This is, however, a strong assumption because the real passengers' weights along every edge strongly depend on the line concept which is to be designed, leading to a chicken-egg-problem. Despite that, line planning methods typically disregard this dependence or solve the LPP as a *bi-level* problem where an optimal line concept is derived for a given passenger assignment, a passenger assignment is performed again based on the paths and frequencies of this line concept, and the process continues iteratively until convergence.

To describe in more detail the passenger convenience metrics, we call *riding time* the time a passenger is in a public transport vehicle on his/her way between his/her origin and destination, without considering the required time for transfers. We also call *travel time of a passenger* the riding time plus some penalty time for every transfer.

To calculate the travel time of a passenger, we need to know the timetable of each line. Because this is not known at the line planning stage, we can use a penalty for each transfer as an approximation. Common ways to represent the convenience of passengers are the sum of all riding times or the sum of all travel times, which need to be as small as possible.

If we use a cost-oriented model, the convenience of passengers is treated as a constraint. For instance, we can apply capacity constraints:

$$\sum_{l \in L: e \in l} d_l f_l \geq q_e \quad \forall e \in E \qquad (8.26)$$

which ensure that a line concept $(L, \mathbf{f})$ can transport q_e passengers along edge e where d_l denotes the capacity of a vehicle operating in line l and $l \in L : e \in l$ denotes that only lines that serve edge e are considered. It is worth noting that the capacity of every vehicle operating a line $l \in L$ is known in advance and the vehicles of every line are considered to be homogeneous. It is also worth noting that the passenger load q_e at each edge $e \in E$ is known in advance based on the origin-destination matrix **W** and a passenger assignment.

Another typical constraint is to ensure that a line concept satisfies the lower edge frequency requirement expressed as:

$$\sum_{l \in L: e \in l} f_l \geq f_e^{\min} \quad \forall e \in E \qquad (8.27)$$

where $f_e^{\min}$ is the minimal number of vehicles needed within the planning period I to transport all passengers in edge e. If the vehicles of all lines have the same capacity, d, then $f_e^{\min} := \lceil \frac{q_e}{d} \rceil$. Typically, there is also an upper edge frequency requirement because it is not possible too many vehicles to use the same edge due to safety reasons:

$$\sum_{l \in L: e \in l} f_l \leq f_e^{\max} \quad \forall e \in E \tag{8.28}$$

The lower and upper edge frequency requirements were introduced by Wegel, 1974 [20]. Similarly, one can use lower and upper node frequency requirements which restrict the number of vehicles stopping at particular stations:

$$\sum_{l \in L: v \in l} f_l \geq f_v^{\min} \quad \forall v \in V \tag{8.29}$$

$$\sum_{l \in L: v \in l} f_l \leq f_v^{\max} \quad \forall v \in V \tag{8.30}$$

Another passenger-related constraint is that if a set of OD-pairs is given, the selected line concept should be feasible, meaning that every OD-pair should be connected by the lines in the line concept ensuring that there is no passenger that cannot travel from his/her origin to his/her destination.

A strategy for determining a line concept is to start first with the generation of a broader *line pool* L_0, where $L \subseteq L_0$ for any line plan L. Following this approach, the selected line plan is a subset of the line pool L_0 (see Ceder and Wilson [21]). Another way of deriving the routes of the lines is to construct them from scratch without considering a line pool. There are heuristics approaches for this, but also exact methods. Borndörfer et al. [22] proposed an exact method that we will discuss in the following section.

Considering a set L_0 of potential lines, the definition of the *basic* line planning problem (LPP) is provided as follows.

Definition 8.5 *Basic Line Planning Problem (LPP) with a given line pool (Schöbel* [16]). Given a transit network $G = (V, E)$, a line pool L_0, and lower and upper frequencies $f_e^{\min}$, $f_e^{\max}$ for all $e \in E$, find a *line concept* $(L, \mathbf{f})$ with $L \subseteq L_0$, $f_l \in \mathbb{N}_0 \ \forall l \in L$, and $f_e^{\min} \leq \sum_{l \in L: e \in l} f_l \leq f_e^{\max}$ for all $e \in E$.

The basic LPP decision problem described above is already NP-complete as shown in [23] by reduction to *exact cover by 3-sets (X3C)*, even for the special case that $f_e^{\min} = f_e^{\max} = 1$ for all $e \in E$. This indicates that most LPPs which contain extensions of the basic LPP are NP-complete decision problems and their optimization counterparts are NP-Hard.

There is a special case of the basic LPP that can be easily solved. This is when the line pool L_0 contains all potential paths of G. Then, the basic LPP is always feasible and can be solved in polynomial time by taking each edge as a line with frequency $f_e^{\min}$. This trivial solution to the decision problem can be used as a starting point and this process can be seen as a primal heuristic when solving an optimization version of the LPP. We can turn this decision problem into an optimization problem by considering the *cost-oriented* version, the *passenger-oriented* version, and the *multi-objective* version, as described below.

Definition 8.6 *Basic cost-oriented Line Planning Problem (LPP) with a given line pool (Schöbel [16]).* Given a transit network $G = (V, E)$, a line pool L_0, lower and upper frequencies $f_e^{\min}$, $f_e^{\max}$ for all $e \in E$, and cost c_l of operating a vehicle of line l, find a *line concept* $(L, \mathbf{f})$ such that:

$$\min_{L, \mathbf{f}} \sum_{l \in L} c_l f_l \tag{8.31}$$

$$\text{s.t. } f_e^{\min} \leq \sum_{l \in L : e \in l} f_l \leq f_e^{\max} \qquad \forall e \in E \tag{8.32}$$

$$L \subseteq L_0 \tag{8.33}$$

$$f_l \in \mathbb{N}_0 \qquad \forall l \in L \tag{8.34}$$

In contrast to the basic LPP decision problem, the basic cost-oriented LPP optimization problem described above is NP-Hard even in the special case when the line pool L_0 contains all potential paths of G. This can be proven by reduction to the *Hamiltonian path* [16], where a Hamiltonian path in a graph G is a *simple path* which visits each vertex exactly once.

Definition 8.7 *Basic passenger-oriented Line Planning Problem (LPP) with a given line pool (Schöbel [16]).* Given a transit network $G = (V, E)$, a line pool L_0, lower and upper frequencies $f_e^{\min}$, $f_e^{\max}$ for all $e \in E$, a budget b, an OD-matrix $\mathbf{W}$, and cost c_l of operating a vehicle of line l, find a *line concept* $(L, \mathbf{f})$ such that the passenger travel times are minimized subject to the following constraints:

$$f_e^{\min} \leq \sum_{l \in L : e \in l} f_l \leq f_e^{\max} \qquad \forall e \in E \tag{8.35}$$

$$\sum_{l \in L} c_l f_l \leq b \tag{8.36}$$

$$L \subseteq L_0 \tag{8.37}$$

$$f_l \in \mathbb{N}_0 \qquad \forall l \in L \tag{8.38}$$

This optimization problem is also NP-Hard, as proven in Schöbel and Scholl [24]. Finally, there is a multi-objective optimization version of the basic LPP, as described below.

Definition 8.8 *Basic multi-objective Line Planning Problem (LPP) with a given line pool.* Given a transit network $G = (V, E)$, a line pool L_0, lower and upper frequencies $f_e^{\min}$, $f_e^{\max}$ for all $e \in E$, an OD-matrix $\mathbf{W}$, and cost c_l of operating a vehicle of line l, find a *line concept* $(L, \mathbf{f})$ such that the passenger travel times are minimized and the costs $\sum_{l \in L} c_l f_l$ are minimized subject to the following constraints:

$$f_e^{\min} \leq \sum_{l \in L : e \in l} f_l \leq f_e^{\max} \qquad\qquad \forall e \in E \qquad (8.39)$$

$$L \subseteq L_0 \qquad\qquad\qquad\qquad (8.40)$$

$$f_l \in \mathbb{N}_0 \qquad\qquad\qquad \forall l \in L \qquad (8.41)$$

8.4.2 Line Planning Problem with a Multi-commodity Flow Formulation

Borndörfer et al. [22] proposed an extended multi-commodity flow formulation for LPP which minimizes the combination of total passenger travel times and operating costs. It generates line routes dynamically, handles frequencies by means of continuous frequency variables, and allows passengers to change their routes according to the computed line system. Until now, we have treated line frequencies as positive integers. However, Borndörfer et al. [22] treat them as continuous variables to reduce the computational complexity of the LPP. The multi-commodity flow formulation of Borndörfer et al. [22] is presented below.

Sets: Consider a set M of different public transport modes (bus, tram, etc.). Consider also an undirected multigraph $G = (V, E) = (V, E_1 \cup E_2 \cup \ldots \cup E_{|M|})$ representing a multi-modal transit network. Set E_1 is the set of all edges corresponding to mode 1, E_2 to mode 2, and so forth. In addition, let $V_1, \ldots, V_i, \ldots, V_{|M|} \subseteq V$ be the terminal sets corresponding to each mode $i \in M$ where lines can start and end. Let L be the set of all feasible lines and $L_e := \{l \in L : e \in l\}$ the set of all feasible lines that use edge $e \in E$. Set L can be seen as a *line pool* from which we need to determine the optimal lines that will be part of the *line plan*. The set of OD-pairs is D. We can also denote the sub-graph of mode $i \in M$ as $G_i = (V_i, E_i)$.

From the undirected graph $G = (V, E)$ we can derive a directed graph $G' = (V, A)$ where each edge $e \in E$ is replaced by two antiparallel arcs a_e and $\bar{a}_e$. Conversely, let $\rho_a \in E$ be the undirected edge corresponding to arc $a \in A$. For an OD-pair $(u, v) \in D$ an (u, v)-passenger path is a directed path in (V, A) from u to v. Let P_{uv} be the set of all (u, v)-passenger paths, $P := \{p \in P_{uv} : (u, v) \in D\}$ the set of all passenger paths, and $P_a := \{p \in P : a \in p\}$ the set of all passenger paths that use arc a.

Parameters: Let $g_1, \ldots, g_{|M|} \in \mathbb{R}_{\geq 0}$ be the fixed costs for the setup of a line for each transport mode. In addition, let $\mathbf{c}^1 = [c_1^1, c_2^1, \ldots, c_{|E_1|}^1]^{\mathsf{T}}$ be the operating costs of the edges of mode 1, $\mathbf{c}^2$ the operating costs of the edges of mode 2, and so forth. We have also vehicle capacities $\kappa_1, \ldots, \kappa_{|M|} \in \mathbb{R}_{\geq 0}$ for each mode, assuming that the operating vehicles of each mode are homogeneous, and edge capacities $\boldsymbol{\lambda} \in \mathbb{R}_{\geq 0}^{|E|}$.

A line of mode i is a path in G_i that connects two terminals $v \in V_i$, $u \in V_i$ where $v \neq u$. The paths of the lines are *simple*, meaning that a repetition of vertices is not allowed. If $\mathbf{r}$ is an indicator vector where $r_l \in M$ indicates the transport mode line l belongs to, then g_{r_l} is the fixed cost and κ_{r_l} the vehicle capacity of line l.

In addition, $c_l = \sum_{e \in l} c_e^{r_l}$ is the cost of line l operated by mode r_l. We have also an origin-destination matrix $\mathbf{W} \in \mathbb{R}_{\geq 0}^{|V| \times |V|}$ which can be used to define the set of OD-pairs as $D := \{(u, v) \in V \times V : w_{uv} > 0\}$.

The travel time of every edge $a \in A$ in the directed graph $G' = (V, A)$ is $\tau_a \in \mathbb{R}_{\geq 0}$. The travel time of a passenger path $p \in P$ is $\tau_p := \sum_{a \in p} \tau_a$. Finally, consider ϕ as an upper bound on the frequency of any line.

Variables: We consider the following variables:

- $\mathbf{y} = [y_1, \dots, y_{|P|}]^\mathsf{T}$, where $y_p \in \mathbb{R}_{\geq 0}$ is the flow of passengers traveling on path p.
- $\mathbf{f} = [f_1, \dots f_{|L|}]^\mathsf{T}$, where $f_l \in \mathbb{R}_{\geq 0}$ is the frequency of line $l \in L$.
- $\mathbf{x} = [x_1, \dots, x_{|L|}]^\mathsf{T}$, where $x_l \in \{0, 1\}$ is a binary decision variable of using line $l \in L$ or not.

With this notation, the LPP is modeled as follows.

LPP with multi-commodity flow formulation, Borndörfer et al. [22]

$$\min_{\mathbf{x}, \mathbf{y}, \mathbf{f}} \sum_{p \in P} \tau_p y_p + \sum_{l \in L} g_{r_l} x_l + \sum_{l \in L} c_l f_l \tag{8.42}$$

$$\text{s.t.} \sum_{p \in P_{uv}} y_p = w_{uv} \qquad \forall (u, v) \in D \tag{8.43}$$

$$\sum_{p \in P_a} y_p \leq \sum_{l : \rho_a \in l} \kappa_{r_l} f_l \qquad \forall a \in A \tag{8.44}$$

$$\sum_{l \in L_e} f_l \leq \lambda_e \qquad \forall e \in E \tag{8.45}$$

$$f_l \leq \phi x_l \qquad \forall l \in L \tag{8.46}$$

$$x_l \in \{0, 1\} \qquad \forall l \in L \tag{8.47}$$

$$f_l \in \mathbb{R}_{\geq 0} \qquad \forall l \in L \tag{8.48}$$

$$y_p \in \mathbb{R}_{\geq 0} \qquad \forall p \in P \tag{8.49}$$

The passenger flow constraints (8.43) and the non-negativity constraints (8.49) model a multicommodity flow problem for the passenger flow. In the LPP case, the commodities correspond to OD-pairs $(u, v) \in D$. Constraints (8.43) ensure that the passenger flow in all paths p that connect the OD-pair (u, v) and thus belong to set P_{uv} is equal to the passenger demand from u to v, w_{uv}. The capacity constraints (8.44) link the passenger paths with the line paths to ensure sufficient transportation capacity on each arc. The frequency constraints (8.45) bound the total frequency of lines using an edge. Constraints (8.46) link the frequencies with the decision variables for the use of lines; they guarantee that the frequency of a line is zero ($f_l = 0$) whenever it

Fig. 8.10 Example
undirected graph
$G = (V, E)$

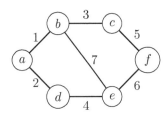

is not used ($x_l = 0$). Here, ϕ is an upper bound on the frequency of a line allowing f_l to take values in the range $[0, \phi]$.

The proposed model is not using passenger-oriented or cost-oriented modeling. Instead, it considers both objectives in the objective function and uses linear scalarization to turn a bi-objective optimization problem into a single-objective one. The two conflicting objectives are the minimization of the total passenger travel times $\tau^\top y$ and the minimization of the costs $g^\top x + c^\top f$ of setting up lines and operating them at frequencies f. The relative importance of these two objectives can be adjusted by an appropriate scaling of the objective function coefficients. We note that the entire model refers to a time horizon (i.e., day or period of the day), and its output is an optimal line concept represented by solution (x^*, f^*). In detail, the components of solution x^* that are equal to 1 correspond to the selected line plan L^* which is a subset of the line pool ($L^* \subseteq L$).

One important aspect is the derivation of feasible lines L of network G that constitute the line pool. This can be a tedious task considering the vast number of potential options. To limit the possible lines, one can impose practical constraints related to the length of a line, which should not deviate too much from a shortest path between its endpoints. There can also be bounds on the number of lines using an edge.

One more point is that transfers between lines are not considered in the multi-commodity LPP formulation because constraints (8.46) only control the total capacity on edges and not the assignment of passengers to l. Finally, as previously discussed, frequencies are modeled as continuous variables. In practice, however, line frequencies should produce a regular timetable and are not always allowed to take arbitrary real values.

Consider the example of the undirected graph in Fig. 8.10. This undirected graph has 7 edges, with their identification numbers presented in the graph.

There is passenger demand from a to f equal to $w_{af} = 200$ and from a to e equal to $w_{ae} = 30$. In this example we have two modes. We assume a line pool where mode 1 has two lines. The first line serves edges 1, 3 and 5. The second line serves edges 1, 7 and 6. Mode 2 has also a line that serves edges 2, 4 and 6. In total, $|L| = 3$ and $r = [1, 1, 2]^\top$. Notice that all three lines are bi-directional because serving an edge means that we cover two directions.

From this line pool, we want to determine a line concept. To present the travel times, we need first to generate from the undirected graph $G = (V, E)$ a directed graph $G' = (V, A)$ where each edge $e \in E$ is replaced by two antiparallel arcs.

Fig. 8.11 Example digraph
$G' = (V, A)$ with arc travel
times τ_a

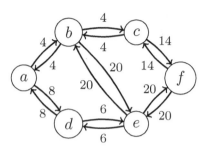

From the set of edges $E = \{1, 2, 3, 4, 5, 6, 7\}$ we generate the set of arcs $A = \{(a, b), (b, a), (a, d), (d, a), (b, c), (c, b), (d, e), (e, d), (c, f), (f, c), (e, f), (f, e), (b, e), (e, b)\}$ where arc (a, b) connects vertices a and b, and so forth. We also have vector $\boldsymbol{\rho}$ where ρ_a returns the edge associated to arc $a \in A$. For instance, $\rho_{cb} = 3$. Using this representation, the digraph $G' = (V, A)$ with the associated travel times τ_a to each arc $a \in A$ is presented in Fig. 8.11.

The costs $\mathbf{c}^1$ of traversing the edges with vehicles of mode 1 are $\mathbf{c}^1 = [8, 16, 8, 12, 28, 40, 40]^\mathsf{T}$ and the costs of traversing the edges with vehicles of mode 2 are $\mathbf{c}^2 = [12, 24, 12, 18, 42, 60, 60]^\mathsf{T}$. The fixed cost to setup a line is 100 units ($g_1 = 100$) for a line of mode 1 and 120 units ($g_2 = 120$) for a line of mode 2. All vehicles of mode 1 have a capacity of 20 passengers ($k_1 = 20$) and all vehicles of mode 2 a capacity of 50 passengers ($k_2 = 50$). In addition, the upper bound of the frequency of every line is $\phi = 8$ trips during the planned period. Finally, the edge capacity is $\lambda_e = 8$ for any edge in the graph.

Regarding paths, the (a, f)-passenger paths that can serve the demand from a to f given our line pool L are:

- path 1: $a \rightarrow b \rightarrow c \rightarrow f$ that can be served by line 1
- path 2: $a \rightarrow b \rightarrow e \rightarrow f$ that can be served by line 2
- path 3: $a \rightarrow d \rightarrow e \rightarrow f$ that can be served by line 3

That is, P_{af} has three paths. The (a, e)-passenger paths that can serve the demand from a to e are paths 4 and 5 expressed as $a \rightarrow b \rightarrow e$ and $a \rightarrow d \rightarrow e$, respectively. Thus, we have 5 paths in total. The travel times of the paths are $\boldsymbol{\tau} = [22, 44, 34, 24, 14]^\mathsf{T}$. The costs of the lines are $\mathbf{c} = [44, 88, 102]^\mathsf{T}$, and the paths corresponding to every arc are presented in Table 8.13.

The solution of this problem is $\mathbf{x}^* = [1, 0, 1]^\mathsf{T}$, $\mathbf{f}^* = [8, 0, 1.4]^\mathsf{T}$, and $\mathbf{y}^* = [160, 0, 40, 0, 30]^\mathsf{T}$. This means that only lines 1 and 3 are required and they have frequencies of 8 and 1.4 trips during the planning period, respectively. Finally, the demand of 200 passengers from a to f is distributed as follows: 160 passengers to path 1 corresponding to line 1 and 40 passengers to path 3 corresponding to line 3. The demand of passengers from a to e is solely served by path 5 corresponding to line 3, meaning that all 30 passengers will be served by line 3.

Table 8.13 Paths associated with each arc

Arc, a	Path(s), P_a
(a, b)	1, 2, 4
(a, d)	3, 5
(b, c)	1
(d, e)	3, 5
(c, f)	1
(e, f)	2, 3
(b, e)	2, 4
(b, a)	none
(d, a)	none
(c, b)	none
(e, d)	none
(f, c)	none
(f, e)	none
(e, b)	none

Practitioner's Corner
Solving LPP in Python 3

One can use `PuLP` to program the LPP model in Python and solve it with solvers that can apply branch-and-cut (CPLEX, Gurobi, CBC). Below follows an example of modeling the aforementioned problem in `PuLP` and solving it with the CBC solver.

```python
import pulp
from pulp import *
V={'a','b','c','d','e','f'}
E={1,2,3,4,5,6,7} #edges
A={'ab','ba','ad','da','bc','cb','de','ed','cf','fc','ef','fe','be','
    eb'} #arcs (directed edges)
rho={'ab':1,'ba':1,'ad':2,'da':2,'bc':3,'cb':3,'de':4,'ed':4,'cf':5,'
    fc':5,'ef':6,'fe':6,'be':7,'eb':7}
M=(1,2)
lambda_g={e:8 for e in E}
r={1:1,2:1,3:2}; k={1:20,2:50}; g={1:100,2:120}
phi=8; L=(1,2,3)
Line_Edges={1:(1,3,5),2:(1,7,6),3:(2,4,6)}
D={'af','ae'}; w={'af':200,'ae':30}
Puv={'af':(1,2,3),'ae':(4,5)}
Pa={'ab':(1,2,4),'bc':1,'cf':1,'ad':(3,5),'de':(3,5),'ef':(2,3),'be'
    :(2,4),'ba':0,'cb':0,'fc':0,'da':0,'ed':0,'fe':0,'eb':0} #if an
    arc's value is 0, it is not part any path
P=(1,2,3,4,5)
tau={1:22,2:44,3:34,4:24,5:14}; c={1:44,2:88,3:102}

model = LpProblem("LPP_ILP", LpMinimize)
# variables
```

```
x = LpVariable.dicts("x", (1 for 1 in L), cat='Binary')
f = LpVariable.dicts("f", (1 for 1 in L), 0, None, LpContinuous)
y = LpVariable.dicts("y", (p for p in P), 0, None, LpContinuous)
# objective function
model += sum(tau[p]*y[p] for p in P) + sum(g[r[l]]*x[l] for l in L) +
    sum(c[l]*f[l] for l in L)
# constraints:
for uv in D:
    model += sum(y[p] for p in Puv[uv])==w[uv]
for a in A:
    if isinstance(Pa[a], Iterable):
        model += sum(y[p] for p in Pa[a]) <= sum(k[r[l]]*f[l] for l
    in L if rho[a] in Line_Edges[l])
    else:
        if Pa[a] != 0:
            model += sum(y[p] for p in [Pa[a]]) <= sum(k[l] * f[l]
        for l in L if rho[a] in Line_Edges[l])
for e in E:
    model += sum(f[l] for l in L if e in Line_Edges[l]) <= lambda_g[e
    ]
for l in L:
    model += f[l] <= phi*x[l]
# optimizing
solver = pulp.PULP_CBC_CMD(mip=True, msg=True, timeLimit=None,
    warmStart=False, path=None, mip_start=False, timeMode='elapsed',
    options=["passCuts=0","preprocess=off","presolve=off","cuts=on","
    heuristics=off","greedyHeuristic=off"])
model.solve(solver)
print ("Status:", LpStatus[model.status])
for v in model.variables():
    if v.varValue>0: print (v.name, "=", v.varValue)
print ("Optimal Solution = ", value(model.objective))
```

The LPP formulation presented in (8.42)–(8.49) is a mixed-integer linear programming formulation. It can be simplified to a linear program by removing the binary variables $\mathbf{x}$. Since LPP minimizes over nonnegative costs, we can assume that inequalities (8.46) are satisfied with equalities, i.e., there is an optimal LPP solution such that $\phi x_l = f_l \Leftrightarrow x_l = \frac{f_l}{\phi}$ for all lines l. Substituting for $\mathbf{x}$, inequalities $f_l \leq \phi$ are dominated by inequalities (8.45) because we assume that $\phi \geq \lambda_e$ for all $e \in E$. Setting $\gamma_l = \frac{g_l}{\phi} + c_l$ we have the following program.

LPP reformulation to a linear program

$$\min_{\mathbf{y},\mathbf{f}} \ \boldsymbol{\tau}^{\mathsf{T}}\mathbf{y} + \boldsymbol{\gamma}^{\mathsf{T}}\mathbf{f} \tag{8.50}$$

$$\text{s.t.} \ \sum_{p \in P_{uv}} y_p = w_{uv} \qquad\qquad \forall (u,v) \in D \tag{8.51}$$

$$\sum_{p \in P_a} y_p \leq \sum_{l:\rho_a \in l} \kappa_{r_l} f_l \qquad \forall a \in A \tag{8.52}$$

$$\sum_{l \in L_e} f_l \leq \lambda_e \qquad\qquad\qquad \forall e \in E \tag{8.53}$$

$$f_l \in \mathbb{R}_{\geq 0} \qquad\qquad\qquad \forall l \in L \tag{8.54}$$

$$y_p \in \mathbb{R}_{\geq 0} \qquad\qquad\qquad \forall p \in P \tag{8.55}$$

Solving this linear programming reformulation for the example of Fig. 8.11 will also give us optimal solution $\mathbf{f}^* = [8, 0, 1.4]^{\mathsf{T}}$ and $\mathbf{y}^* = [160, 0, 40, 0, 30]^{\mathsf{T}}$ resulting in using lines 1 and 3.

8.4.3 Skeleton Heuristic

To generate service lines and derive line concepts, one can also use heuristics that have less computational complexity. One approach is to start with a minimal set of lines that do not cover all vertices with passenger demand and then insert lines in a reasonable way to come up with a, hopefully good, line plan. An approach in this direction is the *skeleton* heuristic, proposed by Lampkin and Saalmans [1] in 1967 and extended by Silman et al. [25] in 1974.

The objective is to generate a desirable set of lines L. The line plan L is evaluated based on the total travel time of passengers:

$$T(L) := \sum_{(u,v) \in D} d_{uv} t_{uv}^L$$

where d_{uv} is the passenger demand of each OD-pair $(u, v) \in V \times V : w_{uv} > 0$, $\mathbf{W}$ the OD-matrix, and t_{uv}^L the shortest travel time from u to v for line plan L. In the skeleton heuristic, the line generation and the setting of frequencies are solved sequentially. Initially, the lines are generated without considering the frequencies and the passenger waiting times. The input is the travel time between every pair of adjacent vertices in the network and OD-matrix. Given a line plan L, $T(L)$ is computed as follows.

Step 1: initially set $t_{uv}^L = r_{uv}$ where r_{uv} is the shortest walking time fom u to v.

Step 2: for every $l \in L$ and every OD-pair $(u, v) \in D$ such that $u \in l$, $v \in l$ calculate the travel time from u to v when using line l. If it is less than t_{uv}^L, then replace t_{uv}^L by this smaller value. After the end of this step, t_{uv}^L represents the minimum travel time on the system for demand pairs (u, v) that lie on one of the lines of the line plan or the walking time for demand pairs (u, v) that are not connected by a line.

Step 3: for any $(u, v) \in D$ for which there exists zone k such that $t_{uk}^L + t_{kv}^L + b \leq t_{uv}^L$ where b is a transfer penalty, set $t_{uv}^L = t_{uk}^L + t_{kv}^L + b$. Note that only one transfer is allowed.

By performing steps 1–3 for a line plan L, we have computed the total passenger travel time $T(L)$ for that line plan. To apply the skeleton heuristic, we can start with a line plan L which is empty. Lines are subsequently added to the line plan. To generate potential line candidates, we need to generate an immense number of potential lines. Because this is not computationally feasible, an approximation is provided with the use of the *skeleton* concept.

A skeleton is a sequence of four vertices with passenger demand (i.e., centroids of zones). With each skeleton is associated a line which consists of the shortest paths between the consecutive demand vertices in the skeleton. We can further reduce the number of skeletons by selecting vertices at the end of our network and considering them as terminals (i.e., they can be the first or last vertex of a skeleton). Our pool of candidate lines L_0 is then the set of all lines derived by the skeletons in our network. This can, of course, result in a sub-optimal line concept, but this heuristic reduces greatly the computational complexity of the LPP.

From the pool of candidate lines, we can add one line at a time to the line plan L. At each iteration, we seek to find the line l' which, when added to L, results in line plan L' such that:

$$\frac{T(L') - T(L)}{t_{l'}} \geq \frac{T(L'') - T(L)}{t_{l''}} \quad \forall l'' \in L_0$$

where $t_{l'}$ is the travel time of line l'. We select a line in this way because we seek to find a line that not only reduces the total travel time of passengers, $T(L') - T(L)$, but also has a small travel time $t_{l'}$ resulting in reduced operational costs. To select such a line l' and add it in the line plan L, we need to evaluate the total travel times of all lines $l'' \in L_0$ using steps 1–3 described above. This requires performing steps 1–3 $|L_0|$ times. In the next iteration, we can add another line to L following the same process. Notice that because the new line plan contains line l', we will have to compute again the values of $T(L)$ considering the addition of every line in L_0 except l'. That is, we would need to perform steps 1–3 $|L_0| - 1$ times. The process can continue with adding new lines.

As we previously mentioned, the LPP is split in two stages. After generating a line plan, we examine its performance by setting the frequency of each line considering the capacity of vehicles, potential restrictions in the number of vehicles, and the demand at each edge. From this, we can evaluate the riding and waiting times of passengers resulting in a more accurate representation of total travel times. At this stage, the objective function to be minimized is the sum of the journey time (which includes allowance for transfer times) plus discomfort penalties proportional to the number of passengers who cannot find seats [25].

To summarize, the skeleton LPP heuristic has two phases. In the first phase, desirable lines, which may be added to an initial set of lines L, are generated. The lines to be added are selected from a set of lines based on a line skeleton consisting of four vertices (zones) with the extremes restricted to a set of terminal zones. In the second phase, optimal frequencies for a set of lines, under the constraint of a given number of available public transport vehicles, are determined.

8.4.4 Further Reading

The LPP is oftentimes referred to as the Transit Network Design and Frequency Setting (TNDFSP) problem or the Transit Route Network Design Problem (TRNDP). For further reading, we refer the interested reader to the following survey papers in this area.

> **Further Reading—survey papers**

- Schöbel, Line planning in public transportation: models and methods, OR spectrum, 2012 [16].
- Kepaptsoglou and Karlaftis, Transit route network design problem: review, Journal of transportation engineering, 2009 [26].
- Guihaire and Hao, Transit network design and scheduling: A global review, Transportation Research Part A: Policy and Practice, 2008 [27].
- Ibarra-Rojas et al., Planning, operation, and control of bus transport systems: A literature review, Transportation Research Part B: Methodological, 2015 [28].
- Iliopoulou et al., Metaheuristics for the transit route network design problem: a review and comparative analysis, Public Transport, 2019 [29].

8.5 Network Complexity and Connectivity

In this section, we will present common indicators used to evaluate the structure and connectedness of already defined public transport networks. In the previous sections, we showed that we can design the stations/stops and line routes of a public transport network based on specific objectives. Once a public transport network is designed, its form and structure can be assessed with the use of graph theory-based indicators that focus more on the topological attributes of the network [30]. Derrible and Kennedy [31] present a set of indicators that were initially developed for general graphs and some of them were later extended to public transport networks.

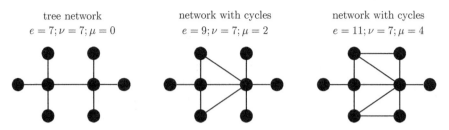

Fig. 8.12 Cyclomatic numbers in different networks represented as undirected graphs

To understand these indicators, we have to start first from graph theory. Graph theory dates back to 1741 when Euler presented the *Seven Bridges of Königsberg* problem. In 1962, Berge introduced the first indicator to evaluate networks, the 1st-Betti number, or *cyclomatic* number, which counts the number of cycles in a graph:

$$\mu := e - (v - p) \quad \text{(cyclomatic number)} \tag{8.56}$$

where e is the number of edges, v the number of vertices, and p the number of subgraphs of a network $G = (V, E)$. Because public transport networks are typically not disjoint, $p = 1$ in most cases. To understand this more, $v - 1$ is the number of edges in a tree graph with v vertices and $e - (v - 1)$ returns the number of extra edges in the network that create cycles. If we have more cycles in the network, we offer more alternative paths to public transport passengers, so we are in favor of higher values of the cyclomatic number μ. In Fig. 8.12 we present two networks with the same number of vertices, but different cyclomatic numbers. It is evident that the network with the higher μ value offers more alternatives to passengers.

Another important indicator for the evaluation of a public transport network is the α-index, also known as the degree of cyclicity, proposed by Garrison and Marble [32] in 1962. This index is expressed as:

$$\alpha := \frac{e - v + 1}{\frac{1}{2}v(v - 1) - (v - 1)} \quad \text{(degree of cyclicity)} \tag{8.57}$$

Its numerator is the cyclomatic number and its denominator is the maximum cyclomatic number possible. This index returns the ratio of actual to potential number of cycles in a graph and we would need it to be as high as possible. Because in many public transport networks we have a single transfer stop in the locations where lines intersect, the maximum number of possible edges is not $\frac{1}{2}v(v - 1)$, but $3(v - 2)$. This can result in the adaptation of the α-index as follows:

$$\alpha := \frac{e - v + 1}{3(v - 2) - (v - 1)} = \frac{e - v + 1}{2v - 5} \tag{8.58}$$

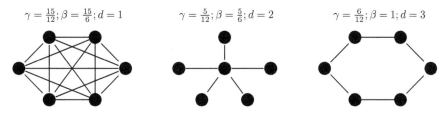

$\gamma = \frac{15}{12}; \beta = \frac{15}{6}; d = 1$ $\gamma = \frac{5}{12}; \beta = \frac{5}{6}; d = 2$ $\gamma = \frac{6}{12}; \beta = 1; d = 3$

Fig. 8.13 Examples of Network Indicators for networks with edge lengths equal to 1 km

Another informative network indicator defined by Garrison and Marble (1962) is the γ-index, also known as connectivity, which examines the ratio of actual to potential edges:

$$\gamma := \frac{e}{\frac{1}{2}\nu(\nu - 1)} \quad \text{(connectivity)} \tag{8.59}$$

If we have a single stop at the locations where lines intersect, then the maximum number of potential edges is $3(\nu - 2)$ and this indicator is calculated as:

$$\gamma := \frac{e}{3(\nu - 2)} = \frac{e}{3\nu - 6} \tag{8.60}$$

To indicate the average number of connections per vertex, there is also the complexity index defined as:

$$\beta := \frac{e}{\nu} \quad \text{(complexity)} \tag{8.61}$$

In practice, we would like our network to be as well connected and as less complex as possible. The above indicators are the most commonly used ones to get an understanding of the topological efficiency of a public transport network. For instance, Derrible and Kennedy [33] used them in a 2010 study to indicate the complexity and robustness of already existing metro networks in several cities and draw general conclusions that are independent of the sizes of the cities.

Another indicator is the *network diameter*, d, which is the longest shortest path connecting a pair of vertices. If l_{ij} is the shortest path between vertices $i \in V$ and $j \in V$, then:

$$d := \max_{i \in V, j \in V} l_{ij} \quad \text{(network diameter)} \tag{8.62}$$

In Fig. 8.13 we present an example of the connectivity, complexity, and network diameter indicators for public transport networks with different structures.

Using Garrison and Marble's indicators, Lam and Schuler [34] introduced a public transport-specific indicator for connectivity as the ratio of potential to actual reciprocal harmonic means of trip times. Let n be the number of trips in the public transport network and T_i the travel time of every trip in the ideal case where the public transport network is completely connected. Then, the harmonic mean of the ideal trip times is:

$$\bar{T} := \frac{1}{\frac{1}{n}\sum_{i=1}^{n}\frac{1}{T_i}} \tag{8.63}$$

If t_i is the real trip time of each trip $i \in \{1, \ldots, n\}$ in the actual network where $t_i = +\infty$ if there are no connections in the public transport network for the origin-destination pair of trip i, then:

$$\bar{t} := \frac{1}{\frac{1}{n}\sum_{i=1}^{n}\frac{1}{t_i}} \tag{8.64}$$

The public transport-related connectivity indicator of Lam and Schuler [34] is then defined as the ratio of the harmonic mean of the ideal trip times and the harmonic mean of the actual trip times:

$$R := \frac{\bar{T}}{\bar{t}} \tag{8.65}$$

The public transport-related connectivity indicator R can take a value within the range [0,1], where a value closer to 1 is better.

Musso and Vuchic [30] introduced two additional public transport network-related indicators in 1998. The first is the *average inter-station spacing*, S, which is the ratio of the route length of the network L and the number of inter-station spacings N_s. The number of inter-station spacings N_s is the total number of vertices (stops) per line minus one terminal to account for the number of edges of the line. If lines have overlapping edges, the edges are counted once:

$$S := \frac{L}{N_s} \quad \text{(average inter-station spacing)} \tag{8.66}$$

The second is the ratio of total lengths of the lines by the length of the network as a whole:

$$\lambda := \frac{\sum_i L_i}{L} \quad \text{(line overlapping index)} \tag{8.67}$$

For transit systems without overlapping lines, $\lambda = 1$.

Table 8.14 Base-year trip distribution matrix, $\mathbf{T}' = \{t'_{ij}\}$

Origins	Destinations				$o_i = \sum_{j \in I} t_{ij}$	Future o_i
	$j = 1$	$j = 2$	$j = 3$	$j = 4$		
$i = 1$	5	50	100	200	$o_1 = 355$	$o_1 = 400$
$i = 2$	50	5	100	300	$o_2 = 455$	$o_2 = 460$
$i = 3$	50	100	5	100	$o_3 = 255$	$o_3 = 400$
$i = 4$	100	200	250	20	$o_4 = 570$	$o_4 = 702$
$d_j = \sum_{i \in I} t_{ij}$	205	355	455	620	$T = 1635$	
Future d_j	260	400	500	802		$T = 1962$

Table 8.15 Cost matrix, c_{ij}

Origins	Destinations				Target o_i
	$j = 1$	$j = 2$	$j = 3$	$j = 4$	
$i = 1$	3	11	18	22	400
$i = 2$	12	3	12	19	460
$i = 3$	15.5	13	5	7	400
$i = 4$	24	18	8	5	702
Target d_j	260	400	500	802	$T = 1962$

Exercises

8.1 Trip distribution—growth factor
Write a computer program and estimate the trip distribution with the doubly-constrained growth factor method using the input of Table 8.14.

8.2 Trip distribution—entropy maximization
Write an equality-constrained mathematical program and estimate the trip distribution with the entropy maximization method by setting the gradient of the Lagrangian function to 0.

8.3 Trip distribution—gravity
Consider the cost matrix of Table 8.15.

For a distance decay function $f(c_{ij}) = \exp(-\beta c_{ij})$ with $\beta = 0.1$, produce the trip distribution matrix with the doubly constrained gravity model.

8.4 MCLP
Devise a primal heuristic or a metaheuristic (i.e., genetic algorithm) to obtain a feasible initial solution guess for the MCLP problem using the input data of Tables 8.11 and 8.12.

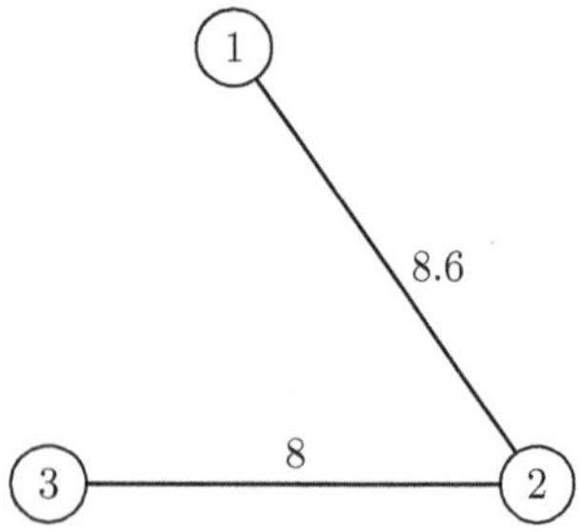

Fig. 8.14 Demand points
and edge distances of the
undirected graph

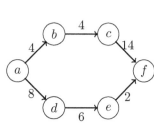

Fig. 8.15 Example network
$G = (V, A)$ with
non-negative edge weights

8.5 m-center Problem

(a) Write a formal algorithm that can solve the m-center problem.
(b) How many points $j \in P'$ we have in the undirected graph of Fig. 8.14 where $j \in P'$ if, and only if, $d(i, j) = d(y, j)$ for two points $i, y \in X : i \neq y$? Consider that X is the set of vertices in the undirected graph.

8.6 Single-pair Shortest Path

Consider the network of Fig. 8.15.
Consider also that traversing each one of the edges (a, d), (d, e) and (e, f) has an energy consumption cost of 15 units and traversing each one of the edges (a, b), (b, c) and (c, f) has an energy consumption cost of 10 units. The maximum allowed energy consumption cost is 35 units. Formulate an integer linear program that can find the shortest path from a to f without consuming more than 35 units of energy.

8.7 All-pairs Shortest Path

Solve the all-pairs shortest path problem for the digraph of Fig. 8.15.

Fig. 8.16 Network vertices and edges

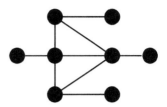

8.8 Line Planning Problem

Consider the network presented in the example of Fig. 8.11 and its corresponding parameter values. Write a computer program to compute its optimal line concept with the use of linear programming considering the LPP's reformulation to a linear program presented in Eqs. (8.50)–(8.55).

8.9 Network indicators

Compute the connectivity, complexity and degrees of cyclicity, and diameter of the network of Fig. 8.16.

References

1. W. Lampkin, P. Saalmans, J. Oper. Res. Soc. **18**(4), 375 (1967)
2. A.G. Wilson, J. Trans. Econ. Policy 108–126 (1969)
3. K. Furness, Traffic Eng. Control **7**(7), 458 (1965)
4. C. Toregas, R. Swain, C. ReVelle, L. Bergman, Oper. Res. **19**(6), 1363 (1971)
5. A.T. Murray, Ann. Oper. Res. **123**(1), 143 (2003)
6. C. Toregas, C. Revelle, Geogr. Anal. **5**(2), 145 (1973)
7. R. Church, C. ReVelle, *Papers of the Regional Science Association*, vol. 32 (Springer-Verlag, 1974), pp. 101–118
8. R. Garfinkel, A. Neebe, M. Rao, Manage. Sci. **23**(10), 1133 (1977)
9. E. Minieka, Siam Rev. **12**(1), 138 (1970)
10. E.W. Dijkstra et al., Numer. Math. **1**(1), 269 (1959)
11. J. Williams, Commun. ACM **7**, 6 (1964)
12. M.L. Fredman, R.E. Tarjan, J. ACM (JACM) **34**(3), 596 (1987)
13. R.W. Floyd, Commun. ACM **5**(6), 345 (1962)
14. J.Y. Yen, Manage. Sci. **17**(11), 712 (1971)
15. F. López-Ramos, SORT-Stat. Oper. Res. Trans. 181–214 (2014)
16. A. Schöbel, OR Spectrum **34**(3), 491 (2012)
17. M. Patriksson, *The Traffic Assignment Problem: Models and Methods* (Courier Dover Publications, 2015)
18. P. Carraresi, F. Malucelli, S. Pallottino, Optim. Ind. **3**, 19 (1995)
19. S. Peeta, A.K. Ziliaskopoulos, Netw. Spat. Econ. **1**(3), 233 (2001)
20. H. Wegel, *Fahrplangestaltung für taktbetriebene Nahverkehrsnetze* (Inst. für Verkehr, Eisenbahnwesen u. Verkehrssicherung, Techn. Univ, 1974)
21. A. Ceder, N.H. Wilson, Transp. Res. B: Methodol. **20**(4), 331 (1986)
22. R. Borndörfer, M. Grötschel, M.E. Pfetsch, Transp. Sci. **41**(1), 123 (2007)
23. M. Bussieck, Optimal lines in public rail transport. Ph.D. thesis, Citeseer (1998)

24. A. Schöbel, S. Scholl, *5th Workshop on Algorithmic Methods and Models for Optimization of Railways (ATMOS'05)* (Schloss Dagstuhl-Leibniz-Zentrum für Informatik, 2006)
25. L.A. Silman, Z. Barzily, U. Passy, Comput. Oper. Res. **1**(2), 201 (1974)
26. K. Kepaptsoglou, M. Karlaftis, J. Transp. Eng. **135**(8), 491 (2009)
27. V. Guihaire, J.K. Hao, Transp. Res. A: Policy and Pract. **42**(10), 1251 (2008)
28. O. Ibarra-Rojas, F. Delgado, R. Giesen, J. Muñoz, Transp. Res. B: Methodol. **77**, 38 (2015)
29. C. Iliopoulou, K. Kepaptsoglou, E. Vlahogianni, Pub. Transport **11**(3), 487 (2019)
30. A. Musso, V.R. Vuchic, *Characteristics of Metro Networks and Methodology for Their Evaluation* (National Research Council, Transportation Research Board Washington, DC, USA, 1988)
31. S. Derrible, C. Kennedy, Transport Rev. **31**(4), 495 (2011)
32. W.L. Garrison, D.F. Marble, *The Structure of Transportation Networks* (Northwestern Univ Evanston Il, Tech. rep., 1962)
33. S. Derrible, C. Kennedy, Physica A: Stat. Mech. Appl. **389**(17), 3678 (2010)
34. T.N. Lam, H.J. Schuler, Transp. Res. Rec. (854) (1982)

Chapter 9
Tactical Planning of Public Transport Services: Frequencies and Timetables

Abstract In the previous chapter we concentrated on the strategic planning phase of public transport services, including the estimation of origin-destination matrices, the selection of station/stop locations, the definition of line routes, and the graph-based evaluation of the resulting public transport networks. Once a public transport network is formed, a new series of actions need to take place to make it operational, including the determination of line frequencies, timetables, vehicle and crew schedules. This chapter focuses on the tactical planning decision level and addresses the frequency settings and timetabling problems.

9.1 Frequency Settings

The frequency settings problem typically splits the day into time periods based on the observed demand patterns (morning peak, morning non-peak, afternoon peak, and so on) and determines the *number of trips* per hour needed to satisfy the passenger demand for each public transport line at each one of the determined daily periods. We have already shown that many methods address the frequency setting problem when tackling the line planning problem (LPP). Nonetheless, these preliminary frequencies can be later updated to better adjust to the passenger demand of the system. Generally, the frequency settings problem can be defined as follows.

Frequency Settings Problem [1]
Frequency Setting is the problem of determining the number of trips for a given set of lines L to provide a high level of service in a specific planning period.

Setting the frequencies of public transport lines seeks to find a balance between reducing passenger waiting times by increasing the frequencies and reducing the fleet size and the vehicle running costs by decreasing the frequencies. Typical data that is used to set the frequencies of service lines when the public transport network is fixed includes:

- Automated Passenger Counting data (APC) typically collected with infrared lights above the doorways of a vehicle.
- Automated Fare Collection data (AFC) typically collected with the use of digital payment cards.
- Automated Vehicle Location data (AVL) typically collected by an in-vehicle tracking system that updates its position within a short interval (Google has developed a feed specification for sharing AVL data, known as GTFS-RT).

To proceed further, we provide the following set of important definitions.

Relevant definitions

- *Frequency*: The number of vehicles operating at a pre-defined time period. It is typically expressed in vehicles per hour.
- *Time Headway* or *Headway*: The time difference between two vehicles, measured at a particular location.
- *Vehicle Load*: The total number of on-board passengers.
- *Dwell time*: The time spent by the time a vehicle arrives at a stop until it departs from it. The dwell time is at least equal to the time needed for completing the passenger boardings and alightings.
- *Deadheading* time: The time period within which a vehicle is running without operating a service and generating revenue.
- *Layover* time: Short period of recovery time built into a schedule (i.e., driver break after completing a trip).

In the remainder of this section, we will provide common closed-form (analytical) methods and mathematical programming methods for the frequency settings problem.

9.1.1 Closed-form Methods

There are a number of closed-form (analytical) methods for determining the frequencies of service lines without resorting to mathematical modeling. Furth and Wilson [2] listed a number of analytical methods, such as the maximum loading point, the load profile, and the revenue/cost ratio methods.

9.1.1.1 Maximum Loading Point

The *Maximum Loading Point* or *Peak Load* method determines the service frequency during a period of the day according to the following rule.

> **Maximum Loading Point** [3]
> Consider a time period j for which we need to determine the frequency of a public transport service with stops $S = \{1, \ldots i, \ldots\}$. Consider also the average passenger load p_{ij} at each stop $i \in S$ of the line. The available vehicles are homogeneous and have the same vehicle capacity c. In addition, the desired in-vehicle load factor in period j is γ_j. The service frequency is the maximum passenger load observed at a stop $i' \in S$ such that $p_{i'j} \geq p_{ij} \ \forall i \in S$, divided by the desired load factor multiplied by the vehicle capacity, $\gamma_j c$.

Mathematically, the service frequency with the maximum loading point method is determined as the number of vehicle trips required for period j:

$$f_j := \frac{\max\limits_{i \in S} p_{ij}}{\gamma_j c}$$

where:

- p_{ij} is the average number of passengers (load) observed on-board when departing from stop $i \in S$ in period j. Note that $\max\limits_{i \in S} p_{ij}$ stands for the maximum observed load (across all stops) in period j and is the value measured at the *peak-load* point.
- c represents the capacity of any vehicle (number of seats plus the maximum allowable standees).
- γ_j is the desired load factor during period j, $0 < \gamma_j \leq 1$. This load factor can be used, for instance, to ensure that every passenger can have a seat or there is no overcrowding.

Let us consider the example of the service in Table 9.1. In this example, we need to set the service frequencies at three time periods, namely 6:00–8:00, 8:00–10:00 and 10:00–12:00. The average passenger load refers to the timeframe of each planning period (2 hrs) where we need to define the service frequency.

Consider the first time period from 6:00 until 8:00. Consider also a desired load factor $\gamma_j = 5/6$ and a fleet of homogeneous vehicles with capacity $c = 60$. Then, the number of required vehicle trips in the time period 6:00–8:00 according to the maximum loading point method is:

$$f_j := \frac{\max\limits_{i \in S} p_{ij}}{\gamma_j c} = \frac{290}{(5/6)60} = 5.8$$

Table 9.1 Average passenger load upon departure from each stop at each 2-h time period

Stop no	Distance to next stop (km)	Passenger load			Total load
		6:00–8:00	8:00–10:00	10:00–12:00	
1	2	210	150	120	480
2	3	110	130	180	420
3	1.5	150	180	140	470
4	4.5	100	180	220	500
5	2	290	120	180	590

where the 5th stop is the peak-load stop. We thus have a frequency of 5.8 vehicle trips in this 2-h period, or $5.8/2 = 2.9$ vehicle trips per hour. This results in a time headway of $60/2.9 = 20.69$ min between the passing of successive vehicle trips.

Another example is the time period from 8:00 to 10:00. In this period both stops 3 and 4 can be considered as peak-load points with a passenger load of 180. The service frequency in this 2-h period is $\frac{180}{(5/6)60} = 3.6$, or 1.8 vehicle trips per hour.

9.1.1.2   Load Profile (Ride Check) Method

An alternative analytical approach to the maximum load point method that can reduce the operational costs is the load profile method (see [3]).

Load Profile (ride check)

Consider a time period j for which we need to determine the frequency of a public transport service with stops $S = \{1, \ldots i, \ldots\}$. Consider also the average passenger load p_{ij} at each stop $i \in S$ of the line. The available vehicles are homogeneous and have the same vehicle capacity c. In addition, the desired in-vehicle load factor in period j is γ_j. The distance between stops i and $i + 1$ is l_i and $L = \sum_{i \in S} l_i$ is the length of the line. The service frequency is the maximum of the following two terms:

- The maximum passenger load observed at a stop $i' \in S$ such that $p_{i'j} \geq p_{ij} \; \forall i \in S$, divided by the the vehicle capacity c.
- The passenger-km traveled, $\sum_{i \in S} p_{ij} l_i$, divided by $\gamma_j cL$.

Mathematically, the load profile method calculates the frequency of a line over a period j as follows:

$$f_j = \max \left[\frac{\max\limits_{i \in S} p_{ij}}{c}, \frac{\sum\limits_{i \in S} p_{ij} l_i}{\gamma_j cL} \right]$$

where:

- p_{ij} is the *average number of passengers (load)* observed on-board when departing from stop $i \in S$ in period j.
- $\sum_{i \in S} p_{ij} l_i$ are the *passenger-km traveled* during time period j.
- $L = \sum_{i \in S} l_i$ is the route length of the line.

The load profile method guarantees, on the average basis of p_{ij}, that the on-board passengers at the max load route segment will not experience crowding above the given vehicle capacity c. The desired occupancy requirement $\gamma_j c$ is therefore relaxed because we might use the full capacity of vehicles at some stops. This allows using fewer trips per hour to perform the service compared to the maximum loading point method. Considering the input from Table 9.1, we have the route length of the line $L = (2 + 3 + 1.5 + 4.5 + 2) = 13$ km. We also have $\sum_{i \in S} p_{ij} l_i = 210 \cdot 2 + 110 \cdot 3 + 150 \cdot 1.5 + 100 \cdot 4.5 + 290 \cdot 2 = 2005$. The selected line frequency when applying the load factor method in the time period 6:00–8:00 would be:

$$
f_j = \max \left[\frac{\max_{i \in S} p_{ij}}{c}, \frac{\sum_{i \in S} p_{ij} l_i}{\gamma_j c L} \right] = \max \left[\frac{290}{60}, \frac{2005}{(5/6) \cdot 60 \cdot 13} \right] = \frac{290}{60} \simeq 4.83
$$

expressed in vehicle trips per 2 hrs, or $4.83/2 \simeq 2.42$ vehicle trips per hour. Notice that for the same time period we required more vehicle trips when we applied the maximum loading point method (namely, 2.9 vehicle trips per hour).

Over the 2-h period, the maximum loading point method provides a total capacity of 2.9 vehicle trips multiplied by 60 passengers per vehicle, resulting in 348 passengers. This capacity is never used because the peak-load point has a passenger load of 290 passengers. However, we are able to maintain the desired occupancy level of 290 passengers even at the peak load point because $348 \cdot (5/6) = 290$. In contrast, the load profile method offers a capacity of $4.83 \cdot 60 \simeq 290$ passengers resulting in maximum crowding at the peak load point. From this, a trade-off emerges: with the load profile method we accept crowding levels beyond the desired occupancy at some stops in exchange for using fewer vehicle trips and having fewer empty vehicles at stops with lower passenger loads; in contrast, with the maximum loading point method we ensure that we never exceed the desired occupancy level, but vehicles might run almost empty at stops with lower in-vehicle passenger load levels. This is made evident in Fig. 9.1.

The biggest disadvantage of the maximum loading point and load profile methods is that they both (naively) assume that the passenger demand (load) will *remain the same* if the frequencies are modified. That is, the assignment of passengers to a line is fixed. In addition, they do not take into consideration the passenger arrival rates at stations/stops (they both assume uniformly distributed passenger arrivals) and they do not take into consideration the available resources (vehicles) at the network level when setting the frequencies of individual lines. The latter might result in setting

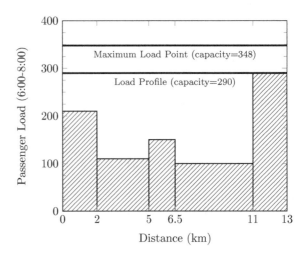

Fig. 9.1 Offered capacity with the maximum loading point and the load profile methods, and passenger load upon departure from each stop

frequencies that cannot be implemented in practice because of the lack of public transport vehicles to operate the required trips.

9.1.2   Model-based Frequency Setting

In addition to mathematical expressions, there have also been several mathematical models for setting up line frequencies in a more elaborate manner. A simple version of the frequency settings problem presented in the literature review of [1] is:

$$\min_{\mathbf{f}} \sum_{i \in V} \sum_{j \in V} d_{ij} t_{ij}^l \qquad (9.1)$$

$$\text{s.t.} \sum_{l \in L} \lceil \rho_l f_l \rceil \leq m \qquad (9.2)$$

where t_{ij}^l is the travel time between i and j via line l, d_{ij} is the passenger demand at the origin-destination pair (i, j), ρ_l is the cycle (round-trip) time of line l, and m is the maximum fleet size. Notice that the above formulation is not complete because there should be a relation between the travel time between i and j via line l, t_{ij}^l, and the frequency of line l, f_l.

Han and Wilson [4] developed a general model to set the frequencies of a set of lines in a directed graph $G = (V, A)$ where the number of available public transport vehicles is limited. This is expressed as follows, using a high-level representation.

$$\min \ J(q_{ij}^k, f_k, A_k) \tag{9.3}$$

$$\text{s.t. } cf_k \geq q_{ij}^k \qquad \qquad \forall k \in R, (i, j) \in L_k \tag{9.4}$$

$$q_{ij}^k = g_{ij}^k(v_{ab}, f_r, A_r) \quad \forall k \in R, (i, j) \in L_k, r \in X_{ij}, a \in V, b \in V \tag{9.5}$$

$$\sum_{k \in R} t_k f_k \leq m \tag{9.6}$$

where f_k is the service frequency on line $k \in R$, A_k is the set of attributes associated to line k, c is the passenger capacity for each public transport vehicle, q_{ij}^k is the passenger flow in arc (i, j) on route k, g_{ij}^k the function that determines the passenger flow assignment in arc (i, j) of line k, v_{ab} the origin-destination flow between vertices a and b, m the total number of available public transport vehicles, V is the set of vertices (stations/stops) of the public transport network, L_k the set of directed edges (arcs) on public transport line k, R the set of lines, t_k the round trip travel time of line k, and X_{ij} the set of routes offering duplicate service between vertices i and j. The objective function J is typically a function of the passenger waiting times at stops and in-vehicle crowding levels. Constraints (9.4) ensure that the in-vehicle passenger load does not exceed the capacity, constraints (9.5) perform a passenger flow assignment, and constraints (9.6) are fleet size constraints ensuring that we will not use more vehicles than available.

As we will later see, solving the passenger assignment problem for a given frequency can be perceived as a separate problem leading to a *bi-level* formulation of the frequency settings problem. Let us, for now, consider the trivial case where the passenger flow from i to j for line k is already known and it is not affected by the selected service frequency. Although this assumption will rarely hold in practice, this can greatly simplify the problem expressed in (9.3)–(9.6) by considering that the passenger flow in any arc (i, j) on any route k, q_{ij}^k, is fixed (parameter). This results in:

$$\min_{\mathbf{f}} \ J(q_{ij}^k, f_k, A_k) \tag{9.7}$$

$$\text{s.t. } cf_k \geq q_{ij}^k \qquad \qquad \forall k \in R, (i, j) \in L_k \tag{9.8}$$

$$\sum_{k \in R} t_k f_k \leq m \tag{9.9}$$

Let us also consider an objective function J that strives to minimize the total number of running vehicles. This will result in replacing the general function $J(q_{ij}^k, f_k, A_k)$ by:

$$J(f_k) := \sum_{k \in R} t_k f_k$$

Suppose also that we want all our passengers to wait less than θ minutes at each station/stop. This will result in adding constraint $f_k \geq \frac{1}{\theta}$. With these modifications,

Fig. 9.2 Example digraph
$G = (V, A)$ where V
represents the stops of the
lines

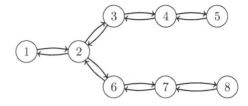

we have the following frequency settings formulation that strives to minimize the total number of running vehicles.

Running vehicle-oriented Frequency Settings formulation

$$\min_{\mathbf{f} \in \mathbb{Z}_{\geq 0}^{|R|}} \sum_{k \in R} t_k f_k$$

$$\text{s.t. } cf_k \geq q_{ij}^k \qquad \forall k \in R, (i, j) \in L_k$$

$$\sum_{k \in R} t_k f_k \leq m \qquad\qquad (9.10)$$

$$f_k \geq \frac{1}{\theta} \qquad \forall k \in R$$

Considering that the values of frequencies are positive integer numbers, this problem can be solved as an integer linear program with $f_k \in \mathbb{Z}_{\geq 0}$ for any $k \in R$.

Consider an example of solving this problem for two lines $R = \{1, 2\}$, vehicle capacity $c = 30$ passengers, round-trip travel times $t_1 = 110$ and $t_2 = 100$ min (or 110/60 and 100/60 hrs), $\theta = 30$ min, and total number of available vehicles $m = 15$. The digraph $G = (V, A)$ of this example is presented in Fig. 9.2.

The set of arcs of line 1 is $L_1 = \{(1, 2), (2, 3), (3, 4), (4, 5), (5, 4), (4, 3), (3, 2), (2, 1)\}$. The set of arcs of line 2 is $L_2 = \{(1, 2), (2, 6), (6, 7), (7, 8), (8, 7), (7, 6), (6, 2), (2, 1)\}$. The passenger flows at each arc for line 1 are $q_{12}^1 = 30, q_{23}^1 = 40, q_{34}^1 = 60, q_{45}^1 = 70$ and for line 2 are $q_{12}^2 = 35, q_{26}^2 = 45, q_{67}^2 = 65, q_{78}^2 = 150$, expressed in passengers per hour. The two lines $R = \{1, 2\}$ can attain a positive integer frequency value expressed in trips per hour and the planning period is one hour. The solution of this problem is $\mathbf{f}^* = [3, 5]^\mathsf{T}$ indicating that line $k = 1$ has an optimal frequency of 3 trips per hour and line $k = 2$ of 5 trips per hour. An example of solving this integer linear programming problem is presented below.

Practitioner's Corner
Frequency Settings example in Python 3

One can use PuLP to program and the CBC solver to solve the example frequency settings problem in (9.10) as follows.

```python
import pulp
from pulp import *
L={1:(12,23,34,45,54,43,32,21),2:(12,26,67,78,87,76,62,21)} #
    line arcs
t={1:110/60,2:100/60} # round trip travel times
c=30; m=15; theta=30/60
q={(12,1):30, (23,1):40, (34,1):60, (45,1):70, (21,1):0, (32,1)
    :0, (43,1):0, (54,1):00,
    (12,2):35, (26,2):45, (67,2):65, (78,2):150, (21,2):0, (62,2)
    :0, (76,2):0, (87,2):0,
    }
model = LpProblem("FS", LpMinimize)
R=(1,2) #lines
# variables
f = LpVariable.dicts("f", (k for k in R), 0, None, LpInteger)
# objective function
model += (sum(t[k]*f[k] for k in R))
# constraints:
for k in R:
    for ij in L[k]:
        model += c*f[k] >= q[ij,k]
model += sum(t[k]*f[k] for k in R) <= m
for k in R: model += f[k] >= 1/theta
# optimizing
solver = pulp.PULP_CBC_CMD(mip=True, msg=True, timeLimit=None,
    warmStart=False, path=None, mip_start=False, timeMode='
    elapsed',options=["passCuts=0","preprocess=off","presolve=off
    ","cuts=on","heuristics=off","greedyHeuristic=off"])
model.solve(solver)
print ("Status:", LpStatus[model.status])
for v in model.variables():
    if v.varValue>0: print (v.name, "=", v.varValue)
print ("Optimal Solution = ", value(model.objective))
```

The optimal solution $\mathbf{f}^* = [3, 5]^\mathsf{T}$ assigns 3 trips per hour to line 1 and 5 trips per hour to line 2. This will require $t_1 f_1 = 5.5$ vehicles to operate line 1 and $t_2 f_2 \approx 8.33$ vehicles to operate line 2. If we want to assign whole vehicles to every line, we can add integer variables $\mathbf{x}$, where x_k indicates the number of whole vehicles assigned to line $k \in R$. This will result in a slight reformulation of program (9.10), as follows.

$$
\begin{aligned}
\min_{\mathbf{f},\mathbf{x}} \quad & \sum_{k \in R} t_k f_k \\
\text{s.t.} \quad & cf_k \geq q_{ij}^k && \forall k \in R, (i, j) \in L_k \\
& t_k f_k \leq x_k && \forall k \in R \\
& \sum_{k \in R} x_k \leq m && (9.11)
\end{aligned}
$$

$$f_k \geq \frac{1}{\theta} \qquad\qquad \forall k \in R$$

$$\mathbf{f} \in \mathbb{Z}_{\geq 0}^{|R|}, \mathbf{x} \in \mathbb{Z}_{\geq 0}^{|R|}$$

This formulation will result in optimal solution $\mathbf{f}^* = [3, 5]^\mathsf{T}$ and $\mathbf{x}^* = [6, 9]^\mathsf{T}$ when solving the problem of Fig. 9.2.

It is worth noting that we can have several different formulations for the frequency settings problem, with different advantages and disadvantages. For instance, we might not be interested in reducing the vehicle running costs as long as the available vehicles suffice to perform the service. Instead, we might be interested in reducing the waiting times of passengers at stations/stops. This will result in adopting another objective function that is *passenger waiting time*-oriented. Given that the waiting time of passengers is related to the time headways of service lines, we can seek to minimize the function:

$$\sum_{k \in R} \sum_{(i,j) \in L_k} \frac{q_{ij}^k}{f_k}$$

where $1/f_k$ is the time headway of line $k \in R$. This formulation will prioritize the increase of frequencies to reduce the passengers' waiting times at stations/stops. Higher frequencies will be provided to lines with more passengers that are affected more by high waiting times. Note that this new objective function is fractional (non-linear). We can also consider a constraint $\frac{1}{f_k} \leq \theta_{\min}$ enforcing every line to have at least a minimum time headway of $\theta_{\min}$ for safety reasons or for avoiding vehicle bunching. This new frequency settings formulation that strives to minimize the passenger waiting times is expressed below.

Passenger waiting times-oriented Frequency Settings formulation

$$\min_{\mathbf{f},\mathbf{x}} \sum_{k \in R} \sum_{(i,j) \in L_k} \frac{q_{ij}^k}{f_k}$$

$$\text{s.t. } cf_k \geq q_{ij}^k \qquad\qquad \forall k \in R, (i,j) \in L_k$$

$$t_k f_k \leq x_k \qquad\qquad \forall k \in R$$

$$\sum_{k \in R} x_k \leq m \qquad\qquad\qquad\qquad (9.12)$$

$$f_k \geq \frac{1}{\theta} \qquad\qquad \forall k \in R$$

$$f_k \leq \frac{1}{\theta_{\min}} \qquad\qquad \forall k \in R$$

$$\mathbf{f} \in \mathbb{Z}_{\geq 0}^{|R|}, \mathbf{x} \in \mathbb{Z}_{\geq 0}^{|R|}$$

The main issue with the passenger waiting times-oriented formulation is the non-linear (fractional) objective function. We can linearize the objective function by taking advantage of the restricted set of frequency values that can be assigned to a line. In practical operations, line frequencies rarely exceed 30 trips per hour, which results in 1 trip every two minutes. If we introduce a discrete set $F = \{1, 2, \dots, 30\}$ containing all possible discrete frequency values of a line in a practical environment, then we can also introduce binary variables $z_{fk} \in \{0, 1\}$ where $z_{fk} = 1$ if line $k \in R$ operates with frequency $f \in F$ and $z_{fk} = 0$ if not. These binary variables can be expressed via an $|F| \times |R|$ matrix $\mathbf{Z}$.

Because we must assign one, and only one, frequency value to each line $k \in R$, we need to add constraints:

$$\sum_{f \in F} z_{fk} = 1 \quad \forall k \in R$$

Importantly, f_k can be now replaced by $\sum_{f \in F} f z_{fk}$ and $\frac{1}{f_k}$ can be replaced by $\sum_{f \in F} \frac{z_{fk}}{f}$. Thus, the objective function can be rewritten as:

$$\sum_{k \in R} \sum_{(i,j) \in L_k} q_{ij}^k \sum_{f \in F} \frac{z_{fk}}{f}$$

which has a linear form since variables z_{fk} appear at the numerator. Replacing variables f_k by $\sum_{f \in F} f z_{fk}$ in the remaining constraints of (9.12) yields the mixed-integer linear problem presented below.

Passenger waiting times-oriented Frequency Settings as a MILP

$$\min_{\mathbf{Z}, \mathbf{x}} \sum_{k \in R} \sum_{(i,j) \in L_k} q_{ij}^k \sum_{f \in F} \frac{z_{fk}}{f}$$

$$\text{s.t. } c \sum_{f \in F} f z_{fk} \geq q_{ij}^k \qquad \forall k \in R, (i,j) \in L_k$$

$$t_k \sum_{f \in F} f z_{fk} \leq x_k \qquad \forall k \in R$$

$$\sum_{k \in R} x_k \leq m \qquad\qquad\qquad\qquad (9.13)$$

$$\sum_{f \in F} f z_{fk} \geq \frac{1}{\theta} \qquad \forall k \in R$$

$$\sum_{f \in F} f z_{fk} \leq \frac{1}{\theta_{\min}} \qquad \forall k \in R$$

$$\sum_{f \in F} z_{fk} = 1 \qquad \forall k \in R$$

$$\mathbf{Z} \in \{0, 1\}^{|F| \times |R|}, \mathbf{x} \in \mathbb{Z}_{\geq 0}^{|R|}$$

Practitioner's Corner
Frequency Settings example in Python 3

One can use `PuLP` to program and the `CBC` solver to solve the frequency settings problem in (9.13) considering the data from the example in Fig. 9.2 and $\theta_{min} = 2$ min, as follows.

```python
import pulp
from pulp import *
L={1:(12,23,34,45,54,43,32,21),2:(12,26,67,78,87,76,62,21)} #
    line arcs
t={1:110/60,2:100/60} # round trip travel times
c=30; m=15; theta=30/60; theta_min=2/60
q={(12,1):30, (23,1):40, (34,1):60, (45,1):70, (21,1):0, (32,1)
    :0, (43,1):0, (54,1):00,
    (12,2):35, (26,2):45, (67,2):65, (78,2):150, (21,2):0, (62,2)
    :0, (76,2):0, (87,2):0,
    }
model = LpProblem("FS", LpMinimize)
R=(1,2) #lines
F=[i for i in range(1,31)] #allowed frequency values
# variables
z = LpVariable.dicts("z", ((f,k) for f in F for k in R), 0, None,
    LpBinary)
x = LpVariable.dicts("x", (k for k in R), 0, None, LpInteger)
# objective function
model += sum(sum(q[ij,k]*sum(float(1/f)*z[f,k] for f in F) for ij
    in L[k]) for k in R)
# constraints:
for k in R:
    for ij in L[k]:
        model += c*sum(f*z[f,k] for f in F) >= q[ij,k]
for k in R: model += t[k]*sum(f*z[f,k] for f in F) <= x[k]
model += sum(x[k] for k in R) <= m
for k in R: model += sum(f*z[f,k] for f in F) >= 1/theta
for k in R: model += sum(f*z[f,k] for f in F) <= 1/theta_min
# optimizing
solver = pulp.PULP_CBC_CMD(mip=True, msg=True, timeLimit=None,
    warmStart=False, path=None, mip_start=False, timeMode='
    elapsed',options=["passCuts=0","preprocess=off","presolve=off
    ","cuts=on","heuristics=off","greedyHeuristic=off"])
model.solve(solver)
print ("Status:", LpStatus[model.status])
for v in model.variables():
    if v.varValue>0: print (v.name, "=", v.varValue)
print ("Optimal Solution = ", value(model.objective))
```

Solving this problem yields optimal solution $z_{3,1}^* = 1$ and $z_{5,2}^* = 1$, and $z_{f,k}^* = 0$ for any other $f \in F$, $k \in R$. That is, the optimal frequency of line 1 is 3 trips per hour and of line 2 is 5 trips per hour. The optimal solution also contains $\mathbf{x}^* = [6, 9]^\mathsf{T}$.

The formulation in (9.13) requires $|F| \times |R|$ binary variables and, given that the problem is a MILP, this high number of discrete variables might create computational issues in large-scale problem instances.

9.1.3 Passenger Assignment based on Line Frequencies

9.1.3.1 Assignment with Fixed Travel Times

One key issue that is not addressed yet is the effect of line frequencies to the route selection of passengers. Passengers would naturally change their plans and use alternatives if their lines become less frequent, requiring more waiting time at stops. For this reason, many frequency settings models combine the frequency setting and the passenger assignment problem resulting in a *bi-level* formulation.

The most prevalent frequency-based passenger assignment formulation is provided by Spiess and Florian [5] in 1989. This model splits passenger trips into trip components that are represented by arcs $a \in A$ in a directed graph $G = (V, A)$. The trip components $a \in A$ are not necessarily inter-station arcs. Trip components can be boarding trip components, alighting trip components, and in-vehicle traveling trip components. Each trip component $a \in A$ has:

- An arc travel time c_a which is equal to 0 if this trip component is a boarding or alighting one since the vehicle is idle.
- A service frequency f_a which can be set to $+\infty$ or a very large positive number M for alighting and in-vehicle traveling trip components since passengers do not have to wait when they alight or the vehicle is running.

In practice, three types of trip components are considered:

1. Boarding trip component ($c_a = 0$, $f_a > 0$);
2. In-vehicle riding trip component ($c_a > 0$, $f_a = +\infty$);
3. Alighting trip component ($c_a = 0$, $f_a = +\infty$).

The waiting time of trip component $a \in A$ can be linked to the service frequency f_a related to this trip component. In this network representation, $i \in V$ are vertices that have incoming and outgoing trip components (arcs). Note that we might have multiple vertices in the same physical location because at that location there might be different travel options for a passenger that seeks to arrive at his/her final destination. For instance, in Fig. 9.3 a network with 4 public transport stations is represented by a digraph with 14 vertices.

Because each arc a is a trip component, it is associated with one, and only one, service line l. For instance, if arc a is part of line l, then the frequency of this arc is $f_a = f_l$ if arc a is related to the boarding trip component to line l. If, however, arc a is an alighting or an in-vehicle riding arc, then $f_a = +\infty$ because passengers do not experience any waiting time when using this trip component. An example of how a physical network with 4 lines and 4 stations is translated to a digraph where each arc is a trip component is provided in Fig. 9.3. In this representation, we use a very large positive number M instead of $+\infty$ to declare the frequencies of alighting and in-vehicle riding trip components.

The representation of the trip components in Fig. 9.3 can be made more compact resulting in a digraph with less arcs. In particular, each vertex $i \in V$ that has only one

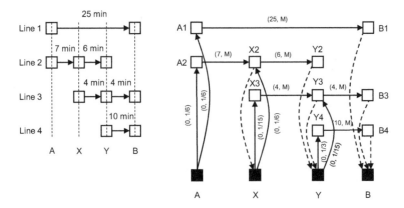

Fig. 9.3 Left: Example of a physical network with 4 lines and stops A, X, Y and B. Right: Representation of the network with a digraph where every arc has a value (c_a, f_a) expressing its travel time and frequency. Arcs that represent alighting trip components are shown with trimmed lines and they have travel times and frequencies (0,M)

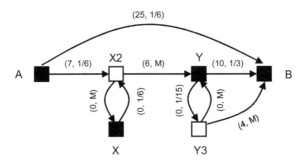

Fig. 9.4 Digraph's transformation to a more compact representation

incoming arc a_1 with (c_{a_1}, f_{a_1}) and only one outgoing arc a_2 with (c_{a_2}, f_{a_2}) where $f_{a_2} = M$ can be replaced by a single arc a with $(c_a, f_a) = (c_{a_1} + c_{a_2}, f_{a_1})$. Thus, the network of Fig. 9.3 may be simplified to the one shown in Fig. 9.4.

In this implementation we assume that the arc travel times c_a are not dependent on the passenger flow (that is, they are fixed). Because of that, we can assign the passenger demand that is expressed in the form of origin-destination matrices in a sequential order by assigning at each iteration the demand from all origin points to a destination vertex $r \in V$. In other words, we do not need to assign the passenger demand from all origin vertices to all destination vertices at the same time.

Let us consider that we want to assign all passenger demand that has vertex r as its final destination. To do so, we seek to define an assignment strategy $\bar{A}$ that will use only arcs $\bar{A} \subseteq A$ to assign all demand to vertex r. Arcs not included in $\bar{A}$ will not be used and a strategy $\bar{A}$ is feasible if it contains at least one path from each origin vertex to the destination vertex r and does not contain any cycles. Let A_i^+ be the set of outgoing arcs from vertex i and A_i^- the set of incoming arcs to vertex i. Let also $\bar{A}_i^+ = A_i^+ \cap \bar{A}$. It follows that $\bar{A}_r^+ = \emptyset$ because r is the destination vertex of this strategy.

We can then define the expected waiting time for the arrival of the 1st vehicle serving arc $a \in \bar{A}_i^+$ known as the *combined waiting time* of arc a:

$$w(\bar{A}_i^+) := \frac{\theta}{\sum\limits_{a \in A_i^+} f_a} \quad \forall i \in V \tag{9.14}$$

where $\theta \geq 0$ is equal to 0.5 for high frequency services where passengers cannot coordinate their arrivals with the arrival times of vehicles and 1 for low frequency and unreliable services. We can also define the probability that arc a is served first among all outgoing arcs in $\bar{A}_i^+$:

$$\mathcal{P}_a(\bar{A}_i^+) := \frac{f_a}{\sum\limits_{a' \in \bar{A}_i^+} f_{a'}} \quad \forall a \in \bar{A}_i^+ \tag{9.15}$$

To assign passenger trips to destination vertex r, we use g_i^r for $i \in V - \{r\}$, where g_i^r is the passenger demand from origin vertex i to the destination vertex r. Because all this demand will end up to destination r, we have that:

$$g_r^r = - \sum\limits_{i \in V - \{r\}} g_i^r$$

The demand g_i^r is assigned to the network yielding arc volumes ν_a and vertex volumes ν_i representing the passenger volume traversing vertex i:

$$\nu_i := \sum\limits_{a \in A_i^-} \nu_a + g_i^r \quad \forall i \in V \tag{9.16}$$

where g_i^r is the demand in vertex i and ν_a is the passenger volume of arc $a \in A_i^-$. We can also express the arc volumes as:

$$\nu_a := \nu_i \mathcal{P}_a(\bar{A}_i^+) \quad \forall a \in A_i^+, \forall i \in V \tag{9.17}$$

We can then cast the passenger assignment problem as finding the optimal assignment strategy $\bar{A}$. For modeling purposes, we introduce binary variables:

$$x_a := \begin{cases} 0 \text{ if } a \notin \bar{A} \\ 1 \text{ if } a \in \bar{A} \end{cases} \tag{9.18}$$

and we cast the problem as:

$$(\mathcal{L}_1) \quad \min_{\mathbf{x},\nu} \quad \sum_{a \in A} c_a \nu_a + \sum_{i \in V} \frac{\nu_i}{\sum_{a \in A_i^+} f_a x_a} \tag{9.19}$$

$$\text{subject to:} \quad \nu_a = \nu_i \frac{f_a x_a}{\sum_{a' \in A_i^+} f_{a'} x_{a'}} \qquad \forall a \in A_i^+, \forall i \in V \tag{9.20}$$

$$\nu_i = \sum_{a \in A_i^-} \nu_a + g_i^r \qquad \forall i \in V \tag{9.21}$$

$$\nu_i \geq 0 \qquad \forall i \in V \tag{9.22}$$

$$x_a \in \{0, 1\} \qquad \forall a \in A \tag{9.23}$$

The objective function of this model is nonlinear (fractional) and it seeks to minimize the in-vehicle travel times and the waiting times.

Because $\sum_{a \in A_i^+} P_a(\bar{A}_i^+)\nu_i = 1$, Eqs. (9.16)–(9.17) result in the *flow conservation* equation:

$$\sum_{a \in A_i^+} \nu_a - \sum_{a \in A_i^-} \nu_a = g_i^r \quad \forall i \in V \tag{9.24}$$

Let us also introduce variables

$$w_i := \frac{\nu_i}{\sum_{a \in A_i^+} f_a x_a} \quad \forall i \in V$$

where $\mathbf{w} = [w_1, \ldots, w_{|I|}]^{\mathsf{T}}$ and w_i is the total waiting time at vertex $i \in V$. Then, formulation $\mathcal{L}_1$ can be replaced by the following.

$$(\mathcal{L}_2) \quad \min_{\mathbf{w},\mathbf{x},\nu} \quad \sum_{a \in A} c_a \nu_a + \sum_{i \in V} w_i \tag{9.25}$$

$$\text{subject to:} \quad \sum_{a \in A_i^+} \nu_a - \sum_{a \in A_i^-} \nu_a = g_i^r \qquad \forall i \in V \tag{9.26}$$

$$\nu_a = x_a f_a w_i \qquad \forall a \in A_i^+, \forall i \in V \tag{9.27}$$

$$\nu_a \geq 0, x_a \in \{0, 1\} \qquad \forall a \in A \tag{9.28}$$

$$w_i \in \mathbb{R} \qquad \forall i \in V \tag{9.29}$$

Formulation $\mathcal{L}_2$ has the nonlinear constraints (9.27). These can be linearized by removing x_a and replacing the equality by inequality:

$$\nu_a \leq f_a w_i \quad \forall a \in A_i^+, \forall i \in V$$

This will result in the final, linear programming formulation of $\mathcal{L}_3$.

Frequency-based Passenger Assignment LP of Spiess and Florian [5]

$$(\mathcal{L}_3) \qquad \min_{\nu \in \mathbb{R}_+^{|A|}, w \in \mathbb{R}_+^{|V|}} \quad \sum_{a \in A} c_a \nu_a + \sum_{i \in V} w_i \tag{9.30}$$

$$\text{subject to:} \quad \sum_{a \in A_i^+} \nu_a - \sum_{a \in A_i^-} \nu_a = g_i^r \quad \forall i \in V \tag{9.31}$$

$$\nu_a \leq f_a w_i \qquad \forall a \in A_i^+, \forall i \in V \tag{9.32}$$

Program $(\mathcal{L}_3)$ assigns passengers from all origins to the destination vertex r to minimize the total travel times at the arcs and the waiting times at vertices. It is a linear program with $|V| + |A|$ variables (ν and $\mathbf{w}$). For in-vehicle riding and alighting arcs that have frequencies $f_a = +\infty$ we set $f_a = M$, where M is a very large positive number. This avoids multiplications by $+\infty$. Constraints (9.31) are the passenger *flow conservation* constraints which ensure that the outgoing flow from all outgoing arcs from vertex i is equal to the incoming flow, $\sum_{a \in A_i^-} \nu_a$, plus the passenger demand, g_i^r, at vertex i. Constraints (9.32) ensure that the passenger volume ν_a in the outgoing arc a of vertex i is lower than or equal to the frequency of that arc multiplied by the total waiting time for all trips at vertex i.

To simplify the notation of the outgoing arcs A_i^+ and the incoming arcs A_i^- from/to every vertex, we can use indicator parameters:

- $\mu_{a,i}$ where $\mu_{a,i} = 1$ if arc a is an outgoing arc of vertex i and $\mu_{a,i} = 0$ otherwise;
- $\lambda_{a,i}$ where $\lambda_{a,i} = 1$ if arc a is an incoming arc of vertex i and 0 otherwise.

This will result in a reformulation of constraints (9.31) to:

$$\sum_{a \in A} \mu_{a,i} \nu_a - \sum_{a \in A} \lambda_{a,i} \nu_a = g_i^r \quad \forall i \in V \tag{9.33}$$

and a reformulation of constraints (9.32) to:

$$\nu_a \mu_{a,i} \leq f_a w_i \quad \forall a \in A, i \in V \tag{9.34}$$

Let us consider the implementation of this model to the example network of Fig. 9.4 when having 1 passenger who wants to travel from 'A' to 'B'. We can set destination vertex $r = $'B'. Because 'A' is the origin, we have passenger demand $g_{\text{'A'}}^r = 1$ at vertex 'A'. Because 'B' is the destination, this demand is exiting there resulting in $g_{\text{'B'}}^r = -1$. For every other $i \in V - \{\text{'A'}, \text{'B'}\}$ we have $g_i^r = 0$.

Practitioner's Corner
Frequency-based Passenger Assignment in Python 3

One can use PuLP to program and the CBC solver to solve the linear program ($\mathcal{L}_3$) of Spiess and Florian as follows.

```python
import pulp
from pulp import *
import pandas as pd
model = LpProblem("Spiess_Florian_Assignment", LpMinimize)
V=['A','X2','X','Y','Y3','B'] #vertices
M=100000; A=[]
df=pd.DataFrame({
 "a_start": ['A','A','X','X2','X2','Y','Y3','Y3','Y'],
 "a_end": ['X2','B','X2','X','Y','Y3','Y','B','B'],
 "c_a": [7,25,0,0,6,0,0,4,10],
 "f_a": [1/6,1/6,1/6,M,M,1/15,M,M,1/3]})
for a in range(0,len(df)):
    A.append((df.iat[a,0],df.iat[a,1]))
c={a:0 for a in A}; f={a:0 for a in A}; mu={(a,i):0 for a in A
    for i in V}; lu={(a,i):0 for a in A for i in V}
g={i:0 for i in V}; g['A']=1; g['B']=-1; i=0
for a in A:
    c[a] = df.iat[i,2]; f[a] = df.iat[i,3]
    mu[a,df.iat[i,0]]=1; lu[a,df.iat[i,1]]=1
    i=i+1
# variables
v = LpVariable.dicts("v", A, 0, None, LpContinuous)
w = LpVariable.dicts("w", V, None, None, LpContinuous)
# objective function
model += sum(c[a]*v[a] for a in A) + sum(w[i] for i in V)
# constraints:
for i in V:
    model += sum(mu[a,i]*v[a] for a in A) - sum(lu[a,i]*v[a] for
    a in A) == g[i]
for i in V:
    for a in A:
        model += v[a]*mu[a,i] <= f[a]*w[i]
# optimizing
solver = pulp.PULP_CBC_CMD(mip=True, msg=True, timeLimit=None)
model.solve(solver)
print ("Status:", LpStatus[model.status])
for v in model.variables():
    print (v.name, "=", v.varValue)
print ("Optimal Solution = ", value(model.objective))
```

Solving this problem yields the following optimal solution.

It is important to stretch again that the problem is separable. If, for instance, we have also passenger demand from 'A' to 'X' we can assign this demand on top of the passenger flows we already presented in Table 9.2. To do so, we would need to solve again the same problem by setting final destination $r =$ 'X'.

Table 9.2 Solution

Arcs		Vertices		
a	v_a	i	v_i	w_i
(A,B)	0.5	A	1	3
(A,X2)	0.5	B	0	0
(X,X2)	0	X	0	0
(X2,X)	0	X2	0.5	5.00E-06
(X2,Y)	0.5	Y	0.5	1.25
(Y,B)	0.417	Y3	0.083	8.33E-07
(Y,Y3)	0.083			
(Y3,B)	0.083			
(Y3,Y)	0			

9.1.3.2 Assignment with Varying Travel Times

Formulation ($\mathcal{L}_3$) assigns the passenger demand probabilistically to different paths, but it does not consider the effect of in-vehicle congestion on the passengers' path selection. As discussed in more recent passenger assignment models (Nielsen [6], 2000), this is a shortcoming because passengers might perceive the in-vehicle travel times in crowded vehicles to be higher compared to the ones in less crowded vehicles. To rectify this, let the arc travel times c_a be no longer constant, but continuous non-decreasing functions of the corresponding arc flows $c_a(v_a)$. By doing so, the assignment problem is no longer separable by destination vertex $r \in R$, where $R \subseteq V$ is the subset of destination vertices. As a result, the optimal assignment strategies can be defined by Wardrop's second principle, implying that only strategies with minimal expected cost will be used by the travelers [7]. Let $\tilde{v}_a^r$ denote the volume on arc $a \in A$ associated with destination vertex $r \in R$. We can then write:

$$v_a := \sum_{r \in R} \tilde{v}_a^r \quad \forall a \in A \tag{9.35}$$

and the problem cannot be separated by destination. To solve the passenger assignment problem from all origins to all destinations $r \in R$, we set first subsets of vertices $V_r \subseteq V$ that contain all vertices before destination r. We also use variables $\tilde{v}_a^r$ and w_i^r indicating the flow in arc a from all passengers traveling to destination r and the waiting time at vertex i for passengers traveling to destination r. Let also g_i^r be the demand in vertex i for passengers that travel to destination $r \in R$. If c_a^0 is the fixed free-flow travel time of arc a without considering any congestion, then the passenger assignment problem that considers in-vehicle crowding can be cast as follows.

$$(\mathcal{L}_4) \quad \min \quad \underbrace{\sum_{a \in A} c_a^0 \nu_a + \sum_{r \in R} \sum_{i \in V} w_i^r}_{\text{actual travel cost}} + \underbrace{\sum_{a \in A} c_a^0 \int_0^{\nu_a} d_a(z)\, dz}_{\text{perceived cost due to congestion}} \tag{9.36}$$

$$\nu_a = \sum_{r \in R} \tilde{\nu}_a^r \quad \forall a \in A \tag{9.37}$$

$$\sum_{a \in A} \tilde{\nu}_a^r \mu_{a,i} - \sum_{a \in A} \tilde{\nu}_a^r \lambda_{a,i} = g_i^r \quad \forall i \in V_r, r \in R \tag{9.38}$$

$$\tilde{\nu}_a^r \mu_{a,i} \le f_a w_i^r \quad \forall a \in A, i \in V, r \in R \tag{9.39}$$

$$\tilde{\nu}_a^r \in \mathbb{R}_+ \;\; \forall a \in A, r \in R, \;\; w_i^r \in \mathbb{R}_+ \;\; \forall i \in V, r \in R \tag{9.40}$$

$$\nu_a \in \mathbb{R}_+ \;\; \forall a \in A \tag{9.41}$$

This passenger assignment model is developed in the PhD thesis of Spiess in 1984 and it satisfies Wardrop's second principle for the traffic network equilibrium [8]. Note that now the objective function includes the nonlinear term of the additional perceived travel cost due to congestion, where $d_a(z)$ is a continuous non-decreasing function of the corresponding arc flows (see Spiess [7]). In 1990, Spiess [9] proposed a series of conical volume-delay functions for passenger assignment, although one of the most common ones is an adaptation of the Bureau of Public Roads (BPR) function that will result in a $\int_0^{\nu_a} d_a(z)\, dz$ equal to:

$$\int_0^{\nu_a} d_a(z)\, dz = \left(\frac{\nu_a}{f_a \gamma_a} \right)^\beta \tag{9.42}$$

where ν_a is the passenger volume of trip component a, γ_a is the nominal capacity of the line that serves arc a, and β a BPR function parameter that typically takes values in the range $2 \le \beta \le 12$. We note that if we set $\beta = 2$, the objective function has a quadratic form.

One additional important point is the inclusion of an additional waiting penalty for passenger transfers, on top of the actual waiting time for the transfer. This penalty can be perceived as extra discomfort due to transferring between lines. To model this, one could provide a frequency penalty $f_a = \epsilon$, where $0 < \epsilon < M$, to alighting trip components because any alighting arc that does not end up to the final destination of a passenger will force the passenger to use a subsequent boarding arc to make a transfer.

9.1.3.3 Frequency Setting with Bi-level Optimization

The frequency settings problem is solved so far in two steps: first, assign the frequencies to service lines assuming a pre-defined passenger demand for each line; second, assign the passengers to lines considering fixed line frequencies. These two steps can be intertwined by following a bi-level optimization approach. In bi-level optimization, one problem is embedded (nested) within another problem. The outer optimization problem is called *upper-level* optimization, whereas the inner optimization problem is called *lower-level* optimization. Bilevel problems have two sets of variables: the upper-level variables and the lower-level variables. Bilevel optimization was introduced by Stackelberg in 1934 and offers a hierarchical structure in solving problems. It can be seen as an interaction game expressed in the bi-level programming problem:

$$(\mathcal{U}_0) \quad \min_{\mathbf{x}} \ F(\mathbf{x}, \mathbf{y}) \tag{9.43}$$

$$\text{s.t.} \ G(\mathbf{x}, \mathbf{y}) \leq \mathbf{0} \tag{9.44}$$

where $\mathbf{y} = \mathbf{y}(\mathbf{x})$ is implicitly defined by:

$$(\mathcal{L}_0) \quad \min_{\mathbf{y}} \ f(\mathbf{x}, \mathbf{y}) \tag{9.45}$$

$$\text{s.t.} \ g(\mathbf{x}, \mathbf{y}) \leq \mathbf{0} \tag{9.46}$$

The bi-level programming model consists of two sub-models: $\mathcal{U}_0$ which is the upper-level problem and $\mathcal{L}_0$ which is the lower level. $F(\mathbf{x}, \mathbf{y})$ is the objective function of the upper level decision-maker that has $\mathbf{x}$ as a decision vector. $G(\mathbf{x}, \mathbf{y})$ is the constraint set of the upper-level decision vector. $f(\mathbf{x}, \mathbf{y})$ is the objective function of the lower-level model, $\mathbf{y}$ is its decision vector, and $g(\mathbf{x}, \mathbf{y})$ is its set of constraints. $\mathbf{y} = \mathbf{y}(\mathbf{x})$ is usually called a *reaction* or *response function*. The bi-level programming model is solved by obtaining the response function through solving the lower-level problem and replacing the variable $\mathbf{y}$ in the upper-level problem with the relationship between $\mathbf{x}$ and $\mathbf{y}$—the response function. This response function connects the upper and lower-level decision variables by solving the models iteratively until convergence.

Let us discuss how bi-level optimization can be applied to the frequency settings problem. We present below the frequency setting formulation of Eq. (9.10) that can take the place of our upper level model.

$$
\begin{aligned}
\text{Upper-level:} \quad &\min_{\mathbf{f} \in \mathbb{Z}_{\geq 0}^{|R|}} \sum_{k \in R} t_k f_k \\
&\text{s.t.} \ c f_k \geq q_{ij}^k \qquad \forall k \in R, (i, j) \in L_k \\
&\qquad \sum_{k \in R} t_k f_k \leq m \\
&\qquad f_k \geq \frac{1}{\theta} \qquad \forall k \in R
\end{aligned}
\tag{9.47}
$$

The decision variables of this upper level model are the service frequencies $\mathbf{f}$. Solving this upper level model we presume that we know the values of the passenger flow in each arc (i, j) on any route k, q_{ij}^k. The values of q_{ij}^k play the role of $\mathbf{y}$ in $\mathcal{U}_0$ and are obtained by solving the lower level passenger assignment model, i.e., the passenger assignment model:

$$\text{Lower-level: } \min \underbrace{\sum_{a \in A} c_a^0 v_a + \sum_{r \in R} \sum_{i \in V} w_i^r +}_{\text{actual travel cost}} \underbrace{\sum_{a \in A} c_a^0 \left(\frac{v_a}{f_a \gamma_a} \right)^{\beta}}_{\text{perceived cost due to congestion}}$$

$$v_a = \sum_{r \in R} \tilde{v}_a^r \quad \forall a \in A$$

$$\sum_{a \in A} \tilde{v}_a^r \mu_{a,i} - \sum_{a \in A} \tilde{v}_a^r \lambda_{a,i} = g_i^r \quad \forall i \in V_r, r \in R$$

$$\tilde{v}_a^r \mu_{a,i} \leq f_a w_i^r \quad \forall a \in A, i \in V, r \in R$$

$$\tilde{v}_a^r \in \mathbb{R}_+ \ \forall a \in A, r \in R, \ w_i^r \in \mathbb{R}_+ \ \forall i \in V, r \in R, v_a \in \mathbb{R}_+ \ \forall a \in A$$

$$(9.48)$$

To solve this lower level model we will use the frequencies $\mathbf{f}$ obtained by the upper level model and we will compute the optimal values of the arc volumes v_a which correspond to the values q_{ij}^k of our upper level model because each arc a corresponds to a specific arc (i, j) and a specific line k in the physical network. Using these newly obtained q_{ij}^k values from the solution of the lower level model, we solve again the upper level model to obtain new values of $\mathbf{f}$ and we continue until convergence. In essence, this requires to repeatedly solve the upper and lower level problems using the solution (output) of the one problem as input to the other problem.

For further reading, we note that common bilevel frequency settings models were developed by [10–14].

9.1.4 Subline Frequency Setting

As of now, we have covered the frequency settings problem for fixed lines. However, if specific line segments have low passenger demand, we can use lower frequencies on these segments by creating sublines. This allows exploiting the available vehicle resources more efficiently. This can be achieved by splitting an *original* public transport line into a number of generated sublines that serve segments of the original line. The subline frequency setting problem can be expressed as follows.

Subline Frequency Setting Problem [15]
Given a fixed public transport line with stops $S = \{1, \ldots, \frac{|S|}{2}, \frac{|S|}{2} + 1, \ldots, |S|\}$
where stops $1, \ldots, |S|/2$ are the ordered stops of the outbound direction and
$|S|/2 + 1, \ldots, |S|$ the ordered stops of the inbound direction, generate all
potential sublines that start from stop 1 and return to stop $|S|$ or start from stop
$|S|/2 + 1$ and return to stop $|S|/2$. Given an objective function, determine
which of these sublines will be operational and which will be their service
frequencies.

When generating sublines, a typical assumption is that vehicles cannot park at
intermediate stops as these stops do not have the necessary parking infrastructure.
Therefore, a requirement is to start a subline service at one of the two terminals,
where the first terminal is the depot (stop 1 in the digraph presented in Fig. 9.5) and
the second terminal is the stop that starts the trip in the opposite direction (stop 5 in
Fig. 9.5).

Sublines are then obtained by short-turning vehicles at intermediate stops. For
instance, a subline in Fig. 9.5 that starts from stop 1 and performs a short-turn at stop
3 will serve stops $1 \rightarrow 2 \rightarrow 3 \rightarrow 6 \rightarrow 7 \rightarrow 8$. Similarly, starting from the terminal
at stop $|S|/2 + 1 = 5$ and performing a short-turn at stop 6 will result in a subline
serving stops $5 \rightarrow 6 \rightarrow 3 \rightarrow 4$. It becomes evident that the number of generated
sublines starting at the same terminal is equal to the number of stops that can be
used for short-turning. For example, in Fig. 9.5 we have 4 sublines if we use all
intermediate stops for short-turning. We can then introduce set $R = \{1, \ldots r, \ldots\}$
where 1 is the original line and $2, \ldots, r, \ldots$ are all potential sublines that can be
generated from line 1. This set is presented in Fig. 9.5 where R consists of the
original line and four sublines.

Even if we develop the set of all potential lines R that can be generated from
the original line 1, we do not necessarily need to assign a frequency to each one
of them. In an analogy to the line planning problem, the set R of all potential lines
that can be generated from an originally planned line can be seen as a line pool. In
reality, some of the sublines in R might be inefficient because the passenger demand
in their line segments is low. It is the task of the subline frequency setting models to
introduce frequencies to sublines that contribute to the improvement of the service
and exclude (i.e, do not assign vehicles to) sublines that are counterproductive. That
is, not all members of set R will be assigned a service frequency. To proceed with
the formulation, we declare the following sets, parameters and variables.

Sets: We first have the set R of all potential lines, where 1 is the original line and
$R \setminus \{1\}$ the set that contains all sublines that can be generated from the originally
planned line. We also have set O that contains all origin-destination pairs (s, y) that
have a passenger demand. In addition, we have set S of the ordered stops of the line.
Finally, set F is a discrete set of potential frequencies that can be assigned to line
services.

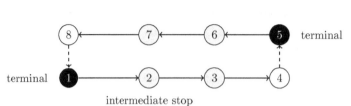

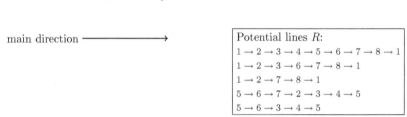

Fig. 9.5 Generation of sublines in digraph $G = (V, A)$ representing the vertices and directed edges (arcs) of an originally planned, bi-rectional line starting from the depot located at stop 1

Parameters: We consider t_r to be the expected *round-trip* travel time of each line $r \in R$. We also require that each subline $r \in R \setminus \{1\}$ should operate with at least a minimum frequency of f_0 in order to be deemed operational. Consider θ to be the minimum allowed service frequency between any origin-destination pair $(s, y) \in O$. This parameter is used to ensure a minimum quality of service to all passengers irrespective of their OD-pair. We also have n available vehicles, which are homogeneous with a capacity c. Importantly, each vehicle is exclusively assigned to one of the possible (sub)lines. To ensure that the original line remains operational, at least k vehicles should be assigned to $r = 1$, where $k \leq 1$. The cost of using a vehicle is w_1 and the cost per time unit driven is w_2. We also have d_{sy} indicating the passenger demand between each OD-pair $(s, y) \in O$ during period T where the demand pattern remains homogeneous. Parameters $\delta_{r,sy}$ denote whether line r connects the OD-pair $(s, y) \in O$ or not (in the former case $\delta_{r,sy} = 1$ and in the later $\delta_{r,sy} = 0$).

Variables: We first have variables $\mathbf{f} = [f_1, \ldots, f_r, \ldots, f_{|R|}]^{\mathsf{T}}$ indicating the frequency of each potential (sub)line expressed in vehicle-trips per hour. We also have variables $\mathbf{x}$, where x_r is the number of vehicles assigned to line $r \in R$ and it is a positive integer. A line is not deemed operational if $x_r = 0$. In addition, for every OD-pair $(s, y) \in O$ we have variables f_{sy} indicating the service frequency for this OD-pair. Additional variables are:

- Binary variables a_r, where $a_r = 1$ if subline $r \in R \setminus \{1\}$ is deemed operational and $a_r = 0$ otherwise,

- Continuous variables b_{rs} indicating the number of passengers that board line r at stop s
- Continuous variables ν_{rs} indicating the number of passengers that alight from line r at stop s
- Continuous variables l_{rs} indicating the in-vehicle passenger load of line r when departing from stop s

Using this notation, the formulation of the subline frequency setting problem is expressed as follows.

Subline frequency setting formulation

$$\min \sum_{r \in R} (x_r w_1 + w_2 t_r T f_r) + \sum_{(s,y) \in O} d_{sy} \frac{1}{f_{sy} + 1} \tag{9.49}$$

subject to:

$$f_r \leq \frac{x_r}{t_r} \qquad\qquad \forall r \in R \tag{9.50}$$

$$f_{sy} \leq \sum_{r \in R} \delta_{r,sy} f_r \qquad\qquad \forall (s,y) \in O \tag{9.51}$$

$$f_{sy} \geq \theta \qquad\qquad \forall (s,y) \in O \tag{9.52}$$

$$x_r \leq a_r M \qquad\qquad \forall r \in R \setminus \{1\} \tag{9.53}$$

$$x_r \geq a_r t_r f_0 \qquad\qquad \forall r \in R \setminus \{1\} \tag{9.54}$$

$$\sum_{r \in R} x_r \leq n \tag{9.55}$$

$$x_1 \geq k \tag{9.56}$$

$$b_{rs} = \sum_{y>s} d_{sy} \frac{f_r}{f_{sy}} \delta_{r,sy} \qquad\qquad \forall r \in R, \forall s \in S \setminus \{|S|\} \tag{9.57}$$

$$\nu_{ry} = \sum_{s<y} d_{sy} \frac{f_r}{f_{sy}} \delta_{r,sy} \qquad\qquad \forall r \in R, \forall y \in S \setminus \{1\} \tag{9.58}$$

$$l_{rs} = l_{r,s-1} + b_{rs} - \nu_{rs} \qquad\qquad \forall r \in R, \forall s \in S \setminus \{1\} \tag{9.59}$$

$$l_{r1} = b_{r1} \qquad\qquad \forall r \in R \tag{9.60}$$

$$l_{rs} \leq c f_r \qquad\qquad \forall r \in R, s \in S \tag{9.61}$$

$$x_r \in \mathbb{Z}_{\geq 0} \qquad\qquad \forall r \in R \tag{9.62}$$

$$f_r \in F \qquad\qquad \forall r \in R \tag{9.63}$$

$$a_r \in \{0, 1\} \qquad\qquad \forall r \in R \setminus \{1\} \tag{9.64}$$

$$b_{rs}, \nu_{rs}, l_{rs} \in \mathbb{R}_{\geq 0} \qquad\qquad \forall r \in R, \forall s \in S \tag{9.65}$$

The objective function (9.49) consists of two terms. The first term strives to minimize the operational costs, where $\sum_{r \in R} x_r w_1$ represents the cost of using $\sum_{r \in R} x_r$ vehicles and $\sum_{r \in R} w_2 t_r T f_r$ is the cost of the total vehicle running times. In more detail, multiplying the planning horizon T with the service frequency f_r expressed in trips per hour gives us the number of vehicle trips of line r during the planning horizon T. Multiplying $T f_r$ with the round-trip travel time t_r of line r returns the total travel time of all vehicle-trips of line r operating in the planning horizon T. The second term of the objective function, $\sum_{(s,y) \in O} d_{sy} \frac{1}{f_{sy}+1}$, is an estimate of the total waiting time of passengers. In more detail, $d_{sy} \frac{1}{f_{sy}+1}$ is the estimated waiting time of passengers traveling from s to y over the planning horizon T. Assuming that passengers arrive randomly at stops, their waiting time is equal to half the headway which is equal to $\frac{1}{2f_{sy}}$ if the relevant departures of all vehicle trips that serve the OD pair $(s, y) \in O$ are perfectly synchronized. This will give us $d_{sy} \frac{1}{2f_{sy}}$ passenger waiting hours during the planning horizon T. At the other extreme, if all vehicle trips that serve the OD-pair (s, y) arrive at s at the same point in time at each 1-h period, then the effective frequency of (s, y) will be 1 vehicle trip per hour resulting in passenger waiting times $d_{sy} \frac{1}{2}$. Because we have several sublines and we cannot expect the arrival times of all trips at s that serve (s, y) to be evenly spread across time, Gkiotsalitis et al. [15] proved that an upper bound of the expected waiting time of a passenger is $\frac{1}{f_{sy}+1}$, and this is the reason behind using this term in the objective function.

Constraints (9.50) ensure that the round-trip travel time t_r of each potential line together with the number of its assigned vehicles, x_r, provides an upper bound on the subline frequency f_r, namely $f_r \leq \frac{x_r}{t_r}$. Constraints (9.51) set the service frequency f_{sy} of each OD-pair $(s, y) \in O$ to be no larger than the total frequency assigned to all sublines r that serve OD-pair (s, y). Note that the 0–1 parameters $\delta_{r,sy}$ force the model to only consider the vehicles of sublines that serve the OD-pair (s, y). Because the original line is always operational, $\delta_{1,sy} = 1$ for any OD-pair (s, y). Constraints (9.52) ensure that each OD-pair (s, y) is served at least with minimum frequency θ, thus guaranteeing a minimum level of service. Constraints (9.53) use a very big positive number M and enforce that when subline $r \in R \setminus \{1\}$ is operational, that is $x_r > 0$, then a_r should be equal to one. Otherwise, $a_r = 0$. Constraints (9.54) state that every subline $r \in R \setminus \{1\}$ which is deemed operational ($a_r = 1$) should have at least a minimum frequency of f_0. Constraint (9.55) is the fleet size constraint ensuring that we will not use more vehicles than available. Constraint (9.56) ensures that at least k vehicles will be assigned to the original line serving all stops. Constraints (9.62) restrict x_r to positive integer values, and constraints (9.63) restrict frequency f_r to take values from a discrete set of feasible frequencies F, thus allowing to require a minimum frequency if a subline is selected for operation.

Constraints (9.57) return the total number of passengers that board vehicles of potential line r at stop s, by splitting the passengers of each OD-pair (s, y) equally over all relevant potential lines for (s, y). In a similar way, constraints (9.58) estimate the number of alighting passengers per stop and potential line. Constraints (9.59)–(9.60) keep track of the in-vehicle load per stop and per potential line. Constraints (9.61) ensure that the capacity restrictions are met per subline.

Note that the mathematical formulation of the subline frequency setting problem is a mixed-integer nonlinear program (MINLP). It is mixed-integer because variables a_r are binary and variables x_r, f_r are restricted to integer/discrete values. It is nonlinear because the objective function (9.49) as well as constraints (9.57)–(9.58) are fractional since they contain a division by one of the variables.

9.2 Transit Network Timetabling

Transit lines can operate either based on their *frequencies* or based on detailed *timetables*. In the latter case, passengers do not expect the arrival of a trip periodically (i.e., with a time headway of $\frac{1}{\text{line frequency}}$). Instead, passengers are aware of the expected arrival time of every public transport trip at every station/stop because this is provided in a timetable. To create this timetable, we need to solve the Transit Network Timetabling Problem (TNT). TNT is solved after we have set the frequencies of the service lines. Solving this problem is particularly important in transit lines with known demand patterns and lines that have several transfer stations/stops and need to synchronize their arrival times with the arrival times of other lines. Although the timetabling problem can be expressed in many forms, one common problem description is the following.

> **Timetabling problem** [1]
> Suppose there is a public transport line operating in a pre-defined frequency. Determine the exact dispatching time of each trip of the line and its arrival/departure time to/from each stop so that an objective function is optimized.

Common objectives for the TNT problem include the even distribution of the in-vehicle passenger load across all trips of the line, the maximization of transfer synchronization events, and the minimization of the passenger waiting times [1, 16, 17].

When developing the timetable of a line, the time headways of consecutive trips can be different (see Table 9.3), and thus the transport service might not be regular. For instance, the time headway between the first two trips is 10 min and between trips 2 and 3 is 8 min. This difference is justified in case the passenger demand patterns change very often or there is a need to synchronize the service of a line with the services of other lines at transfer stops.

The remainder of this section presents closed-form (analytical) methods and common mathematical models for the TNT problem.

Table 9.3 Example of timetable with four trips operating in a line with 3 stops

Trip	Stop		
	1	2	3
1	8:00	8:20	8:40
2	8:10	8:30	8:50
3	8:18	8:38	8:58
4	8:25	8:45	9:05

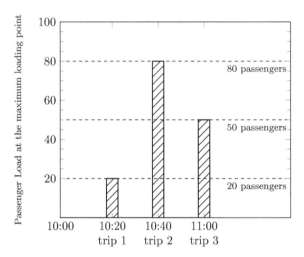

Fig. 9.6 Passenger load at the maximum loading point for each one of the vehicle trips operating from 10:00 until 11:00

9.2.1 Closed-form Methods

A common analytical method for the TNT is the *Even (Average) Load method*. This method determines the timetable of a transit line by striving to even the in-vehicle passenger load of all trips of the line at the maximum loading point. Let us, for example, consider that a line operates with a frequency of 3 vehicle trips per hour from 10 am until 11 am and the average in-vehicle passenger load at the maximum loading point for each one of its trips is presented in Fig. 9.6.

To apply the even load method, we assume *uniform passenger arrival rates* between successive trips that have a 20-min headway. This will give us the following passenger arrival rates expressed in minutes:

- From 10:00–10:20 we have 20/20 = 1 passenger per minute
- From 10:20–10:40 we have 80/20 = 4 passengers per minute
- From 10:40–11:00 we have 50/20 = 2.5 passengers per minute

If our target is to even the in-vehicle load at the maximum loading point, we would like all our trips to have a load of $\frac{20+80+50}{3} = 50$ passengers at that point. To achieve that for the first vehicle trip, we need to change its departure time from 10:20 to 10:20+x such that the delay of x minutes will result in an in-vehicle load of 50 passengers at the maximum loading point. From 10:00 until 10:20 we have a

Fig. 9.7 Finding the dispatching times of trips to achieve an even load with a cumulative plot of the passenger load at the maximum loading point. Note that trip 1 is moved from 10:20 to 27.5 min after 10:00 to attaim a passenger load of 50 passengers at its maximum loading point

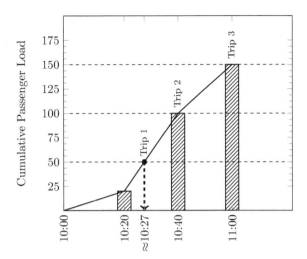

passenger arrival rate of 1 passenger per minute resulting in 20 passengers. For the next x minutes the arrival rate is 4 passengers per minute, and thus $20 + 4x = 50 \Rightarrow x = 30/4 = 7.5\,\text{min}$. Thus, trip 1 should be dispatched 27.5 min after 10:00.

Let us now consider the second trip. The waiting passengers at the maximum loading point of this trip from 10:27 and 30 seconds until 10:40 will be (40–27.5) min multiplied by 4 passengers per minute resulting in 50 passengers. Thus, we do not have to change the dispatching time of the second trip. Since the second trip will be dispatched as planned and the third trip has a passenger load of 50 passengers at its maximum loading point, we do not need to change the dispatching of the third trip as well.

This problem can be also solved graphically by using a cumulative plot and trying to find the time instance when we have a passenger load at the maximum loading point of 50, 50+50, and 50+50+50 passengers. This is presented in Fig. 9.7.

Additional analytical and graphical methods that determine a timetable based on rules governing the desired bus-loads and headways are presented at [18–20].

9.2.2 Model-based Timetabling

Timetabling models differ based on their problem description and their objective function. In the works of [21–23] the emphasis is placed on meeting the passenger demand patterns. There are several other models that try to minimize the passenger waiting times at transfer stops [24–27].

In other approaches, timetabling aims to minimize the passenger waiting times subject to resource limitations (e.g., the availability of public transport vehicles, the fixed dispatching times of the first and last trip of the day, the fixed number of performed trips per day). These constraints are typically a result of the frequency settings stage that precedes timetabling and determines the exact number of trips at different time periods of the day.

9.2.2.1   Waiting Time-based Timetabling

One of the timetabling formulations that use the waiting time of passengers of the line as objective function is presented in [28]. In [28], the timetabling problem is formulated as follows using the notation in Table 9.4. This formulation corresponds to a single-direction public transport line. Stops 1 and $|M|$ are the first and last stops of the line and they can be in close proximity, or even identical stops. At the end of a trip, all remaining passengers alight at stop $|M|$. Then, the public transport vehicle drives to stop 1 to be ready to start a new trip. We also have a set of passengers P waiting to board the public transport line, an ordered set K of trips performed by the transit line during the day, and a set Z of time periods (intervals) of the day with homogeneous travel times.

The dispatching time of the first trip of the day, d_f, and the dispatching time of the last trip of the day, d_l, are fixed to avoid *schedule sliding*. We also know the public transport stop $b_p \in M$ where every passenger $p \in P$ is waiting and the expected arrival time a_p of the p-th passenger at b_p. We note that estimating the arrival time of the p-th passenger, a_p, is not a trivial task. Even though the fare payment records may contain all transactions of passengers, extracting the arrival time distribution of passengers at each stop is not easy. The reason behind this is that payment records provide information about the boarding and alighting times of passengers, and not their waiting times. To overcome this issue, [28] assumed that the passenger arrivals follow roughly the same distribution among different stops of a particular bus service, and estimated the arrival time distribution at only one stop of the public transport service to derive the values of a_p.

The expected travel time of a vehicle of the line from stop m to $m + 1$ is time-dependent and it is provided by parameter λ_{mz} where $z \in Z$ is the time interval within which the trip arrives at stop m. The decision variables of the problem are the dispatching times of trips $\mathbf{x} = [x_1, \ldots, x_{|K|}]^\mathsf{T}$ which will determine also the values of the following variables:

- The arrival time of the k-th trip at stop m, v_{mk}.
- The vehicle onto which the p-th passenger boards, τ_p.
- The waiting time of the p-th passenger.

The resulting mathematical formulation is presented below.

Table 9.4 Nomenclature of waiting time-based timetabling for a single line

Sets	
M	Set of ordered stops of the transit line
P	Set of passengers waiting to board the public transport line
K	Set of ordered trips performed by the transit line during the day
Z	Set of peak/off-peak time intervals with homogeneous travel times
Parameters	
d_f	Fixed dispatching time of the first trip of the planning period
d_l	Fixed dispatching time of the last trip of the planning period
λ_{mz}	Expected travel time from stop m to $m+1$ at the z-th time interval
b_p	Stop at which the p-th passenger is waiting
a_p	Arrival time of the p-th passenger
Variables	
x_k	Dispatching time of the k-th trip
v_{mk}	Arrival time of the k-th trip at stop m
τ_p	Vehicle trip onto which the p-th passenger boards
w_p	Waiting time of the p-th passenger

Waiting time-based Timetabling for a single line

$$\min_{\mathbf{x},\mathbf{w},\tau,\mathbf{V}} \sum_{p \in P} w_p \tag{9.66}$$

$$\text{s.t.} \quad x_1 = d_f \tag{9.67}$$

$$x_{|K|} = d_l \tag{9.68}$$

$$x_k \le x_{k+1} \qquad\qquad\qquad \forall k \in K \setminus \{|K|\} \tag{9.69}$$

$$v_{1k} = x_k \qquad\qquad\qquad \forall k \in K \tag{9.70}$$

$$v_{mk} = v_{m-1,k} + \lambda_{m-1,z} \ \text{ where } v_{m-1,k} \in z \quad \forall m \in M \setminus \{1\}, \forall k \in K \tag{9.71}$$

$$\tau_p = \min_k(v_{b_p,k} : v_{b_p,k} > a_p) \qquad \forall p \in P \tag{9.72}$$

$$w_p = v_{b_p,\tau_p} - a_p \qquad\qquad \forall p \in P \tag{9.73}$$

$$x_k \in \mathbb{Z} \qquad\qquad\qquad \forall k \in K \tag{9.74}$$

$$\tau_p \in K \qquad\qquad\qquad \forall p \in P \tag{9.75}$$

$$w_p \in \mathbb{R} \qquad\qquad\qquad \forall p \in P \tag{9.76}$$

$$v_{mk} \in \mathbb{R} \qquad\qquad\qquad \forall m \in M, k \in K \tag{9.77}$$

Equation (9.66) is the objective function that strives to minimize the waiting times of all passengers over the course of the day. Constraints (9.67)–(9.68) ensure that the first and the last trip of the day will be dispatched according to their planned dispatching times. Constraints (9.69) ensure that the order of trips is maintained. That is, our decisions about the new dispatching times of trips, x_k, $\forall k \in K$, cannot affect the pre-determined dispatching order. Constraints (9.70) set $v_{1k} = x_k$. Hence, v_{1k} refers to the dispatching time of each trip $k \in K$. Constraints (9.71) ensure that the arrival time of every trip at every stop $m \in M \setminus \{1\}$ is equal to its arrival time at the previous stop plus the expected travel time from stop $m - 1$ to m at the time period $z \mid V_{m-1,k} \in z$. Constraints (9.72) return the k-th vehicle trip onto which the p-th passenger boards, and constraints (9.73) return the waiting time of each passenger. Because of constraints (9.71) and (9.72) the problem is not convex and it should either be reformulated or solved with (meta)heuristics.

9.2.2.2 Synchronization-based Timetabling

Another type of timetabling formulation strives to increase the number of synchronizations at the transfer stops between two or more lines while maintaining the time headways of each line as even as possible. This type of formulation has been used in many studies. Here we present the model of [29] presented in 2016.

Sets: Consider a transit network with a set of lines I and a set of stations/stops B. In addition, J_i is the set of lines that share synchronization stops with line $i \in I$ and B_{ij} the set of stops that represent the synchronization (i.e., transfer) stops between lines i and j, where $j \in J_i$. Consider also S_i to be the set of planning periods for line $i \in I$, P_i the ordered set of trips of line $i \in I$, and P_i^s the ordered set of trips of line $i \in I$ during a period $s \in S_i$.

Parameters: Consider parameter d_s^i representing the end of planning period $s \in S_i$ and the beginning of planning period $s + 1 \in S_i$. The total number of trips of line i during the day is known, f_i, and f_i^s is the frequency of trips during planning period s. The p-th trip of line i is denoted as i_p and its travel time from its depot to node b is denoted as t_p^{ib}. Because every trip i_p belongs to a planning period $s \in S_i$ its travel time is $t_p^{ib} = t_s^{ib}$. A synchronization between two trips i_p and j_q where $j \in J_i$ is said to occur at node $b \in B_{ij}$ if the waiting times of passengers that transfer between these two lines is within the minimum, w_{pq}^{ijb}, and maximum, $\bar{w}_{pq}^{ijb}$ waiting times, respectively.

The additional requirement is that the timetables of lines maintain a headway which is close to the even headway. At period s the minimum and maximum headway times for line $i \in I$ are $h_s^i := \frac{d_s^i - d_{s-1}^i}{f_s^i} - \delta_s^i$ and $\bar{h}_s^i := \frac{d_s^i - d_{s-1}^i}{f_i^s} + \delta_s^i$ where δ_s^i is a flexibility parameter of the headway in period s. Note that the term $\frac{d_s^i - d_{s-1}^i}{f_i^s}$ is the *even headway* which requires that all vehicles being evenly distributed in planning period s. Last, let β_{is} be the first trip of line $i \in I$ in planning period $s \in S_i$ and $\bar{\beta}_{is}$ the last trip of line $i \in I$ in planning period $s \in S_i$.

Variables: The variables of the problem are the dispatching time of the p-th trip of line i, x_p^i, and binary variable y_{pq}^{ijb} which is equal to 1 if trip i_p arrives first at stop $b \in B_{ij}$ and trip j_q synchronizes with it, and 0 otherwise. The MILP formulation of the synchronization-based timetabling problem of [29] is provided below.

Synchronization-based timetabling

$$\max \sum_{i \in I} \sum_{j \in J_i} \sum_{b \in B_{ij}} \sum_{p=1}^{f_i} \sum_{q=1}^{f_j} y_{pq}^{ijb} \tag{9.78}$$

$$\text{s.t. } h_s^i \leq x_{p+1}^i - x_p^i \leq \bar{h}_s^i \qquad \forall i \in I, \forall p \in P_i^s \setminus \{|P_i^s|\}, \forall s \in S_i \tag{9.79}$$

$$d_{s-1}^i + \frac{h_s^i}{2} \leq x_{\beta_{is}}^i \leq d_{s-1}^i + \frac{\bar{h}_s^i}{2} \qquad \forall i \in I, s \in S_i \tag{9.80}$$

$$d_s^i - \frac{\bar{h}_s^i}{2} \leq x_{\bar{\beta}_{is}}^i \leq d_s^i - \frac{h_s^i}{2} \qquad \forall i \in I, \forall s \in S_i \tag{9.81}$$

$$x_q^j + t_q^{jb} - (x_p^i + t_p^{ib}) \geq w_{pq}^{ijb} - M(1 - y_{pq}^{ijb}) \qquad \forall i \in I, j \in J_i, b \in B_{ij}, p \in P_i, q \in P_j \tag{9.82}$$

$$x_q^j + t_q^{jb} - (x_p^i + t_p^{ib}) \leq \bar{w}_{pq}^{ijb} + M(1 - y_{pq}^{ijb}) \qquad \forall i \in I, j \in J_I, b \in B_{ij}, p \in P_i, q \in P_j \tag{9.83}$$

$$x_p^i \in \mathbb{R} \qquad \forall i \in I, \forall p \in P_i \tag{9.84}$$

$$y_{pq}^{ijb} \in \{0, 1\} \qquad \forall i \in I, \forall j \in J, \forall p \in P_i, \forall q \in P_j, \forall b \in B_{ij} \tag{9.85}$$

The objective function (9.78) maximizes the number of synchronizations. Constraints (9.79) ensure that the trips of every line are almost evenly spaced within each time period s. Constraints (9.80) and (9.81) ensure that there is a separation time between arrivals of the last trip of period $s - 1$ and the first trip of period s, i.e., the separation between trips β_{is} and $\bar{\beta}_{i,s-1}$ corresponding to the first trip of line i in planning period s and the last trip of line i in planning period $s - 1$ must be within $[\frac{h_{s-1}^i + h_s^i}{2}, \frac{\bar{h}_{s-1}^i + \bar{h}_s^i}{2}]$. Constraints (9.82) and (9.83) are synchronization constraints allowing variable y_{pq}^{ijb} to be equal to 1 if the arrival times of i_p and j_q at node b are within $[w_{pq}^{ijb}, \bar{w}_{pq}^{ijb}]$. If $y_{pq}^{ijb} = 0$ the synchronization constraints are redundant since M is a very big positive number.

Exercises

9.1 Frequency Settings—Maximum Loading Point and Load Profile
The average passenger load in Table 9.5 refers to the timeframe of each planning period (2 hrs) where we need to define the service frequency.

Table 9.5 Average passenger load upon departure from each stop at each 2-h time period

Stop no	Distance to next stop (km)	Passenger load			Total load
		6:00–8:00	8:00–10:00	10:00–12:00	
1	2	210	150	120	480
2	3	110	130	180	420
3	1.5	150	180	140	470
4	4.5	100	180	220	500
5	2	290	120	180	590

Fig. 9.8 Example digraph $G = (V, A)$ where V represents the stops of the lines

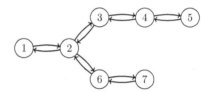

Consider the time period from 10:00 until 12:00. Consider also a desired load factor $\gamma_j = 5/6$ and a fleet of homogeneous vehicles with capacity $c = 60$. Compute the frequency of the line during this 2-hour period with the *maximum loading point* method and the *load profile* method. Determine how many seats will be empty upon the departure from each public transport stop when using each one of the two methods.

9.2 Cost-oriented Frequency Settings
Consider two lines $R = \{1, 2\}$, vehicle capacity $c = 30$ passengers, round-trip travel times $t_1 = 110$ and $t_2 = 100$ min, maximum allowed waiting time $\theta = 30$ min, and total number of available vehicles $m = 15$. The digraph $G = (V, A)$ of this example is presented in the following figure (Fig. 9.8).

The passenger flows at each arc for line 1 are $q_{12}^1 = 30$, $q_{23}^1 = 40$, $q_{34}^1 = 60$, $q_{45}^1 = 70$ and for line 2 are $q_{12}^2 = 35$, $q_{26}^2 = 45$, $q_{67}^2 = 65$, expressed in passengers per hour. Solve this problem to derive the optimal frequency using the MILP formulation of (9.10).

9.3 Passenger-oriented Frequency Settings
Consider the example in the previous exercise with minimum time headway $\theta_{min} = 2$ min. Solve this problem to derive the optimal frequency using the mixed-integer linear programming formulation of (9.13).

9.4 Timetabling with the even load method
Consider a line that has the passenger load profile of Fig. 9.9 at its maximum loading point. Use the Even Load method to determine the dispatching times of the line.

Fig. 9.9 Passenger load at the maximum loading point for each one of the vehicle trips within an 1-h period

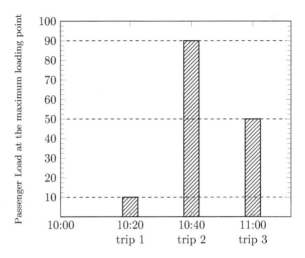

Fig. 9.10 Digraph $G = (V, A)$ with arc travel times and arc frequencies

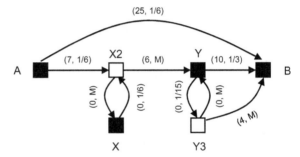

9.5 Passenger Assignment

Consider the digraph in Fig. 9.10. Solve the linear program (9.30)–(9.32) for assigning passengers. Consider as input the origin-destination matrix in Table 9.6 and Fig. 9.10.

9.6 Subline Frequency Setting

Consider the mixed-integer nonlinear subline frequency setting formulation presented in (9.49)–(9.65). Linearize the nonlinear terms in the objective function and the constraints to introduce a mixed-integer linear formulation.

Table 9.6 OD Matrix presenting the number of passengers traveling from an origin vertex to a destination vertex during the hourly planning period

	A	X	Y	B
A	–	0	80	100
X	0	–	20	0
Y	0	0	–	0
B	0	0	0	–

References

1. O. Ibarra-Rojas, F. Delgado, R. Giesen, J. Muñoz, Transp. Res. B: Methodol. **77**, 38 (2015)
2. P.G. Furth, N.H. Wilson, Transp. Res. Record **818**(1981), 1 (1981)
3. A. Ceder, Transp. Res. A: General **18**(5–6), 439 (1984)
4. A.F. Han, N.H. Wilson, Transp. Res. B: Methodol. **16**(3), 221 (1982)
5. H. Spiess, M. Florian, Transp. Res. B: Methodol. **23**(2), 83 (1989)
6. O.A. Nielsen, Transp. Res. B: Methodol. **34**(5), 377 (2000)
7. H. Spiess, *EMME/2 Support Center* (1993)
8. J.G. Wardrop, J.I. Whitehead, Proceed. Inst. Civil Eng. **1**(5), 767 (1952)
9. H. Spiess, Transp. Sci. **24**(2), 153 (1990)
10. I. Constantin, M. Florian, Int. Trans. Oper. Res. **2**(2), 149 (1995)
11. G.S. Yoo, D.K. Kim, K.S. Chon, KSCE J. Civil Eng. **14**(3), 403 (2010)
12. B. Yu, Z. Yang, J. Yao, J. Transp. Eng. **136**(6), 576 (2010)
13. Z. Huang, G. Ren, H. Liu, Math. Probl. Eng. **2013** (2013)
14. İ.Ö. Verbas, H.S. Mahmassani, Transp. Res. Record **2498**(1), 37 (2015)
15. K. Gkiotsalitis, M. Schmidt, E. van der Hurk, Transp. Res. C: Emer. Technol. **135**, 103492 (2022)
16. G. Desaulniers, M.D. Hickman, Handbooks. Oper. Res. Manag. Sci. **14**, 69 (2007)
17. V. Guihaire, J.K. Hao, Transp. Res. A: Policy. Pract. **42**(10), 1251 (2008)
18. A. Ceder, J. Adv. Transp. **25**(2), 137 (1991)
19. A.A. Ceder, Procedia-Soc. Behav. Sci. **20**, 19 (2011)
20. A.A. Ceder, S. Hassold, B. Dano, Publ. Transp. **5**(3), 193 (2013)
21. A. de Palma, R. Lindsey, Transp. Res. B: Methodol. **35**(8), 789 (2001)
22. Z.C. Li, W.H. Lam, S. Wong, A. Sumalee, Transportation **37**(5), 751 (2010)
23. J.A. Mesa, F.A. Ortega, M.A. Pozo, Anna. Oper. Res. **222**(1), 439 (2014)
24. W.D. Klemt, W. Stemme, in *Computer-Aided Transit Scheduling* (Springer, 1988), pp. 327–335
25. P. Chakroborty, K. Deb, P. Subrahmanyam, J. Transp. Eng. **121**(6), 544 (1995)
26. J.R. Daduna, S. Voß, in *Computer-Aided Transit Scheduling* (Springer, 1995), pp. 39–55
27. V. Guihaire, J.K. Hao, Comput. Indus. Eng. **59**(1), 16 (2010)
28. Y. Wang, D. Zhang, L. Hu, Y. Yang, L.H. Lee, IEEE Trans. Intel. Transp. Syst. **18**(9), 2443 (2017)
29. O.J. Ibarra-Rojas, F. López-Irarragorri, Y.A. Rios-Solis, Transp. Sci. **50**(3), 805 (2016)

Chapter 10
Tactical Planning of Public Transport Services: Multi-modal Synchronization

Abstract We have already analyzed the tactical planning stages of setting the service frequencies and timetables. One important aspect in the tactical planning of dense, multi-modal public transport networks is the synchronization of different services so that the services of the lines are coordinated and the passenger transfer times are minimized. This chapter focuses on this tactical planning issue, which has extensions at the operational level. The aim is to provide model formulations for synchronizing the transfers at public transport networks with no hierarchy and in networks with hierarchy, where some lines act as feeder (first/last-mile) and some other lines as collector lines.

10.1 Multi-modal Synchronization Without Hierarchy

The multi-modal synchronization problem has many different perspectives. At a high level, if we seek to synchronize the services of two lines that share common transport stops we can either consider the timetable of one line to be fixed and adapt the timetable of the other line so it is synchronized with the timetable of the first, or adapt the timetables of both lines. In the former case, we have a multi-modal synchronization with hierarchy because the second line needs to adapt its timetable to the fixed timetable of the first line. This implies that the first line is of higher importance (*collector* line) and its timetable cannot be easily modified. It can be, for instance, that the first line is a train line and the second line is a bus line. In this case, we have a distinct line hierarchy during the synchronization process. In the latter case, there is no hierarchy and both lines can change their timetables considering a common synchronization objective (i.e., we have two bus lines of the same importance).

The distinction between multi-modal synchronization with and without hierarchy is critical because it results in different problem definitions and different formulations. We will start first with the description of multi-modal synchronization problems *without hierarchy*. In this category, we have two different sets of problems as presented below.

10.1.1   Maximal Multi-modal Synchronization Without Hierarchy

A practical definition of the maximal multi-modal synchronization problem without hierarchy was provided by Ceder et al. [1] in 2001, and it is expressed as follows.

Maximal multi-modal synchronization without hierarchy
Consider a set of public transport lines and a set of transfer stops at which we seek to synchronize their operations. Adjust the dispatching times of the timetabled trips of the lines so that the number of simultaneous vehicle arrivals at the transfer stops/stations is maximized.

To formulate this problem, one can use the MILP formulation of Ceder et al. [1]. This formulation considers a directed graph $G = (V, A)$ where V is the set of transfer nodes in the network and A is the set of arcs representing the traveling path of the service lines.

Consider a planning horizon T where the dispatching time of each vehicle trip should be within the discrete interval $[0, T]$. The interval is discrete to permit discrete dispatching times in the timetables. The number of lines in the network that need to be synchronized with each other is $|M|$, where M is the set of all these lines. In addition, each line $m \in M$ should have a dispatching time headway that lies within $H_m^{\min}$ and $H_m^{\max}$ where the values of $H_m^{\min}$ and $H_m^{\max}$ reflect the operator's requirements. Let also F_m be the set of scheduled trips for line m during the interval T. This set is defined during the frequency settings phase. Let also t_{mj} be the traveling time from the starting point of line $m \in M$ to vertex $j \in V$, which is allowed to take positive integer values. We also note that the dispatching of the first trip at every line should take place within the time interval $[0, H_m^{\max}]$. We can also pre-define sets V_{mq} which contain the common stops where lines $m \in M$ and $q \in M$ intersect since we require to synchronize their services at these stops. An example with three lines is presented in Fig. 10.1. For this example we have $V_{12} = \{2, 3\}$ and $V_{13} = \{4\}$.

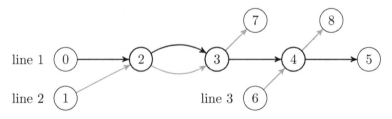

Fig. 10.1 Line 1 with path $0 \to 2 \to 3 \to 4 \to 5$ intersecting with line 2 with path $1 \to 2 \to 3 \to 7$ at stops 2 and 3 ($V_{12} = \{2, 3\}$). Line 1 also intersects with line 3 with path $6 \to 4 \to 8$ at stop 4 ($V_{13} = \{4\}$)

When defining the values of the travel time parameters, we set $t_{mj} = -1$ if the trips of line m do not serve vertex $j \in V$. The decision variables of the problem are the positive integer variables x_{im} representing the dispatching time of the i-th trip of the m-th line, where $i \in F_m$ and $m \in M$. The problem is then cast as follows.

Maximal multi-modal synchronization formulation [1]

$$\max_{\mathbf{x}} \sum_{m=1}^{|M|-1} \sum_{i \in F_m} \sum_{q=m+1}^{|M|} \sum_{j \in F_q} \sum_{n \in V_{mq}} \max[1 - |(x_{im} + t_{mn}) - (x_{jq} + t_{qn})|, 0]$$

$$(10.1)$$

$$\text{s.t. } x_{1m} \leq H_m^{\max} \qquad \qquad \forall m \in M \qquad\qquad\qquad (10.2)$$

$$x_{|F_m|,m} \leq T \qquad \qquad \forall m \in M \qquad\qquad\qquad (10.3)$$

$$H_m^{\min} \leq x_{i+1,m} - x_{im} \leq H_m^{\max} \quad \forall m \in M, \ \forall i : 1 \leq i \leq |F_m| - 1 \qquad (10.4)$$

$$x_{im} \in \mathbb{Z}_{\geq 0} \qquad \qquad \forall i \in V, \ \forall m \in M \qquad\qquad (10.5)$$

Constraints (10.2) ensure that the dispatching time of the first trip of every line will not be beyond the maximum allowed headway. Constraints (10.3) ensure that the last trip of every line is dispatched within the planning horizon. Constraints (10.4) enforce the headway limits to the successive dispatching times of the trips belonging to the same line. The objective function maximizes the number of times vehicles from different lines arrive at the same time at their common transfer stop. Note that because the values of the dispatching time variables and the values of the travel time parameters are forced to be non-negative integers, the objective function can only attain integer values.

One can restate the model by defining variable Y_{mq} representing the overall number of simultaneous arrivals of vehicles in line m and q. This will result in changing the objective function to:

$$\max \sum_{m=1}^{|M|-1} \sum_{q=m+1}^{|M|} Y_{mq} \qquad\qquad\qquad (10.6)$$

and adding constraints:

$$Y_{mq} = \sum_{n \in V_{mq}} \sum_{i \in F_m} \sum_{j \in F_q} \max[1 - |(x_{im} + t_{mn}) - (x_{jq} + t_{qn})|, 0]$$

$$\forall m = 1, \ldots, |M| - 1, \ q = m + 1, \ldots, |M| \quad (10.7)$$

The issue with the model is that equality constraints (10.7) are nonlinear and the feasible region is not a convex set. To linearize them, consider binary variables D_{nijmq} and a very large positive number $\bar{M}$. Then, constraints (10.7) can be replaced by the following set of equisatisfiable constraints for all $m = 1, \ldots, |M| - 1, q = m + 1, \ldots, |M|$:

$$D_{nijmq} \bar{M} \geq x_{im} + t_{mn} - (x_{jq} + t_{qn}) \qquad \forall i \in F_m, j \in F_q, n \in V_{mq} \qquad (10.8)$$

$$D_{nijmq} \bar{M} \geq x_{jq} + t_{qn} - (x_{im} + t_{mn}) \qquad \forall i \in F_m, j \in F_q, n \in V_{mq} \qquad (10.9)$$

$$Y_{mq} \leq \sum_{n \in V_{mq}} \sum_{i \in F_m} \sum_{j \in F_q} (1 - D_{nijmq}) \qquad (10.10)$$

Notice that if $x_{im} + t_{mn} = x_{jq} + t_{qn}$, then there is a simultaneous arrival of the i-th vehicle in line m with the j-th vehicle in line q at vertex n. The variable D_{nijmq} can then yield the value of 0 and Y_{mq} is increased by one according to (10.10) and the maximization objective. If, however, $x_{im} + t_{mn} \neq x_{jq} + t_{qn}$ then the arrivals do not coincide and D_{nijmq} must be equal to 1 to satisfy both (10.8) and (10.9). In that case, Y_{mq} is not increased in (10.10). This results in a mixed-integer linear programming model.

10.1.2 Multi-modal Synchronization Without Hierarchy and Time Windows at Transfers

We previously considered the case of maximizing the simultaneous arrivals of the trips of different lines at common transfer stops. In practice, however, the request of having simultaneous arrivals might be relaxed by allowing a small time window between the arrivals of different lines at the same station/stop. This time window can allow for a small walking distance that might be required when transferring between modes. We have already discussed the recent model of Ibarra et al. [2] presented in 2016 that considers time windows at transfers. Consider a transit network with a set of lines I and a set of stations/stops B. In addition, J_i is the set of lines that share synchronization stops with line $i \in I$ and B_{ij} the set of stops that represent the synchronization (i.e., transfer) stops between lines i and j, where $j \in J_i$. Consider also S_i to be the set of planning periods for line $i \in I$, P_i the ordered set of trips of line $i \in I$, and P_i^s the ordered set of trips of line $i \in I$ during a period $s \in S_i$. We have parameter d_s^i representing the end of planning period $s \in S_i$ and the beginning of planning period $s + 1 \in S_i$. The total number of trips of line i during the day is known, f_i, and f_i^s is the frequency of trips during planning period s.

The p-th trip of line i is denoted as i_p and its travel time from its depot to node b is denoted as t_p^{ib}. Because every trip i_p belongs to a planning period $s \in S_i$ its travel time is $t_p^{ib} = t_s^{ib}$. A synchronization between two trips i_p and j_q where $j \in J_i$ is said to occur at node $b \in B_{ij}$ if the waiting times of passengers that transfer between these two lines is within the minimum, w_{pq}^{ijb}, and maximum, $\bar{w}_{pq}^{ijb}$ waiting times, respectively.

The additional requirement is that the timetables of lines maintain a headway which is close to the even headway. At period s the minimum and maximum headway times for line $i \in I$ are $h_s^i := \frac{d_s^i - d_{s-1}^i}{f_i^s} - \delta_s^i$ and $\bar{h}_s^i := \frac{d_s^i - d_{s-1}^i}{f_i^s} + \delta_s^i$ where δ_s^i is a flexibility parameter of the headway in period s. Note that the term $\frac{d_s^i - d_{s-1}^i}{f_i^s}$ is the *even headway* which requires that all vehicles being evenly distributed in planning period s. Last, let β_{is} be the first trip of line $i \in I$ in planning period $s \in S_i$ and $\bar{\beta}_{is}$ the last trip of line $i \in I$ in planning period $s \in S_i$. The variables of the problem are the dispatching time of the p-th trip of line i, x_p^i, and binary variable y_{pq}^{ijb} which is equal to 1 if trip i_p arrives first at stop $b \in B_{ij}$ and trip j_q synchronizes with it, and 0 otherwise. The MILP formulation of the synchronization-based timetabling problem of [2] is provided below.

Synchronization-based timetabling

$$\max \sum_{i \in I} \sum_{j \in J_i} \sum_{b \in B_{ij}} \sum_{p=1}^{f_i} \sum_{q=1}^{f_j} y_{pq}^{ijb} \tag{10.11}$$

$$\text{s.t. } h_s^i \le x_{p+1}^i - x_p^i \le \bar{h}_s^i \qquad \forall i \in I, \forall p \in P_i^s \setminus \{|P_i^s|\}, \forall s \in S_i \tag{10.12}$$

$$d_{s-1}^i + \frac{h_s^i}{2} \le x_{\beta_{is}}^i \le d_{s-1}^i + \frac{\bar{h}_s^i}{2} \qquad \forall i \in I, s \in S_i \tag{10.13}$$

$$d_s^i - \frac{\bar{h}_s^i}{2} \le x_{\bar{\beta}_{is}}^i \le d_s^i - \frac{h_s^i}{2} \qquad \forall i \in I, \forall s \in S_i \tag{10.14}$$

$$x_q^j + t_q^{jb} - (x_p^i + t_p^{ib}) \ge w_{pq}^{ijb} - M(1 - y_{pq}^{ijb}) \qquad \forall i \in I, j \in J_i, b \in B_{ij}, p \in P_i, q \in P_j \tag{10.15}$$

$$x_q^j + t_q^{jb} - (x_p^i + t_p^{ib}) \le \bar{w}_{pq}^{ijb} + M(1 - y_{pq}^{ijb}) \qquad \forall i \in I, j \in J_I, b \in B_{ij}, p \in P_i, q \in P_j \tag{10.16}$$

$$x_p^i \in \mathbb{R} \qquad \forall i \in I, \forall p \in P_i \tag{10.17}$$

$$y_{pq}^{ijb} \in \{0, 1\} \qquad \forall i \in I, \forall j \in J, \forall p \in P_i, \forall q \in P_j, \forall b \in B_{ij} \tag{10.18}$$

The objective function (10.11) maximizes the number of synchronizations. Constraints (10.12) ensure that the trips of every line are almost evenly spaced within each time period s. Constraints (10.13) and (10.14) ensure that there is a separation time between arrivals of the last trip of period $s - 1$ and the first trip of period s, i.e., the separation between trips β_{is} and $\bar{\beta}_{i,s-1}$ corresponding to the first trip of line i in planning period s and the last trip of line i in planning period $s - 1$ must be within $[\frac{h_{s-1}^i + h_s^i}{2}, \frac{\bar{h}_{s-1}^i + \bar{h}_s^i}{2}]$. Constraints (10.15) and (10.16) are synchronization constraints allowing variable y_{pq}^{ijb} to be equal to 1 if the arrival times of i_p and j_q at node b are within $[w_{pq}^{ijb}, \bar{w}_{pq}^{ijb}]$. If $y_{pq}^{ijb} = 0$ the synchronization constraints are redundant since M is a very big positive number.

10.2 Multi-modal Synchronization with Hierarchy: Feeder and Collector Lines

Until now we considered that synchronized lines are of equal importance and all of them have to adapt the dispatching times of their timetabled trips. There are cases, however, where one of the two lines is a fixed, collector line that cannot modify the dispatching times of its trips. In that case, the feeder line which shares transfer stops should adapt its timetable accordingly to improve the synchronization of the services. A typical example of feeder-collector line synchronization is the synchronization of a feeder bus line to a train service [3–7].

The problem of transfer synchronization between feeder and collector transit lines has been mainly studied at the tactical planning level [1, 8]. There have also been studies that consider stochastic travel times and passenger demand resulting in robust optimization models [9].

We will discuss here the convex model of [10] for the case of multi-modal synchronization with hierarchy. This model considers the regularity of the feeder lines and the synchronization of feeder and collector lines that can reduce the transfer times of passengers. The problem description is provided below.

Multi-modal synchronization with hierarchy

Consider a collector line with a fixed schedule and a feeder line that has some common stops with the collector line where its service should be synchronized. Adjust the dispatching times of the trips of the feeder line so as to minimize the waiting times of passengers at its stops while synchronizing its arrival times at the transfer stops with the arrival times of trips of the collector line.

To present the model, we introduce the nomenclature of Table 10.1.

The decision variables of the problem are the rescheduled dispatching times of the feeder line trips, x_n. Changing the dispatching times of trips modifies their expected arrival times at stops as follows:

$$a_{ns} := x_n + \sum_{z=2}^{s} t_{nz} \quad \forall n \in N_f, \ \forall s \in S_f \setminus \{1\} \tag{10.19}$$

where t_{nz} is the expected travel time of trip n from stop $z - 1$ to stop z, including the dwell time for boardings/alightings. Because overtaking is not allowed, the arrival time of each feeder line trip $n \in N_f \setminus \{1\}$ at stop s should be greater than the arrival time of its previously dispatched trip $a_{n-1,s}$. To achieve this, Eq. (10.19) is modified to:

Table 10.1 Nomenclature

Sets	
N_f	Set of all planned daily trips of the feeder line
N_m	Set of all planned daily trips of the main (collector) line
S_f	Ordered set of public transport stops of the feeder line
S_m	Ordered set of public transport stops of the main line
$B = S_f \cap S_m$	Set denoting all transfer stops between the feeder and the main line
Parameters	
h^*	The ideal headway of the feeder line that should be maintained at all stops for attaining a regular service. Note that this might differ at different times of the day (peak, off-peak)
t_{ns}	The expected inter-station travel time of the feeder line trip $n \in N_f$ between stops $s - 1$ and s, including the dwell time at stop $s - 1$
$\delta_{\min}$	The pre-determined dispatching time of the first trip of the feeder line that prevents starting the daily operations before the drivers (crew members) are available
$\delta_{\max}$	The pre-determined dispatching time of the last trip of the feeder line that prevents schedule sliding
ψ	The required layover time for the feeder line for performing two successive trips with the same vehicle
$\gamma_{n,s}$	The arrival time of trip $n \in N_m$ at stop s which cannot be rescheduled because it belongs to the main line
Y_{bnm}	0–1 parameter, where $Y_{bnm} = 1$ if trip $n \in N_f$ needs to synchronize its arrival times with trip $m \in N_m$ at stop $b \in B$, and 0 otherwise
$\Phi_{n,n'}$	0–1 parameter, where $\Phi_{n,n'} = 1$ if trips $n, n' \in N_f$ of the feeder line are operated by the same vehicle in a sequential order (e.g., n' after n), and 0 otherwise
M	A very large positive number
Variables	
x_n	The (re)scheduled dispatching time of the nth trip of the feeder line
a_{ns}	Arrival time of trip $n \in N_f$ at stop s
h_{ns}	Inter-arrival headway between successive feeder line trips $n - 1$ and n at stop s

$$a_{ns} := \max \left(a_{n-1,s} \; ; \; x_n + \sum_{z=2}^{s} t_{nz} \right) \qquad \forall n \in N_f \setminus \{1\}, \; \forall s \in S_f \setminus \{1\} \quad (10.20)$$

which ensures that the dispatching order of feeder line trips is maintained. There is also a boundary condition for the first trip of the day which does not have a preceding trip,

$$a_{1s} := x_1 + \sum_{z=2}^{s} t_{1z} \qquad \forall s \in S_f \setminus \{1\} \quad (10.21)$$

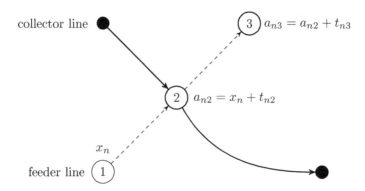

Fig. 10.2 Illustration of the varying arrival times of trip n that belongs to the feeder line based on the dispatching time decision x_n

This results in the varying arrival times of the feeder line at its stops, depicted in Fig. 10.2. These arrival times vary subject to the rescheduled dispatching time, x_n, since the travel and dwell times t_{ns} are considered to be fixed.

The inter-arrival headways of the trips of the *feeder* line are linked to the rescheduling of their trip dispatching times as follows:

$$h_{ns} := a_{ns} - a_{n-1,s} \quad \forall n \in N_f \setminus \{1\}, \ \forall s \in S_f \setminus \{1\} \tag{10.22}$$

To increase the regularity of the feeder line service, the actual inter-arrival time headways at the stops of the feeder line should be as close as possible to their scheduled values, h^*, that were determined during the frequency settings stage. To achieve this, one should minimize the sum of the squared difference between the actual and the ideal headways:

$$f(\mathbf{h}) := \sum_{s \in S_f \setminus \{1\}} \sum_{n \in N_f \setminus \{1\}} \left(\frac{h_{ns}}{2} - \frac{h^*}{2} \right)^2 \tag{10.23}$$

Now, let us consider the waiting times of passengers when transferring from a feeder to a collector line. We have $Y_{bnm} = 1$ if trip $n \in N_f$ needs to synchronize its arrival time with trip $m \in N_m$ at the transfer stop $b \in B$, and $Y_{bnm} = 0$ otherwise. As already discussed, [1] considers a perfect synchronization when trip $n \in N_f$ arrives at the transfer stop $b \in B$ exactly at the same time as trip $m \in N_m$. In this way, the transfer times of passengers are minimized when $a_{nb} - \gamma_{mb} = 0$, where γ_{mb} is the arrival time of the trip of the collector line at stop s. Later studies by [11] and [12] consider a trip n to be synchronized if it arrives within a time window $[0, \Delta t]$ before the arrival of trip m at the transfer stop b. We can consider the flexible synchronization scheme with time windows by modeling the required synchronizations at transfer stops as problem constraints:

$$0 \leq Y_{bnm}(\gamma_{mb} - a_{nb}) \leq \Delta t \quad \forall n \in N_f \setminus \{1\}, \ \forall m \in N_m \setminus \{1\}, \ \forall b \in B \tag{10.24}$$

Obviously, when $Y_{bnm} = 0$ the inequalities in Eq. (10.24) hold for any value of the arrival times a_{nb} and γ_{mb} because it is not required to synchronize the arrival times of those two trips.

Typically, if one vehicle operates two successive trips of the same line, there should be a layover time from the time the first trip is finished until the next trip is dispatched. When rescheduling the dispatching times of our feeder line trips, one should factor in this layover time. This yields the inequality constraints:

$$\Phi_{n,n'}\left(x_{n'} - a_{n,|S_f|}\right) \geq \Phi_{n,n'}\psi \qquad \forall n \in N_f, n' \in N_f \qquad (10.25)$$

That is, if trip n' is the next trip of trip n that is operated by the same vehicle (e.g., $\Phi_{n,n'} = 1$), then the dispatching time of trip n', $x_{n'}$, should be greater than the arrival time of trip n at the last stop, $a_{n,|S_f|}$, plus the required layover time ψ. If trips n, n' are not operated by the same vehicle in successive order, inequality (10.25) is satisfied because $\Phi_{n,n'} = 0$.

The following mathematical program summarizes the proposed feeder line synchronization problem.

Multi-modal synchronization with hierarchy

$$\min_{\mathbf{x},\mathbf{h},\mathbf{a}} \sum_{s \in S_f \setminus \{1\}} \sum_{n \in N_f \setminus \{1\}} \left(\frac{h_{ns}}{2} - \frac{h^*}{2}\right)^2 \qquad (10.26)$$

s.t.: Eqs. (10.20) − (10.25)

$$x_1 \geq \delta_{\min} \qquad (10.27)$$

$$x_{|N_f|} \leq \delta_{\max} \qquad (10.28)$$

$$x_n \in \mathbb{R}_{\geq 0} \qquad \forall n \in N_f \qquad (10.29)$$

$$a_{ns} \in \mathbb{R}_{\geq 0} \qquad \forall n \in N_f, s \in S_f \qquad (10.30)$$

$$h_{ns} \in \mathbb{R}_{\geq 0} \qquad \forall n \in N_f, s \in S_f \setminus \{1\} \qquad (10.31)$$

Inequality constraint (10.27) ensures that we dispatch the first feeder trip of the day when drivers and vehicles are available. Inequality constraint (10.28) ensures that we dispatch the last trip of the day before the latest possible dispatching time, $\delta_{\max}$, to avoid schedule sliding. This formulation results in a nonlinear mathematical program which is not convex because the constraints (10.20) include the non-convex term $\max(a_{n-1,s}; x_n + \sum_{z=2}^{s} t_{nz})$. To rectify this, one can:

- Replace the equality constraint $a_{ns} := \max(a_{n-1,s}; x_n + \sum_{z=2}^{s} t_{nz})$ with the inequality constraints $a_{ns} \geq a_{n-1,s}$ and $a_{ns} \geq x_n + \sum_{z=2}^{s} t_{nz}$

- Introduce the penalty term $M\left(a_{ns} - (x_n + \sum_{z=2}^{s} t_{nz})\right)^2$ to the objective function which will force the value of a_{ns} to be equal to $\max(a_{n-1,s}; x_n + \sum_{z=2}^{s} t_{nz})$ at the solution of the mathematical program. This is because a_{ns} is forced to be equal to $x_n + \sum_{z=2}^{s} t_{nz}$ if $x_n + \sum_{z=2}^{s} t_{nz} \geq a_{n-1,s}$ to avoid a major cost increase in the objective function because of the large value of M.

With this transformation, we have the following quadratic program.

Multi-modal synchronization with hierarchy as a convex program

$$\min \sum_{s \in S_f \setminus \{1\}} \sum_{n \in N_f \setminus \{1\}} \left(\frac{h_{ns}}{2} - \frac{h^*}{2}\right)^2 + \sum_{n \in N_f \setminus \{1\}} \sum_{s \in S_f \setminus \{1\}} M\left(a_{ns} - (x_n + \sum_{z=2}^{s} t_{nz})\right)^2$$

subject to:

$$\text{Eqs. } (10.21) - (10.25)$$

$$a_{ns} \geq a_{n-1,s} \qquad \forall n \in N_f \setminus \{1\}, \forall s \in S_f \setminus \{1\}$$

$$a_{ns} \geq x_n + \sum_{z=2}^{s} t_{nz} \qquad \forall n \in N_f \setminus \{1\}, \forall s \in S_f \setminus \{1\}$$

$$x_1 \geq \delta_{min}$$

$$x_{|N_f|} \leq \delta_{max}$$

$$x_n \in \mathbb{R}_{\geq 0} \qquad \forall n \in N_f$$

$$a_{ns} \in \mathbb{R}_{\geq 0} \qquad \forall n \in N_f, s \in S_f$$

$$h_{ns} \in \mathbb{R}_{\geq 0} \qquad \forall n \in N_f, s \in S_f \setminus \{1\}$$

This quadratic program is convex and it can be solved to global optimality. In more detail, its feasible region is defined by affine/linear inequalities and it is a polyhedron (convex set). The objective function is also convex, as it is proven below.

Let us denote the reformulated objective function as:

$$\tilde{f}(\mathbf{h}, \mathbf{a}, \mathbf{x}) := \sum_{s \in S_f \setminus \{1\}} \sum_{n \in N_f \setminus \{1\}} \left(\frac{h_{ns}}{2} - \frac{h^*}{2}\right)^2 + \sum_{n \in N_f \setminus \{1\}} \sum_{s \in S_f \setminus \{1\}} M\left(a_{ns} - (x_n + \sum_{z=2}^{s} t_{nz})\right)^2$$

We then introduce functions $g_{n,s}(\mathbf{h}) := \left(\frac{h_{ns}}{2} - \frac{h^*}{2}\right)^2$ and $k_{n,s}(\mathbf{a}, \mathbf{x}) := M(a_{ns} - (x_n + \sum_{z=2}^{s} t_{nz}))^2$ so that we can write $\tilde{f}(\mathbf{h}, \mathbf{a}, \mathbf{x})$ as:

$$\tilde{f}(\mathbf{h}, \mathbf{a}, \mathbf{x}) := \sum_{s \in S_f \setminus \{1\}} \sum_{n \in N_f \setminus \{1\}} g_{n,s}(\mathbf{h}) + \sum_{n \in N_f \setminus \{1\}} \sum_{s \in S_f \setminus \{1\}} k_{n,s}(\mathbf{a}, \mathbf{x})$$

Function $g_{n,s}(\mathbf{h})$ is convex with respect to $\mathbf{h}$ for any $n \in N_f \setminus \{1\}$, $s \in S_f \setminus \{1\}$ because

$$\frac{\partial^2 g_{n,s}}{\partial h_{ns}} = \frac{1}{2} > 0 \qquad \forall n \in N_f \setminus \{1\}, \ \forall s \in S_f \setminus \{1\}$$

Similarly, function $k_{n,s}(\mathbf{a}, \mathbf{x})$ is convex for any $n \in N_f \setminus \{1\}, s \in S_f \setminus \{1\}$ because its Hessian matrix is positive semi-definite:

$$\mathbf{H} = \begin{bmatrix} \frac{\partial^2 k_{ns}}{\partial a_{ns}^2} & \frac{\partial^2 k_{ns}}{\partial a_{ns} \partial x_{ns}} \\ \frac{\partial^2 k_{ns}}{\partial x_{ns} \partial a_{ns}} & \frac{\partial^2 k_{ns}}{\partial x_{ns}^2} \end{bmatrix} = \begin{bmatrix} 2M & -2M \\ -2M & 2M \end{bmatrix}$$

since the leading principal minor $\det(\mathbf{H}) = 4M - 4M = 0$ is non-negative. Therefore, $\tilde{f}(\mathbf{h}, \mathbf{a}, \mathbf{x})$ is convex as the sum of the convex functions $g_{n,s}(\mathbf{h})$ and $k_{n,s}(\mathbf{a}, \mathbf{x})$.

Let us consider the example of Fig. 10.2 where we have one common transfer stop between the feeder and the collector line. The collector line has two trips with arrival times at the transfer stop equal to 310 s and 610 s, respectively. The ideal headway of the feeder line is $h^* = 300$ s. The feeder line has two trips. Trip 1 has inter-station travel times equal to 300 s. Trip 2 has an inter-station travel time of 320 s when traveling from the first to the second stop and 290 s when traveling from the second to the third stop. We also have $\delta_{\min} = 0$ s and $\delta_{\max} = 900$ s. The required layover time for the feeder line for performing two successive trips with the same vehicle is $\psi = 15$ s. The trips of the feeder line are performed by different vehicles and a synchronized transfer should be within the time window $[0, \Delta t]$, where $\Delta_t = 20$ s. Solving this problem results in solution $\mathbf{x}^* = [5, 290]^\mathsf{T}$ expressed in seconds,

$$\mathbf{a}^* = \begin{bmatrix} 305 & 605 \\ 610 & 900 \end{bmatrix} \text{ expressed in seconds,}$$

and $\mathbf{h}^* = [305, 295]^\mathsf{T}$ expressed in seconds.

Closing this chapter, we present selected works that use mathematical programs to solve transfer synchronization problems in Table 10.2.

> **Further Reading—survey papers**

- Liu et al., A review of public transport transfer coordination at the tactical planning phase, Transportation Research Part C: Emerging Technologies, 2021 [13].
- Gkiotsalitis et al., A review of public transport transfer synchronisation at the real-time control phase, Transport Reviews, 2022 [23].

Table 10.2 Summary of selected mathematical programming approaches for the transfer synchronization problem [13]

Study	Objective	Decision variable	Formulation	Solution approach
[14]	Min total transfer waiting time	Offset time	Binary program	Branch and Bound
[15]	Min total transfer penalty	Offset time	Integer linear program	CPLEX
[12]	Max number of synchronizations	Departure time	Integer program	Branch and Bound
[16]	Min total transfer waiting time	Departure time	Mixed-integer program	CPLEX
[17]	Max number of smooth transfers	Offset time, travel time	Integer program	CPLEX
[18]	Max weighted sum of synchronized transfers	Departure time	Integer program	CPLEX
[19]	Max number of successful connections, Min connection slack time	Departure/arrival times, dwell time	Integer nonlinear program	Dynamic Programming
[20]	Max number of passengers making successful transfers	Departure/arrival times	Mixed-integer program	Branch and Bound
[21]	Max gain of transfer passengers minus the loss of direct passengers	Departure/arrival times	Mixed-integer program	CPLEX
[22]	Min user and agency costs	Offset time, headway	Integer program	Grid search, enumeration

Exercises

10.1 Multi-modal Synchronization without Hierarchy

Consider the example of Fig. 10.3 with a single transfer stop and inter-station travel times as presented in every arc of the figure. We also have 3 trips in line 1 and 3 trips in line 2. The minimum and maximum time headways of line 1 are $H_1^{\min} = 2$ and $H_1^{\max} = 6$ time units, respectively. For line 2 we have $H_2^{\min} = 3$ and $H_2^{\max} = 6$, respectively. Let also $T = 35$ be the latest possible dispatching time of any trip.

Use the MILP reformulation of the nonlinear maximal multi-modal synchronization problem expressed in (10.1)–(10.5) to derive the timetables of lines 1 and 2. Write a branch and bound algorithm to derive a solution and explain why this solution is a globally optimal one.

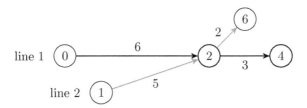

Fig. 10.3 Digraph of lines 1 and 2 with their stop identification numbers and their inter-station travel times

10.2 Multi-modal Synchronization with Hierarchy
Consider the example of Fig. 10.2 presented in the Multi-modal Synchronization with Hierarchy. Program and solve this problem to global optimality.

10.3 Synchronization
Describe the differences between synchronization problems with and without hierarchy. Provide some potential cases where synchronization problems without hierarchy are more suitable.

References

1. A. Ceder, B. Golany, O. Tal, Transp. Res. A: Policy Pract. **35**(10), 913 (2001)
2. O.J. Ibarra-Rojas, F. López-Irarragorri, Y.A. Rios-Solis, Transp. Sci. **50**(3), 805 (2016)
3. S.M. Chowdhury, S. I-Jy Chien, Transp. Planning Technol. **25**(4), 257 (2002)
4. K. Sivakumaran, Y. Li, M.J. Cassidy, S. Madanat, Transp. Res. A: Policy Pract. **46**(1), 131 (2012)
5. M.H. Almasi, A. Sadollah, S.M. Mounes, M.R. Karim, J. Comput. Civil Eng. **29**(6), 04014090 (2015)
6. J. Wu, R. Song, Y. Wang, F. Chen, S. Li, Math. Probl. Eng. **2015** (2015)
7. Y. Yang, X. Jiang, W. Fan, Y. Yan, L. Xia, IEEE Access **8**, 96391 (2020)
8. A. Ceder, Y. Yim, (2003)
9. O.A. Eikenbroek, K. Gkiotsalitis, in *2020 IEEE 23rd International Conference on Intelligent Transportation Systems (ITSC)* (IEEE, 2020), pp. 1–7
10. K. Gkiotsalitis, EURO, J. Transp. Logist. **11**, 100057 (2022)
11. A. Eranki, (2004)
12. O.J. Ibarra-Rojas, Y.A. Rios-Solis, Transp. Res. B: Methodol. **46**(5), 599 (2012)
13. T. Liu, O. Cats, K. Gkiotsalitis, Transp. Res. C: Emerg. Technol. **133**, 103450 (2021)
14. W. Domschke, Operations-Research-Spektrum **11**(1), 17 (1989)
15. M. Schröder, I. Solchenbach (2006)
16. G.K. Saharidis, C. Dimitropoulos, E. Skordilis, Oper. Res. **14**(3), 341 (2014)
17. X. Dou, Q. Meng, X. Guo, Transp. Res. C: Emerg. Technol. **60**, 360 (2015)
18. P. Fouilhoux, O.J. Ibarra-Rojas, S. Kedad-Sidhoum, Y.A. Rios-Solis, Eur. J. Oper. Res. **251**(2), 442 (2016)
19. X. Tian, H. Niu, Adv. Mech. Eng. **9**(6), 1687814017712364 (2017)
20. Y. Wu, Y. Zhu, T. Cao, Math. Probl. Eng. **2018** (2018)
21. M. Takamatsu, A. Taguchi, Transp. Sci. **54**(5), 1238 (2020)
22. M. Estrada, J. Mension, M. Salicrú, Transp. Res. C: Emerg. Technol. **130**, 103283 (2021)
23. K. Gkiotsalitis, O. Cats, T. Liu, *Transport Reviews* (2022), pp. 1–20

Chapter 11
Tactical Planning of Public Transport Services: Vehicle and Crew Schedules

Abstract In the previous chapters on the tactical planning of public transport services we concentrated on the setting of the frequencies and timetables of service lines. Once the timetables of the service lines are fixed, a new series of actions need to take place to make them operational, including the determination of vehicle and crew schedules. This chapter focuses on the tactical planning decision level and addresses these problems in a sequential order starting from the vehicle scheduling problem and moving to the crew scheduling problem.

11.1 Vehicle Scheduling

We have already used an estimate of the required vehicles in the frequency settings formulations. In these formulations, i.e., formulation (9.10), the fleet size m that is required to operate a set of lines $k \in R$ is expected to be:

$$m := \sum_{k \in R} \lceil t_k f_k \rceil$$

where t_k is the round-trip (cycle) time of line k and f_k its frequency. With this approach, it is very simple to estimate the required number of vehicles. However, this approach is very restrictive. First, it requires the lines to operate based on *frequencies*, not on *timetables* with potentially varying dispatching headways between successive trips. Second, the round-trip travel times of lines should be stable across long time periods and the size of the vehicles should be homogeneous. To rectify this, vehicle scheduling is typically performed by solving a separate Vehicle Scheduling Problem (VSP). The first attempts to formulate and solve the VSP were around 1970. Bunte and Kliewer [1] provided an overview of VSP models in 2009 and defined the VSP as follows.

© The Author(s), under exclusive license to Springer Nature Switzerland AG 2022   485
K. Gkiotsalitis, *Public Transport Optimization*,
https://doi.org/10.1007/978-3-031-12444-0_11

Vehicle Scheduling Problem

Consider a set of lines with a set of timetabled trips with fixed travel (departure and arrival) times and start and end locations as well as traveling times between all pairs of end stations. Consider also a set of vehicles with (possibly different) capacities that are located at depot locations. Find an assignment of trips to vehicles such that:

- each trip is covered exactly once,
- each vehicle performs a feasible sequence of trips,
- the overall costs are minimized.

VSP can take different forms, depending on several decisions related to the problem's formulation. Different versions of common VSP models are provided below.

11.1.1 Single-Depot Vehicle Scheduling Problem (SD-VSP)

11.1.1.1 SD-VSP: Minimum Decomposition Formulation

The Single-Depot Vehicle Scheduling Problem (SD-VSP) was first formulated by Saha in 1972 [2] and it is the simplest VSP problem because all vehicles belong to a single depot. The formulation of Saha results in a *Minimal Decomposition Model*. In this formulation we consider a set $T = \{1, 2, \ldots, n\}$ of timetabled trips. An ordered relationship β is proposed that admits the service of a trip j after trip i by the same vehicle, if:

- trip j starts at the end station of trip i
- the dispatching time of trip j is greater than or equal to the arrival time of trip i

The ordered relationship β can be used to characterize any pair of trips i and j. We can introduce parameters $\mathbf{C} = \{c_{ij}\}$ where $c_{ij} = 1$ if trip j can be admitted after trip i (that is, if $i \; \beta \; j$) and $c_{ij} = -\infty$ otherwise. For computer programming purposes, we can set $c_{ij} = -M$ instead of $-\infty$ where M is a very large positive number.

Let us also consider binary decision variables $\mathbf{X} = \{x_{ij}\}$ where $x_{ij} = 1$ if trips i and j are connected (meaning that trip j is performed after trip i by the same vehicle) and $x_{ij} = 0$ if they are not connected. Then, the SD-VSP formulation of Saha is expressed as follows.

SD-VSP—Minimal Decomposition Model formulation

$$\max_{\mathbf{X}} \sum_{i \in T} \sum_{j \in T} c_{ij} x_{ij} \tag{11.1}$$

$$\text{subject to:} \sum_{j \in T} x_{ij} \leq 1 \qquad \forall i \in T \tag{11.2}$$

$$\sum_{i \in T} x_{ij} \leq 1 \qquad \forall j \in T \tag{11.3}$$

$$x_{ij} \in \{0, 1\} \qquad \forall i \in T, j \in T \tag{11.4}$$

The objective function (11.1) strives to maximize the number of connections among trips, resulting in using fewer vehicles. The operational costs related to running the trips are not taken into consideration but the trip travel times are implicitly considered as problem constraints when defining the values of parameters c_{ij}. Constraints (11.2) ensure that every trip i can be connected to at most one other trip j. Constraints (11.3) ensure that from all trips i that can be connected to trip j, at most one of them can make the connection.

Let us consider, for example, the following 7 trips of a bi-directional line that have fixed dispatching times and travel times (see Table 11.1).

From Table 11.1 we can obtain the values of c_{ij}. For instance, trips that operate in the same direction cannot be operated in successive order by the same vehicle because the end station of one trip is not the start station of the other trip. Instead, the end station of trips operating in direction 1 is the start station of trips operating in direction 2, and vice versa. From these trip pairs, only trips 1–6, 1–7, 2–6, 2–7, 3–7, and 4–7 can be operated one after the other by the same vehicle given their dispatching and travel times. Thus, we can produce Table 11.2 of the parameter values c_{ij}.

This example can be solved as a MILP with branch-and-cut resulting in solution $x_{2,6} = 1$, $x_{3,7} = 1$ and $x_{ij} = 0$ for any other $i \in T$ and $j \in T$.

Table 11.1 Pre-determined dispatching times of trips at each direction. The travel time of trips in direction 1 is 30 min and the travel time of trips in direction 2 is 25 min

Direction 1		Direction 2	
Trip	Dispatching time	Trip	Dispatching time
1	08:00	5	08:15
2	08:10	6	08:42
3	08:20	7	09:00
4	08:30		

Table 11.2 c_{ij} parameter values, where M is a very large positive number

Trips	1	2	3	4	5	6	7
1	$-M$	$-M$	$-M$	$-M$	$-M$	1	1
2	$-M$	$-M$	$-M$	$-M$	$-M$	1	1
3	$-M$	$-M$	$-M$	$-M$	$-M$	$-M$	1
4	$-M$	$-M$	$-M$	$-M$	$-M$	$-M$	1
5	$-M$	$-M$	$-M$	$-M$	$-M$	$-M$	$-M$
6	$-M$	$-M$	$-M$	$-M$	$-M$	$-M$	$-M$
7	$-M$	$-M$	$-M$	$-M$	$-M$	$-M$	$-M$

Practitioner's Corner
SD-VSP Minimal Decomposition Model in Python 3

One can use PuLP to program and the CBC solver to solve the Minimal Decomposition Model formulation of the SD-VSP as follows.

```
import pulp
from pulp import *
model = LpProblem("SD_VSP_Minimal_Decomposition_Model", LpMaximize)
T=(1,2,3,4,5,6,7) #trips
M=100000
c={(i,j):-M for i in T for j in T}
c[1,6]=1; c[1,7]=1; c[2,6]=1; c[2,7]=1; c[3,7]=1; c[4,7]=1
# variables
x = LpVariable.dicts("x", ((i, j) for i in T for j in T),lowBound=0,
    upBound=1,cat='Binary')
# objective function
model += sum(sum(c[i,j]*x[i,j] for j in T) for i in T)
# constraints:
for i in T:
    model += sum(x[i,j] for j in T) <= 1
for j in T:
    model += sum(x[i,j] for i in T) <= 1
# optimizing
solver = pulp.PULP_CBC_CMD(mip=True, msg=True, timeLimit=None)
model.solve(solver)
print ("Status:", LpStatus[model.status])
for v in model.variables():
    print (v.name, "=", v.varValue)
print ("Optimal Solution = ", value(model.objective))
```

Solving this problem yields solution $x_{2,6} = 1$, $x_{3,7} = 1$ and $x_{ij} = 0$ for any other $i \in T$ and $j \in T$. Thus, the optimal objective function score is equal to 2. This indicates that trips 2 and 6 will be performed by the same vehicle and trips 3 and 7 will be also performed by the same vehicle. All other trips (that is, trips 1, 4, 5)

Fig. 11.1 Optimal vehicle
assignment. We have
$x_{16} = 1, x_{37} = 1$ and $x_{ij} = 0$
for all other i, j meaning that
trips 1 and 6 are performed
by the same vehicle (dashed
lines), trips 3 and 7 are
performed by the same
vehicle (dotted lines), and all
other trips are performed by
different vehicles (normal
lines). Thus, we need 5
vehicles in total

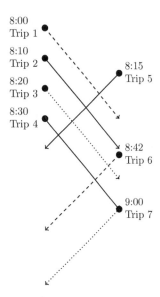

will be performed by other vehicles. This means that we would require 5 vehicles
to operate these 7 trips. The validity of this solution can be graphically checked in
Fig. 11.1.

The *Minimal Decomposition Model* of Saha solves only the minimum fleet size
problem without respecting the operational costs. In addition, no upper bound for
the fleet size can be set.

11.1.1.2 SD-VSP: Assignment Formulation

Another SD-VSP formulation is the *Assignment Problem* formulation provided by
Orloff in 1976 [3]. We will present first the easier-to-follow *Asymmetric Assignment*
formulation provided by Freling et al. in 2001 [4]. In this formulation, trips that can
be served by the same vehicle are linked by *deadheading trips*.

Let T be the set of trips and b_i and e_i the starting and ending locations of each
trip $i \in T$. Let also b_i^t and e_i^t be the starting and ending times of each trip $i \in T$,
respectively. If d_{ij} is the deadheading (and possibly idle) time between two locations
e_i and b_j, two trips $i \in T$ and $j \in T$ are a *compatible* pair if they can be covered by
the same vehicle in a sequence, meaning that $e_i^t + d_{ij} \le b_j^t$.

This formulation builds a *trip precedence network*, where a sequence of trips is
feasible if each consecutive pair of trips in the sequence is compatible. To start forming the precedence network, we consider set of arcs $\bar{A} = \{(i, j) \mid i, j \text{ are compatible}\}$.
The cost of each arc $a_{ij} \in \bar{A}$ is c_{ij} and it can be set equal to the deadhead time d_{ij}.

The *trip precedence network* is a digraph $G = (V, A)$ and it is obtained by introducing artificial vertex $|T| + i$ and a zero cost arc $(i, |T| + i)$ to each vertex $i \in T$. This network has vertices $V = T \cup \{|T| + 1, \ldots, 2|T|\}$ and arcs $A = \bar{A} \cup \{(1, |T| + 1), (2, |T| + 2), \ldots, (|T|, |T| + |T|)\}$. A path of vertices of the form $\{i_1, i_2, \ldots, i_k, |T| + i_k\}$ corresponds to a *feasible vehicle schedule* leaving the deport to perform trips $i_1, i_2, \ldots, i_k$ and returning to the depot. That is, the departure from the depot is not explicitly modeled while the return to the depot is modeled by an arc of type $(i, |T| + i)$. Let c_{sj} be the deadhead cost of going from the depot s to the start location of trip j, b_j, and c_{it} be the deadhead cost of going from the end location of trip i, e_i, to the location of the depot (note that s and t are the same physical location in the single-depot problem). Then, we introduce travel costs $\bar{c}_{ij}$ for every arc (i, j) in the precedence network $G = (V, A)$ as follows:

$$\bar{c}_{ij} := \begin{cases} c_{ij} - c_{sj} \text{ for } (i, j) \in \bar{A} \\ c_{it} \text{ for } (i, j) \in A \setminus \bar{A} \end{cases}$$

The reason for using these costs is that if our vehicle serves arc $(i, j) \in \bar{A}$, then it will serve trips i and j in a sequence and we will not have to go from the depot to the start location of trip j. If, however, our vehicle serves arc $(i, j) \in A \setminus \bar{A}$ we have that arc (i, j) is actually arc $(i, |T| + i)$, meaning that our vehicle moves from the end location of trip i to the depot.

The decision variables are x_{ij}, with $x_{ij} = 1$ denoting whether a vehicle covers arc $a \in A$ and $x_{ij} = 0$ otherwise. With these definitions, we have the following *Asymmetric Assignment Problem* formulation of the SD-VSP.

SD-VSP—Asymmetric Assignment Problem formulation

$$\min_{\mathbf{x}} \sum_{(i,j) \in A} \bar{c}_{ij} x_{ij} + \sum_{j \in T} c_{sj} \tag{11.5}$$

$$\text{subject to: } \sum_{j:(i,j) \in A} x_{ij} = 1 \qquad \forall i \in T \tag{11.6}$$

$$\sum_{i:(i,j) \in A} x_{ij} \leq 1 \qquad \forall j \in V \tag{11.7}$$

$$x_{ij} \in \{0, 1\} \qquad \forall (i, j) \in A \tag{11.8}$$

In this MILP, constraints (11.6) ensure that each vertex $i \in T$ will be matched with exactly one arc in A. If this arc is of the form $(i, |T| + i)$, then our vehicle ends its schedule after visiting trip i. If the arc is of the form $(i, j) \in \bar{A}$, then our vehicle will perform trip $j \in T$ after i. Constraints (11.7) are the asymmetric constraints of

Table 11.3 Deadhead d_{ij} values in minutes

Trips	1	2	3	4	5	6	7
1	20	20	20	20	0	0	0
2	20	20	20	20	0	0	0
3	20	20	20	20	0	0	0
4	20	20	20	20	0	0	0
5	0	0	0	0	25	25	25
6	0	0	0	0	25	25	25
7	0	0	0	0	25	25	25

Fig. 11.2 Digraph $G = (V, A)$ of the trip precedence network for the example's problem

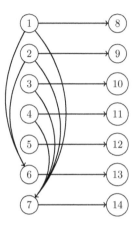

the assignment problem and ensure that each vertex of the digraph will be visited by at most one incoming arc.

Let us consider the implementation of this formulation to the example of Table 11.1 that has $|T| = 7$ trips. We consider the deadhead times of Table 11.3.

Given these deadhead times, the arc set $\bar{A}$ that represents compatible vertex pairs (i, j) for which $e_i^t + d_{ij} \leq b_j^t$ is $\bar{A} = \{(1, 6), (1, 7), (2, 6), (2, 7), (3, 7), (4, 7)\}$ and the set of arcs A is $A = \bar{A} \cup \{(1, 7 + 1), \ldots, (7, 7 + 7)\}$. Considering also that $V = T \cup \{|T| + 1, \ldots, 2|T|\}$, this results in the precedence network of digraph G presented in Fig. 11.2.

Considering deadhead times to/from the depot as $c_{sj} = 17$ min for any $j \in T$ and $c_{it} = 15$ min for any $i \in T$, we can program and solve the asymmetric assignment problem as follows.

Practitioner's Corner

SD-VSP Asymmetric Assignment Model in Python 3

One can use PuLP to program and the CBC solver to solve the Asymmetric Assignment Model formulation of the SD-VSP as follows.

```python
import pulp
from pulp import *
model = LpProblem("SD_VSP_Asymmetric_Assignment_Problem", LpMinimize)
T=(1,2,3,4,5,6,7) #trips
b={1:"1'",2:"2'",3:"3'",4:"4'",5:"5'",6:"6'",7:"7'"}
e={1:"1'''",2:"2'''",3:"3'''",4:"4'''",5:"5'''",6:"6'''",7:"7'''"}
bt={1:0,2:10,3:20,4:30,5:15,6:42,7:60}
et={1:bt[1]+30,2:bt[2]+30,3:bt[3]+30,4:bt[4]+30,5:bt[5]+25,6:bt
    [6]+25,7:bt[7]+25}
d={(i,j):0 for i in T for j in T}
for i in T:
    if i in [1,2,3,4]:
        for j in [1,2,3,4]: d[i,j]=20
        for j in [5,6,7]: d[i,j]=0
    if i in [5,6,7]:
        for j in [1,2,3,4]: d[i,j]=0
        for j in [5,6,7]: d[i,j]=25
csj={j:17 for j in T}
cit={i:15 for i in T}
A_bar=[];A=[]
for i in T:
    for j in T:
        if et[i]+d[i,j]<=bt[j]: A_bar.append((i,j)); A.append((i,j))
V=(1,2,3,4,5,6,7,8,9,10,11,12,13,14)
for i in T:
    A.append((i,len(T)+i))
c_bar={(i,j):0 for (i,j) in A}
for (i,j) in A_bar:
    c_bar[i,j]=d[i,j]-csj[j]
for (i,j) in A:
    if (i,j) not in A_bar: c_bar[i,j]=cit[i]
# variables
x = LpVariable.dicts("x", ((i, j) for (i,j) in A),lowBound=0,upBound
    =1,cat='Binary')
# objective function
model += sum(c_bar[i,j]*x[i,j] for (i,j) in A)+sum(csj[j] for j in T)
# constraints:
for i in T:
    model += sum(x[i,j] for j in V if (i,j) in A) == 1
for j in V:
    model += sum(x[i,j] for i in V if (i,j) in A) <= 1
# optimizing
solver = pulp.PULP_CBC_CMD(mip=True, msg=True, timeLimit=None)
model.solve(solver)
print ("Status:", LpStatus[model.status])
for v in model.variables():
    print (v.name, "=", v.varValue)
print ("Optimal Solution = ", value(model.objective))
```

Solving this problem yields the solution of Table 11.4 with objective function score 160. This solution means that a vehicle performs trips 1 and 7 and then goes to the depot (vertex 14), a vehicle performs trips 2 and 6 and then goes to the depot

Table 11.4 Optimal
solution **x**

Arc $a \in A$	x_a
(1, 6)	0
(1, 7)	1
(1, 8)	0
(2, 6)	1
(2, 7)	0
(2, 9)	0
(3, 10)	1
(3, 7)	0
(4, 11)	1
(4, 7)	0
(5, 12)	1
(6, 13)	1
(7, 14)	1

(vertex 13), a vehicle performs trip 3 and then goes to the depot (vertex 10), a vehicle performs trip 4 and then goes to the depot (vertex 11), a vehicle performs trip 5 and then goes to the depot (vertex 12). Thus, we need 5 vehicles in total to cover all trips.

Let us now turn our attention to the *Assignment Problem* formulation of Orloff [3] for the SD-VSP. Reckon that an Assignment Problem can be expressed as a MILP of the following form.

$$\min \sum_{(i,j) \in N \times T} c_{ij} x_{ij}$$

$$\text{s.t.} \sum_{j \in T} x_{ij} = 1 \qquad \forall i \in N$$

$$\sum_{i \in N} x_{ij} = 1 \qquad \forall j \in T \qquad (11.9)$$

$$x_{ij} \in \mathbb{Z}_+ \qquad \forall (i, j) \in N \times T$$

In our SD-VSP problem, let us redefine the network of the problem as digraph $G = (V, A)$ where $V = T \cup \{s, t\}$ and $A = \bar{A} \cup \{s \times T\} \cup \{T \times t\}$. In this network, vertices s and t are the start depot and the end depot of every trip (both of them are in the same physical location at the SD-VSP). In addition, the set of trips T is ordered according to the increasing starting times of trips.

A path from s to t in the network G represents a *feasible* vehicle schedule and it should include at least one trip from the set T in between s and t. The cost of arc $(i, j) \in \bar{A}$ is $\bar{c}_{ij} = d_{ij}$ where d_{ij} can be either the deadhead time from the end location of trip i to the start location of trip j, the deadhead time plus the possible idle time spent before starting trip j, or the deadhead time plus the possible idle time plus the travel time to complete trip i. The cost of any arc $(s, j) \in A$ is $\bar{c}_{sj} := d_{sj} + c_0$ where c_0 is the cost of adding an extra vehicle. The cost of any arc $(i, t) \in A$ is $\bar{c}_{it} := d_{it}$. The decision variables are x_{ij} for all $(i, j) \in A$ where $x_{ij} = 0$ if arc (i, j) is not part of the solution and $x_{ij} = 1$ otherwise. With these definitions, we proceed to the *Assignment Problem* MILP formulation.

SD-VSP—Assignment Problem formulation

$$\min_{x} \quad \sum_{(i,j) \in A} \bar{c}_{ij} x_{ij} \tag{11.10}$$

$$\text{s.t.} \quad \sum_{j:(i,j) \in A} x_{ij} = 1 \qquad \forall i \in T \tag{11.11}$$

$$\sum_{i:(i,j) \in A} x_{ij} = 1 \qquad \forall j \in T \tag{11.12}$$

$$x_{ij} \in \{0, 1\} \qquad \forall (i, j) \in A \tag{11.13}$$

Constraints (11.11) and (11.12) ensure that each trip is assigned to exactly one predecessor vertex and one successor vertex, and the network is partitioned in a set of disjoint paths from s to t, where each path represents a vehicle schedule.

Because constraints (11.11) and (11.12) and the fact that x_{ij} is integer do not allow x_{ij} to take a value greater than 1, the requirement $x_{ij} \leq 1$ is redundant and the binary variables $x_{ij} \in \{0, 1\}$ can be replaced by positive integer variables $x_{ij} \in \mathbb{Z}_+$ replicating the formulation of the general assignment problem presented in (11.9).

Let us consider the implementation of this formulation to the example of Table 11.1 that has $|T| = 7$ trips. We consider the deadhead times of Table 11.3. The set of trips ordered in ascending start times is $T = \{1, 2, 5, 3, 4, 6, 7\}$ and the vertex set of digraph G is $V = \{s, 1, 2, 5, 3, 4, 6, 7, t\}$. Given these deadhead times, the arc set $\bar{E}$ that represents compatible vertex pairs (i, j) for which $e_i' + d_{ij} \leq b_j'$ is $\bar{A} = \{(1, 6), (1, 7), (2, 6), (2, 7), (3, 7), (4, 7)\}$ and the set of arcs A of digraph G is $A = \bar{A} \cup \{(s, 1), (s, 2), \ldots, (s, 7), (1, t), (2, t), \ldots, (7, t)\}$. This results in the network G of Fig. 11.3.

We consider deadhead times to/from the depot as $d_{sj} = 17$ min for all $j : (s, j) \in A$ and $d_{it} = 15$ min for all $i : (i, t) \in A$. In addition, the cost of using a vehicle is $c_0 = 200$ units. Idle times are not considered and the cost of completing a trip $i \in T$ is also not considered because it is not necessary since all trips are required to be covered by a bus. Given this input, we can program and solve the SD-VSP as an Assignment Problem as follows.

Fig. 11.3 Digraph
$G = (V, A)$ of the example's
problem when using an
assignment formulation

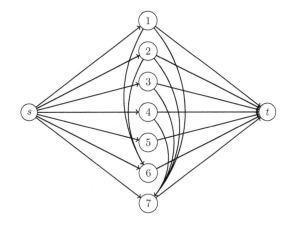

Practitioner's Corner
SD-VSP Assignment Model in Python 3

One can use PuLP to program and the CBC solver to solve the Assignment Model
formulation of the SD-VSP as follows.

```python
import pulp
from pulp import *
model = LpProblem("SD_VSP_Assignment_Problem", LpMinimize)
T=(1,2,5,3,4,6,7) #trips
V=('s',1,2,5,3,4,6,7,'t')
bt={1:0,2:10,3:20,4:30,5:15,6:42,7:60}
et={1:bt[1]+30,2:bt[2]+30,3:bt[3]+30,4:bt[4]+30,5:bt[5]+25,6:bt
    [6]+25,7:bt[7]+25}
d={(i,j):0 for i in V for j in V}
for i in T:
    if i in [1,2,3,4]:
        for j in [1,2,3,4]: d[i,j]=20
        for j in [5,6,7]: d[i,j]=0
    if i in [5,6,7]:
        for j in [1,2,3,4]: d[i,j]=0
        for j in [5,6,7]: d[i,j]=25
A_bar=[];A=[]
for i in T:
    for j in T:
        if et[i]+d[i,j]<=bt[j]: A_bar.append((i,j)); A.append((i,j))
for j in T: A.append(('s',j))
for i in T: A.append((i,'t'))
for j in V:
    if ('s',j) in A: d['s',j]=17
for i in V:
    if (i,'t') in A: d[i,'t']=15
c_bar={(i,j):0 for (i,j) in A}; c_0=200
for (i,j) in A_bar:
    c_bar[i,j]=d[i,j]
for (i,j) in A:
```

```
    if i=='s': c_bar[i,j]=d[i,j]+c_0
    if j=='t': c_bar[i,j]=d[i,j]
# variables
x = LpVariable.dicts("x", ((i, j) for (i,j) in A),lowBound=0,upBound=
    None,cat='Integer')
# objective function
model += sum(c_bar[i,j]*x[i,j] for (i,j) in A)
# constraints:
for i in T:
    model += sum(x[i,j] for j in V if (i,j) in A) == 1
for j in T:
    model += sum(x[i,j] for i in V if (i,j) in A) == 1
# optimizing
solver = pulp.PULP_CBC_CMD(mip=True, msg=True, timeLimit=None)
model.solve(solver)
print ("Status:", LpStatus[model.status])
for v in model.variables():
    print (v.name, "=", v.varValue)
print ("Optimal Solution = ", value(model.objective))
```

Solving this problem yields solution of Table 11.5 with objective function score 1160. Note that the objective function score is 1000 units higher than the objective function score of the Asymmetric Assignment formulation because now we consider an extra cost $c_0 = 200$ units for every deployed vehicle. This already tells us that our optimal solution will require $1000/200 = 5$ vehicles to cover the 7 trips. The schedule of each vehicle as obtained by Table 11.5 is:

- $s \to 1 \to t$
- $s \to 2 \to 6 \to t$
- $s \to 3 \to t$
- $s \to 4 \to 7 \to t$
- $s \to 5 \to t$

The advantage of the *Assignment Problem* formulation compared to the *Minimal Decomposition* formulation is that we can explicitly model the operational costs of vehicles. More importantly, the *Assignment Problem* formulation of SD-VSP in (11.10)–(11.13) has a *totally unimodular* constraint matrix and its optimal solution is equivalent to the solution of the linear program where the integrality constraints (11.13), that were also expressed as $\mathbf{x} \in \mathbb{Z}_{\geq 0}^{|A|}$, are replaced by $\mathbf{x} \in \mathbb{R}_{\geq 0}^{|A|}$.

11.1.1.3 SD-VSP: Network Flow Formulation for Bounded Fleet Size

In the *Minimal Decomposition* formulation and the *Assignment Problem* formulation of SD-VSP the objective is to reduce the number of vehicles required to cover the scheduled trips in T. Both formulations, however, cannot deal with problems where the number of available vehicles is fixed or there is an upper bound, $p \in \mathbb{Z}_+$, reflecting

Arc $a \in A$	x_a
$(s, 1)$	1
$(s, 2)$	1
$(s, 3)$	1
$(s, 4)$	1
$(s, 5)$	1
$(s, 6)$	0
$(s, 7)$	0
$(1, t)$	1
$(1, 6)$	0
$(1, 7)$	0
$(2, t)$	0
$(2, 6)$	1
$(2, 7)$	0
$(3, t)$	1
$(3, 7)$	0
$(4, t)$	0
$(4, 7)$	1
$(5, t)$	1
$(6, t)$	1
$(7, t)$	1

Table 11.5 Optimal solution **x**

the maximum number of available vehicles. To address these cases, one can use a *Network Flow* formulation for the SD-VSP. This formulation was used in the work of Bodin [5] for fixed fleet sizes and it can be adapted to consider an upper bound on the fleet size, such as in the work of Daduna and Paixão [6]. It was first devised by Dantzig and Fulkerson [7] with the aim of minimizing the number of tankers required to meet a fixed transportation schedule.

We present here the Network Flow formulation of Bodin adapted to consider an upper bound on the fleet size. This formulation considers a digraph $G = (V, A)$ with a set of vertices V representing each trip task $i \in T$ and the *source* and *sink* vertices s and t, where both s and t are at the same physical location (the location of the depot) and represent the start and the end of a vehicle schedule. That is, $V = \{T \cup \{s, t\}\}$. The set of directed edges of the digraph A is defined as follows:

$$A = \{t, s\} \cup \{(s, j) \text{ for all } j \in T\} \cup \{(i, t) \text{ for all } i \in T\} \cup$$
$$\{(i, j) \text{ for all } i \in T, j \in T : e_i^t + d_{ij} \leq b_j^t\}$$

That is, set A contains all *pull-out* arcs (s, j) that start from the depot and connect it to a trip $j \in T$, all *pull-in* arcs (i, t) that represent the return to the depot t after performing trip $i \in T$, all *compatible* arcs $\bar{A} = \{(i, j) \text{ for all } i \in T, j \in T : e_i^t + d_{ij} \leq b_j^t\}$ which indicate that tasks (trips) i and j can be performed one after the other, and arc (t, s) which indicates the use of an extra vehicle and makes network G cyclic. The cost of each arc $(i, j) \in A$ is:

- $c_{ij} = d_{ij}$ if $(i, j) \in \bar{A}$, where d_{ij} is the deadhead time from i to j
- $c_{sj} = d_{sj}$ for all $j \in T$, where d_{sj} is the deadhead time from s to j
- $c_{it} = d_{it}$ for all $i \in T$, where d_{it} is the deadhead time from i to t
- $c_{ts} = c_0$ where c_0 is the cost of adding an extra vehicle.

Each path that starts from s and ends at t in network G represents the schedule of a single vehicle. Because we can also use arc (t, s) we will eventually generate a number of circuits and the objective of the network flow formulation is to find the minimum cost set of circuits in G that cover each trip. The decision variables are x_{ij} for all $(i, j) \in A$ where x_{ij} is a positive integer number indicating the flow in arc (i, j). The upper bound to the arc leading back, (t, s), is set to the upper bound p that represents the number of available vehicles. Thus, $x_{ts} \leq p$ and we cannot generate more than p circuits. The minimum cost flow problem is then defined as follows.

SD-VSP—Network Flow formulation

$$\min_{x} \sum_{(i,j)\in A} c_{ij} x_{ij} \tag{11.14}$$

$$\text{s.t.} \sum_{i:(i,j)\in A} x_{ij} - \sum_{i:(j,i)\in A} x_{ji} = 0 \qquad \forall j \in V \tag{11.15}$$

$$\sum_{i:(i,j)\in A} x_{ij} = 1 \qquad \forall j \in T \tag{11.16}$$

$$x_{ts} \leq p \tag{11.17}$$

$$x_{ij} \in \mathbb{Z}_+ \qquad \forall (i, j) \in A \tag{11.18}$$

In the network flow formulation, constraints (11.16) ensure that each trip $j \in T$ is covered exactly once and constraints (11.17) that we will not deploy more than p vehicles.

Let us consider the implementation of the Network Flow formulation to the example of Table 11.1 that has $|T| = 7$ trips. We consider a maximum number of available vehicles $p = 6$. Using the network flow formulation for this problem results in the network G of Fig. 11.4.

Consider costs $c_0 = 200$, $c_{it} = 15$ for any $i \in V$, $c_{sj} = 17$ for any $j \in V$, and c_{ij} values equal to the deadhead costs presented in Table 11.3. The optimal solution is

Fig. 11.4 Resulting digraph
$G = (V, A)$ when using the
network flow formulation

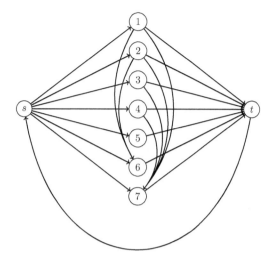

the following five vehicle schedules with a cost of 1160 units expressed in the form
of five circuits:

- $s \rightarrow 1 \rightarrow t \rightarrow s$
- $s \rightarrow 2 \rightarrow 6 \rightarrow t \rightarrow s$
- $s \rightarrow 3 \rightarrow 7 \rightarrow t \rightarrow s$
- $s \rightarrow 4 \rightarrow t \rightarrow s$
- $s \rightarrow 5 \rightarrow t \rightarrow s$

Notice that the vertex sequence at each circuit does not include repeated vertices,
except for the first and the last vertex in the sequence. Thus, they are simple circuits
(cycles).

Practitioner's Corner
SD-VSP Network Flow formulation in Python 3

One can use PuLP to program and the CBC solver to solve the Network Flow formulation of the SD-VSP as follows.

```
import pulp
from pulp import *
model = LpProblem("SD_VSP_Network_Flow_formulation", LpMinimize)
T=(1,2,3,4,5,6,7) #trips
V=('s',1,2,3,4,5,6,7,'t')
bt={1:0,2:10,3:20,4:30,5:15,6:42,7:60}
et={1:bt[1]+30,2:bt[2]+30,3:bt[3]+30,4:bt[4]+30,5:bt[5]+25,6:bt
    [6]+25,7:bt[7]+25}
d={(i,j):0 for i in V for j in V}
for i in T:
    if i in [1,2,3,4]:
        for j in [1,2,3,4]: d[i,j]=20
        for j in [5,6,7]: d[i,j]=0
```

```
    if i in [5,6,7]:
        for j in [1,2,3,4]: d[i,j]=0
        for j in [5,6,7]: d[i,j]=25
for j in T: d['s',j]=17
for i in T: d[i,'t']=15
A_bar=[];A=[]
for i in T:
    for j in T:
        if et[i]+d[i,j]<=bt[j]: A_bar.append((i,j)); A.append((i,j))
for j in T: A.append(('s',j))
for i in T: A.append((i,'t'))
A.append(('t','s'))
c={(i,j):0 for (i,j) in A}
p=6; c_0=200
for (i,j) in A:
    if (i,j) in A_bar: c[i,j]=d[i,j]
    if i=='s': c[i,j]=d[i,j]
    if j=='t': c[i,j]=d[i,j]
c['t','s']=c_0
# variables
x = LpVariable.dicts("x", ((i, j) for (i,j) in A),lowBound=0,upBound=
    None,cat='Integer')
# objective function
model += sum(c[i,j]*x[i,j] for (i,j) in A)
# constraints:
for j_it in V:
    model += sum(x[i,j] for (i,j) in A if j==j_it) - sum(x[j,i] for (
    j,i) in A if j==j_it) == 0
for j_it in T:
    model += sum(x[i,j] for (i,j) in A if j==j_it) == 1
model += x['t','s'] <= p
# optimizing
solver = pulp.PULP_CBC_CMD(mip=True, msg=True, timeLimit=None)
model.solve(solver)
print ("Status:", LpStatus[model.status])
for v in model.variables():
    print (v.name, "=", v.varValue)
print ("Optimal Solution = ", value(model.objective))
```

Solving this problem yields solution with objective function score 1160 and flow $x_{t,s}^* = 5$ indicating that we will need to deploy 5 vehicles. The solution produces the following five vehicle schedules:

- $s \to 1 \to t$
- $s \to 2 \to 6 \to t$
- $s \to 3 \to 7 \to t$
- $s \to 4 \to t$
- $s \to 5 \to t$

Before closing, we note that the SD-VSP with a fixed number of vehicles, known in literature as p-VSP, can be solved by setting the lower and upper bound for the flow of arc (t, s) to p.

11.1.2 Multiple-Depot Vehicle Scheduling Problem

11.1.2.1 MDVSP with Fixed Trip Times

The Multiple-Depot Vehicle Scheduling Problem (MDVSP) is an extension of the SD-VSP in the case of having more than one depot. In MDVSP we have a set M of depots, where at each depot $m \in M$ is stationed a fixed number of vehicles. The MDVSP with fixed trip start times can be defined as follows.

Multiple-Depot Vehicle Scheduling Problem
Consider a set of lines with a set of timetabled trips with fixed travel (departure and arrival) times and start and end locations as well as traveling times between all pairs of end stations. Consider also a set of vehicles with (possibly different) capacities that are located at M different depot locations. Find an assignment of trips to vehicles such that:

- each trip is covered exactly once,
- each vehicle performs a feasible sequence of trips,
- each vehicle is stationed to a specific depot $m \in M$ and it starts/returns to that depot when completing its vehicle schedule,
- the overall costs are minimized.

We can use the models developed for the SD-VSP also for the MDVSP after extending them. Because the MDVSP is NP-Hard [8], there are no polynomial time optimal algorithms for this problem. A very common formulation for the MDVSP is the *Multi-commodity Flow* formulation which is an extension of the *Network Flow* formulation for the SD-VSP. This formulation has been used in several studies (see [5, 8–10]). We present here the multi-commodity flow formulation of Bodin [5] presented in 1983. In this formulation, we use $|K|$ directed graphs $G_k = (V_k, A_k)$, where each digraph G_k is associated to a depot $k \in K$. In more detail, the set of vertices $V_k = T \cup \{s_k\} \cup \{t_k\}$ refers to a specific depot $k \in K$ where s_k is the source and t_k is the sink. We also define arcs A_k for each depot k:

$$A_k = \{(s_k, j) \text{ for all } j \in T\} \cup \{(i, t_k) \text{ for all } i \in T\} \cup \{(t_k, s_k)\} \cup \bar{A}$$

where $\bar{A} = \{(i, j) \text{ for all } i \in T, j \in T : e_i^t + d_{ij} \leq b_j^t\}$ is the set of compatible trips and it is independent of the depot. Reckon that pair (i, j) is compatible if the end time of trip i, e^t_i, plus the deadhead time from i to j, d_{ij}, is less than or equal to the start time of trip j, b^t_j.

The number of vehicles p_k stationed at the k-th depot is the upper bound of available vehicles for that depot. We use a three-index formulation with a set of flow variables x^k_{ij} where x^k_{ij} is a positive integer indicating the flow in arc $(i, j) \in A_k$. The arc costs are defined as:

- $c_{ij} = d_{ij}$ if $(i, j) \in \bar{A}$, where d_{ij} is the deadhead time from i to j
- $c_{s_k j} = d_{s_k j}$ for all $j \in T$, where $d_{s_k j}$ is the deadhead time from s_k to j
- $c_{i t_k} = d_{i t_k}$ for all $i \in T$, where $d_{i t_k}$ is the deadhead time from i to t_k
- $c_{t_k s_k} = c_0$ where c_0 is the cost of adding an extra vehicle

The multi-commodity flow formulation of the MDVSP is provided below.

MDVSP—Multi-commodity Flow three-index formulation

$$\min_{\mathbf{x}} \sum_{k \in K} \sum_{(i,j) \in A_k} c^k_{ij} x^k_{ij} \tag{11.19}$$

$$\text{s.t.} \sum_{i:(i,j) \in A_k} x^k_{ij} - \sum_{i:(j,i) \in A_k} x^k_{ji} = 0 \qquad \forall j \in V_k, \forall k \in K \tag{11.20}$$

$$\sum_{j:(s_k,j) \in A_k} x^k_{s_k,j} \leq p_k \qquad \forall k \in K \tag{11.21}$$

$$\sum_{k \in K} \sum_{i:(i,j) \in A_k} x^k_{ij} = 1 \qquad \forall j \in T \tag{11.22}$$

$$x^k_{ij} \in \mathbb{Z}_+ \qquad \forall k \in K, \forall (i, j) \in A_k \tag{11.23}$$

Constraints (11.20) are the flow conservation constraints that create circuits in each digraph G_k. Constraints (11.21) ensure that we will not use more than p_k vehicles from each depot. Importantly, constraints (11.22) ensure that each trip $j \in T$ will be covered exactly once. Constraints (11.22) make the critical difference between the multi-commodity flow formulation and the network flow formulation of the SD-VSP.

Let us consider the implementation of the Multi-commodity Flow formulation in the example of Table 11.1 that has $|T| = 7$ trips. We consider three depots $K = \{1, 2, 3\}$ where each depot has a maximum number of 2 available vehicles; $p_k = 2$ for all $k \in K$. In addition, the deadhead times from depot $k = 1$ to/from all $i \in T$ are 15 min, from depot $k = 2$ to/from all $i \in T$ are 17 min and from depot $k = 3$ to/from all $i \in T$ are 19 min. The representation of this problem can be done with three digraphs. The solution to this problem is five vehicle schedules:

- vehicle of depot 1: $s_1 \rightarrow 2 \rightarrow t_1$
- vehicle of depot 1: $s_1 \rightarrow 4 \rightarrow t_1$

- vehicle of depot 2: $s_2 \to 1 \to 6 \to t_2$
- vehicle of depot 2: $s_2 \to 3 \to 7 \to t_2$
- vehicle of depot 3: $s_3 \to 5 \to t_3$

Practitioner's Corner
MDVSP Multi-commodity Flow formulation in Python 3

One can use PuLP to program and the CBC solver to solve the Multi-commodity
Flow formulation of the MDVSP as follows.

```
import pulp
from pulp import *
model = LpProblem("MDVSP_Multicommodity_Flow_formulation", LpMinimize
    )
T=(1,2,3,4,5,6,7) #trips
K=(1,2,3)
V={k:('s_1',1,2,3,4,5,6,7,'t_1') for k in K}
V[2]=('s_2',1,2,3,4,5,6,7,'t_2')
V[3]=('s_3',1,2,3,4,5,6,7,'t_3')
bt={1:0,2:10,3:20,4:30,5:15,6:42,7:60}
et={1:bt[1]+30,2:bt[2]+30,3:bt[3]+30,4:bt[4]+30,5:bt[5]+25,6:bt
    [6]+25,7:bt[7]+25}
d_E={(i,j):0 for i in T for j in T}
for i in T:
    if i in [1,2,3,4]:
        for j in [1,2,3,4]: d_E[i,j]=20
        for j in [5,6,7]: d_E[i,j]=0
    if i in [5,6,7]:
        for j in [1,2,3,4]: d_E[i,j]=0
        for j in [5,6,7]: d_E[i,j]=25
d={(i,j,k):0 for k in K for i in V[k] for j in V[k]}
for j in T: d['s_1',j,1]=15; d['s_2',j,2]=17; d['s_3',j,3]=19
for i in T: d[i,'t_1',1]=15; d[i,'t_2',2]=17; d[i,'t_3',3]=19
for i in T:
    for j in T:
        for k in K: d[i,j,k]=d_E[i,j]
A_bar=[]
for i in T:
    for j in T:
        if et[i]+d_E[i,j]<=bt[j]: A_bar.append((i,j))
A1_a=[(V[1][0],j) for j in T]; A1_b=[(i,V[1][8]) for i in T]; A1_c=[(
    V[1][8],V[1][0])]
A2_a=[(V[2][0],j) for j in T]; A2_b=[(i,V[2][8]) for i in T]; A2_c=[(
    V[2][8],V[2][0])]
A3_a=[(V[3][0],j) for j in T]; A3_b=[(i,V[3][8]) for i in T]; A3_c=[(
    V[3][8],V[3][0])]
A={1:A1_a+A1_b+A_bar+A1_c,2:A2_a+A2_b+A_bar+A2_c,3:A3_a+A3_b+A_bar+
    A3_c}
c={(i,j,k):0 for k in K for (i,j) in A[k]}
p={1:2,2:2,3:2}; c_0=200
for k in K:
```

```
    for (i,j) in A[k]:
        if (i,j) in A_bar: c[i,j,k]=d[i,j,k]
        if i==V[k][0]: c[i,j,k]=d[i,j,k]
        if j==V[k][8]: c[i,j,k]=d[i,j,k]
        if i==V[k][8] and j==V[k][0]: c[i,j,k]=c_0
# variables
x = LpVariable.dicts("x", ((i, j, k) for k in K for (i,j) in A[k]),
    lowBound=0,upBound=None,cat='Integer')
# objective function
model += sum(sum(c[i,j,k]*x[i,j,k] for (i,j) in A[k]) for k in K)
# constraints:
for k in K:
    for j_it in V[k]: model += sum(x[i,j,k] for (i,j) in A[k] if j==
    j_it) - sum(x[j,i,k] for (j,i) in A[k] if j==j_it) == 0
for j_it in T: model += sum(sum(x[i,j,k] for (i,j) in A[k] if j==j_it
    ) for k in K) == 1
for k in K: model += sum(x[i,j,k] for (i,j) in A[k] if i==V[k][0]) <=
    p[k]
# optimizing
solver = pulp.PULP_CBC_CMD(mip=True, msg=True, timeLimit=None)
model.solve(solver)
print ("Status:", LpStatus[model.status])
for v in model.variables():
    if v.varValue>0:
        print (v.name, "=", v.varValue)
print ("Optimal Solution = ", value(model.objective))
```

Solving this problem yields solution with objective function score 1166. The solution produces the following five vehicle schedules:

- vehicle of depot 1: $s_1 \to 2 \to t_1$
- vehicle of depot 1: $s_1 \to 4 \to t_1$
- vehicle of depot 2: $s_2 \to 1 \to 6 \to t_2$
- vehicle of depot 2: $s_2 \to 3 \to 7 \to t_2$
- vehicle of depot 3: $s_3 \to 5 \to t_3$

It is important to note that the same formulation can be used for solving the SD-VSP with multiple vehicle types (SD-VSPMVT). SD-VSPMVT has the same problem definition as the SD-VSP but we now have a set of vehicle types K where each vehicle type $k \in K$ has a different capacity. For each task (trip) the set of vehicles that may service it is specified. In SD-VSPMVT each pair $\{s_k, t_k\}$ corresponds to a vehicle type k and flow variables x^k_{ij} refer to covering arc (i, j) with vehicle type k. Similarly, p_k is the maximum number of available vehicles of vehicle type k.

11.1.2.2 MDVSP with Time Windows

The MDVSP with time windows is an extension of the MDVSP that considers trip start and end times within time intervals, instead of fixed trip start and end times. This consideration is favorable in cases of travel time uncertainty.

MDVSPTW is NP-Hard as it includes MDVSP as a special case. In MDVSPTW each task is restricted to begin within a prescribed time interval and vehicles are supplied by different depots. In more detail:

1. Every trip has to be assigned to exactly one vehicle.
2. Every vehicle is associated with a single depot.
3. Each vehicle starts from the depot and returns to it only once, e.g., at the end of its daily schedule.
4. The starting times and ending times of trips are not fixed, but they can take values within a time range (time window). Note that this becomes particularly important when accounting for travel time uncertainty.

The main addition of the MDVSPTW is that starting times and ending times of trips are not fixed, but they can take values within a time range (time window). This problem is formulated in Desaulniers et al. [11] in 1998 and was initially proposed by Bianco et al. [12] in 1995.

We will present here the MDVSPTW formulation of Desaulniers et al. [11] as a multi-commodity flow, nonlinear mixed-integer programming problem. This problem seeks to find the minimum cost exact cover of all scheduled trips during the daily operations.

Sets: Let K be the set of available vehicles and $G_k = (V_k, A_k)$ the digraph associated with vehicle k, where V_k are the vertices and A_k the arcs. The source and sink vertices associated with the depot housing vehicle k are denoted as s_k and t_k, respectively. These vertices indicate the start and the end of the schedule assigned to vehicle $k \in K$. In our network representation, each vertex corresponds to a *scheduled trip* with a specific start and end stop and each arc indicates a trip-to-trip or trip-to-depot transition. The set of all possible trips (task vertices) is T, where task $i \in T$ corresponds to a scheduled trip with a specific start and end stop that needs to be served by a vehicle $k \in K$.

Parameters: Each vertex $i \in T$ is associated with a time window $[l_i, u_i]$ indicating the time period within which we should perform task i. We have also a set of *empty, start, end* and *inter-task* arcs A_k. Arc (s_k, i), where $i \in V_k$, is a start arc. Arc (i, t_k) is an end arc, arc (s_k, t_k) is an empty arc, and arc (i, j), where $i \in V_k \setminus \{s_k, t_k\}$ and $j \in V_k \setminus \{s_k, t_k\}$ and $l_i + t_{ij} \le u_j$, is an inter-task arc. The empty arc (s_k, t_k) is used when vehicle k is not required in the solution. The elapsed time on arc (i, j) that connects vertices i and j is denoted as t_{ij}. This time is equal to the duration of task i plus the travel time between the end location of task i and the start location of task j. Clearly, $t_{s_k t_k} = 0$. In addition, $t_{s_k, j}$ corresponds to the travel time between s_k and the start location of $j \in V_k \setminus \{s_k, t_k\}$ and t_{i, t_k} corresponds to the travel time between the end location of $i \in V_k \setminus \{s_k, t_k\}$ and depot t_k. It is also worth noting that this elapsed time is not vehicle-specific.

The unit waiting cost of a vehicle is λ. We have an indicator parameter I_{ij}^k, where $I_{ij}^k = 1$ if (i, j) is an *inter-task* arc and 0 otherwise. Finally, the cost constant component of performing task j after task i without considering any potential delay (task j starts immediately after i without time delays) is b_{ij}^k.

Variables: The variables of the MDVSPTW problem are:

- x_{ij}^k: binary flow variables, where $x_{ij}^k = 1$ if vehicle k uses arc $(i, j) \in A_k$ and 0 otherwise.
- T_i^k: a time variable associated with each vertex $i \in V_k$. $T_{s_k}^k$ indicates the departure time from the depot, $T_{t_k}^k$ the arrival time at the depot, and T_i^k with $i \in V^k \setminus \{s_k, t_k\}$ the time that service begins at node i.
- c_{ij}^k: cost of performing task j immediately after performing task i. For instance, $c_{s_k, j}^k$ is the cost of performing task j after starting from the depot and c_{i, t_k}^k is the cost of returning to the depot after performing task i.

The cost of performing task j immediately after task i by vehicle k is:

$$c_{ij}^k = b_{ij}^k + I_{ij}^k \lambda (T_j^k - T_i^k - t_{ij}) \tag{11.24}$$

where the waiting cost $I_{ij}^k \lambda (T_j^k - T_i^k - t_{ij})$ is added to the cost constant component b_{ij}^k of performing task j after task i without considering any potential delay. The MDVSPTW can be formulated as the following three-index, mixed-integer nonlinear, multi-commodity network flow model.

MDVSPTW—Multi-commodity Flow three-index formulation

$$\min \sum_{k \in K} \sum_{(i,j) \in A_k} \left(b_{ij}^k + I_{ij}^k \lambda (T_j^k - T_i^k - t_{ij}) \right) x_{ij}^k \tag{11.25}$$

$$\text{subject to: } \sum_{k \in K} \sum_{j:(i,j) \in A_k} x_{ij}^k = 1 \qquad \forall i \in T \tag{11.26}$$

$$\sum_{j:(s_k, j) \in A^k} x_{s_k, j}^k = \sum_{i:(i, t_k) \in A^k} x_{i, t_k}^k = 1 \qquad \forall k \in K \tag{11.27}$$

$$\sum_{i:(i,j) \in A^k} x_{ij}^k - \sum_{i:(j,i) \in A^k} x_{ji}^k = 0 \qquad \forall k \in K \ \forall j \in T \tag{11.28}$$

$$x_{ij}^k (T_i^k + t_{ij}) \leq x_{ij}^k T_j^k \qquad \forall k \in K \ \forall (i, j) \in A_k \tag{11.29}$$

$$l_i \leq T_i^k \leq u_i \qquad \forall k \in K \ \forall i \in V_k \tag{11.30}$$

$$x_{ij}^k \in \{0, 1\} \qquad \forall k \in K \ \forall (i, j) \in A_k \tag{11.31}$$

The objective function is nonlinear and seeks to minimize the total cost, including potential delays when waiting between activities i and j. Constraints (11.26) ensure that each task (trip) is covered exactly once. Satisfying constraints (11.26) returns an *exact cover* of the scheduled trips—where each trip is assigned to exactly one vehicle schedule. Constraints (11.27) and (11.28) are network constraints and indicate that one unit of flow must be sent from the source node to the sink node in each network G_k. Therefore, one schedule (empty or not) will be assigned to each vehicle. Constraints (11.29) ensure compatibility between flow and travel time variables. This relation states that the start of service time at vertex j (or the arrival time at the depot if $j = s_k$) must be greater than or equal to the start of service time at node i (or the departure time from the depot if $i = t_k$) plus the minimal time elapsed on arc (i, j) whenever the flow on this arc is positive. Finally, the time window restrictions are defined by constraints (11.30).

The mixed-integer nonlinear program in Eqs. (11.25)–(11.31) seeks to find the minimum cost *exact cover* S^*, where an *exact cover* of the trip set T is the set of feasible vehicle schedules that contain each trip $i \in T$ in exactly one of the vehicle schedules. The formulation in (11.25)–(11.31) is nonlinear because of the nonlinearity of the objective function and the nonlinear constraints (11.29) that include multiplication of variables. As discussed in Desaulniers et al. [11], the MDVSPTW can be transformed into a MILP by making the following changes:

- consider the cost variables $c_{ij}^k = b_{ij}^k + I_{ij}^k \lambda (T_j^k - T_i^k - t_{ij})$ as parameters $c_{ij}^k = b_{ij}^k + I_{ij}^k \lambda \max\{0, l_j - u_i - t_{ij}\}$ by replacing the exact waiting costs $T_j^k - T_i^k - t_{ij}$ with the minimum waiting costs $\max\{0, l_j - u_i - t_{ij}\}$
- replace time compatibility constraints $x_{ij}^k (T_i^k + t_{ij}) \le x_{ij}^k T_j^k$ by the more conservative requirement $x_{ij}^k (u_i + t_{ij}) \le x_{ij}^k l_j$ or introduce variables $\sigma_{ij}^k \in \mathbb{R}$ and a very big positive number M and request that:

$$T_i^k + t_{ij} - T_j^k + \sigma_{ij}^k \le 0 \tag{11.32}$$

$$\sigma_{ij}^k \le M(1 - x_{ij}^k) \tag{11.33}$$

$$\sigma_{ij}^k \ge -M(1 - x_{ij}^k) \tag{11.34}$$

The mixed-integer linear version of the MDVSPTW is presented below.

MDVSPTW—MILP formulation

$$\min \sum_{k \in K} \sum_{(i,j) \in A_k} \left(b_{ij}^k + I_{ij}^k \lambda \max\{0, l_j - u_i - t_{ij}\} \right) \tag{11.35}$$

$$\text{subject to:} \sum_{k \in K} \sum_{j:(i,j) \in A_k} x_{ij}^k = 1 \qquad \forall i \in T \tag{11.36}$$

$$\sum_{j:(s_k,j) \in A^k} x_{s_k,j}^k = \sum_{i:(i,t_k) \in A^k} x_{i,t_k}^k = 1 \qquad \forall k \in K \tag{11.37}$$

$$\sum_{i:(i,j)\in A^k} x_{ij}^k - \sum_{i:(j,i)\in A^k} x_{ji}^k = 0 \qquad \forall k \in K \ \forall j \in T$$

$$\tag{11.38}$$

$$T_i^k + t_{ij} - T_j^k + \sigma_{ij}^k \le 0 \qquad k \in K \ \forall (i, j) \in A_k$$

$$\tag{11.39}$$

$$\sigma_{ij}^k \le M(1 - x_{ij}^k) \qquad k \in K \ \forall (i, j) \in A_k$$

$$\tag{11.40}$$

$$\sigma_{ij}^k \ge -M(1 - x_{ij}^k) \qquad k \in K \ \forall (i, j) \in A_k$$

$$\tag{11.41}$$

$$l_i \le T_i^k \le u_i \qquad \forall k \in K \ \forall i \in V_k$$

$$\tag{11.42}$$

$$x_{ij}^k \in \{0, 1\} \qquad \forall k \in K \ \forall (i, j) \in A_k$$

$$\tag{11.43}$$

11.2   Crew Scheduling

Crew scheduling (also known as rostering) is the last planning activity in tactical planning and follows the frequency settings, timetabling, and vehicle scheduling steps. In crew scheduling, we need to split the vehicle schedules into pieces (duties) that form tasks which can be performed by crew members given constraints on the number of available drivers and the restrictions of labor contracts.

There are several approaches to solve the crew scheduling problem and they greatly differ based on the crew scheduling problem's definition, which is not standardized. Unlike vehicle scheduling problems, the crew scheduling problem requirements can change vastly based on the requirements of the service provider and the needs of real-life applications [13]. Thus, it requires more effort from modelers to produce a model that can translate the expressed problem definition into a mathematical formulation.

Typically, crew scheduling problems can be formulated as graph problems where we need to cover the vertices of an acyclic graph by means of a number of *minimum cost vertex disjoint paths* that satisfy a set of constraints [13]. In this graph, the existence of an arc (i, j) means that task j can be performed immediately after task i by the same crew.

11.2.1 Crew Scheduling as a Fixed Job Scheduling Problem with Spread-Time Constraints

A special case of many crew scheduling problems is the NP-Hard fixed job scheduling problem with spread-time constraints introduced by Fischetti et al. [14, 15] in 1987 and 1989. The use of spread-time constraints makes the fixed job scheduling problem similar to the crew scheduling problem of public transport drivers. The fixed job scheduling problem with spread-time constraints is defined as follows.

Fixed job scheduling problem with spread-time constraints [14]
Consider a set T of n tasks $T_1, \ldots, T_n$. Each task T_j requires processing without interruption from a given *release time* r_j to a given deadline d_j. Consider an unlimited number of *identical parallel processors* (drivers) each of which can process at most one task at a time. Each driver is only available for s time units from the release time of the earliest task assigned to him/her. That is, any pair of tasks (T_j, T_k) assigned to the same driver must satisfy $d_j - r_k \leq s$. Find a crew schedule to execute all tasks using the minimum number of drivers.

Before proceeding further, we present the broadly acceptable terminology definitions of Desrochers and Soumis [16] presented in 1989.

Definition 11.1 *Vehicle block* or simply *block* is a vehicle schedule starting at the depot and going back to the depot.

Definition 11.2 *Relief point* is a location along a block where a new driver can replace the previous driver in a vehicle.

Definition 11.3 *Relief time* is a time during the day when a public transport vehicle passes a relief point.

Definition 11.4 *Task* is the portion of a block determined by two consecutive relief points of the same public transport vehicle.

Definition 11.5 *Increment of work* is a portion of work between two adjacent relief times of the same public transport vehicle.

From the above definitions, it follows that the j-th increment of works corresponds to task T_j and the associated relief times of the j-th task correspond to r_j and d_j, respectively. Let us consider, without loss of generality, that $r_1 \leq r_2 \leq \cdots \leq r_n$. A driver performing the j-th increment of work will work in time interval $(r_j, d_j]$. For every increment of work j, we can define set I_j with all incompatible increments of work $k > j$ to j:

$$I_j := \{k > j \ : \ r_k < d_j \text{ or } d_k - r_j > s\}$$

Fig. 11.5 Schematic of
release time, r_j, and
deadline, d_j, of each task T_j
ordered such that
$r_1 \leq r_2 \leq \cdots \leq r_5$

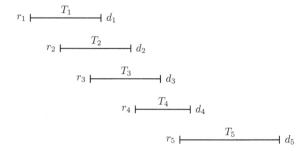

Two increments of work j and k cannot be assigned to the same driver if they overlap or violate the spread time constraint ($j \in I_k$ or $k \in I_j$). For instance, in the example of Fig. 11.5 task T_1 that corresponds to the 1st increment of work has incompatible increments of work $I_1 = \{2, 3\}$ corresponding to tasks T_2 and T_3.

Let m be the upper bound of the number of available drivers. Let binary variables $\mathbf{y} = [y_1, \ldots, y_m]^\mathsf{T}$ where $y_i = 1$ if driver i is used and $y_i = 0$ if not. Let also binary variables x_{ij} where $x_{ij} = 1$ if the j-th increment of work is assigned to driver i and 0 otherwise. Then, the fixed job scheduling problem with spread-time constraints is formulated as the following Vertex Coloring Problem.

Fixed job scheduling problem with spread-time constraints

$$\min_{\mathbf{y}} \sum_{i=1}^{m} y_i \tag{11.44}$$

$$\text{s.t. } x_{ij} \leq y_i \qquad \forall i \in \{1, \ldots, m\}, \forall j \in \{1, \ldots, n\} \tag{11.45}$$

$$\sum_{i=1}^{m} x_{ij} = 1 \qquad \forall j \in \{1, \ldots, n\} \tag{11.46}$$

$$x_{ij} + x_{ik} \leq 1 \qquad \forall i \in \{1, \ldots, m\}, \forall j \in \{1, \ldots, n-1\}, \forall k \in I_j \tag{11.47}$$

$$x_{ij} \in \{0, 1\} \qquad \forall i \in \{1, \ldots, m\}, \forall j \in \{1, \ldots, n\} \tag{11.48}$$

$$y_i \in \{0, 1\} \qquad \forall i \in \{1, \ldots, m\} \tag{11.49}$$

Constraints (11.45) ensure that if there exists even a single increment of work j for which the i-th driver is assigned to it ($x_{ij} = 1$), then y_i should be equal to 1 because we will use this driver during the workday. Constraints (11.46) ensure that each increment of work j is covered by exactly one driver. Constraints (11.47) do not allow the same driver to cover incompatible increments of work. In addition, the objective function minimizes the number of drivers and the formulation is a MILP.

We note that if a driver i is used, then this driver might perform more than one increment of work. All these increments of work (i.e., $(i, j), (j, k), \ldots$) performed by the same driver result in the so-called *duty* of the driver.

To see why the formulation corresponds to a Vertex Coloring Problem, we can introduce network $G = (V, A)$ with vertex set $V = \{1, \ldots, n\}$ and arc set $A = \{(j, k) : j \in V, k \in I_j\}$. Then, the optimal number of drivers is the minimum number of colors needed to color the vertices of G so that no two adjacent vertices are of the same color.

Practitioner's Corner
Fixed job scheduling formulation in Python 3

One can use PuLP to program and the CBC solver to solve the Fixed job scheduling problem. Consider the example of Fig. 11.5 with release times $\mathbf{r} = [0, 1, 2, 3.5, 5]^\mathsf{T}$ and deadline times $\mathbf{d} = [3, 4, 5, 6, 9]^\mathsf{T}$. There is also an upper bound of $m = 20$ drivers and a driver availability of $s = 8$ time units. Then, the crew scheduling problem is solved as follows.

```python
import pulp
from pulp import *
model = LpProblem("Fixed_Job_Scheduling_formulation", LpMinimize)
m=20 #available drivers
n=5 #number of increments of work (tasks)
r={1:0,2:1,3:2,4:3.5,5:5}
d={1:3,2:4,3:5,4:6,5:9}
s=8 #availability of each driver in time units

I=[[]];I[0].append('');I.append([])
for j in range(1,n+1):
    for k in range(1,n+1):
        if k>j and (r[k]<d[j] or d[k]-r[j]>s):
            I[j].append(k)
    I.append([])

# variables
x = LpVariable.dicts("x", ((i, j) for i in range(1,m+1) for j in
    range (1,n+1)),cat='Binary')
y = LpVariable.dicts("y", (i for i in range(1,m+1)),cat='Binary')
# objective function
model += sum(y[i] for i in range(1,m+1))
# constraints:
for i in range(1,m+1):
    for j in range(1,n+1):
        model += x[i,j] <= y[i]
for j in range(1,n+1):
    model += sum(x[i,j] for i in range(1,m+1)) == 1
for i in range(1,m+1):
    for j in range(1,n):
        for k in I[j]:
            model += x[i,j]+x[i,k]<=1
```

```
# optimizing
solver = pulp.PULP_CBC_CMD(mip=True, msg=True, timeLimit=None)
model.solve(solver)
print ("Status:", LpStatus[model.status])
for v in model.variables():
    if v.varValue >= 10E-7:
        print (v.name, "=", v.varValue)
print ("Optimal Solution = ", value(model.objective))
```

Based on the values of $\mathbf{r}$, $\mathbf{d}$ and s, we have the incompatibility sets $I_1 = \{2, 3, 5\}$, $I_2 = \{3, 4\}$, $I_4 = \{3\}$, $I_4 = \{5\}$, $I_5 = \emptyset$. Solving this problem yields solution with objective function score 3, resulting in using 3 crew members (drivers). The solution produces the following three driver duties:

- crew member 1 to serve work increments $2 \rightarrow 5$.
- crew member 2 to serve work increments $1 \rightarrow 4$.
- crew member 3 to serve work increment 3.

11.2.2 Crew Scheduling as a Set Partitioning Problem

The basic crew scheduling problem that seeks to minimize the total cost of driver duties is typically formulated as a set partitioning problem. We first provide a reminder of minimum weight set partitioning by considering the formulation:

$$\min \mathbf{c}^\mathsf{T}\mathbf{x}$$
$$\text{s.t. } \mathbf{Ax} = \mathbf{1} \qquad (11.50)$$
$$\mathbf{x} \in \{0, 1\}^n$$

where $\mathbf{A}$ is an $m \times n$ matrix of zeros and ones, $\mathbf{1}$ is an m-valued column vector and $\mathbf{c}$ is an n-valued column vector. This pure 0–1 linear programming problem is the *set partitioning problem*. If constraints $\mathbf{Ax} = \mathbf{1}$ are replaced by $\mathbf{Ax} \geq \mathbf{1}$ then this problem is the minimum weight *set covering problem* and if they are replaced by $\mathbf{Ax} \leq \mathbf{1}$ then this problem is the minimum weight *set packing problem*.

A commonly used set partitioning formulation was provided by Mingozzi et al. [13]. In this formulation, we consider a graph $G = (V, A)$ with vertices $V = \{0, 1 \dots, n, n+1\}$ where $|V| = n + 2$, the total number of tasks (increments of work) is n, and vertices 0 and $n + 1$ represent the same physical location: the depot where the crew members (drivers) start and finish their duties. We can express the set of tasks as $V' = V \setminus \{0, n + 1\}$. Each task has a start time r_i and end time d_i.

The set of arcs A contains all arcs $(0, j) \; \forall j \in V'$ and $(i, n+1) \; \forall i \in V'$. It also contains all compatible arcs (i, j) where j can be performed immediately after i. To determine the compatible arcs, let t_{ij} be the travel time from the end of task i to the start of task j, t_{0j} the travel time from the depot to the start of task j and t_{in+1} the travel time from the end of task i to the depot $n + 1$. Let also w_i be the working time to perform task $i \in V'$. Then, (i, j) is a compatible arc if $r_j \geq d_i + t_{ij}$.

Let us consider the example of Fig. 11.5 that has 5 tasks with $\mathbf{r} = [0, 1, 2, 3.5, 5]^\mathsf{T}$ and $\mathbf{d} = [3, 4, 5, 6, 9]^\mathsf{T}$. If we consider travel times t_{ij} equal to 0, this will result in the digraph representation G of Fig. 11.6.

For modeling purposes, we can set parameter values $r_0 = d_0 = 0$ and $r_{n+1} = d_{n+1} \geq d_j + t_{jn+1}$ for all $(j, n + 1) \in A$. We can also pre-order the tasks such that $r_1 \leq r_2 \leq \cdots \leq r_{n+1}$. Given the above, we can present the definition of a feasible duty for a crew member.

Definition 11.6 *Feasible duty* is a path $R = \{0, i_1, i_2, \ldots, i_h, n + 1\}$ in graph G with $i_1, \ldots, i_h \in V'$ for which:

- the total working time required by a driver to perform this duty is less that s, where s are the time units corresponding to his/her daily working time. That is, $\sum_{i \in R} w_i \leq s$.
- the total duration of duty R is less than $\bar{s}$, where $\bar{s}$ are the time units which correspond to his/her daily availability time: $t_{0i_1} + (d_{i_h} - r_{i1}) + t_{i_h n+1} \leq \bar{s}$.

The cost of a feasible duty is $c_R = t_{0i_1} + t_{i_1 i_2} + \cdots + t_{i_h n+1}$. We can also add simple constraints, such that introducing a *rest time*, ρ, for drivers between successive tasks. This will result in deleting arcs (i, j) from A if $r_j - d_i < \rho$ because they will not be compatible any longer.

Let m be the number of available drivers and $\bar{R}$ the set of all feasible duties. Let also $\bar{R}_i \subseteq \bar{R}$ be the set of indices of all feasible duties that cover task $i \in V'$. Then, we can introduce binary decision variables $\mathbf{x}$ where $x_l = 1$ if feasible duty $l \in \bar{R}$ is part of the optimal solution, and 0 otherwise. The crew scheduling problem can be expressed as a *set partitioning problem* where we strive to minimize the cost of duties required to cover all tasks. If we add the requirement that we want to use all available drivers, m, then the formulation is expressed as follows.

Fig. 11.6 Example of digraph $G = (V, A)$ for five tasks (increments of work) with $\mathbf{r} = [0, 1, 2, 3.5, 5]^\mathsf{T}$, $\mathbf{d} = [3, 4, 5, 6, 9]^\mathsf{T}$ and inter-task travel times equal to 0. Vertices 0 and 6 are the same physical location (depot)

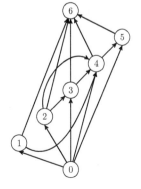

Crew Scheduling as a minimum weight Set Partitioning Problem

$$\min_{x} \sum_{l \in \bar{R}} c_l x_l \tag{11.51}$$

$$\text{s.t.} \sum_{l \in \bar{R}_i} x_l = 1 \qquad\qquad \forall i \in V' \tag{11.52}$$

$$x_l \in \{0, 1\} \qquad\qquad \forall l \in \bar{R} \tag{11.53}$$

Note that this formulation is the general formulation of the set partitioning problem described in (11.50). Constraints (11.52) ensure that every task is covered by exactly one duty. The objective function strives to minimize the duty costs.

If we do not strictly want each task to be assigned to exactly one duty but we might have more duties performing the same task, then we can replace constraints (11.52) by the inequality constraints:

$$\sum_{l \in \bar{R}_i} x_l \geq 1 \quad \forall i \in V' \tag{11.54}$$

This modification will turn our crew scheduling problem to a minimum weight *Set Covering Problem*. There are models that formulate the crew scheduling problem as a set covering problem, such as the one of Desrochers and Soumis [16].

We can additionally add constraint:

$$\sum_{l \in \bar{R}} x_l = m \tag{11.55}$$

This will ensure that we assign all crew members and we generate m duties. By replacing this constraint with the inequality constraint $\sum_{l \in \bar{R}} x_l \leq m$ we will not require to use all available crew members, but consider their availability as an upper bound.

The set partitioning and set covering formulations are pure binary linear programs. However, these problems are impractical to solve because of the large number of variables emerging from the possible number of feasible duties $\bar{R}$ related to graph G and the fact that both set partitioning and set covering are NP-Hard because their decision problem counterparts belong to the list of Karp's 21 NP-complete decision problems.

Fig. 11.7 Digraph
$G = (V, A)$ of the example's
problem when using an
assignment formulation

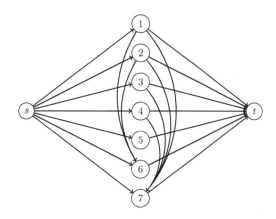

Table 11.6 Trip times

Trip	Start time	End time
1	08:00	08:15
2	08:15	08:30
3	08:40	08:50
4	08:50	09:00

Exercises

11.1 Single-depot Vehicle Scheduling Problem as a linear program

Consider the trip times presented in Table 11.1 that has $|T| = 7$ trips and the deadhead times of Table 11.3. The set of trips ordered in ascending start times is $T = \{1, 2, 5, 3, 4, 6, 7\}$. Consider deadhead times to/from the depot as $d_{sj} = 17$ min for all $j : (s, j) \in A$ and $d_{it} = 15$ min for all $i : (i, t) \in A$. In addition, the cost of using a vehicle is $c_0 = 200$ units. Idle times are not considered and the problem can be represented by the digraph in Fig. 11.7.

Prove that the Assignment Problem formulation of the SD-VSP can be written as a linear program and solve the SD-VSP of this exercise with simplex.

11.2 Single-depot Vehicle Scheduling with asymmetric assignment

Write a branch and bound algorithm and solve the asymmetric assignment problem using the input of the previous exercise.

11.3 SD-VSP digraphs

Construct the digraphs of the *asymmetric assignment* formulation, the *assignment* formulation, and the *network flow formulation* of the SD-VSP with trip start times and end times presented in Table 11.6. Consider that the deadhead time from the depot to the start location of every trip is 15 min and the deadhead time from the end location of every trip to the depot is 20 min. The deadhead time between any two trips is 5 min.

11.4 Single-depot vehicle scheduling problem with different vehicle types

Formulate a single-depot vehicle scheduling problem with different vehicle types using a multi-commodity flow formulation.

11.5 Multi-depot Vehicle Scheduling Problem with Time Windows

Consider the example of Table 11.1 that has $|T| = 7$ trips. Consider three depots $K = \{1, 2, 3\}$ where each depot has 2 available vehicles; $p_k = 2$ for all $k \in K$. In addition, the deadhead times from depot $k = 1$ to/from all $i \in T$ are 15 min, from depot $k = 2$ to/from all $i \in T$ are 17 min and from depot $k = 3$ to/from all $i \in T$ are 19 min.

(a) Solve this problem as a multi-depot vehicle scheduling problem without time windows.

(b) Describe which steps should be taken to solve it as a multi-depot vehicle problem with time windows, where each task (trip) i can be served within the time range [0, 300].

11.6 Crew Scheduling

Consider the release times $\mathbf{r} = [0, 1, 2, 3.5, 5]^\mathsf{T}$ and deadline times $\mathbf{d} = [3, 4, 5, 6, 9]^\mathsf{T}$ of each task in Fig. 11.5. Consider an upper bound of $m = 8$ drivers and a driver availability of $s = 8$ time units.

(a) Produce a digraph G that can represent this problem as a Vertex Coloring problem when using the fixed-job scheduling formulation and solve this problem to optimality.

(b) Justify why the fixed-job scheduling formulation is NP-Hard.

References

1. S. Bunte, N. Kliewer, Pub. Transp. **1**(4), 299 (2009)
2. J. Saha, J. Oper. Res. Soc. **21**(4), 463 (1970)
3. C.S. Orloff, Transp. Sci. **10**(2), 149 (1976)
4. R. Freling, A.P. Wagelmans, J.M.P. Paixão, Transp. Sci. **35**(2), 165 (2001)
5. L. Bodin, Comput. Oper. Res. **10**(2), 63 (1983)
6. J.R. Daduna, J.M. Pinto Paixão, Comput.-Aided Trans. Sched. 76–90 (1995)
7. G.B. Dantzig, D.R. Fulkerson, Naval Res. Logist. Quart. **1**(3), 217 (1954)
8. A.A. Bertossi, P. Carraresi, G. Gallo, Networks **17**(3), 271 (1987)
9. M. Forbes, J. Holt, A. Watts, Eur. J. Oper. Res. **72**(1), 115 (1994)
10. C.C. Ribeiro, F. Soumis, Oper. Res. **42**(1), 41 (1994)
11. G. Desaulniers, J. Lavigne, F. Soumis, Eur. J. Oper. Res. **111**(3), 479 (1998)
12. A. Mingozzi, L. Bianco, S. Ricciardelli, in *Computer-Aided Transit Scheduling* (Springer, 1995), pp. 145–172
13. A. Mingozzi, M.A. Boschetti, S. Ricciardelli, L. Bianco, Oper. Res. **47**(6), 873 (1999)
14. M. Fischetti, S. Martello, P. Toth, Oper. Res. **35**(6), 849 (1987)
15. M. Fischetti, S. Martello, P. Toth, Oper. Res. **37**(3), 395 (1989)
16. M. Desrochers, F. Soumis, Transp. Sci. **23**(1), 1 (1989)

Chapter 12
Tactical Planning of On-Demand and Shared Mobility Services

Abstract We have already explored common tactical planning approaches for setting the frequencies, timetables, vehicle and crew schedules of fixed-line services that follow a pre-determined set of stations/stops and lines (routes). Decisions at the tactical planning stage can be made even for lines that do not follow a fixed route (i.e., in case of demand responsive services or car pooling schemes). This chapter focuses on the tactical planning decision level of on-demand and shared mode services considering the problems of on-demand buses, organizing shared mobility services, the school bus problem, carpooling, and vehicle routing. These problems are special cases of the Traveling Salesman Problem, the Vehicle Routing Problem, and the Dial-a-Ride Problem presented in this chapter.

12.1 Introduction

Until now we have considered the tactical planning problems of fixed-line public transport services, which include the setting of frequencies, timetables, vehicle and crew schedules. The public transport ecosystem is undergoing, however, a rapid change where demand responsive public transport services and shared mobility request the development of different planning tools which can schedule vehicles operating on routes that are not fixed. Common problems in this category are:

- the *Traveling Salesman Problem* which plans the route of a single vehicle in such a way that the vehicle can start from a starting location, pick up passengers from different locations, and return them back to the starting point in the most cost-efficient manner.
- the *Capacitated Vehicle Routing Problem* which plans the routes of multiple vehicles with specific capacities in the most cost-efficient manner such that all passengers are picked up by exactly one vehicle and all vehicles start from and return to the same location.
- the *Dial-a-Ride Problem* which plans the most cost-efficient pickup and delivery routes of multiple vehicles where passenger pickup and delivery points are known in advance and there are time windows associated with the passengers' pickup and deliveries in addition to maximum ride time constraints.

These problems are at the heart of modern applications, such as carpooling and on-demand bus services. In the remainder of this section, we present their problem descriptions and common formulations.

12.2 Planning the Route of a Single Vehicle: Traveling Salesman Problem

We have already addressed the Traveling Salesman Problem (TSP) as the problem of finding the shortest tour of a salesman who visits different cities. The traveling salesman wishes to visit the cities and return to the starting point after visiting each city exactly once. In the context of public transportation, this problem is equivalent to the problem of finding the optimal route of a shared vehicle (i.e., van, bus) that starts from a specific location, picks up a set of passengers from different locations, and returns them back to its start location. This application emerges naturally when solving school bus scheduling problems or when using on-demand vehicles that collect passengers and return them to the station of a collector line (i.e., metro or train station). For instance, the bus school scheduling problem where a bus starts from the school, picks up students, and returns to the school is a TSP. The same holds for the on-demand public transport scheduling problem when a bus/van starts from a central station (metro station), collects passengers from their origins, and transports them back to the central station.

We have already expressed the TSP as a MILP using the MTZ formulation of the sub-tour elimination constraints. Here, we will discuss first its graph theory-related aspects. Let us consider a network $G = (N, A)$ where $N = \{1, 2, ..., n\}$ are the vertices with 1 being the start and end node of the tour and $2, 3, ...n$ the pickup locations of passengers. We also have set A of arcs, where each arc (i, j) has a cost c_{ij} related to the distance traveled from i to j. The TSP requires to find the *Hamiltonian cycle* in G with the minimum total cost, where the Hamiltonian cycle is defined below.

Definition 12.1 *Hamiltonian cycle* in an undirected or directed graph G is any cycle passing through each node $i \in N$ exactly once.

We have previously discussed the asymmetric TSP. Let us now discuss the symmetric TSP in an undirected graph with $c_{ij} = c_{ji}$. We set $c_{ii} = +\infty$. We also introduce binary variables where $x_{ij} = 1$ if arc (i, j) is part of the optimal tour and $x_{ij} = 0$ otherwise.

We can introduce then an *assignment-based* formulation where x_{ij} values will result in exactly one arc (i, j) outgoing from each vertex i and exactly one arc (i, j) incoming to each vertex j. Reckon that the *assignment problem* is a (minimum) weight bipartite perfect matching problem and it is expressed as follows.

$$\min_{\mathbf{x}} \mathbf{c}^{\mathsf{T}}\mathbf{x}$$

$$\text{s.t. } \sum_{i=1}^{n} x_{ij} = 1 \quad \forall j = 1, ..., n$$

$$\sum_{j=1}^{n} x_{ij} = 1 \quad \forall i = 1, ..., n \tag{12.1}$$

$$\mathbf{x} \in \{0, 1\}^{n \times n}$$

which has a totally unimodular constraint matrix and its optimal solution is equivalent to the solution of the following linear program.

$$\min_{\mathbf{x} \in \mathbb{R}^{n \times n}} \mathbf{c}^{\mathsf{T}}\mathbf{x}$$

$$\text{s.t. } \sum_{i=1}^{n} x_{ij} = 1 \quad \forall j = 1, ..., n$$

$$\sum_{j=1}^{n} x_{ij} = 1 \quad \forall i = 1, ..., n \tag{12.2}$$

$$\mathbf{x} \geq \mathbf{0}$$

The TSP formulation is assignment-based, but not exactly the same as the assignment problem described above, because the assignment problem formulation cannot ensure that solution $\mathbf{x}$ corresponds to a tour since it may be a set of sub-tours instead. For instance, consider the example of the symmetric TSP with four vertices presented in the undirected graph of Fig. 12.1. A possible solution that meets the aforementioned constraints is $x_{12} = x_{21} = x_{34} = x_{43}$. However, this solution will not create a tour (cycle) that starts from location 1 and returns back to location 1. Instead, it forms two sub-tours which do not meet the definition of a Hamiltonian cycle.

In Fig. 12.1 we have two sub-tours (1-2-1) and (3-4-3) that do not result in a tour which visits all pickup locations and returns back to the 1st one. For this reason, additional constraints are needed to eliminate sub-tours. To address this, TSP is defined as follows where the sub-tour elimination constraints are expressed in (12.6).

Traveling Salesman Problem

$$\min \sum_{i \in N} \sum_{j \in N} c_{ij} x_{ij} \tag{12.3}$$

$$\text{subject to: } \sum_{j \in N \setminus \{i\}} x_{ij} = 1 \qquad \forall i \in N \tag{12.4}$$

$$\sum_{i \in N \setminus \{j\}} x_{ij} = 1 \qquad \forall j \in N \tag{12.5}$$

$$\{x_{ij}\} \in S \tag{12.6}$$

Fig. 12.1 Solution that satisfies the assignment problem's constraints but does not result in a complete tour because there is no tour that returns back to location 1

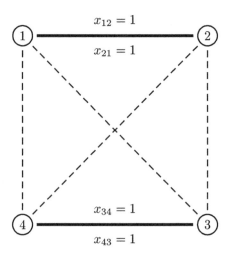

$$x_{ij} \in \{0, 1\} \qquad\qquad \forall i \in N, j \in N \qquad (12.7)$$

The sub-tour elimination constraints (12.6) are defined abstractly by requesting that x_{ij} values belong to a set S that prohibits potential sub-tour solutions. Examples of possible choices of set S in past literature include [1]:

1. $S = \{(x_{ij}) : \sum_{i \in Q} \sum_{j \notin Q} x_{ij} \geq 1$ for every nonempty proper subset Q of $N\}$
2. $S = \{(x_{ij}) : \sum_{i \in R} \sum_{j \in R} x_{ij} \leq |R| - 1$ for every nonempty subset R of $\{2, 3, ..., n\}\}$
3. $S = \{(x_{ij}) : n x_{ij} + y_i - y_j \leq n - 1$ for $2 \leq i \neq j \leq n$ for some $y_i \in \mathbb{R}$ where $i \in \{2, 3, ..., n\}\}$

The first and second examples of set S result in a *non-compact* formulation where the number of constraints increases exponentially with the number of vertices. In more detail, the first and second examples of set S contain nearly 2^n sub-tour breaking constraints. In reverse, the third example of set S results in a *compact* formulation with $n^2 - 3n + 2$ constraints that increase polynomially with the number of vertices. The third example of set S has already been presented in the formulation of (5.66). Using any example set S from the above will have the same effect on the problem's optimality because they will all exclude sub-tours.

The first example of S states that every proper subset Q of vertices must be connected to the other vertices in the network G in the solution $\mathbf{x}$. The second example of S requests that the arcs selected in solution $\mathbf{x}$ contain no cycle, since a cycle on vertices R must contain $|R|$ arcs. The third example of S has the same effect on sub-tour elimination and results in a compact MILP model.

12.3 Planning the Routes of Multiple Vehicles: Capacitated Vehicle Routing Problem

The Capacitated Vehicle Routing Problem (CVRP) is typically modeled with the use of the *vehicle flow formulations* or as a *set partitioning problem* that requires the determination of vehicle routes (circuits) with minimum cost to serve each customer once. The set partitioning problem formulation suffers from the same issues as the set partitioning formulation used for crew scheduling. In more detail, it requires an exponentially increasing number of binary variables, each associated with a different feasible circuit. Although the set partitioning formulation is tighter, it is not preferred in most cases because it requires dealing with a very large number of variables since we have to use a binary variable for every feasible circuit [2].

We proceed with the formal definition of CVRP according to Dantzig [3] in 1959.

Capacitated Vehicle Routing Problem [3]
Consider a fleet of capacitated vehicles located at a depot and a set of customers (passengers) that need to be served by that fleet. Find a set of minimum total cost routes for the fleet of capacitated vehicles to serve a set of customers (passengers located at different spots) under the following constraints:
- each route begins and ends at the depot,
- each customer (passenger) is visited exactly once,
- the total demand of each route does not exceed the capacity of the associated vehicle.

In addition to the requirements of the CVRP's problem statement, a typical extra requirement is that vehicles have maximum route time constraints. However, in the standard CVRP problem statement this requirement is not considered [1].

Below, we provide the two-index *vehicle flow formulation* of the CVRP (see the literature review of Toth and Vigo [2]). Let us consider a graph $G = (V, A)$ with vertex set $V = \{0, 1, ..., n\}$ and arc set A. Vertices $j = \{1, 2, ..., n\}$ correspond to customers (pickup passenger locations) and vertex 0 corresponds to the depot. Each vertex $j \in V \setminus \{0\}$ has an associated passenger demand $d_i > 0$ and the depot 0 has a fictitious demand $d_0 = 0$. Given a subset of passenger pickup points $S \subseteq V$, the total demand of that subset is:

$$d(S) := \sum_{j \in S} d_j$$

Each arc $(i, j) \in A$ has a travel cost c_{ij} related to the cost of traveling from the end of vertex i to the start of vertex j. To not allow loop arcs (i, i), we set $c_{i,i} = +\infty$ for all $i \in V$. Clearly if $c_{ij} = c_{ji}$ for every $(i, j) \in A$, the CVRP is symmetric and graph $G = (V, A)$ becomes undirected. Otherwise, it is asymmetric and $G = (V, A)$ is a digraph.

Let us consider again the subset S of V. We can define $\delta(S)$ as the set of arcs $(i, j) \in A$ for which $i \in S$ or $j \in S$ or $i \in S$ and $j \in S$. We also have a set of K vehicles that we assume having the same capacity. Given a subset $S \subseteq V \setminus \{0\}$ we can denote $\gamma(S)$ as the minimum number of vehicles required to serve all passengers in S. Considering the variant of the CVRP, we might need to use exactly K vehicles, resulting in K circuits, or at most K vehicles. In the latter case, we can add fixed costs associated with the use of an extra vehicle to steer the solution method towards finding a solution with less than K vehicles, if such solution exists. As described in the vehicle scheduling formulations, the fixed cost of adding an extra vehicle can be added to the cost of living the depot c_{0j} for any $j \in V \setminus \{0\}$.

Given the above definitions, it is clear that TSP is a special case of the CVRP problem for the case of a single vehicle, $|K| = 1$, the capacity of which, C, is greater than $d(V)$. Considering two-index decision variables x_{ij} where $x_{ij} = 1$ if arc (i, j) is traversed by a vehicle and $x_{ij} = 0$ if not, the two-index vehicle flow formulation of the CVRP is provided below. In this formulation we require to use all available vehicles in set K.

Two-index vehicle flow formulation of CVRP

$$\min_{\mathbf{x}} \sum_{i \in V} \sum_{j \in V} c_{ij} x_{ij} \tag{12.8}$$

$$\text{s.t.} \sum_{i \in V} x_{ij} = 1 \qquad \forall j \in V \setminus \{0\} \tag{12.9}$$

$$\sum_{j \in V} x_{ij} = 1 \qquad \forall i \in V \setminus \{0\} \tag{12.10}$$

$$\sum_{i \in V} x_{i0} = |K| \tag{12.11}$$

$$\sum_{j \in V} x_{0j} = |K| \tag{12.12}$$

$$\sum_{i \in V \setminus S} \sum_{j \in S} x_{ij} \geq \gamma(S) \qquad \forall S \subseteq V \setminus \{0\} \text{ and } S \neq \emptyset \tag{12.13}$$

$$x_{ij} \in \{0, 1\} \qquad \forall i \in V, j \in V \tag{12.14}$$

Constraints (12.9) and (12.10) are *indegree* and *outdegree* constraints and ensure that exactly one arc enters and leaves each passenger pickup location. Constraints (12.11) and (12.12) ensure that there is a route (circuit) for each vehicle in K which starts and ends at the depot. Constraints (12.13) ensure the connectivity of the solution and impose the vehicle capacity requirements. These constraints increase exponentially with the number of vertices n resulting in a non-compact formulation.

Constraints (12.13) are also expressed in literature as *generalized sub-tour elimination constraints* (GSEC):

$$\sum_{i \in S} \sum_{j \in S} x_{ij} \leq |S| - \gamma(S) \quad \forall S \subseteq V \setminus \{0\} \text{ and } S \neq \emptyset \tag{12.15}$$

The GSEC constrains increase also exponentially with the number of vertices n. Alternatively, we can replace constraints (12.13) with constraints that increase polynomially with the number of vertices. To do so, we can use the *sub-tour elimination constraints* of Miller et al. [4] proposed for the TSP in 1960 after extending them for the CVRP as follows:

$$u_i - u_j + C x_{ij} \leq C - d_j \quad \forall i \in V \setminus \{0\}, j \in V \setminus \{0\} \text{ such that } i \neq j \text{ and } d_i + d_j \leq C \tag{12.16}$$

$$d_i \leq u_i \leq C \qquad \forall i \in V \setminus \{0\} \tag{12.17}$$

where $u_i \in \mathbb{R}$ for $i \in V \setminus \{0\}$ are additional variables representing the load of the vehicle after visiting the pickup location i and C is the capacity of each vehicle assuming a homogeneous fleet. These polynomially increasing constraints impose the capacity constraints of CVRP because if $x_{ij} = 0$ they are not binding (u_i, u_j can take arbitrary values in $\mathbb{R}$ satisfying $u_i \leq C$ and $u_j \geq d_j$), whereas if $x_{ij} = 1$ they impose that $u_j \geq u_i + d_j$.

Let us now consider the case where we want to use at most $|K|$ vehicles and we associate a fixed cost into c_{0j} representing the cost of using an extra vehicle. Then, the above formulation of the CVRP can be modified as follows. Note that we use the polynomially increasing constraints this time.

Two-index vehicle flow formulation of CVRP with an upper bound on the number of available vehicles

$$\min_{\mathbf{x}, \mathbf{u}} \sum_{i \in V} \sum_{j \in V} c_{ij} x_{ij} \tag{12.18}$$

$$\text{s.t.} \sum_{i \in V} x_{ij} = 1 \qquad \forall j \in V \setminus \{0\} \tag{12.19}$$

$$\sum_{j \in V} x_{ij} = 1 \qquad \forall i \in V \setminus \{0\} \tag{12.20}$$

$$\sum_{i \in V} x_{i0} = \sum_{j \in V} x_{0j} \tag{12.21}$$

$$\sum_{j \in V} x_{0j} \leq |K| \tag{12.22}$$

$$u_i - u_j + C x_{ij} \leq C - d_j \quad \forall i \in V \setminus \{0\}, j \in V \setminus \{0\} \text{ such that } i \neq j \text{ and } d_i + d_j \leq C \tag{12.23}$$

$$d_i \leq u_i \leq C \qquad \forall i \in V \setminus \{0\} \tag{12.24}$$

Fig. 12.2 Locations of the depot and the pickup vertices in two dimensions

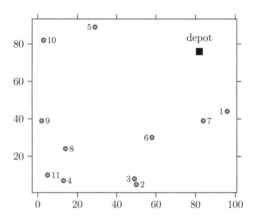

$$x_{ij} \in \{0, 1\} \qquad \forall i \in V, j \in V \qquad (12.25)$$

$$u_i \in \mathbb{R} \qquad \forall i \in V \setminus \{0\} \qquad (12.26)$$

The main difference in this new formulation is that we impose constraint (12.21) which requires that the number of vehicles leaving the depot is the same as the number of vehicles returning to the depot and (12.22) which ensures that we will not use more than $|K|$ vehicles. These two modifications allow the computation of solutions with less than $|K|$ vehicles, if there are feasible CVRP solutions in the case of using less than $|K|$ vehicles. The solution of this problem will be a number of less than $|K|$ cycles in graph G, where the vertex sequence of each cycle will represent the route of a particular vehicle.

Let us consider an example with 11 pickup passenger locations (customers) resulting in vertices $V = \{0, 1, 2, ..., 11\}$. The passenger demand associated to each vertex is $\mathbf{d} = [19, 21, 6, 19, 7, 12, 16, 6, 16, 8, 14]^{\mathsf{T}}$. The locations of the depot and the pickup passenger locations at the 2-dimensional space are presented in Fig. 12.2.

The travel cost of traveling from the end of vertex $i \in V$ to the start of vertex $j \in V$ is presented in Table 12.1.

Given a capacity of $C = 50$ passengers per vehicle and a maximum number of $|K| = 20$ vehicles, CVRP can be solved as follows.

Table 12.1 Cost of traveling c_{ij} from the end of vertex i to the start of vertex j

	0	1	2	3	4	5	6	7	8	9	10	11
0	$+\infty$	34.9285	77.8781	75.5844	97.5807	54.5711	51.8845	37.054	85.6037	88.1419	79.2275	101.415
1	34.9285	$+\infty$	60.3075	59.203	90.8735	80.7094	40.4969	13	84.4038	94.1329	100.464	97.1442
2	77.8781	60.3075	$+\infty$	3.16228	37.054	86.5852	26.2488	48.0833	40.7063	58.8218	90.2109	45.2769
3	75.5844	59.203	3.16228	$+\infty$	36.0139	83.4326	23.7697	46.7547	38.4838	56.3028	87.1321	44.0454
4	97.5807	90.8735	37.054	36.0139	$+\infty$	83.5464	50.5371	77.8781	17.0294	33.8378	75.6637	8.544
5	54.5711	80.7094	86.5852	83.4326	83.5464	$+\infty$	65.7419	74.3303	66.7083	56.8243	26.9258	82.5651
6	51.8845	40.4969	26.2488	23.7697	50.5371	65.7419	$+\infty$	27.5136	44.4072	56.7186	75.6902	56.648
7	37.054	13	48.0833	46.7547	77.8781	74.3303	27.5136	$+\infty$	71.5891	82	91.7061	84.1546
8	85.6037	84.4038	40.7063	38.4838	17.0294	66.7083	44.4072	71.5891	$+\infty$	19.2094	59.0339	16.6433
9	88.1419	94.1329	58.8218	56.3028	33.8378	56.8243	56.7186	82	19.2094	$+\infty$	43.0116	29.1548
10	79.2275	100.464	90.2109	87.1321	75.6637	26.9258	75.6902	91.7061	59.0339	43.0116	$+\infty$	72.0278
11	101.415	97.1442	45.2769	44.0454	8.544	82.5651	56.648	84.1546	16.6433	29.1548	72.0278	$+\infty$

Practitioner's Corner
Fixed job scheduling formulation in Python 3

One can use PuLP to program and the CBC solver to solve the two-index CVRP formulation with an upper bound on the number of available vehicles as follows.

```python
import pulp
from pulp import *
import numpy as np

model = LpProblem("CVRP_formulation", LpMinimize)

n=11 #number of passenger pickup points without the depot
xc = [82,96,50,49,13,29,58,84,14,2,3,5]
yc = [76,44,5,8,7,89,30,39,24,39,82,10]
V=[i for i in range(0,n+1)]
A=[(i,j) for i in V for j in V if i!=j]
c = {(i, j): np.hypot(xc[i]-xc[j], yc[i]-yc[j]) for i in V for j
    in V}
M=1000000 #big value to approximate +infinity
for i in V:
    for j in V:
        if i==j: c[i,j]=M
print('c',c)
d=[0,19,21,6,19,7,12,16,6,16,8,14] #passenger demand at each
    vertex
K=20 #number of available vehicles
C=50 #capacity of each vehicle
# variables
x = LpVariable.dicts("x", ((i, j) for i in V for j in V),cat='
    Binary')
u = LpVariable.dicts("u", (i for i in V),cat='Continuous')
# objective function
model += sum(sum(c[i,j]*x[i,j] for j in V) for i in V)
# constraints:
for j in V:
    if j!=0:
        model += sum(x[i,j] for i in V)==1
for i in V:
    if i!=0:
        model += sum(x[i,j] for j in V)==1
model += sum(x[i,0] for i in V) == sum(x[0,j] for j in V)
model += sum(x[0,j] for j in V) <= K
for i in V:
    if i!=0:
        for j in V:
            if j!=0 and i!=j and d[i]+d[j] <= C:
                model += u[i]-u[j]+C*x[i,j] <= C-d[j]
for i in V:
    if i!=0:
        model += u[i] <= C
        model += d[i] <= u[i]
```

```python
# optimizing
solver = pulp.PULP_CBC_CMD(mip=True, msg=True, timeLimit=None)
model.solve(solver)
print ("Status:", LpStatus[model.status])
for v in model.variables():
    if v.varValue >= 10E-7:
        print (v.name, "=", v.varValue)
print ("Optimal Solution = ", value(model.objective))
```

The optimal solution has objective function score 609.524 and it is presented below.

$x_{0,1} = 1$	$u_1 = 19.0$
$x_{0,3} = 1$	$u_{10} = 15.0$
$x_{0,5} = 1$	$u_{11} = 44.0$
$x_{1,7} = 1$	$u_2 = 27.0$
$x_{10,9} = 1$	$u_3 = 6.0$
$x_{11,8} = 1$	$u_4 = 50.0$
$x_{2,11} = 1$	$u_5 = 7.0$
$x_{3,2} = 1$	$u_6 = 50.0$
$x_{4,0} = 1$	$u_7 = 35.0$
$x_{5,10} = 1$	$u_8 = 50.0$
$x_{6,0} = 1$	$u_9 = 31.0$
$x_{7,6} = 1$	
$x_{8,0} = 1$	
$x_{9,4} = 1$	

This solution indicates that we need only 3 vehicles to serve the passenger demand. These vehicles will perform the following routes (cycles), presented also in Fig. 12.3.

- vehicle 1: $0 \rightarrow 1 \rightarrow 7 \rightarrow 6 \rightarrow 0$.
- vehicle 2: $0 \rightarrow 3 \rightarrow 2 \rightarrow 11 \rightarrow 8 \rightarrow 0$.
- vehicle 3: $0 \rightarrow 5 \rightarrow 10 \rightarrow 9 \rightarrow 4 \rightarrow 0$.

Fig. 12.3 Optimal routes of the three vehicles

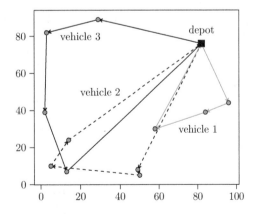

Closing, we note that to inflate the cost of adding extra vehicles and force the solution method to use as few vehicles as possible, each cost c_{0j} with $j \in V$ can be artificially increased by adding to it a very big number (i.e., 10000 units for this particular example).

> **Further Reading - survey papers**

- Baldacci et al., Recent advances in vehicle routing exact algorithms, 4or, 2007 [5].
- Caceres-Cruz et al., Rich vehicle routing problem: Survey, ACM Computing Surveys (CSUR), 2014, [6].
- Vidal et al., Heuristics for multi-attribute vehicle routing problems: A survey and synthesis, European Journal of Operational Research, 2013 [7].

12.4 Dial-a-Ride Problem

Until now we have described the TSP and CVRP problems which are problems where vehicles start from a depot, *pickup* passengers, and return back to the depot. Note that the TSP and CVRP formulations can be equivalently used for problems when vehicles start from the depot with a set of passengers onboard, *deliver* these passengers to a pre-defined set of delivery locations, and return back to the depot. What will happen though if vehicles start from the depot and they need to pickup and deliver passengers before returning back to it? This will result in combining the pickup up and delivery requirements in the scheduling of vehicle routes and cannot be addressed by TSP or CVRP formulations. Instead, we need to provide a new problem description which is embedded in the Dial-a-Ride Problem (DARP).

The first DARP models were developed in the early 1980s to plan the routes of buses based on telephone requests from passengers [8]. Over the years, the applicability of DARP models has been extended to on-demand services including minibuses, shared modes, and last-mile transport modes playing an integral part on the planning of on-demand services [9]. There are both static and dynamic DARP formulations. In static formulations, the transportation requests of passengers are known in advance [10, 11]. The fleet availability and size are also known in advance, and vehicles start from and return to a pre-defined depot. Prospective passengers can impose a time window on both their departure and arrival times from/to their pickup and delivery locations [12]. Typically, DARP formulations impose a maximum limit on the ride times of passengers and the route times of vehicles [13]. The simplest version of DARP considers a single vehicle and can be solved as a dynamic program [8]. DARP resembles the pickup and delivery problem (PDP) in logistics, because its main difference is imposing an additional constraint regarding the maximum ride time of passengers.

12.4.1 Classic DARP

The classic problem description of DARP seeks to define a set of minimum cost routes in order to satisfy a set of transportation requests. Each request involves transporting a set of passengers from a set of origins (*pickup points*), to a set of destinations (*delivery points*). Each passenger is associated with a distinct request. Each request has also a maximum ride time corresponding to the maximum duration of the trip between the pickup and delivery point. In addition, passengers can share the same vehicle subject to vehicle capacity constraints.

> **Classic Dial-a-Ride Problem**
> Consider a fleet of m identical vehicles based at the same depot. Consider also a set of passengers with specific pickup and delivery requests and, possibly, pickup and delivery time windows. Find a set of minimum cost vehicle routes that cover all requests and satisfy additional constraints (i.e., maximum ride time constraints). All vehicle routes should start from and return to the depot and each vehicle cannot perform more than one route.

Below we present the three-index DARP formulation of Cordeau [14].

Sets: Consider a directed graph $G = (V, A)$ where the vertex set V is partitioned into $\{P, D, \{0, 2n + 1\}\}$ where 0 and $2n + 1$ are two copies of the depot, $P = \{1, ..., n\}$ is the set of pickup vertices and $D = \{n + 1, ..., 2n\}$ the set of delivery vertices. We order the sets of pickup and delivery vertices in such a way that each request is a couple $(i, n + i)$ with i being the pickup and $n + i$ the associated delivery request. If two passenger requests have a common pickup but different delivery location, the corresponding pickup vertex is duplicated resulting in two requests. The same applies if some passengers have a common delivery location but different pickup locations. On the other side, if the origin-destination pair of several passengers is the same, the request of these passengers can be represented by a pickup and delivery pair. That is, each pickup i and corresponding delivery vertex $(n + i)$ are associated with exactly one origin-destination pair.

The arc set A of directed edges in the digraph G consists of:

- arcs $\{(0, j) : j \in P\}$ which are all outgoing arcs from the depot.
- arcs $\{(i, j)$ where $i \in P \cup D$ and $j \in P \cup D$ such that $i \neq j$ and $i \neq n + j\}$.
- arcs $\{(i, 2n + 1) : i \in D\}$ which are all returning arcs to the depot.

Fig. 12.4 Digraph
$G = (V, A)$ of DARP in the
case of two requests

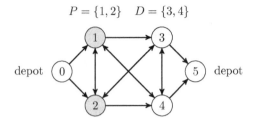

$P = \{1, 2\} \quad D = \{3, 4\}$

Let us consider, for instance, an example with two requests (two pickup and two delivery points). The graph representation, $G = (V, A)$, of this example is presented in Fig. 12.4. Note that vertices 0 and 5 are the same physical location (the location of the depot).

In addition to vertices V and arcs A, we also consider a set K representing the available vehicles.

Parameters: Each vertex $i \in P$ has a passenger load q_i, where $q_i \geq 0$. For the depot, we have $q_0 = q_{2n+1} = 0$. For the delivery vertices we have $q_{n+i} = -q_i$ because q_i passengers will disembark at vertex $n + i$. Each vertex has also a service duration $d_i \geq 0$ with $d_0 = d_{2n+1} = 0$. Each vehicle $k \in K$ has a capacity of Q_k and a maximum route duration T_k. The cost of traversing arcs $(i, j) \in A$ with vehicle k is c_{ij}^k and the travel time t_{ij}. Each vertex i has a time window $[e_i, l_i]$ and the maximum allowed ride time of a passenger is L.

Variables: Consider three-index binary variables x_{ij}^k where $x_{ij}^k = 1$ if arc $(i, j) \in A$ is traversed by vehicle $k \in K$ and 0 otherwise. Consider also continuous variables u_i^k denoting the time at which vehicle k starts servicing vertex i, w_i^k the load of vehicle k when it departs from vertex i, and r_i^k the ride time of passenger i corresponding to request $(i, n + i)$ on vehicle k. The three-index DARP formulation is then presented as follows.

Three-index DARP formulation

$$\min_{\mathbf{x,u,w,r}} \sum_{k \in K} \sum_{(i,j) \in A} c_{ij}^k x_{ij}^k \tag{12.27}$$

$$\text{s.t.} \sum_{k \in K} \sum_{j:(i,j) \in A} x_{ij}^k = 1 \qquad \forall i \in P \tag{12.28}$$

$$\sum_{j \in P} x_{0j}^k = \sum_{i \in D} x_{i,2n+1}^k = 1 \qquad \forall k \in K \tag{12.29}$$

$$\sum_{j:(i,j) \in A} x_{ij}^k - \sum_{j:(i,j) \in A} x_{n+i,j}^k = 0 \qquad i \in P, k \in K \tag{12.30}$$

$$\sum_{j:(j,i)\in A} x_{ji}^k - \sum_{j:(i,j)\in A} x_{ij}^k = 0 \qquad \forall i \in P \cup D, k \in K \qquad (12.31)$$

$$u_j^k \geq (u_i^k + d_i + t_{ij})x_{ij}^k \qquad \forall (i,j) \in A, k \in K \qquad (12.32)$$

$$w_j^k \geq (w_i^k + q_j)x_{ij}^k \qquad \forall (i,j) \in A, k \in K \qquad (12.33)$$

$$r_i^k = u_{n+i}^k - (u_i^k + d_i) \qquad i \in P, k \in K \qquad (12.34)$$

$$u_{2n+1}^k - u_0^k \leq T_k \qquad \forall k \in K \qquad (12.35)$$

$$e_i \leq u_i^k \leq l_i \qquad \forall i \in V, k \in K \qquad (12.36)$$

$$t_{i,n+i} \leq r_i^k \leq L \qquad \forall i \in P, k \in K \qquad (12.37)$$

$$\max\{0, q_i\} \leq w_i^k \leq \min\{Q_k, Q_k + q_i\} \qquad \forall i \in V, k \in K \qquad (12.38)$$

$$x_{ij}^k \in \{0, 1\} \qquad \forall (i,j) \in A, k \in K \qquad (12.39)$$

Constraints (12.28) ensure that each pickup point is covered exactly once. In more detail, it ensures that each pickup point $i \in P$ will be directly connected with exactly one other vertex $j : (i, j) \in A$ by exactly one vehicle k. Constraints (12.29) ensure that each vehicle will move from the depot to exactly one pickup point and it will return to the depot from exactly one delivery point. Constraints (12.30) ensure that if a vehicle k serves pickup point $i \in P$ it will also serve its associated delivery point $n + i$. In essence, constraints (12.28) and (12.30) ensure that each request is covered exactly once.

Constraints (12.29) ensure that each vehicle $k \in K$ will move from the depot to exactly one vertex $i \in V$ and it will also return to the depot from exactly one vertex i. In combination with the flow conservation constraints (12.31), they ensure that each vehicle $k \in K$ will start its route from the depot and return to the depot.

Constraints (12.32) ensure that if arc (i, j) is served by vehicle k, $x_{ij}^k = 1$, then the time at which vehicle k starts servicing vertex j (denoted as u_j^k) is greater than or equal to u_i^k plus the travel time from i to j (denoted as t_{ij}) plus the service time at i (d_i). Constraints (12.35) ensure that vehicle k will return to the depot before T_k. Constraints (12.36) ensure that each vertex $i \in V$ will be served within its predefined time window.

Constraints (12.34) ensure that the ride time of the passenger who uses vehicle k to travel from origin i to destination $n + i$ (denoted as r_i^k) is equal to $u_{n+i}^k - (u_i^k + d_i)$. Constraints (12.37) ensure that the ride time of a passenger from i to $n + i$ who uses vehicle k is less than the maximum allowed ride time of a passenger, L. It is also greater than or equal to the minimum travel time $t_{i,n+1}$ corresponding to a direct connection between i and $n + 1$.

Constraints (12.33) ensure that if a vehicle k serves arc (i, j) the passenger load upon leaving vertex j is greater than or equal to the load at vertex i plus q_j. Constraints (12.38) ensure that the in-vehicle passenger load w_i^k upon leaving vertex i is at least max$\{0, q_i\}$ and, at the same time, does not exceed the vehicle capacity min$\{Q_k, Q_k + q_i\}$.

It is worth noting that constraints (12.32)–(12.34) ensure the consistency of the service time, vehicle load, and ride time variables. In addition, the three-index formulation is mixed-integer and nonlinear. The nonlinearity is caused by the nonlinear, consistency-related constraints (12.32) and (12.33). Cordeau [14] proposed the linearization of these constraints by adding parameters M_{ij}^k and W_{ij}^k such that $M_{ij}^k \geq \max\{0, l_i + d_i + t_{ij} - e_j\}$ and $W_{ij}^k \geq \min\{Q_k, Q_k + q_i\}$ and replacing constraints (12.32) and (12.33) by:

$$u_j^k \geq u_i^k + d_i + t_{ij} - M_{ij}^k(1 - x_{ij}^k) \qquad \forall (i, j) \in A, k \in K \qquad (12.40)$$

$$w_j^k \geq w_i^k + q_j - W_{ij}^k(1 - x_{ij}^k) \qquad \forall (i, j) \in A, k \in K \qquad (12.41)$$

In constraints (12.40) and (12.41) we can simply use a very large positive value M instead of M_{ij}^k and W_{ij}^k to reduce the number of parameters.

The classic DARP is a standardized problem and benchmark data instances are available. We hereby consider the b2-16 benchmark instance with 2 vehicles and 16 requests from http://neumann.hec.ca/chairedistributique/data/darp/branch-and-cut. The coordinates of the vertices are provided in Table 12.2, where vertices 0 and 33 refer to the depot, vertices 1–16 to the pickup points, and vertices 17–32 to the delivery points. Column 4 presents the service duration at each vertex, d_i. Column 5 refers to the number of passengers at each vertex, q_i, having a negative sign for passenger deliveries. Column 6 indicates the earliest time we can visit vertex i, e_i, and column 7 indicates the latest time we can visit vertex i, l_i. The maximum allowed ride time of any passenger, L, is 45 time units. Each vehicle has capacity $Q_k = 6$ passengers, and the maximum route duration for each vehicle is $T_k = 480$ time units.

Below we provide an example of solving this DARP problem with branch and cut.

Practitioner's Corner
DARP in Python 3

One can use PuLP to program and the GUROBI (or CBC or CPLEX) solver to solve the three-index DARP MILP formulation as follows.

Table 12.2 Dataset b2-16 with 16 requests and 2 vehicles

Vertex	x coordinate	y coordinate	d_i	q_i	e_i	l_i
0	0.000	0.000	0	0	0	480
1	5.525	6.750	6	6	0	1440
2	0.366	6.249	1	1	0	1440
3	−3.611	−2.115	5	5	0	1440
4	−9.704	−1.154	6	6	0	1440
5	−0.145	−2.659	6	6	0	1440
6	6.602	−0.700	1	1	0	1440
7	−9.417	1.315	3	3	0	1440
8	−8.108	−0.165	1	1	0	1440
9	9.652	−6.043	2	2	24	39
10	−4.214	−6.219	1	1	25	40
11	−6.312	4.187	4	4	307	322
12	9.505	−4.496	3	3	320	335
13	−5.078	1.371	1	1	264	279
14	−9.106	4.125	3	3	402	417
15	−3.981	4.082	3	3	345	360
16	1.023	5.440	2	2	163	178
17	8.740	−2.490	6	−6	196	211
18	−5.203	8.897	1	−1	151	166
19	5.734	−4.899	5	−5	252	267
20	7.821	−6.475	6	−6	408	423
21	−7.387	−2.528	6	−6	266	281
22	2.965	3.351	1	−1	181	196
23	0.841	6.832	3	−3	341	356
24	6.128	−0.224	1	−1	173	188
25	−8.475	−1.503	2	−2	0	1440
26	4.906	3.128	1	−1	0	1440
27	8.980	2.988	4	−4	0	1440
28	2.914	0.820	3	−3	0	1440
29	1.638	4.757	1	−1	0	1440
30	−8.934	−5.149	3	−3	0	1440
31	5.350	−0.199	3	−3	0	1440
32	2.165	−4.790	2	−2	0	1440
33	0.000	0.000	0	0	0	480

```
import pulp
from pulp import *
import numpy as np
import math
model = LpProblem("DARP_formulation", LpMinimize)
n=16 #number of requests
xc =
    [0,5.525,0.366,-3.611,-9.704,-0.145,6.602,-9.417,-8.108,9.652,

    -4.214,-6.312,9.505,-5.078,-9.106,-3.981,1.023,8.74,-5.203,5.734,

    7.821,-7.387,2.965,0.841,6.128,-8.475,4.906,8.98,2.914,1.638,-8.934,5.35,2.165,0]
yc =
    [0,6.75,6.249,-2.115,-1.154,-2.659,-0.7,1.315,-0.165,-6.043,-6.219,

    4.187,-4.496,1.371,4.125,4.082,5.44,-2.49,8.897,-4.899,-6.475,-2.528,

    3.351,6.832,-0.224,-1.503,3.128,2.988,0.82,4.757,-5.149,-0.199,-4.79,0]

d =
    [0,6,1,5,6,6,1,3,1,2,1,4,3,1,3,3,2,6,1,5,6,6,1,3,1,2,1,4,3,1,3,3,2,0]

q =
    [0,6,1,5,6,6,1,3,1,2,1,4,3,1,3,3,2,-6,-1,-5,-6,-6,-1,-3,-1,-2,
    -1,-4,-3,-1,-3,-3,-2,0]
e =
    [0,0,0,0,0,0,0,0,0,24,25,307,320,264,402,345,163,196,151,252,408,
    266,181,341,173,0,0,0,0,0,0,0,0,0]
l=[480,1440,1440,1440,1440,1440,1440,1440,1440,39,40,322,335,279,

    417,360,178,211,166,267,423,281,196,356,188,1440,1440,1440,1440,1440,1440,1440,1440,480]

V = [i for i in range(0,2*n+2)]
K = (1,2) #vehicles
L = 45 #maximum ride time of every user
Q = {k:6 for k in K} #vehicle capacity
T = {k:480 for k in K} #maximum route duration
P = [i for i in range(1,n+1)] #pickup vertices
D = [i for i in range(n+1,2*n+1)] #delivery vertices
A = [(0,j) for j in P] + [(i,j) for i in P+D for j in P+D if i!=j
    if i!=n+j] + [(i,2*n+1) for i in D]
c = {(i, j): math.sqrt((xc[j]-xc[i])**2+(yc[j]-yc[i])**2) for (i,
    j) in A}
t = {(i, j): math.sqrt((xc[j]-xc[i])**2+(yc[j]-yc[i])**2) for (i,
    j) in A}
M=1000000
# variables
x = LpVariable.dicts("x", ((i, j, k) for (i,j) in A for k in K),
    cat='Binary')
u = LpVariable.dicts("u", ((i, k) for i in V for k in K),cat='
    Continuous')
w = LpVariable.dicts("w", ((i, k) for i in V for k in K),cat='
    Continuous')
r = LpVariable.dicts("r", ((i, k) for i in V for k in K),cat='
    Continuous')
# objective function
model += sum(sum(c[i,j]*x[i,j,k] for (i,j) in A) for k in K)
# constraints:
for i in P:
    model += sum(sum(x[i_t,j,k] for (i_t,j) in A if i_t==i) for k
    in K)==1
for k in K:
    model += sum(x[0,j,k] for j in P)==1
    model += sum(x[i,2*n+1,k] for i in D)==1
for i in P:
    for k in K:
        model += sum(x[i_t,j,k] for (i_t,j) in A if i_t==i) - sum
    (x[i_t,j,k] for (i_t,j) in A if i_t==n+i)==0
```

```
for i in P+D:
    for k in K:
        model += sum(x[j,i_t,k] for (j,i_t) in A if i_t==i) - sum
(x[i_t,j,k] for (i_t,j) in A if i_t==i)==0
for (i,j) in A:
    for k in K:
        model += u[j,k] >= u[i,k]+d[i]+t[i,j] - M*(1-x[i,j,k])
        model += w[j,k] >= w[i,k]+q[j] - M*(1-x[i,j,k])
for i in P:
    for k in K:
        model += r[i,k] == u[n+i,k]-(u[i,k]+d[i])
for k in K:
    model += u[2*n+1,k]-u[0,k] <= T[k]
for i in V:
    for k in K:
        model += e[i] <= u[i,k]
        model += u[i,k] <= l[i]
        model += max(0,q[i]) <= w[i,k]
        model += w[i,k] <= min(Q[k],Q[k]+q[i])
for i in P:
    for k in K:
        model += r[i,k] >= t[i,n+i]
        model += r[i,k] <= L

# optimizing
model.solve(solver = GUROBI_CMD())
print ("Status:", LpStatus[model.status])
for v in model.variables():
    if v.varValue >= 10E-7:
        print (v.name, "=", v.varValue)
print ("Optimal Solution = ", value(model.objective))
```

The optimal solution of this problem instance has objective function score 309.4057. Vehicles 1 and 2 will serve the following vertices.

- vehicle 1: $0 \rightarrow 9 \rightarrow 25 \rightarrow 8 \rightarrow 2 \rightarrow 18 \rightarrow 16 \rightarrow 24 \rightarrow 6 \rightarrow 22 \rightarrow 32 \rightarrow 3 \rightarrow 19 \rightarrow 5 \rightarrow 21 \rightarrow 7 \rightarrow 12 \rightarrow 23 \rightarrow 15 \rightarrow 28 \rightarrow 31 \rightarrow 4 \rightarrow 20 \rightarrow 33$
- vehicle 2: $0 \rightarrow 10 \rightarrow 26 \rightarrow 1 \rightarrow 17 \rightarrow 13 \rightarrow 11 \rightarrow 29 \rightarrow 27 \rightarrow 14 \rightarrow 30 \rightarrow 33$

12.4.2 Dial-a-Ride with Interchange Point

We have covered the classic formulation of DARP. One can apply several modifications to that formulation by adding constraints, using a heterogeneous fleet of vehicles with different capacities, or using multiple depots. Doing so results in straightforward extensions of the classic DARP formulation. Herein, we will discuss more fundamental changes to the structure of the problem, such as allowing an interchange (transfer) point where all picked up passengers can be exchanged among vehicles before proceeding to the delivery phase. This can be beneficial on many occasions because some origin-destination pairs of passengers might be too far and enforcing a vehicle making that connection might be inefficient.

Sets: The Dial-a-Ride problem with interchange (DARPi) lifts the restriction that a passenger picked by a vehicle must be delivered to his/her destination by the same vehicle. Instead, vehicles are allowed to exchange passengers at an interchange point. Let $G = (V, A)$ be a directed graph. The vertex set V is partitioned into $\{O \cup P \cup D\}$ where:

- $P = \{1, ..., n\}$ the set of pickup vertices and $D = \{n + 1, ..., 2n\}$ the set of delivery vertices, exactly as in the classic DARP formulation.
- $O = \{o_1, o_2, o_3, o_4\}$ are four copies of the depot, where o_1 symbolizes the start of the vehicle to pickup passengers, o_2 the return of the vehicle to the interchange point, o_3 the departure of the vehicle from the interchange point to deliver passengers, and o_4 the end of the trip of the vehicle after it has delivered all passengers.

The locations of o_1, o_2, o_3, o_4 are the same because they all refer to the depot location. As in the classic DARP formulation, if two passenger requests have a common pickup but different delivery locations the corresponding pickup vertex is duplicated. The same applies if two requests have a common delivery location but different pickup locations. It follows that we have n requests, where each request is a couple $(i, n + i)$ with $i \in P$ being the pickup point and $n + i \in D$ the associated delivery point for this origin-destination pair.

The feasible arc set A consists of the union of the following arc sets:

- $\{(o_1, j) : j \in P\}$
- $\{(i, j) : i \in P, j \in P, i \neq j\}$
- $\{(i, o_2) : i \in P\}$
- $\{(o_3, j) : j \in D\}$
- $\{(i, j) : i \in D, j \in D, i \neq j\}$
- $\{(i, o_4) : i \in D\}$

Note that a vehicle cannot go directly from a pickup point to a delivery point without passing from the interchange point, denoted by o_2, o_3. In essence, the feasible arc set A of the directed graph G forces a vehicle to start from o_1, serve pickup vertices belonging to P, return to the interchange point o_2 where it exchanges its passengers, and then depart from the interchange point o_3 delivering its newly assigned passengers to their delivery points in D. Finally, it returns back to the depot o_4. An example of this is presented in Fig. 12.5 where 0 is the location of the depot and corresponds to o_1, o_2, o_3, o_4.

Parameters: Similarly to the classic DARP formulation, to each vertex $i \in V$ there is an associated pickup or delivery demand q_i with $q_i \geq 0 \; \forall i \in P$, and $q_i = q_{i-n}$ $\forall i \in D$. This demand represents the number of passengers of the origin-destination pair $(i, n + i)$. Note that vehicles start without passengers (empty) from o_1 and return to o_4 empty. That is, $q_{o_1} = q_{o_4} = 0$. There is also a minimum service duration for boarding/alighting every passenger. If parameter β is the fixed time requirement for handling a single passenger, then this duration is $\beta \sum_{i \in P} q_i$ at the interchange location. For the interchange location, we also assume an additional fixed time for unloading and reloading, $a \in \mathbb{R}_+$.

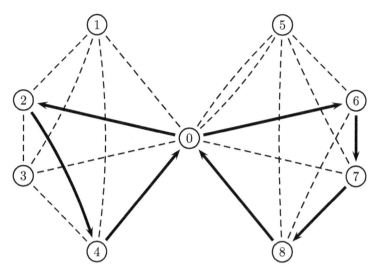

Fig. 12.5 Example of a vehicle trip starting from $o_1 = 0$, serving pickup vertices 2 and 4, returning to $o_2 = 0$ to unload the picked up passengers, reloading passengers and departing from $o_3 = 0$ to deliver the reloaded passengers to 6, 7 and 8, while finally returning to $o_4 = 0$

Let K be the set of available vehicles. The capacity of vehicle $k \in K$ is denoted as Q_k and the maximum allowed duration of route k as T_k. The cost and travel time of traversing a feasible arc $(i, j) \in A$ without performing intermediate stops is c_{ij}^k and t_{ij}, respectively. Note that the triangular inequality holds because both the costs and the travel times are non-negative. Let L be the maximum allowed ride time of any traveler and $[e_i, l_i]$ the time window within which we should serve vertex i.

Variables: Consider variables $u_i^k \in \mathbb{R}_+$ as the time at which vehicle k starts serving vertex $i \in P \cup D$ and $r_i \in \mathbb{R}_+$ the ride time of passenger i corresponding to request $(i, n + 1)$. We also introduce binary variables x_{ij}^k, η_i^k and θ_i^k. x_{ij}^k is equal to 1 if vehicle k serves arc $(i, j) \in A$. $\eta_i^k = 1$ if vehicle k unloads request $i \in P$ to the interchange point. $\theta_i^k = 1$ if vehicle k reloads request $i \in P$ from the interchange point. Binary variables $\tilde{\eta}_k$ and $\tilde{\theta}_k$ indicate also whether vehicle k unloads or reloads at the interchange point, respectively. Finally, additional continuous variables include:

- τ_k referring to the time at which vehicle $k \in K$ finishes unloading at the interchange point,
- w_k referring to the time at which vehicle $k \in K$ starts reloading at the interchange point,
- z_i referring to the time at which request $i \in P$ is unloaded at the interchange point.

The compact, three-index formulation of the DARPi model is presented below. Note that a very large positive number M is used for modeling purposes.

DARP with interchange point

$$\min \sum_{k \in K} \sum_{(i,j) \in A} c_{ij}^k x_{ij}^k \tag{12.42}$$

subject to:

$$\sum_{k \in K} \sum_{j:(i,j) \in A} x_{ij}^k = 1 \qquad\qquad \forall i \in P \cup D \quad (12.43)$$

$$\sum_{i \in P} \sum_{j:(i,j) \in A} q_i x_{ij}^k \leq Q_k \qquad\qquad \forall k \in K \quad (12.44)$$

$$\sum_{i \in D} \sum_{j:(i,j) \in A} q_i x_{ij}^k \leq Q_k \qquad\qquad \forall k \in K \quad (12.45)$$

$$\sum_{j:(o_1,j) \in A} x_{o_1 j}^k = \sum_{j:(o_3,j) \in A} x_{o_3 j}^k = 1 \qquad\qquad \forall k \in K \quad (12.46)$$

$$\sum_{j:(j,o_2) \in A} x_{j o_2}^k = \sum_{j:(j,o_4) \in A} x_{j o_4}^k = 1 \qquad\qquad \forall k \in K \quad (12.47)$$

$$\sum_{i:(i,h) \in A} x_{ih}^k - \sum_{j:(h,j) \in A} x_{hj}^k = 0 \qquad\qquad \forall h \in P \cup D, k \in K \quad (12.48)$$

$$u_j^k \geq u_i^k + t_{ij} - M(1 - x_{ij}^k) \qquad\qquad \forall (i,j) \in A, k \in K \quad (12.49)$$

$$e_i \leq u_i^k \leq l_i \qquad\qquad \forall i \in V, k \in K \quad (12.50)$$

$$\eta_i^k - \theta_i^k = \sum_{j \in P \cup \{o_2\}: j \neq i} x_{ij}^k - \sum_{j \in D \cup \{o_4\}: j \neq i+n} x_{i+n,j}^k \qquad\qquad \forall i \in P, k \in K \quad (12.51)$$

$$\eta_i^k + \theta_i^k \leq 1 \qquad\qquad \forall i \in P, k \in K \quad (12.52)$$

$$\frac{1}{M} \sum_{i \in P} \eta_i^k \leq \tilde{\eta}_k \leq \sum_{i \in P} \eta_i^k \qquad\qquad \forall k \in K \quad (12.53)$$

$$\frac{1}{M} \sum_{i \in P} \theta_i^k \leq \tilde{\theta}_k \leq \sum_{i \in P} \theta_i^k \qquad\qquad \forall k \in K \quad (12.54)$$

$$\tau_k = u_{o_2}^k + a\tilde{\eta}_k + \beta \sum_{i \in P} q_i \eta_i^k \qquad\qquad \forall k \in K \quad (12.55)$$

$$w_k \geq \tau_k \qquad\qquad \forall k \in K \quad (12.56)$$

$$u_{o_3}^k = w_k + a\tilde{\theta}_k + \beta \sum_{i \in P} q_i \theta_i^k \qquad\qquad \forall k \in K \quad (12.57)$$

$$w_k \geq z_i - M(1 - \theta_i^k) \qquad\qquad \forall i \in P, k \in K \quad (12.58)$$

$$z_i \geq \tau_k - M(1 - \eta_i^k) \qquad\qquad \forall i \in P, k \in K \quad (12.59)$$

$$u_{o_2}^k - u_{o_1}^k \leq T_k \qquad\qquad \forall k \in K \quad (12.60)$$

$$u_{o_4}^k - u_{o_3}^k \leq T_k \qquad\qquad \forall k \in K \quad (12.61)$$

$$r_i = \sum_{k \in K} \sum_{j:(j,n+i) \in A} x^k_{j,n+i} u^k_{n+i} - \sum_{k \in K} \sum_{j:(j,i) \in A} x^k_{j,i} u^k_i \qquad \forall i \in P \quad (12.62)$$

$$r_i \leq L \qquad \forall i \in P \quad (12.63)$$

The objective function (12.42) strives to minimize the total vehicle running costs. Constraints (12.43) ensure that each request-related vertex is visited exactly once. Constraints (12.44) and (12.45) ensure that the vehicle capacity at the pickup and delivery route is not exceeded. Constraints (12.46) ensure that the pickup route of each vehicle will start from o_1 and the delivery route of each vehicle will start from o_3. Similarly, constraints (12.47) ensure that every pickup route of a vehicle will return to o_2 and every delivery route will return to o_4. Constraints (12.48) ensure that when a vehicle arrives at a pickup or delivery vertex, it should depart from that vertex (flow conservation). Constraints (12.49) ensure that if arc $(i, j) \in A$ is served by vehicle k, then the time at which vehicle k starts servicing vertex j is greater than or equal to u^k_i plus the travel time from i to j (time consistency constraints). Constraints (12.50) ensure that any vertex $i \in V$ will be served within its time window. We explain now constraints (12.51). When

$$\sum_{j \in P \cup \{o_2\}: j \neq i} x^k_{ij} = 1$$

vehicle k will pickup request i. Similarly, when

$$\sum_{j \in D \cup \{o_4\}: j \neq i+n} x^k_{i+n,j} = 1$$

vehicle k will deliver request i. Constraints (12.51) result in the following four cases: (a) if request i is picked up but not delivered by vehicle k, then $\eta^k_i - \theta^k_i = 1$ and because $\eta^k_i + \theta^k_i \leq 1$ (see constraints (12.52)) we have that $\eta^k_i = 1$ and $\theta^k_i = 0$; (b) if request i is not picked up, but it is delivered by vehicle k, then $\eta^k_i - \theta^k_i = -1$ and because $\eta^k_i + \theta^k_i \leq 1$ we have that $\eta^k_i = 0$ and $\theta^k_i = 1$; (c) if request i is not picked up and not delivered by vehicle k, then $\eta^k_i - \theta^k_i = 0$ and because $\eta^k_i + \theta^k_i \leq 1$ we have that $\eta^k_i = 0$ and $\theta^k_i = 0$; (d) if request i is picked up and delivered by vehicle k, then $\eta^k_i - \theta^k_i = 0$ and because $\eta^k_i + \theta^k_i \leq 1$ we have that $\eta^k_i = 0$ and $\theta^k_i = 0$. Note that cases (c) and (d) have the same result because even if a vehicle picks up a request i it will not unload it to the interchange point ($\eta^k_i = 0$) and will not reload it from the interchange point ($\theta^k_i = 0$) if it is delivered to $i + n$ by the same vehicle.

Constraints (12.53) and (12.54) determine whether vehicle k unloads and/or reloads at the interchange point. Constraints (12.55) determine the time at which vehicle k finishes unloading at the interchange point, which depends on whether it performs any unloading ($\tilde{\eta}_k$) and on the amount of unloading requests ($\sum_{i \in P} q_i \eta^k_i$). Constraints (12.56) ensure that a vehicle starts reloading at an interchange point after it has finished the unloading process. Constraints (12.57) determine the time at which vehicle k has finished its unloading/reloading operations at the interchange point and is ready to leave. Constraints (12.58) ensure that reloading of a request i can start after this request has been unloaded at the interchange point. Constraints (12.59) ensure that the time a request i is unloaded at the interchange point is later than the time vehicle k, which unloads request i, finished its unloading process. Constraints (12.58) and (12.59) combined ensure that if a request is unloaded and reloaded by two different vehicles at the interchange point, the unloading vehicle should have already completed its unloading process before the reloading vehicle can reload this request. Constraints (12.60) and (12.61) ensure that pickup routes from the start of a trip o_1 to the interchange point o_2 and delivery routes from the interchange point o_3 to the end of the trip o_4 do not exceed the maximum allowed route duration T_k. Constraints (12.62) and (12.63) compute the ride time of each request i and ensure that this ride time is less than the maximum allowed ride time L.

In this model, constraints (12.62) are nonlinear. To linearize them, one can introduce variables $\tilde{u}_i$ which indicate the time at which vertex $i \in P \cup D$ starts to get serviced. Each vertex $i \in P \cup D$ is served by exactly one vehicle and one can exploit this fact to linearize constraints (12.62). The idea is to replace constraint (12.62) by $r_i = \tilde{u}_{n+i} - \tilde{u}_i$, $\forall i \in P$. To do so, we need to force $\tilde{u}_i$ to take the value of $u_i^{k^*}$, where k^* is the vehicle that actually serves vertex i. Let σ_{ij}^k be continuous slack variables. Then, we can write $\tilde{u}_i$ as:

$$
\begin{aligned}
\tilde{u}_i + \sigma_{ji}^k = u_i^k \qquad & \forall (j,i) \in A : i \in P \cup D, \forall k \in K \\
\sigma_{ji}^k \leq M(1 - x_{ji}^k) \qquad & \forall (j,i) \in A : i \in P \cup D, k \in K \qquad (12.64) \\
\sigma_{ji}^k \geq -M(1 - x_{ji}^k) \qquad & \forall (j,i) \in A : i \in P \cup D, k \in K
\end{aligned}
$$

Constraints (12.65) presented below together with constraints (12.64) can replace the nonlinear constraints (12.62) resulting in a MILP formulation of the dial-a-ride problem with interchange.

$$
r_i = \tilde{u}_{n+i} - \tilde{u}_i \qquad\qquad \forall i \in P \qquad (12.65)
$$

> Further Reading - survey papers

- Cordeau and Laporte, The dial-a-ride problem: models and algorithms, Annals of operations research, 2007 [11].
- Ho et al., A survey of dial-a-ride problems: Literature review and recent developments, Transportation Research Part B: Methodological, 2018 [9].
- Molenbruch et al., Typology and literature review for dial-a-ride problems, Annals of Operations Research, 2017 [15].

- Cordeau et al., The dial-a-ride problem (DARP): Variants, modeling issues and algorithms, Quarterly Journal of the Belgian, French and Italian Operations Research Societies, 2003 [16].

Exercises

12.1 CVRP for electric vehicles

Consider the CVRP formulation that uses all available vehicles. The fleet size is $|K|$ and the fleet is homogeneous.

(a) transform the two-index vehicle flow formulation of the CVRP to a three-index vehicle flow formulation.
(b) formulate the CVRP considering the additional request that every vehicle should spend up to ϕ units of energy before returning back to the depot for recharging. Consider also that traversing each directed edge (i, j) results in an energy consumption e_{ij}.

12.2 DARP

Solve the dial-a-ride problem for the a2-16 instance of Neumann provided at http://neumann.hec.ca/chairedistributique/data/darp/branch-and-cut using branch and cut. Solve the same instance using a metaheuristic of your choice.

12.3 DARP with interchange point

Consider the network presented in Fig. 12.6 which contains $n = 4$ requests. The inter-station travel times are equal to the travel costs and are presented in Fig. 12.6. The vehicle capacity is $Q_k = 20$ and we have 2 vehicles. The earliest possible start time of each vehicle is $e_{o_1} = 360$ and the latest possible end time is $l_{o_4} = 1320$. The fixed time for unloading and reloading at the interchange point is $a = 10$. In addition, the fixed time for handling a single passenger at the interchange point is $\beta = 1$. The maximum allowed duration of a route is $T_k = 480 \ \forall k \in \{1, 2\}$. The maximum ride time of any traveler is $L = 550$. The passenger demand for pickup/delivery at each vertex is $q_1 = q_5 = 16$, $q_2 = q_6 = 10$, $q_3 = q_7 = 4$, $q_4 = q_8 = 4$.

Finally, Table 12.3 presents the time window for serving each vertex.

Solve this problem using the linearized version of the DARP formulation with interchange point presented in (12.42)–(12.61), (12.63)–(12.65).

12.4 Traveling Salesman Problem

Devise a heuristic for the traveling salesman problem exploiting the fact that an assignment problem can be solved with linear programming.

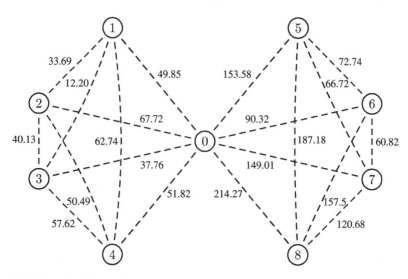

Fig. 12.6 Network where vertices o_1, o_2, o_3, o_4 are at location 0 (depot)

Table 12.3 Lower and upper time for visiting any vertex $i \in P \cup D$. For the depot we impose a lower time of 360 and an upper time of 1320

i	e_i	l_i	e_{n+i}	l_{n+i}
1	442	562	823	943
2	455	575	852	972
3	360	471	793	913
4	475	595	1007	1127

References

1. L. Bodin, Comput. Oper. Res. **10**(2), 63 (1983)
2. P. Toth, D. Vigo, Discrete Appl. Math. **123**(1–3), 487 (2002)
3. G.B. Dantzig, J.H. Ramser, Manag. Sci. **6**(1), 80 (1959)
4. C.E. Miller, A.W. Tucker, R.A. Zemlin, J. ACM (JACM) **7**(4), 326 (1960)
5. R. Baldacci, P. Toth, D. Vigo, 4or **5**(4), 269 (2007)
6. J. Caceres-Cruz, P. Arias, D. Guimarans, D. Riera, A.A. Juan, ACM Comput. Surv. (CSUR) **47**(2), 1 (2014)
7. T. Vidal, T.G. Crainic, M. Gendreau, C. Prins, Eur. J. Oper. Res. **231**(1), 1 (2013)
8. H.N. Psaraftis, Transp. Sci. **14**(2), 130 (1980)
9. S.C. Ho, W.Y. Szeto, Y.H. Kuo, J.M. Leung, M. Petering, T.W. Tou, Transp. Res. Part B: Methodol. **111**, 395 (2018)
10. R. Borndörfer, M. Grötschel, F. Klostermeier, C. Küttner, in *Computer-Aided Transit Scheduling* (Springer, 1999), pp. 391–422
11. J.F. Cordeau, G. Laporte, Ann. Oper. Res. **153**(1), 29 (2007)
12. J.J. Jaw, A.R. Odoni, H.N. Psaraftis, N.H. Wilson, Transp. Res. Part B: Methodol. **20**(3), 243 (1986)
13. S.N. Parragh, Transp. Res. Part C: Emerg. Technol. **19**(5), 912 (2011)

14. J.F. Cordeau, Oper. Res. **54**(3), 573 (2006)
15. Y. Molenbruch, K. Braekers, A. Caris, Ann. Oper. Res. **259**(1), 295 (2017)
16. J.F. Cordeau, G. Laporte, Q. J. Belg. Fr. Ital. Oper. Res. Soc. **1**(2), 89 (2003)

Chapter 13
Operational Planning and Control

Abstract The planning of public transit services takes place at the strategic and tactical level. Planning public transit services requires making the best available use of resources given a known input, where this input typically refers to passenger demand and vehicle travel times. During the actual operations, however, there can be several disruptions both on the passenger demand side (i.e., change of travel patterns) and on the public transport network (i.e., traffic congestion, roadworks, accidents, vehicle breakdowns or absence of drivers). To adapt to unexpected changes that might occur during a day of operations, public transport planning at the operational level is also required. This adaptive planning has the form of implementing control measures, such as rescheduling, stop-skipping, vehicle holding or speed control. This area is more recent and this chapter provides information on new developments at the operational planning and control stages.

13.1 General Overview of Operational Planning

Common corrective actions during daily operations include stop-skipping [1–4], vehicle holding at specific stops [5–9] or rescheduling the dispatching times of trips [10, 11]. Operational planning measures can also be applied in combination. For instance, stop-skipping and vehicle holding [12–15], short-turning and interlining [16], speed control and vehicle holding [17] or short-turning and deadheading [18]. In general, however, operational control approaches are typically applied in isolation because of the need to make decisions in quasi-real-time and the computational complexity of solving several problems.

Apart from holding, rescheduling and stop-skipping that are applied at the locations of public transport stations/stops, inter-station control can also be an option. Typical inter-station control strategies are traffic signal priority [19–22] and speed control [17, 23–25]. Because increasing the speed of vehicles is typically not possible given the speed limits on the road network and the traffic conditions, speed

© The Author(s), under exclusive license to Springer Nature Switzerland AG 2022   545
K. Gkiotsalitis, *Public Transport Optimization*,
https://doi.org/10.1007/978-3-031-12444-0_13

control mainly involves suggestions about speed reductions and it can be seen as an equivalent problem to vehicle holding.

Considering holding, stop-skipping and rescheduling, we note that each control method has its own adverse effects. Stop-skipping increases the inconvenience of passengers who cannot board the vehicle that skips their stop [4, 26]. Holding increases the trip's travel time and the inconvenience of onboard passengers who wait while the vehicle is held at the stop(s) [27]. Finally, rescheduling can affect the crew/vehicle schedules and can result in schedule sliding. Inter-station control strategies have also adverse effects. For instance, providing traffic signal priority can penalize the general traffic in the transport network.

For any control strategy to be effective, it needs to be acted upon by drivers. Simulation studies have shown that cooperative fleet control strategies such as even-headway holding (i.e. equalizing the headway from the backward and forward vehicles) are highly robust to the impacts of imperfect driver compliance [28, 29] as well as imprecise communication with the driver display [28].

In addition to technical considerations, experiences gained in the aforementioned cases also highlight the importance of the human factor in designing control schemes. Empirical analysis of vehicle movement data has shown that drivers' behavior and heterogeneity have a considerable impact on service reliability [30, 31]. Moreover, there is evidence that drivers adjust their speeds in response to real-time guidance and that this response depends on where performance is measured [32]. This means that to be effective, it is essential to devise incentives that are easy to understand and communicate, consistent and transparent throughout the service chain—from drivers to control center dispatchers and management [33].

In terms of key performance indicators when applying control measures, the main challenge in high-frequency services is to maintain the planned headways between vehicles at each stop, whereas in low-frequency services it is to adhere to the scheduled arrival/departure times to/from stops [33–36]. If the demand and the travel times of all vehicle trips that operate in a service line are equal and stable, vehicle trips will maintain an even headway at all downstream stops. This will result in a regular service where the actual waiting times at stops meet the passengers' expectations. Nevertheless, travel time and passenger demand variations during the actual operations result in unreliable and inconsistent services [37, 38]. Knoppers and Berrebi [39, 40] have shown that the fixed service intervals cannot be maintained at all stops. Indeed, even if vehicles are dispatched according to their planned headways, their headways are expected to deviate from their scheduled values as they move towards downstream stops [41].

To address the adverse effects of the demand and travel time variability, several periodic optimization approaches have emerged over the past 40 years. Periodic optimization approaches of fleet operations consider multiple decision variables and are based on iterative, finite-horizon optimization(s) of fleet operations. At time t, the current state of the fleet operations (i.e., current positions of running trips) is used as input and, together with the expected travel times within a relatively short time horizon $t + T$, the control measures (i.e., holding, stop-skipping, rescheduling) of multiple trips are determined. We note here that there are two main issues with the periodic optimization methods:

- If the number of vehicle trips, $i = \{1, 2, \ldots, \}$, that belong to the periodic optimization time period $t + T$ is too large, determining the appropriate control measures of all those trips results in multi-variable optimization problems that cannot be solved in real-time [42]. Additionally, as in model predictive control (MPC), the control measures of multiple trips will be updated before we even have the chance to implement them in practice because of the continuous updates of the operational status [43];
- If the optimization horizon is too short resulting in controlling only one trip at a time, the decisions are myopic. Hence, the performance of one selected trip might be improved, but the performance of the overall operations might deteriorate [44].

When defining the control measures of fleet operations in real-time, a balance should be established between considering multiple future trips or focusing on one trip at a time. This is prevalent in holding, rescheduling, and stop-skipping methods.

13.2 Rescheduling Approaches

Rescheduling approaches try to adapt to changes in passenger demand or service supply in order to improve the disrupted operations of public transport services. Depending on the implemented control actions, we can have different levels of rescheduling. We might have, for instance, the rescheduling of vehicles when we observe a change in passenger demand or when there is a disruption in the supply (i.e., some vehicles are delayed or one of the planned vehicles is not available). Rescheduling can also occur by changing the dispatching times of planned trips leading to a modified timetable. In the former case, we have to adapt the solution of the *Vehicle Scheduling Problem*. In the latter case, we have to adapt the solution of the *Timetabling Problem*. The two rescheduling aspects are presented in the following sub-sections.

13.2.1 Rescheduling Vehicles

We have already presented the Vehicle Scheduling Problem (VSP) and the Dial-a-Ride Problem (DARP) at the tactical planning stage. These problems can be extended to the operational phase by leveraging real-time information from telematics and passenger requests. These new problems are called *dynamic VSP* and *dynamic DARP* to emphasize the real-time decisions [45].

The classic VSP and DARP problems consider that all input is known beforehand and vehicle routes do not change once they are in execution. In contrast, dynamic VSP and DARP formulations can adapt to changes in the input during the actual operations. Below we provide the definition of dynamic vehicle scheduling problems by Pillac et al. [45].

> **Dynamic and Deterministic Vehicle Scheduling Problems**
> In dynamic and deterministic VSP problems, part or all of the input is unknown and revealed dynamically during the design or execution of the routes. For these problems, vehicle routes are redefined in an ongoing fashion, requiring technological support for real-time communication between the vehicles and the decision maker.

The first reference to dynamic routing was on the dial-a-ride problem in the work of Wilson and Colvin [46]. Psaraftis [47] in 1980 introduced also the dynamic DARP requiring immediate replanning of the current vehicle route when a new customer request is made.

Dynamic routing problems make additional considerations compared to classic routing problems. For instance, a new passenger request can be rejected if it is not possible to serve it or if the cost of serving it is too high. This allows dynamic routing formulations to explicitly consider the possibility of acceptance/denial of passenger requests [48, 49].

Dynamic routing implies that a moving vehicle can be redirected using real-time knowledge of the vehicle's position and online passenger requests. Dynamic routing problems have also different objective functions compared to classic routing problems [50]. For instance, dynamic routing problems might not only strive to minimize the routing cost, but also maximize the number of serviced passenger requests or the revenue. In addition, online passenger requests might require to be serviced as soon as possible adding another objective to the objective function of dynamic routing problems. For example, if a passenger makes an online request for a shared vehicle, the delay between the time of the request and the arrival of the vehicle can be considered as an additional objective.

In 1980, Psaraftis [47] presented the first periodic reoptimization approach for the DARP problem with a single vehicle making use of a dynamic programming approach. The objective was to find the optimal route each time a new passenger request is known. Periodic reoptimization approaches start at the beginning of the day with a first optimization that produces an initial set of routes. Then, an optimization procedure periodically solves a *static problem* corresponding to the current state, either whenever the available data changes, or at fixed intervals of time—known as *decision epochs*. Because periodic reoptimization approaches solve static routing problems at each epoch, their advantage is that they can use well-known solution methods that are developed for classic routing problems described in the previous chapters. The disadvantage is that the reoptimization should be performed before updating the routing plan, and this might result in computation time delays.

> **Further Reading—survey papers**

- Pillac et al., A review of dynamic vehicle routing problems, European Journal of Operational Research, 2013 [45].
- Psaraftis, Dynamic vehicle routing: Status and prospects, Annals of operations research, 1995 [50].

13.2.2 Rescheduling the Dispatching Times of High-frequency Services

Another dynamic control action is the rescheduling of trip dispatching times. This can modify the schedules of the planned timetables. Potential stop-skipping and/or holding actions can be easily applied on top of dispatching time rescheduling as soon as the vehicles are en-route [51, 52].

Rescheduling solutions in train operations commonly adopt local re-timing to adjust their timetables [53–56]. Most works on dispatching time rescheduling model the problem as an integer mathematical program where the decision variables are the dispatching times of trips. Due to their discrete nature, rescheduling problems cannot be easily solved. Several works have employed rescheduling to adjust the dispatching times of trips to the travel time and passenger demand variations. Bly [57] used rescheduling of depleted services to provide equal headways for the available fleet in the schedule. Gkiotsalitis [58] proposed periodic rescheduling that does not focus only on the running buses, but reschedules the dispatching times of all remaining daily trips while considering operational constraints related to layover times and capacity limits. Li et al. [59] modeled and solved the single depot rescheduling problem in pseudo-polynomial time using a parallel auction algorithm. In a follow-up work, [60] showed that the rescheduling problem with discrete decision variables is NP-hard, and used a Lagrangian relaxation-based insertion heuristic for its solution.

Gkiotsalitis and van Berkum [61] proved that the rescheduling problem with continuous variables is not convex when the vehicle capacity is considered in the optimization process. If, however, the vehicle capacity is not considered, the rescheduling problem can be solved to global optimality. This approach was later extended in [62], where an analytic solution for the rescheduling problem in rolling horizons was introduced.

Finally, we should note that dispatching time rescheduling is also used for the synchronization of services to reduce the transfer waiting times of passengers [63–65]. A summary of selected rescheduling methods is provided in Table 13.1.

Table 13.1 Summary of selected rescheduling works

Study	Problem	Mathematical program	Stocha-sticity	Solution method
[66]	Rescheduling	Integer nonlinear	No	Genetic algorithm
[59]	Rescheduling	Integer linear	No	Parallel action algorithm
[60]	Rescheduling	Integer linear	No	Lagrangian relaxation-based heuristic
[67]	Rescheduling and signal priority	Macroscopic model	No	Heuristic
[58]	Rescheduling	Integer nonlinear	No	Sequential hill climbing
[63]	Rescheduling considering passenger transfers	Integer program	No	Genetic algorithm
[64]	Rescheduling considering passenger transfers	Linear program	No	CPLEX
[61]	Rescheduling	Nonlinear convex program	No	CPLEX
[65]	Rescheduling considering passenger transfers	Mixed-integer linear program	No	CPLEX
[62]	Rescheduling	Quadratic convex program	No	Analytic solution
[68]	Rescheduling and stop-skipping	Integer nonlinear program	No	CPLEX

It is worth noting that rescheduling is typically coupled with other control measures, such as vehicle holding, and it is applied as a standalone measure in limited cases. This can be justified because rescheduling only impacts the dispatching times of trips and cannot mitigate potential irregularities appearing at intermediate stops. This is also a reason why rescheduling is mostly applied for synchronizing services provided by different lines and it has limited applicability for improving the regularity of a single line.

In terms of practical implementation, rescheduling models that consider the synchronization of lines are hard-to-solve and past literature typically resorts to heuristics or metaheuristics. Because of this, it is not always guaranteed that an efficient solution can be computed in near real-time. The same applies for models that couple rescheduling with vehicle holding. Standalone rescheduling models that only consider the regularity of services have the most preferable model formulations that return a globally optimal solution in real-time.

The formulation of [62] for rescheduling the dispatching times of trips is provided below. This formulation applies to high-frequency public transport services which aim to minimize the difference between the actual and planned arrivals times of trips at stations/stops.

Sets: Consider a public transport line with an ordered set of stops $S = \{1, 2, \ldots, |S|\}$. Following a periodic rescheduling approach, we can split the day into $M = \{1, 2, \ldots, m, \ldots\}$ decision epochs (also known as rolling horizons). At the m-th *decision epoch* we have an ordered set of future trips $N_m = \{j, j + 1, \ldots, |N|\}$ for which we need to decide their modified dispatching times. We can decide how many future trips N_m we want to include in our reoptimization. We can consider all trips until the end of the day, however this will increase the variables of the reoptimization problem without tangible gains since another reoptimization will occur in the next epoch and the conditions in the network will most probably change forcing us to update our decisions.

To demonstrate the implementation of the periodic reoptimization, consider that the pre-determined length of every set N_m is n. At the beginning of the day, we need to modify the dispatching times of trips $1, 2, \ldots, n$. A new decision epoch might start when a new trip is about to be dispatched. For instance, if the second trip of the day is about to be dispatched, we need to modify the dispatching times of trips $2, 3, \ldots, n + 1$ given the realized dispatching time of trip 1. For simplification, if $j \in N$ is the first trip that needs to be dispatched at a decision epoch m, it can be re-indexed as trip 1. Similarly, trips $j + 1, \ldots, j + n - 1$ can be re-indexed as $2, 3, \ldots, n$. If trip j is not the first trip of the day, it has a previously dispatched trip $j - 1$ which is re-indexed as 0. In addition, trip $j + n$, which is outside of the decision epoch, is re-indexed as $n + 1$. Following this re-indexing, trips 0 and $n + 1$ do not belong to N_m and are the *boundaries* of decision epoch m because their dispatching times cannot be modified at this epoch.

Parameters: Let us now consider the parameters of the problem. First, each trip has an already planned dispatching time δ_j which can be modified at a decision epoch. In addition, γ_s is the marginal increase in the dwell time of a vehicle trip at stop s arising from a unit increase in the time headway. When modifying the dispatching times of trips, the time difference of the dispatching times of adjacent trips should be within the range $H_{\min}, H_{\max}$ where $H_{\min}$ is the minimum possible time headway and $H_{\max}$ the maximum possible time headway. The minimum possible time headway requirement can be used to avoid bunching (i.e., two vehicles being dispatched at the same time), and the maximum possible time headway requirement can be used to avoid vehicles being dispatched with a considerable time difference with each other resulting in excessive passenger waiting times. From the frequency settings stage, there is also a scheduled (target) headway h_{js}^* that we wish trips $j - 1$ and j to achieve upon their arrival at any stop $s \in S$. The interstation travel time from stop s to $s + 1$ is t_{js}. We also consider ζ to be the maximum dispatching time delay of the last trip in the epoch in order to avoid schedule sliding (that is, postponing the

dispatching of trips in the next epoch). Let us also denote as 0 the latest trip which
was dispatched before decision epoch m. The dispatching time of this boundary trip
cannot be modified, so we can consider its arrival times at stops $s \in S \setminus \{1\}$ as a_{0s}
because these values are known or can be estimated.

Variables: We will now describe the problem's variables. First, we have the arrival
time of every trip $j = \{1, \ldots, n\}$ at stops $s \in \{2, \ldots, |S|\}$. This results in $n \times (|S| - 1)$
continuous variables, denoted as a_{js}. We also have the time headway h_{js} between
trips j and $j - 1$ upon their arrival at stop s, which results in a set of continuous
variables. In addition, continuous variables k_{js} indicate the dwell time of any trip
j at any stop s. Finally, we have n continuous decision variables $\mathbf{x} = [x_1, \ldots, x_n]^{\mathsf{T}}$
representing the dispatching time modification of trip j. Each $x_j \in \mathbf{x}$ can be seen as
a dispatching time offset from the planned dispatching time δ_j. The mathematical
program is then defined as follows.

Dispatching Time Rescheduling formulation at decision epoch m

$$\min_{\mathbf{x}, \mathbf{h}, \mathbf{a}, \mathbf{k}} \sum_{s=2}^{|S|-1} \sum_{j=1}^{n} \left(h_{j,s} - h_{j,s}^* \right)^2 \tag{13.1}$$

$$\text{s.t. } H_{\min} \leq (\delta_j + x_j) - (\delta_{j-1} + x_{j-1}) \leq H_{\max} \quad \forall j \in N_m \setminus \{1\} \tag{13.2}$$

$$H_{\min} \leq (\delta_1 + x_1) - \delta_0 \leq H_{\max} \tag{13.3}$$

$$a_{js} = (\delta_j + x_j) + \sum_{\phi=1}^{s-1} t_{j\phi} + \sum_{\phi=2}^{s-1} k_{j\phi} \quad \forall j \in N_m, \forall s \in S \setminus \{1\} \tag{13.4}$$

$$h_{js} = a_{js} - a_{j-1,s} \quad \forall j \in N_m, \forall s \in S \setminus \{1\} \tag{13.5}$$

$$k_{js} = \gamma_s h_{js} \quad \forall j \in N_m, \forall s \in S \setminus \{1\} \tag{13.6}$$

$$x_n \leq \zeta \tag{13.7}$$

$$\mathbf{x} \in \mathbb{R}^n, \mathbf{h}, \mathbf{a}, \mathbf{k} \in \mathbb{R}^{n \times (|S|-1)} \tag{13.8}$$

The objective function (13.1) minimizes the squared deviation between the actual
headways h_{js} and the scheduled (target) headways h_{js}^* at all stops, except for the first
and the last. This is a typical objective in high-frequency service lines. Constraints
(13.2) ensure that the dispatching time differences of consecutive trips are between
the minimum and maximum allowed headway. Constraint (13.3) imposes the same
condition to trips 0 and 1 since trip 0 has been already dispatched at time δ_0. Equality
constraints (13.4)–(13.6) set the values of the arrival time, time headway, and dwell

Table 13.2 Trip travel times τ_{js} from stop s to stop $s + 1$ per bus trip j in seconds

j	Stop	
	$s = 1$	$s = 2$
1	900	720
2	920	700
3	880	640

time variables. The dwell time is proportional to the time headway since it is typically considered to be the product of h_{js} and γ_s, where parameter $\gamma_s > 0$ (see Daganzo [38]). Finally, constraint (13.7) ensures that the last trip of the epoch will not be dispatched with a delay of more than ζ to avoid schedule sliding.

All equality and inequality constraints are affine functions resulting in a feasible region which is a *convex set*. The objective function is quadratic and strictly convex. Thus, the mathematical program can be solved to global optimality. In detail, it is a convex inequality-constrained quadratic program (IQP).

Let us consider an example with a decision epoch m where three future public transport trips (namely 1, 2 and 3) are expected to be dispatched. Trip 0 that precedes trip 1 has been dispatched when our periodic reoptimization in epoch m starts. The dispatching time of trip 0 is $\delta_0 = 0$ s. Its expected arrival times to stops $s = 2$ and $s = 3$ are $a_{0,2} = 900$ s and $a_{0,3} = 1600$ s.

The originally planned dispatching times of trips 1, 2 and 3 are $\delta_1 = 600$ s, $\delta_2 = 1200$ s and $\delta_3 = 1800$ s. The predicted inter-station travel times of trips 1, 2 and 3 are given in Table 13.2.

In addition, $\gamma_s = 0.035$ for stop $s = 2$. The target time headways are $h_{j,s}^* = 600$ s, $\forall j \in \{1, 2, 3\}$, $\forall s \in \{2, 3\}$ and the minimum and maximum allowed headways at the dispatching stop are $H_{\min} = 300$ and $H_{\max} = 900$.

As an inequality-constrained quadratic program, this problem can be solved with the active set method. If we consider $\zeta = 20$ and the objective function $\sum_{s=2}^{|S|} \sum_{j=1}^{n} (h_{j,s} - h_{j,s}^*)^2$ we actually have to minimize the squared deviation between the actual and target time headways at stop 2. This has a trivial solution $\mathbf{x}^* = [0, -20, 20]^\mathsf{T}$ that results in an objective function score equal to 0 since all actual headways at stop 2 will meet their target values.

Let us now consider a more complicated scenario where we also consider the headway deviation at the last stop in the objective function. That is, the objective function is reformulated as: $\sum_{s=2}^{|S|} \sum_{j=1}^{n} (h_{j,s} - h_{j,s}^*)^2$. In that case, the optimal solution becomes $\mathbf{x}^* = [-26.8269, -43.9654, 20]^\mathsf{T}$ with objective function value 2982.35. The remaining variable values are presented in the following table (Table 13.3).

Table 13.3 Rescheduling solution for $\zeta = 20$

x_1	−26.8269	$a_{0,2}$	900	$k_{1,2}$	20.0611	$h_{1,2}$	573.173
x_2	−43.9654	$a_{0,3}$	1600	$k_{1,3}$	21.4632	$h_{1,3}$	613.234
x_3	20	$a_{1,2}$	1473.17	$k_{2,2}$	21.1002	$h_{2,2}$	602.862
		$a_{1,3}$	2213.23	$k_{2,3}$	20.4365	$h_{2,3}$	583.901
		$a_{2,2}$	2076.03	$k_{3,2}$	21.8388	$h_{3,2}$	623.965
		$a_{2,3}$	2797.13	$k_{3,3}$	19.7646	$h_{3,3}$	564.704
		$a_{3,2}$	2700				
		$a_{3,3}$	3361.84				

13.3 Vehicle Holding

Control methods for vehicle holding have been studied since the early 1970s [5, 69]. If the arrival times of a vehicle at particular stops are not as expected, that vehicle can be held for a short time at a stop (typically called *control point* or *time point*) in order to normalize its time headway with its preceding and following vehicles and avoid *bunching*. We can formally define the vehicle holding problem as follows.

> **Vehicle Holding**
> Vehicle holding is the problem of deciding *whether to hold* and for *how long to hold* vehicles at specific stations/stops in order to meet a specific objective (i.e., improve the regularity of the service [38, 52, 70], reduce the waiting times of passengers [15, 71, 72], avoid overcrowding, adhere to the planned schedule [73, 74]).

There are two main vehicle holding approaches:

- Myopic approaches that decide about the holding of a single vehicle when it arrives at a control point stop based on its current time headway with its following and/or preceding vehicle that operates at the same line [5].
- Periodic re-optimization approaches which decide about the holding times of running vehicles at particular time instances (decision epochs) solving mathematical programs with multiple variables.

The problems in the family of myopic approaches have a single decision variable: the decision about the holding time of the vehicle in question at its next stop. The problems in the family of multivariate periodic re-optimization try to suggest holding times for all running vehicles. In the latter case, some of these holding time suggestions will be implemented in practice while others might be updated at the next decision epoch.

The two vehicle holding approaches differ in the nature of the decision variables (single-variate optimization vs multivariate optimization). There are further sub-categorizations, however, including formulations that consider the vehicle capacity as a constraint (or not), formulations that consider stochasticity in the vehicle travel times (or not), formulations that consider the potential of schedule sliding due to holding (or not), and formulations that consider the implications of charging in case of having a fleet of electric vehicles. In the following, we present common holding approaches based on this sub-categorization.

13.3.1 Single-variable Vehicle Holding Problems

In single-variable holding problems, we decide the holding time of a single trip when it arrives at a time point stop. Single-variable optimization problems lead to closed-form expressions that can determine the holding time of a trip in real-time without the need of solving a complex mathematical program [27, 75, 76]. With such approaches, the holding time decision of a trip n is typically made when it arrives at a time point stop s based on:

- Its actual time headway with its preceding trip, $n - 1$, (one-headway-based control logic)
- Or its actual time headway with its preceding trip, $n - 1$, and its expected time headway with its following trip, $n + 1$, (two-headway-based control logic)

This can be understood with the use of the time-space diagram of Fig. 13.1, where:

- H_s is the target headway (i.e., scheduled headway that we need to adhere to),
- t the time when trip n has completed all its boardings/alightings at stop s and is ready to depart (in case no holding is applied),
- $d_{n-1,s}$ the actual departure time of trip $n - 1$ from stop s,
- $d_{n,s}$ the optimized departure time of trip n from stop s, which includes the holding time of $d_{n,s} - t$,
- $\tilde{d}_{n+1,s}$ the expected departure time of trip $n + 1$ from stop s,
- $(d_{n,s} - t)$ the decided holding time of trip n at stop s.

13.3.1.1 One-headway-based Logic

In the one-headway-based control logic, the holding time of trip n when it arrives at stop s depends on the target headway, H_s, the time when it finishes its boardings/alightings and is ready to depart, t, and the departure time of its preceding trip from stop s, $d_{n-1,s}$. A simple closed-form expression that can determine the holding time $h_{n,s}$ of trip n at stop s is provided below.

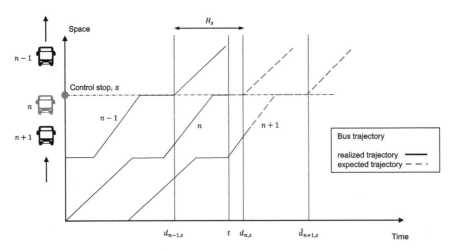

Fig. 13.1 Time-space diagram of the realized and expected trajectories of three successive vehicle trips

One-headway-based vehicle holding logic

$$
h_{n,s} = \begin{cases} 0 & \text{if } t - d_{n-1,s} \geq H_s \\ H_s - (t - d_{n-1,s}) & \text{otherwise.} \end{cases} \tag{13.9}
$$

This simple closed-form expression allows vehicle n to depart immediately after finishing its boardings/alightings (no holding) if $t - d_{n-1,s} \geq H_s$ which indicates that trip n is behind schedule and needs to catch up with its preceding trip. If it is not behind schedule, it is held at stop s for time $H_s - (t - d_{n-1,s})$ to meet the target headway. Given that holding trip n for time $H_s - (t - d_{n-1,s})$ might not always be desirable because it increases the travel time of trip n, a weight $0 \leq a \leq 1$ can be applied to $H_s - (t - d_{n-1,s})$ resulting in $a\big(H_s - (t - d_{n-1,s})\big)$. Intuitively, if $a = 0$ vehicle n is never held, whereas if $a = 1$ vehicle n is held for time $H_s - (t - d_{n-1,s})$ to adhere again to the target headway. Any other value of a in the range $(0, 1)$ determines the strength of the holding control. The value of a depends on the preferences of the service operator because some operators might be more willing to hold their vehicles at stops for a prolonged time in order to adhere to the scheduled headways, whereas other operators might not. Fu and Yang [27] performed a sensitivity analysis testing different values of a and concluded that tighter control can reduce the average passenger waiting time. At the same time, tighter control increases the total trip travel time and triggers a holding action more often.

13.3.1.2 One-headway-based Logic that Avoids Schedule Sliding

Let us restate the one-headway-based vehicle holding logic which defines the holding time of vehicle n at stop s as follows:

$$h_{n,s} = \begin{cases} 0 & \text{if } t - d_{n-1,s} \geq aH_s \\ H_s - (t - d_{n-1,s}) & \text{otherwise.} \end{cases} \tag{13.10}$$

where $a \in [0, 1]$ is the parameter indicating the holding strength. Because the departure time of trip n from stop s is $d_{ns} = t + h_{ns}$ we can express this departure time as:

One-headway-based holding without schedule sliding

$$d_{ns} = \begin{cases} t & \text{if } t - d_{n-1,s} \geq aH_s \\ t + H_s - (t - d_{n-1,s}) = H_s + d_{n-1,s} & \text{otherwise.} \end{cases} \tag{13.11}$$

When holding a vehicle at a stop we add an additional time to the total travel time of its trip. This will delay the termination of this trip and might also delay the dispatching of the next trip that will be performed by the same vehicle resulting in schedule sliding. If ρ_n is the latest possible arrival time of trip n at the last stop of the line to avoid schedule sliding and t_s is the expected travel time of trip n from stop s to the last stop, then we would need the vehicle holding time at stop s to be less than $\rho_n - t_s$ in order to avoid schedule sliding. This means that instead of departing at time $d_{n-1,s} + H_s$ we might need to depart earlier (at time $\rho_n - t_s$) to avoid schedule sliding. The necessary time $\rho_n - t_s$ to arrive at the final destination while avoiding schedule sliding might be even lower than the current time t, meaning that the trip is already delayed even if we do not apply holding. In that case, the trip should depart immediately. This will result in the amended formula:

One-headway-based holding with schedule sliding [77]

$$d_{ns} = \begin{cases} t & \text{if } t - d_{n-1,s} \geq aH_s \\ \max\{t, \min\{\rho_n - t_s, H_s + d_{n-1,s}\}\} & \text{otherwise.} \end{cases} \tag{13.12}$$

Consider the example where the departure time of the previous trip from stop s is $d_{n-1,s} = 1000$ s, the time when trip n is ready to depart from stop s is $t = 1500$ s, the scheduled (ideal) time headway between trips $n - 1$ and n at stop s is $H_s = 600$ s, the expected travel time of vehicle n from stop s to the last stop of the line is $t_s = 3000$ s, and the holding strength parameter a has value $a = 0.9$. Consider also

Table 13.4 Departure times, d_{ns}, of trip n from stop s for different values of ρ_n

ρ_n (s)	One-headway-based holding (s)	One-headway-based holding with schedule sliding (s)
4800	1600	1600
4600	1600	1600
4550	1600	1550
4500	1600	1500
4200	1600	1500

that the latest possible arrival time at the last stop of the line to avoid schedule sliding is $\rho_n = 4200$ s. Using the one-headway control logic without considering schedule sliding, trip n will depart from stop s at time:

$$d_{ns} = \begin{cases} t & \text{if } t - d_{n-1,s} \geq aH_s \\ H_s + d_{n-1,s} & \text{otherwise.} \end{cases}$$

Because $t - d_{n-1,s} = 500$ s is smaller than $aH_s = 0.9 \cdot 600 = 540$ s we have that $d_{ns} = H_s + d_{n-1,s} = 600 + 1000 = 1600$ s.

Let us consider now the case where we schedule sliding. In that case, we have:

$$d_{ns} = \begin{cases} t & \text{if } t - d_{n-1,s} \geq aH_s \\ \max\{t, \min\{\rho_n - t_s, H_s + d_{n-1,s}\}\} & \text{otherwise.} \end{cases}$$

and because $t - d_{n-1,s}$ is smaller than aH_s the optimal departure time is $d_{ns} = \max\{t, \min\{\rho_n - t_s, H_s + d_{n-1,s}\}\} = \max\{1500, \min\{1200, 1600\}\} = 1500$ s. In Table 13.4 we provide more examples with different values of ρ_s to demonstrate the impact of considering schedule sliding when deciding the holding time of a vehicle.

13.3.1.3   Two-headway-based Logic

The *one-headway-based* control logic can be expanded to a *two-headway-based* control logic that considers also the expected departure time of the following trip $n + 1$ at stop s, $\tilde{d}_{n+1,s}$. For instance, a two-headway-based control logic can be expressed as follows.

Algorithm 44 Two-headway-based holding control logic [27]

0: If $t < d_{n-1,s} + H_s$, then:
1:     If $\frac{1}{2}(\tilde{d}_{n+1,s} - d_{n-1,s}) < H_s$, then:
2:        $d_{n,s} = d_{n-1,s} + H_s$
3:     Else:
4:        $d_{n,s} = d_{n-1,s} + [\frac{1}{2}(\tilde{d}_{n+1,s} - d_{n-1,s}) + H_s]/2$
5: Else:
6:     $d_{n,s} = t$

Algorithm 44 states that if $d_{n-1,s} + H_s \geq t$, trip n is already delayed and it will depart stop s immediately ($d_{n,s} = t$). If $d_{n-1,s} + H_s < t$, then if the time interval between the departures of the preceding and following vehicle trips, $\tilde{d}_{n+1,s} - d_{n-1,s}$, is less than $2H_s$ we have that $d_{n,s} = d_{n-1,s} + H_s$. This implies that if we cannot adhere to both target headways between vehicle trip n and its preceding and following vehicle trips, we strive to maintain a headway H_s with its preceding vehicle trip. Because trips n and $n + 1$ will be then close to each other, we can subsequently hold vehicle trip $n + 1$ when it arrives at stop s to hopefully maintain a time headway of H_s between n and $n + 1$. Finally, if $\tilde{d}_{n+1,s} - d_{n-1,s} \geq 2H_s$ we have a large time interval between trips $n - 1$ and $n + 1$ and we want the headway between trip n and its preceding and following trips to be as close as possible to H_s. For this, we set $d_{n,s} = d_{n-1,s} + [\frac{1}{2}(\tilde{d}_{n+1,s} - d_{n-1,s}) + H_s]/2$.

13.3.1.4 Two-headway-based Logic Considering Vehicle Capacity

The one-headway-based and two-headway-based vehicle holding logics presented above do not consider the capacity of vehicles. To consider the capacity of vehicles, a reformulation is needed as suggested in Gkiotsalitis and van Berkum [78]. To present the formulation of [78] that includes vehicle capacity limits, we will use the notation of Table 13.5.

Considering a regularity-based service, the objective function strives to minimize the *squared deviation* between the realized/expected headways of trip n with its adjacent trips, $n - 1, n + 1$, and the ideal headway, H_s [5, 36]:

$$f(x, \tilde{d}_{n+1,s}) := \left((t + x) - d_{n-1,s} - H_s\right)^2 + \left(\tilde{d}_{n+1,s} - (t + x) - H_s\right)^2 \quad (13.13)$$

Because trip n cannot serve more passengers than its capacity, $\phi_n + x\lambda_s \leq c_n$ where $x\lambda_s$ is the number of additional passengers who are willing to board trip n if it is held at stop s for time x after it completes its boardings/alightings. Additionally, ϕ_n is the sum of the in-vehicle passenger load of trip n and the number of (potentially) stranded passengers when it has completed its boardings/alightings at stop s. Because there is no holding time $x \in \mathbb{R}_{\geq 0}$ which can always guarantee that the capacity of trip

Table 13.5 Notation

Indices	
n	Index of the trip for which a holding decision needs to be made at the current time instance.
$n - 1$	Index of the preceding trip of trip n.
$n + 1$	Index of the following trip of trip n.
s	Index of the specific stop at which a holding decision for trip n needs to be made.
Parameters	
t	Time when trip n has completed its boardings/alightings at stop s and is ready to depart if there is no further holding.
$d_{n-1,s}$	Realized departure time of trip $n - 1$ from stop s.
λ_s	Arrival rate of passengers at stop s (i.e., passengers per second).
c_j	Capacity of trip j, where $j \in \{n - 1, n, n + 1\}$.
ϕ_n	Observed in-vehicle passenger load of trip n at time t including the number of passengers who are refused to board trip n at stop s due to overcrowding. By definition, ϕ_n can be greater than c_n.
$\tilde{l}_{n+1}$	Expected in-vehicle load of trip $n + 1$ at the time of its arrival at stop s.
$\tilde{\beta}_{n+1}$	Expected passenger alightings of trip $n + 1$ at stop s.
$\tilde{a}_{n+1,s}$	Expected arrival time of trip $n + 1$ at stop s.
H_s	Target (ideal) headway of adjacent trips at stop s.
t_b	Required time for each passenger boarding.
t_a	Required time for each passenger alighting.
ζ	Maximum allowed holding time.
M_1, M_2	Very large positive numbers, where $M_1 \gg M_2 \gg 0$.
Decision variable	
x	Holding time of trip n at stop s. Note that $x \in \mathbb{R} \mid 0 \leq x \leq \zeta$.
Variables	
$d_{n,s}$	Departure time of trip n from stop s, where $d_{n,s} := t + x$.
$\tilde{d}_{n+1,s}$	Expected departure time of trip $n + 1$ from stop s.
l_n	Stranded passengers who are refused to board on trip n at stop s.

n suffices, the number of stranded passengers, l_n, by trip n at stop s can be expressed as the following equality constraint:

$$l_n := \max(0, \phi_n + x\lambda_s - c_n) \tag{13.14}$$

Since constraint $\phi_n + x\lambda_s \leq c_n$ cannot be always satisfied, it can be perceived as a *soft* constraint which is allowed to be violated if, and only if, the holding time x cannot ensure that there are no stranded passengers at stop s. This soft constraint is added to the objective function as a penalty term $M_1 \max(0, \phi_n + x\lambda_s - c_n)$ resulting in:

$$f(x, \tilde{d}_{n+1,s}, l_n) \triangleq \left((t+x) - d_{n-1,s} - H_s \right)^2 + \left(\tilde{d}_{n+1,s} - (t+x) - H_s \right)^2 + M_1 l_n \tag{13.15}$$

Note that the very large positive number M_1 in the penalty term $M_1 \max(0, \phi_n + x\lambda_s - c_n)$ ensures that the satisfaction of constraint $\phi_n + x\lambda_s \leq c_n$ is prioritized over $\left((t+x) - d_{n-1,s} - H_s \right)^2 + \left(\tilde{d}_{n+1,s} - (t+x) - H_s \right)^2$.

Another constraint is related to the vehicle capacity limit of the following trip, $n+1$. Note that the vehicle capacity limit of the preceding trip, $n-1$, is not considered because our decision variable, x, cannot affect its value. When trip $n+1$ arrives at stop s it has an in-vehicle load $\tilde{l}_{n+1}$ and is expected to alight $\tilde{\beta}_{n+1}$ passengers. Because of the time needed for the alightings, $\tilde{\beta}_{n+1} t_a$, we get $\tilde{\beta}_{n+1} t_a \lambda_s$ more passenger boardings. In addition, the stranded passengers by trip n, l_n, are willing to board trip $n+1$. Furthermore, by the time trip n departs stop s, $(t+x)$, until trip $n+1$ arrives there, we have $(\tilde{a}_{n+1,s} - (t+x))\lambda_s$ more passengers willing to board trip $n+1$. Thus, the expected in-vehicle load of trip $n+1$ when it departs from stop s is $\tilde{l}_{n+1} - \tilde{\beta}_{n+1} + \tilde{\beta}_{n+1} t_a \lambda_s + l_n + (\tilde{a}_{n+1,s} - (t+x))\lambda_s$. This is the lowest possible in-vehicle load of trip $n+1$ when it departs from stop s because the holding time of trip $n+1$ at stop s is not factored in since it is not a decision variable at this time instance.

Consider only the passengers that will arrive while boarding passengers $\tilde{\beta}_{n+1} t_a \lambda_s + l_n + (\tilde{a}_{n+1,s} - (t+x))\lambda_s$ and assume that the number of passenger arrivals during subsequent boardings is negligibly small. That is to say, while boarding passengers $(\tilde{\beta}_{n+1} t_a \lambda_s + l_n + (\tilde{a}_{n+1,s} - (t+x))\lambda_s)t_b \lambda_s$ the number of new passengers arriving at the stop is insignificant because the time duration of $(\tilde{\beta}_{n+1} t_a \lambda_s + l_n + (\tilde{a}_{n+1,s} - (t+x))\lambda_s)t_b^2 \lambda_s$ is infinitesimal and $(\tilde{\beta}_{n+1} t_a \lambda_s + l_n + (\tilde{a}_{n+1,s} - (t+x))\lambda_s)t_b^2 \lambda_s^2 \approx 0$. This assumption results in a closed-form expression of the expected in-vehicle load of trip $n+1$ from stop s. This in-vehicle load should be lower or equal to the capacity of the vehicle that operates trip $n+1$. This is expressed in the inequality constraint of Eq. (13.16).

$$\tilde{l}_{n+1} - \tilde{\beta}_{n+1} + \left(\tilde{\beta}_{n+1} t_a \lambda_s + l_n + (\tilde{a}_{n+1,s} - (t+x))\lambda_s \right)(1 + t_b \lambda_s) \leq c_{n+1} \tag{13.16}$$

Considering the capacity limit of trip $n+1$, it is conjectured that the inequality constraint (13.16) cannot be always satisfied for $x \in \mathbb{R} \mid 0 \leq x \leq \zeta$. Similarly to the capacity constraint of trip n, the capacity constraint of trip $n+1$ expressed in Eq. (13.16) can be perceived as a *soft* constraint which is allowed to be violated if, and only if, our holding time x cannot ensure that there are no stranded passengers by trip $n+1$ at stop s. This soft constraint is added to the objective function as a penalty term $M_2 \max \left[0, \tilde{l}_{n+1} - \tilde{\beta}_{n+1} + \left(\tilde{\beta}_{n+1} t_a \lambda_s + l_n + (\tilde{a}_{n+1,s} - t - x)\lambda_s \right)(1 + t_b \lambda_s) - c_{n+1} \right]$:

$$f(x, \tilde{d}_{n+1,s}, l_n) := \left((t+x) - d_{n-1,s} - H_s\right)^2 + \left(\tilde{d}_{n+1,s} - (t+x) - H_s\right)^2 + M_1 l_n$$
$$+ M_2 \max \Big[0, \tilde{l}_{n+1} - \tilde{\beta}_{n+1} + \left(\tilde{\beta}_{n+1} t_a \lambda_s + l_n + (\tilde{a}_{n+1,s} - t - x)\lambda_s\right)$$
$$(1 + t_b \lambda_s) - c_{n+1} \Big] \tag{13.17}$$

The expected departure time of trip $n+1$ from stop s, $\tilde{d}_{n+1,s}$, is equal to the expected arrival time at stop s, $\tilde{a}_{n+1,s}$, plus the required time for boardings/alightings (dwell time). The required time for boardings/alightings is $\tilde{\beta}_{n+1} t_a$ for passenger alightings and $\left(\tilde{\beta}_{n+1} t_a \lambda_s + l_n + (\tilde{a}_{n+1,s} - (t+x))\lambda_s\right)(1 + t_b \lambda_s) t_b$ for passenger boardings. Note that all $\left(\tilde{\beta}_{n+1} t_a \lambda_s + l_n + (\tilde{a}_{n+1,s} - (t+x))\lambda_s\right)(1 + t_b \lambda_s)$ passengers might not be able to board trip $n+1$ at stop s if its capacity limit is reached. Hence, the required time for passenger boardings is:

$$\min \Big[\left(\tilde{\beta}_{n+1} t_a \lambda_s + l_n + (\tilde{a}_{n+1,s} - (t+x))\lambda_s\right)(1 + t_b \lambda_s) t_b, (c_{n+1} + \tilde{\beta}_{n+1} - \tilde{l}_{n+1}) t_b \Big]$$

This results in the *expected* departure time of trip $n+1$ from stop s:

$$\tilde{d}_{n+1,s} := \tilde{a}_{n+1,s} + \tilde{\beta}_{n+1} t_a + \min \Big[\left(\tilde{\beta}_{n+1} t_a \lambda_s + l_n + (\tilde{a}_{n+1,s} - t - x)\lambda_s\right)(1 + t_b \lambda_s) t_b,$$
$$(c_{n+1} + \tilde{\beta}_{n+1} - \tilde{l}_{n+1}) t_b \Big] \tag{13.18}$$

The above-mentioned constraints form the following regularity-based vehicle holding program, (Q), that determines the holding time x of trip n at stop s.

(Q)

$$\min_{x, \tilde{d}_{n+1,s}, l_n} f(x, \tilde{d}_{n+1,s}, l_n)$$

$$\text{s.t. } l_n = \max(0, \phi_n + x\lambda_s - c_n)$$

$$\tilde{d}_{n+1,s} = \tilde{a}_{n+1,s} + \tilde{\beta}_{n+1} t_a + \min \Big[\left(\tilde{\beta}_{n+1} t_a \lambda_s + l_n + (\tilde{a}_{n+1,s} - t - x)\lambda_s\right)$$
$$\times (1 + t_b \lambda_s) t_b, (c_{n+1} + \tilde{\beta}_{n+1} - \tilde{l}_{n+1}) t_b \Big]$$

$$0 \le x \le \zeta$$

$$l_n \in \mathbb{R}, \tilde{d}_{n+1,s} \in \mathbb{R}$$

$$\tag{13.19}$$

Program (Q) is a nonlinear programming problem (NLP). It is also non-convex because of the non-smooth max and min terms in the first two equality constraints that set the values of the variables $l_n, \tilde{d}_{n+1,s}$ and are not affine functions. Consider the nonlinear constraint $l_n = \max(0, \phi_n + x\lambda_s - c_n)$ where the max term introduces non-smoothness. To rectify this, we replace the equality constraint with the following two inequality constraints:

$$l_n \geq 0$$
$$l_n \geq \phi_n + x\lambda_s - c_n \qquad (13.20)$$

Note that the term $M_1 l_n$ in the objective function forces l_n to receive its lowest possible value which is always greater than or equal to zero and has the equivalent effect of term $M_1 \max(0, \phi_n + x\lambda_s - c_n)$.

The objective function of program $(\bar{Q})$ has another non-smooth term: $M_2 \max\left[0, \tilde{l}_{n+1} - \tilde{\beta}_{n+1} + (\tilde{\beta}_{n+1} t_a \lambda_s + l_n + (\tilde{a}_{n+1,s} - t - x)\lambda_s)(1 + t_b \lambda_s) - c_{n+1}\right]$. With the introduction of another slack variable ν_2 that takes the value of the above term at the solution of the program, the objective function becomes:

$$f(x, \tilde{d}_{n+1,s}, l_n, \nu_2) := \left(t + x - d_{n-1,s} - H_s\right)^2 + \left(\tilde{d}_{n+1,s} - t - x - H_s\right)^2 + M_1 l_n + M_2 \nu_2 \qquad (13.21)$$

and program (Q) is reformulated to

$$
\begin{aligned}
(\hat{Q}) \ \min \ & f(x, \tilde{d}_{n+1,s}, l_n, \nu_2) \\
\text{s.t.} \ & l_n \geq 0 \\
& l_n \geq \phi_n + x\lambda_s - c_n \\
& \nu_2 \geq 0 \\
& \nu_2 \geq \tilde{l}_{n+1} - \tilde{\beta}_{n+1} - c_{n+1} + (\tilde{\beta}_{n+1} t_a \lambda_s + l_n + (\tilde{a}_{n+1,s} - t - x)\lambda_s)(1 + t_b \lambda_s) \\
& \tilde{d}_{n+1,s} = \tilde{a}_{n+1,s} + \tilde{\beta}_{n+1} t_a + \min\left[(\tilde{\beta}_{n+1} t_a \lambda_s + l_n + (\tilde{a}_{n+1,s} - t - x)\lambda_s)(1 + t_b \lambda_s)t_b, (c_{n+1} + \tilde{\beta}_{n+1} - \tilde{l}_{n+1})t_b\right] \\
& 0 \leq x \leq \zeta \\
& l_n \in \mathbb{R}, \tilde{d}_{n+1,s} \in \mathbb{R}, \nu_2 \in \mathbb{R}
\end{aligned}
$$
$$(13.22)$$

The equality constraint that defines the value of variable $\tilde{d}_{n+1,s}$ is the last non-smooth term due to the nonlinear term $\min\left[(\tilde{\beta}_{n+1} t_a \lambda_s + \nu_1 + (\tilde{a}_{n+1,s} - t - x)\lambda_s)(1 + t_b \lambda_s)t_b, (c_{n+1} + \tilde{\beta}_{n+1} - \tilde{l}_{n+1})t_b\right]$. One can re-write $\tilde{d}_{n+1,s}$ as:

$$\tilde{d}_{n+1,s} = \tilde{a}_{n+1,s} + \tilde{\beta}_{n+1} t_a + (\tilde{\beta}_{n+1} t_a \lambda_s + l_n + (\tilde{a}_{n+1,s} - t - x)\lambda_s)(1 + t_b \lambda_s)t_b - \nu_2 t_b \qquad (13.23)$$

because $\tilde{a}_{n+1,s} + \tilde{\beta}_{n+1} t_a + (\tilde{\beta}_{n+1} t_a \lambda_s + l_n + (\tilde{a}_{n+1,s} - t - x)\lambda_s)(1 + t_b \lambda_s)$ $t_b - \nu_2 t_b = \tilde{a}_{n+1,s} + \tilde{\beta}_{n+1} t_a + \min\left[(\tilde{\beta}_{n+1} t_a \lambda_s + l_n + (\tilde{a}_{n+1,s} - t - x)\lambda_s)(1 + t_b \lambda_s)t_b, (c_{n+1} + \tilde{\beta}_{n+1} - \tilde{l}_{n+1})t_b\right]$ at the solution of program $(\hat{Q})$.

To simplify the notation, let $k := 1 + t_b \lambda_s$, where k is a parameter. Then, we can replace $\tilde{d}_{n+1,s}$ in the objective function by $\tilde{a}_{n+1,s} + \tilde{\beta}_{n+1} t_a + (\tilde{\beta}_{n+1} t_a \lambda_s + l_n + (\tilde{a}_{n+1,s} - t - x)\lambda_s)(1 + t_b \lambda_s)t_b - \nu_2 t_b$ and remove the respective equality constraint related to $\tilde{d}_{n+1,s}$. This will result in re-writing the objective function as:

$$f(x, l_n, \nu_2) := \left(t + x - d_{n-1,s} - H_s\right)^2 + \Big[\tilde{a}_{n+1,s} + \tilde{\beta}_{n+1}t_a + \big(\tilde{\beta}_{n+1}t_a\lambda_s + l_n + (\tilde{a}_{n+1,s} - t - x)\lambda_s\big)kt_b$$

$$- \nu_2 t_b - t - x - H_s\Big]^2 + M_1 l_n + M_2 \nu_2$$

$$(13.24)$$

and this leads to the reformulation of program $(\hat{Q})$ to:

$$(\tilde{Q}) \quad \min_{x, l_n, \nu_2} f(x, l_n, \nu_2)$$
$$\text{s.t.} \qquad l_n \geq 0$$
$$\qquad\qquad l_n \geq \phi_n + x\lambda_s - c_n$$
$$\qquad\qquad \nu_2 \geq 0$$
$$\qquad\qquad \nu_2 \geq \tilde{l}_{n+1} - \tilde{\beta}_{n+1} - c_{n+1} + \big(\tilde{\beta}_{n+1}t_a\lambda_s + l_n + (\tilde{a}_{n+1,s} - t - x)\lambda_s\big)k$$
$$\qquad\qquad 0 \leq x \leq \zeta$$
$$\qquad\qquad l_n \in \mathbb{R}, \nu_2 \in \mathbb{R}$$

$$(13.25)$$

This reformulated program $(\tilde{Q})$ has a quadratic objective function and linear inequality constraints and attains an equivalent solution to (Q). Any locally optimal solution of program $(\tilde{Q})$ is also a globally optimal one because the feasible region is a convex set and the objective function is convex. To prove that the objective function is convex, we compute the first-order partial derivatives of $f(x, l_n, \nu_2)$:

$$\frac{\partial f}{\partial x} = 2x + 2(t - d_{n-1,s} - H_s) + 2x(\lambda_s kt_b + 1)^2 - 2(\lambda_s kt_b + 1)\Big[\tilde{a}_{n+1,s} + \tilde{\beta}_{n+1}t_a$$

$$+ \big(\tilde{\beta}_{n+1}t_a\lambda_s + l_n + (\tilde{a}_{n+1,s} - t)\lambda_s\big)kt_b - \nu_2 t_b - t - H_s\Big]$$

$$\frac{\partial f}{\partial l_n} = 2k^2 t_b^2 l_n + 2kt_b\Big[\tilde{a}_{n+1,s} + \tilde{\beta}_{n+1}t_a +$$

$$\big(\tilde{\beta}_{n+1}t_a\lambda_s + (\tilde{a}_{n+1,s} - t - x)\lambda_s\big)kt_b - \nu_2 t_b - t - x - H_s\Big] + M_1$$

$$\frac{\partial f}{\partial \nu_2} = 2t_b^2 \nu_2 - 2t_b\Big[\tilde{a}_{n+1,s} + \tilde{\beta}_{n+1}t_a +$$

$$\big(\tilde{\beta}_{n+1}t_a\lambda_s + l_n + (\tilde{a}_{n+1,s} - t - x)\lambda_s\big)kt_b - t - x - H_s\Big] + M_2$$

Therefore, the *Hessian* matrix of f reads:

$$\mathbf{H} = \begin{bmatrix} \frac{\partial^2 f}{\partial x^2} & \frac{\partial^2 f}{\partial x \partial v_1} & \frac{\partial^2 f}{\partial x \partial v_2} \\[2mm] \frac{\partial^2 f}{\partial v_1 \partial x} & \frac{\partial^2 f}{\partial v_1^2} & \frac{\partial^2 f}{\partial v_1 \partial v_2} \\[2mm] \frac{\partial^2 f}{\partial v_2 \partial x} & \frac{\partial^2 f}{\partial v_2 \partial v_1} & \frac{\partial^2 f}{\partial v_2^2} \end{bmatrix} = \begin{bmatrix} 2 + 2(\lambda_s k t_b + 1)^2 & -2(\lambda_s k t_b + 1)k t_b & 2(\lambda_s k t_b + 1)t_b \\[2mm] -2(\lambda_s k t_b + 1)k t_b & 2k^2 t_b^2 & -2k t_b^2 \\[2mm] 2(\lambda_s k t_b + 1)t_b & -2k t_b^2 & 2t_b^2 \end{bmatrix}$$

To show that the Hessian matrix, $\mathbf{H}$, with elements $H_{ij} \in \mathbf{H}$, is positive semi-definite, it suffices to prove that all the leading principal minors are non-negative:

$$\mathbf{H} \text{ is positive semi-definite} \Leftrightarrow H_{11} \geq 0, \begin{vmatrix} H_{11} & H_{12} \\ H_{21} & H_{22} \end{vmatrix} \geq 0, \ \det(\mathbf{H}) \geq 0.$$

In our case, we have $H_{11} = 2 + 2(\lambda_s k t_b + 1)^2 > 0$.

In   addition, $\begin{vmatrix} H_{11} & H_{12} \\ H_{21} & H_{22} \end{vmatrix} = (2 + 2(\lambda_s k t_b + 1)^2)2k^2 t_b^2 - 4(\lambda_s k t_b + 1)^2 k^2 t_b^2 =$
$4k^2 t_b^2 > 0$.

Furthermore,

$$\det(\mathbf{H}) = (2 + 2(\lambda_s k t_b + 1)^2) \begin{vmatrix} H_{22} & H_{23} \\ H_{32} & H_{33} \end{vmatrix}$$

$$+ 2(\lambda_s k t_b + 1)k t_b \begin{vmatrix} H_{21} & H_{23} \\ H_{31} & H_{33} \end{vmatrix}$$

$$+ 2(\lambda_s k t_b + 1)t_b \begin{vmatrix} H_{21} & H_{22} \\ H_{31} & H_{32} \end{vmatrix}$$

$$= (2 + 2(\lambda_s k t_b + 1)^2) \cdot 0 + 2(\lambda_s k t_b + 1)k t_b \cdot 0$$
$$+ 2(\lambda_s k t_b + 1)t_b \cdot 0 = 0.$$

Thus, f is convex. Using the KKT conditions, Gkiotsalitis and van Berkum [78] proved that the solution of this problem when simplifying the notation by setting $\eta := k t_b$ and $\theta := \tilde{a}_{n+1,s} + \tilde{\beta}_{n+1}t_a + \left(\tilde{\beta}_{n+1}t_a\lambda_s + (\tilde{a}_{n+1,s} - t)\lambda_s\right)k t_b - t - H_s$ is the following holding time.

Two-headway-based holding when considering the capacity of vehicles

$$x^* = \left\{ \max \left(0, \min \left[\zeta, \frac{c_n - \phi_n}{\lambda_s}, \frac{(\lambda_s \eta + 1)\theta - (t - d_{n-1,s} - H_s)}{1 + (\lambda_s \eta + 1)^2} \right] \right) \right\} \qquad (13.26)$$

Table 13.6 Parameter values of the example scenario

Parameter	Value	Unit	Parameter	Value	Unit
$d_{i-1,s}$	1000	s	t_a	1.5	s
t	1500	s	t_b	4	s
H_s	600	s	$a_{n+1,s}$	2500	s
ϕ_n	40	Passengers	ζ	300	s
c_n, c_{n+1}	60	Passengers	M_1	10E+14	-
$\tilde{\beta}_{n+1}$	10	Passengers	M_2	10E+12	-
$\tilde{l}_{n+1}$	50	Passengers	λ_s	0.02	passengers / s

Consider, for instance, an example with trip n that arrives at control point stop s and completes its boardings/alightings at time $t = 1500$ s. The parameters of our scenario are presented in Table 13.6.

Given this input, we have:

- $\eta = kt_b = (1 + t_b\lambda_s)t_b = (1 + 4 \cdot 0.02)4 = 4.32$
- $\theta = \tilde{a}_{n+1,s} + \tilde{\beta}_{n+1}t_a + \left(\tilde{\beta}_{n+1}t_a\lambda_s + (\tilde{a}_{n+1,s} - t)\lambda_s\right)kt_b - t - H_s = 502.696$

and the optimal holding time is:

$$x^* = \left\{ \max\left(0, \min\left[\zeta, \frac{c_n - \phi_n}{\lambda_s}, \frac{(\lambda_s\eta + 1)\theta - (t - d_{n-1,s} - H_s}{1 + (\lambda_s\eta + 1)^2}\right]\right)\right\}$$

$$= \left\{ \max\left(0, \min\left[300, 1000, 296.3534\right]\right)\right\}$$

$$= 296.3534 \text{ seconds.}$$

Closing this section, a summary of selected holding methods for single-variable (myopic) vehicle holding problems is presented in Table 13.7.

Even if their focus was on speed control, it is also worth metnioning the works of [23, 81] since speed control resembles the vehicle holding problem in terms of mathematical formulation.

13.3.2 Multi-variable Vehicle Holding Problems

Examples of multi-variable holding optimization methods are the periodic optimization mathematical programs of [12, 42, 44, 82]. Such mathematical programs simultaneously determine the holding times of all vehicles that are expected to operate within a rolling horizon (*decision epoch*). The optimized holding times are updated in *rolled rolling horizons* when new information becomes available.

We will present here the multi-variable vehicle holding formulation of Eberlein et al. [44] which considers real-time information and assumes that travel times and

Table 13.7 Summary of selected single-variable holding methods [79]

Study	Objective function	Mathematical program	Stochasticity	Solution method	Capacity
[75]	Waiting time and in-vehicle delay	Convex	Considered	Line search	Ignored
[51]	Waiting time and in-vehicle delay	Non-convex	Considered	Multiagent negotiation heuristic	Ignored
[8]	Waiting time	Non-convex	Ignored	First-depart-first-hold rule	Considered
[70]	Equalize headways	No program	Ignored	Closed-form expression	Ignored
[27]	Waiting time	No program	Ignored	Closed-form expression	Ignored
[80]	Guaranteeing a maximum standard deviation from the schedule	Non-convex	Considered	Closed-form expression	Ignored
[78]	Squared deviation from target headway	Convex	Ignored	Closed-form expression	Considered

passenger arrival rates remain constant in rolling horizons with short time duration. This formulation is a quadratic, non-convex program that minimizes the total passenger waiting times.

Consider a bi-directional line with stops $1, 2, \ldots, N/2$ in the *outbound* direction and $N/2 + 1, \ldots, N$ in the *inbound* direction. The inter-station travel time is t_k when traveling from stop $k - 1$ to k. Headway h_{ik} is the time headway between trips $i - 1$ and i at stop k. We also require a minimal safe headway of h_0 between consecutive trips of the same line. In addition, we have a_{ik}, s_{ik} and d_{ik} representing the arrival, dwell, and departure time of trip i at stop k. The in-vehicle passenger load when arriving at stop k is $L_{i,k-1}$ and we assume a constant rate of passenger arrivals at stop k denoted as r_k.

Eberlein et al. [44] assume that passengers alighting at stop k are proportional to the in-vehicle passenger load when arriving at this stop, yielding:

$$A_{ik} := q_k L_{i,k-1}$$

where q_k is the alighting fraction (parameter) of the in-vehicle passenger load of trip i when arriving at stop k. The dwell time of trip i at stop k is an affine function of the boarding and alighting passengers:

$$s_{ik} := c_0 + c_1 B_{ik} + c_2 A_{ik}$$

where $c_0 \geq 0$ is a fixed value indicating the required time to stop even if there are no passenger boardings/alightings, c_1 a parameter indicating the time spent for each boarding, and c_2 a parameter indicating the time spent for each alighting. The number of passenger boardings at stop k is:

$$B_{ik} := r_k h_{ik}$$

and the in-vehicle passenger load:

$$L_{ik} := L_{i,k-1} + B_{ik} - A_{ik} = L_{i,k-1}(1 - q_k) + r_k h_{ik}$$

The departure time of trip i from each stop is:

$$d_{ik} := a_{ik} + s_{ik}$$

and the arrival time of the first trip at each stop $k > 1$:

$$a_{1k} = d_{1,k-1} + t_k$$

for any other trip $i > 1$ its arrival time is:

$$a_{ik} = \max\{d_{i,k-1} + t_k, d_{i-1,k} + h_0\}$$

in order to ensure that we maintain a minimum safety headway h_0 between the departure time of trip $i - 1$ and the arrival time of trip i at stop k. This is important in the case of train operations that share the same track.

When we implement a vehicle holding decision at stop k this will affect the departure times at the downstream stops $k + 1, \ldots, k_t$ where $k_t = N/2$ if stop k is at the outbound direction and $k_t = N$ if it is at the inbound direction. Let us now consider a *decision epoch* where we perform re-optimization by deciding on holding vehicle i and its consecutive trips in I_m at stop k. The impact of this decision will affect consecutive trips $I_m = \{i, i + 1, \ldots, i + m - 1\}$ with trip $i + m$ being the boundary vehicle trip which is not controlled. The choice of the length m of the rolling horizon can impact the optimization phase. For instance, considering all trips until the end of the day might be impractical since this will increase significantly the number of decision variables, whereas the travel times will change in the near future making any decisions about vehicles operating later in the day to be of little use. On the other side, if we only consider a very small number of trips (i.e., only the following trip) we make myopic decisions (see the previously discussed single-variable vehicle holding problems).

The vehicle holding problem is solved dynamically, where at each decision epoch we decide the departure times of trips $i, i + 1, \ldots, i + m - 1$ at control stop k. These new departure times are adjusted to include vehicle holding. It is worth noting that although we decide the departure times of trips $i, \ldots, i + m - 1$ it does not necessarily mean that we will apply the optimal holding times in all trips since we might move to the next decision epoch before all trips $i, \ldots, i + m - 1$ manage to arrive at control stop k. The mathematical program of vehicle holding at a decision epoch considering trip i, which is the first trip expected to arrive at the control stop k, and its subsequent trips $i + 1, \ldots, i + m - 1$ is presented below.

Multi-variate vehicle holding [44]

$$\min \sum_{j \in I_m} \sum_{k'=k}^{k_t} r_{k'}(d_{jk'} - d_{j-1,k'})^2 \tag{13.27}$$

$$\text{s.t. } d_{jk} - a_{jk} - s_{jk} \geq 0 \qquad\qquad \forall j \in I_m \tag{13.28}$$

$$d_{jk'} - a_{jk'} - s_{jk'} = 0 \qquad\qquad \forall j \in I_m, k' : k' > k \tag{13.29}$$

$$d_{i+m,k} - a_{i+m,k} - s_{i+m,k} = 0 \tag{13.30}$$

$$a_{jk'} - d_{j-1,k'} \geq h_0 \qquad\qquad \forall j \in I_m, k' : k' > k \tag{13.31}$$

$$d_{j,k_c} \leq \max(\tau_{j,c}, d^0_{j,k_c}) \qquad\qquad \forall j \in I_m \tag{13.32}$$

$$a_{jk'} = \max(d_{j,k'-1} + t_{k'}, d_{j-1,k'} + h_0) \quad \forall j \in I_m, k' : k' \geq k \tag{13.33}$$

$$s_{jk'} = c_0 + c_1 r_{k'} h_{jk'} + c_2 q_{k'} L_{j,k'-1} \qquad \forall j \in I_m, k' : k' \geq k \tag{13.34}$$

$$L_{jk'} = L_{j,k'-1} + r_{k'} h_{jk'} - q_{k'} L_{j,k'-1} \qquad \forall j \in I_m, k' : k' \geq k \tag{13.35}$$

$$d_{j,k'}, a_{j,k'}, s_{j,k'}, L_{j,k'} \in \mathbb{R}_{\geq 0} \qquad \forall j \in I_m, k' \in \{k, \ldots, k_t\} \tag{13.36}$$

The objective function (13.27) strives to minimize the total waiting time of passengers arriving at stops $k' = k, k + 1, \ldots, k_t$ by minimizing the squared differences between the departure times of consecutive trips. Constraints (13.28) ensure that each trip $j \in I_m$ can leave the holding station k either when it finishes its passenger alighting and boarding, or later, i.e., in case a holding is applied to that trip. Constraints (13.29) ensure that trips $j \in I_m$ will leave any other stop $k' > k$ immediately after they finish their boardings and alightings. Constraint (13.30) ensures that we will not apply holding to the boundary trip $i + m$. Constraints (13.31) enforce the safety headway requirement. Constraints (13.32) ensure that if a vehicle is already delayed for its next trip, it will not be held to avoid schedule sliding. In particular, $\tau_{j,c}$ is the scheduled departure time at the next dispatching terminal k_c minus the minimal layover time. Constraints (13.33)–(13.35) set the values of arrival times, dwell times and in-vehicle passenger loads.

The decision variables of this problem are the departure times d_{jk} of each trip $j \in I_m$ at the control stop k. The holding times of trips $j \in I_m$ at control stop k can be derived as $d_{jk} - (a_{jk} + s_{jk})$. This formulation is considered to be a multi-variate vehicle holding formulation because at a specific decision epoch we determine the holding times of multiple trips at a control stop. In contrast, single-variable vehicle holding formulations determined the vehicle holding time of one trip at a time. The mathematical formulation is nonlinear. The feasible region is not convex due to equality constraints (13.33) that are not affine functions.

Table 13.8 Summary of selected multi-variable holding methods [79]

Study	Objective function	Mathematical program	Stochasticity	Solution method	Capacity
[42]	Waiting time and in-vehicle delay	Non-convex	Ignored	Algorithm of [83]	Considered
[72]	Waiting time, in-vehicle delay and extra waiting time of stranded passengers	Non-convex	Ignored	MINOS solver	Considered
[44]	Waiting time	Non-convex	Ignored	Purpose-built heuristic	Ignored
[84]	Waiting time and in-vehicle delay	Convex	Ignored	Decomposition-based heuristic	Ignored
[85]	Waiting time and extra waiting time of stranded passengers	Convex	Ignored	Simulated annealing	Considered
[86]	Waiting time, in-vehicle delay and extra waiting time of stranded passengers	Non-convex	Ignored	Decomposition-based heuristic	Considered
[6]	Waiting time	No program	Considered	Dynamic programming	Ignored
[74]	Waiting time and in-vehicle delay	Non-convex	Ignored	Branch and bound	Ignored
[15]	Waiting time, in-vehicle delay and extra waiting time of stranded passengers	Non-convex	Considered	Genetic algorithm	Considered
[7]	Waiting time, in-vehicle delay and extra waiting time of stranded passengers	Non-convex	Ignored	MINOS solver	Considered

Closing this section, a summary of selected holding methods for multi-variable vehicle holding problems is presented in Table 13.8.

13.4   Stop-skipping

Stop-skipping (also known as *expressing*) refers to skipping a stop (or a series of stops) to achieve a specific objective. Typical objectives are the minimization of one or a combination of the following:

– Waiting time of passengers;
– In-vehicle time of passengers;
– Vehicle travel time;
– Reduction of control actions;

– Deviation from the planned schedule;
– Unsatisfied demand;
– Inconvenience of passengers that need to alight when skipping their stop.

Stop-skipping strategies can be devised at either the *tactical* planning level or the *operational* level (dynamic stop skipping). Depending on the level of control, the objectives of a stop-skipping strategy might differ. At the tactical planning stage, the focus is on developing reliable, resilient or robust strategies that will maintain a good performance in case of disruptions during actual operations. Stop-skipping can be addressed at the tactical planning stage where a stop-skipping plan is devised before the start of the daily operations and is not updated ever since [87, 88]. The advantage of a fixed stop-skipping plan is that it can be communicated to the drivers and passengers well in advance. However, a fixed stop-skipping plan is inflexible and cannot be modified during actual operations.

In contrast to stop-skipping at the tactical planning level, dynamic stop-skipping strategies at the operational level are reactionary and oftentimes less sophisticated in order to reduce the computational costs. In dynamic stop skipping, the skipped stops of a vehicle trip are typically determined just before its dispatching from the first stop [13, 26, 89, 90].

Stop-skipping can correct service inconsistencies due to the inherent travel time and passenger demand variations, but might result in increased waiting times for passengers waiting at the skipped stops [2]. Thus, most stop-skipping approaches address the problem holistically considering the waiting times of passengers, their in-vehicle times, and the total vehicle trip travel times. The two former objectives concern the passenger-related costs, whereas the last objective concerns the cost of the operator.

Unlike rescheduling and holding, stop-skipping impacts the composition of the service line since many stops might not be served. That is, a number of origin-destination pairs might not be accommodated when applying stop-skipping. This has major implications for the service of passenger demand. Potential issues worthy of investigation are the locations of stops that can be skipped to ensure that critical stops with significant passenger demand are not skipped. The stops that can be skipped might also differ from vehicle trip to vehicle trip to ensure that a stop cannot be skipped by consecutive vehicle trips leading to excessive waiting times for passengers waiting at that stop.

There are two main approaches for determining the skipped stops of vehicle trips at the operational level:

• Approaches that determine the skipped stops of one vehicle trip at a time [1, 26]. In these approaches, the decisions are made when the vehicle trip is about to be dispatched.
• Approaches that determine the skipped stops of a set of vehicle trips I_m in a rolling horizon [87].

The aforementioned approaches differ in their decision mechanisms, where the former focuses on one trip i at a time and the latter on a set of trips $I_m = \{i, i + 1, \ldots\}$

in a rolling horizon. This split is similar to the single-variate and multi-variate vehicle holding decision strategies. In the remainder, we present formulations of the single-trip stop-skipping approaches and the stop-skipping approaches in rolling horizons.

13.4.1 Single-trip Stop-skipping

If one focuses on the stop-skipping decisions of a single vehicle, the solution space comprises $2^{|S|}$ different options, where $|S|$ is the total number of stops that can be optionally skipped. Despite the exponential increase of the solution space, it is possible to explore the solution space of all available stop-skipping options in medium-sized service lines with up to 20 stops [1, 26].

Addressing the stop-skipping problem at the operational level requires computing a stop-skipping solution in near real-time. Given the computational complexity of the stop-skipping problem, we typically devise a stop-skipping strategy for one trip at a time to reduce the size of the solution space [4].

Inherently, stop-skipping is a binary, 0–1 optimization problem where a stop can be skipped (0) or served (1). If we consider a single trip that operates in a service line with $S = \{1, 2, \ldots, |S|\}$ stops and we can skip/serve every stop, the solution space of potential skip/serve options is $2^{|S|}$ (see Fig. 13.2). This exponential increase of the solution space with the number of stops allows evaluating all solution space options with the use of *brute-force* for lines that typically do not exceed 20 stop-skipping candidate stops [26].

Fu et al. [26] presented a stop-skipping formulation for the single-trip stop-skipping problem at the operational level. This formulation determines the skipped stops of a vehicle trip i when it is about to be dispatched. This formulation assumes that the previous vehicle $i - 1$ and the following vehicle trip $i + 1$ will serve all their stops. This is enforced to avoid skipping stops by consecutive vehicles because that could lead to increased passenger waiting times and inconvenience. The formulation of Fu et al. [26] is expressed below.

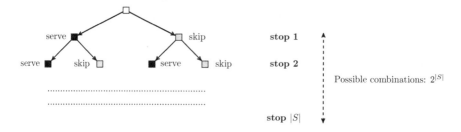

Fig. 13.2 Potential serve/skip options for a trip operating in a line with $|S|$ stops

Sets: Consider a vehicle trip i that is about to be dispatched. We use set $S = \{1, \ldots, s, \ldots, |S|\}$ containing all stops of the service line ordered from the first to the last. We also consider a set $I = \{i - 1, i, i + 1\}$ of the preceding and following vehicle trips of trip i.

Parameters: Consider $\mathbf{t} = [t_1, \ldots, t_{|S|-1}]^{\mathsf{T}}$ to be the inter-station travel times between stop $s - 1$ and s, where $s \in S \setminus \{|S|\}$. The average boarding and alighting times per passenger are r_1 and r_2, respectively. The departure times of vehicle trip $i - 1$ that is already operational when deciding about the skipped stops of vehicle trip i are $d_{i-1,s}$ for any $s \in S$. We also know the dispatching times d_{n1} of all vehicles $n \in I$ which are considered to be fixed. Every time the trip serves a stops it also needs a fixed time δ for acceleration and deceleration purposes ($\frac{\delta}{2}$ for each). The average passenger arrival rate at stop s for passengers whose destination is stop $y > s$ is denoted as λ_{sy}. Obviously, $\lambda_{sy} = 0$ for any $y \in S : y \leq s$. Because the problem has multiple objectives, we use weight factors c_1, c_2 and c_3 indicating the unit time value associated with the passenger waiting times, the passenger in-vehicle travel times, and the vehicle operation times, respectively. This allows to employ the weighted-sum method and use a single objective function instead of solving a multi-objective optimization problem.

Variables: Consider d_{ns}, a_{ns} and k_{ns} to be the departure, arrival and dwell time of vehicle trip $n \in \{i, i + 1\}$ at stop s. In addition, h_{ns} is the departure headway between trip $n - 1$ and n at stop s for $n \in I \setminus \{i - 1\}$. The number of passengers waiting for vehicle trip $n \in I$ and traveling from stop s to y is $w_{n,sy}$, where $w_{n,sy} = 0$ for any $y \in S : y \leq s$. The number of passengers traveling from stop s to stop y skipped by vehicle trip n is denoted by $l_{n,sy}$, where $l_{n,sy} = 0$ for any $y \in S : y \leq s$. In addition, because vehicle trip $i - 1$ is not allowed to skip any stops and because the capacity is assumed not to be a restrictive factor, $l_{i-1,sy} = 0$ for any $s \in S$, $y \in S$. The number of passengers at stop s skipped by vehicle trip n is denoted by m_{ns}, where $m_{ns} = \sum_{y=s+1}^{|S|} l_{n,sy}$. The number of passengers boarding vehicle trip n at stop s is denoted by u_{ns}. Because there are no boardings at the last stop, $u_{n|S|} = 0$ for any n. The number of passengers boarding vehicle trip n at stop s whose destination is y is denoted by $b_{n,sy}$. Note that $b_{n,sy} = 0$ for $y \in S : y \leq s$. The number of passengers alighting vehicle trip n at stop s is denoted by ν_{ns}. Because there are no alightings at the first stop, $\nu_{n1} = 0$ for any n. Finally, the decision variables are binary $x_{ns} \in \{0, 1\}$, where $x_{ns} = 1$ if vehicle trip serves stop s and $x_{ns} = 0$ if vehicle trip s does not serve stop s for any $n \in \{i, i + 1\}$ and $s \in S$. Note that $x_{i+1,s} = 1$ because vehicle trip $i + 1$ is our boundary trip and we assume that it will serve all stops. That is, we only decide about the values of x_{is} for all $s \in S$. In addition, the first and the last stop of the line cannot be skipped resulting in $x_{n1} = x_{n|S|} = 1$ for all $n \in \{i, i + 1\}$.

Using the aforementioned sets, parameters and variables, the formulation of the single-trip stop skipping problem when deciding about the skipped stops of a trip i is provided below. We note that the objective function is:

$$Z := c_1 \sum_{n=i}^{i+1} \sum_{s=1}^{|S|} \left[(u_{n,s} - m_{n-1,s}) \frac{h_{n,s}}{2} + m_{n-1,s} \left(\frac{h_{n-1,s}}{2} + h_{n,s} \right) \right]$$

$$+ c_2 \sum_{n=i}^{i+1} \sum_{s=1}^{|S|-1} \sum_{y=s+1}^{|S|} \left[b_{n,sy} \sum_{z=s+1}^{y} (t_z + (k_{nz} + \delta) x_{n,z}) \right] \qquad (13.37)$$

$$+ c_3 \sum_{n=i}^{i+1} \sum_{s=2}^{|S|} (t_s + (k_{ns} + \delta) x_{n,s})$$

where the generalized cost of the objective function includes three terms multiplied by c_1, c_2 and c_3, respectively. The first term includes two components. The first component, $(u_{n,s} - m_{n-1,s}) \frac{h_{n,s}}{2}$, computes the total waiting time of the passengers arriving after the departure of vehicle trip $n-1$ at stop s, assuming random arrivals with an average passenger waiting time equal to half the headway. The second component, $m_{n-1,s} (\frac{h_{n-1,s}}{2} + h_{n,s})$, represents the total waiting time of those passengers who have been stranded by vehicle trip $n-1$ ($m_{n-1,s}$) and have to wait for an average amount of time equal to $m_{n-1,s} \left(\frac{h_{n-1,s}}{2} + h_{n,s} \right)$ because we do not allow two consecutive vehicle trips to skip the same origin-destination pair. The second term of the objective function calculates the total in-vehicle time of passengers summed over all origin-destination pairs and the final term computes the total vehicle trip times. The complete formulation is provided in the following mathematical program.

Single-trip stop skipping problem [26]

min Z (13.38)

s.t. Eq. (13.37) (13.39)

$$a_{ns} = d_{n,s-1} + t_s + \frac{\delta}{2} x_{n,s-1} + \frac{\delta}{2} x_{ns} \qquad \forall n \in \{i, i+1\}, \forall s \in S \setminus \{1\} \qquad (13.40)$$

$$d_{ns} = a_{ns} + k_{ns} \qquad \forall n \in \{i, i+1\}, \forall s \in S \qquad (13.41)$$

$$h_{ns} = d_{ns} - d_{n-1,s} \qquad \forall n \in \{i, i+1\}, \forall s \in S \setminus \{1\} \qquad (13.42)$$

$$k_{ns} = r_1 u_{ns} + r_2 \nu_{ns} \qquad \forall n \in \{i, i+1\}, \forall s \in S \qquad (13.43)$$

$$b_{n,sy} = x_{ns} w_{n,sy} x_{ny} \qquad \forall n \in \{i, i+1\}, \forall s \in S, \forall y \in S : y > s \qquad (13.44)$$

$$u_{ns} = x_{ns} \sum_{y=s+1}^{|S|} w_{n,sy} x_{ny} \qquad \forall n \in \{i, i+1\}, \forall s \in S \setminus \{|S|\} \qquad (13.45)$$

$$\nu_{ns} = x_{ns} \sum_{y=1}^{s-1} w_{n,ys} x_{ny} \qquad \forall n \in \{i, i+1\}, \forall s \in S \setminus \{1\} \qquad (13.46)$$

$$w_{n,sy} = l_{n-1,sy} + \lambda_{sy} h_{ns} \qquad \forall n \in \{i, i+1\}, \forall s \in S, \forall y \in S : y > s \qquad (13.47)$$

$$l_{n,sy} = w_{n,sy} - w_{n,sy}x_{ns}x_{ny} \qquad \forall n \in \{i, i+1\}, \forall s \in S, \forall y \in S : y > s \tag{13.48}$$

$$m_{ns} = \sum_{y=s+1}^{|S|} l_{n,sy} \qquad \forall n \in \{i, i+1\}, s \in S \setminus \{|S|\} \tag{13.49}$$

$$b_{n,sy} = w_{n,sy} = l_{n,sy} = 0 \qquad \forall n \in \{i, i+1\}, \forall s \in S, \forall y \in S : y \leq s \tag{13.50}$$

$$x_{i+1,s} = 1 \qquad \forall s \in S \tag{13.51}$$

$$x_{is} \in \{0, 1\} \qquad \forall s \in S \tag{13.52}$$

$$x_{i1} = x_{i|S|} = 1 \tag{13.53}$$

Constraints (13.40) set the arrival time of vehicle trip $n = i$ or $i + 1$ equal to the departure time at the previous stop $d_{n,s-1}$ plus the inter-station travel time t_s plus the time lost in acceleration and deceleration depending on whether vehicle trip n served stop $s - 1$ and/or s. If $n = i + 1$ we can write $a_{i+1,s}$ as $d_{i+1,s-1} + t_s$ because trip $i + 1$ will serve all stops. Constraints (13.41) set the departure time of each vehicle trip n at stop s equal to the arrival time at s plus the dwell time. Constraints (13.42) return the time headways upon departure at stop s between vehicle trips $n - 1$ and n. Constraints (13.43) return the dwell time of vehicle trip n at stop s based on the number of passengers who will board, u_{ns}, and alight, v_{ns}, at s.

Constraints (13.44) return the number of passengers boarding vehicle trip n at stop s whose destination is y. Note that if vehicle trip n skips wither stop s or y, then $b_{n,sy} = 0$ because these passengers will wait for the next trip. Constraints (13.45) return the number of passengers boarding vehicle trip n at stop s. Constraints (13.46) return the number of passengers alighting vehicle trip n at stop s. Constraints (13.47) return the number of passengers waiting for vehicle trip n and traveling from stop s to stop $y > s$. This is equal to the number of stranded (skipped) passengers from the previous vehicle trips $n - 1$ that travel from s to y, $l_{n-1,sy}$, plus the headway h_{ns} multiplied by the passenger arrival rate λ_{sy} for the origin-destination pair (s, y). Constraints (13.48) return the number of skipped passengers by vehicle trip n whose origin-destination pair is (s, y) assuming that if vehicle trip n skips either s or y all passengers from s to y will wait for the next trip. Constraints (13.48) return the total number of skipped passengers at stop s by vehicle n.

Constraints (13.49) set the values of variables $b_{n,sy}$, $w_{n,sy}$, $l_{n,sy}$ to 0 if the destination stop y is before the origin stop s because such option is invalid. Constraints (13.50) ensure that the following trip $i + 1$ will serve all stops and constraints (13.52) ensure that the first and the last stop of the line cannot be skipped by any vehicle trip.

The resulting mathematical program is mixed-integer with binary and continuous decision variables. In addition, it is nonlinear with nonlinearities both in the objective function and the constraints. Solving this model to global optimality is achieved with brute-force by evaluating all possible combinations of the decision variables x_{is} resulting in $2^{|S|}$ evaluations. Considering that the first and the last stop of the line cannot be skipped, the number of combinations can be reduced to $2^{|S|-2}$. In addition, if the service provider allows us to skip only a small number of stops belonging to a subset S' of S the possible combinations can be reduced even further to $2^{|S'|}$. The latter is very common since it is typically not possible to skip any stop of the line indiscriminately (i.e., skipping stops with high passenger demand or stops with line transfers might be prohibited for practical reasons). Taking advantage of the fact that we decide about the skipped stops of only one trip, we can compute globally optimal solutions with brute-force using the aforementioned model for networks with, approximately, $|S'| = 20$ stops [26].

13.4.2 Stop-skipping of Multiple Trips in Rolling Horizons

We discussed about the skipped stops of a single trip at the operational level. Other approaches calculate a stop-skipping plan for several trips in a rolling horizon [87] that might even cover all remaining trips until the end of the day in extreme cases. Such approaches cannot return a globally optimal solution within a reasonable computation time because of the exponential increase of the problem size when considering more than one trip. To this end, these problems are typically solved with metaheuristics that try to return a sub-optimal solution within a reasonable time. In more detail, if we consider that vehicle trip i is about to be dispatched and we want to compute the optimal skipped stops of vehicle trip i and its following vehicle trips $i + 1, i + 2, \ldots, |N|$, devising an optimal stop-skipping plan requires evaluating all the solutions from the solution space $2^{|N||S|}$ which grows exponentially with the number of trips and stops of the line.

Typical assumptions of model formulations for the stop-skipping problem in rolling horizons are (see [79]):

- Vehicle trips that serve the same line do not overtake each other. This is a common assumption in past studies [80, 91, 92];
- Passenger arrivals at stops are random [35, 93];
- Passengers traveling between any origin-destination pair cannot be skipped by two consecutive vehicle trips of the same line [1, 4, 26].

These assumptions were also used at the single-trip stop-skipping problem. There are some differences, however. First, the assumption that vehicle trips do not overtake each other might not always be realistic if we have multiple trips in a rolling horizon. Second, there are no hard assumptions about forcing a vehicle trip to serve all stops if its previous trip skipped some of them. The only requirement is to not allow the same origin-destination pair to be skipped twice.

Before proceeding to the modeling, we introduce the nomenclature of Table 13.9. Similarly to the single-trip stop-skipping formulation, we have that:

$$l_{n,sy} := \begin{cases} 0, & \text{if } y \leq s \\ w_{n,sy} - w_{n,sy}x_{n,s}x_{n,y}, & \text{if } y > s \end{cases} \tag{13.54}$$

for any $n \in N, s \in S \setminus \{|S|\}, y \in S$. The number of passengers at stop s skipped by vehicle trip n is:

$$m_{n,s} := \sum_{y=s+1}^{|S|} l_{n,sy}, \quad \forall n \in N, s \in S \setminus \{|S|\} \tag{13.55}$$

The number of passengers waiting for bus n at stop s whose destination is stop $y > s$ is:

$$w_{n,sy} := \begin{cases} l_{n-1,sy} + \lambda_{sy}h_{n,s}, & \forall n \in N \setminus \{1\}, \forall s \in S \setminus \{|S|\} \\ \tilde{w}_{1,sy}, & \text{for } n = 1, \forall s \in S \setminus \{|S|\} \end{cases} \tag{13.56}$$

Note that $w_{n,sy} = \tilde{w}_{1,sy}$, for $n = 1, \forall s \in S \setminus \{|S|\}$ reflects the boundary condition which is imposed at the first trip of the rolling horizon. The value of $\tilde{w}_{1,sy}$ does not depend on the decisions in this rolling horizon. Hence, in the current rolling horizon $\tilde{w}_{1,sy}$ is a parameter.

The expected number of passengers who will board trip n at stop s (assuming trip n stops at stop s) is:

$$u_{n,s} := x_{n,s} \sum_{y=s+1}^{|S|} w_{n,sy}x_{n,y}, \quad \forall n \in N, s \in S \setminus \{|S|\} \tag{13.57}$$

and we also have the boundary condition:

$$u_{n,|S|} = 0, \quad \forall n \in N \tag{13.58}$$

From the total amount of passengers boarding vehicle trip n at stop s ($u_{n,s}$), the number of passengers boarding trip n at stop $s \in S \setminus \{|S|\}$ whose destination is stop y is:

Table 13.9 Nomenclature

Sets	
N	Ordered set of vehicle trips in a rolling horizon, $N = \{1, \ldots n, \ldots, \lvert N \rvert\}$. Trip 1 is the first trip to be dispatched in this rolling horizon;
S	Set of the ordered public transport stops of the line, $S = \{1, \ldots, s, \ldots, \lvert S \rvert\}$;
Parameters	
$\mathbf{T}$	Is a $\lvert N \rvert \times (\lvert S \rvert - 1)$ matrix of inter-station travel times. $t_{n,s}$ is the expected inter-station travel time of the n-th trip between stop $s-1$ and s, where $s \in S \setminus \{1\}$;
$\mathbf{g}$	Is an $\lvert N \rvert$-valued vector indicating the capacity of each vehicle trip $n \in N$;
r_1	Average boarding time per passenger;
r_2	Average alighting time per passenger;
δ	Average bus acceleration plus deceleration time for serving a bus stop;
$\boldsymbol{\Lambda}$	$\lvert S \rvert \times \lvert S \rvert$ matrix where each element λ_{sy} denotes the average passenger arrival rate at stop s whose destination is stop y (note: $\lambda_{sy} = 0$, $\forall 1 \leq y \leq s$);
c_1	Unit time value associated with the passenger waiting times;
c_2	Unit time value associated with the passenger in-vehicle travel time;
c_3	Unit time value associated with the vehicle operation time;
$\tilde{d}_{n,1}$	Planned departure time of every trip $n \in N$ from the first stop;
$\tilde{w}_{1,sy}$	Number of passengers waiting for trip 1, which is the first trip of the rolling horizon, and traveling from stop $s \in S$ to stop $y \in S$;
$\tilde{x}_s$	$\tilde{x}_s = 0$ if the already dispatched trip that precedes trip 1 will skip stop $s \in S$, and $\tilde{x}_s = 1$ otherwise;
Decision variables	
$\mathbf{x}$	$\lvert N \rvert \times \lvert S \rvert$-dimensional matrix of the decision variables where each $x_{n,s} \in \mathbf{x}$ can take a binary value $\{0, 1\}$ with $x_{n,s} = 1$ denoting that the n-th bus trip will serve stop s.
Variables	
$\mathbf{D}$	$\lvert N \rvert \times \lvert S \rvert$ matrix of departure times where $d_{n,s} \in \mathbb{R}_{\geq 0}$ is the departure time of trip n from stop s;
$\mathbf{A}$	$\lvert N \rvert \times \lvert S \rvert$ matrix of arrival times where $a_{n,s} \in \mathbb{R}_{\geq 0}$ is the arrival time of trip n at stop s;
$\mathbf{K}$	$\lvert N \rvert \times \lvert S \rvert$ matrix of dwell times where $k_{n,s} \in \mathbb{R}_{\geq 0}$ is the dwell time of trip n at stop s;
$\mathbf{H}$	$(\lvert N \rvert - 1) \times \lvert S \rvert$ matrix of headways where $h_{n,s} \in \mathbb{R}_{\geq 0}$ is the time headway between the departure of trip $n-1$ and the arrival of trip n at stop s, where $n \in N \setminus \{1\}$ and $s \in S$;
$\mathbf{W}$	$\lvert N \rvert \times \lvert S \rvert \times \lvert S \rvert$ matrix where each $w_{n,sy} \in \mathbf{W}$ denotes the number of passengers waiting for trip n and traveling from stop s to y (note: $w_{n,sy} = 0$, $\forall y \leq s$);
$\mathbf{L}$	$\lvert N \rvert \times \lvert S \rvert \times \lvert S \rvert$ matrix where each $l_{n,sy} \in \mathbb{R}_{\geq 0}$ denotes the number of passengers traveling from stop s to stop y skipped by trip n (note: $l_{n,sy} = 0$, $\forall y \leq s$);
$\mathbf{M}$	$\lvert N \rvert \times \lvert S \rvert$ matrix where each $m_{n,s} \in \mathbb{R}_{\geq 0}$ denotes the number of passengers at stop s skipped by trip n, where $n \in N$, $s \in S$ (note: $m_{n,s} = \sum\limits_{y=s+1}^{\lvert S \rvert} l_{n,sy}$);
$\mathbf{U}$	$\lvert N \rvert \times \lvert S \rvert$ matrix where each $u_{n,s} \in \mathbb{R}_{\geq 0}$ denotes the number of passengers boarding trip n at stop s, where $n \in N$, $s \in S$ (note: $u_{n,\lvert S \rvert} = 0$, $\forall n \in N$);
$\mathbf{B}$	$\lvert N \rvert \times \lvert S \rvert \times \lvert S \rvert$ matrix where each $b_{n,sy} \in \mathbb{R}_{\geq 0}$ denotes the number of passengers boarding trip n at stop s whose destination is stop y (note: $b_{n,sy} = 0$, $\forall y \leq s$);
$\mathbf{V}$	$\lvert N \rvert \times \lvert S \rvert$ matrix where each $\nu_{n,s} \in \mathbb{R}_{\geq 0}$ denotes the number of passengers alighting trip n at stop s, where $n \in N$, $s \in S$ (note: $\nu_{n,1} = 0$, $\forall n \in N$);
$\boldsymbol{\mu}$	$\lvert S \rvert$-valued vector, where each $\mu_s \in \mathbb{R}_{\geq 0}$ denotes the average passenger arrival rate at stop s (note: $\mu_s = \sum\limits_{y=s+1}^{\lvert S \rvert} \lambda_{sy}$).
$\boldsymbol{\Gamma}$	$\lvert N \rvert \times \lvert S \rvert - 1$ matrix where each $\gamma_{n,s} \in \mathbb{R}_{\geq 0}$ denotes the in-vehicle passenger load of trip n when traveling from stop s to stop $s+1$.

$$b_{n,sy} := \begin{cases} x_{n,s} w_{n,sy} x_{n,y}, & \text{if } y > s \\ 0, & \text{if } y \le s \end{cases} \tag{13.59}$$

The expected number of alighting passengers for trip n at stop s depends on the number of passengers traveling between stops y and s ($y < s$) and whether the trip will serve stop y:

$$\nu_{n,s} := x_{n,s} \sum_{y=1}^{s-1} w_{n,ys} x_{n,y}, \quad \forall n \in N, s \in S \setminus \{1\} \tag{13.60}$$

A special case is the first stop which requires the boundary condition:

$$\nu_{n,1} = 0, \quad \forall n \in N \tag{13.61}$$

The dwell time of each vehicle trip n at each stop s is:

$$k_{n,s} := r_1 u_{n,s} + r_2 \nu_{n,s}, \quad \forall n \in N, s \in S \setminus \{1\} \tag{13.62}$$

Note that if passengers use different door channels for boardings/alightings; then, the dwell time can be expressed as $k_{n,s} := \max\left(r_1 u_{n,s}; r_2 \nu_{n,s}\right)$.

The in-vehicle passenger load of any trip $n \in N$ when it is traveling from stop s to stop $s + 1$ is also derived by:

$$\gamma_{n,s} := \begin{cases} u_{n,1}, & \text{if } s = 1 \\ \gamma_{n,s-1} + u_{n,s} - \nu_{n,s}, & \text{if } 1 < s < |S| \end{cases} \tag{13.63}$$

When deciding about the skipped stops, it will be beneficial if the in-vehicle passenger load of any trip n when traveling from stop s to $s + 1$, $\gamma_{n,s}$, does not exceed its capacity g_n. This can be achieved by adding the inequality constraint:

$$\gamma_{n,s} \le g_n, \quad \forall n \in N, \ \forall s \in S \setminus \{|S|\} \tag{13.64}$$

The arrival time of bus trip n at stop s is:

$$a_{n,s} := d_{n,s-1} + t_{n,s} + \frac{\delta}{2}(x_{n,s-1} + x_{n,s}), \quad \forall n \in N, s \in S \setminus \{1, 2\} \tag{13.65}$$

Equation (13.65) requires a boundary condition for the arrival time at the second stop, $a_{n,2}$, $\forall n \in N$. This is provided by the originally planned dispatching time $\tilde{d}_{n,1}$ of every trip $n \in N$:

$$a_{n,2} := \tilde{d}_{n,1} + t_{n,2} + \frac{\delta}{2}(x_{n,1} + x_{n,2}), \quad \forall n \in N \tag{13.66}$$

In addition, the departure time of vehicle trip n from stop $s \in S \setminus \{1\}$ is equal to its arrival time at that stop plus the dwell time $k_{n,s}$:

$$d_{n,s} := a_{n,s} + k_{n,s}, \ \forall n \in N, s \in S \setminus \{1\} \tag{13.67}$$

Assuming that overtaking is not allowed, the time headway between the arrival of trip n at stop s and the departure of its preceding one from stop s is:

$$h_{n,s} := a_{n,s} - d_{n-1,s}, \ \forall n \in N \setminus \{1\}, s \in S \setminus \{1\} \tag{13.68}$$

Finally, note that the time headway at the first stop is calculated based on the boundary condition that considers planned departure times of the respective trips:

$$h_{n,1} := \tilde{d}_{n,1} - \tilde{d}_{n-1,1}, \ \forall n \in N \setminus \{1\} \tag{13.69}$$

Let us now consider the objective function:

$$Z := c_1 \sum_{n=2}^{|N|} \sum_{s=1}^{|S|-1} \left[(u_{n,s} - m_{n-1,s}) \frac{h_{n,s}}{2} + m_{n-1,s} \left(\frac{h_{n-1,s}}{2} + k_{n-1,s} + h_{n,s} \right) \right]$$
$$+ c_2 \sum_{n=2}^{|N|} \sum_{s=1}^{|S|-1} \sum_{y=s+1}^{|S|} \left[b_{n,sy} \sum_{z=s+1}^{y} (t_{n,z} + (k_{n,z} + \delta)x_{n,z}) \right] \tag{13.70}$$
$$+ c_3 \sum_{n=2}^{|N|} \sum_{s=2}^{|S|} (t_{n,s} + (k_{n,s} + \delta)x_{n,s})$$

where the generalized cost of the objective function includes three terms. The first term includes two components. The first component, $(u_{n,s} - m_{n-1,s})\left(\frac{h_{n,s}}{2} \right)$, computes the total waiting time of the passengers who arrive after the departure (or passing) of bus $n-1$ at stop s, assuming random arrivals with an average passenger waiting time equal to half the headway. The second component represents the total waiting time of those passengers who have been stranded by vehicle $n-1$ ($m_{n-1,s}$) and have to wait for an average amount of time equal to $m_{n-1,s}\left(\frac{h_{n-1,s}}{2} + k_{n-1,s} + h_{n,s} \right)$ because we do not allow two consecutive bus trips to skip the same origin-destination pair. The second term of the objective function calculates the total in-vehicle time of passengers summed over all O-D pairs and the final term computes the total vehicle trip time.

Incorporating the previously formulated vehicle movement equations yields the following mathematical program.

Stop-skipping of multiple vehicle trips in rolling horizons considering vehicle capacities [79]

$$\min \ Z \tag{13.71}$$

$$\text{s.t. Eqs. (13.54)} - (13.70) \tag{13.72}$$

$$x_{n,1} = x_{n,|S|} = 1 \qquad\qquad \forall n \in N \tag{13.73}$$

$$(x_{n-1,s} x_{n-1,y}) + (x_{n,s} x_{n,y}) \geq 1 \quad \forall n \in N \setminus \{1\}, \forall s \in S, \forall y \geq s \tag{13.74}$$

$$(\tilde{x}_s \tilde{x}_y) + (x_{1,s} x_{1,y}) \geq 1 \qquad \forall s \in S, \ \forall y \geq s \tag{13.75}$$

$$x_{n,s} \in \{0, 1\} \qquad\qquad \forall n \in N, \ \forall s \in S \tag{13.76}$$

The equality constraints (13.73) ensure that the first and last stop of each trip $n \in N$ cannot be skipped. The inequality constraints (13.74)–(13.75) ensure that if an origin-destination pair is skipped by one trip, it will be served by its next one.

The mathematical program is a mixed-integer nonlinear programming problem with binary and continuous variables and it is solved every time a new rolling horizon starts. The problem can be solved to global optimality with brute-force (exhaustive search of the solution space) for small instances or with metaheuristics that can return a suboptimal solution in larger instances.

Let us consider an example with a circular public transport line with 5 stops. We consider $|N| = 4$ trips in the rolling horizon and fixed travel times $t_{n,s} = 60$ s, $\forall n \in N$, $\forall s \in S \setminus \{1\}$. The passenger arrival rate at stop $s \in S$ with destination $y \in S$ is set to $\lambda_{n,sy} = 1$ passenger per minute if $y > s$ and 0 if $y \leq s$. The number of passengers waiting for trip 1, which is the first in this rolling horizon, and traveling from stop $s \in S$ to stop $y \in S$ is set as $\tilde{w}_{1,sy} = 12$ if $y > s$ and 0 otherwise. The capacity of each vehicle $n \in N$ is $g_n = 75$ passengers and the previous trip that precedes the first trip in our rolling horizon does not skip any stops ($\tilde{x}_s = 1$, $\forall s \in S$). The values of the other parameters of the example are presented in Table 13.10.

Finally, the planned dispatching times of the 4 trips are $\tilde{d}_{n,1} = 600(n - 1)$, $\forall n \in \{1, 2, \ldots, 4\}$. That is, trip $n = 1$ is dispatched at $\tilde{d}_{n,1} = 0$ s, trip $n = 2$ at $\tilde{d}_{n,1} = 600$ s and so forth. This indicates a 10-min dispatching headway among trips. We then evaluate the cost of the objective function for any potential stop-skipping combination when we have $|N| = 4$ trips in the rolling horizon. This requires $2^{|N| \times |S|} = 2^{20} = 1,048,576$ evaluations of the objective function with the brute-force method. The

Table 13.10 Parameter values of the example scenario

Parameter	Value (s)	Parameter	Value ($/h)
r_1	2	c_1	10
r_2	1	c_2	5
δ	20	c_3	7

globally optimal solution in this rolling horizon when evaluating all possible stop-skipping options with brute force is:

$$\mathbf{x}^* = \begin{bmatrix} 1 & 1 & 1 & 1 & 1 \\ 1 & 1 & 1 & 1 & 1 \\ 1 & 1 & 1 & 1 & 1 \\ 1 & 0 & 0 & 0 & 1 \end{bmatrix}$$

with a generalized objective function cost,

$$Z^* = 563,491 \ \$$$

Closing this section, and concluding this book, we present a summary of selected stop-skipping studies in Table 13.11. For abbreviation, we use the following identifiers to represent different problem objectives:

– O1: waiting time of passengers;
– O2: in-vehicle time of passengers;
– O3: vehicle travel time;
– O4: reduction of control actions;
– O5: deviation from the planned schedule;
– O6: unsatisfied demand;
– O7: inconvenience of passengers that need to alight when skipping their stop.

> **Further Reading—survey papers**

• Gkiotsalitis and Cats, At-stop control measures in public transport: Literature review and research agenda, Transportation Research Part E: Logistics and Transportation Review, 2021 [79].
• Ibarra-Rojas et al., Planning, operation, and control of bus transport systems: A literature review, Transportation Research Part B: Methodological, 2015 [100].

Exercises

13.1 Rescheduling

(a) Write the KKT conditions of the rescheduling formulation in (13.1)–(13.8), considering also the headway deviation at the last stop as part of the objective function.

Table 13.11 Summary of selected stop-skipping methods [79]

Study	Problem	Trips considered	Real-time	Objective function	Mathematical program	Solution method
[26]	Stop-skipping	Single	Yes	O1+O2+O3	Integer nonlinear	Brute-force
[14]	Stop-skipping and holding	Multiple	Yes	O1+O4	Mixed-integer nonlinear	Genetic algorithm
[94]	Stop-skipping	Multiple	No	O1+O2+O3	Integer linear	Genetic algorithm
[4]	Stop-skipping	Single	Yes	O1+O2+O3	Integer non-linear	Genetic algorithm
[89]	Stop-skipping and short-turning	Single	Yes	O5+O6	Integer nonlinear	Heuristic
[13]	Stop-skipping and holding	Single	Yes	O1+O2+O3 +O5	Mixed-integer	–
[95]	Stop-skipping	Multiple	Yes	O1	Integer non-linear	Analytic solution for a simplified model
[15]	Stop-skipping and holding	Multiple	Yes	O1+O2	Mixed integer non-linear	Genetic algorithm
[1]	Stop-skipping	Single	Yes	O1+O7	Integer non-linear	Brute-force
[96]	Stop-skipping	Multiple	No	O3	Integer non-linear	Decomposition and simulated annealing
[97]	Stop-skipping	Multiple	No	O1+O2+O3	Integer non-linear	Artificial bee colony
[98]	Stop-skipping	Multiple	No	O1+O3	Integer non-linear	Response surface methodology
[99]	Stop-skipping	Multiple	Yes	O1+O2+O3	Integer non-linear	Brute-force and heuristics

(b) Consider an example with 3 trips in a decision epoch m where all times are expressed in seconds. The last trip that was dispatched before the start of this decision epoch had dispatching time $\delta_0 = 0$ and arrival times at stops 2 and 3 of the line equal to 850 and 1600, respectively. The originally planned dispatching times of trips 1, 2 and 3 are 600, 1150, and 1820, respectively. Their inter-station travel times are 900 from stop 1 to 2 and 730 from stop 2 to 3. The target headway at each stop is 600, the maximum dispatching time delay of the last trip is 20, and the marginal increase in dwell time at each station arising from a unit increase in the time headway is 0.035. The minimum and maximum allowed headways at the dispatching stop are $H_{\min} = 300$ and $H_{\max} = 900$. Solve this rescheduling problem using the Active Set method for inequality-constrained quadratic programming.

13.2 One-headway-based vehicle holding

Consider trip n that has completed its boardings/alightings at stop s at time $t = 1500$. The departure time of its previous trip from stop s is $d_{n-1,s} = 1050$. The scheduled (ideal) time headway between trips $n - 1$ and n at stop s is $H_s = 600$, the expected travel time of vehicle n from stop s to the last stop of the line is $t_s = 3000$, and the holding strength parameter a has value $a = 0.9$. Consider also that the latest possible arrival time at the last stop of the line to avoid schedule sliding is $\rho_n = 4250$. Using the one-headway control logic with and without considering schedule sliding, compute the optimal holding time of trip n with the one-headway-based vehicle holding method.

13.3 Vehicle holding in rolling horizons

Turn the multivariate vehicle holding problem formulated in (13.27)–(13.35) into a mathematical program with a convex feasible region by reformulating constraints (13.33).

13.4 Speed control

Provide a formulation for speed control aiming at minimizing the squared deviation between the planned and the actual time headway of a vehicle and its preceding vehicle at a particular stop. Consider that the preceding vehicle has already arrived at that stop and its arrival time is known. The planned time headway is also known and symbolized as h^*. In addition, the speed limit is s_{max} and the vehicle has to at least drive with a minimum speed of s_{min} for safety reasons.

13.5 Stop-skipping in rolling horizons

Consider the stop-skipping model in rolling horizons. Use the input parameters of the example presented in Table 13.10 and solve this problem with a metaheuristic.

References

1. A. Sun, M. Hickman, J. Intell. Transp. Syst. **9**(2), 91 (2005)
2. X. Chen, B. Hellinga, C. Chang, L. Fu, J. Adv. Transp. **49**(3), 385 (2015)
3. Y. Yu, Z. Ye, C. Wang, in *CICTP 2015* (2015), pp. 2397–2409
4. Z. Liu, Y. Yan, X. Qu, Y. Zhang, Transp. Res. C: Emerg. Technol. **35**, 46 (2013)
5. G.F. Newell, Transp. Sci. **8**(3), 248 (1974)
6. S.J. Berrebi, K.E. Watkins, J.A. Laval, Transp. Res. B: Methodol. **81**, 377 (2015)
7. D. Hernández, J.C. Muñoz, R. Giesen, F. Delgado, Transp. Res. B: Methodol. **78**, 83 (2015)
8. W. Wu, R. Liu, W. Jin, Transp. Res. B: Methodol. **104**, 175 (2017)
9. A. Gavriilidou, O. Cats, Transp. A: Transp. Sci. 1–29 (2018)
10. A. Adamski, A. Turnau, Transp. Res. A: Policy Pract. **32**(2), 73 (1998)
11. J. Strathman, K. Dueker, T. Kimpel, R. Gerhart, K. Turner, P. Taylor, S. Callas, D. Griffin, J. Hopper, Transp. Res. Rec.: J. Transp. Res. Board **1666**, 28 (1999)
12. X.J. Eberlein, Real-time control strategies in transit operations: Models and analysis. Ph.D. thesis, Massachusetts Institute of Technology, Department of Civil and Environmental Engineering (1995)

13. G. Lin, P. Liang, P. Schonfeld, R. Larson, Adaptive control of transit operations, Final Report. Tech. rep., MD-26-7002. (US Department of Transportation, 1995)
14. C.E. Cortés, D. Sáez, F. Milla, A. Núñez, M. Riquelme, Transp. Res. C: Emerg. Technol. **18**(5), 757 (2010)
15. D. Sáez, C.E. Cortés, F. Milla, A. Núñez, A. Tirachini, M. Riquelme, Transportmetrica **8**(1), 61 (2012)
16. K. Gkiotsalitis, Z. Wu, O. Cats, Transp. Res. C: Emerg. Technol. **98**, 14 (2019)
17. J.C. Muñoz, C.E. Cortés, R. Giesen, D. Sáez, F. Delgado, F. Valencia, A. Cipriano, Transp. Res. C: Emerg. Technol. **28**, 101 (2013)
18. C.E. Cortés, S. Jara-Díaz, A. Tirachini, Transp. Res. A: Policy Pract. **45**(5), 419 (2011)
19. A. Skabardonis, Transp. Res. Rec. **1727**(1), 20 (2000)
20. H. Liu, A. Skabardonis, W.b. Zhang, in *82nd Transportation Research Board Annual Meeting, Washington, DC* (2003)
21. L.A. Koehler, W. Kraus Jr., Transp. Res. C: Emerg. Technol. **18**(3), 288 (2010)
22. N. van Oort, J. Boterman, R. van Nes, Publ. Transp. **4**(1), 39 (2012)
23. C.F. Daganzo, J. Pilachowski, Transp. Res. B: Methodol. **45**(1), 267 (2011)
24. Y. Wang, W. Ma, W. Yin, X. Yang, Transp. Res. Rec. **2424**(1), 48 (2014)
25. K. Ampountolas, M. Kring, in *2015 IEEE 18th International Conference on Intelligent Transportation Systems* (IEEE, 2015), pp. 60–65
26. L. Fu, Q. Liu, P. Calamai, Transp. Res. Rec.: J. Transp. Res. Board (1857), 48 (2003)
27. L. Fu, X. Yang, Transp. Res. Rec.: J. Transp. Res. Board **1791**(1), 6 (2002)
28. O. Cats, A.N. Larijani, Á. Ólafsdóttir, W. Burghout, I.J. Andréasson, H.N. Koutsopoulos, Transp. Res. Rec. **2274**(1), 100 (2012)
29. W. Phillips, A. del Rio, J.C. Muñoz, F. Delgado, R. Giesen, Transp. Res. A: Policy Pract. **78**, 463 (2015)
30. J.G. Strathman, T.J. Kimpel, K.J. Dueker, R.L. Gerhart, S. Callas, Transportation **29**(3), 321 (2002)
31. R.G. Mishalani, M.R. McCord, S. Forman, in *Computer-Aided Systems in Public Transport* (Springer, 2008), pp. 301–317
32. O. Cats, J. Transp. Eng. A: Syst. **145**(1), 04018078 (2019)
33. O. Cats, Transp. Policy **36**, 223 (2014)
34. M. Trompet, R.J. Anderson, D.J. Graham, Transp. Res. Rec. **2111**(1), 177 (2009)
35. E.R. Randall, B.J. Condry, M. Trompet, S.K. Campus, in *Transportation Research Board 86th Annual Meeting, 21st–25th january* (2007)
36. M. Trompet, X. Liu, D. Graham, Transp. Res. Rec.: J. Transp. Res. Board (2216), 33 (2011)
37. X. Chen, L. Yu, Y. Zhang, J. Guo, Transp. Res. A: Policy Pract. **43**(8), 722 (2009)
38. C.F. Daganzo, Transp. Res. B: Methodol. **43**(10), 913 (2009)
39. P. Knoppers, T. Muller, Transp. Sci. **29**(1), 101 (1995)
40. S.J. Berrebi, E. Hans, N. Chiabaut, J.A. Laval, L. Leclercq, K.E. Watkins, Transp. Res. C: Emerg. Technol. **87**, 197 (2018)
41. E. Hans, N. Chiabaut, L. Leclercq, R.L. Bertini, Transp. Res. C: Emerg. Technol. **61**, 121 (2015)
42. G. Sánchez-Martínez, H. Koutsopoulos, N. Wilson, Transp. Res. B: Methodol. **83**, 1 (2016)
43. M. Nikolaou, Adv. Chem. Eng. **26**, 131 (2001)
44. X.J. Eberlein, N.H. Wilson, D. Bernstein, Transp. Sci. **35**(1), 1 (2001)
45. V. Pillac, M. Gendreau, C. Guéret, A.L. Medaglia, Eur. J. Oper. Res. **225**(1), 1 (2013)
46. N.H.M. Wilson, N.J. Colvin, *Computer Control of the Rochester Dial-a-ride System.* 77 (Massachusetts Institute of Technology, Center for Transportation Studies, 1977)
47. H.N. Psaraftis, Transp. Sci. **14**(2), 130 (1980)
48. A. Attanasio, J.F. Cordeau, G. Ghiani, G. Laporte, Parallel Comput. **30**(3), 377 (2004)
49. M. Gendreau, F. Guertin, J.Y. Potvin, É. Taillard, Transp. Sci. **33**(4), 381 (1999)
50. H.N. Psaraftis, Ann. Oper. Res. **61**(1), 143 (1995)
51. J. Zhao, S. Bukkapatnam, M.M. Dessouky, Intell. Transp. Syst. IEEE Trans. **4**(1), 43 (2003)
52. O. Cats, A.N. Larijani, H.N. Koutsopoulos, W. Burghout, Transp. Res. Rec. **2216**(1), 51 (2011)

53. A. D'Ariano, F. Corman, D. Pacciarelli, M. Pranzo, Transp. Sci. **42**(4), 405 (2008)
54. F. Corman, A. D'Ariano, D. Pacciarelli, M. Pranzo, Transp. Res. B: Methodol. **44**(1), 175 (2010)
55. L. Meng, X. Zhou, Transp. Res. B: Methodol. **67**, 208 (2014)
56. Y. Zhu, R.M. Goverde, J. Rail Transp. Plan. Manag. 100196 (2020)
57. P. Bly, *Depleted bus services: the effect of rescheduling* (Tech, Rep, 1976)
58. K. Gkiotsalitis, J. Intell. Transp. Syst. 1–20 (2019)
59. J.Q. Li, P.B. Mirchandani, D. Borenstein, in *Computer-Aided Systems in Public Transport* (Springer, 2008), pp. 281–299
60. J.Q. Li, P.B. Mirchandani, D. Borenstein, Transp. Res. E: Logistics. Transp. Rev. **45**(3), 419 (2009)
61. K. Gkiotsalitis, E. van Berkum, Transp. Res. C: Emerg. Technol. **110**, 143 (2020)
62. K. Gkiotsalitis, Appl. Math. Model. **82**, 785 (2020)
63. F. Cevallos, F. Zhao, in *Applications of Advanced Technology in Transportation* (2006), pp. 737–742
64. C. Coffey, R. Nair, F. Pinelli, A. Pozdnoukhov, F. Calabrese, in *Proceedings of the 5th ACM SIGSPATIAL International Workshop on Computational Transportation Science* (ACM, 2012), pp. 26–32
65. D. Rizopoulos, G.K. Saharidis, *Energy Systems*. pp. 1–30 (2020)
66. K. Gkiotsalitis, A. Stathopoulos, Int. J. Transp. Sci. Technol. **5**(2), 68 (2016)
67. P.B. Mirchandani, J.Q. Li, M. Hickman, Publ. Transp. **2**(3), 159 (2010)
68. E. Altazin, S. Dauzère-Pérès, F. Ramond, S. Tréfond, Transp. Res. C: Emerg. Technol. **79**, 73 (2017)
69. E. Osuna, G. Newell, Transp. Sci. **6**(1), 52 (1972)
70. J.J. Bartholdi, D.D. Eisenstein, Transp. Res. B: Methodol. **46**(4), 481 (2012)
71. F. Delgado, J.C. Muñoz, R. Giesen, A. Cipriano, Transp. Res. Rec.: J. Transp. Res. Board **2090**(1), 59 (2009)
72. F. Delgado, J.C. Munoz, R. Giesen, Transp. Res. B: Methodol. **46**(9), 1202 (2012)
73. M.D. Rossetti, T. Turitto, Transp. Res. A: Policy Pract. **32**(8), 607 (1998)
74. K. Gkiotsalitis, O. Cats, Transp. B: Transp. Dyn. **7**(1), 1258 (2019)
75. M.D. Hickman, Transp. Sci. **35**(3), 215 (2001)
76. N. Van Oort, N. Wilson, R. Van Nes, Transp. Res. Rec.: J. Transp. Res. Board (2143), 67 (2010)
77. K. Gkiotsalitis, Intell. Transp. Syst. (2020)
78. K. Gkiotsalitis, E. Van Berkum, Transp. Res. C: Emerg. Technol. **121**, 102815 (2020)
79. K. Gkiotsalitis, O. Cats, Transp. Res. E: Logistics. Transp. Rev. **145**, 102176 (2021)
80. Y. Xuan, J. Argote, C.F. Daganzo, Transp. Res. B: Methodol. **45**(10), 1831 (2011)
81. K. Ampountolas, M. Kring, Intell. Transp. Syst. (2020)
82. S. Shen, N.H. Wilson, in *Computer-aided Scheduling of Public Transport* (Springer, 2001), pp. 335–363
83. M.J. Powell, Cambridge NA Report NA2009/06 (University of Cambridge, Cambridge) (2009) pp. 26–46
84. A. Sun, M. Hickman, in *Computer-aided Systems in Public Transport* (Springer, 2008), pp. 339–359
85. S. Zolfaghari, N. Azizi, M.Y. Jaber, Int. J. Transp. Manag. **2**(2), 99 (2004)
86. A. Puong, N.H. Wilson, in *Computer-Aided Systems in Public Transport* (Springer, 2008), pp. 319–337
87. W.C. Jordan, M.A. Turnquist, Transp. Sci. **13**(3), 242 (1979)
88. P.G. Furth, Transp. Sci. **20**(1), 1 (1986)
89. Y. Li, M. Gendreau, et al., (1991)
90. X.J. Eberlein, Transp. Res. A **1**(31), 69 (1997)
91. Q. Chen, E. Adida, J. Lin, Publ. Transp. **4**(3), 165 (2013)
92. K. Gkiotsalitis, O.A. Eikenbroek, O. Cats, in *Transportation Research Board (TRB) 99th Annual Meeting* (2020)

93. P. Welding, J. Oper. Res. Soc. **8**(3), 133 (1957)
94. K. Gkiotsalitis, Transp. Res. Rec. **2673**(3), 611 (2019)
95. X.J. Eberlein, N.H. Wilson, C. Barnhart, D. Bernstein, Transp. Res. B: Methodol. **32**(2), 77 (1998)
96. A. Jamili, M.P. Aghaee, Transp. Res. C: Emerg. Technol. **61**, 63 (2015)
97. J. Chen, Z. Liu, S. Zhu, W. Wang, Transp. Res. E: Logist. Transp. Rev. **83**, 1 (2015)
98. W. Wu, R. Liu, W. Jin, C. Ma, Transp. Res. E: Logistics. Transp. Rev. **130**, 61 (2019)
99. K. Gkiotsalitis, *Transportmetrica A: Transport Science* (2020), pp. 1–29
100. O. Ibarra-Rojas, F. Delgado, R. Giesen, J. Muñoz, Transp. Res. B: Methodol. **77**, 38 (2015)

Appendix A
Solutions

Problems of Chapter 1

1.1 $a^x \ln a \cos(a^x)$.

1.2

$$\begin{bmatrix} 10x_1 + \sin x_4 \\ 9x_2^2 \\ 6^{x_3} \ln 6 \\ x_1 \cos x_4 \end{bmatrix}$$

1.3 For the disjunctive clause, $x_1 + (1 - x_2) + x_3 + (1 - x_4) \geq 1$ where $\mathbf{x} \in \{0, 1\}^4$. For the conjunctive clause, $x_1(1 - x_2)x_3(1 - x_4) \geq 1$ where $\mathbf{x} \in \{0, 1\}^4$.

1.4 No because in the objective function we have multiplications between variables x_1 and x_2.

1.5 It is a mixed-integer program because variables $\boldsymbol{\pi}$ are continuous and variables $\mathbf{z}$ are integer.

1.6 The affine hull of two different points is the line passing through them. The affine hull of three different points that are not colinear is a plane that includes these three points.

1.7 It has a unique globally optimal solution because the objective function of the unconstrained minimization problem is strictly convex. This becomes evident from the Hessian matrix of the objective function:

$$\begin{bmatrix} 2 & 0 \\ 0 & 12 \end{bmatrix}$$

which has positive eigenvalues 2 and 12, and thus it is positive definite.

1.8 All inequality constraints are affine functions, and thus they form a polyhedron. To satisfy all inequality constraints, x_2 cannot receive values outside of the range

Fig. A.1 Plot of $x^2 + 5x$ in the feasible region $\mathcal{F}$ and example of a line connecting two points in the graph

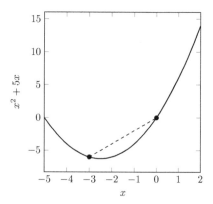

$[0, 40]$ and x_1 cannot receive values outside of the range $[-153/5, 58/3]$. Thus, there exists a finite number $M > 0$ such that $|x_1| + |x_2| \leq M$ for any values of x_1 and x_2 that satisfy the inequality constraints, and our polyhedron is a polytope.

1.9 The Hessian matrix of any linear or affine function with n variables is an $n \times n$ matrix $\mathbf{H}$ of all zeros. We thus have $\mathbf{c}^\mathsf{T}\mathbf{H}\mathbf{c} \geq 0$ for any $\mathbf{c} \in \mathbb{R}^n$ meaning that the Hessian is positive semi-definite and the objective function is convex. We also have $\mathbf{c}^\mathsf{T}\mathbf{H}\mathbf{c} \leq 0$ for any $\mathbf{c} \in \mathbb{R}^n$ meaning that the Hessian is negative semi-definite and the objective function is concave.

1.10 The feasible region $\mathcal{F}$ consists of two inequality constraints of affine functions, and thus it is a convex set. In addition, a line segment drawn from any point $(x_1, x_1^2 + 5x_1)$ to another point $(x_2, x_2^2 + 5x_2)$, where $x_1, x_2 \in \mathcal{F} \subseteq \mathbb{R}$ with $x_1 \neq x_2$, lies above the graph of $x^2 + 5x$ presented in Fig. A.1. Thus, the objective function is strictly convex. Summarizing, the problem is convex.

1.11 (a) First, observe that $(15x_1 - 12x_2 + 7)^4 + e^{x_2}$ is the non-negative weighted sum of functions $(15x_1 - 12x_2 + 7)^4$ and e^{x_2}. Function e^{x_2} is convex, so we have to prove that $(15x_1 - 12x_2 + 7)^4$ is also convex. From the composition with an affine mapping, if $f(\mathbf{x}) : \mathbb{R}^n \to \mathbb{R}$ is convex, then for matrix $\mathbf{A} \in \mathbb{R}^{m \times n}$ and vector $\mathbf{b} \in \mathbb{R}^m$ function $f(\mathbf{A}\mathbf{x} + \mathbf{b})$ is also convex. Let us set $f(y) = y^4$. Then, f is obviously convex. Because f is convex, $f(15x_1 - 12x_2 + 7)$ is also convex since $15x_1 - 12x_2 + 7$ is an affine mapping $\mathbf{A}\mathbf{x} + \mathbf{b}$ with $\mathbf{A} = [15, -12]$ and $b = 7$.
(b) We can write e^{5x^2} as $g(f(x))$ where $g(f(x)) = e^{f(x)}$ and $f(x) = 5x^2$. Using the general composition property, because e^y is convex and non-decreasing with respect to $y = f(x)$, and $f(x) = 5x^2$ is convex, e^{5x^2} is a convex function.

1.12 Introduce additional variables $r \in \mathbb{R}$, $\delta \in \{0, 1\}$ and a parameter M, where M is a very large positive number. Then,

$$\min\ r$$
$$\text{s.t.: } d_i \geq t$$
$$r \geq (d_{i+1} + h - d_i)$$
$$r \geq -(d_{i+1} + h - d_i)$$
$$r \leq (d_{i+1} + h - d_i) + \delta M$$
$$r \leq -(d_{i+1} + h - d_i) + (1 - \delta)M$$
$$d_i \in \mathbb{R}_{\geq 0}, r \in \mathbb{R}, \delta \in \{0, 1\}$$

1.13 We need to solve:

$$\max\ \sum_{i \in I} y_i$$
$$\text{s.t. } y_i \leq x_i \beta_i \qquad\qquad\qquad \forall i \in I$$
$$\sum_{i \in I} x_i \leq 1$$
$$\mathbf{x} \in \{0, 1\}^{|I|}, \mathbf{y} \in \mathbb{R}_{\geq 0}^{|I|}$$

We actually seek to find which $x_i \in \mathbf{x}$ will be selected to maximize $\sum_{i \in I} y_i$. This gives solution $\mathbf{x} = [0, 0, 0, 0, 1]^\mathsf{T}$ with $\mathbf{y} = [0, 0, 0, 0, 12]^\mathsf{T}$.

1.14

$$\tilde{f}(x) := \lambda_1 f(x_1) + \lambda_2 f(x_2) + \cdots + \lambda_{15} f(x_{15})$$
$$\text{s.t. } x = \lambda_1 x_1 + \lambda_2 x_2 + \cdots + \lambda_{15} x_{15}$$
$$\sum_{i=1}^{15} \lambda_i = 1$$
$$\text{SOS2}(\lambda_1, \lambda_2, \ldots, \lambda_{15})$$

with $(x_1, x_2, \ldots, x_{15}) = (1, 2, \ldots, 15)$ and $(f(x_1), f(x_2), \ldots, f(x_{15})) = (\frac{1}{1^2} + 0.05, \frac{1}{2^2} + 0.05, \ldots, \frac{1}{15^2} + 0.05)$.

1.15 The eigenvalues are $10.3852, -0.3852$, and -6. The matrix is indefinite because it has at least one positive and at least one negative eigenvalue.

Problems of Chapter 2

2.1 $\Theta(3^n)$

2.2 The possible combinations of two integer numbers that can take any value in the range $[1, n]$ is n^2. We can perform two operations for each pair of numbers: one to compute its product and one to compute its sum. If their values are 15 and 8, respectively, we have found our answer. This requires $2n^2$ algorithmic steps. Thus, $O(n^2)$.

2.3 The *subset sum* decision problem asks whether given a set S of n integers $\{s_1, s_2, \ldots, s_n\}$, there is a subset $T \subseteq S$ such that the integers in that subset have a sum equal to zero. To answer this question for a specific set S, we need to create all possible subsets of that set and sum their integer elements to check if there is a subset with integers that sum to zero. If we declare variables $t_1, t_2, \ldots, t_n$, where $t_i = 1$ if integer i belongs to subset T and $t_i = 0$ if not, then to create all possible subsets $T \subseteq S$ we have to explore 2^n potential solutions (subsets). Thus, the computational complexity of the subset sum problem is exponential, $O(2^n)$.

2.4 There is no known algorithm that can decide the Traveling Salesman Problem in polynomial time. Given that an oracle provides a tour with vertex sequence $\{v_1, \ldots, v_n\}$ in a graph G, we can develop a polynomial-time certifier to check whether this tour has a total distance of less than or equal to a prescribed cost k. The certifier requires polynomial time because it simply adds the distances of all edges to check if the total distance is less than or equal to k. This results in complexity $O(|E|)$, where E is the set of edges in graph G. Thus, the decision problem is in NP.

2.5 The answer can be found within the chapter itself with the use of poly-time reductions.

2.6 The answer can be found within the chapter itself with the use of poly-time reductions.

2.7 The answer can be found within the chapter itself with the use of poly-time reductions.

2.8 NP-complete problems are in NP and they are all equally hard because if one of them is solved, all of them can be solved using reductions. NP-Hard problems, however, might not be equally hard because some of them might belong in NP while others might not.

Problems of Chapter 3

3.1 Function f is *not* continuous at $x_0 = 1$ because, even though there exists $f(x_0) = 0$, we have:
$$\lim_{x \to x_0^+} f(x) \neq \lim_{x \to x_0^-} f(x)$$
since $\lim_{x \to x_0^+} f(x) = 0$ and $\lim_{x \to x_0^-} f(x) = -5$.

3.2 We can write $f(x) = |x|$ as:

$$f(x) = \begin{cases} x & \text{if } x \geq 0 \\ -x & \text{if } x < 0 \end{cases}$$

f is continuous because it is continuous at any point in $x \in \mathbb{R}$ (even at $x_0 = 0$ since $\lim\limits_{x \to x_0^+} f(x) = \lim\limits_{x \to x_0^-} f(x) = f(x_0) = 0$). However, it is not C^1 because it is not differentiable at $x_0 = 0$ since:

$$\lim_{h \to 0^-} \frac{f(0+h) - f(0)}{h} \neq \lim_{h \to 0^+} \frac{f(0+h) - f(0)}{h}$$

given that $\lim\limits_{h \to 0^-} \frac{f(0+h)-f(0)}{h} = \frac{-h}{h} = -1$ and $\lim\limits_{h \to 0^+} \frac{f(0+h)-f(0)}{h} = \frac{h}{h} = 1$.

Thus, its differentiability class is C^0.

3.3 $f(\mathbf{x}) = |x_1| + 5|x_2|$ is $\mathbb{R}^2 \to \mathbb{R}$. Let us consider any two points $\mathbf{x}, \mathbf{y}$ in $\mathbb{R}^2$. Then, f is Lipschitz continuous if there exists some L such that:

$$|f(\mathbf{x}) - f(\mathbf{y})| \leq L\|\mathbf{x} - \mathbf{y}\| \Rightarrow$$
$$|(|x_1| + 5|x_2|) - (|y_1| + 5|y_2|)| \leq L(|x_1 - y_1| + |x_2 - y_2|) \Leftrightarrow$$
$$|(|x_1| - |y_1|) + 5(|x_2| - |y_2|)| \leq L(|x_1 - y_1| + |x_2 - y_2|)$$

for which there exists $L > 0$, i.e., $L = 5$, satisfying the inequality for any $\mathbf{x}, \mathbf{y}$ in $\mathbb{R}^2$.

3.4 $f(x) = 12x^3 + 6$ is C^∞. We have $f''(x) = 72x$, which yields $f''(0) = 0$. In addition, $f''(0 + h) = 72h$ and $f''(0 - h) = -72h$ which have opposite signs for some $h > 0$ in the neighborhood of 0. Thus, $x^* = 0$ is an inflection point of f. Equivalently, this could have been shown by observing that $f'''(x) = 72 \neq 0$.

3.5 We have to find all points of $f(x) = x^3 - 3x$ at $[-20, 20]$ which are end points, stationary points, or points where the first-order derivative does not exist. The derivative of f exists everywhere in $[-20, 20]$. The endpoints are $x_1 = -20$ with $f(x_1) = (-20)^3 - 3 \cdot (-20) = -7940$ and $x_2 = 20$ with $f(x_2) = 20^3 - 3 \cdot 20 = 7940$. We also have $f'(x) = 3x^2 - 3$ which is equal to zero for points $x_3 = 1$ and $x_4 = -1$ with $f(x_3) = -2$ and $f(x_4) = 2$. For both x_3 and x_4 we have a non-zero second-order derivative, thus they are stationary points and not inflection points. Thus, the global minimum in $[-20, 20]$ is $x_1 = -20$. For completeness, there exists an inflection point (which is not a local extrema) at $x_5 = 0$.

3.6 (a) Yes. The Hessian matrix of the objective function is:

$$\begin{bmatrix} 4 & 0 \\ 0 & 2 \end{bmatrix}$$

which has positive eigenvalues 2 and 4, and thus it is positive definite.

(b) g has a unique global minimizer because its Hessian is positive definite. This is $\mathbf{x}^* = [0, -1]^\mathsf{T}$.

3.7 Based on Algorithm 3 we have order of convergence $q = 1$ and asymptotic error $\mu = \frac{1}{2}$ for the Bisection method. Newton's method has quadratic convergence $q \approx 2$ close to the root, thus it is preferable.

3.8 The answer can be found after implementing Armijo's rule or strong Wolfe conditions in gradient descent's algorithm presented in Algorithm 8. The answer is $x^* \simeq 6$ for both methods. When applying Armijo's rule we require 13 iterations, whereas when applying strong Wolfe conditions we require 9 iterations.

3.9 Implement Algorithm 11 that performs Hessian modification with Cholesky's decomposition. Starting from initial solution guess $[0.5, 0]^T$, the solution is $\mathbf{x}^* \approx [0.22224, -0.07408]^T$ after 3 iterations.

3.10 Using the Conjugate Gradient line search of Fletcher-Reeves presented in Algorithm 13, the solution is $\mathbf{x}^* = [1.000036121.00009026]^T$ after 11 iterations.

Problems of Chapter 4

4.1 The *Dual Problem* is:

$$\min_{\lambda,\mu} \sup_{\mathbf{x}} \ \mathcal{L}(\mathbf{x}, \lambda, \mu)$$

$$\text{s.t.: } \mu \geq 0$$

with Lagrangian function:

$$\mathcal{L}(\mathbf{x}, \lambda, \mu) := f(\mathbf{x}) - \lambda^T \mathbf{c}(\mathbf{x}) + \mu^T \mathbf{g}(\mathbf{x})$$

4.2 The objective function and the constraints are C^∞ because they are polynomials. We have Lagrangian function:

$$\mathcal{L}(\mathbf{x}, \lambda) := 2x_1^2 + 6x_2 + x_3 + \lambda_1(x_1 - 5x_2) + \lambda_2(x_1 - 9x_2)$$

the gradient of which is equal to vector $\mathbf{0}$ at:

$$[x_1^*, x_2^*, x_3^*, \lambda_1^*, \lambda_2^*]^T = [-59/180, -59/900, -59/1620, 6/5, 1/9]^T$$

Point $\mathbf{x}^* = [-59/180, -59/900, -59/1620]^T$ is a regular point because $a_1 \nabla c_1(\mathbf{x}^*) + a_2 \nabla c_2(\mathbf{x}^*) = \mathbf{0}$ if, and only if, $a_1 = a_2 = 0$ where $c_1(\mathbf{x}) = x_1 - 5x_2$ and $c_2(\mathbf{x}) = x_1 - 9x_3$. Finally, we check if the projected Hessian $\mathbf{Z}^T \nabla_{\mathbf{xx}}^2 \mathcal{L}(\mathbf{x}^*, \lambda^*)\mathbf{Z}$ is positive semi-definite, where $\mathbf{Z}$ is an orthonormal basis for the null space of $\nabla \mathbf{c}(\mathbf{x})^*$:

$$\mathbf{Z} = \begin{bmatrix} 0.97481183 \\ 0.19496237 \\ 0.10831243 \end{bmatrix}$$

The projected Hessian $\mathbf{Z}^T \nabla_{\mathbf{xx}}^2 \mathcal{L}(\mathbf{x}^*, \lambda^*)\mathbf{Z} = 3.801$. Thus, it is positive definite and solution $\mathbf{x}^*, \lambda^*$ satisfies the second-order necessary conditions. In fact, it also satisfies the second-order sufficient conditions and this point is a strict local minimizer.

4.3 We have Lagrangian:

$$\mathcal{L}(\mathbf{x}, \boldsymbol{\mu}) := (x_1^4 - 4x_1 + 2x_2^2 - 4x_2) - \mu_1 x_1 + \mu_2(x_2 - 3) + \mu_3(x_1 + x_2 - 2)$$

with KKT point $[x_1^*, x_2^*, \mu_1^*, \mu_2^*, \mu_3^*] = [1, 1, 0, 0, 0]^\mathsf{T}$. For this solution, only inequality constraint $g_3(x) := x_1 + x_2 - 2 \le 0$ is active; thus, we have linear independence on the active constraints and the KKT point is a regular point. Because $\mu_3^* = 0$, we have the critical cone:

$$C(\mathbf{x}^*) := (\nabla g_3(\mathbf{x}^*))^\mathsf{T} \mathbf{d} \le 0 \Rightarrow [1, 1] \begin{bmatrix} d_1 \\ d_2 \end{bmatrix} \le 0 \Rightarrow d_1 + d_2 \le 0$$

For $\mathbf{d} \in \mathbb{R}^2 \setminus \{\mathbf{0}\} : d_1 + d_2 \le 0$ we request that: $\mathbf{d}^\mathsf{T} \nabla_{\mathbf{xx}}^2 \mathcal{L}(\mathbf{x}^*, \boldsymbol{\mu}^*) \mathbf{d} > 0$. Because

$$\nabla_{\mathbf{xx}}^2 \mathcal{L}(\mathbf{x}, \boldsymbol{\mu}) = \begin{bmatrix} 12x_1^2 & 0 \\ 0 & 4 \end{bmatrix},$$

we have $\mathbf{d}^\mathsf{T} \nabla_{\mathbf{xx}}^2 \mathcal{L}(\mathbf{x}^*, \boldsymbol{\mu}^*) \mathbf{d} = 12d_1^2 + 4d_2^2$, which is greater than 0 for any $\mathbf{d} \in \mathbb{R}^2 \setminus \{\mathbf{0}\} : d_1 + d_2 \le 0$. Hence, the KKT point satisfies the second-order sufficient conditions for local optimality and it is a strict local minimizer.

4.4 (a) In a linear program in standard form, a *basic solution* satisfies only the equality constraints $\mathbf{Ax} = \mathbf{b}$. Thus, it might be infeasible since $\mathbf{x} \ge \mathbf{0}$ is not guaranteed. A *basic feasible solution* satisfies also $\mathbf{x} \ge \mathbf{0}$.

(b) For the linear program, we have the following basic solutions presented in Table A.1.

From these basic solutions, only the 3rd, 5th, and 6th are basic feasible solutions. From these, the optimal one is the 3rd.

4.5 Write a software code and solve the Phase I LP and the Phase II LP based on the description of the two-phase simplex.

4.6 The optimal solution is $x_1 = 2/3, x_2 = 1/3, x_3 = 0, x_4 = 0, x_5 = 1$ with objective function score -5.

Table A.1 Basic solutions and objective function scores

basic solution	x_1	x_2	x_3	x_4	$f(\mathbf{x})$
1	-20	100	0	0	-400
2	30	0	-25	0	1050
3	40/3	0	0	100/3	466.7
4	0	60	-10	0	180
5	0	40	0	20	120
6	0	0	20	60	0

4.7 We have $\mathbf{c} = [-4, 0, 0]^\mathsf{T}$, $\mathbf{b} = [5, 5]^\mathsf{T}$ and $\mathbf{A} = \begin{bmatrix} 1 & 2 & 0 \\ 2 & 2 & 1 \end{bmatrix}$. From the Hessian of the objective function, we also have:

$$\mathbf{Q} = \begin{bmatrix} 2 & 0 & 0 \\ 0 & 4 & 0 \\ 0 & 0 & 2 \end{bmatrix}$$

with positive eigenvalues 2, 4, and 2. Thus, the equality-constrained QP is strictly convex. An initial feasible solution to this problem is $\mathbf{y} = [5, 0, -5]^\mathsf{T}$. We also have:

$$\mathbf{Z} = \text{Basis of null}(\mathbf{A}) \simeq \begin{bmatrix} -2/3 \\ 1/3 \\ 2/3 \end{bmatrix}$$

resulting in:

$$\mathbf{z}^* = (\mathbf{Z}^\mathsf{T}\mathbf{Q}\mathbf{Z})^{-1}(-\mathbf{Z}^\mathsf{T}\mathbf{c} - \mathbf{Z}^\mathsf{T}\mathbf{Q}\mathbf{y}) = 4.8$$

Thus, $\mathbf{x}^* = \mathbf{y} + \mathbf{Z}\mathbf{z}^* = [5, 0, -5]^\mathsf{T} + [-2/3, 1/3, 2/3]^\mathsf{T}4.8 = [1.8, 1.6, -1.8]^\mathsf{T}$ with optimal value $f(\mathbf{x}^*) = 4.4$.

4.8 We have no equality constraints, $b = [4, 7, 5, 0]^\mathsf{T}$, $c = [-7, -3]^\mathsf{T}$, the Hessian of the objective function:

$$\mathbf{Q} = \begin{bmatrix} 2 & -1 \\ -1 & 2 \end{bmatrix}$$

and

$$\mathbf{A} = \begin{bmatrix} -1 & 2 \\ 1 & 3 \\ 1 & -2 \\ -1 & 0 \end{bmatrix}$$

Considering initial feasible solution $\mathbf{x}_{k=0} = [0, 2]$, the 1st and 4th inequality constraints become active. We obtain step $\mathbf{p}_{k=0} = [0, 0]^\mathsf{T}$ and $\boldsymbol{\mu}_{k=0} = [-0.5, -8.5]^\mathsf{T}$. Thus, we remove the 4th inequality constraint from the active set and we perform the next iteration which yields $\mathbf{p}_{k=1} = [5.667, 2.833]^\mathsf{T}$ with $\mu_{k=1} = -0.5$. Since $\mathbf{p}_{k=1} \neq \mathbf{0}$ we compute step length $\alpha_{k=1} = 0.070588$ resulting in $\mathbf{x}_{k=2} = \mathbf{x}_{k=1} + \alpha_{k=1}\mathbf{p}_{k=1} = [0.4, 2.2]^\mathsf{T}$, where the 2nd constraint is a blocking constraint and it is added to the active set. In the next iteration we have $\mathbf{p}_{k=2} = [0, 0]^\mathsf{T}$ and $\boldsymbol{\mu}_{k=2} = [-5.24, 3.16]^\mathsf{T}$. Thus, we remove the 1st constraint from the active set and we proceed to the next iteration. Then, we have $\mathbf{p}_{k=3} \simeq [3.0231, -1.0077]^\mathsf{T}$ with $\mu_{k=3} \simeq 1.346$. This results in full step $\alpha_{k=3} = 1$ and $\mathbf{x}_{k=4} \simeq \mathbf{x}_{k=3} + \alpha_{k=3}\mathbf{p}_{k=3} = [3.42308, 1.19231]^\mathsf{T}$. In the next iteration, we have $\mathbf{p}_{k=4} \simeq [0, 0]^\mathsf{T}$ with $\mu_{k=4} \simeq 1.346$. Thus, we terminate the algorithm with optimal solution $\mathbf{x}^* = [3.42308, 1.19231]^\mathsf{T}$ with $f(\mathbf{x}^*) = -18.4808$.

Table A.2 Iterations

k	$\mathbf{x}_k$	μ_k
0	$[12, 2]^{\mathsf{T}}$	0.2
1	$[3.51127862, 12.33220781]^{\mathsf{T}}$	2
2	$[3.52879258, 12.45551544]^{\mathsf{T}}$	$2 \cdot 10$
3	$[3.53054478, 12.46788562]^{\mathsf{T}}$	$2 \cdot 10^2$
4	$[3.53072001, 12.46912303]^{\mathsf{T}}$	$2 \cdot 10^3$
5	$[3.53073753, 12.46924678]^{\mathsf{T}}$	$2 \cdot 10^4$
6	$[3.53073928, 12.46925915]^{\mathsf{T}}$	$2 \cdot 10^5$
7	$[3.53073945, 12.46926039]^{\mathsf{T}}$	$2 \cdot 10^6$
8	$[3.53073947, 12.46926051]^{\mathsf{T}}$	$2 \cdot 10^7$
9	$[3.53073947, 12.46926053]^{\mathsf{T}}$	$2 \cdot 10^8$
10	$[3.53073947, 12.46926053]^{\mathsf{T}}$	

Table A.3 Iterations

k	$\mathbf{x}_k$	μ_k	λ_k
0	$[12, 2]^{\mathsf{T}}$	0.2	2
1	$[2.65229873, 7.03730934]^{\mathsf{T}}$	2	0.737922
2	$[3.48828603, 12.17125892]^{\mathsf{T}}$	$2 \cdot 10$	0.057012
3	$[3.52861574, 12.45426738]^{\mathsf{T}}$	$2 \cdot 10^2$	−0.285326
4	$[3.53063326, 12.46851043]^{\mathsf{T}}$	$2 \cdot 10^3$	−0.456588
5	$[3.53073416, 12.46922302]^{\mathsf{T}}$	$2 \cdot 10^4$	−0.542224
6	$[3.53073921, 12.46925865]^{\mathsf{T}}$	$2 \cdot 10^5$	−0.585042
7	$[3.53073946, 12.46926043]^{\mathsf{T}}$	$2 \cdot 10^6$	−0.606451
8	$[3.53073947, 12.46926052]^{\mathsf{T}}$	$2 \cdot 10^7$	−0.617155
9	$[3.53073947, 12.46926053]^{\mathsf{T}}$	$2 \cdot 10^8$	−0.622507
10	$[3.53073947, 12.46926053]^{\mathsf{T}}$		

4.9 Implement Algorithm 19 and check if your solution is $\mathbf{x}^* = [3.42308, 1.19231]^{\mathsf{T}}$.

4.10 Starting from initial solution guess $\mathbf{x}_0 = [0.5, 0]^{\mathsf{T}}$, the result is $\mathbf{x}^* \simeq [0.4714, -0.7778]^{\mathsf{T}}$ with objective function value -0.62854. The solutions at each iteration always remain within the feasible region because we start from a feasible initial solution guess and we implement a barrier method.

4.11 (a) Starting from initial solution guess $[12, 2]^{\mathsf{T}}$ and $\mu_{k=0} = 0.2$, the Quadratic Penalty method converges to solution $\mathbf{x}^* = [3.53073947, 12.46926053]^{\mathsf{T}}$ after performing the iterations of Table A.2.

(b) With the Augmented Lagrangian method we get the same solution as before after performing the iterations of Table A.3.

Table A.4 Simplex Tableau of the optimal solution to the continuously relaxed LP

Row	z	x_1	x_2	x_3	x_4	x_5	Rhs
0	1	1.5	12.25	0	0	0	31.5
1	0	2.3	−17.4	0	1	0	2.2
2	0	1	4.5	1	0	0	9
3	0	2.5	1	0	0	1	12

Problems of Chapter 5

5.1 Yes, because each lower bound will be the objective function value of a globally optimal solution to a continuously relaxed problem. At some point, the difference between the lower bound and the upper bound will be less than a precision error ϵ resulting in an optimal solution. However, even if this is theoretically guaranteed, we might not be able to find such a solution in large problem instances due to computational complexity issues (i.e., the exploration tree becomes too large or solving a continuous relaxation of the problem becomes computationally expensive).

5.2 The continuous relaxation of the mixed-integer program is an inequality-constrained quadratic program. The Hessian of the objective function is:

$$\mathbf{Q} = \begin{bmatrix} 0 & 0 & 0 \\ 0 & 40 & 0 \\ 0 & 0 & 0 \end{bmatrix}$$

which is positive semi-definite with eigenvalues 0, 0, 40. Thus, any locally optimal solution of our continuously relaxed quadratic program is also a globally optimal one. Relaxing the integerality constraints by allowing $x_1, x_2 \in \mathbb{R}_+$ gives us solution $\mathbf{x}^* = [0, 5, -1]^\mathsf{T}$ with $f(\mathbf{x}^*) = 495.0$. Thus, we set the current upper bound of our maximization problem to $U = 495$. Because this solution is feasible to our original mixed-integer program, we also have lower bound $L = 495$ and we can terminate branch and bound with solution $\mathbf{x}^* = [0, 5, -1]^\mathsf{T}$.

5.3 Solving the continuous relaxation of the problem with simplex will produce solution $[0, 0, 0, 9, 2.2, 12]^\mathsf{T}$ with objective function score equal to 35. This solution has the tableau presented in Table A.4.

This solution is infeasible to the integer linear program. We can produce a valid inequality using a Gomory cut. We can use row 1 of the tableau with right-hand-side value equal to 2.2. Then, we add slack variable x_6 and the following valid inequality constraint:

$$x_6 - \sum_{i=1}^{5} \left(\bar{a}_{1j} - \lfloor \bar{a}_{1j} \rfloor \right) x_j = -(\bar{b}_1 - \lfloor \bar{b}_1 \rfloor)$$

where $\bar{a}_{1j}$ are the coefficients of the previously provided tableau at row 1 and b_1 the right-hand-side value.

5.4 Let $x_i = 1$ if item i belongs to the knapsack and $x_i = 0$ if not. We have formulation:

$$\max_{\mathbf{x}} \; \mathbf{p}^\mathsf{T}\mathbf{x}$$
$$\text{s.t. } \mathbf{w}^\mathsf{T}\mathbf{x} \leq c$$
$$x_i \in \{0, 1\} \qquad\qquad \forall i \in \{1, 2, \dots, 6\}$$

which is a pure binary linear program with solution $\mathbf{x}^* = [1, 1, 1, 0, 0, 0]^\mathsf{T}$.

5.5 Let $x_i = 1$ if vertex $i \in V$ belongs to the independent set and $x_i = 0$ if not. We have formulation:

$$\max_{\mathbf{x}} \; \sum_{i \in V} x_i$$
$$\text{s.t. } x_i + x_j \leq 1 \qquad\qquad \forall (i, j) \in E$$
$$x_i \in \{0, 1\} \qquad\qquad \forall i \in V$$

which is a pure binary linear program.

The steps of solving this problem with branch and bound can be found in this chapter, where we discuss the use of branch and bound for binary optimization problems.

5.6 Let M be the set of factories and N the set of cities. The produced products by each factory are $\mathbf{a} = [11, 6, 5]^\mathsf{T}$, and the products that must be supplied to every city are $\mathbf{b} = [6, 8, 4, 4]^\mathsf{T}$. Hitchcock's Transportation Problem can be formulated as an integer linear program:

$$\min \sum_{i \in M} \sum_{j \in N} c_{ij} x_{ij}$$
$$\text{s.t.: } \sum_{j \in N} x_{ij} = a_i \qquad\qquad \forall i \in M$$
$$\sum_{i \in M} x_{ij} = b_j \qquad\qquad \forall j \in N$$
$$x_{ij} \geq 0 \qquad\qquad \forall i \in M, j \in N$$
$$x_{ij} \in \mathbb{Z} \qquad\qquad \forall i \in M, j \in N$$

Solving its continuous relaxation we get solution $x_{1,1} = 3$, $x_{1,3} = 4$, $x_{1,4} = 4$, $x_{2,2} = 6$, $x_{3,1} = 3$, $x_{3,2} = 2$, and $x_{ij} = 0$ for any other $i \in M$, $j \in N$. This has an objective function value of 131. Notice that this solution is feasible to the integer linear program, thus it is both a lower and upper bound and the branch and bound algorithm terminates.

5.7 We have the following potential vertex sequences:

- Simple path: a, b, e
- Hamiltonian path: a, b, d, e, c, f
- Cycle: a, b, e, d, a
- Hamiltonian cycle: a, b, c, f, e, d, a

Problems of Chapter 6

6.1 Implement the steps of differential evolution described in this chapter. Note that your results might change between different runs, thus it is advised to run your algorithm multiple times and experiment with different parameter settings (i.e., population size, mutation strategy). Your target is to find a solution close to the globally optimal solution $\mathbf{x}^* = [0, -2]^\mathsf{T}$ with $f(\mathbf{x}^*) = 10$.

6.2 Implement the particle swarm optimization metaheuristic described in Algorithm 31. Note that your results might change between different runs, thus it is advised to run your algorithm multiple times and experiment with different parameter settings.

When solving the problem with the active set method, the result is $\mathbf{x}^* = [5, 0]^\mathsf{T}$ with $f(\mathbf{x}^*) = 25$. The problem is a quadratic program and the objective function has the positive semi-definite Hessian:

$$\mathbf{Q} = \begin{bmatrix} 2 & 0 \\ 0 & 0 \end{bmatrix}$$

with eigenvalues 2 and 0. Thus, the solution of the active set method is a globally optimal one.

6.3 This is equivalent to a maximum profit knapsack problem, which is formulated as follows:

$$\max_{\mathbf{x}} \ \mathbf{c}^\mathsf{T}\mathbf{x}$$

$$\text{s.t. } \mathbf{w}^\mathsf{T}\mathbf{x} \leq d$$

$$x_i \in \{0, 1\} \qquad\qquad \forall i \in \{1, 2, \ldots, 6\}$$

where $x_i = 1$ if object i is selected to be part of the subset. The globally optimal solution to this problem when applying branch and cut is $\mathbf{x}^* = [1, 1, 1, 0, 0, 0]^\mathsf{T}$ with objective function score equal to 41. Implement the genetic algorithm metaheuristic described in this chapter and compare your solution to the globally optimal one. Note that your genetic algorithm results might change between different runs, thus it is advised to run your algorithm multiple times and experiment with different parameter settings.

6.4 Follow the Tabu search and Hill Climbing steps described in this chapter to implement the two algorithms and solve the knapsack problem. Check if your solutions are close to $\mathbf{x}^* = [1, 1, 1, 0, 0, 0]^\mathsf{T}$.

The difference between the two is that the Hill Climbing method does not involve stochastic search processes and it always returns the same solution when provided with the same input.

6.5 The reason is that a metaheuristic follows a stochastic search process. Changing its parameter values (population size, termination criteria, etc.) may provide improved solutions when solving a problem.

Problems of Chapter 7

7.1 Let $f_1(x) = x^2 - x$ and $f_2(x) = x$. We have $f_1(x_1) = 0$ and $f_2(x_1) = 0$. We also have $f_1(x_2) = 2$ and $f_2(x_2) = 2$. Thus, x_1 dominates x_2.

7.2 We have the feasible region:

$$X = \left\{ \mathbf{x} \in \mathbb{R}^2 \; : \; \begin{matrix} x_1 + 2x_2 \leq 15 \\ x_1 + x_2 \geq 0 \end{matrix} \right\}$$

We first have:

$$\mathbf{x}_1^* = \underset{\mathbf{x} \in X}{\text{argmin}} \; f_1(\mathbf{x}) = [-15, \, 15]^\mathsf{T}$$

with $f_1(\mathbf{x}^*) = -105$. We then proceed with:

$$\mathbf{x}_2^* = \underset{\mathbf{x} \in X}{\text{argmin}} \; \{f_2(\mathbf{x}) \text{ s.t.: } f_1(\mathbf{x}) \leq -105\} = [-15, \, 15]^\mathsf{T}$$

that has $f_2(\mathbf{x}_2^*) = -45$. Note that $\mathbf{x}_1^*$ and $\mathbf{x}_2^*$ are identical. Thus, we can terminate the algorithm obtaining solution $[-15, \, 15]^\mathsf{T}$.

7.3 We have the following optimal values for each pair (w_1, w_2), where $f(\mathbf{x}) = w_1 f_1(\mathbf{x}) + w_2 f_2(\mathbf{x})$.

w_1	w_2	f	x_1^*	x_2^*	$f_1(x_1^*, x_2^*)$	$f_2(x_1^*, x_2^*)$
1	2	-10.29	-0.857	0.857	1.47	-5.88
1	3	-16.2	-0.9	0.9	1.62	-5.94
2	1	-3.6	-0.6	0.6	0.72	-5.04
3	1	-3	-0.5	0.5	0.50	-4.50
4	2	-7.2	-0.6	0.6	0.72	-5.04

Plot the values of f_1 and f_2 in a 2-dimensional plot. These values belong to the Pareto Front because all weight values (w_1, w_2) are non-negative (see Theorem 7.1).

7.4 Our objective function is composed of two scalar objective functions, $f_1(\mathbf{x}) = x_1^2 + 2x_2$ and $f_2(\mathbf{x}) = x_1 + 3x_2$, where f_2 is linear. Thus, it is convenient to add f_2 to the constraints resulting in the formulation:

$$\min_{x \in \mathbb{R}^2} x_1^2 + 2x_2$$

$$\text{s.t. } x_1 + x_2 \geq 10$$

$$x_1 + 3x_2 \leq \epsilon_2$$

Using multiple values of ϵ_2, we can get the results of the following table and we can plot the values of functions f_1 and f_2 in a 2-dimensional plot.

ϵ_2	x_1^*	x_2^*	$f_1(\mathbf{x}^*)$	$f_2(\mathbf{x}^*)$
$+\infty$	1	9	19	28
1000	1	9	19	28
100	1	9	19	28
25	2.5	7.5	21.25	25
20	5	5	35	20
10	10	0	100	10
5	12.5	-2.5	151.25	5
2	14	-4	188	2
0	15	-5	215	0
-2	16	-6	244	-2
-5	17.5	-7.5	291.25	-5
-10	20	-10	380	-10

7.5 Implement the NSGA-II and the MOEA/D algorithms and compare your results against the Pareto Fronts in Figs. 7.3 and 7.4, respectively. Note that NSGA-II and MOEA/D are metaheuristics and their results might differ between different algorithmic runs.

7.6 The answer to this question is presented in this chapter, where the differences between the two methods are explained.

Problems of Chapter 8

8.1 Compare your results with the results of the following table.

Origins	Destinations				$\sum_{j \in I} t_{ij}$	Future o_i
	$j = 1$	$j = 2$	$j = 3$	$j = 4$		
$i = 1$	5.21	43.77	97.54	254.10	400.6	$o_1 = 400$
$i = 2$	44.90	3.77	84.01	328.30	460.9	$o_2 = 460$
$i = 3$	76.80	128.97	7.18	187.17	400.1	$o_3 = 400$
$i = 4$	133.09	223.49	311.27	32.44	700.3	$o_4 = 702$
$\sum_{i \in I} t_{ij}$	260	400	500	802.01	$T \simeq 1962$	
Future d_j	260	400	500	802		$T = 1962$

8.2 The entropy maximization problem is formulated as:

$$\max_{\mathbf{T} \in \mathbb{R}_{\geq 0}^{|I| \times |I|}} \quad -\sum_{i \in I} \sum_{j \in I} (t_{ij} \ln t_{ij} - t_{ij})$$

$$\text{s.t.} \sum_{j \in I} t_{ij} = o_i, \ \forall i \in I$$

$$\sum_{i \in I} t_{ij} = d_j, \ \forall j \in I$$

This maximization problem has *Lagrangian* function:

$$\mathcal{L}(\mathbf{T}, \boldsymbol{\lambda}', \boldsymbol{\lambda}'') = -\sum_{i \in I} \sum_{j \in I} (t_{ij} \ln t_{ij} - t_{ij}) - \sum_{i \in I} \lambda_i' \left(\sum_{j \in I} t_{ij} - o_i \right) - \sum_{j \in I} \lambda_j'' \left(\sum_{i \in I} t_{ij} - d_j \right)$$

where $\boldsymbol{\lambda}' = [\lambda_1', \lambda_2', \ldots \lambda_{|I|}']^\mathsf{T}$ and $\boldsymbol{\lambda}'' = [\lambda_1'', \lambda_2'', \ldots \lambda_{|I|}'']^\mathsf{T}$ are the Lagrange multipliers. Notice that:

$$\frac{\partial \mathcal{L}(\mathbf{T}, \boldsymbol{\lambda}', \boldsymbol{\lambda}'')}{\partial t_{ij}} = -(\ln t_{ij} + \frac{t_{ij}}{t_{ij}} - 1) - \lambda_i' - \lambda_i'' = -\ln t_{ij} - \lambda_i' - \lambda_i''$$

Therefore, setting the partial derivatives to 0,

$$t_{ij} = \exp(-\lambda_i' - \lambda_j'') \quad \forall i \in I, j \in I$$

8.3 Compare your results with the results of the following table.

Origins	Destinations				
	$j = 1$	$j = 2$	$j = 3$	$j = 4$	Target o_i
$i = 1$	157.01	100.33	66.10	76.41	$399.9 \simeq 400$
$i = 2$	57.48	201.07	108.46	92.88	$459.9 \simeq 460$
$i = 3$	25.27	46.15	136.26	192.38	$400.06 \simeq 400$
$i = 4$	20.24	52.45	189.17	440.33	$702.2 \simeq 702$
Target d_j	260	400	500	802	$T = 1962$

8.4 Consider the MCLP formulation:

$$\max_{\mathbf{x}, \mathbf{y}} \sum_{i \in M} c_i y_i$$

$$\text{s.t.:} \sum_{j \in N} \beta_{ij} x_j \geq 1 \qquad \forall i \in M$$

$$\sum_{j \in N} a_{ij} x_j \geq y_i \qquad \forall i \in M$$

$$\sum_{j \in N} x_j = p$$

$$x_j \in \{0, 1\} \qquad \forall j \in N$$

$$y_i \in \{0, 1\} \qquad \forall i \in M$$

Apply a greedy heuristic that starts with an empty solution set and adds a public transport stop to it at each iteration as follows:

- First, selects the public transport stop which is within a distance r of the most possible demand points $i \in M$, such that $\sum_{j \in N} \beta_{ij} x_j \geq 1$ for as many demand points i as possible.
- Continues this process until adding p stops to the set.

Note that this solution might be feasible, but it is most probably not optimal because we do not consider the number of demand points covered by the selected stops (i.e., we do not ensure that they will be within a distance $l \ll r$). We only strive to ensure that there will be no stranded passengers. This heuristic will evaluate p times the impact of every stop selection to constraints $\sum_{j \in N} \beta_{ij} x_j \geq 1$. Thus, it has a polynomial time complexity of $O(|M||N|p)$.

8.5 (a) The algorithm consists of the following steps.

Step 1: consider undirected graph $G = (X, \Gamma)$ where X is the *finite* set of vertices (passenger demand points) and Γ the *finite* set of undirected edges (road sections). Find all points P' such that $j \in P'$ if, and only if, for some $i \in X$ and $y \in X$ where $i \neq y$, j is the unique point in its edge such that we have equal shortest path distances $d(i, j) = d(y, j)$. Create finite set of points $Q' = P' \cup X$.

Step 2: select m arbitrary points from Q' resulting in subset $U_0 = \{u_1, \ldots, u_m\}$

Step 3: compute the performance of this subset:

$$\tau_{U_0} := \max_{i \in X} \min_{j \in U_0} d_{i,j}$$

where d_{ij} is the shortest distance from i to j.

Step 4: Produce matrix $\mathbf{A} \in \{0, 1\}^{|X| \times |Q'|}$ where $a_{ij} = 0$ if $d_{ij} \geq \tau_{U_0}$ and $a_{ij} = 1$ if $d_{ij} < \tau_{U_0}$.

Step 5: Use matrix $\mathbf{A}$ to solve the minimal set covering problem:

$$\min_{\mathbf{x}} \sum_{j \in Q'} x_j$$
$$\text{s.t. } \mathbf{Ax} \geq \mathbf{1} \tag{13.77}$$
$$\mathbf{x} \in \{0, 1\}^{|Q'|}$$

Step 6: If the solution $\mathbf{x}^*$ of this problem has more than m values equal to 1 or does not exist, set U_0 is indeed a globally optimal solution. If not, then the columns of matrix $\mathbf{D}$ corresponding to the components of $\mathbf{x}^*$ that are equal to 1 represent another set U_1 for which we produce matrix $\mathbf{A} \in \{0, 1\}^{|X| \times |Q'|}$ where $a_{ij} = 0$ if $d_{ij} \geq \tau_{U_1}$ and $a_{ij} = 1$ if $d_{ij} < \tau_{U_1}$.

Step 7: solve again the minimum set covering problem considering the new matrix $\mathbf{A}$ and repeat this process until $\mathbf{x}^*$ has more than m values equal to 1 or does not exist.

(b) We have four points in P'. The first point j is in the middle of the edge connecting vertices 2 and 3 because for vertices $i = 2$ and $j = 3$ we have that $d(i, j) = d(y, j)$ if j is in the middle of the distance between vertices 2 and 3. The same holds for the second point in P' which is in the middle of the edge connecting vertices 1 and 2. The third point $j \in P'$ is located in a distance of 8.3 units from vertex 1 because $d(1, j) = d(3, j)$ for that point. That is, it is in the middle of the distance between vertices 1 and 3. If we denote these points as 4, 5, and 6, we have that $Q' = X \cup P' = \{1, 2, 3, 4, 5, 6\}$.

8.6 Consider binary variables $x_{ij} = 1$ if arc $(i, j) \in A$ belongs to the shortest path and $x_{ij} = 0$ if not. Consider w_{ij} to be the non-negative weight (distance) of each arc and z_{ij} its energy consumption. We have that the shortest path's energy consumption cannot exceed $k = 35$ energy units. The shortest path problem that considers energy consumption limitations is then formulated as:

$$
\min_{\mathbf{x}} \sum_{(i,j)\in A} w_{ij} x_{ij}
$$

$$
\text{s.t.} \quad \sum_{(i,j)\in A: i=a} x_{ij} - \sum_{(i,j)\in A: j=a} x_{ij} = 1
$$

$$
\sum_{(i,j)\in A: i=f} x_{ij} - \sum_{(i,j)\in A: j=f} x_{ij} = -1
$$

$$
\sum_{(i,j)\in A: i=k} x_{ij} - \sum_{(i,j)\in A: j=k} x_{ij} = 0 \qquad \forall k \in V \setminus \{a, f\}
$$

$$
\sum_{(i,j)\in A: i=k} x_{ij} \leq 1 \qquad \forall k \in V
$$

$$
\sum_{(i,j)\in A} z_{ij} x_{ij} \leq k
$$

$$
x_{ij} \in \{0, 1\} \qquad \forall (i, j) \in A
$$

The optimal solution is $a \to b \to c \to f$ with a cost of 22 and an energy consumption of 30.

8.7 The solution is the following.
$(a, b) : 4, (a, c) : 8, (a, d) : 8, (a, e) : 14, (a, f) : 16, (b, c) : 4, (b, f) : 18, (c, f) : 14, (d, e) : 6, (d, f) : 8, (e, f) : 2$.

8.8 Implement simplex to solve the linear program in Eqs. (8.50)–(8.55) and compare your results with the optimal solution: $\mathbf{f}^* = [8, 0, 1.4]^\mathsf{T}$ and $\mathbf{y}^* = [160, 0, 40, 0, 30]^\mathsf{T}$. Hint: to apply simplex, collect all variables $\mathbf{f}, \mathbf{y}$ into $\mathbf{x} = \begin{bmatrix} \mathbf{f} \\ \mathbf{y} \end{bmatrix}$ and all variable coefficients at the constraint functions to matrix $\mathbf{A}$. In addition, collect $\boldsymbol{\tau}$ and $\boldsymbol{\gamma}$ to a cost vector $\mathbf{c} = \begin{bmatrix} \boldsymbol{\tau} \\ \boldsymbol{\gamma} \end{bmatrix}$ and turn the minimization problem to a maximization one by multiplying the objective function by -1. For the two inequality constraints, add slack variables to turn the linear program to its standard form.

8.9 Assuming a single transfer stop in the locations where lines intersect, we have:

- Connectivity: $\frac{e}{3v-6} = \frac{9}{18}$
- Complexity: $\frac{e}{v} = \frac{9}{8}$
- Degree of cyclicity: $\frac{e-v+1}{2v-5} = \frac{2}{11}$
- Diameter: $\max\limits_{i \in V, j \in V} l_{ij} = 4$

Problems of Chapter 9

9.1 With the maximum loading point method we have frequency $\frac{220}{(5/6)60} = 4.4$ vehicles every 2 hours. This yields an offered capacity within the 2-hour period of $4.4 \cdot 60 = 264$ passengers. Thus, we have empty seats: $(264 - 120) + (264 - 180) + (264 - 140) + (264 - 220) + (264 - 180) = 480$ in total upon the departure from each public transport stop.

With the load profile method we have frequency $\max\{\frac{220}{60}, \frac{2340}{650}\} \simeq 3.67$ vehicles every 2 hours. This yields an offered capacity within the 2-hour period of $3.67 \cdot 60 = 220$ passengers. Thus, we have empty seats: $(220 - 120) + (220 - 180) + (220 - 140) + (220 - 220) + (220 - 180) = 260$ in total upon the departure from each public transport stop.

9.2 The optimal frequencies are $f_1 = 3$ and $f_2 = 3$ with objective function score 10.5.

9.3 Use formulation:

$$\min_{\mathbf{Z},\mathbf{x}} \sum_{k \in R} \sum_{(i,j) \in L_k} q_{ij}^k \sum_{f \in F} \frac{z_{fk}}{f}$$

$$\text{s.t. } c \sum_{f \in F} f z_{fk} \geq q_{ij}^k \qquad\qquad \forall k \in R, (i,j) \in L_k$$

$$t_k \sum_{f \in F} f z_{fk} \leq x_k \qquad\qquad \forall k \in R$$

$$\sum_{k \in R} x_k \leq m$$

$$\sum_{f \in F} f z_{fk} \geq \frac{1}{\theta} \qquad\qquad \forall k \in R$$

$$\sum_{f \in F} f z_{fk} \leq \frac{1}{\theta_{\min}} \qquad\qquad \forall k \in R$$

$$\sum_{f \in F} z_{fk} = 1 \qquad\qquad \forall k \in R$$

$$\mathbf{Z} \in \{0,1\}^{|F| \times |R|}, \mathbf{x} \in \mathbb{Z}_{\geq 0}^{|R|}$$

to derive optimal solution $x_1 = 8$, $x_2 = 7$, $z_{4,1} = 1.0$, $z_{4,2} = 1.0$ with objective function score 86.25. That is, the optimal frequency of the first line is $f_1 = 4$ and of the second line is $f_2 = 4$.

9.4 The new dispatching times are:

- 10 and 29 minutes
- 10 and 40 minutes
- 11 and 0 minutes

9.5 The optimal solution is derived by solving the linear program 3 times. One to assign the 100 passengers traveling from A to B. One to assign the 80 passengers from A to Y, and one to assign the 20 passengers from X to Y. This yields the following passenger flows at the arcs of the digraph. The total passenger demand at each arc appears in the last column of the following table.

Arcs		v_a		
a	For A to B	For A to Y	For X to Y	Total
(A,B)	50	0	0	50
(A,X2)	50	80	0	130
(X,X2)	0	0	20	20
(X2,X)	0	0	0	0
(X2,Y)	50	80	20	150
(Y,B)	41.7	0	0	41.7
(Y,Y3)	8.3	0	0	8.3
(Y3,B)	8.3	0	0	8.3
(Y3,Y)	0	0	0	0

9.6 Let us introduce the following binary variables:

- $\zeta_{f,r}$ where $\zeta_{f,r} = 1$ if subline r is operated with frequency $f \in F$
- $u_{f,sy}$ where $u_{f,sy} = 1$ if OD-pair $(s, y) \in O$ is operated with frequency $f \in F \setminus \{0\}$
- $h_{f_1, f_2, r, sy}$ where $h_{f_1, f_2, r, sy} = 1$ if line $r \in R$ operates with frequency f_1 and OD-pair $(s, y) \in O$ is served with frequency $f_2 \in F \setminus \{0\}$

Then, we can reformulate the subline frequency settings problem as a MILP:

$$\min \sum_{r \in R} \left(x_r w_1 + w_2 t_r T \sum_{f \in F} f \zeta_{f,r} \right) + \sum_{(s,y) \in O} d_{sy} \frac{1}{\sum_{i \in F \setminus \{0\}} f + 1} u_{f,sy}$$

subject to:

$$\sum_{f \in F} \zeta_{f,r} = 1 \qquad \forall r \in R$$

$$\sum_{f \in F} f \zeta_{f,r} \leq \frac{x_r}{t_r} \qquad \forall r \in R$$

$$\sum_{f \in F \setminus \{0\}} u_{f,sy} = 1 \qquad \forall (s, y) \in O$$

$$\sum_{f \in F \setminus \{0\}} fu_{f,sy} \leq \sum_{r \in R} \delta_{r,sy} \sum_{f \in F} \zeta_{f,r} \qquad \forall (s,y) \in O$$

$$\sum_{f \in F \setminus \{0\}} fu_{f,sy} \geq \theta \qquad \forall (s,y) \in O$$

$$\sum_{r \in R} x_r \leq n$$

$$x_1 \geq k$$

$$\sum_{f_1 \in F} \sum_{f_1 \in F \setminus \{0\}} h_{f_1,f_2,r,sy} = 1 \qquad \forall r \in R, \forall (s,y) \in O$$

$$2h_{f_1,f_2,r,sy} \leq \zeta_{f_1,r} + u_{f_2,sy} \qquad \forall f_1 \in F, f_2 \in F \setminus \{0\}, r \in R, (s,y) \in O$$

$$b_{r,s} = \sum_{y>s} d_{sy}\delta_{r,sy} \sum_{f_1 \in F} \sum_{f_2 \in F \setminus \{0\}} \frac{f_1}{f_2} h_{f_1,f_2,r,sy} \qquad \forall r \in R, \forall s \in S \setminus \{|S|\}$$

$$v_{r,y} = \sum_{s<y} d_{sy}\delta_{r,sy} \sum_{f_1 \in F} \sum_{f_2 \in F \setminus \{0\}} \frac{f_1}{f_2} h_{f_1,f_2,r,sy} \qquad \forall r \in R, \forall y \in S \setminus \{1\}$$

$$l_{r,s} = l_{r,s-1} + b_{r,s} - v_{r,s} \qquad \forall r \in R, \forall s \in S \setminus \{1\}$$

$$l_{r,1} = b_{r,1} \qquad \forall r \in R$$

$$l_{r,s} \leq c \sum_{f \in F} f\zeta_{f,r} \qquad \forall r \in R, s \in S$$

$$x_r \in \mathbb{Z}_{\geq 0} \qquad \forall r \in R$$

$$u_{f,sy} \in \{0,1\} \qquad \forall f \in F \setminus \{0\}, r \in R$$

$$\zeta_{f,r} \in \{0,1\} \qquad \forall f \in F \setminus \{0\}, (s,y) \in O$$

$$h_{f_1,f_2,r,sy} \in \{0,1\} \qquad \forall f_1 \in F, f_2 \in F \setminus \{0\}, r \in R, (s,y) \in O$$

$$b_{rs}, v_{rs}, l_{rs} \in \mathbb{R}_{\geq 0} \qquad \forall r \in R, \forall s \in S$$

Problems of Chapter 10

10.1 We need to solve the following MILP:

$$\max \sum_{m=1}^{|M|-1} \sum_{q=m+1}^{|M|} Y_{mq} \qquad (13.78)$$

$$\text{s.t. } x_{1m} \leq H_m^{\max} \qquad \forall m \in M \qquad (13.79)$$

$$x_{|F_m|,m} \leq T \qquad \forall m \in M \qquad (13.80)$$

$$H_m^{\min} \leq x_{i+1,m} - x_{im} \leq H_m^{\max} \qquad \forall m \in M, \forall i : 1 \leq i \leq |F_m| - 1 \qquad (13.81)$$

$$D_{nijmq}\bar{M} \geq x_{im} + t_{mn} - (x_{jq} + t_{qn}) \qquad \forall i \in F_m, j \in F_q, n \in V_{mq}, m \in \{1, \dots, |M|-1\}, q \in \{m+1, \dots, |M|\} \qquad (13.82)$$

$$D_{nijmq}\bar{M} \geq x_{jq} + t_{qn} - (x_{im} + t_{mn}) \qquad \forall i \in F_m, j \in F_q, n \in V_{mq}, m \in \{1, \dots, |M|-1\}, q \in \{m+1, \dots, |M|\} \qquad (13.83)$$

$$Y_{mq} \leq \sum_{n \in V_{mq}} \sum_{i \in F_m} \sum_{j \in F_q} (1 - D_{nijmq}) \qquad \forall m \in \{1, \dots, |M|-1\}, q \in \{m+1, \dots, |M|\} \qquad (13.84)$$

$$x_{im} \in \mathbb{Z}_{\geq 0} \qquad \forall i \in V, \forall m \in M \qquad (13.85)$$

$$Y_{mq} \in \mathbb{R}_{\geq 0} \qquad \forall m \in M, \forall q \in M \qquad (13.86)$$

$$D_{nijmq} \in \{0,1\} \qquad \forall m \in M, q \in M, i \in F_m, j \in F_q, n \in V_{mq} \qquad (13.87)$$

To apply branch and bound, we can use Pulp (see below) resulting in dispatching times $x_{11} = 0$, $x_{21} = 3$, $x_{31} = 6$ for line 1 and dispatching times $x_{12} = 1$, $x_{22} = 4$, $x_{32} = 7$ for line 2. This will yield 3 synchronizations, $Y_{12} = 3$. Since we are having only one transfer stop, $V_{mq} = \{2\}$, we can drop index n from variables D_{nijmq} resulting in the following implementation.

```python
import pulp
from pulp import *
model = LpProblem("TransferSync", LpMaximize)
M=(1,2)
H_min={1:2,2:3}
H_max={1:6,2:6}
F={1:(1,2,3),2:(1,2,3)}
t={(1,2):6,(2,2):5}
V={(1,2):2}
T=35 #time interval of scheduled trips
Big_M = 10000


# variables
x = LpVariable.dicts("x", ((i,m) for m in M for i in F[m]), 0, None,
    LpInteger)
Y = LpVariable.dicts("Y", ((i,m) for i in M for m in M), 0, None,
    LpContinuous)
D = LpVariable.dicts("D", ((i,j,m,q) for m in M for q in M for i in F[m]
    for j in F[q]), 0, None, LpBinary)

# constraints:
for m in M: model += x[1,m] <= H_max[m]
for m in M: model += x[len(F[m]),m] <= T
for m in M:
    for i in F[m]:
        if i!=len(F[m]):
            model += H_min[m] <= x[i+1,m]-x[i,m]
            model += x[i + 1, m] - x[i, m] <= H_max[m]
for m in M:
    if m<=len(M)-1:
        for q in M:
            if q>=m+1:
                model += Y[m,q] <= sum(sum( 1-D[i,j,m,q] for j in F[q])
    for i in F[m])
                n=2
                for i in F[m]:
                    for j in F[q]:
                        model += D[i, j, m, q] * Big_M >= x[j, q] + t[q,
    n] - (x[i, m] + t[m, n])
                        model += D[i, j, m, q] * Big_M >= (x[i, m] + t[m,
    n]) - (x[j, q] + t[q, n])

model += sum(sum(Y[m,q] for q in M if q>=m+1) for m in M if m<=len(M)-1)
# optimizing
solver = pulp.PULP_CBC_CMD(mip=True, msg=True, timeLimit=None, warmStart=
    False, path=None, mip_start=False, timeMode='elapsed',options=["
    passCuts=0","preprocess=off","presolve=off","cuts=on","heuristics=off
    ","greedyHeuristic=off"])
model.solve(solver)
print ("Status:", LpStatus[model.status])
```

```
for v in model.variables():
    print (v.name, "=", v.varValue)
print ("Optimal Solution = ", value(model.objective))
```

Finally, the branch and bound solution is a globally optimal one because the continuous relaxation of the mixed-integer program is a linear programming problem.

10.2 The optimal solution is $\mathbf{x}^* = [5, 290]^\mathsf{T}, \mathbf{h}^* = [305, 295]^\mathsf{T}$, and $\mathbf{a}^* = \begin{bmatrix} 305 & 605 \\ 610 & 900 \end{bmatrix}$.

10.3 The main difference is that synchronization problems with hierarchy adjust the schedules of the feeder lines considering the schedules of collector lines as fixed, or not easily adjustable.

Synchronization problems without hierarchy are more suitable in cases of transfers between modes of the same importance that are scheduled by the same service provider or closely collaborating service providers. For instance, in case of synchronizing the schedules of train lines that share common stations.

Problems of Chapter 11

11.1 The assignment problem formulation of the SD-VSP is:

$$\min_{\mathbf{x}} \sum_{(i,j)\in A} \bar{c}_{ij} x_{ij}$$

$$\text{s.t.} \sum_{j:(i,j)\in A} x_{ij} = 1 \qquad \forall i \in T$$

$$\sum_{i:(i,j)\in A} x_{ij} = 1 \qquad \forall j \in T$$

$$x_{ij} \in \{0, 1\} \qquad \forall (i,j) \in A$$

The requirement that $x_{ij} \leq 1$ in $x_{ij} \in \{0, 1\}$ is redundant, and we can thus write $x_{ij} \in \mathbb{Z}_+$. The equality constraints $\sum_{j:(i,j)\in A} x_{ij} = 1 \; \forall i \in T$ and $\sum_{i:(i,j)\in A} x_{ij} = 1 \; \forall j \in T$ are also totally unimodular. Thus, its solution is equivalent to the solution of a linear programming problem where each $x_{ij} \in \{0, 1\}$ is replaced by $x_{ij} \in \mathbb{R}_+$.

Given the above reformulation, solving the linear programming problem returns an optimal value of 1160 units.

11.2 Use the steps of the branch and bound algorithm for binary optimization problems (Algorithm 23).

11.3 We have four trips $T = \{1, 2, 3, 4\}$. The set of compatible arcs for these trips given their start/end times and their deadhead times is $\bar{A} = \{(1, 3), (1, 4), (2, 3), (2, 4)\}$. For the *Asymmetric Assignment Formulation*, we have digraph $G = (V, A)$ with $V = T \cup \{|T| + 1, \ldots, |T| + |T|\}$ and $A = \bar{A} \cup \{(1, |T| + 1), (2, |T| + 2), \ldots, (|T|, 2|T|)\}$. Thus, we have the following figure (Fig. A.2).

For the *Assignment Formulation* we have digraph $G = (V, A)$ where $V = T \cup \{s, t\}$ and $A = \bar{A} \cup \{s \times T\} \cup \{T \times t\}$. In this network, vertices s and t are the start

Fig. A.2 Digraph of the
Asymmetric Assignment
formulation of the SD-VSP

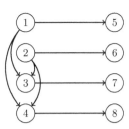

Fig. A.3 Digraph of the
Assignment formulation of
the SD-VSP

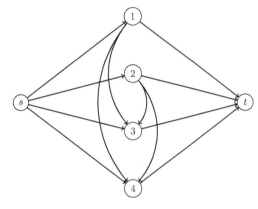

Fig. A.4 Digraph of the
Network Flow formulation
of the SD-VSP

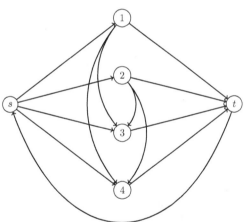

depot and the end depot (same physical location). This results in the following digraph
(Fig. A.3).

Finally, for the *Network Flow* formulation we have $V = T \cup \{s, t\}$ and $A = \{t, s\} \cup \{(s, j)$ for all $j \in T\} \cup \{(i, t)$ for all $i \in T\} \cup \bar{A}$. This results in the following digraph (Fig. A.4).

11.4 Use $|K|$ directed graphs $G_k = (V_k, A_k)$, where each digraph G_k is associated to the vertices (trips) that can be served by vehicle type $k \in K$ given its capacity. The set of vertices $V_k = T \cup \{s_k\} \cup \{t_k\}$ refers to a specific vehicle type $k \in K$ where s_k is the source and t_k is the sink (depot). We also define arcs A_k for each vehicle type k:

$$A_k = \{(s_k, j) \text{ for all } j \in T\} \cup \{(i, t_k) \text{ for all } i \in T\} \cup \{(t_k, s_k)\} \cup \bar{A}$$

where $\bar{A} = \{(i, j) \text{ for all } i \in T, j \in T \ : \ e_i^t + d_{ij} \le b_j^t\}$.

The number of vehicles that belong to the k-th vehicle type is p_k. Use a three-index formulation with a set of flow variables x_{ij}^k, where x_{ij}^k is a positive integer indicating the flow in arc $(i, j) \in A_k$. The arc costs are defined as:

- $c_{ij} = d_{ij}$ if $(i, j) \in \bar{A}$, where d_{ij} is the deadhead time from i to j
- $c_{s_k j} = d_{s_k j}$ for all $j \in T$, where $d_{s_k j}$ is the deadhead time from s_k to j
- $c_{i t_k} = d_{i t_k}$ for all $i \in T$, where $d_{i t_k}$ is the deadhead time from i to t_k
- $c_{t_k s_k} = c_0$ where c_0 is the cost of adding an extra vehicle

The multi-commodity flow formulation of the problem is:

$$\min_{\mathbf{x}} \sum_{k \in K} \sum_{(i,j) \in A_k} c_{ij}^k x_{ij}^k$$

$$\text{s.t.} \sum_{i:(i,j) \in A_k} x_{ij}^k - \sum_{i:(j,i) \in A_k} x_{ji}^k = 0 \qquad \forall j \in V_k, \forall k \in K$$

$$\sum_{j:(s_k,j) \in A_k} x_{s_k,j}^k \le p_k \qquad \forall k \in K$$

$$\sum_{k \in K} \sum_{i:(i,j) \in A_k} x_{ij}^k = 1 \qquad \forall j \in T$$

$$x_{ij}^k \in \mathbb{Z}_+ \qquad \forall k \in K, \forall (i, j) \in A_k$$

11.5 (a) Solving this problem as a multi-depot vehicle scheduling problem without time windows has solution:

- Vehicle of depot 1: $s_1 \to 2 \to t_1$
- Vehicle of depot 1: $s_1 \to 4 \to t_1$
- Vehicle of depot 2: $s_2 \to 1 \to 6 \to t_2$
- Vehicle of depot 2: $s_2 \to 3 \to 7 \to t_2$
- Vehicle of depot 3: $s_3 \to 5 \to t_3$

with an objective function score of 1166.

(b) To solve the problem as a multi-depot vehicle scheduling problem with time windows, we should convert it to the following MILP formulation:

$$\min \sum_{k \in K} \sum_{(i,j) \in A_k} \left(b_{ij}^k + I_{ij}^k \lambda \max\{0, l_j - u_i - t_{ij}\} \right)$$

$$\text{s.t.:} \sum_{k \in K} \sum_{j:(i,j) \in A_k} x_{ij}^k = 1 \qquad \forall i \in T$$

Fig. A.5 Digraph of the fixed job scheduling problem expressed as a minimum vertex coloring problem

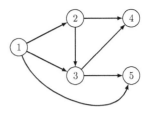

$$\sum_{j:(s_k,j)\in A^k} x^k_{s_k,j} = \sum_{i:(i,t_k)\in A^k} x^k_{i,t_k} = 1 \qquad \forall k \in K$$

$$\sum_{i:(i,j)\in A^k} x^k_{ij} - \sum_{i:(j,i)\in A^k} x^k_{ji} = 0 \qquad \forall k \in K \ \forall j \in T$$

$$T^k_i + t_{ij} - T^k_j + \sigma^k_{ij} \le 0 \qquad k \in K \ \forall(i,j) \in A_k$$

$$\sigma^k_{ij} \le M(1 - x^k_{ij}) \qquad k \in K \ \forall(i,j) \in A_k$$

$$\sigma^k_{ij} \ge -M(1 - x^k_{ij}) \qquad k \in K \ \forall(i,j) \in A_k$$

$$l_i \le T^k_i \le u_i \qquad \forall k \in K \ \forall i \in V_k$$

$$x^k_{ij} \in \{0, 1\} \qquad \forall k \in K \ \forall(i,j) \in A_k$$

11.6 (a) To formulate the fixed job scheduling problem as a minimum Vertex Coloring problem, we create a digraph $G = (V, A)$ where $V = \{1, 2, 3, 4, 5\}$ are the vertices of the graph corresponding to the increments of work (tasks), and $A = \{(j, k) : j \in V, k \in I_j\}$ are the arcs. Based on the release and the deadline times, we have the incompatibility sets:

- $I_1 = \{2, 3, 5\}$
- $I_2 = \{3, 4\}$
- $I_3 = \{4\}$
- $I_4 = \{5\}$
- $I_5 = \emptyset$

Thus, $A = \{(1, 2), (1, 3), (1, 4), (2, 3), (2, 4), (3, 4), (4, 5)\}$. This digraph is presented in Fig. A.5.

We have a set of $M = \{1, 2, \ldots, 8\}$ drivers. Let binary variables **y** represent colors (drivers), where $y_j = 1$ if color $j \in M$ is used. Let also binary variables **x**, where $x_{ij} = 1$ if vertex $i \in V$ is colored by color $j \in M$. Then, we can express the fixed job scheduling problem as a minimum Vertex Coloring problem based on the following formulation:

$$\min_{\mathbf{x,y}} \sum_{j \in M} y_j$$

$$\text{s.t} \sum_{j \in M} x_{ij} = 1 \qquad\qquad \forall i \in V$$

$$x_{ij} + x_{kj} \leq y_j \qquad\qquad \forall (i,k) \in A, j \in M$$

$$x_{ij} \in \{0,1\} \qquad\qquad \forall i \in V, j \in M$$

$$y_j \in \{0,1\} \qquad\qquad \forall j \in M$$

which provides solution $x_{1,4} = 1, x_{2,2} = 1, x_{3,3} = 1, x_{4,4} = 1, x_{5,2} = 1$ and $y_2 = 1, y_3 = 1, y_4 = 1$ indicating that increments of work 1 and 4 will be served by driver 4, increments of work 2 and 5 will be served by driver 2, and increment of work 3 will be served by driver 3. Note that we need 3 drivers in total. These 3 drivers were driver 2, 4 and 5 but, in practice, they could have been any combination of 3 different numbers from 1 to 8.

(b) The decision problem of the fixed job scheduling corresponds to the Vertex Coloring decision problem, which is one of Karp's 21 NP-complete problems. Thus, the fixed job scheduling decision problem is NP-complete and its optimization counterpart is NP-Hard.

Problems of Chapter 12

12.1 (a) Consider the two-index vehicle flow formulation of the CVRP that requests to use all available vehicles $|K|$:

$$\min_{\mathbf{x,u}} \sum_{i \in V} \sum_{j \in V} c_{ij} x_{ij}$$

$$\text{s.t.} \sum_{i \in V} x_{ij} = 1 \qquad\qquad \forall j \in V \setminus \{0\}$$

$$\sum_{j \in V} x_{ij} = 1 \qquad\qquad \forall i \in V \setminus \{0\}$$

$$\sum_{i \in V} x_{i0} = |K|$$

$$\sum_{j \in V} x_{0j} = |K|$$

$$u_i - u_j + C x_{ij} \leq C - d_j \quad \forall i \in V \setminus \{0\}, j \in V \setminus \{0, i\} \text{ such that } d_i + d_j \leq C$$

$$d_i \leq u_i \leq C \qquad\qquad \forall i \in V \setminus \{0\}$$

$$x_{ij} \in \{0,1\} \qquad\qquad \forall i \in V, j \in V$$

$$u_i \in \mathbb{R} \qquad\qquad \forall i \in V \setminus \{0\}$$

We can turn this two-index vehicle flow formulation into a three-index formulation where variable $x_{ij}^k \in \{0,1\}$ indicates whether arc (i,j) will be served by vehicle

$k \in K$ or not, and $u_i^k \in \mathbb{R}$ represents the load of vehicle k after visiting location i. This three-index vehicle flow formulation is presented below.

$$\min_{\mathbf{x},\mathbf{u}} \sum_{k \in K} \sum_{i \in V} \sum_{j \in V} c_{ij} x_{ij}^k$$

$$\text{s.t.} \sum_{k \in K} \sum_{i \in V} x_{ij}^k = 1 \qquad \forall j \in V \setminus \{0\}$$

$$\sum_{k \in K} \sum_{j \in V} x_{ij}^k = 1 \qquad \forall i \in V \setminus \{0\}$$

$$\sum_{k \in K} \sum_{j \in V} x_{0j}^k = |K|$$

$$u_i^k - u_j^k + C x_{ij}^k \le C - d_j \quad \forall i \in V \setminus \{0\}, j \in V \setminus \{0, i\} \text{ such that } d_i + d_j \le C, k \in K$$

$$d_i \le u_i^k \le C \qquad \forall i \in V \setminus \{0\}, k \in K$$

$$x_{ij}^k \in \{0, 1\} \qquad \forall i \in V, j \in V, k \in K$$

$$u_i^k \in \mathbb{R} \qquad \forall i \in V \setminus \{0\}, k \in K$$

(b) Add the following inequality constraints to the three-index vehicle flow formulation:

$$\sum_{i \in V} \sum_{j \in V} e_{ij} x_{ij}^k \le \phi \quad \forall k \in K$$

12.2 The answer to this question can be found in this chapter by implementing the dial-a-ride formulation to the a2-16 instance.

12.3 Let us encode the depot vertices as $o_1 = 0, o_2 = 9, o_3 = 10$, and $o_4 = 11$. Then, we can use branch and bound to solve the mixed-integer linea program resulting in optimal solution:

- Vehicle 1 to perform pick-up route $0 \to 3 \to 1 \to 2 \to 9$
- Vehicle 2 to perform pick-up route $0 \to 4 \to 9$
- Vehicle 1 to perform delivery route $10 \to 5 \to 7 \to 6 \to 11$
- Vehicle 2 to perform delivery route $10 \to 8 \to 11$

The objective function score of the optimal solution is 1054.99 and the route duration of every route is:

- Route duration of pickup route 1: 151.368
- Route duration of pickup route 2: 103.638
- Route duration of delivery route 1: 371.435
- Route duration of delivery route 2: 428.538

The implementation of the branch and bound algorithm using the Pulp modeling framework and the Gurobi solver in Python is provided below.

```
import numpy as np
import math
import pulp
```

```python
from pulp import *

model = LpProblem("DARPi", LpMinimize)
n=4; Qk=20; LL=550; a_par=10; b_par=1
no_vehicles=2; K=[] #vehicle number
for i in range(1,no_vehicles+1): K.append(i)
TT={i:480 for i in K} #maximum route duration
Mbig=10000 #a large positive number
q={1:16,2:10,3:4,4:4,5:16,6:10,7:4,8:4}
e_bound={0:360,1:442,2:455,3:360,4:475,5:823,6:852,7:793,8:1007,
         9:360,10:360,11:360}
l_bound={0:1320,1:562,2:575,3:471,4:595,5:943,6:972,7:913,8:1127,
         9:1320,10:1320,11:1320}
c={(0, 0): 0.0, (0, 1): 49.85, (0, 2): 67.72, (0, 3): 37.76, (0, 4): 51.82, (0,
    5): 153.58, (0, 6): 90.32, (0, 7): 149.01, (0, 8): 214.27, (0, 9): 0.0, (0,
    10): 0.0, (0, 11): 0.0, (1, 0): 49.85, (1, 1): 0.0, (1, 2): 33.69, (1, 3):
    12.2, (1, 4): 62.74, (1, 5): 203.15, (1, 6): 137.24, (1, 7): 192.96, (1, 8):
    239.35, (1, 9): 49.85, (1, 10): 49.85, (1, 11): 49.85, (2, 0): 67.72, (2,
    1): 33.69, (2, 2): 0.0, (2, 3): 40.13, (2, 4): 50.49, (2, 5): 212.41, (2, 6)
    : 141.81, (2, 7): 191.37, (2, 8): 220.33, (2, 9): 67.72, (2, 10): 67.72, (2,
    11): 67.72, (3, 0): 37.76, (3, 1): 12.2, (3, 2): 40.13, (3, 3): 0.0, (3, 4)
    : 57.62, (3, 5): 191.23, (3, 6): 126.07, (3, 7): 182.63, (3, 8): 233.72, (3,
    9): 37.76, (3, 10): 37.76, (3, 11): 37.76, (4, 0): 51.82, (4, 1): 62.74,
    (4, 2): 50.49, (4, 3): 57.62, (4, 4): 0.0, (4, 5): 167.84, (4, 6): 95.43,
    (4, 7): 141.32, (4, 8): 176.63, (4, 9): 51.82, (4, 10): 51.82, (4, 11):
    51.82, (5, 0): 153.58, (5, 1): 203.15, (5, 2): 212.41, (5, 3): 191.23, (5,
    4): 167.84, (5, 5): 0.0, (5, 6): 72.74, (5, 7): 66.72, (5, 8): 187.18, (5,
    9): 153.58, (5, 10): 153.58, (5, 11): 153.58, (6, 0): 90.32, (6, 1): 137.24,
    (6, 2): 141.81, (6, 3): 126.07, (6, 4): 95.43, (6, 5): 72.74, (6, 6): 0.0,
    (6, 7): 60.82, (6, 8): 157.5, (6, 9): 90.32, (6, 10): 90.32, (6, 11): 90.32,
    (7, 0): 149.01, (7, 1): 192.96, (7, 2): 191.37, (7, 3): 182.63, (7, 4):
    141.32, (7, 5): 66.72, (7, 6): 60.82, (7, 7): 0.0, (7, 8): 120.68, (7, 9):
    149.01, (7, 10): 149.01, (7, 11): 149.01, (8, 0): 214.27, (8, 1): 239.35,
    (8, 2): 220.33, (8, 3): 233.72, (8, 4): 176.63, (8, 5): 187.18, (8, 6):
    157.5, (8, 7): 120.68, (8, 8): 0.0, (8, 9): 214.27, (8, 10): 214.27, (8, 11)
    : 214.27, (9, 0): 0.0, (9, 1): 49.85, (9, 2): 67.72, (9, 3): 37.76, (9, 4):
    51.82, (9, 5): 153.58, (9, 6): 90.32, (9, 7): 149.01, (9, 8): 214.27, (9, 9)
    : 0.0, (9, 10): 0.0, (9, 11): 0.0, (10, 0): 0.0, (10, 1): 49.85, (10, 2):
    67.72, (10, 3): 37.76, (10, 4): 51.82, (10, 5): 153.58, (10, 6): 90.32, (10,
    7): 149.01, (10, 8): 214.27, (10, 9): 0.0, (10, 10): 0.0, (10, 11): 0.0,
    (11, 0): 0.0, (11, 1): 49.85, (11, 2): 67.72, (11, 3): 37.76, (11, 4):
    51.82, (11, 5): 153.58, (11, 6): 90.32, (11, 7): 149.01, (11, 8): 214.27,
    (11, 9): 0.0, (11, 10): 0.0, (11, 11): 0.0}
t=c

P=[];D=[];V=[]
for i in range(1,n+1): P.append(i)
for i in range(n+1,2*n+1): D.append(i)
for i in range(0,2*n+4): V.append(i)
PcupD=[]
for i in range(1,2*n+1): PcupD.append(i)
PplusDepot=P+[0,2*n+1]; DplusDepot=D+[2*n+2,2*n+3]

A0=[(0,j) for j in P]; A00=[(j,2*n+1) for j in P]
A000=[(i,j) for i in P for j in P if j!=i]
A1=[(2*n+2,j) for j in D]; A11=[(j,2*n+3) for j in D]
A111=[(i,j) for i in D for j in D if j!=i]
A=A0+A00+A000+A1+A11+A111

#VARIABLES
```

```
x = LpVariable.dicts("x", ((a[0],a[1],k) for a in A for k in K), 0, None,
    LpBinary)
w = LpVariable.dicts("w",K,0,None,LpContinuous)
r = LpVariable.dicts("r",V,0,None,LpContinuous)
u = LpVariable.dicts("u", ((v,k) for v in V for k in K), 0, None, LpContinuous)
eta = LpVariable.dicts("eta",((p,k) for p in P for k in K),None,None, LpBinary)
theta = LpVariable.dicts("theta",((p,k) for p in P for k in K),None,None,
    LpBinary)
eta_k = LpVariable.dicts("eta_k",K, None,None,LpBinary)
theta_k = LpVariable.dicts("theta_k",K, None,None,LpBinary)
tau = LpVariable.dicts("tau",K,0,None, LpContinuous)
z = LpVariable.dicts("z",P,0,None, LpContinuous)
p = LpVariable.dicts("p",((v,k) for v in V for k in K),None,None, LpContinuous)
sigma = LpVariable.dicts("sigma",((a[0],a[1],k) for a in A for k in K),None,None,
    LpContinuous)
p_active = LpVariable.dicts("p_active",V,None,None,LpContinuous)

#OBJECTIVE FUNCTION
model += sum(sum(c[i,j]*x[i,j,kei] for (i,j) in A) for kei in K)

#CONSTRAINTS
for ii in PcupD:
    model += sum(sum(x[i,j,kei] for (i,j) in A if i==ii) for kei in K) == 1
for ii in P:
    model += sum(sum(q[i]*x[i,j,kei] for (i,j) in A if i==ii) for kei in K)<=Qk
for ii in D:
    model += sum(sum(q[i]*x[i,j,kei] for (i,j) in A if i==ii) for kei in K)<=Qk
for kei in K:
    model += sum(x[h,j,kei] for (h,j) in A if h==0 )==1
    model += sum(x[h,j,kei] for (h,j) in A if h==2*n+2 )==1
    model += sum(x[j,h,kei] for (j,h) in A if h==2*n+1 )==1
    model += sum(x[j,h,kei] for (j,h) in A if h==2*n+3 )==1
    for hh in PcupD:
        model += sum(x[i,h,kei] for (i,h) in A if h==hh) - sum(x[h,j,kei] for (h,
    j) in A if h==hh)==0
    for (i,j) in A:
        model += u[j,kei]>=u[i,kei]+t[i,j]-Mbig*(1-x[i,j,kei])
    for i in V:
        model += int(e_bound[i])<=u[i,kei]
        model += u[i,kei]<=l_bound[i]
    for i in P:
        model += eta[i,kei]-theta[i,kei]== sum(x[i,j,kei] for j in P+[2*n+1] if (
    i,j) in A and j!=i ) - sum(x[i+n,j,kei] for j in D+[2*n+3] if (i+n,j) in A
    and j!=i+n)
        model += eta[i,kei]+theta[i,kei]<=1
    model += (1/Mbig) * sum(eta[i,kei] for i in P) <= eta_k[kei]
    model += eta_k[kei] <= sum(eta[i,kei] for i in P)
    model += (1/Mbig) * sum(theta[i,kei] for i in P) <= theta_k[kei]
    model += theta_k[kei] <= sum(theta[i,kei] for i in P)
    model += tau[kei] == u[2*n+1,kei]+a_par*eta_k[kei]+b_par*sum(q[i]*eta[i,kei]
    for i in P)
    model += w[kei]>=tau[kei]
    model += u[2*n+2,kei]==w[kei]+a_par*theta_k[kei]+b_par*sum(q[i]*theta[i,kei]
    for i in P)
    for i in P:
        model += w[kei]>=z[i]-Mbig*(1-theta[i,kei])
        model += z[i]>=tau[kei]-Mbig*(1-eta[i,kei])
    model += u[2*n+1,kei]-u[0,kei] <= TT[kei]
    model += u[2*n+3,kei]-u[2*n+2,kei] <= TT[kei]
for i in P:
```

```
        model += r[i]<=LL
for (j, i) in A:
    if i in P + D :
        for k in K:
            model += p_active[i]+sigma[j,i,k]==p[i,k]
            model += sigma[j,i,k]<=Mbig*(1-x[j,i,k])
            model += sigma[j,i,k]>=-Mbig*(1-x[j,i,k])
for i in P:
    for kei in K:
        model += r[i]==p_active[n+i]-p_active[i]

model.solve(solver = GUROBI_CMD())
print ("Status:", LpStatus[model.status])
for v in model.variables():
    if v.varValue >= 10E-7:
        print (v.name, "=", v.varValue)
print ("Optimal Solution = ", value(model.objective))
```

12.4 Consider the formulation of the Traveling Salesman Problem:

$$\min \sum_{i \in N} \sum_{j \in N} c_{ij} x_{ij}$$

$$\text{s.t.:} \quad \sum_{j \in N \setminus \{i\}} x_{ij} = 1 \qquad \forall i \in N$$

$$\sum_{i \in N \setminus \{j\}} x_{ij} = 1 \qquad \forall j \in N$$

$$\{x_{ij}\} \in S$$

$$x_{ij} \in \{0, 1\} \qquad \forall i \in N, j \in N$$

where $\{x_{ij}\} \in S$ is an abstract requirement representing the sub-tour elimination constraints. If we ignore the sub-tour elimination constraints, we have the minimum weight assignment problem:

$$\min \sum_{i \in N} \sum_{j \in N} c_{ij} x_{ij}$$

$$\text{s.t.:} \quad \sum_{j \in N \setminus \{i\}} x_{ij} = 1 \qquad \forall i \in N$$

$$\sum_{i \in N \setminus \{j\}} x_{ij} = 1 \qquad \forall j \in N$$

$$x_{ij} \in \{0, 1\} \qquad \forall i \in N, j \in N$$

for which $x_{ij} \in \{0, 1\} \; \forall i \in N, j \in N$ can be replaced by $x_{ij} \in \mathbb{R}_{\geq 0} \; \forall i \in N, j \in N$, resulting in a linear program that can be solved in polynomial time. When solving this problem, we receive an optimal solution with $C_1, C_2, \ldots, C_p$ sub-tours. If, by luck, $p = 1$, then our linear programming solution has no sub-tours and it is the optimal solution of the TSP. If $p \geq 2$ we can implement the following heuristic to receive a feasible solution:

1. Select the two sub-tours from $C_1, \ldots, C_p$ that have the most vertices.
2. Merge the two selected sub-tours in a way that results in the smallest possible cost increase. If $p = 1$ after this merge, stop. Otherwise, go back to step 1.

Problems of Chapter 13

13.1 (a) We have Lagrangian function:

$$
\mathcal{L}(\mathbf{x}, \mathbf{h}, \mathbf{a}, \mathbf{k}, \boldsymbol{\lambda}, \boldsymbol{\lambda}', \boldsymbol{\lambda}'', \boldsymbol{\mu}, \boldsymbol{\mu}', \boldsymbol{\mu}'') :=
$$

$$
\sum_{s=2}^{|S|} \sum_{j=1}^{n} \left(h_{j,s} - h_{j,s}^*\right)^2 + \sum_{j \in N_m} \sum_{s \in S \setminus \{1\}} \lambda_{js}\left[a_{js} - (\delta_j + x_j) - \sum_{\phi=1}^{s-1} t_{j\phi} - \sum_{\phi=2}^{s-1} k_{j\phi}\right] +
$$

$$
\sum_{j \in N_m} \sum_{s \in S \setminus \{1\}} \lambda'_{js}\left[h_{js} - (a_{js} - a_{j,s-1})\right] + \sum_{j \in N_m} \sum_{s \in S \setminus \{1\}} \lambda''_{js}\left[k_{js} - \gamma_s h_{j,s}\right] +
$$

$$
\sum_{j \in N_m \setminus \{1\}} \mu_j\left(H_{\min} - [(\delta_j + x_j) - (\delta_{j-1} + x_{j-1})]\right) +
$$

$$
\sum_{j \in N_m \setminus \{1\}} \mu'_j\left[(\delta_j + x_j) - (\delta_{j-1} + x_{j-1}) - H_{\max}\right] + \mu''_1(x_n - \zeta) +
$$

$$
\mu''_2\left(H_{\min} - [(\delta_1 + x_1) - \delta_0]\right) + \mu''_3\left[(\delta_1 + x_1) - \delta_0 - H_{\max}\right]
$$

where $\boldsymbol{\lambda}, \boldsymbol{\lambda}', \boldsymbol{\lambda}'', \boldsymbol{\mu}, \boldsymbol{\mu}', \boldsymbol{\mu}''$ are KKT multipliers related to equality and inequality constraints. We then have the stationarity conditions:

$$
\nabla_{\mathbf{x}, \mathbf{h}, \mathbf{a}, \mathbf{k}} \mathcal{L}(\mathbf{x}, \mathbf{h}, \mathbf{a}, \mathbf{k}, \boldsymbol{\lambda}, \boldsymbol{\lambda}', \boldsymbol{\lambda}'', \boldsymbol{\mu}, \boldsymbol{\mu}', \boldsymbol{\mu}'') = 0
$$

We have primal feasibility conditions:

$$
\begin{align}
H_{\min} \leq (\delta_j + x_j) - (\delta_{j-1} + x_{j-1}) &\qquad \forall j \in N_m \setminus \{1\} \tag{13.88} \\
(\delta_j + x_j) - (\delta_{j-1} + x_{j-1}) \leq H_{\max} &\qquad \forall j \in N_m \setminus \{1\} \tag{13.89} \\
H_{\min} \leq (\delta_1 + x_1) - \delta_0 & \tag{13.90} \\
(\delta_1 + x_1) - \delta_0 \leq H_{\max} & \tag{13.91} \\
a_{js} = (\delta_j + x_j) + \sum_{\phi=1}^{s-1} t_{j\phi} + \sum_{\phi=2}^{s-1} k_{j\phi} &\qquad \forall j \in N_m, \forall s \in S \setminus \{1\} \tag{13.92} \\
h_{js} = a_{js} - a_{j-1,s} &\qquad \forall j \in N_m, \forall s \in S \setminus \{1\} \tag{13.93} \\
k_{js} = \gamma_s h_{js} &\qquad \forall j \in N_m, \forall s \in S \setminus \{1\} \tag{13.94} \\
x_n \leq \zeta & \tag{13.95}
\end{align}
$$

We have dual feasibility conditions:

$$\mu_j \geq 0 \qquad\qquad \forall j \in N_m \setminus \{1\} \qquad\qquad (13.96)$$
$$\mu'_j \geq 0 \qquad\qquad \forall j \in N_m \setminus \{1\} \qquad\qquad (13.97)$$
$$\mu''_1, \mu''_2, \mu''_3 \geq 0 \qquad\qquad\qquad\qquad\qquad\qquad (13.98)$$

Finally, we have complementary slackness conditions:

$$\mu_j \big(H_{\min} - [(\delta_j + x_j) - (\delta_{j-1} + x_{j-1})] \big) = 0 \qquad \forall j \in N_m \setminus \{1\} \qquad (13.99)$$
$$\mu'_j \big((\delta_j + x_j) - (\delta_{j-1} + x_{j-1}) - H_{\max} \big) = 0 \qquad \forall j \in N_m \setminus \{1\} \qquad (13.100)$$
$$\mu''_1 (x_n - \zeta) = 0 \qquad\qquad\qquad\qquad\qquad\qquad (13.101)$$
$$\mu''_2 \big(H_{\min} - [(\delta_1 + x_1) - \delta_0] \big) = 0 \qquad\qquad\qquad (13.102)$$
$$\mu''_3 \big((\delta_1 + x_1) - \delta_0 - H_{\max} \big) = 0 \qquad\qquad\qquad (13.103)$$

(b) We have solution $\mathbf{x} = [-50.4996, -0.508296, -70.5084]^\mathsf{T}$ with objective function score 0.482954. Note that in this solution we also consider the headway deviation at the last stop as part of the objective function.

13.2 Without schedule sliding, we have:

$$d_{ns} = \begin{cases} t & \text{if } t - d_{n-1,s} \geq aH_s \\ H_s + d_{n-1,s} & \text{otherwise.} \end{cases}$$

resulting in new departure time: $d_{ns} = 600 + 1050 = 1650$.
With schedule sliding, we have:

$$d_{ns} = \begin{cases} t & \text{if } t - d_{n-1,s} \geq aH_s \\ \max\{t, \min\{\rho_n - t_s, H_s + d_{n-1,s}\}\} & \text{otherwise.} \end{cases} \qquad (13.104)$$

resulting in new departure time: $d_{ns} = \max\{1500, \min\{4250 - 3000, 600 + 1050\}\} = 1500$.

13.3 Given formulation:

$$\min \sum_{j \in I_m} \sum_{k'=k}^{k_t} r_{k'} (d_{jk'} - d_{j-1,k'})^2$$

$$\begin{aligned}
\text{s.t. } & d_{jk} - a_{jk} - s_{jk} \geq 0 && \forall j \in I_m \\
& d_{jk'} - a_{jk'} - s_{jk'} = 0 && \forall j \in I_m, k' : k' > k \\
& d_{i+m,k} - a_{i+m,k} - s_{i+m,k} = 0 \\
& a_{jk'} - d_{j-1,k'} \geq h_0 && \forall j \in I_m, k' : k' > k \\
& d_{j,k_c} \leq \max(\tau_{j,c}, d^0_{j,k_c}) && \forall j \in I_m
\end{aligned}$$

$$a_{jk'} = \max(d_{j,k'-1} + t_{k'}, d_{j-1,k'} + h_0) \qquad \forall j \in I_m, k' : k' \geq k$$
$$s_{jk'} = c_0 + c_1 r_{k'} h_{jk'} + c_2 q_{k'} L_{j,k'-1} \qquad \forall j \in I_m, k' : k' \geq k$$
$$L_{jk'} = L_{j,k'-1} + r_{k'} h_{jk'} - q_{k'} L_{j,k'-1} \qquad \forall j \in I_m, k' : k' \geq k$$
$$d_{j,k'}, a_{j,k'}, s_{j,k'}, L_{j,k'} \in \mathbb{R}_{\geq 0} \qquad \forall j \in I_m, k' \in \{k, \ldots, k_t\}$$

We can replace $a_{jk'} = \max(d_{j,k'-1} + t_{k'}, d_{j-1,k'} + h_0) \ \forall j \in I_m, k' : k' \geq k$ by:

$$a_{jk'} \geq d_{j,k'-1} + t_{k'} \tag{13.105}$$
$$a_{jk'} \geq d_{j-1,k'} + h_0 \tag{13.106}$$
$$a_{jk'} \leq d_{j,k'-1} + t_{k'} + \delta M \tag{13.107}$$
$$a_{jk'} \leq d_{j-1,k'} + h_0 + (1 - \delta)M \tag{13.108}$$

where M is a very big positive number and δ a binary variable. This will result in a convex feasible region.

13.4 Let a be the known arrival time of the preceding vehicle at the stop. Let also r be the fastest possible arrival time of our vehicle to that stop when our vehicle drives on the maximum allowed speed limit, s_{max}. If t is the current time and the position of our vehicle is x km before the location of our stop, then parameter r is equal to $(r - t)s_{max} = x \Rightarrow r = \frac{x}{s_{max}} + t$.

Suppose now that we have a limitation on the minimum possible speed, s_{min}, since the vehicle cannot completely stop in a mixed-traffic environment. Then, our optimal speed recommendation s is bounded by $s_{min} \leq s \leq s_{max}$.

Using speed recommendation s, our vehicle arrives at the stop at time $r' = \frac{x}{s} + t$, where r' is bounded by $\frac{x}{s_{max}} + t \leq r' \leq \frac{x}{s_{min}} + t$. We seek to minimize the squared deviation between the planned and the actual time headway at this stop. Thus, we can formulate the problem as:

$$\min_{r' \in \mathbb{R}} \ ((r' - a) - h^*)^2$$
$$\text{s.t.} \ \frac{x}{s_{max}} + t \leq r' \leq \frac{x}{s_{min}} + t$$

Retrieving the optimal value of r' by solving this single-variable quadratic program will give us the optimal speed recommendation $s^* = \frac{x}{r'-t}$.

13.5 Implement a metaheuristic and compare the performance of your solution against the performance of the globally optimal solution, $Z^* = 563{,}491$.

Index

© The Editor(s) (if applicable) and The Author(s), under exclusive license to Springer Nature Switzerland AG 2022

K. Gkiotsalitis, *Public Transport Optimization*,

https://doi.org/10.1007/978-3-031-12444-0